The Life of Mammals

THE LIFE OF MAMMALS

THEIR ANATOMY AND PHYSIOLOGY

J. Z. YOUNG

WITH THE ASSISTANCE OF
M. J. HOBBS

SECOND EDITION

CLARENDON PRESS · OXFORD · 1975

Oxford University Press, Ely House, London W.1.

GLASGOW NEW YORK TORONTO MELBOURNE WELLINGTON
CAPE TOWN IBADAN NAIROBI DAR ES SALAAM LUSAKA ADDIS ABABA
DELHI BOMBAY CALCUTTA MADRAS KARACHI LAHORE DACCA
KUALA LUMPUR SINGAPORE HONG KONG TOKYO

CASEBOUND ISBN 0 19 857156 9
PAPERBACK ISBN 0 19 857158 5

© OXFORD UNIVERSITY PRESS 1975

PRINTED IN GREAT BRITAIN
BY J. W. ARROWSMITH LTD., BRISTOL

Preface to the Second Edition

In the seventeen years since this book was first published many of the developments foreshadowed in it have duly occurred. One of these has been a wide increase in use of the concept of the book itself—the study of the life of animals as a whole. When the title was first put forward for *The life of vertebrates* it was so unfamiliar that the publishers questioned it—did I mean 'The lives', 'The life-history', or what? But the idea caught on so well that it is now a commonplace to study the whole life system of animals, which is what I aimed to do. Indeed I have myself helped to edit a whole series of works with this theme, the Weidenfeld and Nicolson Natural History Series, including another *Life of Mammals* by my fellow editor Dr. Leo Harrison Matthews. His book was able to supply the natural history that space does not allow here.

The present work deals mainly with what is more commonly called the structure and function of mammals. In both these aspects of study there have been enormous advances since 1957. Use of the electron microscope was then only just beginning and has since extended to the point where the details of molecular organization can be almost or quite visualized, for instance in muscle. The electron micrographs in this edition allow many such points to be illustrated.

The main theme of this new edition, as of the first, is the consideration of the body as a self-controlling machine. There has been a continual increase of precision in the use of this concept for the study of homeostasis. Perhaps the growth of cybernetics has not been quite so rapid as some of its supporters may have hoped, but the development of the subject has been of use to Biology. It has progressed furthest for those parts of the body in which analogies from man-made machines are most obvious. For instance, more than 300 factors regulating the circulation of the blood have been identified and equations written for their inter-actions (Chapter 21).

Since the first edition there has been an enormous increase of information of all sorts about almost every part of the body. We have tried to include much of this new biochemical and physiological knowledge where it is relevant to the specific function of an organ system. Progress has been so rapid that some organs whose nature was then wholly mysterious are now reasonably well understood—for instance the pineal body, which now merits a chapter to itself. Knowledge of the brain and sense organs has advanced very rapidly and the chapters on the nervous system have been considerably expanded.

For the present edition the book has been revised throughout, but we have tried not to allow a mass of new information to blur the presentation of fundamental outlines of knowledge about each organ system. The book is meant to be used by those not yet very familiar with biology, and also can be used to accompany dissections of, say, the rabbit or the rat. Knowledge of the gross anatomy of these animals has not altered much in seventeen years and the chapters dealing with this subject are the least changed. In particular I have retained the beautiful drawings made from skeletons and dissections of the rabbit by the late Miss E. R. Turlington.

The long section on embryology has now been broken up and the relevant parts incorporated with the accounts of each organ system. This has meant omitting treatment of the development of non-mammalian vertebrates. About this there has been so much new information here that it would be impossible to give a good account in a short space, and excellent text books are now available. Dr. Ruth Bellairs has kindly checked the revision of the accounts of mammalian early embryology and organogenesis.

In the work of revision I have fortunately had very much able help. I am particularly grateful to Dr. Michael Hobbs who has been responsible for organizing a great deal of the work; collecting references; assembling figures; preparing the index; and, above all, providing wise advice.

We have received substantial help from Dr. E. C. R. Hall-Craggs in revising the chapters on the skeletal and muscular systems, from Professor R. D. Harkness on the connective tissues, and from Dr. K. E. Webster on the nervous system. We are also much indebted to Dr. M. Nixon for help in many places. To Mrs. J. Astafiev, who helped with the first edition (as Miss J. de Vere) we are grateful for drawing many of the new figures. Many others have helped with advice or illustrations, including those listed on page viii. Finally it is a pleasure once again to thank the Secretary and staff of the Clarendon Press for their continuing help.

July 1974   J.Z.Y.

Preface to the First Edition

WHEN presenting any subject it has to be decided whether to accept a conventional method, known and widely accepted, or to try to invent a better one. The latter is the more exciting course, but a compromise is safer for the author and kinder to his readers, who want to learn the subject as it is. This book attempts to have the best of both methods. It presents mammalian anatomy, physiology, histology, and embryology as they are, and at the same time tries to show what in the future they may be. It begins with a discussion of biological method, which may seem to be a strange and difficult subject to the scientist, who is apt to leave these matters to philosophers. I recommend even the beginner to read it before he passes to the detailed work of the rest of the book. It may show him how many parts of the subject are imperfect and how they can be illuminated by use of a new language that considers the body as a self-controlling machine. This idea is not new but in developing it I have been pleased to find how interesting it makes the consideration of the structure and function of many parts of the body. I have not been able to apply the methods of information theory fully or exactly, but I hope that others may come to do so and to find in the study of control the clue that unifies many parts of Biology.

As one tries to treat the whole organization of life in this way it seems that a new science is growing with every word. This is very exciting, but the average student of biology or medicine cannot wait for a new science to be born. He must learn now, and the book attempts to give an orthodox presentation that will be acceptable by the most conventional, while yet showing how new approaches can be incorporated.

The book is meant to be used as a systematic aid to the student of mammals and man who has already some general familiarity with biology. It should serve as a companion to a course of practical study of dissection and of histology and embryology. Drawings of dissections and sections are included with the hope that they will be found useful in the laboratory. The book was first drafted some years ago when I was writing the *Life of Vertebrates* and in a sense it forms part of that work. But it became clear that a Life of Mammals of somewhat different scope might be useful not only to zoologists beginning their course but to others as well.

Most students first learn about mammals through dissecting the rabbit or the rat and the book therefore gives special attention to these species. For the medical student these mammals are an introduction to the study of man and I have tried to present their biology in a way that will be useful for this purpose. In particular the sections on the skeletal and muscular systems and on neurology are meant to provide an introduction to those subjects as they are treated by human anatomists and physiologists. Too often the zoologist who trains future medical students is unfamiliar with the methods and terminology that his pupils will meet later. One aim of this book is to bridge the gaps between the basic and the medical aspects of biology. Unfortunately the book ignores altogether many aspects of mammalian life that many would wish to study, for example ecology. It provides only a framework or scheme for the study of mammalian life, within which much else could be incorporated.

It is impossible for any work of this sort to be authoritative and original as well as comprehensive. I have tried to cover the ground but have had no hesitation in lingering longer in the fields that have interested me especially. I doubt whether anyone suffers by learning from a book in which he can discern the personality, interests, and foibles of the author. Perhaps some of the complaints about the aridity of science come from the attempt to produce giant comprehensive textbooks written by collections of depersonalized authors. One of the chief lessons for any intending scientist is that the facts that he is told may be wrong and that points of view differ and change. This is no excuse for making mistakes, as I have no doubt often done; but let a critic realize that the anxiety involved in uttering such a work is formidable. References appropriate to each chapter are given at the end of the book. Recent references have generally been preferred and this has meant omitting classical ones. A contemporary survey, even if of little originality or critical worth, is often the most useful reference from which to work backwards through the literature.

Any system of arrangement of so vast a subject-matter will seem to be in some way illogical. The plan adopted assumes that the student will first dissect the parts of the mammalian body and learn the microscopic appearance of the tissues at the same time. Having dealt with the skeletal and vascular systems he will study the individual organs, and last the nervous system, receptor organs, and endocrines. During this work he will often want to know about development, but in practice the study of embryology is distinct and is here treated separately at the end.

In due time we shall learn how to combine macroscopic and microscopic studies with those of biochemistry and biophysics. All that I have been able to do here is to issue reminders that there is a connexion and to suggest possible means of unification.

Preparation of such a treatise involves collaboration by many people. I should like to thank most warmly all those who have given help, directly or indirectly. Many have read sections dealing with their own specialities and are mentioned separately below. Others have allowed their illustrations to be copied or have provided the material for new pictures. The figures have been drawn from life or redrawn from other figures by Miss E. R. Turlington and Miss J. I. D. de Vere. Their faculties of observation, care, and skill have been disciplined continually by a consideration of the needs of the user. The production of such representations is an integral part of the work of biology, for which we cannot be too grateful. Mr. J. P. Stanier and Dr. E. G. Gray have helped with innumerable tasks of editing, assembling of figures, and of bibliography. Finally, it is a pleasure to record my thanks to the Secretary and Staff of the Clarendon Press for their skilful and willing collaboration in the preparation of the book.

J.Z.Y.

January 1957

Acknowledgements

Our thanks are due to the following who gave advice on various parts of the manuscript.

D. Barker	E. G. Merrill
A. Boyde	N. A. Mitchison
G. Gabella	Patricia M. Preston
E. G. Gray	N. R. Saunders

Contents

1 Communication, control, and the continuation of life

1. Homeostasis: the maintenance of a steady state

THE sciences of anatomy and physiology reveal a huge number of different parts in the body and everywhere intricate chemical operations continually at work. This book will try to show how life depends upon this great variety. If we are to understand organisms we must know the parts of which they are composed, for the very simple reason that life continues only by virtue of this division into many different parts.

Any physical system that differs from its surroundings will tend to merge by diffusion into those surroundings. Living organisms are not exempt from the operation of this tendency; the matter and energy that they contain tend to spread away into the environment. But it is characteristic of organisms that they resist this tendency to dissipation; indeed they take in materials from the environment against the concentration gradients. A man is a watery system and tends to lose water to the drier air around him, but in compensation he has elaborate means for seeking and taking water from an environment that contains little of it. Organisms show a great number of such devices by means of which they maintain themselves intact in spite of the tendency to diffuse into the surroundings. All these activities are so directed that they expend energy in ways that ensure that the organism maintains its independence from the environment.

How can it be that organisms seem able in this way to evade the otherwise universal tendency to ever greater disorder (to increase entropy)? They can only do it if they have access to sources of orderliness that prevent their dissipation. And of course we can recognize now that these sources are indeed provided: by heredity. The secret of the purposefulness of living actions is that they are directed by an outside source. That source is their inherited nucleic acids, containing the information collected by millions of years of trial and error. This information decides which particular chemical reaction is likely to be useful in each situation the animal encounters. The living organism can thus be considered to be continually making predictions as to how to expend energy to prevent its disintegration. It is not in equilibrium with its surroundings but maintains a steady state of interchange. Its sources of information enable it to take in fuel and then to expend energy in just those ways that ensure survival. This process of regulation is called *homeostasis* (a term first used by Cannon in 1932).

In the animals that we call 'higher', such as the mammals, this tendency to make predictions that ensure maintenance of an improbable state appears in a marked degree. These animals differ widely in composition from the medium in which they live. Thus a human being only remains as a distinct entity by virtue of the expenditure of much energy, by an elaborate control system with an extensive information store. In this way he obtains the materials necessary to prevent the dispersal of his system.

Physiologists early paid special attention to one aspect of the process of self-maintenance—namely, the composition of the blood. Claude Bernard noticed that in animals the blood constitutes a kind of 'internal environment' around the separate cells and that in mammals there are elaborate processes of regulation, or homeostasis, which tend to keep the composition of the blood constant. On this fact Bernard based his famous dictum (1878): '*la fixité du milieu intérieur c'est la condition de la vie libre*' (see Langley (1973) for reading on homeostasis). The phrase reads well and is still often quoted, but it only partly meets the need we feel today for a general principle in physiological studies. All organisms keep themselves more or less constant and distinct from their environment; in this sense all animals are free, whether they have an 'internal environment' or not. The concept of a process of homeostasis can be extremely valuable if we recognize that it implies control by the use of an information store and if we generalize this idea to cover all life. It expresses in a word the tendency to self-maintenance that is the characteristic of all living activities.

2. Living communication systems

Living therefore depends upon the transmission of information. We can describe living things by using the analogy of the process of human communication by language. In speech we recognize a transmitter, a communication channel, and a receiver. The transmitter wishes to send a message, that is, he aims to produce some effect in the receiver. To do this he *encodes* the message, by selection of suitable items from a previously chosen set. They may be spoken sounds or written words or the dots and dashes of the Morse code. They are sent along the communication channel and then *decoded* by the receiver. If the latter has been suitably instructed in the pre-established conventions of the set he will 'understand' the message, that is, he will act as the transmitter intended.

For our discussion perhaps the most important feature of information transfer is that it depends upon selection from a set. If the transmitter wishes to produce a pattern of action by the receiver he must break the pattern up into a series of discrete units whose size is comparable to that of the error that can be allowed. Each item is then transmitted as a whole and, of course,

because of inevitable random noise in the channel many errors will be made. But nevertheless the message will get through; the pattern will be reproduced. Any attempt to send an exact copy of a pattern without breaking it up into units is bound to fail because of progressive divergence from the original by fluctuations in the course of transmission. This is why we use words, and this is also why every living organism has so many parts. A mammalian body continues to exist because it consists of a vast number of units—the cells. There are some thousand different sorts of cells in the body— about 50 different sorts of skin alone (p. 24). Within each species of cell there are many further subspecies. For example, the whole of our system of resistance to infection depends upon the presence of numerous varieties of lymphocytes, each sort producing one type of antibody (p. 191).

But the division into sets is much more subtle than this. Each cell is itself very far from being a homogeneous whole. The electron microscope shows that it contains thousands of distinct particles, and these are of many sorts (Figs 1.1 and 1.2). There are mitochondria, ribosomes, endoplasmic reticulum, Golgi

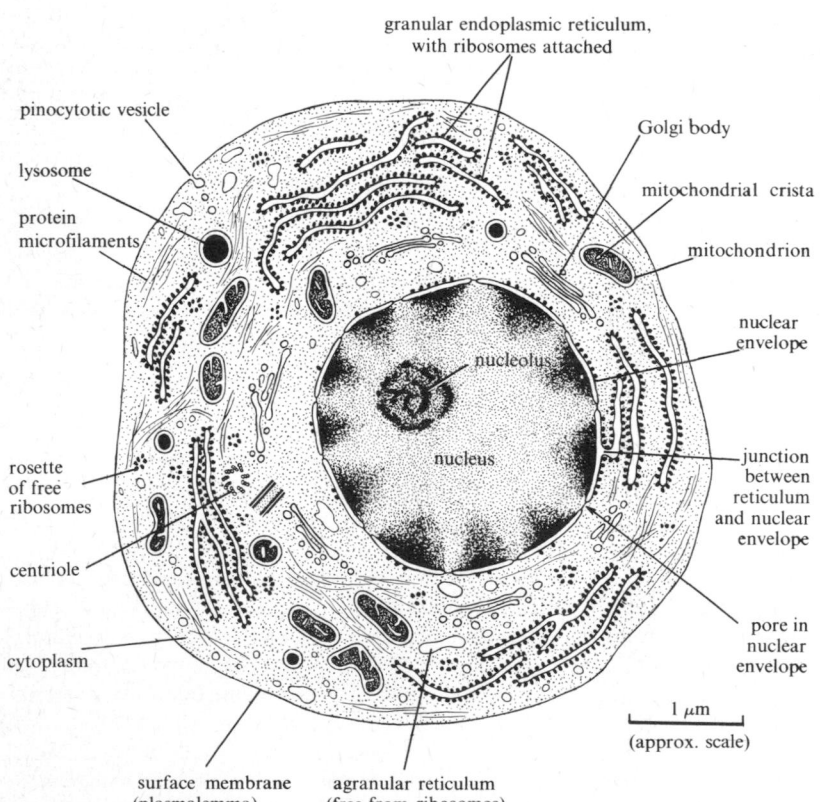

granular endoplasmic reticulum, with ribosomes attached

pinocytotic vesicle

lysosome

protein microfilaments

Golgi body

mitochondrial crista

mitochondrion

nuclear envelope

nucleolus

nucleus

rosette of free ribosomes

centriole

cytoplasm

junction between reticulum and nuclear envelope

pore in nuclear envelope

surface membrane (plasmalemma)

agranular reticulum (free from ribosomes)

1 μm (approx. scale)

FIG. 1.1. Diagram of a cell reconstructed from electron micrographs. (Drawing by Prof. E. G. Gray.)

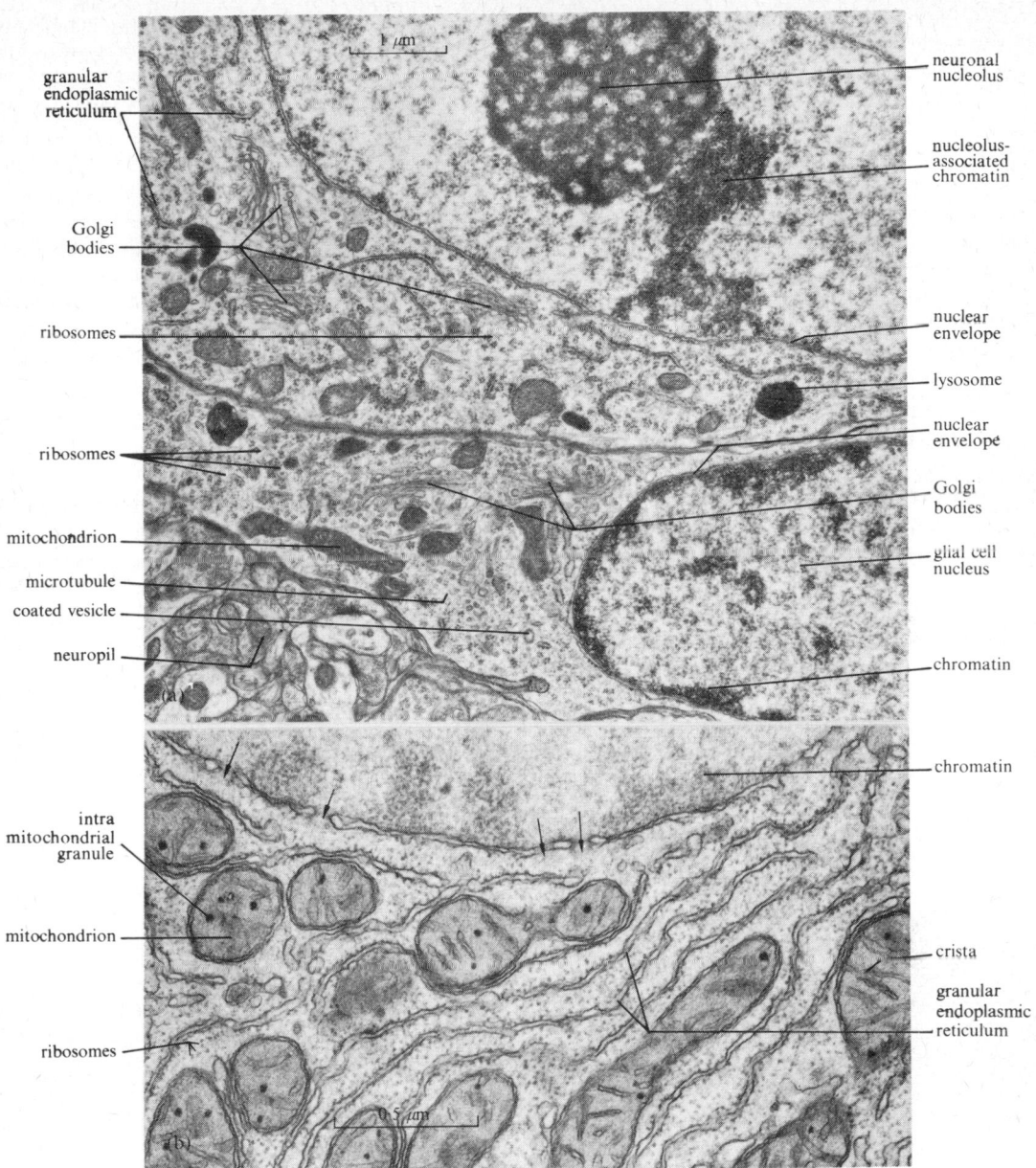

FIG. 1.2. Electron micrographs showing examples of cellular organization.

(a) Parts of a neuron and of a satellite glial cell from the lateral geniculate nucleus of a rat. The nucleus of the neuron contains a large nucleolus with the appearance characteristic of cells active in protein synthesis, including a patch of nucleolus-associated chromatin. In the cytoplasm are three elements of the Golgi apparatus and scattered, individual cisternae of granular endoplasmic reticulum, mitochondria, lyso-some-like dense bodies, clumps of free ribosomes, and microtubules. In the glial cell, clumped chromatin is con-centrated at the nuclear periphery, and a large Golgi apparatus lies in the cytoplasm. Also in the cytoplasm are abundant free ribosomes and a coated vesicle. (b) Part of the nucleus and perinuclear cytoplasm of a rat liver cell. Note the many perpendicularly sectioned nuclear pores (arrows). In the cytoplasm are mitochondria with well-defined cristae showing continuity with the inner mitochondrial membrane. The electron-dense granules in the mitochondria are probably storage sites for divalent cations. The granular endoplasmic reticulum consists of long, parallel cisternae. The ribosomes in this micrograph are clearly not aggregated into the distinctive polyribosomal arrays (rosettes, spirals, etc.) seen in (a). This is because, $3\frac{1}{2}$ hours before sacrifice, the rat was treated with the ethyl analogue of a methionine, which is a potent inhibitor of cytoplasmic protein synthesis and causes rapid (but re-versible) detachment of ribosomes from the granular endo-plasmic reticulum and disaggregation of polysomes into their component monoribosomes. (Micrographs kindly supplied by Dr. A. R. Lieberman.)

vesicles, fibrils, liposomes, microsomes, nucleoli, and, of course, the chromosomes. The application of chemical science by molecular biologists now enables us to give an exact account of many of these components of cells. We can see them as complex aggregates of molecules carrying the enzymes that control metabolism. In our analogy with language they can be considered as signals, that is to say pre-selected units, comparable to words, by means of which the cells operate to transmit a stable pattern of life.

3. The hereditary information store

Above all we can recognize that the hereditary material is itself a most subtle code. Nucleic acids are built of sequences of nucleotides, arranged in pairs. Each successive triplet of bases along the molecule organizes the attachment of one of the 20 amino acids that go to make up proteins. During evolution those sequences have been selected that make proteins that are effective in ensuring the life of that particular type of individual.

The system thus conforms precisely to the communication of information in a code along the channel provided by cell division and reproduction. The code is a set of sequences of nucleotides pre-arranged by natural selection in the past and ensuring appropriate reactions and survival if conditions remain similar. There are also, of course, provisions for change by mutation and recombination (see Young 1971, Chapter 28).

4. Information acquired during life

This method of thought can be applied wherever we find that the organism makes 'choices' between one of several possible paths, and we shall find that it is by repeatedly making such 'decisions' that the organism maintains its integrity. For example, the nucleic acids can be said to carry information that decides which of a large set of characters an individual shall develop. The receptors of the nervous system provide the information by which decision is reached among the numerous possible courses of behaviour provided by the various nerve cells (Chapters 34–6). Again, each molecule that is taken into the body may become incorporated into one of many chemical compounds, and we can try to describe the organization of biochemical events in terms of the 'information' by which decision is made among the various possible courses of metabolism.

In mammals the information that is stored during the individual's own life-time is of special importance in determining the course of action. The store is mainly within the nervous system, but the characteristics acquired in many tissues during life constitute information stores, providing means to meet future eventuali-

ties. When an animal lives in an environment that is poor in oxygen the oxygen-carrying power of its blood is increased (p. 175). If it is called upon to use one set of muscles more than others these undergo hypertrophy (p. 69). The bones may be said to carry stores of information about the stresses imposed upon them (p. 43). When an animal meets a particular set of bacteria or other infective agents, antibodies capable of neutralizing these are produced (p. 191). Such 'adaptations' by the organism can all be regarded as 'memories', stores of information that improve its representation of the environment and provide powers of action likely to be useful if the future resembles the past. Something is known about the means by which these 'adaptive' actions of the organs are produced. They can be regarded as the result of selection, by the influence of the environment, among the set of reactions that are possible within the cells. This set of reactions includes those that are likely to be appropriate for the organism because of the past operation of natural selection upon the genes of the population. The events occurring in the environment then select appropriate combinations of actions from the 'code' that is provided by heredity, and this enables the organism to produce a representation that allows it to take such actions that it remains in a steady state in its environment. There is thus a two-way communication between the hereditary system and the environment, along the channel provided by the body.

5. The double dependence of the organism

Each cell carries receptors appropriate to detect a relevant stimulus or stress at its surface. This then releases a signal within the cell. There is some evidence that this 'second messenger' in many cells is the enzyme adenyl cyclase. This activates the system, using cyclic AMP to release the cell's energy-yielding systems (p. 456). Each cell is, of course, already specialized to use energy for the performance of some particular sort of work—say synthesis of bone or of a digestive enzyme. Upon receipt of the message the synthetic system is activated, and this in turn switches on the actions of the nucleus by which new messenger RNA is synthesized from the template of the DNA. The composition of every part of the body therefore varies continually under a double system of control. The information of heredity, operating from within, provides the system with certain tendencies that have been selected by past history. On the other hand, the events occurring in the environment constantly modify the operations of this system within the tissues. We shall find many examples of this *double dependence* of tissues. The composition of every cell depends upon the interaction of influences from within and from without. No tissue maintains a constant state of development for long; if used more it hypertrophies but if left without use it undergoes atrophy (see Young 1946).

6. Repair and reproduction

The numerous particles within each cell are continually in use, for instance, the mitochondria, for respiration. Many will carry errors, and if they work wrongly can be eliminated, whilst the cell as a whole can continue. Similarly, among the many individual cells of a tissue any that are damaged or in error can be replaced. In tissues especially liable to damage provision is made for continual replacement. The lining of the intestine is renewed every day (p. 145), and the cells of the blood or skin are renewed quite frequently. There are therefore mechanisms for repair and replacement in all cells, and all of these are provided for by the hereditary code of the nucleic acids. But these repair mechanisms are themselves liable to error. Indeed, one strand of the DNA itself will sometimes be damaged during copying, and there are systems by which a faulty section is identified and removed and then replaced by copying the other strand. But this itself involves specific repair enzymes, and these in turn may have faults due to random factors.

It seems that it must be impossible to have a perfect system of repairers of repairers of repairers of repairers of . . . And indeed life does not really defeat this regress. It avoids it by the system of reproduction. No cell or individual animal or plant keeps its integrity for ever. At intervals all reproduce by division, and thus die. The new progeny are like their parents, but not exactly alike. Since they are numerous there will be selection among them, and those that are successful survive. Thus once again we see that the continuity of life depends upon selection among a set.

7. The development of regulation

The various regulators in the body all follow rather similar paths of development (Adolph 1968). Their earliest actions spring from pacemakers within the tissue developed under intrinsic hereditary influence. Thus pieces of future embryo chick hearts transplanted before they beat will start to do so with the proper frequency. Such intrinsic rhythms provide the set points to which many regulated organs will return. No one yet knows how such pacemakers arise. A different example is the rate of replacement of red cells, which have a half-life of 14 days in a new-born human and 30 days in an adult. The shift occurs in the first year, but no known treatment alters either rate.

Pacemakers provide a factor of safety insuring main-

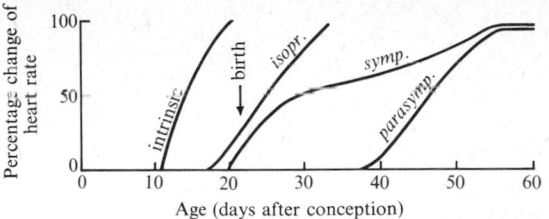

FIG. 1.3. Relative changes of heart-rate due to intrinsic regulation and to three extrinsic effects in foetal and post-natal rats. *Isopr.,* effect of isopropylnoradrenaline (hormone); *symp.,* sympathetic accelerator effect; *parasymp.,* vagal depressor effect. (From Adolph (1968). *Origins of physiological regulations.* Academic Press, New York.)

tenance of the rate between limits. In the cardiac pacemaker this depends on the presence of many nerve cells, usually discharging in synchrony. Later in development extrinsic influences gradually come to play a part in the regulation, this sequence also being genetically programmed. Thus further hormones gradually come to influence the heart (Fig. 1.3). Sensitivities to external influences also change in a serial order and tolerance limits change. Feed-back systems appear after extrinsic regulation has begun. In general, there is a shift from earlier intrinsic regularity to later plasticity as more extrinsic influences become effective and adaptations to environment are added. Higher organisms have great capacity to vary their controls according to past individual experiences.

8. Methods of control

Biological problems thus resolve themselves into studies of *control*, by selection among alternatives in the light of information that depends upon past associations of events. In this way we can speak of the difficult problems of adaptation of an organism to its environment, saying that the former has gradually built up a *representation* of the latter. This leads to the concept of *memories*, or *information stores*, not only in the nervous system (p. 359) but elsewhere also. The whole stability of the organism can be treated in the language that is used by the engineer to describe the control of machines.

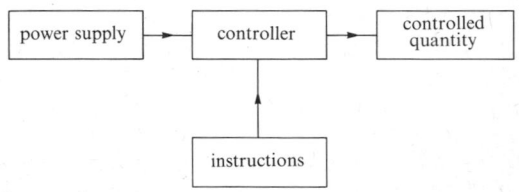

FIG. 1.4. Basic general control system. The controlled quantity is varied by the power supply under the direction of the controller, which is in turn governed by the instructions. (From Wilkins in Kalmus (ed.) (1966). *Regulation and control of living systems.* Wiley, New York.)

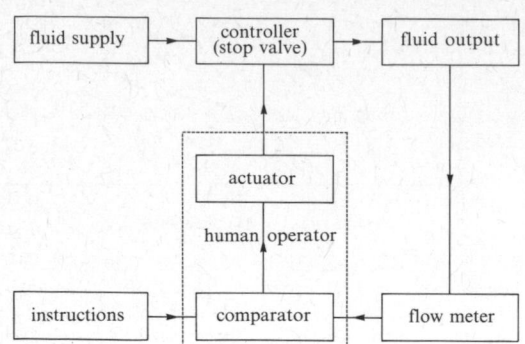

Fig. 1.5. Control system in which the controlled quantity is maintained by feed-back via the agency of a human operator. (From Wilkins, *loc. cit.*)

The making of artefacts to assist in the control of machines has led to considerable clarification of the concept of control and extension of the terminology available for describing its variables. Of course, the engineer and the biologist work in opposite directions. The engineer sees things in terms of synthesis, whereas the biologist looks at problems with an analytical eye. The biologist's problem is to find out how the system has been designed; but he also wants to learn how to control it (see Machin 1964).

The essential features of a system for regulation are shown in Fig. 1.4 (Wilkins 1966). A controller, whose behaviour is itself regulated by instructions, adjusts the flow of energy from the power to the controlled system. The energy required to give the instructions is much less than that supplied from the power; the controller thus acts as an amplifier. To maintain any given rate of flow in spite of variations, the output can be measured

by a comparator and any difference from the required output corrected by change of the setting of the controller, as by the agency of a human operator (Fig. 1.5). The use of information about the output in this way is known as *feed-back*, and that shown in Fig. 1.5 is said to be a closed-loop system, whereas that of Fig. 1.4 is called (misleadingly) an open loop.

In automatic control systems the measurement of the output (say rate of flow of water) must be converted by a *transducer* into whatever form of signal operates the control of the power, for example, a voltage. This is a good example of the process of *encoding* (p. 360). The instruction or *command signal* must also be in the same language, and the output signal and command signal are compared in a summation point or comparator and made to produce an *error signal*, which activates the controller. The output of the comparator, if necessary after amplification, is made to control the power output ('negative feed-back'). The whole device is in effect a monitoring system, which ensures that the actual performance of the machine corresponds to that called for by the instruction.

Obviously the study of such servo-systems involves investigation of the concepts of instruction, information, coding, and representation, briefly considered above. The instruction is a coded representation of a pattern of action for the machine. The receptor provides a representation of the course that the machine output actually follows. The comparator measures the difference between the two representations and provides a controlling output that reduces the mismatch.

Such systems are liable to three main types of error, which are also seen in the control systems of living organisms (Fig. 1.6). If there is a considerable delay or

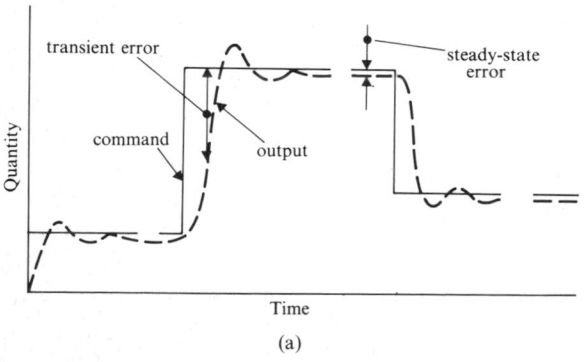

(a)

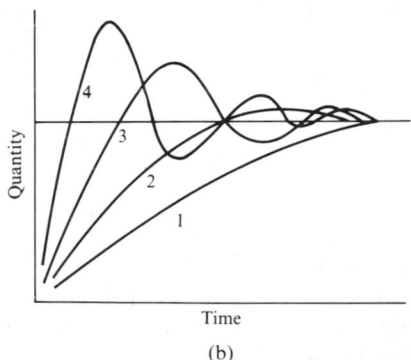

(b)

Fig. 1.6. (a) Diagram to show the errors liable to occur between a command (solid line) and the response (dashed line) during a sudden change in command. There is a lag in initial response (transient error), followed by some oscillation, and then a divergence between the maintained response and the command (steady-state error). (b) This shows that the transient errors and oscillation vary with degree of damping. Curves 1, 2, 3, and 4 show the response in systems with decreasing degrees of damping. Curve 1 is over-damped and shows a large transient error but little oscillation. Curve 4, the least damped, has a small transient error but a large oscillation. (After Brown and Campbell (1948). *Principles of servomechanisms*. Wiley, New York.)

damping in the operation of the system the output may not follow the command closely. It may produce a large *transient error* during adjustment to any change, though a small *steady-state error*. Conversely if the feed-back system is over-sensitive and produces large changes in the servo-motor control there may be over-correction and *oscillation* in response to even the smallest changes. One means of preventing oscillatory behaviour is to make the correction dependent upon rate of change of error as well as upon its magnitude. (For further details see p. 14 and Wilkins (1966).)

9. Control in living systems

The controls operated by living systems are numerous and varied, and they proceed with time scales differing between the rapid adjustments mediated by the nervous system and the very slow ones involved in evolution. Processes analogous to negative feed-back are involved in many of these controls, especially in higher organisms. Some of these adjustments are operated by statistical processes, such as those of natural selection, stability being ensured simply by eliminating those processes that do not match. Organisms are able to proceed in this way because of the great number of individuals and parts that are involved and the presence of minute variations among these. Thus very subtle adjustments can be made in the composition of a population or of a part of the body, but such adjustments are rather slow. Where adjustments are rapid, as in the nervous system, there are means for preventing errors. Thus the cerebellum provides, among other things, a mechanism for the prevention of oscillation (p. 300).

In dealing with any part of the body we have therefore to ask what are the controls that regulate its action. Some tissues adjust very quickly. The nervous system in particular is so organized that, following a small change in the environment, information is communicated to the brain and compared with the appropriate representation stored there. The brain region in question then emits an output that quickly adjusts the activity to suit the new conditions, and a feed-back system ensures that the muscles are brought smoothly into action (p. 269). Control of the operations of the respiratory system is another example of rapid adjustment, the representation used for matching being located in the medulla oblongata (details on p. 156). Instances of such rapid control occur endlessly throughout the animal body. The most elaborate forms of human behaviour depend upon comparing the input from the receptors with a detailed representation of the environment that is gradually built up in the cerebral cortex (p. 341).

The secretions of the endocrine glands provide a set of signals controlling the slower changes in the body, such as those concerned with nutrition, excretion, growth, and reproduction. Presumably chemical agents provide a more effective means for regulation of such long-lasting processes than would the brief signals of the nervous system (p. 455).

Still slower adjustments are made by changes in the composition of parts of the body. No tissue is in equilibrium with its surroundings; all are continually changing their materials. As molecules break down and are replaced by others there are small but important alterations in the fabric of the body. These 'adaptations' ensure that each tissue is suited to represent the conditions that fall upon it. Such growth changes are influenced by the information that is provided by surrounding events; they thus provide a record of this information. This slow control by growth changes is found in all tissues and probably plays a part in the setting up of the information store of the nervous system.

On a still longer time-scale, living organization is regulated by the process of evolutionary change. The representation that controls all the operations of more rapid control that we have been considering is coded into the hereditary material of the genes. This hereditary information store includes elaborate provisions ensuring that it shall itself be subject to continual change. Each new version seeks survival in the environment and control is produced by the slow process of natural selection.

This method of speaking about animals may seem at first to be unnecessarily complicated and laboured. Yet as each aspect of life is considered, the language will be found to be illuminating and stimulating. It provides the possibility of unifying the different treatments that are adopted by anatomy, physiology, genetics, and evolution. Such unification is itself a great gain, and it has the special advantage that it brings into focus important aspects of living organization that are at present neglected or obscure. In particular, it allows description of the control of the interchanges that go on continually in the tissues. Revelation of the extent of this turnover is one of the largest changes produced by recent biological research. If the substance of the body is continually changing then the old analogy of a simple machine having 'structure' and 'function' is no longer adequate. Comparison with machines that control themselves is much more helpful, especially if we can find ways of describing how the control system itself changes with time. The methods used by information theory may be able to do this because they tell us how to deal with situations that are controlled by the accumulated effects of associated events, some of them occurring long ago. By comparing the events in organisms with the selection of words in a language or other code we obtain a means of speaking about 'organization' and 'adaptation', which will at least serve until some better method is found to provide a more exact science of the study of living organization.

10. The mammalian homeostatic system

The approach suggested for the study of living organization can be summarized by describing the characteristics of a homeostatic machine that would be able to maintain a steady state under conditions similar to those encountered by mammals. This account amounts to a description of mammalian organization given in the terms that might be used for describing a man-made machine. The various headings mentioned therefore correspond to the separate parts of the body and its activities as they are usually considered. Such a system of description is not fundamentally novel; its advantage over that usually adopted is that it uses the recent developments in the language and methods of engineers to give a more complete and consistent account of living activities.

The machines are heterogeneous (polyphasic) chemical systems operating by virtue of the properties of carbon compounds, arranged into a hierarchy of subsystems. They have the characteristic that their actions are so controlled that they maintain a steady state of interchange with the surroundings in spite of variations in the environment. The regulation depends upon the fact that the organism contains representations of the features of its environment that are relevant for its existence. These representations are matched against environmental conditions, and when deviations occur processes are set up that so alter the organism (or the environment) that the mismatch is corrected.

The persistence of a stable set of the machines (the 'race') is ensured by elimination (death) of the old and replacement by new ones. This is achieved by the presence of a coded version of the representation, the genes, consisting of nucleo-proteins. These have been selected over a long period to include units that represent various probable environmental conditions. Out of this code at intervals new combinations are chosen and then transmuted into the control systems of a varied set of adults.

The fundamental property of the system is that the nucleic acids replicate themselves when in suitable surroundings. Then by natural selection it is ensured that the code includes elements which when suitably combined allow the organisms to react appropriately if future conditions resemble those of the past. The gene

representations thus enable the organisms to receive information from the environment. In highly developed forms of the machines, such as those of mammals, the long history of past associations of events provides most elaborate control systems that are able to maintain stability in spite of the fact that they differ widely from the environment and that the latter fluctuates considerably.

The detailed controls of each organism are very numerous. They enable it to take in the necessary materials, including sources of free energy, and to maintain itself in surroundings in which they are available, and to eliminate waste. These are the metabolic activities, commonly considered as the subject matter of physiology. The whole organism is surrounded by a covering that separates it from the surroundings but allows regulated exchanges at certain points (Chapters 2 and 3). It is supported by a framework (Chapters 6–10) which permits movement of the parts in relation to each other and of the whole in relation to the surroundings (Chapter 7).

For taking in materials the machine has a number of effector systems such as the mouth, digestive apparatus, and lungs (Chapters 13–15). These are arranged to operate with a periodicity that is regulated by the rate of demand. The system is physically large and transportation of materials between its parts is ensured by a suitable circulatory mechanism (Chapters 17 and 21).

Maintenance of the whole in situations where materials are available is ensured by a special system of effectors that moves the whole structure. Detection of suitable conditions is ensured by a variety of receptors (Chapters 38–44). There is an elaborate communication system by which the information collected by the receptors is made to operate the effectors appropriately (Chapters 25–36). This communication system includes a variety of predictors. Some of these are 'reflex systems' built so that they operate alike in all machines of each type on receipt of certain signals (Chapters 27 and 28). Other predictors operate by storing information as to the relation between certain sets of signals from the receptors and the presence of suitable or unsuitable external conditions. These are the parts that learn and continually change the representation that they contain of surrounding conditions (Chapters 34–6). Stability is partly ensured by feed-backs operated by receptors within the machine, which signal deficiency or excess of some component or activity. A system of chemical signalling by the liberation of substances into the circulatory system is used for the slower regulation of some of the fundamental chemical processes on which the whole system depends (Chapters 45–9). Mechanisms are present that are set into operation to counteract many of the deleterious influences that are likely to destroy the system. Some of these operate through the nervous system to produce attack or retreat. Others provide chemical defence against the intrusion of living parasites and toxic substances. After each defensive action the system is usually so modified as to be better able to meet a similar danger in the future (Chapters 19 and 20).

The activities of the genes also provide that the whole system shall ultimately cease to operate but shall set up a new set of genes (reproduction). This new set then proceeds to provide a system that is similar to but not identical with the parent (embryogenesis, Chapters 51–3). Thus each type of organization exists in a sufficiently large variety of forms to ensure homeostasis in spite of slow changes in the surroundings.

In this manner we can give an account of all the various operations that an organism performs and of the ways in which they are regulated to ensure homeo-stasis. We use as the central concept the description of the sets of nucleotides that make up the DNA able to produce proteins that ensure self-maintenance. This condition has been derived by selection of groups of nucleotides that have proved effective in the past. The account cannot be a simple one, for the living machine performs a range of adjustments vastly greater than that of any man-made machine. We must not expect the study of biology to be simpler than that of engineering. The amateur who wishes to enjoy a superficial study of the actions of living things will not wish to be troubled with so complicated an analysis. But the professional biologist or medical man will find in it the basis of a method by which he can describe and control living processes. The system is far from perfect but we may be amazed that after investigation extending in its present form over a few years it is possible to see the outlines of a unified terminology for description of man, animals, and plants

2 The skin and control of temperature

1. The boundary of the body

THE skin marks the boundary between the organism and its environment and makes a large contribution to the characteristic appearance of the individual. In the fully terrestrial mode of life adopted by most mammals, the properties of the surface layers are obviously of first importance and the skin is an elaborate structure, showing special features not found in lower vertebrates. These features are connected with temperature control and with the formation of an impermeable yet sensitive surface which is able to resist tensile forces and to prevent the access of bacteria.

The skin and its derivatives therefore play a very important part in the life of a mammal (Montagna 1962). Study of the differences between the skins of various mammals shows how closely the activities of the body are related to the conditions of the environment. In this chapter, besides describing the histological structure of the skin, we shall see how some of its many derivatives play their part in the life of the species and how this part may vary in different sorts of mammals.

Although the skin is a resistant, relatively inert material, it does not persist unchanged throughout life but is continually renewed by the activity of the cells below. Even the outer boundary of the organism is, therefore, not in static equilibrium but in a steady state, continually changing its material.

Control of the metabolic interchanges in the skin is largely a matter of the operation of hereditary morphogenetic factors, serving to maintain different sorts of skin on the various parts of the body (p. 24). Yet there are also changes in the skin in relation to changes in the environment. These changes may be fast, such as those concerned with temperature control, or slow, for example, the hardening of the palms of the hands.

Being the boundary of the body the skin 'represents' the environment in a particularly clear sense. It is enabled to conform to the outside conditions by its texture, colour, scent, temperature, and other features. This conformity is maintained by control on all time-scales ranging from the few seconds required for temperature regulation to daily or seasonal changes (p. 15) and the many generations for selection of a new set of genes controlling the coat colour.

2. The epidermis

The skin consists of two chief layers (Fig. 2.1); an outer, nonvascular, stratified squamous epithelium, the *epidermis*; and below this a layer of vascular connective tissue, the *corium* or *dermis*, projecting outwards by papillae, which are vascular and sensitive, into the epidermis. The sweat glands, sebaceous glands, and hair follicles are derivatives of the epidermis that project inwards into the dermis.

The epidermis is a stratified squamous epithelium. It consists of many layers of cells the inner of which are active and protoplasmic and are continually producing cells that pass outwards and gradually undergo a process of cornification, by which they are converted to masses of the protein keratin. This substance is the distinctive component of the epidermis and its derivatives, hairs, hooves, nails, and horns. Keratin contains large amounts of the sulphur-containing amino acids cystine and methionine, and its most interesting property is that it has a very low solubility in water and therefore makes an ideal protective outer covering for the body. The arrangement of the keratin molecules determines the physical properties of the tissue and the various epidermal derivatives show different molecular arrangements with characteristic X-ray diffraction patterns.

The process of keratin formation is continuous from within outwards, but the conditions at certain levels give characteristic appearances to the cells and thus several layers can be named. The innermost part of the epidermis, the *stratum germinativum* or *stratum Malpighii*, contains the active cells, among which mitotic figures appear. The process is so rapid that the living cells of the plantar epidermis of the rat are renewed every 14 days. Outside these cells begins the horny layer, the innermost cells of which (*stratum granulosum*) contain granules of a substance known as

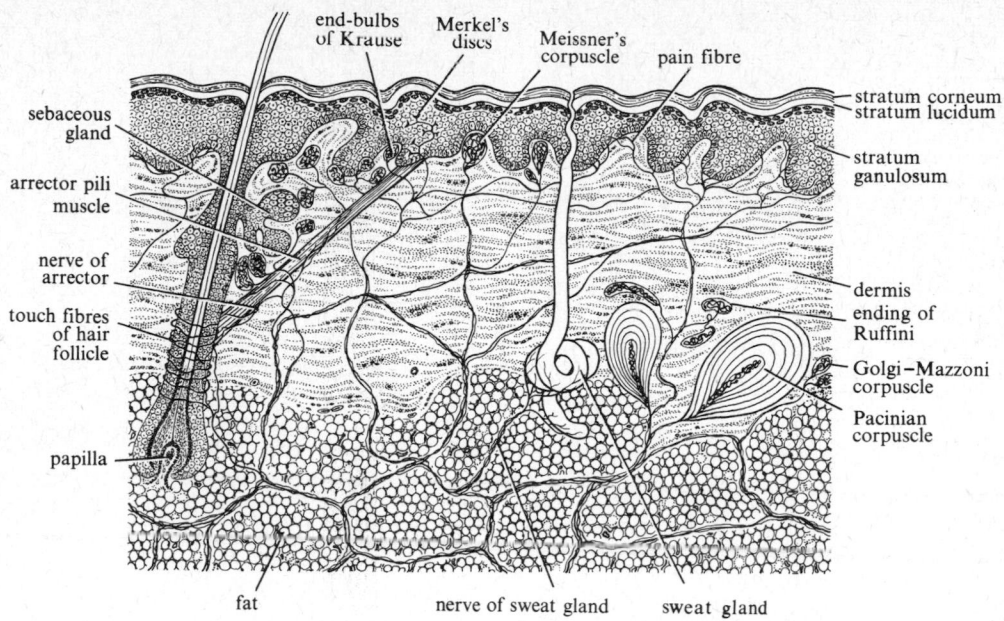

Fig. 2.1. Composite diagrammatic section of the human skin. (After Woollard, Weddell, and Harpman (1940). *J. Anat.* **74**.)

keratohyalin. Keratohyalin granules are complex, and contain calcium and phospholipid but no sulphur amino acids or nucleic acids. Enzymes may be absorbed on the granules. Outside this level the cells come to contain the clear substance eleidin in the *stratum lucidum* and then finally become keratinized and non-nucleated in the many layers of the *stratum corneum.*

The cells of the epidermis are joined to each other and to the underlying dermis by desmosomes (Fig. 2.2). At these sites long filaments within the cells come into contact with a dense plaque lying beneath the surface membrane. The region between the cells is occupied by an aggregation of polysaccharide material. The significance of these various structures is not understood. They presumably serve partly to give the appropriate mechanical properties but may also influence permeability between the cells (Kelly 1966).

The keratinizing process, of course, is not peculiar to mammals, but it is highly developed in them and is specialized for appropriate purposes in various parts of the body. It is suggested that the hairs represent solid outgrowths of the epidermis first evolved in the scale hinge regions between the scales of mammal-like reptiles (Spearman 1966) (Fig. 2.3). The hair is made up of filaments of orientated keratin molecules, about 8 nm across, each containing eleven microfibrils of length 2 nm. It may be that nine of these microfibrils form a ring with two at the centre, recalling the structure of cilia.

The thickness of the epidermis varies, being greatest on regions such as the footpads that suffer special friction. Mechanical influences are known to stimulate mitosis in the Malpighian layer, but the skin of the soles of the feet of man is thick before birth, showing that hereditary influences must be at work in controlling the regional characteristics of the skin. Pieces of skin transplanted to distant parts of the body retain their characteristic properties. Skin of the abdominal wall transplanted to the face of a man does not become more hairy, and skin of the sole of the foot grows thick wherever it is placed. We may say that the genus epidermis contains a number of distinct self-reproducing species, each with characteristics that are determined during development. The production of keratin is also made use of in special parts of the body to form claws (or nails), and in some mammals, horns, spines, or scales.

Dark colour in the skin is due to melanin, produced by a system of branched melanophores (pigmentary dendritic cells), which lie among the cells of the stratum germinativum (p. 23).

Among the many important specializations of the epidermis are the 'touch pads' found on the hands and feet of most mammals, usually seen on the palm and on the tips of the digits. The skin here carries papillary ridges, giving better grip and special sensitivity (p. 368). Rows of sweat glands open upon the crests. The genetically determined pattern of the ridge is called a dermatoglyph or fingerprint.

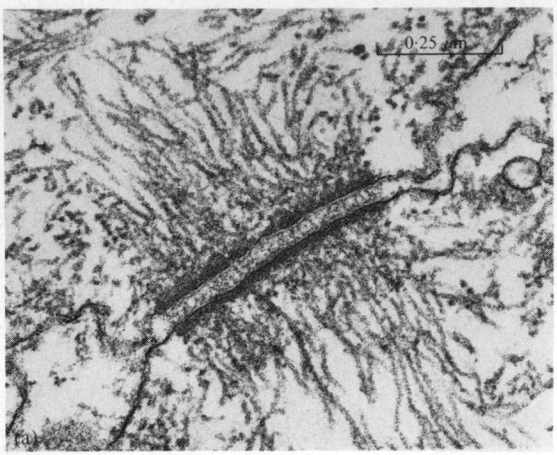
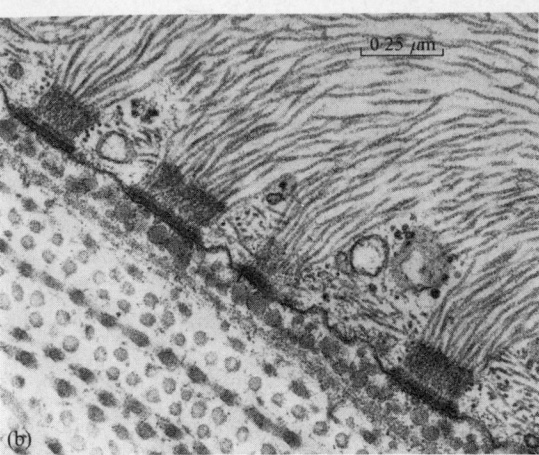

FIG. 2.2. Electron micrographs of the skin of a newt. Fixed in OsO₄ with addition of the dye ruthenium red, which attaches to mucopolysaccharides and makes them electron- dense. (a) Desmosome between two epidermal cells. (b) Hemidesmosomes attaching the epidermal cells to the dermis. (From Kelly 1966.)

3. Hairs

Each hair consists of a rod of elongated keratinized cells forming a thread that is usually cylindrical but sometimes flattened, for example, in the negro races of man. There is usually a central region of each hair, the medulla, consisting of cells that contain air. In the outer cortex there are often granules of pigment. Each hair develops as a thickening of the stratum germinativum of the epidermis, pushing inwards to the corium and becoming differentiated into a central shaft and an outer sheath. The mesenchyme at the base of the ingrowth forms a vascular papilla, and the epidermal cells above this continue to produce keratin, thus providing the constantly growing base of the hair, whose activities soon push the tip out of the follicle and above the surface.

The hairs were probably developed originally to retain heat, but they are associated with sebaceous glands, which keep them and the skin surface oily and hence waterproofed. They are also provided with nerve fibres, which wrap around the base of the hair follicle (Fig. 2.1), and when they are moved provide nerve impulses that are recorded as the sensation of touch (p. 368). In nearly all mammals certain hairs on the snout, the *vibrissae*, have become elongated and stiffened for sensory purposes, and these make the

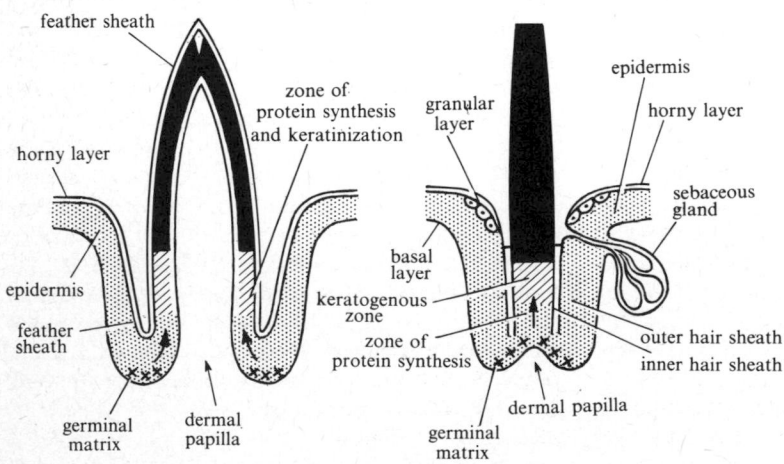

FIG. 2.3. Diagrammatic drawings from sections through (a) a growing feather follicle, and (b) a growing hair follicle. (From Spearman (1966). *Biol. Rev.* **41**.)

animals sensitive to very light touch or even to small changes of air pressure. The rabbit possesses vibrissae above and below the eyes and on the nose.

4. Sebaceous glands and hair muscles

Each *sebaceous gland* consists of a few short sacks, the alveoli, opening into the hair follicle. The secretory cells are derived from the epidermis and at the centre of each alveolus they become converted into the substance *sebum*, formed mainly of esters of cholesterol, which keeps the hair greasy. Attached to each hair is a small bundle of unstriped muscle fibres, the *arrector pili* muscle, whose contraction raises the hair, thus allowing for air movement and greater heat loss. The movements of the hairs probably also serve to press the sebum out of the follicles. The contraction is controlled by sympathetic nerve fibres:

In the production of sebum numerous smooth-surfaced membranes of the Golgi zone are involved. As a droplet grows it is surrounded by the remains of these membranes. The cell gradually becomes filled with granules and finally disintegrates, releasing them (holocrine secretion). New sebaceous cells are continually differentiating, being at first highly basophilic. The glands are not innervated and are hormonally controlled, mainly by the direct effect of androgens. Oestrogens reduce the glands. The pimples of acne are the result of increased production of testosterone, causing excessive development of the glands.

Sebum serves to protect the hair against wetting and gives it greater sleekness and hence better heat insulation. But sebum perhaps plays little part in protecting the non-hairy regions of the skin of man—especially in the child, where there is very little. The sebum of man, therefore, has been said (probably wrongly) to be vestigial and dispensible. Other functions that have been assigned to it are antibacterial and antiseptic effects, as a precursor of vitamin D, prevention of water loss, and a barrier function generally. These properties of the skin certainly also involve other factors, mainly the stratum corneum.

5. Temperature regulation

It is difficult to overestimate the importance of the advance in living organization that is made possible by the maintenance of a high and constant temperature (homothermy) (Bligh 1966). Sir Joseph Barcroft pointed out that many refinements of organization can only operate under constant conditions. For instance, if there is a constant temperature, elaborate patterns of activity can be set up in the cerebral cortex, allowing for persistent and complicated memories (p. 341). Similarly, in various parts of the body, there are intricate sets of biochemical reactions that would be disturbed by large temperature fluctuations. At the same time achievement of a high temperature allows a greatly increased level of activity. The birds and mammals have been experimenting independently with high temperatures for probably more than 100 million years, but it may be that one or both groups will eventually make still more spectacular innovations of organization on this basis, including perhaps the use of still higher temperatures. However, the rate of denaturation of protein increases with temperature much faster than does enzymatic activity, and the limit may be nearly reached.

It is not known how the temperature-regulating mechanism first arose. The egg-laying mammals, platypus and spiny anteater, possess hair, and they probably diverged from the other mammals not later than the early Jurassic Period, nearly 150 million years ago. Therefore it seems that the mammalian line began to be warm-blooded earlier than this date, as also did the line that was leading to the birds. This may have been a response to either cold or warm conditions; reptiles are severely limited in distribution by temperature. It is also possible that the condition did not follow any special climatic change but that the early avian and mammalian stocks were pioneers, driven by the competition of their many reptilian cousins to seek life in colder or hotter land regions, which were not yet inhabited by tetrapods.

Cold will make a reptile dormant unless the animal can be active enough to keep itself warm by heat produced as a by-product of muscular activity (Whittow 1970). This will be made more easy if the animal is large (incidentally this may have been a reason for the great size of many reptiles) and, of course, especially if a heat-insulating mechanism is developed. Many reptiles achieve a temperature above their surroundings by appropriate behaviour, for example, basking in the sun (Young 1962). It is not difficult to understand how a temperature above that of the surroundings could be achieved by sufficiently active reptilian animals. Even in the present day the heat of muscular work remains the chief source of heat in mammals. In the early stages of the evolution of high temperature, alteration of *heat production* was probably the main means of temperature regulation, as it still is today in monotremes and bats. In all mammals a fall of external temperature calls forth extra muscular activity by shivering. The higher mammals also possess a mechanism for the control of *heat loss*, and they maintain a constant temperature largely by this downward regulation.

Control of the temperature-regulating mechanism is centred on the hypothalamic region of the forebrain

(p. 307). especially in the *tuber cinereum*, which is large in birds and mammals. After removal of the tuber an animal no longer regulates properly, and its temperature fluctuates with that of its surroundings.

If cooled blood reaches the hypothalamus, the shivering mechanism is set in action; conversely perfusion of the carotid arteries with overwarm blood causes sweating or panting. Heat-producing systems other than the muscles may be called upon. The liver and other organs give out heat in the course of their work, and it is possible that during periods of sleep they may be stimulated to increased activity and greater heat output. Thyroid secretion increases the basal metabolism and thus the amount of secretion produced by the thyroid affects the temperature; other endocrine organs are probably also involved.

6. The temperature control system

Regulation of temperature at a fixed level is achieved by the presence of a system with a 'set-point', from which deviation of temperature initiates heating or cooling. The mechanism shows many of the features that are used in man-made thermostatic systems. The set-point or reference temperature is determined in the tuber cinereum. Experiments by electrical stimulation and excision show that the posterior part of the hypothalamus controls heat production, initiating shivering, and the anterior part controls heat loss, by sweating. The tuber contains receptors sensitive to changes of a few tenths of a degree Centigrade, perhaps much less, but the details of the neural mechanism that sets the temperature are not known. Injection of 5-hydroxytryptamine into the cerebral ventricle causes a rise of temperature, while the adrenalines cause a fall. It may be that the setting of the control is determined by the relative concentration of these amines. Fevers depend upon a raising of the set-point by toxins or other pyrogenic substances in the blood. The biological 'value' of the fever is less clear.

The maintenance of a nearly constant temperature includes not only the central reference system but also a set of peripheral sensors to detect superficial variations and initiate anticipatory control, before there is any change in deep temperature. This is probably part of the function of the receptors for heat and cold in the skin, muscle, and elsewhere. Such peripheral senses are also used in artificial thermostats, but acting alone they would produce wide fluctuations.

The methods used singly or together to regulate temperature or any other measurable quantity to within a prescribal deviation from the set-point may be considered in four classes (Hardy 1961):

1. On–off regulators. Heating or cooling devices may be switched on or off by signals from the peripheral receptors. Thus a dog exposed to cold, shivers in bursts, and, conversely, pants spasmodically when heated. The length of these bursts of activity depends upon the temperature. Such a system inevitably shows oscillations and these are proportional to the thermal lag in the system.

2. Proportional control. In this there is a continuous linear relation between the difference from the set-point and the magnitude of effect. This produces a very stable control, and there is much evidence for its action. The body temperature of man rises in proportion to the amount of exercise, whatever the external temperature. Thus the cooling system must be driven by the difference between the body temperature produced by exercise and the resting body temperature. This method of control does not, however, apply to exposure to cold since the temperature rises even with mild cold and there is no load error.

3. A third system of regulation used artificially is to have a fixed relation between the deviation from a set-point and the rate of application of restoring action. This is known as integral control but does not seem to be used by the body. Thus a fixed deviation in internal body temperature, say in fever or exercise, does not drive the sweating system to increasing levels because it has persisted for hours, indeed conversely the rate tends to be constant. (But, of course, it may be that the 'set-point' itself may be influenced in conditions of fever.)

4. In a fourth type of control the effect on action is proportional to the rate of change of temperature. This method has the advantage that it quickly produces a rapid correcting action, as when a human operator on a tracking task overcorrects in order to reach position quickly. But there is an obvious liability to overshoot and instability. This may be avoided by combination of this form of control with proportional control.

Rate control of temperature is seen when, after administration of a pyrogen (heat-producing drug), there is an immediate large shift in metabolic rate and heat loss. Anticipatory panting by a dog before exercise or shivering in man before a plunge into cold water may be similar.

By means of such comparisons with artificial thermostatic systems it has been possible to obtain a reasonably satisfactory insight into the mechanism of temperature regulation. Equations can be written to describe responses to heat and cold using various assumptions, and the computed solutions used to foresee actual

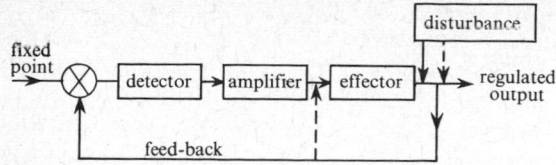

FIG. 2.4. Principle of temperature regulation by a two-stage feed-back. (From Hardy 1961.)

performance in exercise, illness, various types of heat insulation, ventilation, or other conditions.

The control system depends basically upon a closed-loop regulator and feed-back circuit (Fig. 2.4). To estimate the 'gain' of the circuit the feed-back can be 'opened' (p. 6). This is done by the technique known as a thermal clamp, in which several tubes are placed close to the hypothalamus and water circulated through them and kept at a fixed temperature in spite of changes in the blood temperatures. A rise of hypothalamic temperature of 0·2–0·5 °C was found to cause a fall of up to 10 times as much in body temperature. This factor gives us the amplification produced by the feed-back control.

7. Heat conservation

Hair tends to trap a layer of air, which insulates the body, reducing heat exchange with the environment. The air-retaining capacity of the hair is increased in some animals by a roughening of the surfaces of the hairs (for example, in the wool of sheep and the fur of some rabbits), which causes them to stick together to make a *pelt*. The hairs are not of the same kind all over the surface of the body; often long, stiff hairs are found mingled with the softer fur. In many mammals the hair changes according to the time of year so as to provide a thicker covering in winter than in summer.

The hairs do not grow straight out from the skin but are placed obliquely, so that they all 'set' in a particular direction, giving the coat its characteristic reaction to brushing. It has been suggested that the primary direction of the hairs provides a backward slope, which would prevent them from catching in grasses, etc. The reasons for the deviations from this direction, which occur in many mammals, are not understood.

Not all mammals use an air layer for the retention of heat. In whales and sea cows the body is covered with a thick layer of fat, developed in the dermis; the hairs are reduced to a few vibrissae. In man, hairs are present over most parts of the body, but their function is mainly tactile and heat is conserved by a layer of fat in the dermis. Very large mammals, such as the elephant or rhinoceros, have little need to minimize heat loss, since their surface area is small relative to the large heat-producing volume; the hairs are accordingly reduced. Related species that existed during glacial periods had well-developed hair (mammoths and the woolly rhinoceros).

8. Heat loss

Various agencies are responsible for control of heat loss. The convection from the surface of the body can be controlled by varying the position of the hairs. The skin contains devices for losing as well as for retaining heat. The *sweat glands* are derived from the epidermis, but they lie deeply placed in the dermis. They are tubular glands, much coiled at their inner ends and surrounded by *myo-epithelial cells* that contain contractile fibres and serve to expel the secretion. The sweat is a watery solution, whose evaporation serves to cool the body. The amount of liquid produced is controlled by the action of nerve fibres of the sympathetic system (p. 285). Sweat glands are not well developed in all

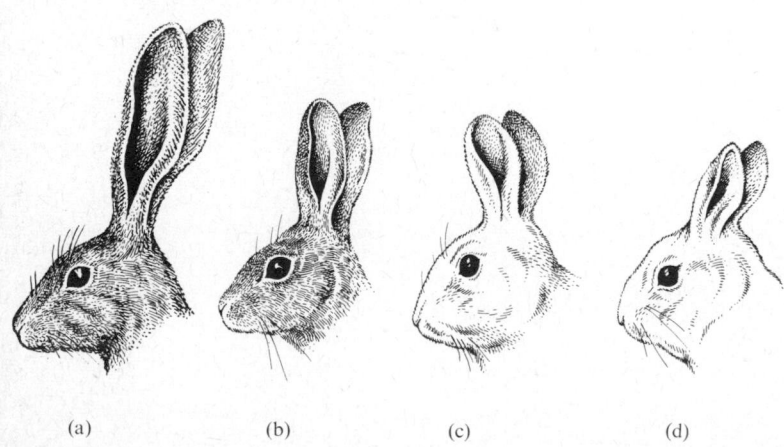

(a)   (b)   (c)   (d)

FIG. 2.5. The ear length of hares in relation to latitude.
(a) Antelope jack rabbit (*Lepus alleni*) from Arizona, (b) Black-tailed jack rabbit (*L. californicus*) from Oregon. (c) Snowshoe hare (*L. americanus*) from British Colombia. (d) Arctic hare (*L. articus*) from tundra region.

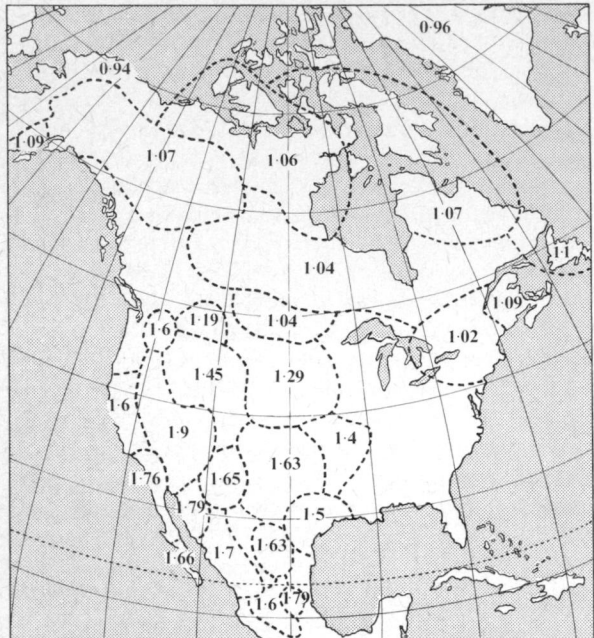

FIG. 2.6. Ear length of North American hares in relation to skull length. (Data of Nelson, after Hesse (1928). *Z. wiss. Zoo.* **132.**)

mammals, for instance, in cats they are found only on the pads of the feet. In man they occur all over the body, most abundantly in the axillae and groin. They are altogether absent from the skin of whales and sea cows and from the burrowing golden mole of South Africa, *Chrysochloris*.

The amount of heat lost from the surface is also controlled by regulating the flow of blood. The action of the sweat gland itself increases the blood supply by stimulating the release of the vasodilator polypeptide *bradykinin*. The sweat contains an enzyme that promotes the release of this substance from the protein in the surrounding tissue spaces. Thus the whole capillary bed is opened, the skin becomes flushed and loses much heat, and the sweat glands receive more blood. Conversely, in cold conditions a set of arterio-venous anastomoses is opened up, so that the blood short-circuits the capillaries and the skin becomes blue or white. These processes are influenced by the hypothalamus. The surface of the lungs also provides a means of heat loss. Animals such as the dog that have few sweat glands resort to panting.

Large animals use special means for losing heat, for instance, the African elephant increases its total surface area by nearly one-sixth when it raises its ears. The flapping movements of the ears are more frequent in large elephants than in small. The flow of blood to the ears is increased in hot weather. The ears are used for heat loss in smaller animals also; in North America the hares (or jack rabbits in the vernacular) have longer ears in the warmer southern regions (Figs 2.5 and 2.6).

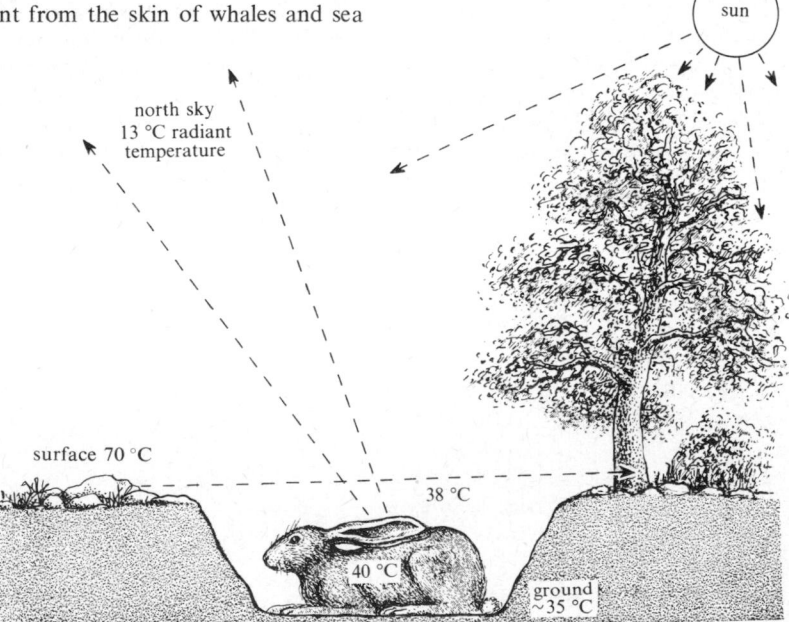

FIG. 2.7. A hypothetical thermal situation for a jack rabbit. When sitting in a shaded depression, radiation to the sky probably constitutes an important means of heat dissipation without water loss. (After Schmidt-Nielsen 1964.)

The jack rabbits of the south-western United States live in areas where there is no free water, and cannot lose heat by evaporation. They probably use the ears as radiators (Fig. 2.7). During the day they always sit in the shade and, if disturbed, run to another shelter. They quickly become prostrated by heat. Probably other medium-sized desert animals use their ears to exploit a microclimate in this way. Larger animals (for example, camels) exposed to standard desert meteorological conditions have small ears (Schmidt-Nielsen 1964).

9. Temperature regulation in various mammals

With such devices for control of heat production and heat loss most mammals maintain a constant temperature of about 39 °C. There are none with a much greater temperature, but several achieve only a lower, inconstant, temperature. In the earlier types of mammals the temperature is lower (Martin 1902). With an air temperature of 15 °C, monotremes average 30 °C, marsupials and hedgehog 35 °C, man 37 °C, and cat and rabbit 39 °C. Moreover, the lower forms are unable to regulate fully, either at high or low temperatures. The temperature of the spiny anteater (*Tachyglossus*) falls to 25 °C at 5 °C external; both it and the platypus rise in temperature with the environment above 30 °C and die of heat apoplexy at about 37 °C. These imperfections of regulation greatly restrict the living conditions, especially of the spiny anteater, which hibernates in winter and burrows in hot weather. The platypus has a few sweat glands, the spiny anteater none; neither of them increases the heat loss by panting at high temperatures or vasodilatation in the skin. Temperature is therefore regulated in these animals mainly by variations of heat production, and they have been observed to become warmer during periods of activity. Nevertheless, the difference from the reptilian condition is striking (Fig. 2.8).

In the armadilloes, and probably in other edentates, the temperature remains between 32 °C and 35 °C for as long as the surroundings are between 16 °C and 28 °C. Above or below these limits some observers have reported variations with the external temperature others good regulation to cold, though not to heat. Armadilloes also show marked diurnal temperature rhythms (2·5 °C). They regulate partly by appropriate behaviour (rolling up). It may be that they are using gradually improving temperature regulation to make life possible in a widening range of conditions. They have spread rapidly northwards in the southern United States in recent decades.

The insectivorous bats are peculiar in that they have no mechanisms for control of the amount of heat produced. Every time a bat suspends itself and goes to sleep its temperature falls and within an hour reaches that of the surroundings. On waking, the bat is at first in a strangely 'reptilian' condition; its fur feels cold like that of a dead mouse. It can crawl, squeak, and bite, but not fly. If left on its 'perch' it will perform a series of physical jerks, shivering and raising itself repeatedly by flexing the legs; since the wings are folded this produces a rise of temperature. Rapid and continuous movements of the head and ears begin at 25 °C, and at about 30 °C flight may be attempted. Its temperature is high immediately after flight, but then drops rapidly, and thereafter varies with activity. It is probable that this condition is peculiar to the bats; the large area of the wings allows such a great heat loss that strict temperature regulation is impossible. Other small animals minimize their surface area and increase their fur thickness. The bat must be small and yet have a large wing area.

There are minor fluctuations in temperature even in the higher mammals, and usually there is a daily rhythm. In man, the temperature is highest in the afternoon and falls to a minimum in the small hours of the night, the extreme variations covering about 1·2 °C. The power of temperature regulation is imperfect in new-born mammals, especially those such as the mouse that are very small. A bath at 27 °C will lower the temperature of a new-born human baby by 2–3 °C in 10 minutes, but with such cooling the heat production and carbon dioxide output rise. Downward temperature

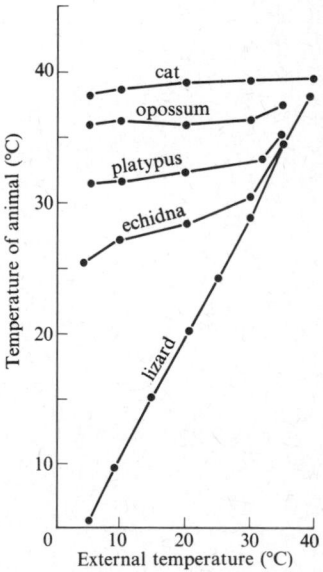

FIG. 2.8. Changes of body temperature with external change in lizards and various mammals. (After Martin 1902.)

regulation by sweating and panting only becomes apparent at a later age.

In the *hibernating mammals*, the temperature regulation is abandoned during the winter (Whittow 1971, 1973; Mrosovsky 1971); for many weeks or months the body remains at a temperature little above that of the surroundings. During this period the animal is inactive and takes no food; the small amount of energy necessary for breathing and circulation is provided from reserves of fat. The animals that behave in this way are mostly small and therefore present a relatively large surface for cooling. They would presumably not be able to produce enough heat to maintain a high temperature in winter. The hibernation is therefore a means for extending their distribution and is indeed a device adopted also by amphibia and reptiles. Hibernation is found among small mammals living either in northerly regions or in high mountains. Familiar mammals that hibernate are the dormouse (*Muscardinus*), marmot (*Cynomys*), hedgehog (*Erinaceus*), and the insectivorous bats.

3   Epidermis and dermis. Appendages and appearances

1. Claws, nails, hooves, and horns

BESIDES producing hair, in various parts of the body the Malpighian layer also gives rise to special structures made of keratin. The tips of the digits of mammals are nearly always covered with hardened plates of keratin. In their simplest form these are *claws*, similar to those of reptiles and birds; they may partly embrace the tip of the digit, as in the rabbit or cat. There are transitions between this condition and the *hoof* of ungulates, which completely covers the digit, and the *nail* of primates, which is a flat plate found on one surface only. These are all products of the Malpighian layer, but it is not known what difference in the physical process of keratin deposition produces them.

The nail of man (Fig. 3.1) is usually considered to be a modified stratum lucidum, and the underlying corium (*nail bed*) is very vascular, giving the pink appearance. At the base the epidermis is invaginated to form the *nail root*, so that the lucidum is here covered by a layer of stratum corneum which projects for a short distance as a whitish fold (the *eponychium*), more extensive in the foetus. Under the nail root the corium is less vascular, producing the whitish *lunula*. The nail grows mainly by proliferation of the cells of the nail root, but its innermost portion is formed directly from the nail bed and has a structure different from the rest. The upper (outer) part of the nail is the hardest, and this consistency is due to a high calcium content and not to

disulphide bonds, which are relatively few. In addition to this *nail plate* of horny epidermal cells, other mammals often have also a *sole plate*, a pad of somewhat less modified epidermis on the opposite side of the terminal phalanx (Fig. 3.1); this is especially developed in the hoof of ungulate animals, where the weight is carried on the tip of the digit and the sensory functions of the leg are reduced.

The *horns* found in various mammals are formed of keratin produced by special areas of skin overlying prolongations of the bones of the skull; the *antlers* of deer, however, are purely bony. Horns are either used for defence (and hence are found in both sexes) or as part of the sexual and social organization of species, usually where the males maintain their dominance by fighting. The horn of the rhinoceros is formed of a compact mass of keratin produced by the aggregation of a number of coarse threads; the process is apparently a modification of hair-formation.

Various other special protective devices of the skin have been developed. Thus the pangolins, *Manis* sp, of Africa and Asia are covered with horny scales. Small scales occur on various mammals, for instance on the tail of rats. In hedgehogs and porcupines, and also in the monotreme *Tachyglossus*, the epidermis produces sharp spines, usually considered to be modified hairs. It may be mentioned here that in armadilloes there is a dermal skeleton of bony plates.

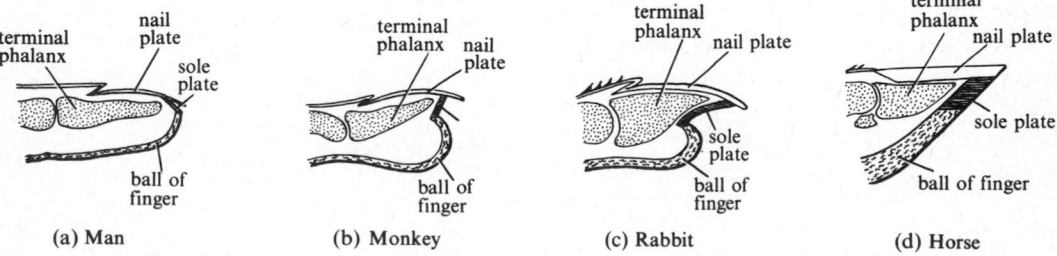

FIG. 3.1. Longitudinal section through the end of the digits of the manus of various mammals. (After Weber (1927). *Die Säugetiere*. Gustav Fischer, Jena.)

2. Glands producing attractive or repellent scents

A further characteristic feature of the skin of mammals is the development of scent-producing glands, derived from sweat or sebaceous glands or sometimes intermediate between the two. These play a large part in the life of the animals by providing the mechanism for recognition of members of the species, especially in social forms, and for bringing the sexes together. Most species of mammals have a characteristic smell, often different in the two sexes. Perhaps this condition has grown up in association with the development of sensitive chemo-receptors for hunting purposes (p. 409). Although many mammals have good eyes, for recognition they rely mainly upon smell. Birds, on the other hand, which are microsmatic, have few scent glands but often show distinctive colours. In mobile animals, such as birds or primates, smell would evidently be of less use than sight for either recognition or hunting.

The characteristic smell of the rabbit is due to a pair of sebaceous glands, the *inguinal glands*, opening between the anus and the penis or vulva. Many other mammals have similar glands in this region, for instance, the *civet glands* of carnivora and the *musk glands* of certain deer. Many antelopes have a sub-orbital gland with which they place olfactory 'marks' upon the trees. Ruminants also often have glands on the feet and legs; this is true especially in the smaller species (for example, duikers). The glands are absent in the large bovids (and in giraffes), which can follow each other by sight or sound. The elephant has a gland on its forehead. The peccary has a large gland opening on its back. Marsupials commonly possess a complex scent gland at the base of the neck. Rhinoceros have scent glands on the feet, and the callosities of the legs of the horse may be of similar nature. It is significant that such glands are less marked in the higher (visual) primates, but the gentle lemur (*Hapalemur*) has a large one on the arm.

Scents have a highly 'emotional' quality in man, and there is no doubt that they act as important signals in both sexual and social organization. In animals with large scent glands there probably are characteristic differences of scent between individuals. These smells may be left upon trees and other objects and serve to mark the territory of an individual. No massive scent glands are present in man, but the characteristic smells of the secretions of the axilla and genital regions are due to modified sweat glands. Those in the axilla change in histological appearance during the menstrual cycle in women, being enlarged during the pre-menstrual period but reduced during menstruation.

Of the specifically repulsive smells developed the best-known is that of the skunk (*Mephitis)* of America,

a carnivore in which the perineal glands can emit a jet of secretion whose unpleasant smell is perceptible far away. Presumably in connexion with this habit the animals have a conspicuous warning coloration of black and white.

3. Mammary glands

Mammary glands giving mammals their characteristic power of nourishing the young are products of the epidermis, resembling sweat glands in some respects (Bloom and Fawcett 1968; Ham 1974). In the resting state they consist of a branching system of ducts. During pregnancy the epithelium of these ducts divides actively, producing secretory end-sacks or alveoli. After birth of the young the alveoli produce the milk; the cells produce carbohydrates, proteins, fat, and water and pass them to the lumen. The growth of the mammary glands is stimulated by the combined action of oestrogens and progesterone, and the release of the milk is stimulated by prolactin and growth hormone, produced by the anterior lobe of the pituitary (p. 439). All these endocrine secretions are necessary, and each influences the secretion of the others. Emptying of the breasts by nursing also stimulates the production of prolactin, probably by a nervous pathway to the anterior pituitary.

The milk varies in composition in different mammals. In cows' milk the dry weight includes about 20 per cent each of protein (mostly casein) and fat, and 60 per cent of sugars (mostly the disaccharide lactose). Salts and vitamins are present in the proportions in which they occur in the blood. It is usual to speak of milk as a 'nearly ideal food' for the young. Evidently the ideal varies in different species, the milk of whales and seals, for example, contains 12 times more fat and 4 times more protein than that of the cow, but no sugars. There is a relationship between the fat and protein content of the milk and the speed of growth after birth (see Table 3.1). Such differences show how complicated the factors are that determine the characteristics of each variety of living organization.

TABLE 3.1

	Man	Horse	Cow	Pig	Sheep	Cat	Dog	Rabbit	Seal
Number of days in which weight of young is doubled after birth	180	60	47	18	10	9	8	6	5
Protein in milk (g/l)	19	20	33	37	70	95	97	104	119

The mammary glands develop by thickening of the epidermis to form '*milk lines*' extending along each side of the abdominal wall. The definitive mammae appear either dispersed along this line, as in the sow or bitch, or mainly at the hind end (ruminants) or front end (primates). The position and number of the mammae and nipples varies with the number of offspring and habit of the animal. The peculiar pectoral position in primates was probably associated first with arboreal life and carrying of the young.

Organs similar to mammary glands are found in the egg-laying monotremes. The gland of the echidna consists of numerous ducts, opening on hairs, but without special nipples, on areas of skin on either side of the pouch in which the eggs and young are placed. The milk is a thick, yellow substance, which is licked rather than sucked by the young and contains proteins and fat but probably no lactose.

The mammary apparatus is well developed in marsupials, where the teat is sucked into the specially shaped mouth of the new-born while it is in the pouch and the milk is pumped in by the mother.

4. The dermis

Many of the special structures derived from the epidermis are actually lodged in the underlying dermis (corium); in addition, the tissues of the latter are of great importance for the properties of the skin as a whole, and nearly all the nerve endings of the skin lie here. The dermis forms a tough, flexible, somewhat elastic covering over the whole surface of the body. Its function is to hold together, protect, and support the body and to carry blood to the surface. It contains the nerve endings and in some mammals (whales and man) carries much fat to serve as a heat-insulating layer. It plays a large part in the defence against bacterial invasion.

5. Musculature of the skin

The skin contains unstriped muscle fibres, and these permit a certain degree of movement, for instance, in the nipples, penis, scrotum, and labia majora of man. In monotremes there is a well-developed layer of striped muscles below the skin over the whole surface of the body, and these are innervated from the ventral roots of spinal nerves. This musculature appears to be a special derivative of the myotomes that is peculiar to the early mammals but is reduced in later forms.

The skin musculature is often considered together with the striped musculature of the face under the title of *panniculus carnosus*. The face muscles, however, found from the monotreme stage onwards, are lateral

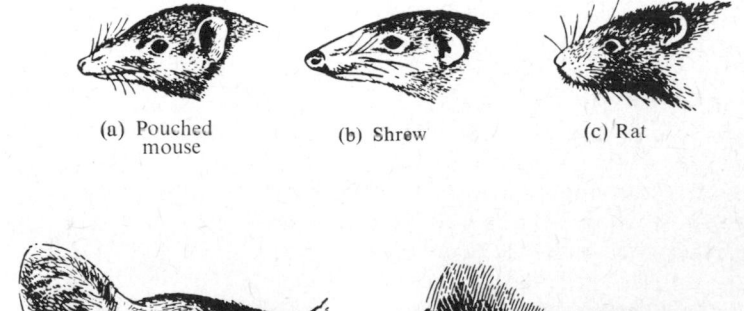

(a) Pouched mouse   (b) Shrew   (c) Rat

FIG. 3.2. The heads of various mammals. (From Portmann (n.d.). *Die Tiergestalt*. Rheinhardt, Basle.)

(d) Puma   (e) Proboscis monkey

plate muscles (see p. 119), innervated from the facial nerve. They are developed in connexion with a variety of functions, including chewing, suckling, and moving the ears, nose, and whiskers. In a few groups with good vision these facial skin muscles are specially developed to provide systems of social signals. This is most conspicuous in primates but present also in carnivores, Equidae, and some artiodactyls.

6. Expression, body shape, and colour in mammals

Mammals are mostly much less highly coloured than birds, the colour serving usually for concealment rather than for recognition or stimulation, which are accomplished by scent (p. 20). This dull colour is probably correlated with the fact that the early mammals were crepuscular or nocturnal animals, whose eyes lacked cones and were ill suited for detailed form discrimination or colour discrimination. Even today the majority of mammals are colour-blind (p. 391).

Bright colours are found chiefly among primates; these, with their arboreal habitat, have developed a predominantly visual sensory organization. Yet the external form of the body is probably significant in many mammals, perhaps more so in those that we recognize as 'higher' (Portmann 1952). The heads of many small insectivores and rodents are almost exactly alike; they are, as we say 'featureless'. On the other hand, in a lion or a monkey the muscles and other parts are developed to provide a means by which the creature exercises its influence on members of the same or other species (Fig. 3.2). The external form therefore does more than provide a 'physiological sack' for the organs; shape itself comes to have a significance for these types of animal organization that survive by cooperative action between individuals. Fig. 3.3 shows examples in which the front or hind end of the animal have characteristic shapes and colour patterns.

In the primates there are a number of compound expressions each serving as a sign to others of the internal motivational state (van Hooff 1962). These include four 'agonistic' expressions serving to indicate the balance of tendencies to flee or attack (Fig. 3.4). The attack face and aggressive threat face are given by dominant animals and are met by the scared threat face and crouch face by the animals lower in the hierarchy. The grin face also has an appeasing function but may be followed by approach if no strong threat is met. Lip-smacking and teeth-chattering are performed by monkeys that are grooming each other but may also be an invitation to approach and groom. During the approach of a male to a female in heat he gives a peculiar 'sniffing' face. The pout face is given by animals looking for comfort or food and is probably

part of the mother–child relation. The play face is given during the more or less ritual encounters of young animals. Obviously this analysis gives only a general outline of a very complicated communication system. Each facial display is accompanied by special sounds, postures, and actions such as urination or defaecation.

The same system of communication seems to be used by platyrrhine and catarrhine monkeys and by apes. Some of the expressions, such as the sexual sniffing or the lip-smacking after grooming and eating bits of the skin removed, are obviously ritualized elements of 'functional' actions. The grin face is perhaps a further modified scared threat face and has a possible relationship to the human smile as a gesture of appeasement.

Colouration is brilliant in only a few mammals; for example, the pink lower lip and pink and blue buttocks and thighs of the male mandrill. Another startling pattern is that of the gelada baboon, whose pink collar

(a)

(b)

FIG. 3.3. Characteristic forms and postures.
(a) Wild mountain ram showing well-developed appearance of both anterior and posterior ends. (b) Postures adopted by wolves; the animal to the left ranks high and that on the right low in the social structure of the group. (From Portmann, after Schenkel.)

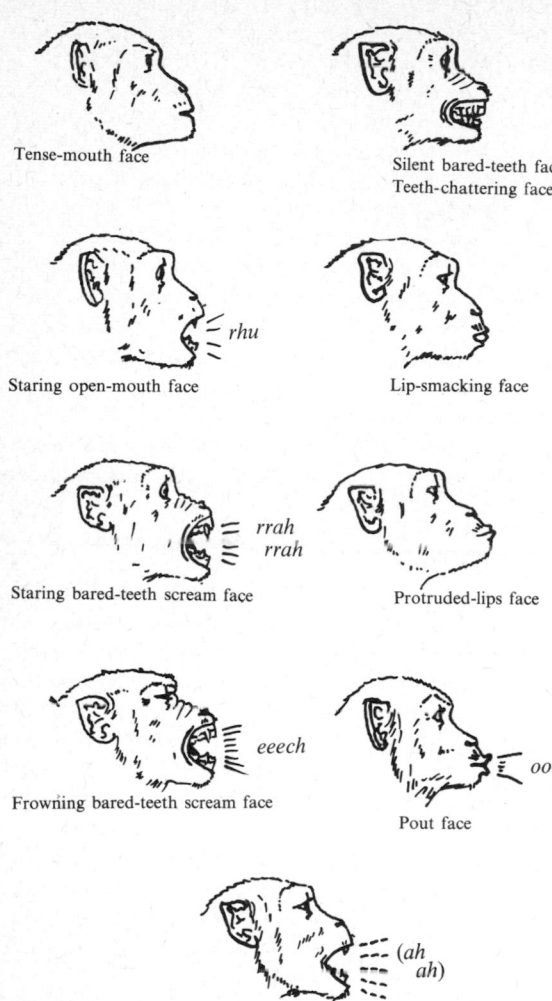

Tense-mouth face

Silent bared-teeth face
Teeth-chattering face

Staring open-mouth face   *rhu*

Lip-smacking face

Staring bared-teeth scream face   *rrah rrah*

Protruded-lips face

Frowning bared-teeth scream face   *eeech*

Pout face   *ooo*

Relaxed open-mouth face   *(ah ah)*

FIG. 3.4. Monkey (*Macaca*) facial expressions which have developed as a means of social communication. (After Van Hoof (1962). *Symp. Zool. Soc. Lond.* **8**.)

and chest, together with his white upper eyelids, give him a ferocious appearance when he bares his fangs by turning back the upper lip and showing the white inside. In this case the colour is made to 'change' suddenly by lowering the lids and turning back the lip, presumably to scare away an attacker by presentation of a startling pattern. Similar startling patterns are produced by many types of animals, and such colouration may be called dymantic (startling) to distinguish it from concealing (cryptic) or warning (aposematic) patterns.

A particular use is made of colour change as a means of communication in man, by blushing. This is a sudden dilatation of the capillaries in the skin when social

conditions become such as to make communication difficult, when the individual, as we say, feels embarrassed. Apart from these few special examples the colours do not change from moment to moment in mammals, as they do in some cyclostomes, fishes, amphibians, and reptiles. Moreover, colour patterns are fairly constant in wild mammals among the various individual members of the species. They vary more in domestication, where selection is less strict or less uniform.

The colours are usually the result of various mixtures of black, due to the pigment melanin, with white, formed by the presence of air in the hairs. The pigment is produced by special cells the *melanophores* (*pigmentary dendritic cells*) located in the basal layer of the epidermis. These are branched cells with long processes (Fig. 3.5), looking so like the neuroglia cells of the central nervous system (p. 277) that they have been called the 'epidermal glial system'. They are formed during development by migration from the neural crest (p. 347). Each branch of a dendritic cell ends as a cup-shaped button in contact either with an ordinary epidermal cell or with another dendritic cell, forming an elaborate branching system among the cells of the base of the epidermis. There is sometimes protoplasmic continuity between the branches of the dendritic cells. The pigment is manufactured in the dendritic cells and transferred to the epidermal cells. The melanin is brown or black and is produced by the action of the enzyme tyrosinase on tyrosine.

The activities of the dendritic cells are influenced by environmental factors. In the familiar case of the white

pigment   epidermal cell

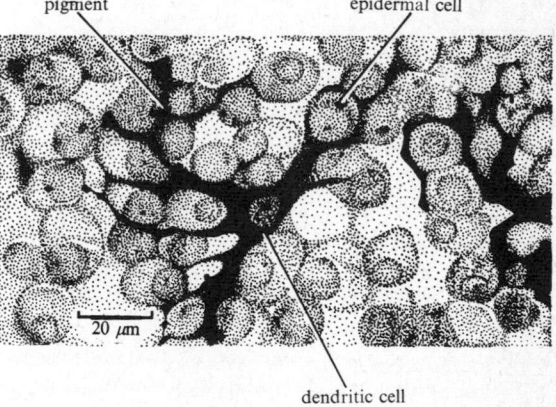

20 μm

dendritic cell

FIG. 3.5. Pigmentary dendritic cells seen in a preparation made by lightly squashing a piece of guinea-pig skin in Ringer's solution. The nucleated dendritic cells have processes that twine around the epidermal cells and transfer pigment to the latter. (After Billingham and Medawar (1948). *Heredity, Lond.* **2**.)

human skin, exposure to light produces increased synthesis of pigment. Exposure to cold will induce pigmentation of ears and other points of albino guinea-pigs, which, like white men, possess dendritic cells that produce melanin only under certain conditions.

The white patches that make up the pattern of spotted guinea-pigs and other mammals are due to the fact that the dendritic cells of these areas lack the power to produce melanin; therefore they are said to be *non-pigmentary dendritic cells* (Billingham and Medawar 1948). The cells of Langerhans are branched cells in the superficial layers of the epidermis. They are sometimes supposed to be of neural or 'neural–humoral' nature but may be worn out pigment cells.

Many mammals wear coats of different colours at different seasons. Some arctic species become white in winter; deer become darker in winter and lighter and dappled in the summer. Different mammals present particular colour patterns for purposes of concealment (cryptic) or warning (aposematic, as in skunks). Sexual colour displays are found mainly in primates. Whitening of the hair in man is due to some failure of the melanocytes, allowing the air-filled medullary cells to show. It is not known whether this is the result of disappearance of melanocytes or their inability to produce melanin. Sudden whitening may be caused by their removal by phagocytosis.

7. Control of colour and expression

The appearance of the skin is maintained in an appropriate condition by regulation with time-scales that vary according to the speed with which change is required. Blushing and the facial and other expressive movements of communication are under the immediate control of the nervous system. The colour of the sexual skin and other secondary sexual characters are controlled by chemical signals from the endocrine glands, producing effects that last for days or weeks. Changes in the whole shape of the body or of the basic colour patterns involve alteration of the hereditary make-up by variation and natural selection.

8. The skin and homeostasis

Evidently the skin is far more than a passive covering for the body. Study of even a few of its actions has led us to consider a variety of means by which homeostasis of the individual and race are maintained. The skin plays a part in rapid changes such as those by which temperature is kept constant, as well as in the slow adjustments by which sexual and social organization are maintained and the race kept in balance with the environment by the production of varied new individuals. All these activities of the skin can be considered within one system of ideas by emphasizing the part that each organ plays in homeostasis.

The many actions of mammalian skin are made possible by the multiplicity of morphogenetic processes that are at work in the embryo and the adult, producing the many types of cell, with their characteristic actions, by differentiation. Transplantation experiments in chick embryos have shown that the character of the epidermis is determined by contact with the underlying dermis. It is probable that the maintenance of the specific characteristics of each area of epidermis in the adult is due to the continuation of this inductive effect. These morphogenetic processes all operate under the instruction provided by the hereditary mechanism, and embryology shows that the genus skin becomes divided into its several species of cells by emphasis on particular types of enzyme action. Thus it is ensured that the skin provides a representation of the types of change in the surroundings that may influence the animal. Its keratin gives it the qualities required to resist deformation, the dermis gives resistance to infection, and hairs and sweat glands provide a representation of the probable temperature changes. The colour and form of the outside of the body represent aspects of the activity of other individuals of the same and other species that the animal may meet. The receptor organs are sensitive to a certain range of the changes likely to occur around the skin.

Evidently the mammalian skin provides a very elaborate representation of conditions in its environment, which is another way of saying that mammals are 'higher' or 'well-developed' animals. The representation that the skin provides at any moment is mainly the result of the instructions of heredity, but information from the nervous system also arrives to enable the representation to be more complete—for example, in relation to temperature. The skin of each individual also carries its own memory of conditions that have been encountered in the past, as we all know from human faces.

Thus the outer layers of the body provide a mirror of the conditions both without and within and of the conditions experienced in the remote and immediate past.

4   The connective tissues

1. Consistency of the tissues

CONNECTIVE tissue, being dispersed throughout the body, is less conspicuous than the special tissues characteristic of the various organs, but its importance for the life of the whole individual is immense (Peacock and van Winkle 1970; Fitton Jackson 1964; Rhodin 1967). Besides holding together and supporting the other cells it also provides a large part of the means of defence against infection.

The physical properties of the dermis are mostly those characteristic not of its muscles, described in the last chapter, but of connective tissue. Similar material is found throughout and around nearly all the organs. The protoplasm of cells is watery and soft; the firm consistency and resistance to stretch of the organs is a property of the connective tissue they contain rather than of their own cells. This is well illustrated by the difference between a peripheral nerve, which contains much connective tissue and is therefore a relatively strong substance, and the matter of the brain or spinal cord, which has a minimum of connective tissue, is almost liquid, and can hardly be picked up with forceps.

2. Mesenchyme and its derivatives

In an embryo the cells of most of the tissues are rather sparsely distributed in a soft jelly-like *ground substance*. This itself is the product of scattered *mesenchyme cells*. These have large nuclei and long fine branches extending through the ground substance (Fig. 4.1). They have the potentiality to develop into all the various types of cell that occur in connective tissue. Some of these, the *fibroblasts*, produce the structural proteins collagen and elastin that give the connective tissue its characteristic strength in resisting tensile deformation. Others form

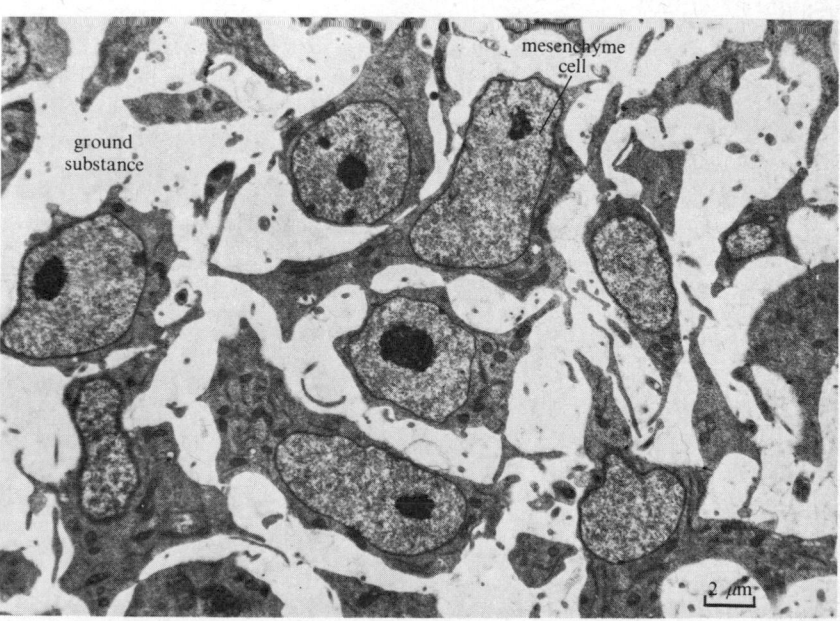

FIG. 4.1. Electron micrograph showing mesenchyme cells in the limb-bud of a chick embryo. (From a preparation and photograph by Dr. R. Bellairs.)

the cells that produce cartilage and bone, able to resist compression as well as tension. Another set, the mast cells, produce (among other things) the polysaccharides that compose the ground substance. A further whole class is concerned with protection either by production of antibodies (the plasma cells) or by ingestion of foreign particles (histiocytes or reticulo-endothelial cells). In these latter respects the cells of the connective tissues overlap in function with those of the blood. Indeed the lines that lead to the blood-forming tissues, the myeloid and lymphatic haemoblasts, are derived also from mesenchymal cells, as are the endothelial cells that line the vessels themselves.

Evidently the mesenchyme and its derivatives play several fundamental parts in maintaining the life of the individual. They are concerned with the physical support of the tissues, with protection against invasion, and with the whole transport system of the body. When we understand more of the systems that control the differentiation of these cells we shall probably be able to make a better classification. It is already an interesting task to consider how the various proteins and polysaccharides that they produce may be inter-related and perhaps formed from partly shared portions of the genotype.

Many of these derivatives of the mesenchyme are formed continuously throughout life. It is still a debated question whether they are produced from a reserve of undifferentiated mesenchyme cells, persisting in the adult. Once differentiated the cells do not change from one type into another. Unfortunately, little is known of the influences that induce differentiation. There is no means for identifying 'undifferentiated mesenchyme cells' in adult tissues.

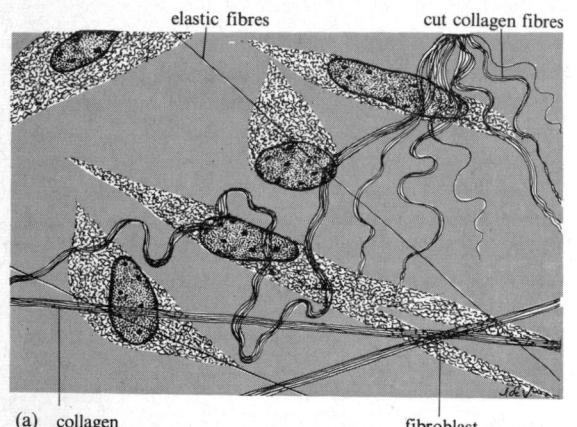

(a) collagen

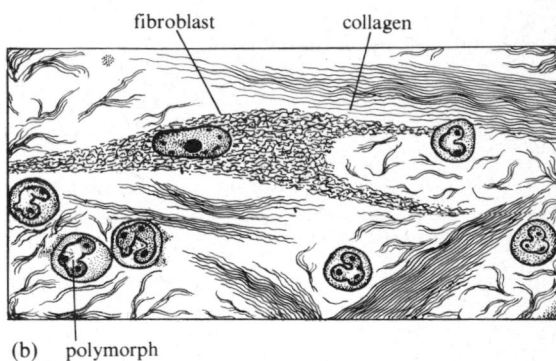

(b) polymorph

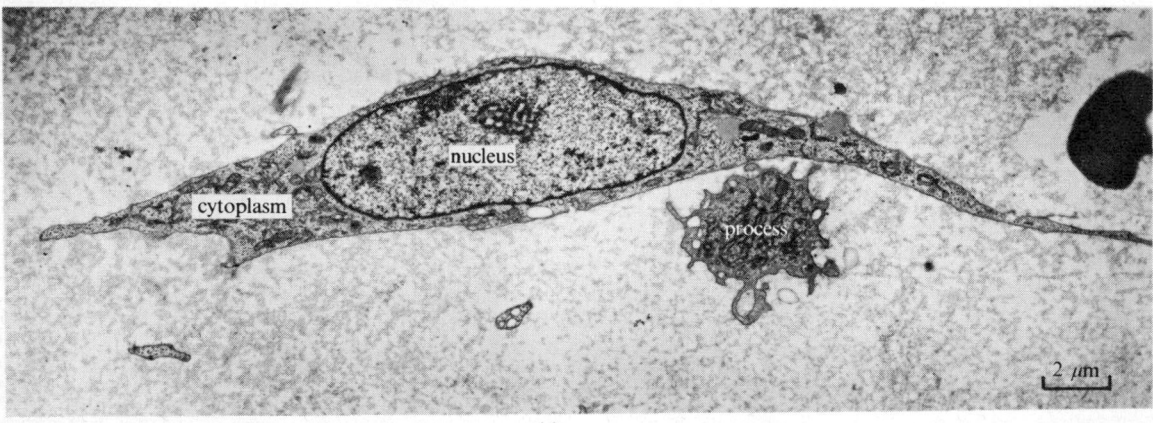

(c)

FIG. 4.2. (a) Areolar tissue from the skin of a rabbit. (b) Fibroblast and young collagen fibres from a section of the skin of a monkey. (Drawn from preparations and photographs by Mr. K. C. Richardson.) (c) Electron micrograph of a fibroblast and the process from an adjacent one in the submucosa of the small intestine of a guinea pig. (Preparation and photograph by Dr. G. Gabella.)

3. Fibroblasts

The characteristic cell of connective tissue is called a *fibroblast* when it is young and active but in the mature tissue it is known as a *fibrocyte* (Kulonen and Pikkarainen 1973). The fibroblast (Fig. 4.2) has fairly extensive cytoplasm with some basophilia, due to a high content of ribosomal endoplasmic reticulum. There is a large nucleolus. These cells are thus in process of active production of protein (collagen), and they perhaps also produce some of the ground substance. In mature connective tissue, the fibrocytes have little obvious cytoplasm. The nuclei have little chromatin and are long, flattened ovoids. After injury fibroblasts multiply rapidly. They migrate into a wound area and rapidly produce collagen there ('fibrosis', formation of 'scar tissue'). It is not certain whether these fibroblasts come from the old connective tissue, from 'undifferentiated cells', or by migration of cells from the blood (perhaps monocytes).

4. Collagen

This structural protein is found throughout the animal kingdom from sponges onwards (Harkness 1961, 1966). It occurs as fibres having great tensile strength but some flexibility because of their composition (Fig. 4.3(a),(b)). One-third of the amino-acid residues are of glycine, which being small with no side chain, allows the chains to pack closely together. The imino-acids, hydroxyproline and hydroxylysine, make up another 25 per cent and are rare except in collagen. These imino acids have flat, rigid, five-membered rings, which ensure that the chains can only fold in certain ways and hence produce the strong rope-like structure.

$$CH_2—COOH$$
$$|$$
$$NH_2$$

Glycine

$$HO—CH—CH_2$$
$$|\ \ \ \ |$$
$$CH_2\ \ CH—COOH$$
$$\backslash\ \ /$$
$$N$$
$$H$$

Hydroxyproline

$$H_2N—CH_2—CH—CH_2—CH_2—CH—COOH$$
$$|\qquad\qquad\qquad |$$
$$OH\qquad\qquad\qquad NH_2$$

Hydroxylysine

From the characteristic X-ray diffraction patterns and other data (Grant and Prockop 1972; Miller and Matukas 1974), the structure of collagen is found to consist of three polypeptide chains, each coiled on itself and the three coiled together like a rope. The molecule usually includes two chains very similar in amino-acid composition $((\alpha_1)_2)$ and a third with a different composition (α_2). Every third amino acid is a glycine, and these are held together by hydrogen bonds between the CO and NH groups. Between the glycines are either a hydroxyproline or proline residue. The amino-acid sequences are arranged as sub-units of which there are five types in two of the chains, and seven in the third. Only a limited amount of genetic information is thus needed to specify the molecule. The arrangement provides tensile strength, and some 'give' and flexibility is provided by uncoiled regions without the imino acids. The chains are linked by both covalent and hydrogen bonds. The molecules are rods about 280 nm long and 1·4 nm wide with a molecular weight of 300000. These molecules of tropocollagen associate spontaneously to form long fibres with a pattern of light and dark bands with a major periodicity of 64 nm. The banding depends on the sequence of acidic and basic amino acids. Some collagen does not form fibrils at all, particularly that in the basement membranes below cells (p. 30).

The details of the length and arrangement of the fibrils vary in different tissues. The first fibrils are formed on the surfaces of the fibroblasts, but they then grow in diameter after they have left the cell. The latter, therefore, must have produced the protein (or its precursor tropocollagen) partly in solution. Labelled glycine injected into young mice appears in the cytoplasm of odontoblasts in 30 minutes, but by 4 hours it is outside the cell in young collagen. After 35 hours it is found in the organized dentine. Once laid down the constituents of the collagen turn over very little, if at all. Much of it probably remains throughout life. Indeed it becomes more cross-linked and insoluble as part of the process of ageing. Other 'degenerative' changes may occur through infection or injury, or in skin exposed to the sun. On heating collagen the three chains come apart and contract at a temperature only a few degrees above that of the body (43 °C). The significance of this 'melting' or 'thermal contraction' is not clear: it has even been suggested that it might provide a repairable molecular reference standard for the heat-regulating mechanism (p. 13).

Collagen is very resistant to most proteolytic enzymes, but is occasionally removed for physiological reasons, for example, in the uterus or during re-moulding of bone. A powerful collagenase has been found in tadpoles at metamorphosis.

5. Fibrous tissue and tendon

The fine collagen fibrils aggregate into bundles, 10 μm or more thick, making the characteristic wavy strands of *white fibrous tissue*, from which it derives its name of

FIG. 4.3. (a) Drawing from an electron micrograph of collagen fibrils from a rat's tail. The fibrils have been obtained by dispersing trypsin-treated material in weak acid. Note that the cleavage plane at the frayed end is longitudinal. (After Schmitt, Hall, and Jakus (1942). *J. Cell Comp. Physiol.* **20.**) (b) High-resolution electron micrograph of rat-tail collagen. (Preparation and photograph made by Professor E. G. Gray.)

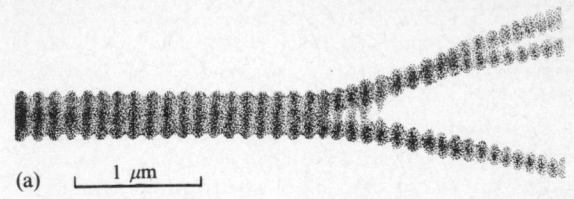

(a) |—— 1 μm ——|

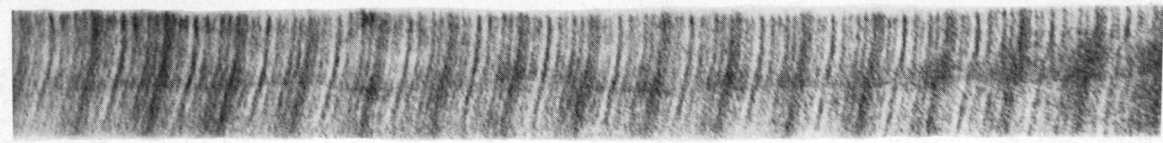

(b) |— 0·1 μm —|

fascia (Latin *fascis*, a bundle) (Fig. 4.4). Collagen is recognized in histological preparations by the appearance of these wavy bundles. It stains readily with the dyes, aniline blue and light green, that are used in the staining methods of Mallory and Masson, respectively, and red with acid fuchsin in van Giesen's stain. The nature of these reactions is not understood chemically, and the colours are not specific evidence of the presence of collagen. Collagen is found in almost pure form in the ligaments uniting the bones and in the tendons that join muscles to bones (Fig. 4.5); also in such membranes as the periosteum around bones or perimysium around muscles. It forms the main component of areolar tissue, the sticky material that occupies the spaces between organs and gets its name from the fact that it readily takes up bubbles of air, assuming a white colour.

Tendons are formed by collections of fibroblasts, later arranged in rows with collagen between. The blood supply disappears in the adult. Where the tendon is subject to friction it acquires inner and outer sheaths of collagen with synovial fluid between (see p. 41). Damaged tendons can regenerate from fibroblasts of

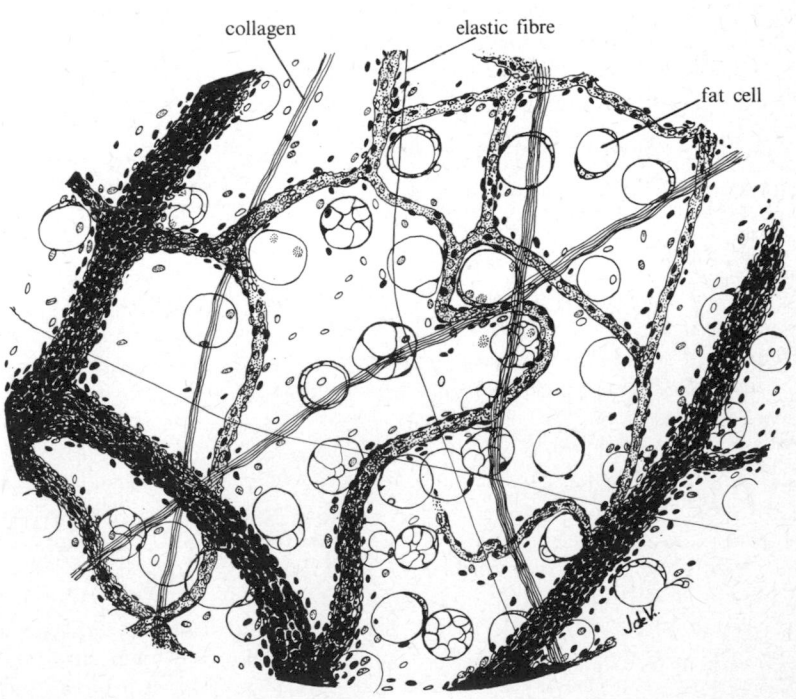

FIG. 4.4. Areolar tissue with fat cells and blood-vessels from the skin of a rabbit. (Drawn from preparation and photograph by Mr. K. C. Richardson.)

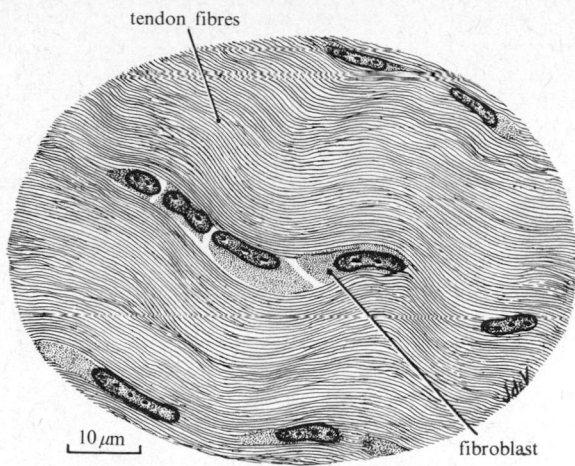

tendon fibres

10 μm

fibroblast

FIG. 4.5. Section of a portion of a tendon from a monkey, *Macacus*. (Drawn from a preparation and photograph by Mr. K. C. Richardson.)

the sheath. Tendons contain up to 90 per cent of their dry weight of collagen. They have a tensile strength of up to 50 kg mm^{-2} and a high coefficient of elasticity.

Collagen fibres are so called because when boiled with water they dissolve to form gelatin or glue (Greek κόλλα, glue). Further characteristic properties are that the fibres swell in weak acids and alkalis and are dissolved by strong acids or alkalis, which are accordingly used in the process of maceration of tissues, when it is desired to allow the cells to become separated. Tannic acid and the salts of some heavy metals convert collagen to the substance leather, which is usually produced by tanning the collagen of the skin. The property that gives collagen its chief value to the organism is that, while highly flexible, it resists tension and thus gives to the tissues their power to resist deformation by stretching (see Viidik 1973).

6. Ground substance

All connective tissues are permeated by an intercellular matrix or ground substance, composed largely of a solution of polymers known as mucopolysaccharides (more strictly called glycosaminoglycans). The commonest, hyaluronic acid, consists of long unbranched chains of disaccharide pairs.

CH_2OH

COO^-

OH

HO

NH_2COCH_3

D-Glucuronic acid   N-Acetyl-D-glucosamine

Hyaluronic acid

Various sulphated mucopolysaccharides, such as chondroitin sulphate and dermatan sulphate, occur in associated states (usually in collagen), but hyaluronic acid, which carries a negative charge due to its carboxyl groups and dissolves in water to form a viscous solution, occurs free, as a space-filler. The molecules of hyaluronic acid are immensely long: with molecular weights up to 10 million the length would be 20 μm. They carry one large charge (COO^-) and therefore repel each other but attract such free ions as Na^+ or K^+. In life, they are randomly coiled around the few protein units and occupy a very large volume. A single molecule would occupy a sphere of 1 μm diameter. In a weak solution 1 g would fill a volume of 5 l of water. The importance of this filling effect of hyaluronic acid is that it prevents the diffusion of other large molecules while allowing that of small ones such as glucose, and ions.

The result of the presence of hyaluronic acid is to prevent the spread of any large molecules, particles, or bacteria through the tissue. If it is broken down by the enzyme hyaluronidase such substances spread rapidly. This 'spreading factor', therefore, is used to improve the spreading of local anaesthetics. Some pathogenic bacteria produce the enzyme and it is a constituent of snake venom and the stings of bees and wasps.

Hyaluronic acid is found in all mammalian tissues though obviously only in small amounts (<0.1 g per 100 g of wet skin). It forms most of the amorphous ground substance that fills up all the interstices of an embryo or young mammal. *Wharton's jelly* of the umbilical cord contains this mucoid in combination with a protein and prevents kinking of the vessels. The hyaluronic acid–protein from synovial fluid forms semi-stiff coils of particle weight 4×10^6 which resist compression and provide a special form of lubrication for joints (p. 41).

Heparin is a mucopolysaccharide that prevents clotting of the blood. It was first isolated from liver but occurs in all connective tissues, being produced by the *mast cells* (Fig. 4.6). This inappropriate name was given by Ehrlich, in allusion to the mast used as pig food, because he thought the cells were in some way 'overfed'. They are indeed full of granules, and these are metachromatic, that is, when stained with some dyes (toluidin blue) they give a colour different from that of other tissue stained with the solution. This is a property of mucopolysaccharides and serves to identify them in sections. Mast cells have various other actions. They contain most of the base histamine (derived from histidine) in the body and release it during anaphylactic shock and other immunity responses. 5-hydroxytryptamine also occurs in these cells (in the rat). The full

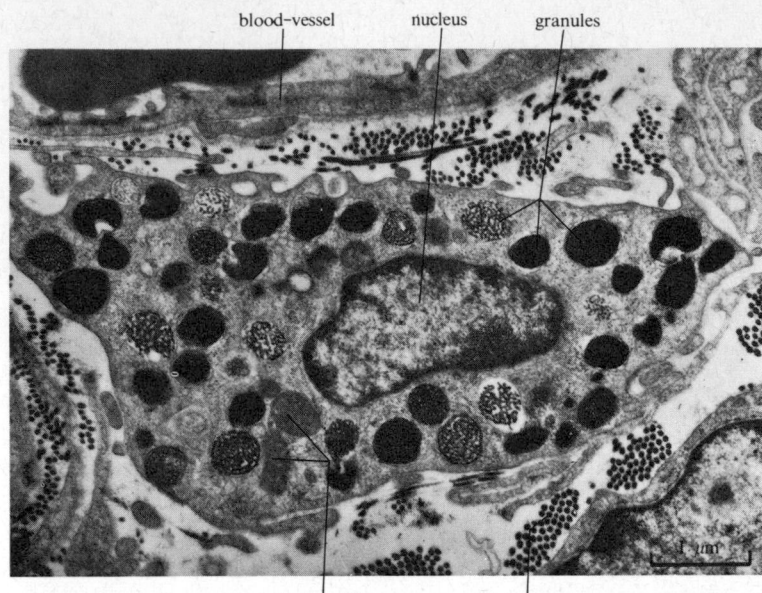

blood-vessel   nucleus   granules

mitochondria   collagen

FIG. 4.6. Electron micrograph of a mast cell from a rat's tongue. The granules are in various stages of maturation. (Preparation and photomicrograph made by Dr. G. Gabella.)

significance of the mast cells remains obscure. The heparin may serve to prevent the clotting of fibrinogen that escapes from the vessels (p. 171). Another suggestion is that it is simply the sulphated remains of the muco-polysaccharides that were first laid down in the embryo and then removed as they were replaced by collagen. Its relation to the altered permeability of the vessels in immunity response would then be an accidental result of the attachment of antibodies to the mast cells. A person suffering from an immunity response such as hay-fever or asthma may find it hard to believe that such marked phenomena are 'accidental'.

7. Reticular tissue

This is formed of fine networks of a protein similar to collagen, but the fibres always branch and are not arranged into bundles. As a result reticulin does not show the birefringence in polarized light that is characteristic of collagen. Moreover, among the fibres of reticulin there is much polysaccharide material, so that the substance forms sheets rather than fibres. These sheets make up the *basement membranes* lying below many epithelia and around many other types of cells, for instance, fat cells. The sarcolemma around muscle fibres and neurilemma around nerve fibres are composed largely of reticulin. It also forms a framework for many organs, for instance, when interspersed with lymph corpuscles in *lymphoid tissue*. Reticulin is best stained by certain special silver techniques and hence is sometimes known as *argyrophil tissue*.

8. Elastic tissue

The other main component of the intercellular substance of connective tissue is *elastin*, the substance of the yellow elastic fibres (Harkness 1964). This is found mixed with white fibres in areolar and other connective tissues (Fig. 4.7). The elastic fibres are thin, optically refractile, and homogenous in appearance (Franzblau 1971). They branch at intervals. They curl up where they are cut or if the tissue is not stretched. Elastin differs from collagen in composition. It contains much glycine but little hydroxyproline. There is much amino acid with hydrophobic non-polar side chains (for example, valine). Hence it is unwettable and chemically inert: for example, it is insoluble in boiling 0·1 N NaOH which dissolves all other proteins. It has only one-tenth the tensile strength of collagen, but will extend to double or more its resting length.

Elastin is probably a product of the fibroblasts, but the factors responsible for producing it are not known. It is always found associated with collagen. Elastic fibres are usually recognized by the fact that they stain with the dye orcein. A few tissues, for instance, the *ligamenta flava* uniting the vertebrae, are composed mainly of elastic fibres, and it is important in the coats of the larger arteries where it acts as an energy reservoir (p. 168).

9. Defensive functions of connective tissue

Connective tissue has a protective as well as a supporting function, and it contains a variety of cells that are concerned in the removal of foreign bodies and in

FIG. 4.7. (a) Elastic fibres and capillaries from the mesentery of a rat. (Drawn from a preparation and photograph made by Mr. K. C. Richardson.) (b) An electron micrograph showing an elastic fibre with an electron-dense amorphous core surrounded by microfilaments. In the centre of the picture the knife has grazed the surface of the elastic fibre and only microfilaments are visible. (From the oesophagus of a rat stained with uranium and lead.) (Preparation and photograph by Dr. G. Gabella.)

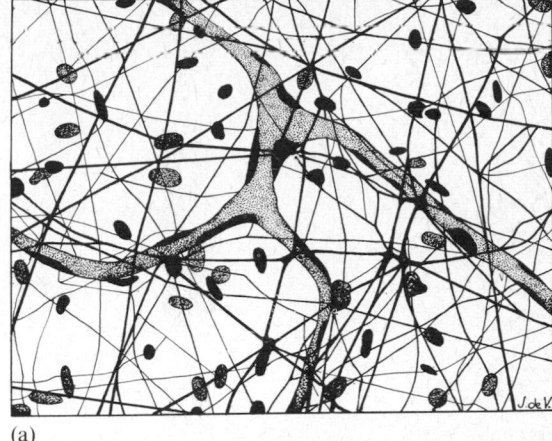

(a)

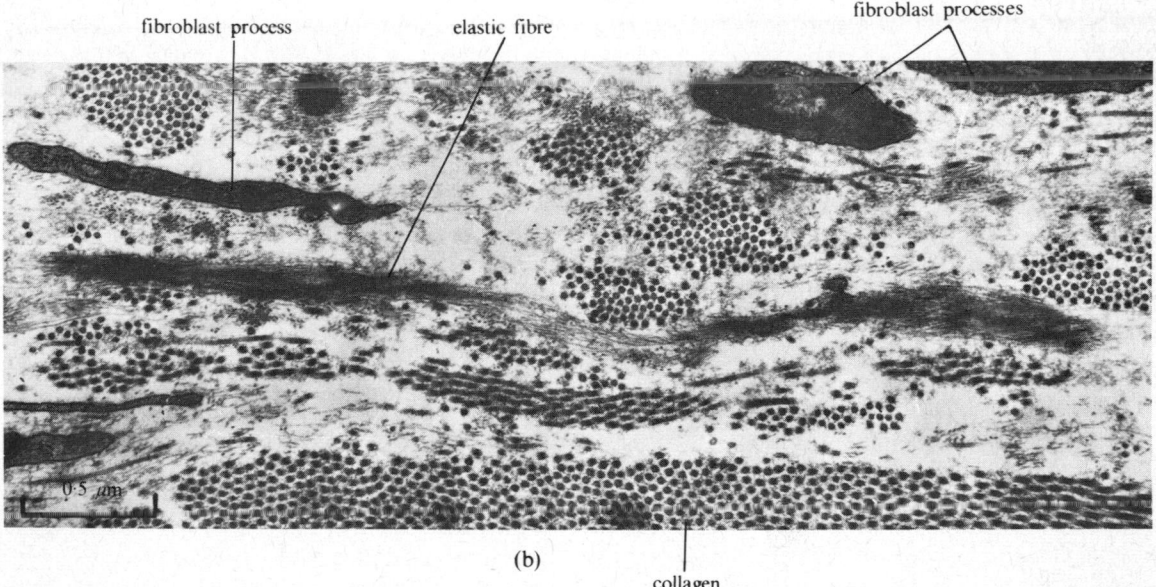

fibroblast process   elastic fibre   fibroblast processes

0·5 μm

(b)

collagen

resistance to infection (see Chapter 19). The most conspicuous of these are the *histiocytes* or *macrophages* (Fig. 4.8) (Vernon-Roberts 1972; Carr 1973), which are often almost as numerous as the fibroblasts. They are flattened or spindle-shaped cells with a more definite outline than the fibroblasts. The nucleus is small, kidney-shaped or irregular in outline, and composed of coarse, dark-staining granules. Great variations occur, however, and it is often not easy to distinguish histiocytes from fibroblasts except by their characteristic property of taking up bacteria or other foreign particles and their power to store certain dyes, such as *trypan blue* or *lithium carmine*, after these have been injected into the animal. In conditions of infection or trauma the histiocytes become migratory and ingest foreign bodies or degeneration products and their

protoplasm is then seen to contain large vacuoles; in this state they are known as *macrophages*. It is not clear whether they act by carrying away the debris or by dissolving it with enzymes on the spot. The macrophages were first so called to distinguish them from the 'microphages', which are now usually called polymorphonuclear leucocytes (p. 180). These latter enter the connective tissue in a few hours after an injury.

If any large pieces of foreign material enters the connective tissue it is attacked by *giant cells*. These contain many nuclei and are formed by the fusion of macrophages or monocytes.

The histiocytes thus provide a defensive system of cells dispersed throughout the tissues and sometimes known collectively as the *reticulo-endothelial system* or better as the *macrophage system*. The former name

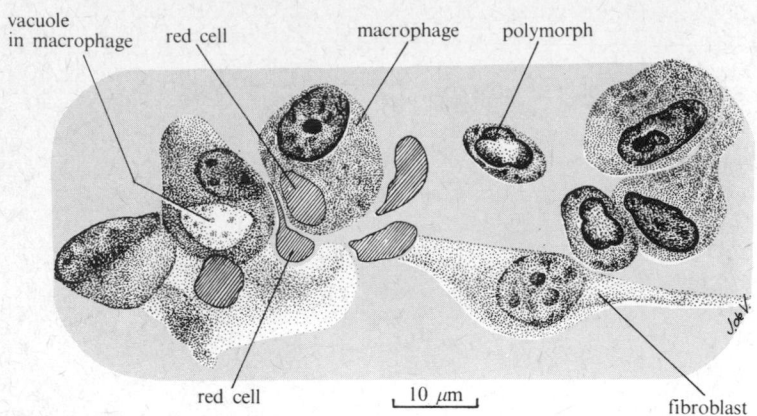

vacuole
in macrophage   red cell   macrophage   polymorph

red cell |⊢ 10 μm ⊣| fibroblast

Fig. 4.8. Macrophages removing extravasated red cells in a wound of guinea-pig skin, made one week previously. (Drawn from a preparation and photograph of Mr. K. C. Richardson.)

recalls the similarity of these cells to the reticular cells that line the sinuses of the lymph nodes and spleen (p. 176), which may be regarded as specialized macrophages. Other members of the macrophage group are the Kupffer cells of the liver (p. 150), the monocytes of the blood (p. 183), and the microglia cells of the central nervous system (p. 277). *Plasma cells* (p. 190), *eosinophil cells,* and *pigment cells* are also to be found in connective tissue; *monocytes* and *lymphocytes*, similar to those of the blood, may also be found there. There is evidently a close community of structure and function between the blood and blood-vessels and the connective tissues (see p. 26). The blood itself, like collagen, is an intercellular matrix, and it is produced in spaces lined by cells that are remarkably similar to some of those in connective tissue. Moreover, substances usually reach the cells from the blood only after transport across the connective-tissue membranes that surround the cells.

In addition to its power of removing foreign bodies, connective tissue, through its plasma cells, also plays an important part in the formation of antibodies after invasion of the body by bacteria and other foreign substances (Chapter 20).

10. The forms of connective tissue

Connective tissue should be thought of not as a mere set of fibres, nor even of membranes, but as a material with varied and complex properties that are of great importance to the organism. The condition of the proteins of the connective tissues, together with those of the blood, probably goes far to determine the *constitution* of the individual, especially as regards resistance to disease.

The connective tissue assumes different forms in various parts of the body, according to the forces acting on it and the chemical conditions to which it is subjected. It makes protective layers around most of the

organs. For example, the capsules of glands, the meninges of the brain, the periosteum of the bones, perimysium of muscles, and perineurium of nerves can all be regarded as sheets of connective tissue, each with its special properties. There is often a distinct sheet of connective tissue around every cell of a tissue, for example, each fibre of a peripheral nerve runs in an endoneurial tube (p. 240). The connective tissue is often in such close relation to the cell surface that nutrient and other substances can only reach the cell after passing through layers of connective tissue. The question of the permeability properties of this substance and the nature of the fluid spaces in connective tissue are therefore obviously of first importance, though unfortunately they are little understood. Connective tissue, like others, contains minute spaces filled with *tissue fluid*, which thus intervenes between the walls of the capillaries and the cells (p. 169). The amount of this fluid varies from moment to moment in any organ. In general it only becomes abundant when irritation or some other abnormality produces the condition of *oedema*. It is not safe to make any assumptions about the composition of this intercellular fluid or about the extent to which the connective tissue forms a barrier to diffusion between the blood stream and the cells.

The fascia forms *loose connective tissue* or *areolar tissue* between the organs, allowing movement between them. It makes fine sheets, the *deep fascia*, between many muscles and bones dividing the limbs into 'compartments'. In situations where tension falls on the connective tissue, the collagen fibres are laid down along the line of pull, producing a *tendon* (Fig. 4.5), or *ligament*. The characteristic cells in a tendon are fibroblasts of a special form known as tendon cells. The strength of a tendon is about twice the force its muscle can exert, allowing therefore a safety factor for say, the end of a jump.

11. Serous membranes

In situations where the sliding of surfaces over each other is of special importance, the connective tissue develops into *serous membranes*, the surface layers being covered with flattened cells known as *mesothelium*, which is not strictly an epithelium but a modified form of connective tissue. Such coverings are found over the abdominal viscera (*peritoneum*), lungs and thoracic cavity (*pleura*), and heart (*pericardium*). It is convenient to call these flattened cells mesothelium, to distinguish them by their position and function from such flattened surface cells as the endothelium and epithelium that are found in other parts of the body.

12. Synovial membranes

The surfaces of the joint cavities are lined by *synovial membranes*. These resemble the mesothelia of the viscera in some ways and may be considered as a modified form of connective tissue. However, they are not covered by a complete epithelium. These membranes produce the synovial fluid that occupies the joint cavity (p. 41). A similar tissue lines the *synovial bursae* or sacks that occur where there is a play between one tissue and another, for instance between the skin and the knee-cap (patella). Again, there are synovial *tendon sheaths* that allow smooth movement of the tendons where they are bound down by sheets of fascia.

13. Fatty tissue

Enzymatic systems capable of synthesizing or breaking down fat are present throughout the body, and in certain cells of the connective tissue they become specially developed for the storage of fat. Adipose tissue is found at many sites in the body, especially in the dermis and mesentery and around the kidneys. It serves mainly as a reserve but also has mechanical functions, forming soft and elastic pads in some places.

Fat cells are formed from fibroblasts and can be recognized by the appearance of small droplets of fat in the cytoplasm. These grow and fuse until the cell becomes a thin-walled bag, enclosing a single large droplet of neutral fat, with a flattened nucleus at one side (Fig. 4.4). Fat cells can synthesize fat from carbohydrate (for example, under the influence of insulin). They can also make it from the fatty acids that come from the breakdown of dietary fat. If these are of vegetable origin they will be largely unsaturated fats, which are held to be less liable to produce fatty degeneration of the arteries (arteriosclerosis) than are the saturated fats that the animal or man form from carbohydrates. The deposition and withdrawal of fat is controlled partly by hormonal signals and partly by variation in the demand by other tissues. The anterior pituitary occupies a central place in the control of fat metabolism, being itself influenced by the nearby nervous centres of the hypothalamus (p. 306). The provision of reserves of fuel is an obvious example of predictive action by the body. The reserve is maintained at a level that is adequate but not excessive by the action of a control system, itself basically laid down by heredity. Environmental factors such as available food supplies and demands for the output of work obviously influence the balance.

14. Control of the connective tissues

The connective tissues provide a part of the relatively stable framework for the body, and control of their organization is therefore little influenced by the quick-acting signals of the nervous system. There are however, powerful hormonal influences, especially those arising from the steroids cortisone and hydrocortisone, which are powerful anti-inflammatory agents, probably acting by inhibiting division of fibroblasts and forming collagen and mucopolysaccharides. ACTH (adreno-corticotropic hormone) has the same effect acting mainly through the adrenal, but partly direct. Somatotropin has the reverse effect, promoting proliferation of fibroblasts and formation of collagen and ground substance. Sex hormones also have direct effects. It is uncertain how these various influences maintain the appropriate constitution for the individual. Oestrogens produce the enormous concentration of hyaluronidase in the sexual skin of some primates at ovulation. Relaxin is a hormone produced during pregnancy, which produces softening of the connective tissue of the symphysis pubis. The differentiation of the various types of cell and fibre is thus produced by hereditary factors at each point. The working of this system is of primary importance in the maintenance of the body, but we understand little of the factors that control the production of the various cell types.

Some investigators claim that certain cells seen around the blood-vessels constitute a reserve of *undifferentiated cells*, capable of developing into most or all of the types seen in connective tissue (p. 26). Unfortunately, the whole question of the nature of the transformations of cell types and the stimuli that produce them is very obscure. There are certainly some similarities between, for instance, fibroblasts and histiocytes, and possible transitional forms between types as distinct as fat cells and fibroblasts can be found. Evidently the genus 'connective tissue', like the genus 'epidermis', contains many different species. The condition of any piece of tissue is the product of the powers of differentiation conferred on it during development and the forces that fall on it by the action of other tissues and the outside world.

An example of such interaction is the control of the deposition of collagen in fibrous tissue or tendon. The material is found to occur along the lines where the tissue is under tension. Weiss (1961) has shown how this state of affairs is reached by study of tissue cultures. In such situations as a healing wound, in which new collagen is being laid down, the fibrin fibres of a blood clot are, at first, randomly orientated, but as the clot is dissolved the parts not subjected to tension disappear first, leaving a web of orientated fibrin fibres. Fibroblasts grow into this web and collagen is then laid down along the tension lines. Readjustment of the earliest network of the collagen proceeds by the same method, fibres being removed when they are not in the line of tension. In this way a piece of connective tissue, tendon, or ligament composed of fibres orientated along the lines of tension and hence well adapted to its position is produced.

This is a good example of the way in which the organism makes the forecasts that ensure its survival (p. 4). The hereditary instruction of the genes ensures that collagen is laid down along lines of tension. The stresses of the environment may be regarded as providing information that selects those sites in which the collagen is produced. The two together provide the organism with a memory in the form of a set of collagen fibres so orientated that the organism can meet the stresses that are likely to fall upon it, assuming that the future stresses are like those of the past. The direction of orientation of the fibres thus provides a forecast based on past experience of the likely direction of tension stress in the future.

Thus the connective tissues, although not under the control of the nervous system, nevertheless carry memories of the situations that have been experienced. Such memories are provided by the arrangement of the collagen and elastin fibres, the condition of the histiocytes, plasma cells, mast cells, and other defence mechanisms, the number of fat cells, and many other features. Moreover, the condition of these tissues is continually changing under the influence of the chemical and physical forces acting upon them and of chemical signals received from the endocrine organs. By these relatively slow methods of control the connective tissue is continually adjusted to suit the environment.

5　Skeletal tissue: cartilage and bone

1. Resistance to tension and compression

COLLAGEN and the other scleroproteins of connective tissue provide resistance to tensile forces, but where parts of the organism are liable to be distorted by compression there develops either the mucopoly-saccharide–protein combination *cartilage* or the system of organic matrix and inorganic particles that makes *bone*. This power of the tissues to produce materials that are appropriate to resist the forces that fall upon them obviously provides forecasts that are of central importance for ensuring the survival of the organization of the animal or man.

Knowledge about the control of bone formation is likely to help greatly with the study of various problems about living organization and its evolution. The investigation of skeletal structures plays an important part in the study of the evolution of mammals, because as fossils the bones constitute the main means of following the course of phylogenesis. The solution of certain clinical problems would also be easier if we knew more about how skeletal material forms normally and why it tends to appear in tendons, arteries, and other tissues that are subjected to powerful forces.

There has been much controversy as to the origin of the cells that give rise to the various forms of connective and skeletal tissues (see Hall 1970). It is of great importance to know whether there is a reserve of 'undifferentiated cells', in connective tissue, capable of developing in any direction. It is now considered that the various cell-types are to some extent interconvertible. Thus the cells of cartilage may become dedifferentiated during ossification of a long bone and converted into bone cells, but this is not the only source of the latter (p. 38). The results of extensive experiments on this question may be summarized in the following diagram showing the possible sources of the various cell types now to be considered (Fig. 5.1).

2. Cartilage

Cartilage is a form of connective tissue in which the general amorphous ground substance has been added to and thickened. Fifty per cent of the dry weight is mucopolysaccharide and only 40 per cent is collagen, whereas for bone, in the 20 per cent that is organic, the proportions are 1·2 per cent and 93 per cent. The characteristic cells are the rounded chondrocytes, which lie in little spaces, surrounded on all sides by matrix (Fig. 5.2). At the margin of the cartilage new cells are

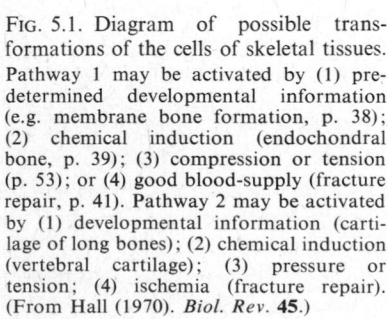

FIG. 5.1. Diagram of possible transformations of the cells of skeletal tissues. Pathway 1 may be activated by (1) predetermined developmental information (e.g. membrane bone formation, p. 38); (2) chemical induction (endochondral bone, p. 39); (3) compression or tension (p. 53); or (4) good blood-supply (fracture repair, p. 41). Pathway 2 may be activated by (1) developmental information (cartilage of long bones); (2) chemical induction (vertebral cartilage); (3) pressure or tension; (4) ischemia (fracture repair). (From Hall (1970). *Biol. Rev.* **45**.)

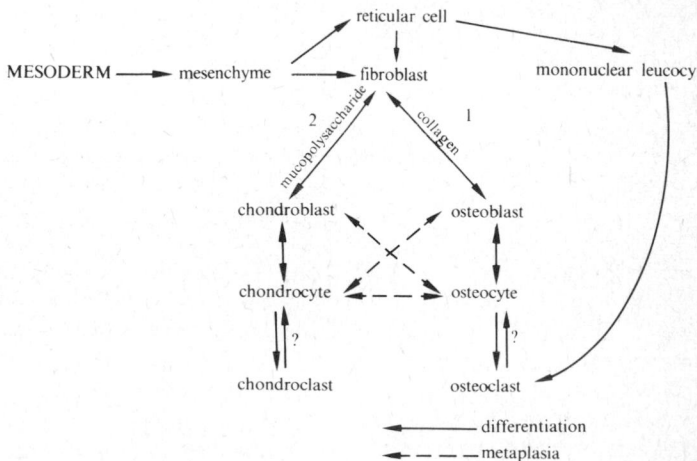

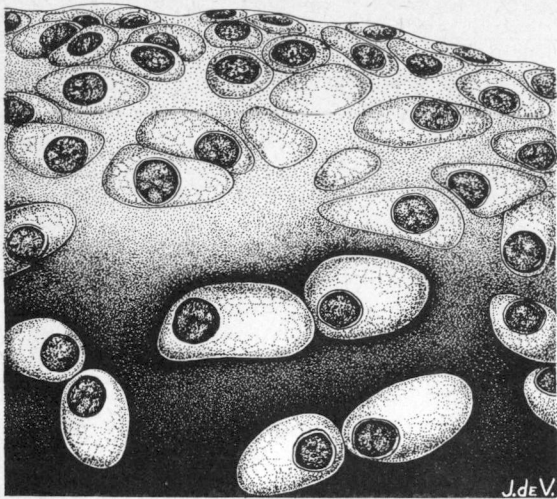

FIG. 5.2. Section of articular cartilage at the end of the femur of a frog. (From a preparation and photograph by Mr. K. C. Richardson.)

continually added from a special superficial layer of connective tissue, the *perichondrium*. The cartilage therefore grows mainly by addition from outside by appositional growth, but there may also be some interstitial growth by division of the cells within the formed or partly formed cartilage.

The characteristic interstitial substance of the cartilage consists of the mucopolysaccharides, chondroitin sulphate and keratan sulphate. In ordinary *hyaline* or *glass-like cartilage* this substance is accompanied by collagen; little of the cartilage is truly glass-like, most of it is *fibrous cartilage*. In the *elastic cartilage* of the external ear, epiglottis, and a few other regions, the matrix contains elastic fibres.

The mechanical properties of cartilage are produced by the negatively charged sulphate ions trapped in the sponge-like mesh of collagen fibrils. The sulphates provide a so-called negative fixed charge density, which attracts positive ions, mainly sodium. These exert a high osmotic pressure which draws in water and keeps the sponge expanded.

Cartilage is the material that forms the first outline of much of the skull, the vertebrae, and the long bones, but in mammals it is mostly replaced in the adult by bone, cartilaginous remains being found only at the ends of the bones, within the joints, at the ventral ends of the ribs, and in a few other situations. As a structural material cartilage stands between connective tissue and bone (p. 45). It is much better able to resist compression than is connective tissue, though not so strong in this respect as is bone. It is also well able to resist tension, a property that is, of course, present also in

bone, though bones may snap under strong muscular effort before rupture of the tendon that transmits the pull. Cartilage is highly adapted to the strains placed upon it, for instance, in the ring-like cartilages of the trachea the outer region contains abundant fibres, since this side meets mostly tension stresses; the inner part of the cartilage, subjected mostly to compression, contains more matrix. The combination of materials in cartilage has been likened to that in reinforced concrete or a rubber tyre.

As cartilage matures it tends to become calcified. This stops the diffusion by which the cells are nourished (cartilage is avascular), so that they die and are replaced by bone. The integrity of the cartilage of the adult is, however, of great importance, especially in the joints. Articular cartilage can be repaired after damage. New chondrocytes are produced by division and produce new intercellular substance. The healed defect may contain masses of cells intermediate between fibrocytes and chondrocytes ('fibrocartilage'). The cartilage is continually broken down by the wear of the joint. In older people it is not fully replaced, and osteoarthritis is in part the result of this.

The development of cartilage has been followed in the regenerating limb buds of *Ambystoma* by autoradiography after intra-peritoneal injection of [^{3}H]proline. After 15 minutes radioactivity appeared in the cytoplasm of the chondroblasts and after 30 minutes in the Golgi zone. After 1–2 hours it was mostly in the matrix and diffused away from the cells to form a band still present after 16 days and presumably mainly in the collagen.

3. Bone

Bone is a tissue especially adapted to meet the compression forces falling upon it, but also able to resist tension. Unlike cartilage, the bones are closely related to the blood system; indeed the central marrow of some bones is occupied by the chief blood-forming tissues. Bone is a plastic substance in the sense that it is readily modified, perhaps because the bone cells are branched structures, whose fine processes penetrate the matrix, every part of which thus remains close to the protoplasm of the cells. Even the hard part of the bone is continually changed. If the radioactive isotope of phosphorus (^{32}P) is ingested with food or injected, it is taken up by the bones within a few hours. About one-third of the material taken up after a single test injection was found to be retained in the bones of a rat after 3 weeks. This plasticity of bone allows it to change its structure quite rapidly if the forces falling upon it change; each bone thus carries a memory of the stressing that has occurred

Haversian
canal

canaliculi

lacuna

(a)

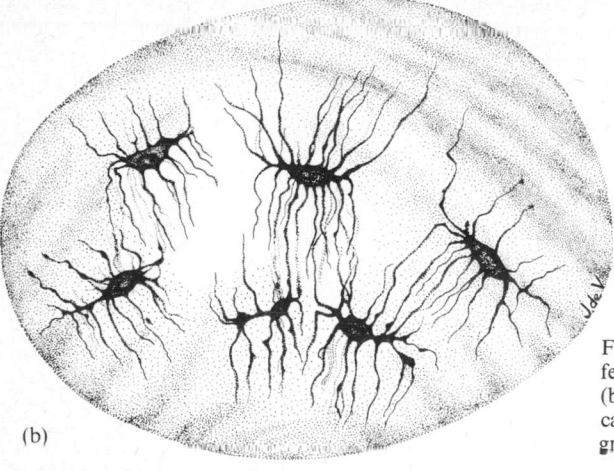

(b)

FIG. 5.3. (a) Portion of a transverse section of a human
femur after decalcification, showing Haversian systems.
(b) Osteocytes from human adult compact bone after decalcification and staining. (From preparations and photographs by Mr. K. C. Richardson.)

in the recent past, providing a forecast to meet the
forces likely to be applied in the future.

Sections of bone reveal a series of spaces, the *lacunae*,
occupied in life by the cell bodies of the *osteocytes*, each
surrounded by a system of *canaliculi*, which contain the
long fibrous processes of the cells (Fig. 5.3). In the
osteoblasts of young bone the processes of the cells
unite, and the canaliculi thus make a continuous communicating system throughout the bone.

The matrix of the bone consists of 80 per cent inorganic calcium phosphate partly in the form of
crystals (hydroxyapatite) and partly amorphous.
Twenty per cent of the bone is made up of organic
material, mainly collagen, highly cross-linked (Hancox
1972). This large protein content gives the bone a
certain flexibility, avoiding the brittleness of a purely
inorganic structure. The arrangement of the spicules of
apatite gives the bone its characteristic hardness, and

the detailed organization differs with the function of
the bone. It is usual to distinguish between *spongy* (or
cancellous) *bone*, made of a network of spicules visible
to the naked eye, and *compact bone*, where the spicules
are closely packed and produce a hard white material;
there are no essential differences, however, and all bone
first passes through a spongy stage.

The shafts of long bones consist of a tube of compact
bone and a central cavity, the marrow. In adults this
marrow is mostly fat and is known as *yellow marrow*.
The ends of the shaft are composed of spongy bone,
among the trabeculae of which there is *red marrow*,
which is a bone-forming tissue that is found also in the
flat bones of the skull, sternum, and ribs (p. 173).

In compact bones the bone cells are arranged in a
regular manner around arteries and veins running along
the bone. Each group of vessels constitutes a *Haversian
canal*, and around it the bone cells are arranged in

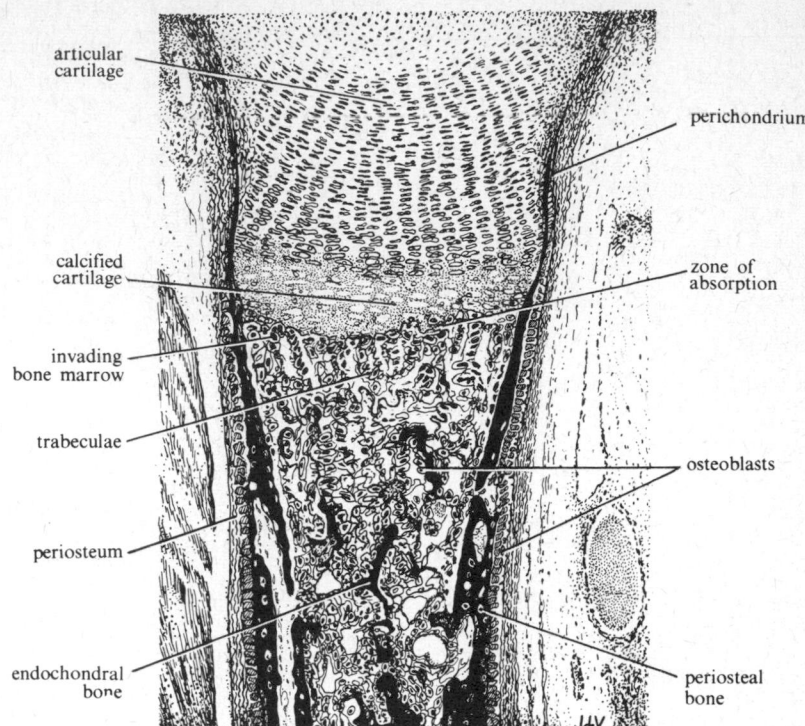

articular cartilage

perichondrium

calcified cartilage

zone of absorption

invading bone marrow

trabeculae

osteoblasts

periosteum

endochondral bone

periosteal bone

J&V.

FIG. 5.4. Section of a developing metatarsal bone from a foetal cat, showing the formation of endochondral bone within the extending cylinder of periosteal bone. (From a preparation and photograph by Mr. K. C. Richardson.)

concentric layers to form a Haversian system (Fig. 5.3). Within the lamellae the fibrils and the particles of inorganic material are placed in various directions, so that the whole constitutes a column of great strength. Both in compact and in spongy bone the direction of the main spicules and lamellae is related in a definite manner to the forces acting upon the material, producing a structure suited to the load that must be carried. The details of the architecture of the bones will be considered later (p. 50).

The Haversian canals communicate at intervals with the vessels at the surface and centre of the bone by means of the *canals of Volkmann*. The outer and inner surfaces of the bones are known as *periosteum* and *endosteum*, respectively; they consist of a special variety of connective tissue able to produce bone. The periosteum is tightly attached to the bone over most of the surface by strands of collagen, which continue into the substance of the bone as the *fibres of Sharpey*. Muscles, tendons, and ligaments are attached to the periosteum, which is here tightly fastened to the underlying bone. Where there is a joint at the end of a bone the periosteum becomes continuous with the joint capsule. The periosteum thus serves three functions: (1) its inner osteogenic layers produce bone; (2) its outer fibrous layers prevent the spread of the bone-

forming tissues; (3) it provides attachment for muscles, tendons, and ligaments.

4. Development of bone

The formation of bone occurs either in pre-existing cartilage (*endochondral ossification*), producing *cartilage bones*, or in a connective tissue membrane, to produce a *dermal* or *intramembranous bone*. The dermal bones originate in phylogeny from scales lying near the surface. The skull bones still occupy this position, other dermal bones are more deeply sunken (clavicle). The conversion of cartilage into bone is preceded by changes in the cartilage itself (Fig. 5.4). The cells divide and become arranged in rows, and their nuclei degenerate. The division always occurs in a direction parallel to the compression deformation. This ensures that the bone continually elongates along the axis that is compressed. The matrix meanwhile becomes calcified by the deposition of lime salts, a process that occurs in other degenerating tissues, for instance the walls of diseased blood vessels. This deposit is not bone, however, and it is soon removed by an ingrowth of blood vessels and connective tissue, which appear to dissolve away the calcified matrix. Certain cells of the connective tissue that are similar to fibroblasts then differentiate into *osteoblasts*. These form layers of cells

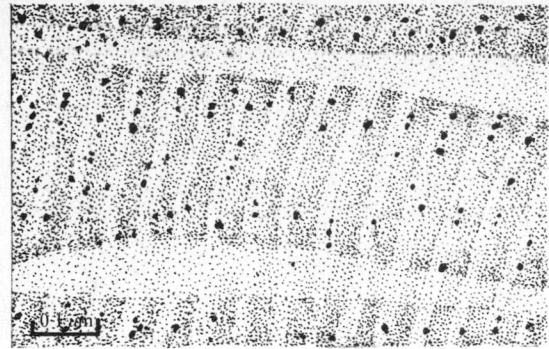

Fig. 5.5. Drawing of collagen fibrils of periosteal bone from a 16-day chick embryo. The dense apatite crystals are embedded within the fibrils in an organized fashion. The diameter of these collagen fibrils was about 60–70 nm. (Drawing made from an electron micrograph of Fitton Jackson (1957). *Proc. R. Soc.* **B146.**)

that produce new bone by the provision of the necessary enzymes.

The first product of the osteoblasts is a mucopolysaccharide matrix containing collagen fibres. Minute crystals of apatite are then formed within the spaces provided by the overlapping molecules in the collagen lattice (Fig. 5.5). The crystals probably remain within the fibres even in adult bone. They are extremely small (20–40 nm by 1·5–3·0 nm) and thus present an enormous area in which the atoms are near the surface. This allows for rapid interchange between the crystals and interstitial fluids, which allows the inner structure of the bone to be continually remodelled to give an internal system of girders (trabeculae) suited to meet the mechanical stresses put upon it (p. 50). Moreover, the

accessibility of the crystals allows the bones to operate as reservoirs of calcium. The accessibility of this reservoir is shown by taking the blood from a dog, removing the calcium, and replacing the blood. The calcium level is rapidly recovered, and by repeating the experiment 3 or 4 times the amount of calcium originally in the blood can be released in 1 hour.

The question of the rate of turnover of calcium in the bones is complicated. Ingested radioactive calcium (^{45}Ca ions) appears in the bones within a few minutes. Some of it remains for weeks or years. Therefore we can recognize 'exchangeable' and 'non-exchangeable' fractions of the bone calcium. The latter is in situations such as the shafts of the long bones, where the tissue may remain unchanged for a long time. On the other hand, the exchange is very active in the ends of the long bones, especially in a young animal. Clearly there is an elaborate relationship between accretion and remodelling and turnover of the calcium of the bones (see Vaughan 1970). It is not known what stimulus initiates the differentiation of the osteoblasts nor how these cause the production of bone. The body fluids are supersaturated with the ions needed for hydroxyapatite precipitation. Pure collagen fibres, reconstituted from solution of various types of connective tissue, are able to cause nucleation of apatite crystals from metastable calcium phosphate solutions. It is uncertain what prevents the mineralization of collagen in soft tissues or promotes it in the hard ones.

The formation of cartilage and bone can be induced by implantation of various materials into soft tissues such as kidney (McLean and Urist 1968). Thus pieces of cartilage devitalized with alcohol, acetone, or heat induced bone formation, whereas devitalized

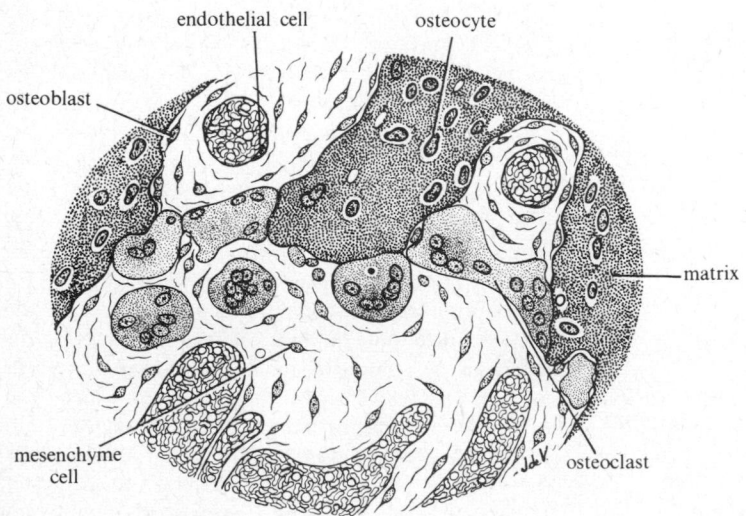

Fig. 5.6. Portion of a section from the mandible of a human foetus of 6 months. (From a preparation and photograph by Mr. K. C. Richardson.)

muscle induced cartilage, followed by bone. The induction is not an effect of the vehicles, which are ineffective by themselves. The untreated tissues will not produce these effects, probably because they are removed too quickly. But implantation of bony callus will induce further bone. It is probable, therefore, that bone formation is induced by specific chemical substances, though the nature of these is not known. Evidently the release of these substances can be triggered by various conditions, including hereditary, mechanical, and pathological factors, leading to the deposition of bone (p. 42) (see Pritchard 1961, 1963).

An elaborate reconstruction is undertaken before the bone reaches its final form, and this is made possible by the presence of *osteoclasts* (Fig. 5.6), multinucleate cells which dissolve away unwanted bone and are abundant throughout the process of bone differentiation.

The differentiation of bone cells has been studied by autoradiography of the periosteum of young rats injected with [³H]thymidine. Cell division takes place in 'pre-osteoblasts' of the connective tissue. Some of the products remain as pre-osteoblasts; others are 'modulated' into osteoblasts, and then after 2–3 days become enfolded into the new bone as osteocytes. The origin of osteoclasts is still uncertain. Some studies suggest that they are formed by the fusion of osteoblasts but more

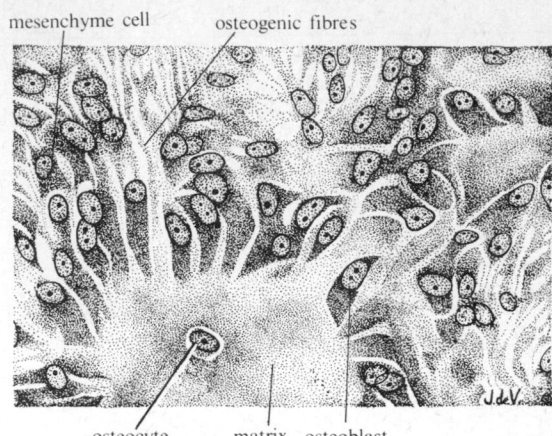

FIG. 5.8. The process of ossification in a membrane bone. (From a preparation and photograph by Mr. K. C. Richardson.)

probably it is by fusion of monocytes or pre-osteoblasts (Fitton Jackson 1964). Little is known of the factors that control the differentiation of these various cells.

The ossification of a long bone proceeds in a characteristic manner. The cartilaginous precursor provides the general shape of the bone. Osteogenesis begins around the middle of the shaft; here the perichondrium produces osteoblasts and is henceforward known as periosteum. A layer of bone, the *periosteal bone collar*, forms around the middle of the shaft. Soon afterward, blood vessels and connective tissue penetrate into the shaft and replace its cartilage with bone. In this way the whole length becomes ossified, forming the *diaphysis* of the bone, with a bone-marrow cavity at its middle. The ends of the bone ossify by separate centres that develop somewhat later and are known as *epiphyses* (Fig. 5.7). There may be one or several such centres at the end of the bone; they remain separate from the diaphysis throughout the growth period. Growth occurs by the addition of new cartilage between the diaphysis and the ends, and its conversion to bone, producing increase in length; growth in thickness comes largely from the periosteum. The epiphyses fuse with the diaphysis only towards the end of development, when the bones are no longer elongating.

The formation of a membrane bone differs from the above account only in that the osteoblasts begin to appear in a connective tissue matrix instead of in cartilage (Fig. 5.8). They form rows of cells, which soon lay down trabeculae of bone. The membrane bones of the skull develop in this way from one or a few centres for each bone, bone formation spreading outwards until the bony area meets that of a neighbouring bone and

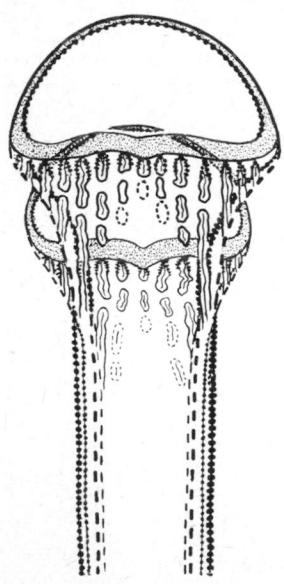

FIG. 5.7. A mammalian long bone cut in longitudinal section to show the head and part of the shaft at two stages of its growth. The growth in length and width is shown and the mode of remodelling of head into shaft. (After A. E. Needham (1964). *The growth process in animals*. Pitman, London.)

the characteristic jagged suture line is formed. Subsequent enlargement of the skull is produced by addition of new bone externally and its removal from the inner side. Bone is also added at the sutures, and the bones remain separate throughout the growth period.

The repair after a bone has been fractured begins, like all repair processes, by the formation of a blood clot. This is then invaded by cells of the periosteum, which produce a form of fibro-cartilage known as *callus*. The cells that do this are probably those of the periosteum that would otherwise have given rise to bone. It is interesting that cartilage is formed even after fracture of the membrane bones of the skull, which, of course, were not pre-formed in cartilage. The stimulus to form cartilage is probably a combination of ischemia and shearing forces. Later bone develops when blood vessels grow in and pressure and tension are applied.

5. Joints

The discontinuities between bones have a very different significance according to whether or not there is movement between them (see Barnett, Davies, and MacConaill 1961). Where little movement occurs, as between the bones of the skull, the bones are separated by *fibrous joints* and the thin connective tissue layers between them remain as a *suture* (Fig. 5.9). Some slightly moveable bones are united by *cartilaginous joints* such as the intervertebral discs, composed of a central mass of mucopolysaccharide, the nucleus pulposus, surrounded by a strong web of collagen fibrils,

thus making a cushion. Where there is free movement there is a *synovial joint* between the bones. The articular surfaces are covered by smooth hard bone, and this is covered by a thin layer of smooth hyaline cartilage containing a high proportion of chondroitin sulphate. The ratio of this to collagen is especially high in the cartilage of joints and areas that are subject to great pressures, such as the knee. The ratio increases with age, predisposing the joint to damage.

The bones are held together by *capsular ligaments*, having fibrous thickenings arranged in directions that prevent separation of the bones. The capsule consists of an outer dense fibrous layer and an inner more cellular *synovial membrane*. The latter has folds projecting between the bones, and it secretes the very thin layer of the viscous *synovial fluid* that separates the bones. There is as little as 0·2 ml of this even in the largest human joint, the knee. This fluid contains the large polysaccharide molecules of hyaluronic acid, combined with protein to give mucin, whose physical properties are suitable for ensuring free movement. This is probably achieved partly by the process known as 'hydrodynamic' or 'full-film' lubrication, which is used for axles continually rotating, but also by 'surface' or 'boundary' lubrication. The lubricant (synovial fluid) becomes attached to the moving surfaces, allowing repeated changes of direction. The lubrication must be very efficient since the coefficient of friction is very low (about 0·003–0·015, as against 0·10 between two artificial lubricated surfaces or 0·03 between ice and steel in skating). The hyaluronic acid also confers upon the synovial fluid the capacity to remain as a continuous

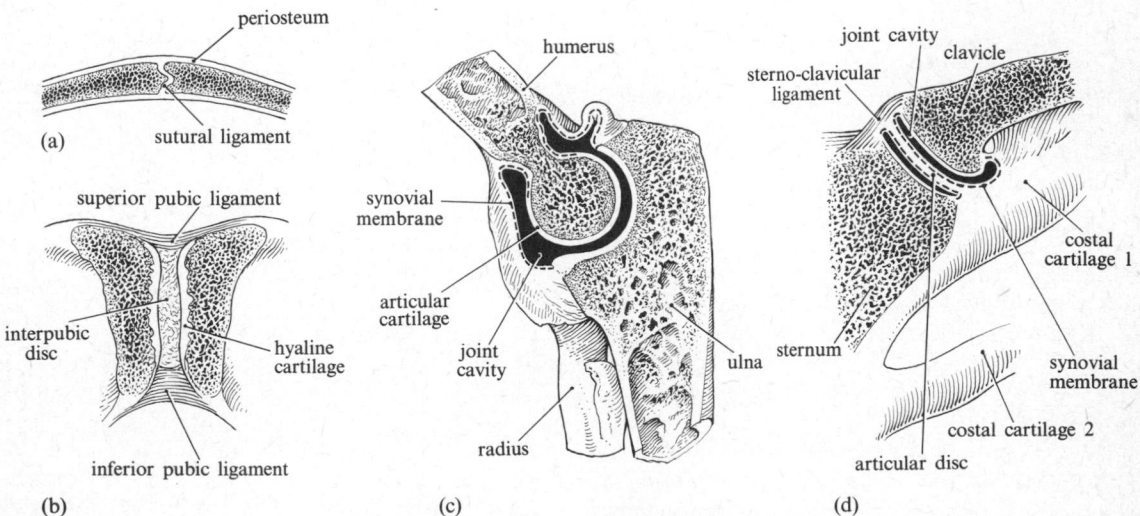

FIG. 5.9. Various types of joints found in man. (a) Sutural joint between bones of the skull. (b) Cartilaginous joint at the pubic symphysis. (c) Simple synovial joint at the elbow. (d) Sterno-clavicular joint, a synovial joint with articular disc.

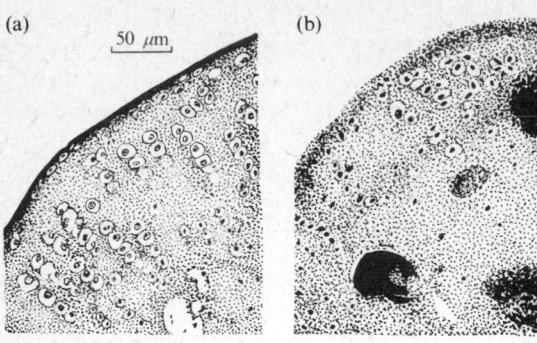

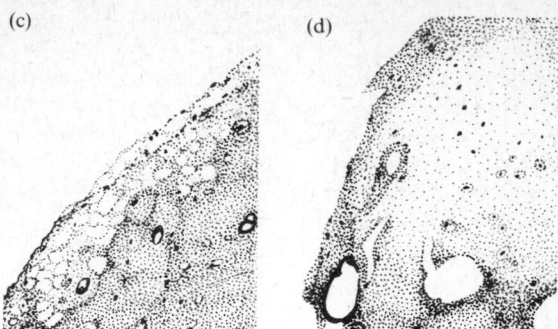

FIG. 5.10. Sections through the cartilage of the ankle joints from four rabbits where the viscosity of the synovial fluid had been reduced, by an injection of hyalase.

(a) Control animal given 48 hours exercise; (b) Hyalase-treated animal given 30 hours exercise; (c) and (d) Hyalase-treated and given 48 hours exercise. (Drawn from Barnett (1956.) *J. Bone Jt. Surg.* **38** B.)

film rather than rupturing under pressure. A proof of the efficiency of the lubrication has been given by injection of hyalase into the ankle joint of rabbits. After 30–48 hours of exercise the injected joints show gross evidence of wear (Fig. 5.10) (Barnett, Davies, and MacConaill 1961).

The incidence of wear in joints is probably greatly reduced by arrangements ensuring that the fit between the surfaces is not exact and no single parts carry the whole thrust. In joints that bear a great deal of pressure the male surface departs greatly from a circle. There is thus translation between the surfaces during movement, and no one part of either surface suffers attrition producing wear such as is seen in the bearing of an axle. Thus the trochlea of the humerus in man is nearly a perfect sphere seen in profile, but is an obvious spiral in a quadruped. The articular cartilage has no nerves or blood-vessels, but these are abundant in the capsule. In some synovial joints there are pads of fibro-cartilage, the *articular discs*, which, being pliable, serve to maintain the congruity of the surfaces during movement. Pads of fatty tissue in the capsule also serve this function.

Synovial bursae are sacks, filled with synovia, which occur where one structure moves upon another. The *tendon sheaths*, which surround tendons where they change direction, are lined by synovial membranes, and these continue over the surface of the fascial tunnels in which the tendons slide.

The control of the structure of joints is largely influenced by hereditary factors. Limb-buds transplanted to atypical positions may produce bones united by synovial joints even though few muscles and no normal strains are present. Yet the details of the form of the joint develop in response to the conditions imposed. Once again we see the double dependence of the tissues (p. 4). For example, the fibres of the ligaments in the capsule that prevent disarticulation develop in the directions in which strain occurs.

6. Factors controlling bone deposition

The determination of part of the developing embryo to produce cartilage and bone occurs very early. Tiny fragments of material taken from a chick embryo at 4 days or even earlier, when they consist only of apparently undifferentiated mesoderm, can be grown on the chorio-allantoic membrane (p. 483) of an older embryo. There they may develop into an almost perfect femur, with recognizable articular facets (Fig. 5.11).

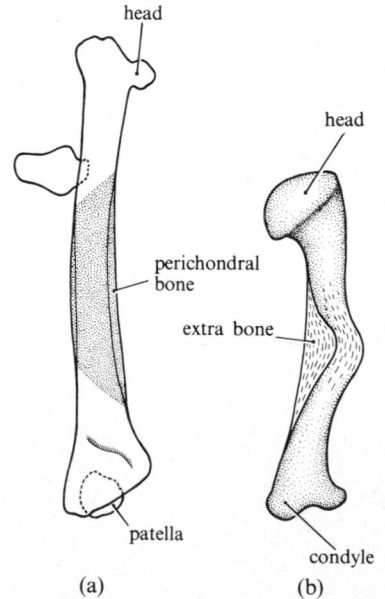

FIG. 5.11. Bones developed by growth of fragments of limb-buds in the chick. Pieces of the limb-buds of 4-day embryos were grafted to the chorio-allantoic membrane. (After Murray (1936). *Bones.* Cambridge University Press.)

Many experiments of this sort could be cited to show that the existence of a 'cartilage' or 'bone' as a separate entity is determined by inherited morphogenetic factors and is not the result of pressure, tension, or other mechanical influences. Nevertheless, the final detailed form of the bone depends on the forces that fall upon it (Glucksmann 1942). Joint surfaces develop on bones that are not stressed, but the surfaces come to conform properly only when there is movement between them. Similarly the sizes and shapes of the various ridges and depressions to which muscles are attached depend upon the use of these muscles. In young rabbits compelled by means of splints to rotate the foot laterally the direction of twist of the femur was found to become markedly abnormal. When there is a deficiency of certain vitamins or other nutritional requirements, the outer shape of the bones may be nearly normal but the walls are thin and there are other structural defects. The shape of a bone is therefore controlled by double dependence, partly by the hereditary forces acting from within, partly by the stresses imposed from without (see p. 51).

In view of the great importance of the shapes of bones for the study of evolutionary history it would be especially valuable to know something of the factors that determine the time and place of bone deposition and thus produce the characteristic patterns of 'bones'. In particular it is not clear how arrangement into a set of irregular bones has any advantage over ossification of the skull in some other pattern. The bones of the skull form in approximately their normal shapes in animals where for some reason the brain is very small, even though the reduced brain cannot give the normal pressure stimulus. When a whole bone, say the parietal, is removed from a young mammal the space is usually filled by growth of those that remain.

These observations do not tell us why bone does not form in uniform sheets in the skull, but they suggest that the division into separate 'bones' is not the result of simple mechanical factors. It is greatly to be hoped that the controlling factors will be found ultimately, and further experiments of the above sort might be interesting. It should be possible to discover experimentally how pressure, tension, or more general morphogenetic factors determine that bone is produced in restricted areas. We shall return to consideration of this subject in the next chapter, after examining the arrangement of the skeleton.

6 The framework of the body

1. Forces acting upon the body

THE skeleto-muscular system provides a means by which the body can be supported in various positions against the force of gravity. It can also move the body as a whole or change the position of its parts so as to move other bodies, especially in the 'handling' actions of primates. The arrangement of the skeleton and muscles differs according to the environment in which each species of animal lives, in other words the hereditary organization provides an apparatus suitable to meet the conditions in which the animal will find itself. In this sense the skeleton and muscles provide a representation of certain features of the environment. Moreover, during the life of the individual, the stresses placed upon the tissues provide information by which this representation is made to conform more exactly to the particular conditions that the animal meets.

Study of the skeleton and muscles must take account of the physical principles that affect any system performing mechanical operations. The strength of the parts must be sufficient to meet the forces that fall upon them, with an adequate margin of safety. The frequency with which bones are broken and muscles and ligaments torn shows the severe limitations imposed upon the actions of the body by the need to carry its own weight under various circumstances. Anything that reduces this load will be of importance in ensuring the survival of the animal.

The loads that the tissues must bear are chiefly those imposed by their own weight and actions (leaving aside the rather unusual load-carrying habits of man). It is important, therefore, that the body should be served by tissues that combine the maximum of strength with the minimum of weight. The forces that tissues must resist are the external forces, especially gravity, and the internal forces brought into play to counteract gravity or to move the body.

When two equal and opposite forces are applied to a body it is said to be under *stress* (force per unit area). The term stress is also applied to the internal forces within a material that prevent its distortion when loaded. We may call this 'internal stress' (p. 46). Changes of shape of a body under loads are called *strains* and are measured as ratio of change of length to original length. The ratio of stress to strain is the modulus of elasticity (Young's modulus). Stress is subdivided according to the direction in which the forces act (Fig. 6.1) into (1) compression, when parts tend to be pushed together, (2) tension, when parts of the body tend to be pulled apart, and (3) shear, when parts tend to be slid over each other. A pair of equal and opposite parallel forces that are not collinear form a couple tending to rotate the body. Two couples of the opposite sense of rotation are said to cause torsion (4). When a body is bent, tension, compression, and shear operate together but in different places.

In considering the properties and arrangements of the tissues that resist external forces we can ask three questions. (a) Are the materials themselves well suited to retain their shape under the forces that they are likely to meet? (b) Are the materials arranged in such a way as to use the minimum amount of material necessary to meet the forces acting upon them? (c) Are these forces themselves kept to a minimum by efficient arrangement of the parts of the whole system? Such a discussion of efficiency of materials is, of course, subject to the limitations that restrict the organization of all living systems. The building materials available are the metabolizable carbon compounds. The engineer

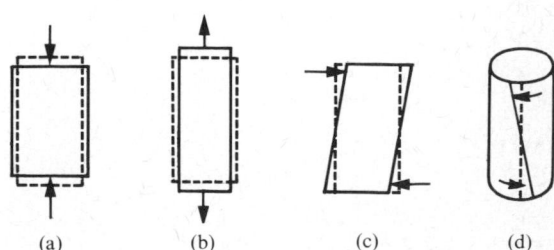

FIG. 6.1. Diagram of the action of external forces on a body. (a) Compression, (b) tension, (c) shearing or bending, (d) torsion.

could make 'better' bones out of, say, alloys of aluminium. Further, the organism must grow in size, function properly at all sizes, and be constantly replaced, even in the adult. Hence bones cannot be used to apply constant tensions, and muscles have to be used instead, although they have a lower strength–weight ratio than bone and use more metabolic energy. Finally, certain types of structure that are very efficient cannot be built by the body at all—notably wheels and axles.

2. The strength of living materials

It is difficult to discuss the 'efficiency' of the materials in a living system quite as an engineer would do. The nature of living systems and their dependence upon past history impose limitations on the materials available. However, we may attempt to answer the first of the above questions by measurement of the strength of the materials under various conditions. Table 6.1 shows figures for various substances and tissues.

These figures show the load at which the substance breaks under the three conditions, and it will be seen that bone and tendon compare favourably with all but the strongest non-living materials. Tendon is an efficient material in resisting tensile forces, and bone

TABLE 6.1

Substance	Ultimate strength (lb in^{-2})		
	Tension	Compression	Shear
Hard steel	90000–400000	—	65000–220000
Cast iron	11000–30000	50000–130000	18000
Granite	—	15000–45000	—
Concrete	0–1000	0–10000	0–2000
Bone	16300–21000	17900–23900	—
Cartilage	2410	21200	3270
Tendon	6400–8500	—	—
Nerve	13000	—	—
Artery	200	—	—
Muscle	80	—	—

TABLE 6.2

Substance	Ultimate strength in tension (lb in^{-2})	Density (lb in^{-3})	$\dfrac{\text{Strength}}{\text{Density}}$(in)
Hard steel	90000–400000	0·28	321400–1428500
Duralumin wire	70000	0·10	700000
Granite	1500	0·10	15000
Bone	16300–21000	0·068	239700–308800

resists both compression and tension, especially the former, but it is rather easily broken under shearing forces.

Table 6.2 shows the relationship between strength and density of various substances. The last column expresses the length of a piece of material that can be hung and support its own weight. Bone evidently combines strength and lightness as well as all but the best non-living materials.

3. Arrangement of materials along trajectories

The question of the arrangement of the material within the bones has attracted much attention since the anatomist Meyer and the engineer Culmann showed in 1880 that the trabeculae of bone in the human femur are arranged in an economical manner similar to that adopted by an engineer who disposes his material in order to make the best use of it in girders (p. 50). To understand this and also the principles on which the

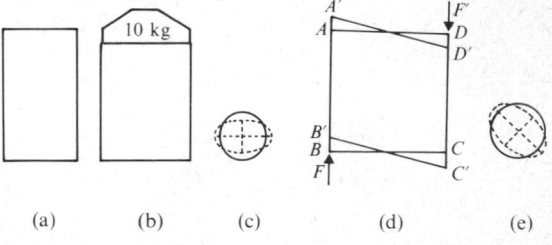

(a)   (b)   (c)   (d)   (e)

FIG. 6.2 The effects produced by the application of a load to a body.
(a) and (b) show how a block when loaded is shortened vertically but not decreased in volume, and is therefore expanded horizontally. This is to say that any small circle imagined on a vertical plane becomes an ellipse; there is maximum compression in the vertical and tension is the horizontal plane and these are called the principal axes of the strains. (c), (d), and (e) show how shearing forces *FF'* also produce two principal axes of strain set at right angles. (After Murray (1936). *Bones*. Cambridge University Press.)

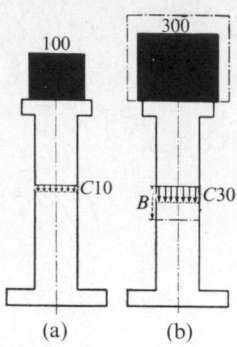

FIG. 6.3. Effect of increased load on stressing of a column. The lengths of the arrows are proportional to the stress. Increase of the load from 100 kg to 300 kg increases the stressing (C) from 10 kg cm^{-2} to 30 kg cm^{-2}. The breaking stress (shown dotted, B) is 60 kg cm^{-2}, and the column therefore has a factor of safety of 6 times in the first case but only twice in the second. (After Pauwels (1948). *Z. Anat. Entw. Gesch.* **114**.)

third of the above questions may be answered, we must consider how an engineer analyses the effects of forces acting upon a body. In a block of material supporting a load any small spherical region will tend to be deformed by the strains into an ellipsoid (Fig. 6.2). In any plane there will be a line of maximum compression strain and at right angles to this a line of maximum tension strain. Therefore we can easily state in a piece of material how the axes of strain will run when force is applied in any given direction. The internal stressing must be arranged to meet these strains. As Murray (1936) puts it, 'in the directions of principal stress there are no shearing stresses, but in any other direction there will be shearing stresses'. Lines indicating the axes of strain are called '*trajectories*', and they can be drawn so that the closeness of the lines corresponds to the degree of stress. Whether the lines are straight or

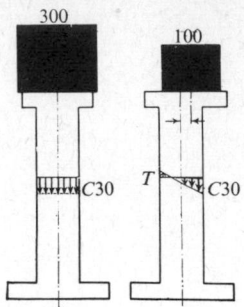

FIG. 6.5. In a column loaded asymmetrically compression strains C develop along one part of the axis and tension strains T along another. The breaking strain is reached with a smaller load. (After Pauwels, *loc. cit.*)

curved they must fall into two sets—*compression lines* and *tension lines*—crossing each other at right angles. The engineer builds his structures so that the main girders lie in these lines of greatest stress.

4. Symmetry of loading

Whether any material will be strong enough to meet a given load depends on the way the load is distributed. Fig. 6.3 shows how a load symmetrically placed on a girder produces compression strains uniformly distributed through it. Increasing the load increases the stress until some breaking stress is reached. If we imagine that the breaking stress is 60 kg cm^{-2}, then in Fig. 6.3(a), where the stress is 10 kg cm^{-2}, there is a margin of safety of 6 times; in Fig. 6.3(b) there is a margin of twice. The only way that a girder of given shape and material can be strengthened is by increasing its diameter. If the diameter of the girder is greater the margin of safety is increased (Fig. 6.4).

If the load is not symmetrically applied the compression and tension stresses will be as in Fig. 6.5. More-

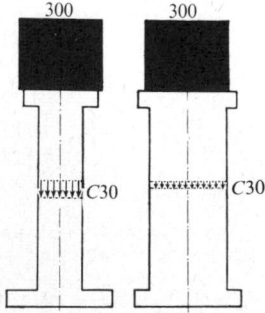

FIG. 6.4. Increase in the dimensions of the column carrying a load decreases the stress, giving a greater margin of safety. (After Pauwels, *loc. cit.*)

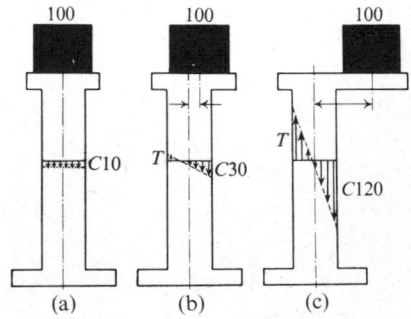

FIG. 6.6. As asymmetry of loading increases, the strain on the supporting column becomes very much greater. (After Pauwels, *loc. cit.*)

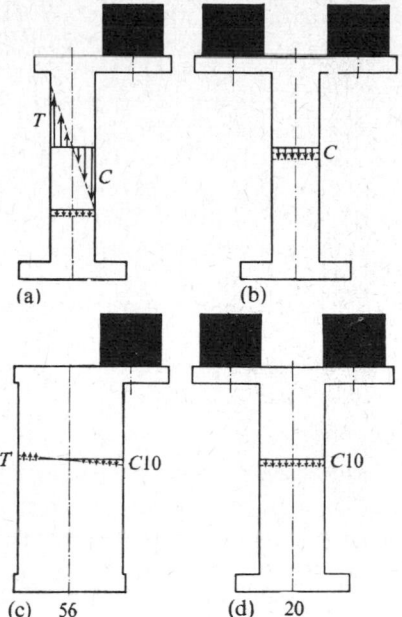

FIG. 6.7. Effect of a counterweight in reducing the strains set up by asymmetrical loading.
The large compression and tension strains shown in (a) are reduced to smaller compression strains in (b), in spite of the greater load carried. This allows a considerable reduction in the size of the column that is needed for a given margin of safety ((c) and (d).) Thus for a breaking limit of 10 kg cm^{-2} the column of a particular material needed can be reduced in weight from 56 kg to 20 kg by counterweighting. (After Pauwels, *loc. cit.*)

over, the stresses are then no longer uniformly distributed, and a given load may set up much higher stresses in some parts than were present with a symmetrical arrangement (Fig. 6.6). The breaking point of the whole structure is limited by that of any one point, and it is therefore clear how important it is for a living body to shorten the length of any lever arms, such as that of Fig. 6.6, in order to avoid having to develop massive structures to meet the stresses that would result.

5. Counterweighting

There are various means by which the stresses set up can be reduced and the structure made lighter without loss of strength. A counterweight has the effect not only of distributing the stress uniformly through the girder (Fig. 6.7) but of greatly reducing the total stress and hence the amount of material necessary to carry the load. Animals sometimes employ this principle of counterweighting even where it adds to the total weight to be moved; thus it has been suggested that the large weight of the tusks of elephants may serve as a counter-

weight to the body (Young 1962, p. 711). In most mammals there is a careful balancing of weight about the main weight-bearing limbs, for instance, by the tail in the kangaroo and the head and neck in the giraffe; no doubt it is only by such balance that the great weight can be supported by limbs of reasonable proportions.

Even more important in animal organization is the reduction of size of compression members by appropriate *bracing*, whose effect is analogous to that of counterweighting. This effect can be shown by use of the photoelastic method illustrated in Fig. 6.8. In this technique a model of the girder or other object to be studied is cut out of a polymerized resin that becomes birefringent where it is stressed and can therefore be made to show the trajectories of strain in the material by placing and viewing it in polarized light. The closeness of the dark lines then indicates the degree of stressing, each line representing a multiple of the stress. Thus in Fig. 6.8 (a)–(d) each parallel thin dark line in the column represents a stress of 10 kg cm^{-2}. With the asymmetrical load the maximum compression stress is 94 kg cm^{-2} and the tension stress 79 kg cm^{-2}. Addition of a brace G of increasing tension reduces the total stress and converts it through an asymmetrical compression stress (Fig. 6.8(e)) to a pure compression stress (Fig. 6.8(f)). There are many examples of such bracing action throughout the skeleto-muscular system (Figs 6.11 and 12). It should be noted that the behaviour of the model used in the photoelastic method is much oversimplified as compared with the living tissue, bone.

6. Economy in use of materials

As an introduction to the application of these principles to the body we may look at the way the amount of material in a crane may be reduced. With the primitive form of Fig. 6.9(a) the weight of a certain metal needed to make a crane able to carry a given load is 218 kg. By counterweighting this can be reduced to 189 kg (Fig. 6.9(b)), by counterweighting and bracing to 164 kg, and with an appropriate bending of the girder to 155 kg. Further reductions are possible because not all parts of the cross-section of the material are equally stressed. A symmetrical load is distributed uniformly over the material (Fig. 6.10), but the bending stresses set up by an asymmetrical load fall mostly at the periphery. Therefore the centre part of the girders can be omitted, the girder being left either circular (as in many bones) or of H shape (as is more usual in engineering practice) (Figs 6.9 and 6.10). Furthermore the stressing is least at the tip of the girder and increases downwards. This factor alone allows the members to be reduced by shaping, as in Fig. 6.9(f). Proper construction allows a reduction of over 4 times in the weight of material employed to support a given load. Comparable saving in animals would be of outstanding selective and evolutionary importance and architecture of the bodies of mammals and man shows abundant signs of the operation of such factors.

7. Bracing of the long bones

If the long bones are subject to bending stresses we can immediately understand that their circular form with hollow centre leads to an effective and economical use of material, for example, in the human femur. When the weight is all on one leg the femur is loaded eccen-

trically. The situation is made less simple by the tilting of the pelvis at this stage of walking, which tends to bring the centre of gravity of the body over the femoral head, but the femur still remains eccentrically loaded. Such an arrangement would put severe compression and tension stresses on the edges of the femur, which would have to be very massive to meet them were it not that the forces are reduced by the action, as braces, of the abductor muscles running from the ilium to the femur (Fig. 6.12).

A further bracing action is performed by the ilio-tibial tract (Fig. 6.11). Fig. 6.12 shows by investigation of the photoelastic properties of a model of the femur

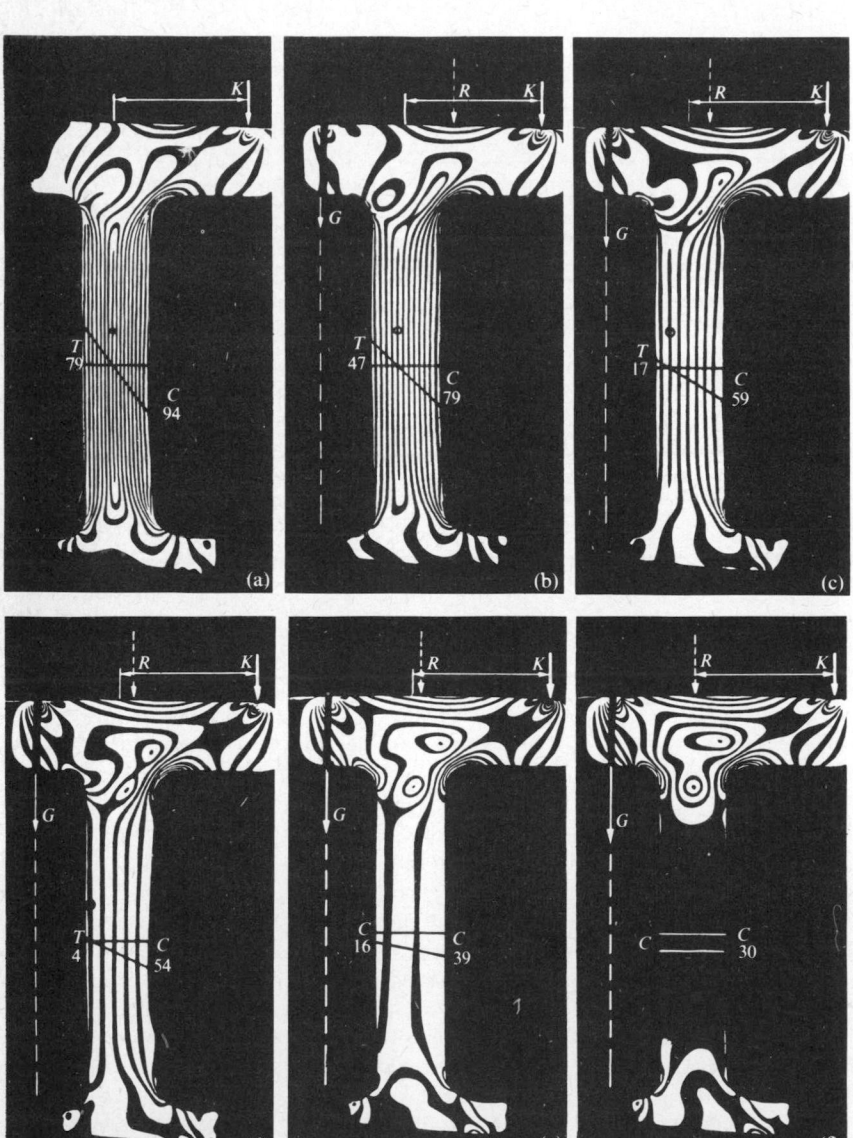

FIG. 6.8. Copies of photographs of a column of polymerized resin loaded and photographed in polarized light. The closer the dark lines, the greater is the stress gradient. In (a) the load K is applied asymmetrically and sets up large compression and tension stresses; in (b) a brace G is added, and the resultant force operates at R, reducing the stress. As the tension of the brace is increased in (c)–(f), the resultant approaches the centre of the column and finally reduces the effect to a small compression stress. (After Pauwels, *loc. cit.*)

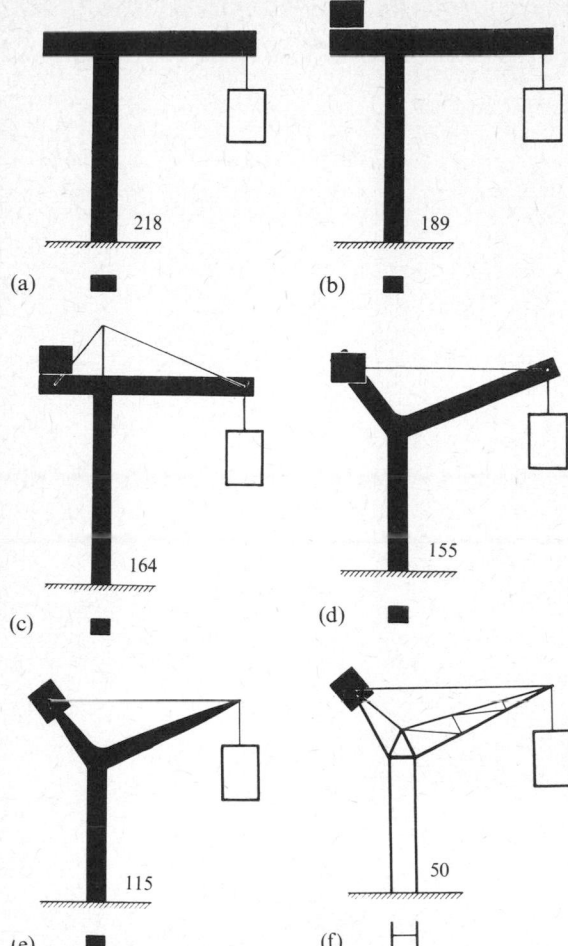

(a)   218

(b)   189

(c)   164

(d)   155

(e)   115

(f)   50

FIG. 6.9. Diagram to show the effect of proper counterweighting, bracing, and arrangement of girders in allowing reduction of weight of a crane of a given material carrying a given load. In the series (a)–(f) the weight of the girder necessary to carry the weight is progressively reduced from 218 kg to 50 kg (see text). (After Pauwels, *loc. cit.*)

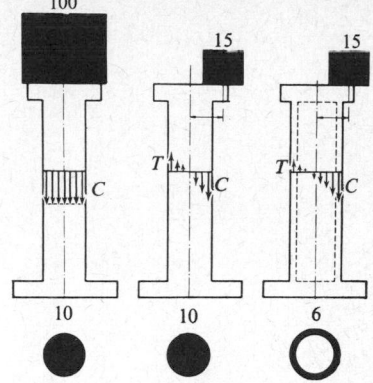

FIG. 6.10. Diagrams of the stressing of a column under asymmetrical loading, showing how the stressing lies at the periphery, allowing a reduction of the weight of the supporting column by use of a hollow construction. (After Pauwels, *loc. cit.*)

the bones of stresses greater than they can bear.

Confirmation of this action of the ilio-tibial tract is found in the fact that the femur is built in such a way that the strongest part of its circumference corresponds at each level to the part that would need to be most strongly stressed to meet the strains that remain after

how the presence of the tract reduces the compression stress from 83 kg cm^{-2} to 48 kg cm^{-2} and the tension stress from 69 kg cm^{-2} to 8 kg cm^{-2}. Such great reduction of tension stress is characteristic of this type of bracing and is especially valuable since bone is less resistant to tension forces than to compression forces. The effect of the ilio-tibial tract is to reduce the stressing of the femur by as much as one-half. The enormous advantage of this in allowing saving of material in the bones is obvious. Interference with the action of such braces as the ilio-tibial tract by severance or loss of tension through damage to muscles or their nerves or by weakening of ligaments may lead to the placing on

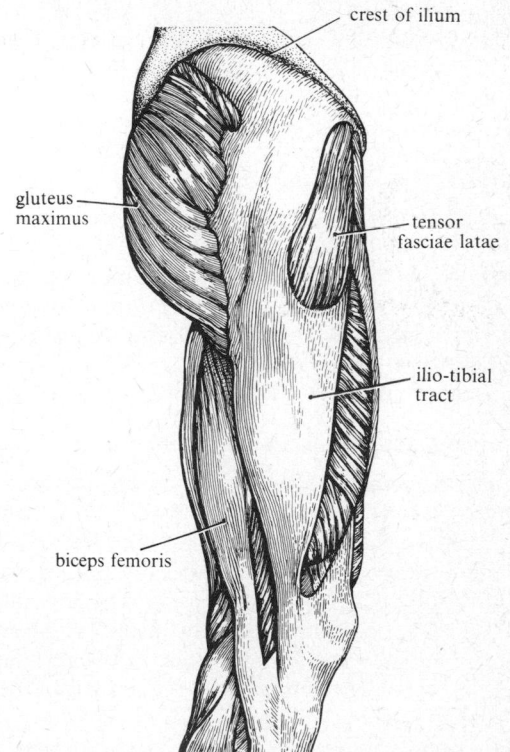

FIG. 6.11. Lateral view of human thigh. (After Pauwels, *loc. cit.*)

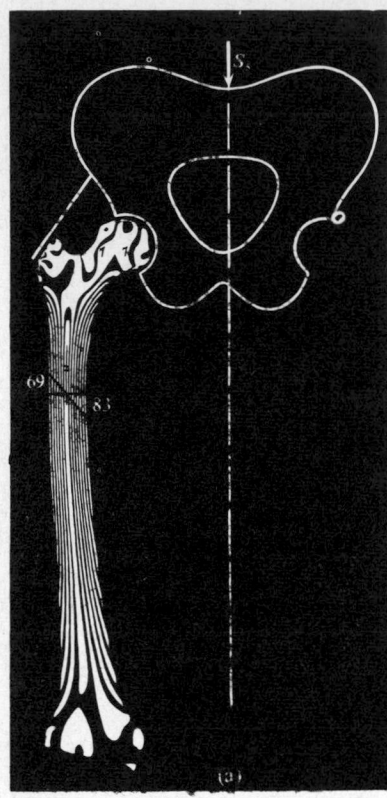
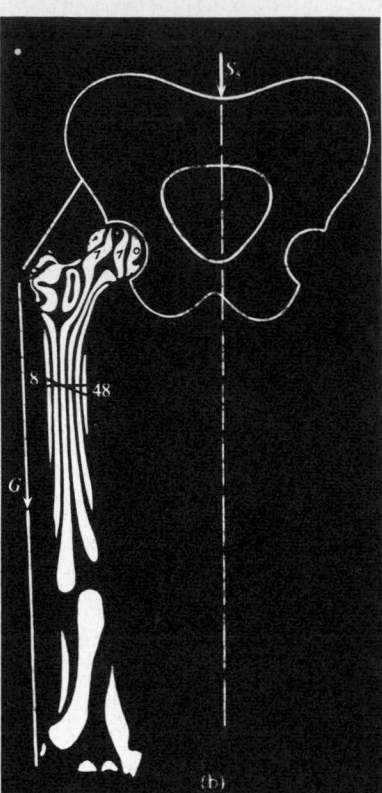

FIG. 6.12. (a) The distribution of stressing in the femur shown by transmission of polarized light by a flat model. (b) The effect of the ilio-tibial tract is to reduce the tension stress greatly and the compression stress considerably. (After Pauwels, *loc. cit.*)

the combined action of the body-weight and the tension of the tract.

Many similar examples of the effect of muscles and ligaments in reducing the strain by counterbalancing can be found throughout the body. The brachialis muscle acts in this way on the lower end of the humerus (p. 93) and the deltoid acts similarly at the upper end. The long head of the biceps muscle, like other two-joint muscles, has a similar action along the whole length of the bone.

8. Arrangement of trabeculae within the bones

Within the bones themselves the trabeculae are so arranged as to lie in the directions in which the material is stressed. In the human femur the trabeculae are seen to follow lines similar to the trajectories or lines of internal stress in a crane (Fig. 6.13). The trabecular lines cross approximately at right angles, suggesting that they are following lines of tension and compression stressing, as would the trajectories set up by the action of bending forces.

The main compression lines (cc^1) start from the surface of the head and run down to join the compacta of the shaft on its medial side. The tension lines (tt^1)

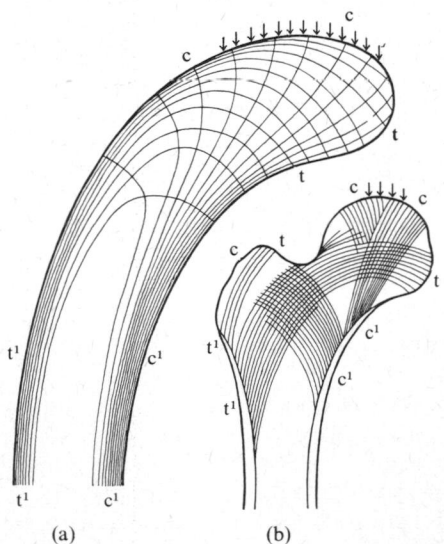

FIG. 6.13. (a) Trajectories in the head of a crane of the type known as a Fairbairn crane loaded as shown by the arrows. cc^1 compression lines; tt^1 tension lines. (b) Diagram of the trabeculae of the head of the femur; the arrows show the approximate method of loading. (After Murray 1936.)

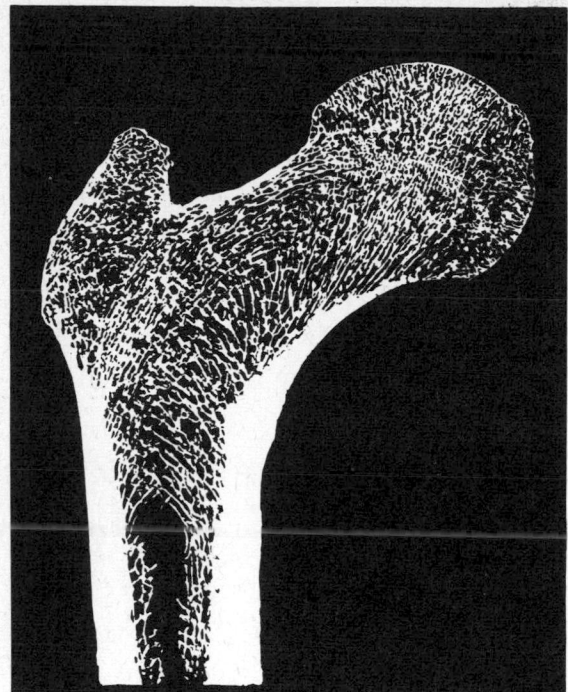

FIG. 6.14. Frontal section of upper end of human femur. (After Koch. (1917). *Am. J. Anat.* **21**.)

run across the underside of the head and neck to the lateral side of the shaft. The situation is complicated by the fact that other forces fall on the bone; for instance, the pull of the muscles that are attached to the greater trochanter (p. 97). Such attachments will themselves produce tension stresses in the bone, especially if the muscles are in frequent use, as in walking.

Similar arrangements of the bone trabeculae occur throughout the body but are often somewhat obscured, so that doubt has been thrown on the whole trajectorial theory. The trabeculae often do not cross at right angles and therefore cannot be simple stress lines. Yet the general agreement with the theory is striking (Fig. 6.14); divergences are only to be expected in a system such as a living body, which, besides its great complexity, is continually varying its activity. The action of muscles must produce varying forces acting on the bones. The fact that animals can jump so well shows that the forces exerted by the muscles may greatly exceed those due to the weight.

There can be no doubt that the internal arrangement of the bony trabeculae provides a stressing for the bone appropriate to the forces that act upon it. But how does it come about that the stress lines become 'materialized'

in the bone? 'Functional' factors play a considerable part in determining bone structure. The trabeculae of the femur, for instance, do not have their characteristic arrangement at birth; this appears as the limb is used for walking (Fig. 6.15). Similarly, study of bones that have been badly set or have become fused together at unusual angles shows the development of a completely new architecture under the influence of the new forces.

Some explanation is to be found in the study of the incidence of stresses following the application to any material of forces that tend to distort it. The resulting strains can be considered as occurring along axes at right angles to each other but inclined to the main axes of the material. The view usually expressed about the morphogenesis of bone is that it is laid down along the lines of maximum stressing by compression and tension. As the loads increase, further lines of bone are added. Any bone that is not stressed is removed. The bony material that remains therefore 'materializes' the stress trajectories, exactly as an engineer would do in his diagrams. It may be that as spicules of bone grow out into the cartilage at the epiphyses (p. 38), their tips become oriented by the stresses present in that region. Whatever the mechanism, the effect is to produce a structure that combines strength with lightness by placing material only where it is stressed.

However, another school of thought holds that deposition of bone is not a result of simple compression or tension but of bending (Frost 1964). Changes of curvature at surfaces are held to provide a signalling system whose feed-back ensures that bone is removed

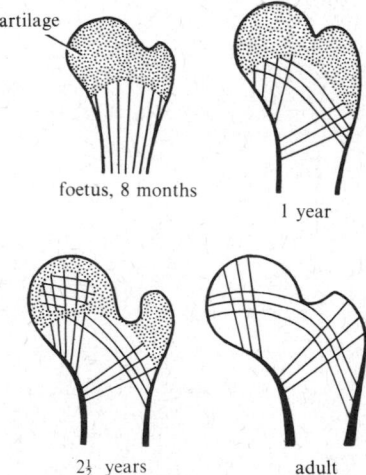

FIG. 6.15. Drawings of frontal sections of the heads of human femora at different ages to show the changes in the arrangement of the trabeculae with use. (After Townsley (1948). *Am. J. Phys. Anthrop.* **6**.)

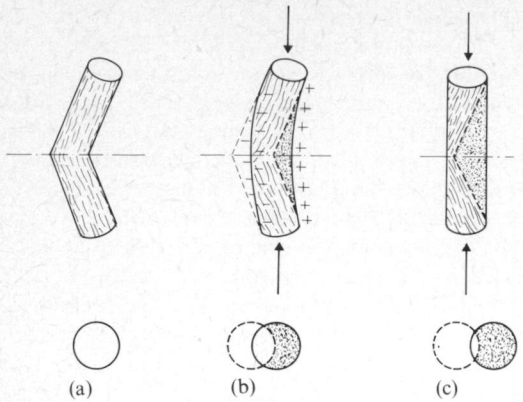

(a)   (b)   (c)

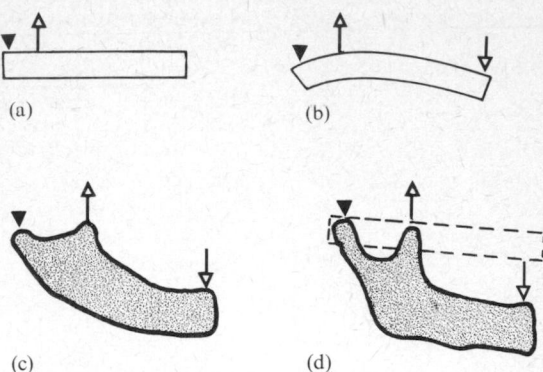

(a)   (b)

(c)   (d)

FIG. 6.18. Derivation of human mandible shape. (a) A straight bone supported by a fulcrum at one end and a muscle nearby will (b) bend when a force is exerted near the other end. It is assumed that the bone is immune to change at the muscle attachment site. (c) Formation is inhibited at the top (convex) surface and resorption at the bottom (concave) surface until the bone acquires a shape not subject to bending (d). (Redrawn from Frost, *loc. cit.*)

FIG. 6.16. Correction of abnormal angulation. Longitudinal sections (upper row): a child's long bone which has mis-set after fracture (a) is subject to compression and bending when loaded. On the bending theory removal of bone is signalled at the convex longitudinal surface (minus signs) and deposition at the concave surface (plus signs); (b) resulting finally in a straightening of the bone (c). The cross-sections at the level of the angulation show the drift of bone substance into the line along which the stress is applied; the eroded part is within the dotted line and the newly deposited part is shaded. Microscopical examination will reveal the original longitudinal axis as a grain in the original fragments. (Redrawn from Frost, H. M. (1964). *Laws of bone structure*. Courtesy of Charles C. Thomas, Publisher, Springfield, Illinois.)

where the curve is convex and deposited where it is concave (Fig. 6.16).

Evidence has been produced that bone functions as a transducer, either by its solid-state or piezo-electric properties. Currents, which are transmitted through the tissues, can be shown to be generated when a bone is bent. Experiments with battery packs implanted in bone show that with a current of 3 μA passing for 14 days, there is massive deposition of bone at the cathode (Bassett, Pawluk, and Becker 1964). According to this scheme therefore, bone growth is controlled by a

precise feed-back system; the theory is not proved and further experiments are needed (see Evans 1973).

No doubt many factors are involved in producing the final form of a bone. Thus growth is prevented if an epiphysial disc is strongly compressed, a fact used clinically to restrain growth of long bones (Sijbrandij 1963). After removal of the epiphysis at one end of a bone the other end produces bone abnormally rapidly (Hall-Craggs 1969).

If the more-or-less liquid contents of a bone exert a hydraulic effect they will tend to make the walls bulge under compression. The resultant bending would stimulate the deposition and removal of bone to produce the familiar waisted shape of vertebrae (Fig. 6.17).

Yet another bone whose shape agrees with the bending theory is the human mandible (Fig. 6.18). The temporalis and masseter muscles tend to bend the bone upwards. Here, however, we should have to assume that removal at the convex surface is inhibited where the muscle is actually attached. Similar special conditions would have to be postulated for other muscle attachments. It remains for further experiment and comparison of bones to elicit the precise operation of applied loads on bone deposition.

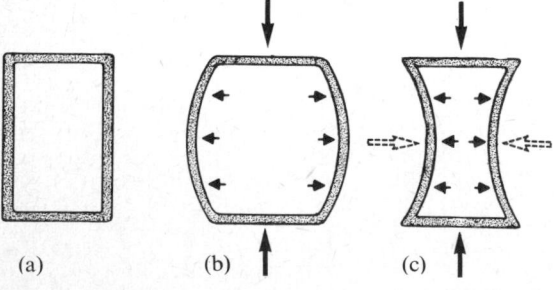

(a)   (b)   (c)

FIG. 6.17. Origin of waisting in vertebrae. (a) The bone is hollow with more or less liquid contents. (b) On loading, the sides bulge due to the internal hydraulic pressure and this stress is relieved by drift in bone substance resulting in (c) a narrow-waisted form. (Redrawn from Frost, *loc. cit.*)

9. Loading and safety margin of the human femur

The effectiveness with which bone fulfils its functions may be assessed by considering the factor of safety that it provides. Fig. 6.19 shows the calculations made by Koch of the distribution of lines of maximum stress in the femur under a load of 100 lb (45 kg) applied to

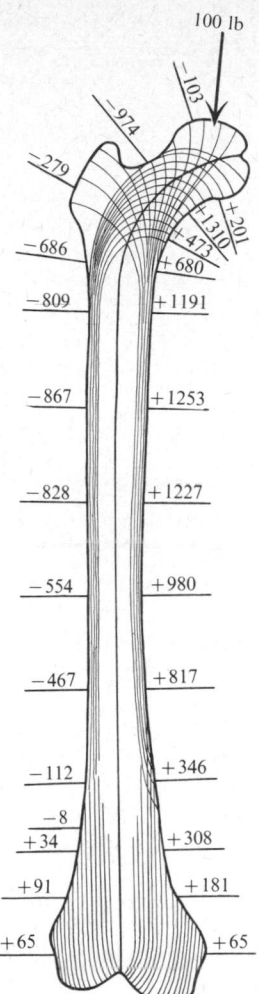

100 lb

−103
−974
−279
−1310
+201
+473
−686
+680
−809   +1191
−867   +1253
−828   +1227
−554   +980
−467   +817
−112   +346
−8
+34   +308
+91   +181
+65   +65

FIG. 6.19. Diagram of the lines of stress and their quantity (lb in^{-2}) in a normal femur for a load of 100 lb (45 kg) on the head. The plus signs show compression- and the minus signs tension-stresses. Note that the greatest stressing is in the neck and upper third of the shaft. The stresses in the standing position are about 0·6 of those shown; when walking × 1·6, and when running × 3·2 (1 lb in^{-2} = 0·07031 kg cm^{-2}). (After Koch (1917). *Am. J. Anat.* **21**.)

the head. For the purpose of the analysis he divided the bone into 75 sections and found that the greatest stresses are set up in the neck. Here they may reach 1340 lb in^{-2} (94·2 kg cm^{-2}) for compression and 974 lb in^{-2} (68·5 kg cm^{-2}) for tension. This is a region in which fractures commonly occur, especially in old people, whose bones have become brittle. Fractures are also common in the shaft, where the stress is nearly as great as it is in the head.

Measurements of the pressure within the hip joint

have recently been made on living people (Rydell 1966). Strain gauges were built into artificial femoral heads before these were inserted into patients, and readings were taken 6 months later. When standing on one leg the load on the hip joint exerts a force of about 2·6 times the body-weight. When walking the load reaches 2·8 times and in running up to 5 times the body-weight. These direct measurements include all incidental forces due to muscles and ligaments. They show that the pressure is applied from above and is directed medially and ventrally. This must produce a moment of the head about the shaft leading to compression in the lower cortex of the neck and tension in the upper. This is the situation assumed to exist in the interpretation of trabeculae on p. 50.

To determine the factor of safety we must estimate the stresses that the bone will bear not under a static load of 100 lb (45 kg) but when the person is walking, running, jumping, or falling. With the strengths for bone given in Table 6.1 (p. 45) we thus have a safety factor of about 5·7 times for both forms of stress during running. These compare with the factors of safety of about 5 times usually allowed in a steel structure. The body evidently builds economically. The above calculations make rather simple assumptions about the way the load falls on the bone, and in practice the action of the muscles would complicate the situation considerably.

10. Control of bone organization

The moulding of bones during later development and throughout life is certainly influenced by mechanical factors, although, as we have seen, the trajectories follow stress lines rather more complicated than those found in simple girders. Three general rules may be stated about the control of bone growth.

(a) Continuous pressure on bone leads to its absorption. This has been proved experimentally by pressing the neural spines of rabbits with bags of mercury and is shown in some brain tumours, whose local pressure may cause a hole to develop right through the skull.

(b) On the other hand, intermittent pressure stimulates bone deposition, as can be proved by rhythmical applications of weight on the neural spines. There is probably a similar effect of intermittent tension; thus the tendons of muscles that are frequently used may become ossified, for example, a 'rider's bone' develops in the tendon of the adductor longus muscle of active horsemen. The formation of ridges and tuberosities at points of muscle attachment can be similarly explained. However, this may involve addition of bone on the endosteal surface beneath the attachment, even if it is being removed from the outer side (Hoyte and Enlow 1966).

(c) Finally, the third rule is that bone that is unstressed undergoes atrophy. This can be seen in the texture of the bones of limbs that have not been used. The hollow form of long bones is probably determined, like their general shape, by hereditary factors, but the fact that no bony tissue is laid down later in life at the centre of the hollow shaft is presumably related to the fact that this central region is not stressed.

With this knowledge we can give some 'explanation' of the arrangement of the trabeculae in spongy bone, assuming that new bone is being continually replaced and is laid down along lines of intermittent compression or tension and removed where there is none (Wolff's law). Presumably the deformation produced by the application of force to the bone is sensed by some receptor component of the bone, perhaps the piezo-electric properties (p. 52). The appropriate signal is then sent to the mesenchymal cells that lay down or absorb bone. The receptors and signals are unknown, but are

perhaps produced partly by the strains at the surface when the bone is bent. The cross-sectional size of each bone is thus adjusted to the level that reduces bone-surface deformation below the level of detectability by the mesenchymal cells. The loads are largely due to the action of the muscles, and the size will therefore be correlated with the usage. To proceed further we should need to understand the underlying physical and chemical factors that control bone formation. Deposition of bone is preceded by a concentration of calcium salts on the site by the action of the enzyme phosphatase, and further study of the factors causing the release of this enzyme or its activation should be helpful.

Bone formation involves a special form of the metabolism of calcium that occurs in all cells. It may be that we shall come to understand the process better when we find out how it is related to the factors that maintain the ionic balance in other tissues. Some modification of the processes that occur in other cells has produced this extraordinary substance which is so hard and tenacious.

7 Muscles

1. The significance of movement for homeostasis

MAMMALIAN life is largely based on the system of expending energy to achieve positions in which appropriate raw materials can be obtained. The elaborate system for receiving information from the environment is allied with correspondingly elaborate effectors, the muscles, triggered and ready to respond according to the signals they receive from the nervous system. In studying muscles we may therefore examine first how they produce their actions, converting supplies of food and oxygen into movement. Investigations of the visible structure and chemical and energy interchanges of muscle tell us much about the nature of the contractile process. Secondly, we can study how the process of contraction is controlled by the signals received from the nervous system. Thirdly, we can investigate the arrangement of the muscle fibres in relation to the bones and other tissues and so reach an understanding of the various movements that are made.

2. Muscles for movement and for holding

Animals are able to keep themselves alive in spite of unfavourable circumstances either by moving to more suitable conditions or by so manipulating the surroundings as to make them favourable. In mammals this is achieved by a framework, composed of the skeletal parts, joined by ties that can be varied in length, the muscles. The mobility ensured in this way enables the body to perform varied feats that as yet can be only roughly paralleled by human tools and constructs. A horse can move rapidly over rough ground and jump accurately over obstacles that no machine can pass so simply. No instrument yet built approaches the human fingers in the number of delicate operations that can be performed.

The parts being freely mobile, the muscles serve to fix as well as to move them; they act as braces of variable length, holding the organs of the body in place. This postural action, important in all animals, is especially so in those that live unsupported by water and must therefore maintain themselves erect against gravitational forces. The recognition of these two distinct functions, of movement and of holding, enables us to reach a proper understanding of the various sorts of muscle. It explains the fact that the tension is developed at different speeds in different muscles, and especially that there is variation in the speed with which muscles contract and relax. We can thus recognize two main classes of muscles, the *holding muscles* and the *movement muscles*; the former contract slowly and relax slowly and can therefore maintain the parts in position for long periods. The latter contract and relax fast and are able to produce quick and quickly changing movements.

3. Prime movers and antagonists

Muscle is a tissue specialized for the conversion of chemical energy into mechanical energy. When the muscle is serving for holding, the energy is expressed as tension. When the muscle produces a movement, the energy is expressed as external work.

It is characteristic of all muscles that they can pull but not push. Before they can do work again they must be elongated. The termination of the action of any muscle is followed by a more or less sudden diminution of tension, thus allowing whatever forces are acting against the muscle to elongate it. This general picture of muscle as acting in one direction is the clue to an understanding of the arrangement of the whole muscular system. It explains the fact that every muscle when it acts as a *prime mover* or *agonist* has an *antagonist*, able to restore it to its 'resting' length. A single muscle acting by itself can accomplish one single movement only and must then wait until it is elongated. Often this elongation will be accomplished by the opposing antagonistic muscle, but it may be brought about by gravity or other forces.

4. The neuro-muscular-skeletal system as a whole

Muscles are thus adjustable ties, able to vary their length and tension upon receipt of impulses from the nervous system. Therefore we cannot deal with the muscles without considering the nerves, sensory and central as well as motor. All these together form a single system, triggered to respond in such ways as will produce the adjustments necessary to maintain and reproduce the organism under whatever circumstances it may encounter. On the other hand, we cannot properly consider muscle without also dealing with the skeleton, since these two make the system of adjustable girders, the muscles being the chief tension members, the braces or ties, while the bones are the compression members or struts. The stability of the whole body must thus be considered as dependent jointly on the nervous, muscular, and skeletal systems. The body collapses when any part fails. Lack of control due to shortage of oxygen in the nervous system (as in a faint) disturbs the posture of a man as effectively as does fracture of his femur.

Although the system is capable of so much movement it can at other times remain steady and still. Moreover, all the movements are smooth and precise. This is made possible by an abundant supply of receptor organs in the muscles, the proprioceptors, which provide information to the nervous system about the movement of every part. The nerve centres themselves are so arranged that this feed-back of information ensures that stability is maintained and that smooth, precise movements are produced.

5. Isometric and isotonic contraction

The function of muscular tissue, therefore, is to exert tension which may or may not result in movement. If the ends of the muscle are fixed, the effect of the initiation of its activity is to exert a force along its length without any actual movement and this is said to be *isometric contraction*. If a muscle is allowed to shorten without increase of tension, for instance, in lifting a weight, the contraction is said to be *isotonic* or, better, *anisometric*. Muscles contract isometrically when they support loads or fix bones relative to each other. For instance, when a man carries a heavy suitcase in his hand the weight is sustained by approximately isometric contraction of the muscles that hold up the shoulder, such as the trapezius (see p. 89) at the back of the neck. During movements, for example, in bending the elbow or raising the heel off the ground in walking, conditions are neither fully isometric nor isotonic. It is obvious that the two types of contraction are not likely to be fully separated, there is seldom shortening without some increase of tension or vice versa.

6. Structure of smooth muscle

The muscles in which the powers of holding are most fully developed are the smooth (unstriped) muscles of

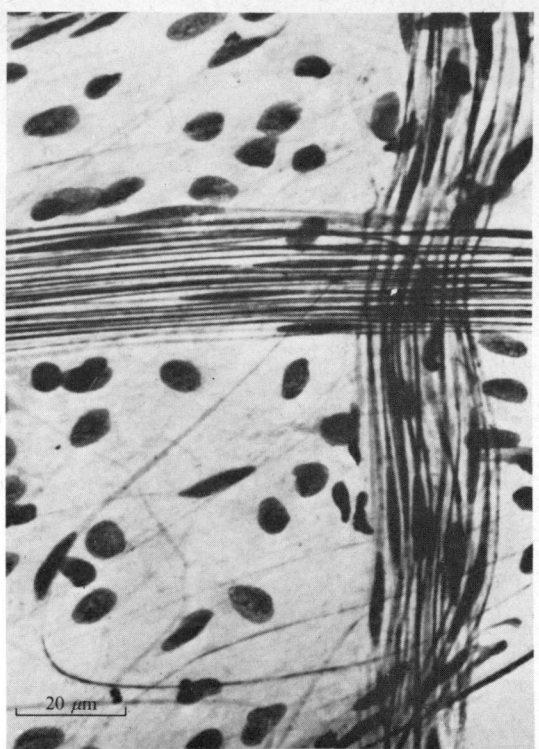

FIG. 7.1. Light micrograph of smooth muscle from cat bladder. The nuclei are elongated indicating that the muscle is in a relaxed condition.

the viscera. These have the functions of holding the organs in place and producing slow movements such as are needed to propel the food along the gut or to squeeze the secretion from a gland. Such movements are effective even if they are not quick or precise and they are usually produced by coats of muscle that are wrapped around the organs but not attached to any skeletal elements. Receptors are present in these abdominal viscera to record the amount of tension and thus provide the information with which the nervous system ensures regulation of the organs, but smooth muscle has no elaborate system of proprioceptors such as is used for the rapid and precise control of the muscles that produce actions of the limbs (p. 265) (see Prosser 1974).

The smooth muscle fibres consist of elongated cells, up to 0·5 mm long and 6 μm wide, each with a single nucleus midway along its length (Fig. 7.1). Part of the protoplasm of these cells shows under suitable conditions a longitudinal striation ('myofibrils') and is birefringent in polarized light. There is, however, no sign of any repetition of structure along the axis of the fibre, which is therefore said to be *smooth*, in distinction from

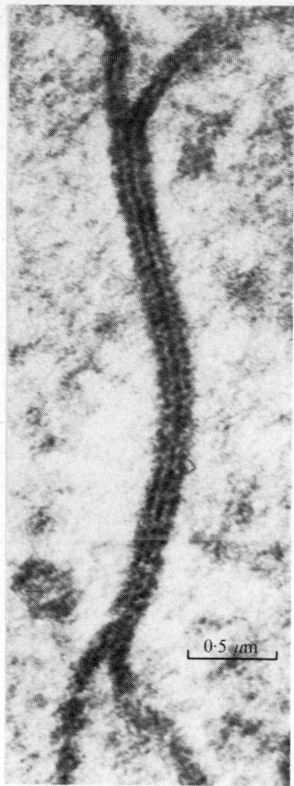

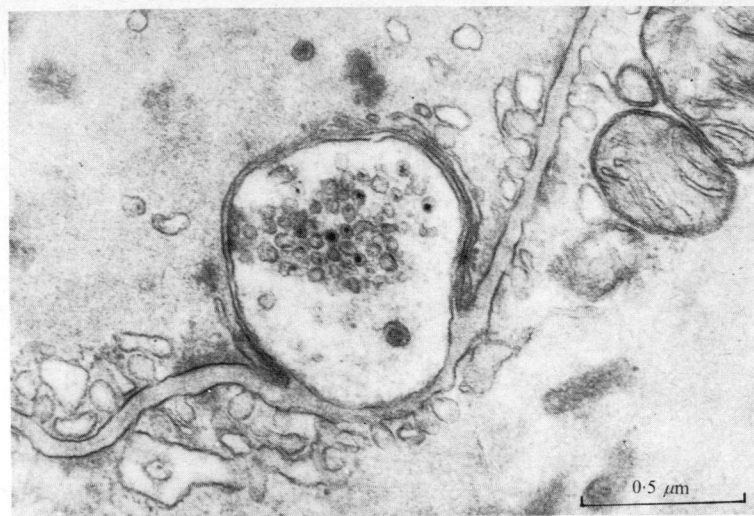

Fig. 7.3. Electron micrograph of adrenergic nerve ending, with granulated vesicles, on smooth muscle. (Figure kindly supplied by Dr. J. Taxi.)

Fig. 7.2. Electron micrograph of the contact region between adjacent smooth muscle cells. (Figure kindly supplied by Dr. J. Taxi.)

the *striped* or striated appearance of other muscles. Smooth muscles are also said to be *involuntary* because in man it is seldom possible to initiate their contraction at will. Nevertheless, by the Pavlov technique of the conditioned reflex it is possible to arrive at a condition first in which ringing a bell causes contraction of the pupil of the eye and then one in which thinking about the bell produces the same result. Moreover, accommodation of the focus of the eye is produced 'at will' by unstriped muscles. The concept of volition is evidently a complicated one and it is better not to introduce it into the nomenclature of muscular structure.

The exact nature of the contractile material of the smooth muscle cell is still uncertain. Actin and myosin are present, but are not regularly arranged as thick and thin fibrils as they are in striped muscle (p. 58). There are bundles of myofilaments embedded in sarcoplasm. Smooth muscle fibres are usually associated with collagen and reticular tissue to make bundles or sheets. Neighbouring muscle fibres may come into very close contact with each other by a zonula occludens (Fig. 7.2). This may allow electrical connexion and hence activation to spread between them. The nerve fibres of smooth muscle do not form discrete end-plates but

branch repeatedly and run along close to the surfaces of the fibres, stimulating them by the release of transmitter substances (Fig. 7.3) (p. 288).

In mammals smooth muscle occurs in the wall of the gut and urino-genital ducts, and around the blood vessels, in the skin, in the iris of the eye, and, mixed with striped fibres, in the levator palpebrae superioris that raises the upper eyelid. Smooth muscle fibres often occur in association with connective tissue and especially with elastic fibres. The individual smooth muscle cells often run in various directions, crossing each other, especially in the investments of hollow organs; in some situations they run in parallel bundles and may make definite 'muscles' composed of parallel fibres, with an insertion, for instance, in the arrectores pilorum muscles of the hairs (p. 13).

7. Structure of striated muscle

The greater part of movement in mammals is performed by *striated* or *voluntary muscles* and these also serve to hold organs and limbs in position without movement. The fibres of striated muscle (Fig. 7.4) are much larger than those of smooth muscle and commonly reach 4 cm in length and 100 μm in diameter. Such masses of protoplasm are too large to be regulated by a single nucleus and the striped muscle fibres are syncytia; each possesses hundreds of nuclei. Initially these nuclei are derived from numerous myoblasts which fuse to form the muscle fibre. In the fully formed condition each fibre is a unit mass of protoplasm, surrounded by a single surface membrane and with the nuclei mostly near the surface. As in most highly specialized cells these nuclei have lost the ability to divide and additional nuclei for growth and possibly also for repair are provided by the division of satellite cells, which lie outside the surface membrane but within the basement membrane of muscle fibres (see Fuchs 1974).

Each striated muscle fibre consists of numerous myofibrils embedded in sarcoplasm, the latter containing numerous mitochondria and two systems of spaces, the transverse (T) system and the sarcoplasmic reticulum (Fig. 7.9, p. 61). The fibre is surrounded by a system of membranes, whose naming presents some difficulties. The term 'sarcolemma' was invented by light microscopists to describe what they could see. The electron microscope shows a typical cell membrane at the surface of the sarcoplasm. This may show the typical three layers and is about 7–10 nm thick and therefore

not visible by light microscopy. It is best called the *sarcoplasmic membrane*. Outside this lies a basement membrane up to 50 nm thick and outside that a layer of reticular fibres (fine fibres of collagen). All these three together probably correspond to the sarcolemma of light microscopy, about 1 μm thick. Outside that again are further layers of collagen constituting the *endomysium*. The contractile substance is probably attached to the sarcolemma by the Z lines (see below). Probably the sarcolemma with the endomysium constitutes an elastic bag. The muscle substance is of a visco-elastic nature and perhaps force is transmitted from the contractile protein to the sarcolemma merely by friction. There may, however, be special means of attachment at the ends of the muscle fibres.

The supporting connective tissue of the endomysium constitutes a packing between the fibres and holds together the various tissues in the muscle, including the blood and lymph vessels, which are very abundant, and the nerve fibres. Bundles of muscle fibres are held together by larger strands of connective tissue, known as the *perimysium,* and the whole muscle is often surrounded by a more definite sheet, the *epimysium*.

Individual muscle fibres usually taper slightly at their ends and are attached either to the periosteum of bone or to a tendon, or, if they end within the substance of the muscle, to the connective tissue of the perimysium. The mode of attachment at the ends has been much discussed; it is probable that the sarcolemma provides the main attachment to the tendinous or other tissue, and it is not likely that myofibrils continue directly into collagen fibrils as has been claimed. The means by which force is exerted from the contractile substance of the fibres through the sarcolemma to the ends of the muscle has never been satisfactorily determined.

8. The contractile mechanism

Each muscle fibre shows light and dark bands as seen with transmitted light. These are seen in the electron microscope to be due to striations of the myofibrils, all in register (Fig. 7.5). The main division is into anisotropic (A) bands and isotropic (I) bands, the former being strongly birefringent in polarized light. Each I band is crossed by a Z line and the A band has a lighter central H zone (Figs 7.5 and 7.6). The length between successive Z lines is called a *sarcomere* and is between 2·3 μm and 2·8 μm long in the relaxed fibre of all vertebrates. During contraction it becomes reduced and in the process the I band disappears, the A bands coming into contact with the Z lines.

These appearances can now be explained and the contractile process is understood in some molecular detail as a result of electron microscopic studies and

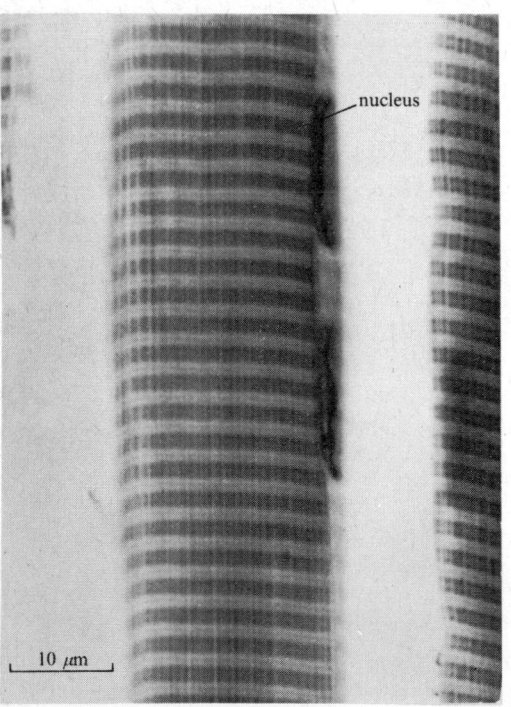

nucleus

10 μm

FIG. 7.4. Light micrograph of rabbit striated muscle in longitudinal section. In addition to the principal striations (A band, dark; I band, light) a thin Z line can be seen bisecting each I band.

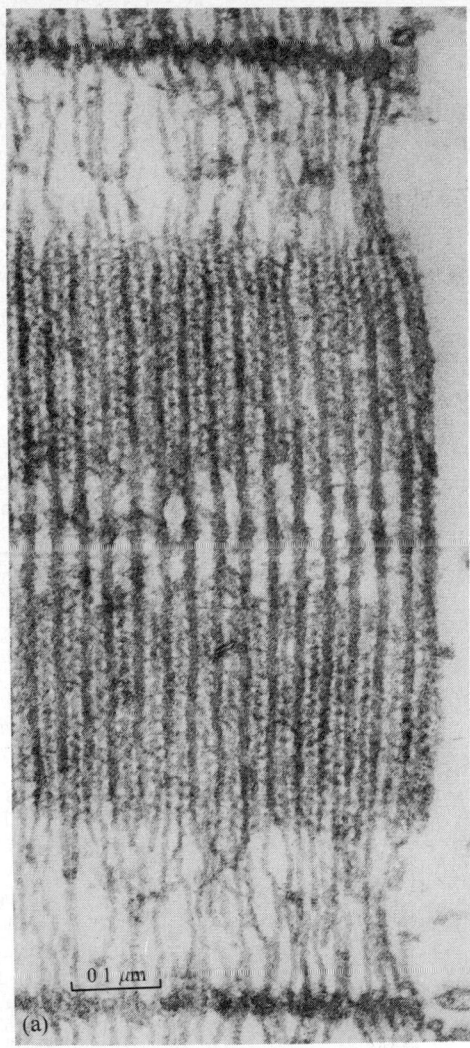

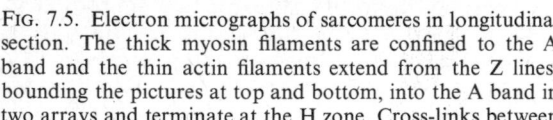

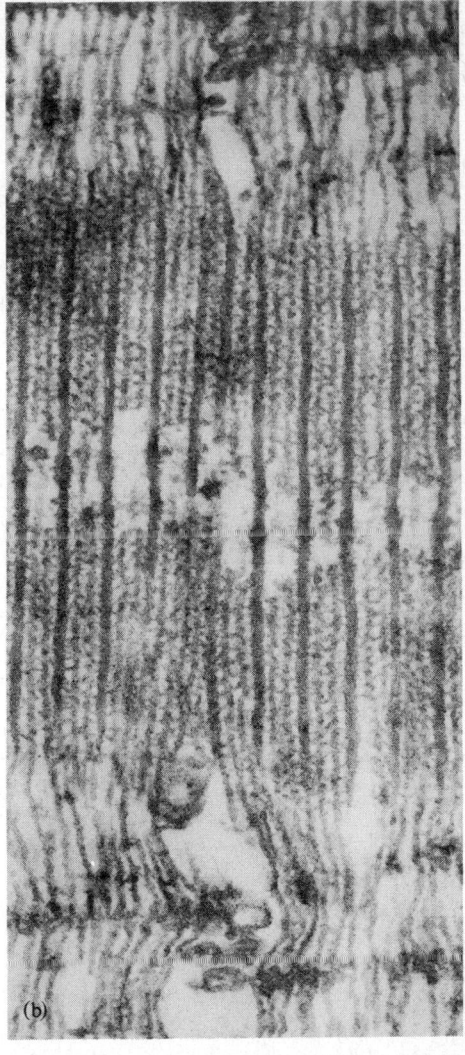

FIG. 7.5. Electron micrographs of sarcomeres in longitudinal section. The thick myosin filaments are confined to the A band and the thin actin filaments extend from the Z lines, bounding the pictures at top and bottom, into the A band in two arrays and terminate at the H zone. Cross-links between thick and thin filaments are visible. The number of actin filaments appearing between the myosin filaments depends on plane of section as explained in Fig. 7.6.(b) (Figures kindly supplied by Dr. H. E. Huxley.)

experiments, especially using X-ray diffraction (Huxley 1972). The myofibril contains two types of myofilaments, thick ones, made of the protein myosin, forming the A band, and more numerous thin ones of actin in the I band, attached to the Z lines. Cross-sections show corresponding appearances in different parts of the sarcomere (Fig. 7.6(a)). In very thin sections cross-bridges are seen between the thick and thin filaments (Fig. 7.5).

The shortening of the sarcomere during contraction is due to a sliding of the two sorts of filaments past each other, the lengths of both remaining constant. This has been proved in several ways, and Fig. 7.7 shows twice the normal number of thin filaments in the double overlap region at the centre of a sarcomere. The energy to produce this sliding comes from splitting off the terminal phosphate from adenosine triphosphate (ATP). This is a result of its interaction with myosin itself, which acts as an ATPase enzyme. This splitting is strongly activated by actin under conditions that allow the two proteins to associate. This is, of course, exactly what is provided for by the existence of the two sorts of filaments in contact. It was first suggested by H. E. Huxley in 1957 that the projections on the thick filaments represent the parts of the myosin molecule that interact with actin. Each cross-bridge could produce only a small movement and the relatively large movement of the filaments must be achieved by each cross-bridge going through many cycles of enzymic change as the muscle shortens. A molecule of ATP (or perhaps two molecules) would be split each time a cross-bridge went through the cycle. The evidence that this model is correct comes from quantitative studies of the proteins, their appearance in the electron microscope after separation and of the changes in X-ray diffraction patterns during contraction. Myosin molecules are long rods with a globular region at one end. On tryptic digestion they break into a shorter terminal heavy meromyosin (HMM) which has all the ATPase

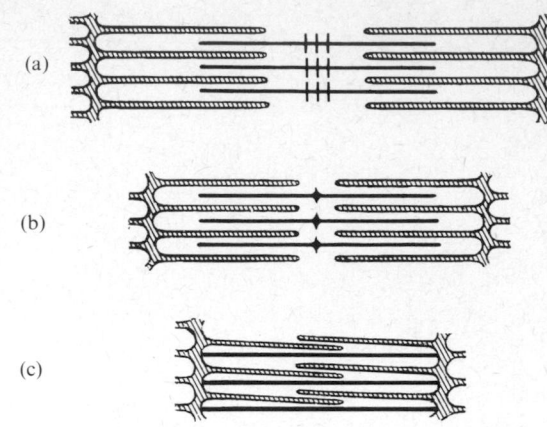

FIG. 7.7. Diagrams of longitudinal sections of muscle during isometric contraction, seen by electron microscopy.

In (a) the H zone is quite wide but as the muscle shortens it is reduced in size and only a small clear band is visible on either side of the M line (b). On the strongest contraction, a zone denser than the rest of the A band as seen in electron micrographs replaces the H zone and this is believed to arise from a double overlap of actin filaments (c). (After Huxley, *loc. cit.*)

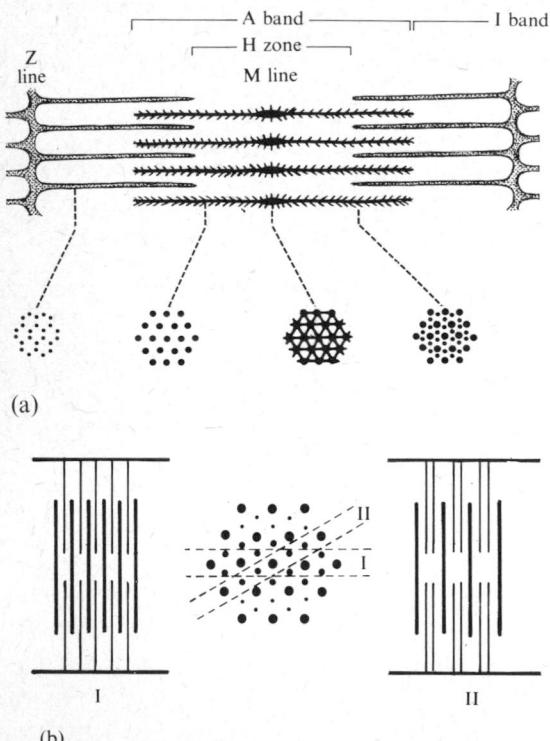

FIG. 7.6. The structure of striated muscle.
(a) Diagrammatic representation of one sarcomere. Under a longitudinal section the appearance of transverse sections at various levels is indicated. Note that the I band comprises elements of adjacent sarcomeres. (b) Illustration showing how plane of section affects the appearance of longitudinal sections. (After Huxley, in Bourne (ed.) (1972). *Structure and function of muscle I*, Part I. Academic Press, New York.)

and actin-binding power and a longer light meromyosin (LMM) which provides the ability to form filaments. The HMM can be further split and is found to have two S_1 subfragments and a short S_2 or rod fragment. The whole myosin molecule is probably made of two helical polypeptide chains, wound together in the rod part and with two separate S_1 units. In the filaments the molecules are packed with LMM parts as a backbone and the HMM parts projecting sideways at intervals to interact with the actin filaments.

Isolated thick filaments in fact show projections along their length but with a bare area at the centre. This suggests that they are made of aggregates of the molecules in anti-parallel arrays (Fig. 7.8). This would allow the two halves of the A band to draw in actin filaments in opposite directions. Similarly, the actin

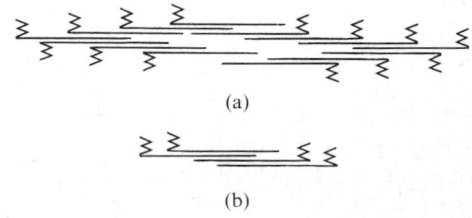

FIG. 7.8. Possible arrangement of myosin molecules with half oriented in one direction and half in the other. The globular regions (the side projections) are absent from the centre to an extent depending on the packing arrangements. (From Huxley. *loc. cit.*)

filaments would also be expected to show reversal at the Z line of the sites able to activate the ATPase. Actin molecules are also formed of two-chain helices. In life they are tightly combined with two regulatory proteins, tropomyosin and troponin (see below). Actin filaments alone do not show the detail of their polarity, but if they are allowed to combine with HMM or S_1 the combined units show a characteristic arrow-head appearance, and this reverses at the Z line.

The mechanism of activation of the contraction involves the action of calcium. At rest the proteins tropomyosin and troponin somehow inhibit the power of actin to activate the ATPase effect of myosin. The presence of calcium releases this inhibition. The normal initiation of contraction is by nerve impulses arriving at the motor end-plate (see below). These set up a propagated muscle action potential over the surface of the fibre. This in turn initiates the shortening of the myofibrils via two systems of channels, the transverse (T) system, opening at the surface of the sarcolemma, and a closed sarcoplasmic reticulum (SR). The openings of the T system, which are only sometimes visible in micrographs, probably lie at the level of the Z lines. Its tubules branch around each myofibril, forming a system of sacs extending across the fibre (Figs 7.9 and

7.10). If a frog muscle is soaked before fixation in a solution containing the electron opaque protein ferritin then the granules of this can later be seen to have entered these channels of the T system. By these and other observations it is clear that the 'spaces' in the muscle fibre are in direct and open communication with the intercellular spaces. Fluorescent dye can also diffuse readily in and out. It is presumed that the surface of this reticulum, being continuous with the plasma membrane of the fibre, allows for the propagation of the effects of the muscle action potential throughout the fibre. The T system contains a high concentration of sodium. Exactly how propagation occurs in it is not known. Almost certainly it is not an all-or-nothing impulse. In any case, the effect is probably to produce the release of calcium from the second system of sarcoplasmic channels (SR). These consist of distended sacks, forming collars around the I bands and extending as longitudinal channels to make a network over the A band, especially at its lighter centre (H band). This system of distended vesicles contains much granular material and is supposed to be responsible for releasing calcium all over the myofibril, to activate it. Probably this system of tubules is not in open communication with the T system.

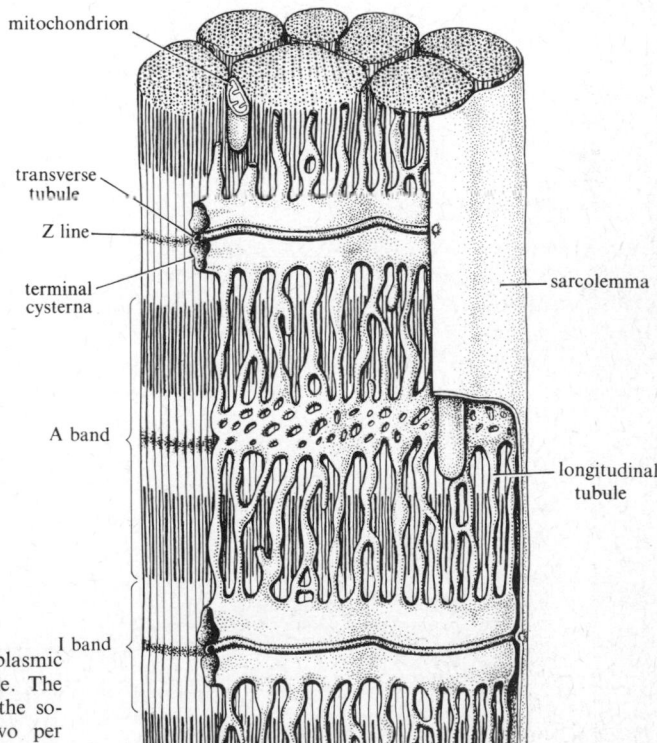

FIG. 7.9. Diagrammatic reconstruction of the sarcoplasmic reticulum around myofibrils of frog skeletal muscle. The terminal cysternae and transverse tubule constitute the so-called triads. In mammalian muscle there are two per sarcomere at the level of the A–I junctions. (After Peachey (1965). *J. Cell. Biol.* **25,** Part II.)

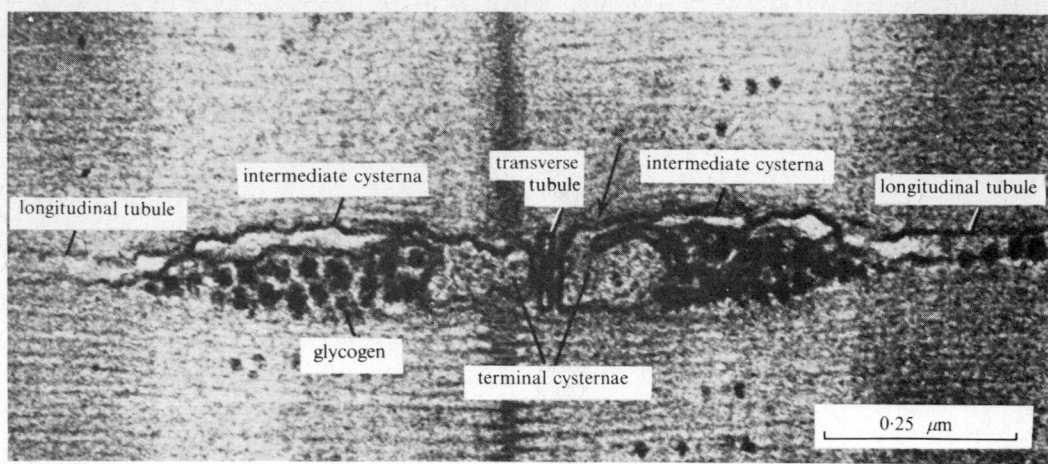

FIG. 7.10. Electron micrographs of triad and intermediate cysternae in longitudinal section. A flattened transverse tubule is seen between terminal cysternae at the level of a Z line. Intermediate cysternae link the longitudinal tubules and terminal cysternae, sometimes joining the latter near the centre (arrow). (From Peachey, *loc. cit.*)

The movement of the calcium during contraction can be followed by autoradiography, with ^{45}Ca, or spectrophotometrically. Free calcium appears in the sarcoplasm of toad muscles 1–5 ms after a stimulus. Its peak is at about 65 ms and half is gone at 125 ms, that is, before the peak of tension development. The calcium is then pumped back into sarcoplasmic reticulum, where there is much ATPase. The calcium moves from the I band to the A band during contraction, and is still mainly found in the latter 20 s after a 5 s tetanus. By 3·5 min later it has returned to the I band, presumably to the longitudinal tubules and collar.

9. Innervation of muscle

Striated muscle fibres normally only produce their contraction when they receive impulses from nerve fibres through the *motor end-plates* (Figs 7.11 and 7.12). These are essentially regions in which the surface membranes of the nerve and muscle cells come into close apposition. At this point the nerve fibre loses its myelin and spreads out into a number of processes in contact with the muscle-fibre surface. These processes vary in number and arrangement, but the essential point is that they allow a large area of apposition between axoplasm and sarcoplasm. They may be inflated into terminal bulbs, but it should be emphasized that the region of apposition and not the terminal bulb is the essential agent of transmission. There has been much dispute as to the relations of nerve and muscle substances at the end-plates. The membranes of both fibres remain complete at the region of contact. Nerve and muscle cells contain much potassium but little sodium (p. 244), and we may express the condition by saying that the potassium spaces of the two tissues are not continuous (Fig. 7.13).

The terminal branches of the nerve fibre spread out over a specialized region of sarcoplasm known as the sole plate (or plasm). The sarcoplasmic membrane is much folded here, dipping in a series of junctional folds. The axonal surface is not correspondingly

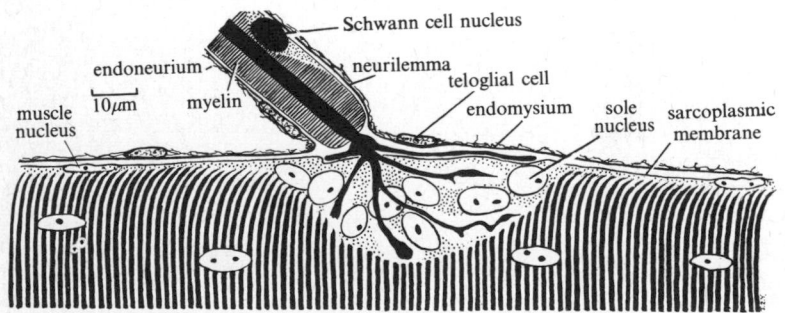

FIG. 7.11. Diagram of the relationship between a nerve-fibre and a muscle fibre at the motor end-plate. The 'sole nuclei' are ordinary muscle nuclei concentrated in the region of the myoneural junction. (From Gutmann and Young (1944). *J. Anat.* **78**.)

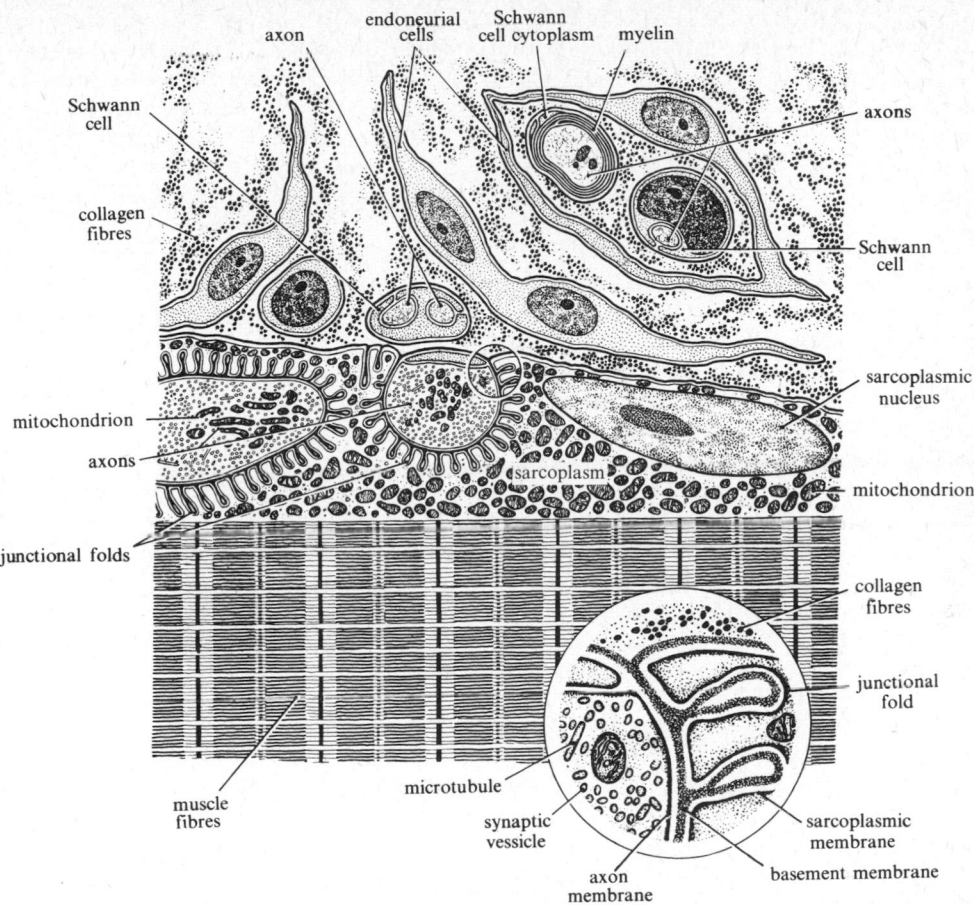

FIG. 7.12. Drawing of electron micrograph of section of a motor end-plate. (From a preparation and photograph by Dr. J. D. Robertson.)

folded but is closely apposed to the sarcoplasm of the sole plate across a 20–60 nm cleft. A layer of basement membrane intervenes and dips into the folds. The upper surface of the axon is covered with terminal Schwann cell processes, but these do not extend between the protoplasm of axon and muscle fibre.

The terminal branches of the axon contain many mitochondria and numerous synaptic vesicles (p. 251). These are about 50 nm in diameter and may be aggregated close to the axon–muscle interface. They are

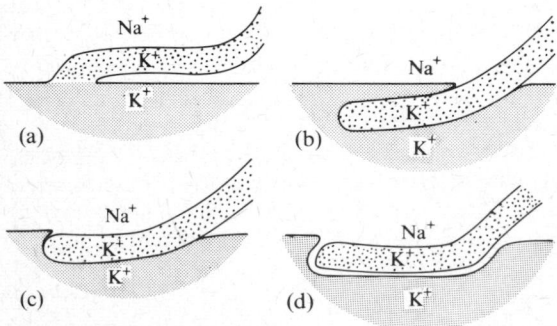

FIG. 7.13. Diagrams showing the various ways in which it has been suggested that the nerve and muscle fibre might be related at the motor end-plate.

K^+ shows the spaces occupied by solutions rich in potassium, Na^+ those rich in sodium. The arrangement shown at (a), with continuity between the potassium spaces of nerve and muscle, is almost certainly never found. If the situation is as in (b) or (c) then one membrane separates the potassium spaces, whereas in (d) there is a sodium space all round the nerve ending. Histological studies have not yet decided whether the arrangement is as (b), (c), or (d). (From Gutmann and Young, *loc. cit.*)

usually presumed to contain the synaptic transmitter, acetylcholine (ACh). Tubules of at least two sizes also occur in the end-plate and have been supposed to form and store ACh.

The enzyme acetylcholinesterase (AChE) also occurs within the axonal terminal and in large amounts in the sole plasm.

10. Transmission from nerve to muscle

The essential feature of the transmission from nerve to muscle is the production of minute amounts of a substance able to depolarize the muscle membrane, probably the ester *acetylcholine* (p. 251). The resulting local change in permeability causes a large ionic flow across the end-plate membrane, and this current stimulates the adjacent regions of the muscle fibre (Fatt and Katz 1953). The end-plate thus acts as an amplifier of the minute electric signal in the nerve fibre so that it excites the muscle fibre. Transmission along a nerve fibre takes place by the spread of electric currents in advance of the active region (p. 243), and finally along the fine terminal branches that are in contact with the surface membrane of the muscle fibre. Muscle fibres themselves produce electrical action potentials ('muscle spikes'), which travel along their surfaces and are essentially like the action potentials that traverse nerve fibres, though they move at a slower rate.

The current produced by the tiny nerve fibre would not be able to stimulate the much larger muscle fibre. Thus the area of the terminal branches at a frog's motor end-plate has been calculated to be $2 \cdot 3 \times 10^{-5}$ cm^2. This at 1 mA cm^{-2} would give a total current of $2 \cdot 3 \times 10^{-8}$A, sufficient to produce depolarization of the muscle fibre of only 1–2 mV even if there was full continuity of the contents of nerve and muscle as in Fig. 7.13. To excite the muscle fibre its potential must be lowered from 90 to 50 mV, which would require 10^{-6} A. This mismatch between nerve and muscle is overcome by the amplification produced at the motor end-plate by chemical means (see Katz 1966).

The terminal nerve fibres contain large quantities of acetylcholine and of the enzyme choline acetyl transferase by which it is synthesized from choline. After stimulation of the motor nerve acetylcholine can be collected by perfusion but is normally rapidly destroyed by the enzyme acetylcholinesterase. Special staining shows that this enzyme is abundant on the postsynaptic membrane. If the action of the enzyme is paralysed, for instance, by application of the drug eserine, the effect of nerve stimulation is to produce prolonged tetanic contraction and excess acetylcholine is obtained from perfusing the muscle. It is estimated that perhaps

5×10^6 molecules are released per impulse per end-plate. By applying acetylcholine through a fine pipette inserted close to the end-plate it is found that excitation of a muscle fibre requires something more than 10^7 molecules. The disagreement is not unreasonable.

The propagated muscle action potential resembles that in nerve (p. 243) and can be excited either directly by electrical stimulation or through the motor end-plate. Only the region of the plate itself is highly sensitive to ACh. Applications further away do not excite. Curiously enough this restriction of sensitivity disappears if the motor nerve is cut and the end-plate allowed to degenerate. The whole surface of the muscle fibre then becomes hyperexcitable to chemical stimulation.

The effect of the transmitter on the muscle is recorded by placing a microelectrode within the muscle fibre close to the end-plate. After arrival of a nerve impulse no change is seen for 0·7 ms and after this a local depolarization, the end-plate potential (e.p.p.) appears and increases to a maximum about 0·5 ms later. At this time a muscle action potential arises and travels away at a constant speed over the surface of the muscle fibre. The e.p.p. itself is a purely local phenomenon and declines passively after the first few milliseconds; it is attenuated with distance by the leaky cable properties of the muscle fibre.

The amplifying effect of the acetylcholine is produced by its reaction with receptor molecules on the outer surface of the postsynaptic membrane, to make the latter highly permeable to small cations. Sodium flows in and produces the depolarization that shows as the e.p.p. and is capable of discharging neighbouring regions of the muscle-fibre membrane. The actual potential changes are complicated by the accompanying outflow of potassium, so that the e.p.p. never moves beyond the zero potential level. However, by suitable experiments it can be shown that the 10^{-17} mol of ACh released causes a flow of at least 10^{-12} mol univalent ions. This amplification is thus the result of a 'short circuit' or 'shunt' placed across the membrane. The acetylcholine is acting as a trigger, releasing the potential energy stored across the muscle-fibre membrane. The membrane can be shown to increase its conductance locally by a transient puncture opening a pathway of $2–3 \times 10^{-5}$ Ω.

Studies with microelectrodes in the end-plate region show that small e.p.p.s of about 0·5 mV are produced even when the fibre is at rest. These miniature e.p.p.s are produced locally at random at different points along the terminal fibres. The effect of the arrival of a nerve impulse is to cause the release of a few hundred of them within 1 ms—enough to activate the muscle membrane.

Similar spontaneous quantal e.p.p.s have been found at synaptic junctions in the CNS. The presumption is that each is caused by the release of a 'packet' of ACh, perhaps one vesicle. But this has never been proved. Indeed it is not known how the events in the pre-synaptic membrane serve to release acetylcholine when an impulse arrives. Calcium is somehow involved, for in its absence no ACh is released by nerve impulses.

11. Motor units

It has long been considered that in mammals each striated muscle fibre receives only a single nerve fibre, and no convincing evidence has been produced for sympathetic or other nerve fibres running to the muscle fibres, such as might produce either tonic or inhibitory effects. Each nerve cell controls a certain number of muscle fibres, the whole constituting a motor (*neuromyal*) unit (Hunt and Kuffler 1954). Sherrington and colleagues (Eccles and Sherrington 1930) determined, for a few different muscles, the number of muscle fibres in the unit and the tension that can be produced. For instance, in the lateral head of the gastrocnemius muscle of the cat each nerve cell controls a unit producing an isometric tension of 30 g. There are 430 such units, so that the whole muscle can produce about 13 kg. More recent work suggests that motor units may vary in size in a single muscle and maximum tetanic tensions ranging from 0·5 g to 120 g have been recorded from units of cat gastrocnemius. In muscles that perform delicate movements the units are smaller, so that the action of the muscle can be more finely graded; thus the ratio of muscle cells to nerve cells varies from upwards of 150:1 in the biceps muscle of the leg to 2:1 in human eye muscles.

In mammals, therefore, all action by striped muscles depends on the arrival of impulses in the motor nerve fibres to set off the contractile process. The tension ceases to be developed when the impulses stop arriving. There are no separate inhibitory nerve fibres able to switch the action off, although these may exist in other vertebrates, as they certainly do in Crustacea. In mammals the balance between excitation and inhibition is struck centrally and the striped muscle itself receives from its nerve only excitor impulses. These impulses are all alike; there are not some that produce 'tonus' whilst others produce movement; the only variable is the frequency with which they arrive at the end-plate, sending the muscle into more or less frequent contraction.

12. Twitch and tetanus

The unit of muscular action is the *twitch*, which consists of a single process of development of tension spreading throughout the muscle fibre following propagation of an impulse over its whole surface. When single impulses are fired along a motor nerve at sufficiently long intervals apart, each will produce a single twitch, followed by complete relaxation (Fig. 7.14). If the impulses are more frequent, the successive contractions of the muscle fibre occur without time for relaxation. The result is more or less complete fusion of the contractions to produce a *tetanus*, whose tension may be much greater than that developed by the single twitches. In most of the ordinary actions of muscles the impulses arrive in such a way that the contraction is tetanic.

13. The energy for contraction

Whatever mechanism is involved in contraction it must somehow be connected with the reactions by which energy is made available in the muscle cell, essentially by the oxidation of foodstuffs. If the contraction

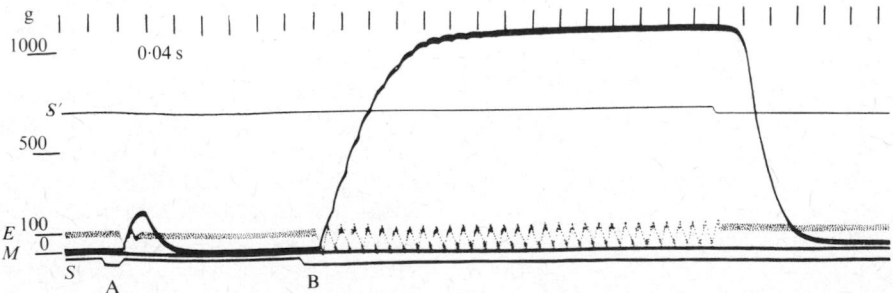

FIG. 7.14. Isometric responses of gastrocnemius muscle of frog to electrical stimulation of its nerve. At A a single break induction shock produces a single twitch response. At B a series of shocks at 50 per second produces a tetanus. Line *E* shows the shadow of the string of a galvanometer connected with electrodes to record the electrical activity of the muscle. Line *M* shows the movement of a stretched wire recording the tension of the muscle. Line *S* shows the onset and *S'* the cessation of tetanization. Time above = 0·04s. Note the steps by which the plateau is reached and that the muscle continues to give discrete electrical responses during tetanization. (From Fulton (1926). *Muscular contraction and the reflex control of movement*. Baillière, Tindall and Cox, London.)

involves the release of stored chemical energy, then much of the chemical work can be done after contraction is over. A muscle is able to contract in the absence of oxygen, and any individual twitch should be considered as an anaerobic change. Indeed the muscle is capable of a series of twitches, or a tetanus, even in the absence of oxygen, but it then becomes fatigued much more quickly than if oxygen is available. During anaerobic contraction lactic acid accumulates and the carbohydrate stores of the muscle (such as glycogen) decrease. If oxygen is then admitted, the lactic acid disappears, some of it being burnt to carbon dioxide and water and another part remaining in the muscle and becoming resynthesized to its precursors, including glycogen, which increases during this period.

14. The carbohydrate cycle

There must be materials available, therefore, from which energy can be derived anaerobically in order to reset the muscle spring, and indeed it is clear that such a mechanism is necessary in normal life to produce effective tension before combustion with oxygen can take place to provide the energy.

A key substance in these changes is *adenosine triphosphate* (ATP), which is a nucleotide containing the heterocyclic adenine radicle combined with a sugar (D-ribose) and three molecules of phosphoric acid. This substance is readily hydrolysed anaerobically to adenosine diphosphate (ADP) yielding 42 kJ (kilojoules) of free energy per mole. Re-phosphorylation occurs rapidly also anaerobically, by breakdown of creatine phosphate (CP) to creatine (C) and phosphoric acid by means of the enzyme creatine phosphotransferase (the Lohmann reaction):

Creatine phosphate   Creatine

ATP is then re-synthesized from the phosphoric acid.

These events take place very rapidly during a twitch, leaving a condition in which C and ADP concentrations are higher than at rest and CP and ATP concentrations lower. The resting concentrations are then restored by breakdown of glycogen, which may be aerobic or anaerobic. This recovery process takes more than 30 minutes even after a single twitch (frog muscle at $0\,°C$). The amount of energy (heat and work) that can be obtained from a muscle in the absence of oxygen is proportional to the quantity of CP that disappears (if glycolysis is prevented), as would be expected if the CP provides the energy for the contraction.

There is no doubt that in a normally contracting muscle the aerobic breakdown of glycogen ultimately provides the energy, and it is possible to imagine how this may be done by using the energy to re-form the ATP and CP that have been used up in the immediate breakdown. As the ATP splits during contraction its products become engaged in the phosphorylation of glycogen, and thus energy is made available for creatine phosphate resynthesis. The lactic acid resulting from the glycogen breakdown is partly burned to carbon dioxide and water to provide energy by which the remainder is resynthesized to glycogen, completing the carbohydrate cycle.

In aerobic functioning, most of the lactic acid is produced after the contraction is over. It is probable that the phosphorylation cycles found in muscle extracts and under anaerobic conditions are not always strictly followed in the normal muscle, which may be able to make short cuts if well supplied with oxygen. It is certain, however, that the adenosine triphosphate and creatine phosphate are normally engaged somehow in the recharging, and it is easy to see that an anaerobic process is necessary to allow the muscle to work properly. As A. V. Hill puts it: 'With muscles, work precedes oxidation; with the internal combustion engine oxidation precedes work' (1927).

Not only the muscle but the body as a whole can work for a short time at much higher rates than respiration would seem to allow. An *oxygen debt* amounting to over 16 l may be accumulated during a short sprint and paid back by increased respiration during the subsequent hour or longer. The debt is largely in the form of lactic acid, which escapes from the muscles to the circulation (it may appear in the urine) and is later oxidized throughout the body. It must not be forgotten, however, that muscles contain considerable quantities of haemoglobin and the red muscles, which are called upon for sustained contractions, are less quickly fatigued than the white.

15. Energy output and efficiency of muscle

Study of the heat produced by muscle also shows that most of the metabolic work is performed when the contraction is already over. We can distinguish an *initial heat* occurring during the contraction and a *recovery heat*, more than half the total and present only under aerobic conditions.

The amount of work that can be done by a man or animal is limited by the rate of oxygen intake and the size of the oxygen debt that can be tolerated. In man the resting oxygen consumption is about 250 ml min^{-1} and this can be increased to $4\frac{1}{2}$ l or more during exercise, the maximum depending on the state of 'training' of the individual. The oxygen debt can reach about 16 l, and it is obvious that the maximum output of the individual can last only for the time during which this debt is incurred. Greater speeds of running can be maintained over shorter distances.

The work done in producing movement of a man over a flat surface is mostly related to overcoming the internal friction or viscosity of the muscles and the force necessary for this purpose is found to vary with the speed of movement; other factors, such as the air resistance, are relatively unimportant. 'An athlete . . . like a raindrop falling through air, rapidly attains a certain speed depending on the ratio of propelling force to resistance, and this speed remains constant until other factors, for example fatigue, cause it to fall off' (Hill). The propelling force demands the expenditure of about 2 horse-power when a man is running at maximum speed. From the oxygen consumption it is possible to estimate the energy used during such an effort and it is found that the mechanical work done accounts for 20–25 per cent of the total energy used. This efficiency can rarely be achieved in the performance of external work, for which there is an optimum rate for any given set of conditions, up to a maximum of about 25 per cent. The work done is approximately inversely proportional to the speed of contraction but in practice there is an optimum speed for any given task.

The organism is so arranged that the force it can exert is generally less than that which would bring about its own downfall by breakage of bones or rupture of ligaments, though such accidents are not infrequent among athletes. The inherent strength of a contracting voluntary muscle fibre is roughly constant (4–5 kg cm^{-2} cross-section) but large animals do not disrupt themselves because the speed varies a thousandfold or more, being greater in smaller muscles and the muscles of smaller animals (Hill 1950). The net result is an approximately equal velocity of movement in relation to the earth on the part of all animals, irrespective of their linear dimensions. Hill points out that a racehorse and

a whippet both move at about 40 m.p.h. and can jump to about the same heights, and that the jump of a grasshopper (or indeed a flea) is of the same order. A blue whale is 5000 times heavier than a dolphin but both swim at a maximum speed of about 20 knots. The factor that limits the speed of contraction of muscles may be their own internal viscosity or the rate at which they can liberate energy (see p. 58). There must be a safety factor that determines that the organism shall not disrupt itself. Such features of muscle action make it possible for organisms of different sizes to exist. 'If a man's muscles could be altered without altering his general design so as to allow him to run 25% faster, athletics would become a highly dangerous pastime; pulled tendons, torn muscles, even damaged bones, would be so frequent as to make it almost prohibitive' (Hill 1950). In practice, if one animal is 1000 times as heavy as another, the movements of its parts will be 10 times slower, while the speed over the ground remains the same. One advantage of the greater size is that the larger animal will take ten times as long to become exhausted with maximum effort.

Similar differences exist between different muscles of one animal. Thus the muscles of the eyelid are small and very fast. If the muscles of the limbs contracted at this speed they would break the bones and, moreover, maintenance of posture would require great expenditure of energy.

From these considerations H. E. Huxley has shown how the sliding filament mechanism serves to meet the following conditions necessary for adequate muscular contraction. (a) Clearly the tension developed per unit area should be maximal, and this depends on the extent of overlap of the filaments at various stages. Evidently the constant lengths of sarcomere and thick filaments found in vertebrates provide the best compromise and indicate that there is some limiting factor in the muscles themselves, perhaps the strength of the actin filaments. In insects where the A bands are longer, there are more thin filaments to each thick one, reducing the tension on each of them. (b) The variation in the velocity of shortening to suit animals of different sizes cannot therefore be produced by varying the sarcomere length and must be a result of variations in the rate of binding of myosin to ATP in the different muscles. (c) The extent of coupling is matched to the work to be done by making the number of active cross-bridges vary with the load. Only those developing tension by combination with actin split ATP. (d) The arrangement ensures that muscles can generate tension over a range of lengths. (e) The system evidently allows for the tension to be rapidly switched off by a fast pump to re-absorb calcium. (f) There is little waste of energy because in such a liquid crystalline system, movement occurs with little viscosity, the filaments being held in place by long-range forces.

16. Red and white muscles

The arrangement of actin and myosin filaments in striated muscles enables them to contract fast and to relax fast. This is obviously necessary for muscles that produce movement. The smooth muscles, on the other hand, are used mainly for holding, that is to say, their function is not so much to shorten as to stay at a given length. In order to maintain a given tension it is obviously efficient to have a system that relaxes only slowly. The tension is thus maintained with a minimal number of acts of shortening and therefore a minimal expenditure of energy. In some bivalve molluscs such as the scallop, *Pecten,* which move by flapping their shells, there are two adductor muscles, a striped one that contracts and relaxes fast and a smooth one that contracts only slowly but, also relaxing slowly, is able to hold the animal closed for long periods.

In mammals the distinction is not so clear. Smooth muscles are used mainly in the viscera; in the limbs holding as well as movement is a function of the striped muscles. These, however, show differences according to whether they are mainly used for movement or for holding. The 'white' muscles contract and relax rapidly and are the muscles of movement, whereas the 'red' muscles, having more sarcoplasm, contract and relax slowly and therefore are better suited to hold tension. The large amount of myoglobin in these 'red' fibres may also assist in the prolonged maintenance of tension. Any maintained 'tonus' of a muscle depends upon a series of contractions, which, if they are sufficiently close together, fuse to a tetanus. To maintain a tension by means of a quickly relaxing fibre would therefore involve great expenditure of energy. On the other hand, quick movements can only be effected if relaxation follows soon after contraction. Red muscles assume a tetanic contraction at a much lower frequency than white.

The two types of muscle are not, in fact, sharply distinct and the properties of individual muscles vary in the different animal species according to their use. All mammalian extrafusal muscle fibres are twitch muscles, that is, they respond to a single motor impulse to give a propagated action potential and a twitch. They normally have a single end-plate on each muscle fibre. Some slow fibres in frog's muscles have a distributed innervation, rather than local end-plates and may show local contractions and hence a graded build-up of tonus. This may also be the condition of some intrafusal fibres of mammals (p. 266).

Mammalian muscles can thus be classified as fast-twitch and slow-twitch muscles with the characters shown in Table 7.1. These characteristics develop in the

TABLE 7.1

	Fast twitch	Slow twitch
Appearance	Pale (white muscle)	Dark (red muscle)
Contraction time	Fast (*ca.* 25 ms)	Slow (*ca.* 70 ms)
Relaxation time	Fast	Slow
Tetanus	High impulse frequency	Low impulse frequency
Metabolic enzymes	Predominantly glycolytic pathway	Predominantly oxidative pathway
Nerve fibres	Larger	Smaller
Functional characteristics	More easily fatigued, phasic	Not easily fatigued, tonic

peri-natal period and are nerve-dependent. Innervation also determines these characteristics in the adult, for example, a substantial transformation from one type to the other occurs when fast and slow muscles are cross-innervated. It has not yet been clearly established whether this neural influence is mediated by impulse patterns, neurotransmitter substances, or a combination of these.

There is no consistent association of speed with fibre diameter, and although there are many reports of histological variations, for instance, in the proportions of myofibrils and sarcoplasm, these have not been successfully correlated with the speed of contraction of the muscle. In the rabbit the distinction of colours is particularly marked and the white muscles are in nearly every case the more superficial and larger ones, often working across more than one joint and therefore well suited to produce movement (see p. 70). The red muscles lie deeper, close to the bones, and they usually work only across one joint, which they serve to fix. A typical example is to be found in the calf muscles, where the gastrocnemius and plantaris muscles (p. 101), lying superficially and working across both knee and ankle, are white, whereas the deeper soleus, arising from the tibia, is red. The duration of twitch of the gastrocnemius muscle of a variety of mammals has been shown to be 40–140 ms, whereas that of the soleus varies from 120 ms to 440 ms. Further, the condition of tetanus is reached at a frequency of 10 per second for the 'red' muscles and 30 per second for the 'white'.

These physiological differences between soleus and gastrocnemius are also found in the cat although here both muscles are white in colour. It is evident that differences of speed of contraction are more important than those of colour. Presumably the fast ('white') muscles are found throughout the mammal in situations where movement is important, whereas the slow ('red') fibres serve to hold or fix the bones, as in the action of maintaining the extensor muscles, which serve to maintain the weight of the body against gravity.

There are undoubtedly similar differences between muscles in man, although the colour difference is not usually so sharp as in the rabbit. Muscles such as the trapezius or the gluteus maximus that are continuously or often in action have redder fibres than those such as biceps or the finger muscles that are used mainly for movement. The factors that make for redness in a muscle are still not well understood. Not all the red muscles of the rabbit are slow: the masseter contracts as fast as the white muscles. Moreover, in birds the muscles of the wings, which contract very fast, are red, especially so in the stronger fliers. It seems that myoglobin is produced in muscle fibres that often build up a large oxygen debt, because of either slow tonic or fast phasic actions. The capacity to react in this way to oxygen deficiency is another example of the process of 'adaptation' by which the organism records its past history and prepares for the future.

17. Effects of use and disuse upon muscles

When white muscle fibres are put out of use, which can be done by cutting either their nerve or their tendon, they undergo a process of *atrophy* in which they shrink and lose some of their cross-striations. The nuclei come to lie centrally and the sarcoplasm increases in amount, so that white fibres come to resemble red fibres. At the same time the contraction of the fibres, which, even if they are without nerves, can still be elicited by electrical, mechanical, or chemical stimulation (indeed they are hyperexcitable to such agents) becomes very slow and is followed by slow relaxation. After long periods the atrophy leads to complete disappearance of the muscle fibres. Conversely, use of muscles leads to their hypertrophy. This may occur quite rapidly. After stimulation of an isolated frog's muscle for several hours the enzyme creatine phosphotransferase is increased in amount, an example of adaptive enzyme synthesis.

The type of hypertrophy that occurs varies with use. Thus the muscles of a weight lifter increase in cross-sectional area, allowing large isotonic tensions. The proportion of the fibres occupied by myofibrils is increased. Exercise requiring sustained output of power (for example, long distance running) leads to increase in number of mitochondria and amount of myoglobin as well as increased power of the heart and better blood supply to the muscles.

Use and disuse, whose effects are profound throughout the body, affect muscles perhaps more rapidly than any other tissue. Muscles may become noticeably wasted after a few days in bed and are developed by a few long walks. The muscles like other tissues thus carry a memory of the demands made upon them in the recent past and are ready to meet similar situations in the future. The details of the chemical system that ensures this hypertrophy with use are not known; presumably it is related to the system that is responsible for ensuring that some muscles develop the fast and others the slow character.

18. The arrangement of muscle fibres

The effect of the contraction of muscle fibres depends on their direction and attachment, which vary greatly in different muscles. Smooth muscle fibres often run in bundles around hollow organs, either in various directions, as in the wall of the bladder, or in regular layers of circular and longitudinal fibres, as in the gut. In a few situations smooth muscle fibres form a distinct muscle, for instance, in the retractor of the penis of the dog. The little muscles of hairs are also of this sort. The pupil of the eye is closed by a sphincter in the iris composed of smooth muscle fibres running round the edge of the pupil and is opened by radially arranged dilator fibres.

Striped muscle fibres are usually organized into rather definite 'muscles', often enclosed in an epimysium and further individualized by the presence of a single tendon at one or both ends. The arrangement of the muscle fibres depends on the nature of the work to be done. Each muscle fibre can contract to about one-half of its maximum length. Muscles that produce much movement but exert little force therefore consist of parallel bundles of long fibres, for example, those of the sterno-hyoid muscles that pull down the hyoid apparatus during swallowing. On the other hand, muscles that produce much force consist of numerous short fibres, ending within the muscle belly with a more or less pennate arrangement (Fig. 7.15). This may mean that their direction of pull is slightly out of line with that of the whole muscle but greater force can be exerted by the presence of many short fibres. These *pennate muscles* may have the tendon along one side (unipennate), down the middle (bipennate), or many small tendons joining (multipennate) (Fig. 7.15). The

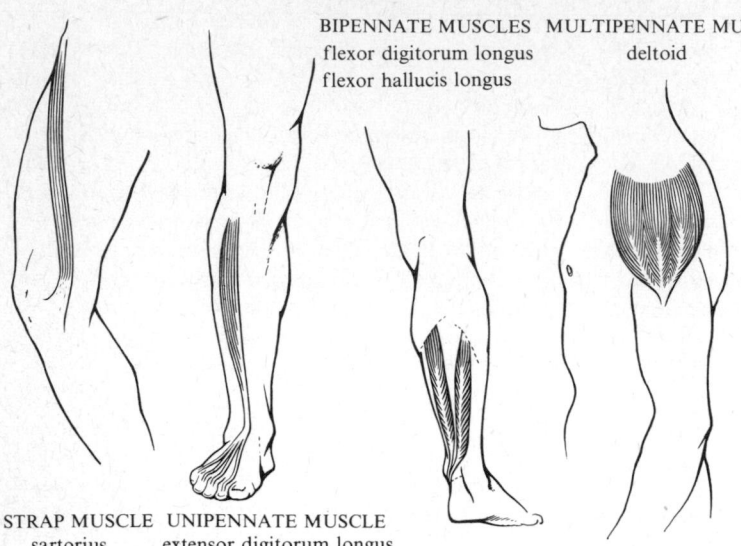

BIPENNATE MUSCLES   MULTIPENNATE MUSCLE
flexor digitorum longus   deltoid
flexor hallucis longus

STRAP MUSCLE   UNIPENNATE MUSCLE
sartorius   extensor digitorum longus

FIG. 7.15. Diagrams of some human muscles to show the different arrangement of fibres within the muscle.

last-named arrangement is found in muscles that are able to exert great force (for example, the human deltoid). Muscles are attached to bones by union of the sarcolemma with the periosteum directly or through a tendon. The attachment on a distal bone is usually placed close to the joint, allowing considerable movement of its distal end quickly, and with little shortening of the muscle. Such a muscle may be called a 'spurt' muscle, for example, biceps brachii. However, in certain muscles the distal attachments are placed a long way from the joint, enabling them to exert a greater effect, even if slowly, for example, the deltoid in man. Other similarly arranged muscles seem to stabilize joints both in flexed and extended conditions and these may be called 'shunt' muscles, for example, brachio-radialis.

Muscles often serve to move a more distal movable part relative to the body, and it is common therefore to distinguish the more fixed proximal 'origin' of the muscle from the more mobile (and usually more distal) 'insertion'. This use is harmless if it is realized that the distinction is arbitrary and that the action is equal at both ends of a muscle; which end is fixed depends solely on the conditions at the time. For instance, the distal end of the biceps muscle is usually regarded as its insertion, and often indeed it is the upper arm that is fixed and the lower which moves, but a man pulling himself up by flexing his arms has the lower arm fixed. Many such cases arise, and it is wise to regard both ends of the muscle equally as points of *attachment*.

19. Levers in the body

The majority of the striped muscles of the body produce their effects by exerting a turning moment at a joint across which they act, that is to say, they tend to move one member of the joint on the other with a force that is proportional to the distance of the line of direction of action of the muscle from the centre of the joint. The muscles are arranged in various ways to produce their actions. Many of them produce tension directly between their points of attachment, and the distribution of weight on the bones will then result in them being used as levers in various ways (Fig. 7.16). *First-order levers*, in which the fulcrum lies between the load and the power, are not common in the body; an instance is the nodding of the head about the atlas as fulcrum. *Second-order levers*, where the pull is applied at a greater distance from the fulcrum than is the load but on the same side, are more usual, and the classic example is the foot, where the fulcrum is the balls of the toes (metatarsal heads), the weight bears on the ankle joint (talus), and the pull (gastrocnemius) is applied at the heel bone (calcaneum), the whole foot acting somewhat as a single lever. However, like so many other attempts to apply mechanical principles to the body this one must be used with caution; the foot is a complicated structure with many parts, not a simple lever, and moreover, since the muscles pull on the leg as well as the heel, the analogy is inappropriate anyway. *Third-order levers*, in which the pull is applied nearer to the fulcrum than the load, are also common—

movement is powerful though slow; it is therefore well adapted for digging. In the horse l/h is small ($\frac{1}{12}$) and the muscle is effective in the production of the rapid movement of running. The shapes of many bones are determined by such factors.

The muscles may be said to act in three ways (Elftman 1941). '1. When movement is not desirable, they must exert forces which will balance the forces present, so that the rotation of the levers does not take place. 2. When movement does take place, the muscles must be able not only to accelerate the movement but also to decelerate it. 3. The muscles must be able to regulate the energy of the system, by contributing energy from their chemical stores, or by removing energy by dissipation into heat, as occasion requires.'

In many situations where the muscles cross joints at which weight is carried, it is convenient to consider them as elastic braces. Thus Gray has considered the legs of a mammal as if supporting the body balanced upon them (Fig. 9.1, p. 85). Without braces the condition would obviously be unstable and the muscles about the joints compensate for the tendency of horizontal forces to upset the equilibrium. In a similar way the human body is balanced upon its two legs by a system of braces at front, back, and sides of the hip joint (p. 97). If the muscular action is cut off, for example, by failure of the nervous system, the body falls down. By modification of their action as braces such muscles move the limbs and produce locomotion. Often they are assisted in this by special longer muscles, exerting greater leverage, and acting either across the same one joint or across several.

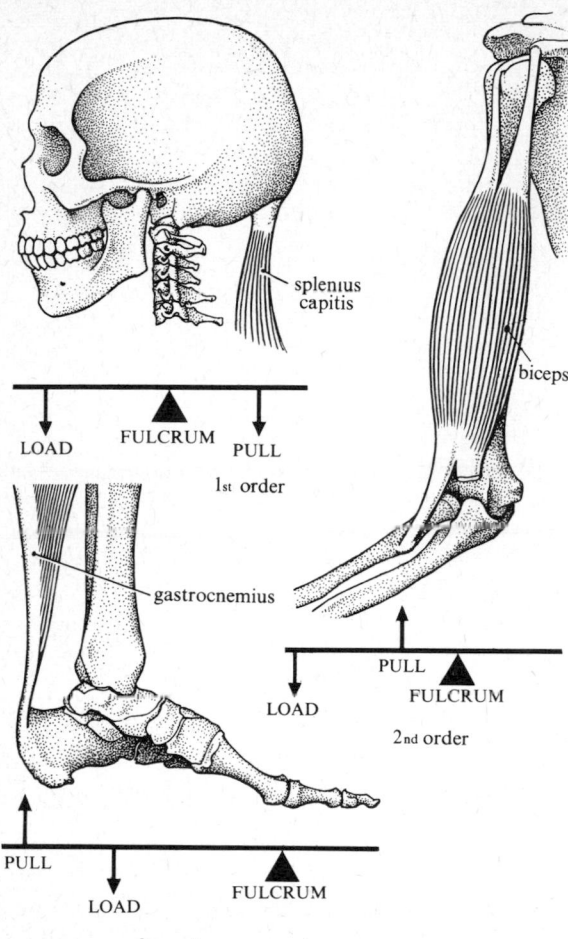

FIG. 7.16. Diagrams of some joints and muscles in man to show arrangements comparable to those of the different orders of lever.

the action of the biceps in lifting a weight held in the hand is a good example.

The significance of many features of the skeleton can be understood by considering the mechanical advantages with which the muscles work. They are differently arranged according to whether speed of action or great force is the main consideration. Thus there is a striking contrast between the forelimb of animals that run at high speeds (such as the horse) and those that dig, for example, the armadillo (Fig. 7.17). We may consider the teres major, one of the main muscles flexing the arm at the shoulder, running from the scapula to the humerus (p. 91). The effectiveness of this muscle will depend upon the ratio between l, the moment of the muscle about the fulcrum, and h, the distance from the fulcrum to the ground. In the armadillo l/h is large ($\frac{1}{4}$) and the

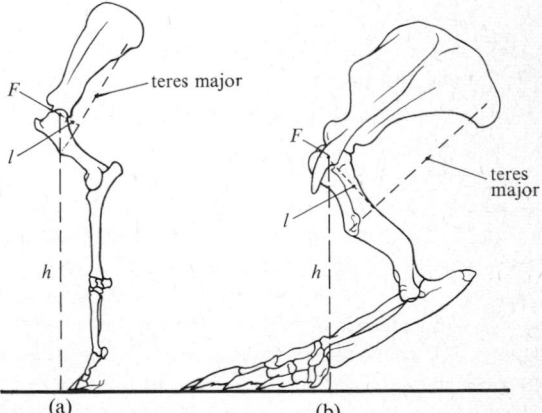

FIG. 7.17. Left forelimbs of (a) horse, and (b) armadillo, showing the line of action of the teres major muscle. l shows the moment arm of the muscle about the fulcrum F, and h the distance from the fulcrum to the ground. (Figure kindly supplied by Dr. J. Maynard Smith.)

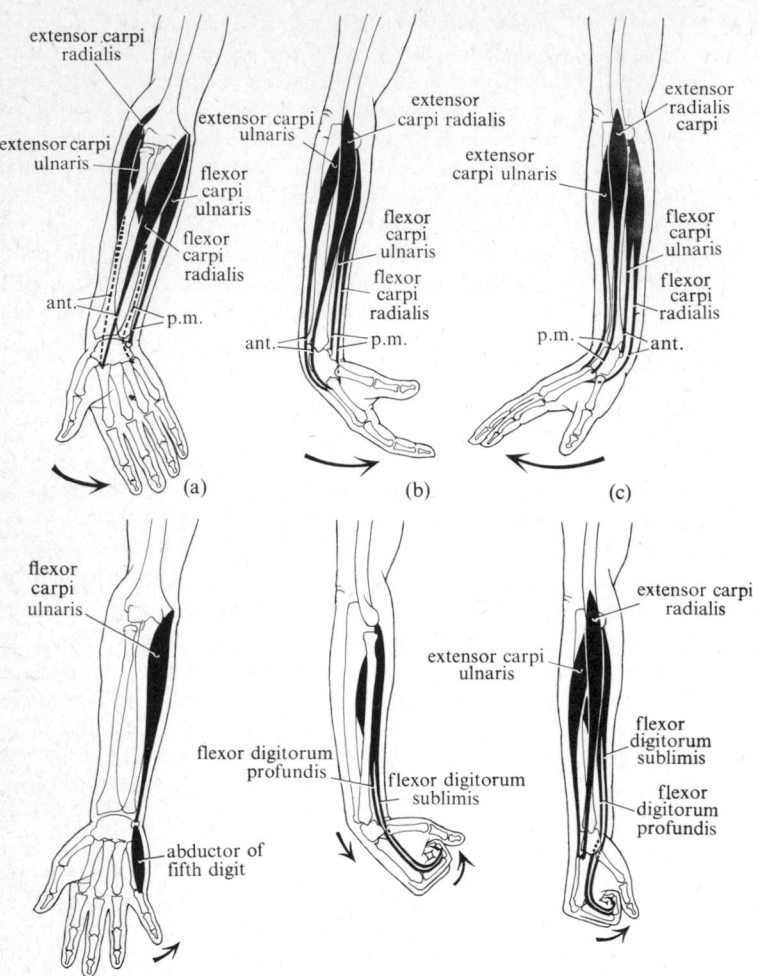

FIG. 7.18. Actions of flexor carpi ulnaris, showing how muscles may act together in different combinations. In each case the prime movers are labelled p.m. and their antagonists ant.

20. Actions of muscles in combination

A single muscle rarely acts alone to produce a movement. The actions that the brain demands are not the contractions of single muscles but the execution of specific movements, and for these the co-operation of various muscles is usually required. Nevertheless, we can often recognize one or more muscles that specifically produce the movement of a joint, and such muscles are called the *prime movers* (agonists) of the action. For instance, in bending the hand towards the little-finger side (ulnar deviation) the flexor carpi ulnaris and extensor carpi ulnaris muscles are the prime movers (Fig. 7.18(a)). Every muscle after contraction has to be stretched again and in most situations, therefore, there are *antagonists* of the prime movers and these must relax during the action. They relax at a rate that ensures a steady movement. In adduction of the

hand the antagonists of flexor and extensor carpi ulnaris are flexor and extensor carpi radialis, on the other side of the wrist. In other movements the muscles work in different combinations; for instance, in flexion of the wrist, flexor carpi ulnaris and flexor carpi radialis act together as prime movers, the extensor muscles being now the antagonists (Fig. 7.18(b)).

Contraction of the prime movers can only produce the required effect if the bone from which they act is fixed. On account of the mobility of the whole skeleto-muscular system this may involve the combined action of a surprisingly large number of muscles, working as *fixation muscles*. A good example of this can also be seen with flexor carpi ulnaris. The little finger is pulled aside ('abducted') by a short muscle (abductor digiti quinti) (Fig. 7.18(d)). This runs from the pisiform bone, which can be felt at the wrist, to the side of the proximal phalanx of the little finger. Contraction along this line will pull the finger aside only if the pisiform bone is fixed by contraction of flexor carpi ulnaris, which is also attached to the pisiform. This muscle can easily be seen to be in action as a fixation muscle during the bending aside of the little finger (Fig. 7.18(d)).

Movements are seldom as simple as they seem, and during many quite small movements a large part of the whole musculature may be brought into action. A special case of this is when a muscle passing over two joints is needed to produce action at only one of these, the other being then fixed by the action of *synergistic muscles*. For example, clenching of the fist is produced by the long flexors of the fingers, which would also bend the wrist forward were it not for the extensor carpi radialis and ulnaris, which can be seen to come into action as synergists (Fig. 7.18(e) and (f)). It will be clear that this is really only a special case of muscles acting as fixators, and many writers do not limit the use of the word 'synergistic muscle' in this way, but make it synonymous with 'fixation muscle', in fact with any muscle whose action assists that of the prime movers. All of these terms signify only arbitrary divisions, useful in the attempt to present an account of the activities of this system of adjustable braces and ties by which the body is supported and moved to the various situations required for its survival.

8 The vertebral column

1. Posture of mammals and man

THE vertebral column and limbs, with their muscles and ligaments, are of particular importance in the life of terrestrial animals. In aquatic creatures the liquid medium supports the weight and the musculo-skeletal system enables the animal to move about and to find surroundings suitable to its needs. In land animals, in the absence of the support of a liquid medium, the skeleton and muscles must carry the entire weight of the body and we have therefore the conception of a *posture*, maintained by the combined action of these parts.

The members of the skeleton serve as struts, while the ligaments and muscles act as ties. It is clearly necessary to consider all these components together; it is senseless to consider compression members of a bridge, or any other structure for resisting deformation, without considering also the ties that go with them.

Discussion of the mechanics of a terrestrial vertebrate is complicated by the fact that the animal is seldom in a situation where it can be considered as a static structure, such as a bridge. Its members are flexible and the ties connecting them include muscles that exert varying tension; therefore we have to think all the time of a system in a steady state rather than in static equilibrium. In nearly all positions work must be done to keep the body up and the energy for this work is obtained from outside the system. If the supplies of energy are not available the body collapses, and the completeness of this collapse affords a measure of the importance of the muscles in maintaining the posture.

The system for the support of the body has become progressively improved throughout the period since the mammal-like reptiles first diverged from other amphibious fish-like stocks. At that early period the muscles of the back were arranged as a series of segmental blocks; the limbs were short and they rarely or never carried the body-weight. In the mammals the locomotor system has changed so much that few traces of the original metameric fish plan remain. The limbs and their muscles carry the whole weight of the body and are able to produce swift motion by acting as long levers. Some mammals can move over rough ground or through the trees in a way that excites the amazement of the engineer, whose machines can produce swift progression only over smooth surfaces. Moreover, many other forms of locomotion have been developed to a high degree in mammals, such as the leaping of some carnivores, the arboreal specializations of primates, the flying of bats, the burrowing of moles, and the swimming of whales, sea-cows, and seals. These developments of the locomotor system show perhaps more than any other part of the body how the exploration of new habitats by the animals has led to the appearance of special types of organization.

2. Some principles for analysis of the skeleton

It has been realized at least since the time of Galileo that there are similarities between the means for distribution of the weight of the body on to the legs and the stressing of bridges and other structures made by engineers. Certain principles are common to all bodies under such conditions. One of the simplest of them is given by Gray (1944): 'although a horse at rest can, by muscular effort, control the degree of support given by an individual leg towards the support of the body, yet the resultant of the thrusts of all four legs must always represent a force equal in magnitude but opposite in direction to the weight of the animal'. As the systems of statics and dynamics have developed, various attempts have been made to apply them to the animal body, and such analyses can be helpful in showing correlations between the positions and physical properties of bones, tendons, and muscles, and the life, posture, and activities of the whole animal. It is not yet possible to make the analysis complete enough to be the main basis for organizing communication of information about the animals: the living system is too complicated for us to be able to see it as a whole and also to name its parts according to any consistent scheme. We still continue to speak of 'the vertebral column' and not of 'the main mammalian compression strut'. Nevertheless, analysis with more exact methods takes us a considerable way towards an understanding of the parts played by the various bones and muscles and it brings us within sight of a general terminology and method of treatment that will make this part of biology both less laborious and more precise.

The basis for attempts made so far in this direction has been that bone is a tissue chiefly suited to resist compression; it also plays a part in resistance to tension, but consideration of the skeleton alone shows mainly the compression members or struts of the living girders. To understand properly how the weight is distributed we must also consider the ligaments and muscles, which resist tension. The principles of such an analysis can be shown by considering the distribution of strain during support of the weight of the body on the hind legs in a short-tailed quadruped such as the rabbit.

The whole back may be considered as a loaded beam and for simplicity we may consider first a beam attached to a wall (Fig. 8.1). In such a structure, as already explained (p. 44), we can recognize (a) compression lines running downwards under the load but then horizontally to the position of attachment of the beam, and (b) symmetrically disposed tension lines which, as it were, support the load from above and cross the compression lines at right angles. Dissection of the back of the rabbit shows at once a muscle, the *sacrospinalis* (Figs 8.3 and 8.8), the largest muscle in the body, attached to the sacrum behind and tapering forwards to be inserted all along the vertebrae of the back. The vertebrae thus form the compression members and the back muscles the tension members of a bracket or cantilever for supporting the weight of the body on the hind legs. The analogy can be pursued further, for the

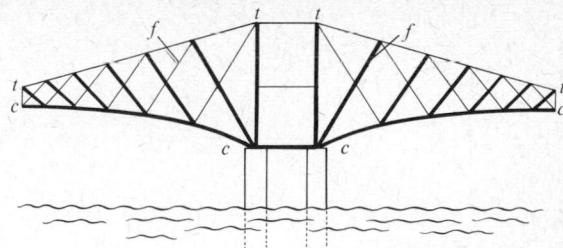

FIG. 8.2. Diagram of cantilever girder, such as is used in the Forth Bridge. *cc*, main compression member; *f*, struts and ties of the filling of the girder (thick lines, compression members; thin lines, tension members); *tt*, main tension member. (After D'Arcy Thompson, *loc. cit.*)

muscle mass is greatest and the bones are largest close to the point of attachment, where, as the stress diagram shows us, the tendency to strain is greatest.

This analogy serves as a simple introduction to the type of method used, but by more careful analysis more fruitful results can be obtained. Some simplification must always be introduced, and no analysis yet devised gives full insight into the mechanics of the animal at rest and in motion. D'Arcy Thompson (1942) has made a suggestive comparison between the backbone of a quadrupedal mammal and two cantilever girders, which rest on the legs rather as the girders of the Forth Bridge rest upon their pillars (Fig. 8.2). A great part of the body-weight in a standing quadruped falls on the front legs and the similarity to a girder is clearly marked in this part of the backbone (Fig. 8.3). The compression member of the girder is represented by the bodies of the vertebrae, arranged in a row to form arches curving upwards from a low point opposite the limb. The tension members are the *ligamentum nuchae*, which runs along the back of the neck, and other ligaments and muscles that run between the vertebral spines, some for long, others for short distances (p. 79). Between the main compression and tension members a cantilever girder is given a 'filling' of struts and ties. These are arranged on the principle that the triangle is the only geometrical figure that cannot be distorted if the lengths of the side remain constant. In the body the vertebral spines are the struts, and the short muscles and ligaments of the back are the ties. The slope of the vertebral spines is approximately forwards in front of the forelegs and backwards behind, as in a cantilever girder, but other forces operate to cause them to point slightly backwards even at the shoulders. As Fig. 8.3 shows, the principle of triangles is used widely, combining strength with lightness.

A significant feature of the construction of girders by an engineer is that the depth and hence strength of

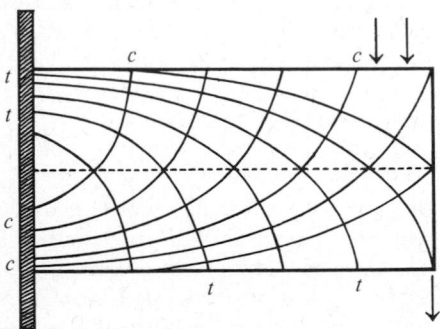

FIG. 8.1. Diagram of stressing in a loaded beam attached to a wall. The closer the lines the greater the stress in that region. *cc*, compression lines; *tt*, tension lines. (After D'Arcy Thompson (1942). *Growth and form*. Cambridge University Press.)

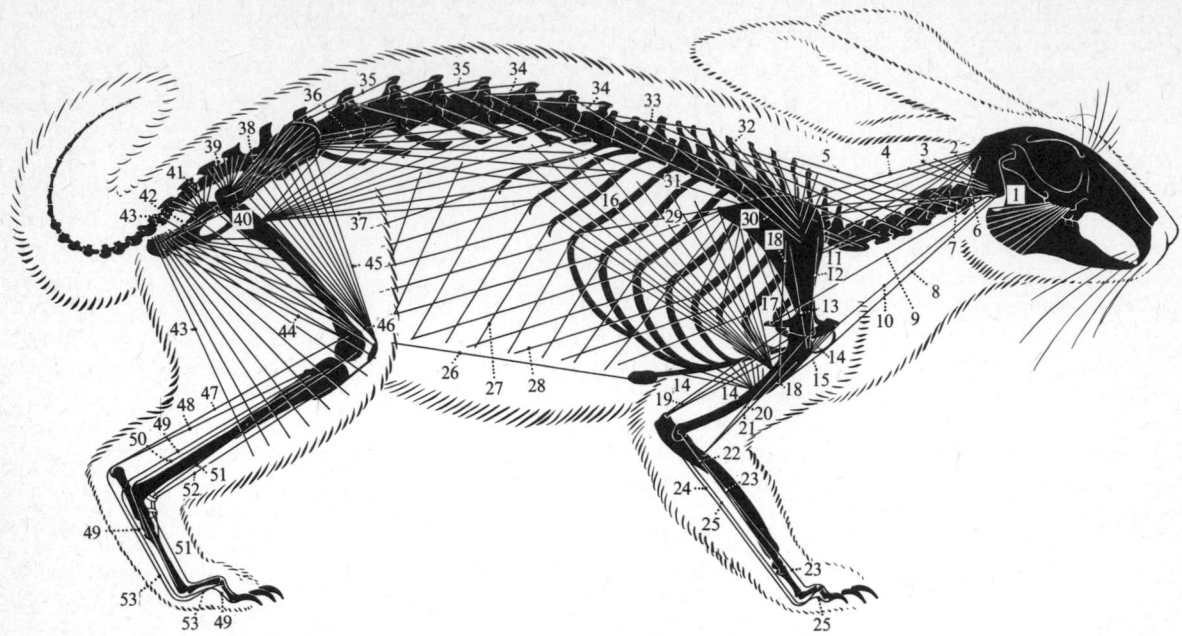

FIG. 8.3. Diagram of the skeleton and muscles of rabbit, to show the general arrangement of struts and ties.
1. masseter; 2. obliquus capitis; 3. splenius capitis; 4. semi-spinalis capitis; 5. longissimus cervicis; 6. longissimus capitis; 7. obliquus capitis inferior; 8. basioclavicularis; 9. levator scapulae; 10. sternomastoid; 11. scalenus; 12. supraspinatus; 13. infraspinatus; 14. pectoralis; 15. cleido-humeralis; 16. latissimus dorsi; 17. subscapularis (displaced caudally); 18. deltoid; 19. triceps; 20. bicep brachii; 21. brachialis; 22. extensor carpi ulnaris; 23. extensor digitorum communis; 24. flexor digitorum sublimis; 25. flexor digitorum profundus; 26. rectus abdominis; 27. transversus abdominis; 28. external oblique; 29. serratus anterior; 30. trapezius; 31. ilio-costalis; 32. longissimus; 33. semispinalis dorsi; 34. longissimus dorsi; 35. multifidus; 36. sacro-spinalis; 37. psoas major; 38. gluteus medius; 39. piriformis; 40. gluteus maximus; 41. abductor caudae; 42. gemellus inferior; 43. biceps; 44. adductors; 45. rectus femoris; 46. vastus intermedius; 47. gastrocnemius and plantaris; 48. soleus; 49. flexor digitorum longus; 50. peroneal muscles; 51. extensor digitorum; 52. tibialis anterior; 53. plantaris.

the structure should be at every point proportional to the bending moments. These will be greatest over the point of attachment of the girder and the vertebral spines are longest above the forelegs. Many such features can be recognized in the skeleton of the rabbit, and they appear even more clearly in those hoofed mammals that are large and rapidly moving, for example, the horse.

Such comparisons are stimulating but there is a danger that they may be accepted uncritically, without recognition of the fact that they give only suggestions and not exact solutions of the statics, still less of the dynamics of the vertebrate body. We can, however, pursue the analysis further, as Gray has done, by determining the bending moments along the vertebral column. Fig. 8.4(a) shows a reptile with fourteen vertebrae, the loading of each vertebra being given near the vertical arrows and the bending moment curve shown by the dotted line, with strain on the dorsal musculature above and on the ventral musculature below the base line. Fig. 8.4 shows the effect of shorten-

ing of the body in relieving the strain on the ventral musculature. One of the most marked features of later reptilian and early mammalian stocks was this decrease in body length. The effect of the absence of a tail is shown in Fig. 8.4(c), where the ventral musculature is again under strain. The animal in Fig. 8.4(d) is so constructed that the loading of the central part of the body is asymptotic to the base line and such a type consists of two exactly balanced cantilevers.

3. The vertebrae

The vertebral column of mammals forms the main compression member of a complicated girder by means of which the weight of the body is carried and can be propelled forwards or stopped. The arrangement of the girder differs considerably in different mammals, but five regions can always be recognized: the cervical, thoracic, lumbar, sacral, and coccygeal. They are developed to varying extents according to the way the weight is distributed. In the rabbit the back is used by the animal in two distinct ways. When springing for-

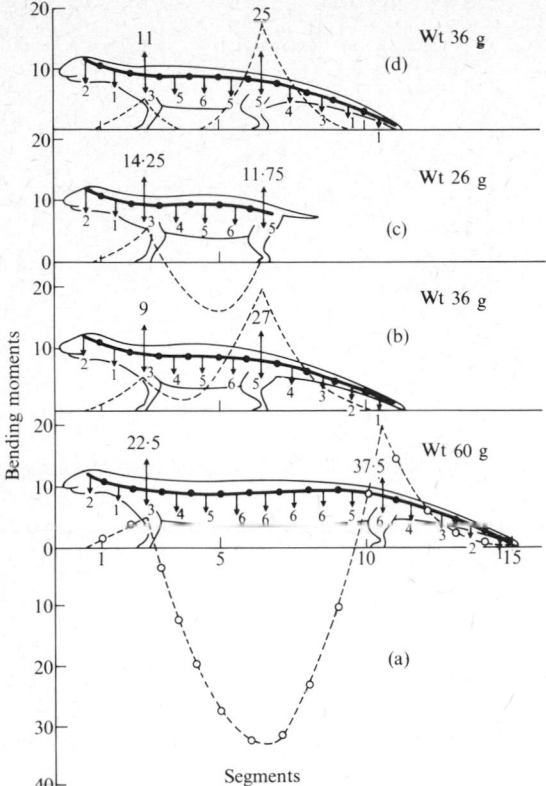

FIG. 8.4. Analysis of the bending moments along the vertebral column of a series of tetrapod forms. The figures near the arrows show the load on each vertebra in grams, the dotted line shows the bending-moment curve, strain on the dorsal muscles above and ventral muscles below the axis. The limbs are shown acting as vertical struts and the proportion of the weight carried on fore- and hind legs is given. (a) Long-bodied reptile; the greatest strain falls on the ventral musculature at the middle of the body. (b) Effect of shortening the body. (c) Same without tail, showing strain on ventral musculature. (d) Central part of back and belly relieved of strain by loading the weight on two balanced cantilevers. (From Gray (1944). *J. exp. Biol.* **20**.)

to the diaphysis of long bones, with smaller *vertebral epiphyses* in front and behind. The surfaces of the centra are usually flat and between them are plates of fibrocartilage, the *intervertebral discs*. The central portion of each of these discs, the *nucleus pulposus*, represents the remains of the notochord (p. 121). The intervertebral discs act as cushions between the centra. Movement between the bodies of the vertebrae is usually rather limited and the surfaces between them are not synovial joints. The arrangement allows some bending and rotation to occur, but this is checked by the interlocking articular facets of the vertebrae and by dorsal and ventral longitudinal ligaments, which run the length of the column.

The regular segmentation found in a fish-like verte-

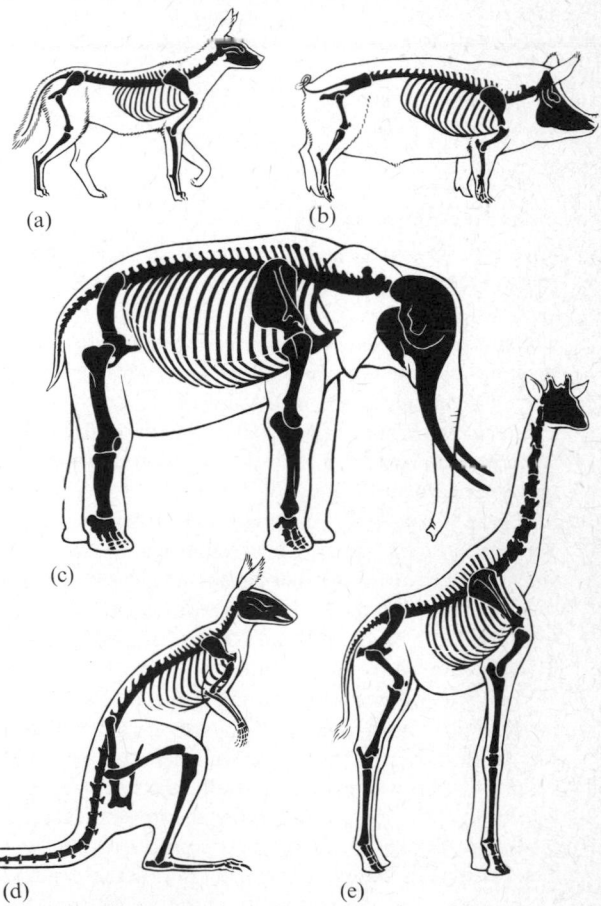

FIG. 8.5. Arrangement of the backbone girder in various mammals. (a) Balanced cantilevers (dog); (b) single girder balanced on forelegs and hind legs (pig); (c) single girder with balance largely about forelegs (elephant); (d) single girder balanced on hind legs (wallaby); (e) single girder balanced on forelegs (giraffe).

wards it forms a single girder, braced on to the hind legs. When standing still it is a double cantilever girder and much of the weight is carried on the forelegs. In the larger quadrupedal mammals this last is the usual arrangement, the weight being balanced on the forelegs, while the hind legs are used for pushing (Fig. 8.5). The head often acts as a counterweight (elephant, rhinoceros, giraffe). In man the situation is completely changed by the use of the vertebral girder as a vertical pillar.

In all mammals the vertebral column contains a number of *centra* and acts as a compression strut. These centra develop from a larger middle part, corresponding

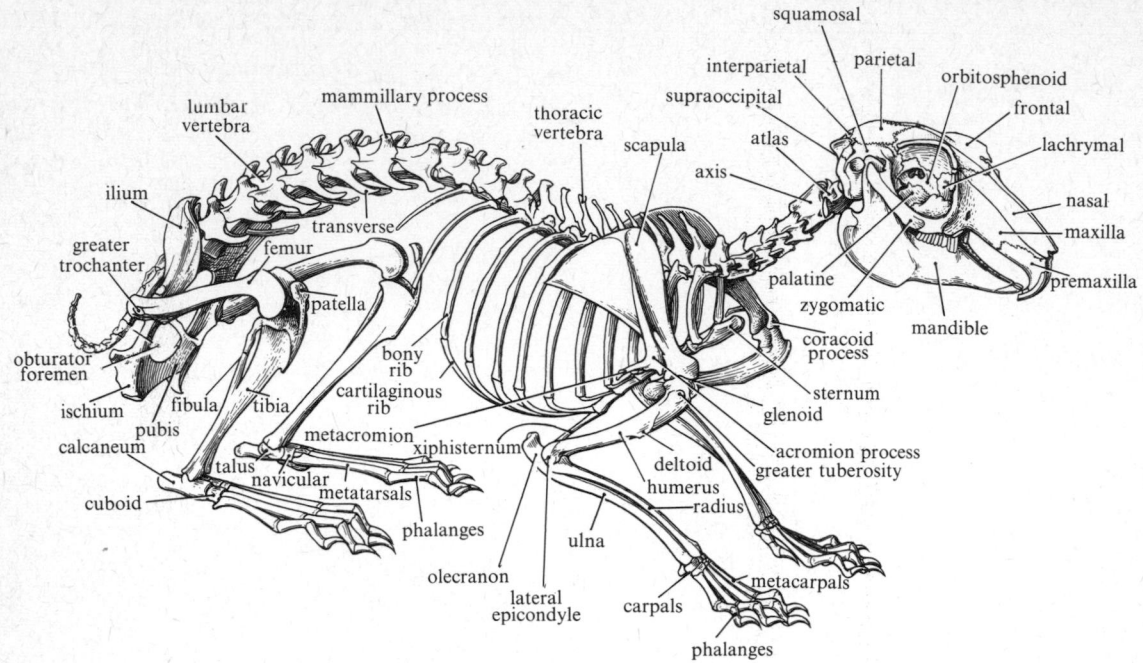

Fig. 8.6. Skeleton of a rabbit.

brate has been modified in mammals so that different types of vertebrae are produced along the length of the back (Figs 8.6 and 8.7). The boundaries between the sets are sometimes fairly sharp, but throughout the vertebral column we can see well illustrated the operation of gradation of morphogenetic processes (p. 120). The neural spines and the various other projections and facets gradually change in direction and in length along the body; evidently the original continuous series of vertebrae has become modified by the interposition in a graded manner of special developmental processes. Whereas one process repeated many times produces the similar members of the backbone of a fish or newt, several diverse processes have been introduced to make the varied vertebrae of a mammal.

The vertebral girder of the rabbit may be considered to consist of two parts: lumbo-sacral, behind, and cervico-thoracic, in front. Each of these has a central main compression member, with struts above and below. In the lumbo-sacral girder the upper struts are the neural spines and mamillary processes and the lower struts are the transverse processes. In the thoracic region the upper struts are the neural spines, and the lower struts are the transverse processes and ribs. The ribs have to be considered as part of the girder in spite of the fact that they are jointed with the vertebrae and move during respiration. The symmetry of this girder can be seen in the skeleton of a large animal such as

an elephant (Fig. 8.5), which carries much of the weight on the forelegs.

The neck and the tail are integrated with the main girders of the back to differing extents in various animals. Many mammals have become short-bodied, and the tail does not then have a balancing or cantilever function as it does in lower animals. The neck retains this condition to a considerable extent in some mammals, as witnessed by the presence of long cervical neural spines and transverse processes. There is a tendency for the head to be supported in such a manner that the neck vertebrae act as a simple compression strut; that is to say, as a separate single-arm girder held in front of the thorax by long braces. In such necks the cervical vertebral processes are short, for example, in the giraffe, rabbit, monkey, and man.

There are pronounced differences in the arrangement of the skeleton and muscles in different mammals (Fig. 8.5), and it is difficult to make generalizations about the girders. Yet the same principles apply throughout and once understood they can be applied to the particular conditions in each species.

We shall deal in detail with the rabbit, where all the points mentioned can be verified by dissection.

4. The lumbo-sacral girder of the rabbit

In describing the vertebral column it is convenient to begin with the more posterior region, where the struc-

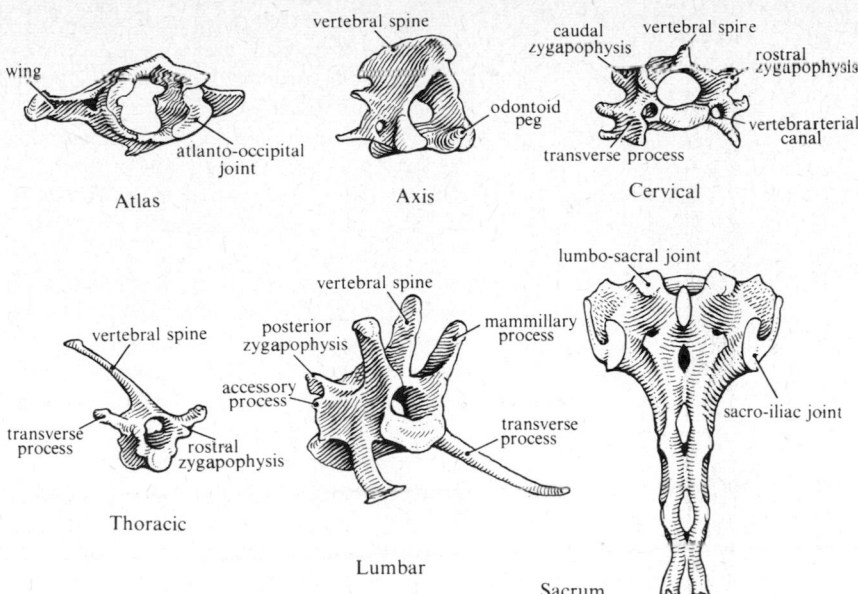

FIG. 8.7. Vertebrae of a rabbit.

ture of the girder is simple. The *pelvic girdle* is firmly attached to the vertebral column in all mammals (except certain aquatic forms). In the rabbit four vertebrae are modified and fused to form the *sacrum*, but of these only the first articulates with the ilium (Fig. 8.7). The surfaces of articulation at this *sacro-iliac joint* will be found to be rather difficult to separate, but the two parts are not actually fused and they permit a small amount of movement. Such small movements are often allowed in the body even at points where great weight is carried. They serve to prevent fracture under sudden forces. The articular surfaces of the sacrum and ilium are partly covered with cartilage and partly roughened for the attachment of the strong *interosseus sacro-iliac ligament*, which, together with bands of fibres above and below the joint, holds the parts together. The amount of movement possible at the sacro-iliac joints becomes greater shortly before parturition, the ligaments being loosened by the action of the hormones circulating at that time (p. 492).

The sacral vertebrae have low spines and expanded upper surfaces for muscle attachment. The pelvic girdle and sacrum thus form a platform, balanced on the hind leg, to the front of which the vertebral column is attached as a projecting bracket. The structure of the bones and muscles of this part of the back can easily be understood by considering the distribution of stress in such a bracket (Fig. 8.1). The compression lines occupy the lower portion of the girder and the tension lines the upper. In the backbone the bodies of the vertebrae are the compression members and the *lumbar vertebrae* have large bodies for this purpose. The movements allowed between them are mainly in the sagittal plane, giving arching or straightening of the back and some rotation. The vertebrae in this region have large surfaces for the attachment of muscles. Besides the broad neural spines and transverse processes there are also large *mamillary processes*, not found elsewhere in the vertebral column. To these vertebrae are attached the tension members of the girder, in the form of the sacrospinalis muscle (Fig. 8.3).

5. The muscles of the back in the rabbit

The essential pattern of the arrangement of the muscle of the back is that close to the midline most of the fibres run cranially and medially, so that caudal lateral parts of the vertebral girder are joined to cranial medial parts. More laterally the muscle fibres run cranially and laterally, so that here caudal medial parts of the vertebral girder are joined to cranial lateral parts (Fig. 8.8).

The medial muscle mass lies over the vertebral column between the spinous processes in the midline and the mamillary and transverse processes laterally. Its fibres originate on the mamillary and transverse processes and insert on to more cranial spinous processes. In the lumbar region, from the sacrum to the more caudal thoracic vertebrae this muscle is called *multifidus* and more cranially it is called *semispinalis dorsi*. Superficially the fibres originate directly on the mamillary and

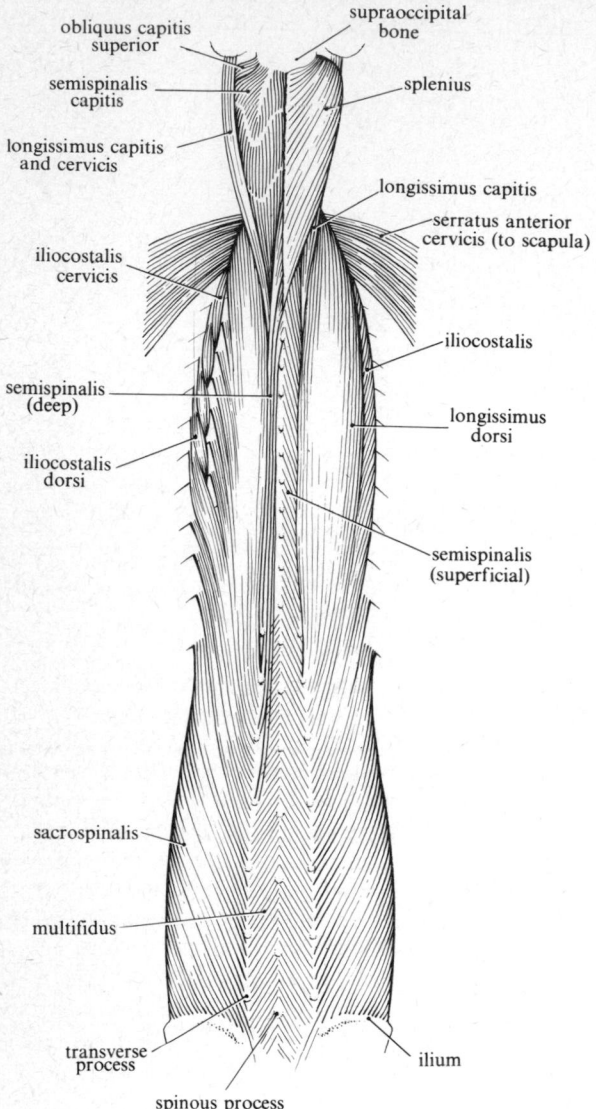

obliquus capitis
superior

supraoccipital
bone

semispinalis
capitis

splenius

longissimus capitis
and cervicis

longissimus capitis

serratus anterior
cervicis (to scapula)

iliocostalis
cervicis

iliocostalis

semispinalis
(deep)

longissimus
dorsi

iliocostalis
dorsi

semispinalis
(superficial)

sacrospinalis

multifidus

transverse
process

ilium

spinous process

Fig. 8.8. Dissection of the back muscles of the rabbit.

transverse processes of the vertebrae and are inserted on more cranial spinous processes. The deeper fibres originate from long tendons that join more caudal mamillary and transverse processes with spinous processes further cranially.

In the neck the medial part of the muscle mass includes the *semispinalis capitis*, arising from the transverse processes of the caudal cervical vertebrae and the cranial thoracic vertebrae. It is inserted on to the lateral surface of the external occipital protuberance but is attached also between the external occipital pro-

tuberance and the transverse process of the atlas. Ventral to the semispinalis capitis is the *semispinalis cervicis*, which originates on the articular processes of the more caudal cervical vertebrae and the first thoracic vertebra. It is inserted on to the spinous processes of the cervical vertebrae, especially on to that of the axis.

The more lateral muscles make up the *sacrospinalis* muscle, which is the dominant feature of the dorsal and lumbar regions after the lumbo-dorsal fascia has been removed. It forms a large mass, lateral to the multifidus, between the crest of the ilium and the thorax. Its fibres originate on the crest of the ilium, the dorsal surface of the sacrum, the mamillary processes of the more caudal lumbar vertebrae, and the lumbo-dorsal fascia. They run cranially and laterally and are inserted on to the transverse processes of the lumbar vertebrae. Fibres belonging to this muscle are originating and inserting all along its length; as some leave it others join it.

In the thoracic region it divides into a more medial *longissimus dorsi* and a more lateral *iliocostalis dorsi*. Both muscles are inserted on to the ribs. The longissimus dorsi receives fibres from the semispinalis muscle in the thoracic region, and these are inserted with it on to the ribs. The iliocostalis receives fibres, originating on the ribs, which run cranially to the seventh cervical vertebra as the *iliocostalis cervicis*. In the neck, longissimus dorsi is inserted on to the transverse processes of the more caudal cervical vertebrae and carries on cranially to the transverse process of the atlas as the *longissimus cervicis*.

From the transverse processes of the second, third, and fourth thoracic vertebrae, a bundle called *longissimus capitis* runs forward along the lateral border of the *splenius* muscle. The latter is a triangular sheet arising from the ligamentum nuchae; its fibres run cranially and laterally to the mastoid part of the skull and the transverse process of the atlas. Cranially the fibres of longissimus turn and run parallel with the fibres of splenius and are inserted with them on the mastoid part of the skull.

6. The tail

The tail varies from the four minute structures making the *coccyx* of man to the fifty or so vertebrae of the scaly anteater *Manis* and the powerful swimming organ of the whales. The muscles (*abductor caudae*) are well developed only in the cranial portion of the tail, the terminal segments being moved by tendons. By these muscles the tail can be moved in all directions, since there are usually no articular facets on the caudal vertebrae. The muscles are developed to different extents in various mammals. In man they are still present in spite of the absence of the tail, being incorporated in the *levator ani* muscle, which forms a floor to the whole pelvis, an important function in the upright position.

The tail is put to various uses, illustrating the adaptability of mammalian organization. It acts as a balancing organ in many species (p. 78), especially in animals that walk or leap mainly on the hind legs, such as kangaroos and squirrels. In the ungulates, which are sufficiently well balanced on their legs without it, we find the tail converted into a fly-whisk. In the extinct *Glyptodon* it formed a defensive mace. In some monkeys it is a prehensile organ, strong enough to carry the whole animal and provided with a sensitive under surface, which has a considerable representation in the cerebral cortex (p. 330). In whales it is a swimming organ of immense strength. In rabbits and many artiodactyls its white underside is shown as a warning of danger to others, marking a degree of social organization and power of communication. Even the pig is said to raise the tail when he is enraged. Man is one of the few mammals who, finding no use for his tail, has lost it.

7. The thoraco-cervical girder of the rabbit

There are usually seven lumbar vertebrae in the rabbit and in front of them there is a sharp change of structure, marking the point where the two portions of the vertebral girder join. The more caudal thoracic vertebrae resemble the lumbars in some respects, but whereas the lumbar transverse processes point headwards, the ribs point caudally and the neural spines also change direction just in front of this level. Evidently the whole direction of the ties and struts reverses at this central point in the girder.

The arrangement in the thoracic region is conditioned by the fact that the forelimb does not articulate directly with the vertebral column. The foreleg and scapula make a pillar on which the weight of the body hangs by means of a sling of muscle, the *serratus anterior* (Fig. 8.3), whose fibres run from the ribs to the border of the scapula, where the bending moment is greatest. Thus in spite of the difference in arrangement at the two ends of the vertebral girder, we can recognize a general similarity. The lumbar transverse processes correspond as lower struts to the ribs and the fibres of sacrospinalis 'hang' the weight on to the ilium, as the fibres of serratus anterior do on the scapula.

8. Ribs and sternum

The ribs, characteristic of the thoracic region, are attached to the vertebrae at two places. A *capitulum* at the end of each articulates with the body of its own vertebra and in the ribs at the head end of the series also with the body of the vertebra in front, while a *tuberculum*, on the side of the rib, articulates with the transverse process. The dorsal part of each rib is, in a rabbit, under compression stress from the weight of the body and the pull of the serratus, and this region is ossified as the *bony rib*. The ventral parts of the ribs turn cephalad and medially; these parts are under tension rather than compression and are *cartilaginous* instead of bony. This is a beautiful example of the way in which the consistency of the tissue varies with its function.

The ventral ends of the ribs are attached to the *sternum*, the more cephalic ones directly by the cartilaginous ventral pieces, the more caudal indirectly (Fig. 8.6). One or two free or floating ribs at the hind end of the series are not attached at all. The sternum is formed by endochondral ossification in a number of paired segmental *sternebrae*, which unite to form the main 'body' or *mesosternum*, to which is attached a posterior *xiphisternum*. In mammals with a well-developed clavicle there is often a large separate plate, the *presternum* ('manubrium' of man). The point of articulation between the manubrium and the body is known in man as the *sternal angle*, and it lies opposite the articulation of the second costal cartilage. Being easily felt through the skin it makes a convenient landmark for counting the ribs.

9. Abdominal muscles

The ribs also serve for the transfer of much of the weight of the abdominal viscera to the vertebral column through the action of the abdominal muscles. These muscles are derived from the hypaxial portion of the myotomes (p. 119) and are innervated from ventral roots. They serve to hold the viscera in place, acting as slings attached to the ribs, lumbar vertebrae, and pelvic girdle. In a well-balanced animal little strain falls on this region (p. 77), and the abdominal muscles are usually thin. The *obliquus externus abdominis* (Fig. 8.3) runs from the ribs and wall of the thorax caudally, ventrally, and medially to form a broad aponeurosis meeting its fellow in the middle line at the *linea alba*. Some of its fibres run to the crest of the ilium and the portion of the lower border of the aponeurosis between the ilium and the pubis forms the *inguinal ligament*, marking the caudal border of the abdomen. The *obliquus internus* consists of fibres running nearly at right angles to those of the externus, from the lumbo-dorsal fascia, ilium, and inguinal ligaments to the linea alba and the caudal ribs. The fibres of *transversus abdominis* lie still deeper and run in the same general direction but more transversely. The *rectus abdominis* consists of fibres close to the midline, running directly from the lower end of the thoracic cage to the pubis. This muscle is crossed by tendinous insertions, which at least approximately mark the segmental divisions. These are also sometimes present in the other abdominal muscles in some animals. In species where this musculature is well developed, seven or eight such segments can be recognized; usually, however, only three or four are present. The abdominal muscles serve to assist in flexing the body in the sagittal plane and the obliqui are also responsible for lateral bending and rotation. All the muscles aid in respiration by fixing the lower ribs and by supporting the viscera.

10. Thoracic vertebrae and muscles

A characteristic feature of the thoracic region is the great length of the neural spines, which in the rabbit point backwards. These upper struts of the girder are united by a ligament, the *supraspinous ligament*, making the upper tension member of the girder. The spines are highest over the withers, where the bending moments are greatest; their length and slope varies greatly in different mammals. In quadrupeds the more cephalic thoracic spines usually point caudally and the lumbar spines headward, with one *anticlinal vertebra* almost vertical between. This represents the point between the two cantilever girders with which the column has been compared (p. 77).

In man, where the column is not constructed as a cantilever girder system, all the spines point caudally, but the thoracic spines more sharply than the lumbar, so that the eleventh thoracic may be considered anticlinal.

The muscles of the thoracic and cervical region are arranged as already explained to form a system of ties bracing the cranial part of the vertebral girder on to the more caudal part. Thus in the rabbit the large *semispinalis dorsi* muscles form a series of powerful ties, running from a more lateral lumbar to a medial thoracic attachment (p. 79). Their more caudal attachments are by long tendons to the lumbar mamillary processes and each divides into conspicuous serial slips inserting on to the more cephalic thoracic spines. Each portion of the muscle has thus an elongated fan shape, with a tendinous caudal base and several muscular cephalic endings. These ties extend from the cephalad to the lower cervical spines. Just as the sacrospinalis serves to carry the front part of the body braced on to the hind legs, so the semispinalis is able to transfer the weight of the hinder part of the body on to the front legs. The girder is so made as to carry weight either mainly on its front or mainly on its hinder portions or on both.

The *serratus posterior* muscles consist of slips running ventrally and caudally from the thoracic spines to the ribs. These fibres correspond functionally to the multi-fidus in the posterior girder, running across the semi-spinalis.

Further muscles attached to the thoracic spines assist in supporting the body on the scapula. The *trapezius* in the rabbit is a broad superficial sheet running from the midline in the neck and thoracic region to the spine of the scapula. Ventral to the trapezius the *rhomboid* muscles are further slips with the same general direction, from the thoracic spines to the vertebral scapular border.

11. The neck muscles

The arrangement for the support of the head consists essentially of long muscles running from the cephalic end of the thoracic region to the back of the skull. The function of the seven cervical vertebrae thus tends to become that of a simple single flexible compression member, without any long projecting struts. The neural spines may be quite small, as in the rabbit, but in animals such as the gorilla, where a heavy head is carried on the neck as a projecting girder, they are long. In man the head is balanced on the vertebral column acting as a vertical girder and the cervical spines are short. The cervical transverse processes, though not long, receive the insertions of many muscles, especially in the posterior neck region, and are divided into dorsal and ventral portions, between which runs the vertebral artery in a foramen characteristic of neck vertebrae.

In the rabbit the more caudal cervical vertebrae serve as a base for the whole neck and receive attachments of *semispinalis* on to their spines and of *longissimus* (p. 80) on to their transverse processes, thus bracing the whole neck and head on to the two main parts of the vertebral girder. The head is held up on the neck mainly by two sets of muscles that run from the back of the skull to the lower cervical and upper thoracic regions. *Semispinalis capitis* is the more dorsal of these, passing between the occipital protuberance and the transverse processes all the way from C_3 to T_4. The second brace, *longissimus capitis*, is attached to the mastoid process and acts below the atlanto-occipital joint, bending the head downwards; it is attached caudally to the transverse processes T_{2-4}. Below these muscles lies a still deeper layer, the *semispinalis cervicis*, which, from the spines of the axis and cervical vertebrae, runs to the articular processes of C_4–T_1 and thus also holds the head back. There are also other small, short muscles between the neck vertebrae.

On the ventral side the musculature of the neck is weaker; *longus cervicis* runs from the ventral side of the atlas and other cervical vertebrae to the vertebral bodies farther back. The *scalene muscles* pass from the lower sides of the transverse processes of C4–7 to the outer side of the first five ribs. These muscles work with the sternomastoid and longissimus capitis and semispinalis capitis in balancing the head on the neck, a function particularly important in types with long or vertically held necks such as the giraffe or man.

12. Cervical vertebrae

The neck is characteristically developed in mammals to allow movement in all directions of the head, with its sense organs and mouth. Whereas the head of a fish or an amphibian can only be turned by movement of the whole body, that of a mammal is like a separate organ and can be directed to almost any point in space by the muscles of the thin neck. The cervical vertebrae allow a considerable degree of movement and provide attachments for the muscles that hold up the head. They have small bodies, with a large neural canal (since the spinal cord is large in the neck, see p. 257), and are broad in proportion to their length. The cervical articular facets are of simple form; the cephalic zygopophyses point dorsally and the caudal ones ventrally on each vertebra. They have flat surfaces allowing a considerable degree of movement in the sagittal plane and some lateral movement (bending the head sideways), though this is always combined with rotation.

13. Atlas and axis

Although there is considerable mobility along the whole neck, the extensive movements of the head are mostly due to the special arrangement of the first two cervical vertebrae, the *atlas* and *axis*, which are larger than the others and form joints that allow much movement. The two occipital condyles rest in large concave facets on the front of the atlas and this *atlanto-occipital joint* allows movement in the sagittal plane, as in nodding the head. The first or *atlas* vertebra is very wide and has a thin neural arch but no centrum; its transverse processes are broad and long, giving good leverage for the muscles that hold and rotate the head and neck.

The *axis* vertebra, on the other hand, is narrow in the transverse plane. It bears on the cranial surface of the centrum a knob, the *odontoid process*, which can be

shown by its development to represent the centrum of the atlas segment. This knob and the articular facets of the axis are so arranged that the atlanto-axial joint allows rotation of the head and atlas on the neck, thus, with the atlanto-occipital joint, allowing movement in all directions. The odontoid articulates with the atlas, and is separated from the neural canal proper by a transverse ligament. The neural spine of the axis presents a large flat surface for the muscles running forwards to the skull and backwards to the other vertebrae; these muscles hold up the head and extend the neck backwards in the sagittal plane.

14. Proprioceptors of the neck

The head and neck are thus supported by an elaborate set of muscles; these have departed a long way from the simple segmental myotomes present in fishes and amphibia, and they allow delicate and controlled movement. The position of the head in space is evidently of great importance to a mammal and especially to a man. It is perhaps not an accident that when our 'personality' is in some way reduced we 'hang our heads' and on recovery again 'hold our heads high'. The extensor muscles of the neck of man are at work continually throughout waking life to maintain the position of the head, but if we fall asleep when sitting they allow it to flop forwards on to the chest. This 'tonic' action of the muscles is produced by stretch receptor organs (p. 265), which are numerous in these as in other 'antigravity' muscles. The proprioceptor organs of the neck muscles have also a wider function to perform in producing adjustments of posture in other parts of the body. Together with the receptors of the eyes and the static organs of the ear (p. 401) they provide the information that determines the position of the eyes and of the muscles controlling the limbs (p. 301). For every position of the head there is an appropriate position of the eyes and limbs, and the whole constitutes a delicate system by which the animal or man directs itself to whatever feature of the environment is claiming attention for the moment. The neck is thus of great importance in controlling the life of a mammal.

9 The forelimb

1. The position of the limbs in a mammal

AFTER the fishes first came on land the limbs changed from flaps projecting laterally to struts with a skeleton on an axial plan. A fin with dorsal and ventral muscles, serving to raise and depress it for purposes of steering, thus became changed into a limb turned under the body, and able to move in various directions and to carry and move the weight of the body. Distinct traces of the original fin plan still remain and they can be made the basis of a simple account of the various muscles of the limbs.

The human arm and hand possess an exceptional mobility and can be placed in positions that serve to illustrate the changes that have taken place during evolution. The relation of a limb to the original position of the fin is shown by stretching the arm out sideways with the thumb pointing towards the head. We can then recognize that the limb has a *pre-axial border*, occupied by the radius bone and the thumb (pollex, digit 1), corresponding to the front of the fin, and a *post-axial border*, with the ulna and little finger.

The muscles of a fish fin serve mainly to raise and lower it and are divided into upper abductor and lower adductor sets. These muscles have become modified to form the muscles of the limbs, which are moved forwards and backwards in locomotion. The more cephalic of the original adductors and abductors now work together as protractors, the more caudal ones act as retractors. Around the shoulder and hip joints there are therefore muscles that move the limb in each of the four chief directions and also produce actions of rotation by which the whole limb is revolved on its axis. At the more distal joints (elbow and knee) movement is mainly in two directions, distinguished by the arbitrary terms flexion and extension. It is not possible to say exactly how the flexor and extensor muscles are related to the original adductors and abductors of the more distal segments of a fin.

The mammalian limbs have therefore changed from a condition in which they projected laterally from the body to one in which they are brought under the body for support, and elongated to act as levers. The first change was the bending of the limb at the elbow and wrist, allowing the ventral surface of the 'hand' to be placed on the ground while the radius and ulna were held vertically. This stage was already reached in mesozoic amphibians and reptiles. To produce a less broad base and more efficient support the humerus then came to be held directed backwards, so that the elbow pointed back and the limb was brought directly under the body. This, however, would have the effect of leaving the hand pointing outwards and backwards. To avoid this the radius early became partly twisted round the ulna, so that in the *prone* position typical of most quadrupedal mammals the radius is lateral at the elbow but passes across the front of the ulna to become medial at the wrist. The hand is thus directed forwards and the first digit is medial. The bones are fixed in this position in many mammals (for example, the rabbit) but in primates the hand can be returned to the more primitive *supine* position, with the radius and ulna parallel and the thumb pointing laterally. In man much use is made of the possibility of changing from one position to the other and thus turning the hand in various directions.

Somewhat simpler changes have been necessary to bring the hind limb into a position underneath the body. The first change from a simple lateral fin was, as in the forelimbs, a bending at knee and ankle, bringing the sole on to the ground and the tibia and fibula to a vertical position, with the tibia pre-axial. The limb was then brought under the body by rotation medially at the hip joint, so that the knee points towards the head. The foot is thus placed under the body, pointing headward, with its sole on the ground and the tibia medially; there is no need for a 'pronation' to direct the foot headward.

This analysis gives us a means of understanding the changes that have taken place during the evolution of the limbs and is a help in understanding the present arrangement of the muscles. There are not sufficient paleontological data to allow a fully satisfactory history

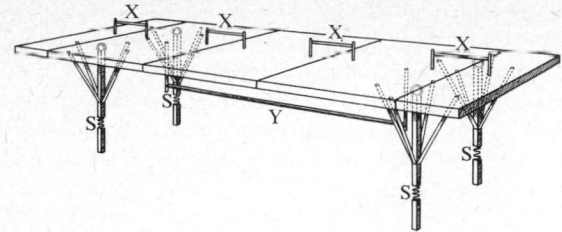

FIG. 9.1. Diagram of a balanced table top for comparison with the balance of a quadruped. The top consists of sections jointed together and supported by elastic braces X above and Y below. The legs articulate with the top by means of universal joints, each protected by four elastic braces (see text). Each leg is itself extensible by a spring S. (From Gray 1944.)

to be given of the detailed sequence of events as they actually occurred.

2. Functioning of the limbs

The limbs provide the chief means by which the body moves about in the world and moves other objects towards or away from itself. In most mammals the limbs support the weight in a standing posture, and it is convenient to consider how the muscles and bones co-operate to maintain this posture and to provide the horizontal forces by which the animal is propelled or stopped. To support the body the limb must exert a vertical thrust against the shoulder or hip, and to drive it forward there must be a thrust along the axis of the vertebral column. To see how these purposes are served Gray (1944) suggests that we may treat the muscles as elastic springs. For an analogy he takes a table whose four legs are attached to the top by universal joints (Fig. 9.1). Each leg contains within it a spring to represent the muscles that extend the legs. Compensation against horizontal forces is produced by four elastic braces running between each leg and the table

top, those in front acting as protractors and those behind as retractors. In the animal these are provided by the muscles attaching the limbs to the body, which serve to brace the leg when it acts as a strut and also to move it as a lever. Many of the muscles lie in positions that can be closely compared with such braces (Fig. 9.2), and this conception gives an insight into the significance of the arrangement of the musculo-skeletal system (Gray 1968).

The table top of Gray's analogy is made further comparable with the segmented animal by imagining it divided into sections, united by hinges and stabilized by braces attached either above or below. These braces can also be recognized in the body, and this part of the analogy could obviously be developed to give some idea of the form and distribution of the ties of the vertebral column. The effect of the analysis is to compare the body to a segmented overhung beam supported on four legs, which can act either as struts or as levers. When they act as struts alone they exert forces only along their mechanical axis, the effect of the muscles acting at the shoulder and hip being zero. When the limbs are acting as levers, on the other hand, the extrinsic muscles are used, so that forces are applied at right angles to the mechanical axes of the limb both to the ground and to the body. This causes the body to be pushed along in a manner that Gray suggests to be analogous to the propulsion of a canoe by a two-handed paddle.

Another method of locomotion is provided by the fact that the struts themselves are extensible by means of their intrinsic muscles, the body then being pushed forward, as in punting a boat with a pole. Analysis of the effects of the limbs in pushing in various ways can be developed in some detail to give a picture of the diagonal co-ordination of the limb movements and the effects produced by the thrust of each limb in tending

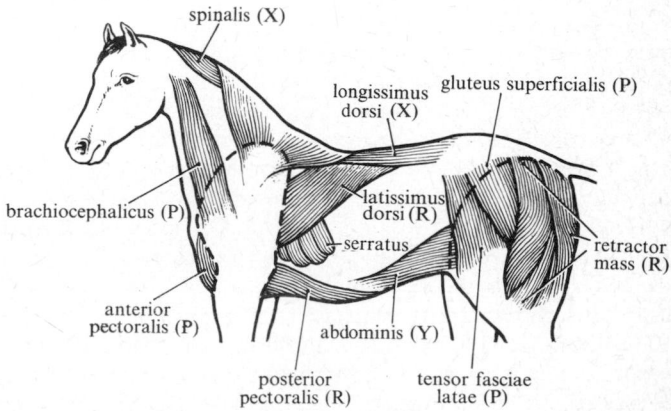

FIG. 9.2. Diagram of the functional relationships of some of the extrinsic limb muscle and axial muscles of a horse. The letters X and Y refer to the braces as shown in Fig. 9.1. P indicates protractors and R retractors. (From Gray 1968.)

to produce forward movement (or braking action), together with unwanted pitching or rolling couples, which must be compensated by contraction of muscles of other limbs or of the vertebral column.

Each type of animal presents a complete, co-ordinated action system, suitable to maintain life in the way appropriate to the species. The long legs of swiftly running animals and the long bodies of climbing types can only work in conjunction with suitable body and legs respectively, and the whole organization has to be considered together. By looking at the anatomy of animals in this way we can see the meaning of the shapes of the various parts. The limbs of terrestrial mammals mostly fall into two classes. (a) In *graviportal* animals, like the elephant or rhinoceros, the great weight is carried by the limbs acting as rigid extensible struts. Locomotion is by the extrinsic muscles. (b) In *cursorial* animals, adapted for fast speeds, the necessary forces are produced largely by the intrinsic muscles acting across the joints of the limb, and the foot is long.

The greatest possible speed is again related to the mode of life. A cheetah can give a burst of speed of 60 m.p.h. but only for 15 seconds. The maximum for a horse is 43 m.p.h. but it is adapted for long distances and can keep up 20 m.p.h. for an hour. A man can sprint at 24 m.p.h. and keep up 3 m.p.h. all day. He exerts 2·5 horse-power (84 kW) in the sprint. The powers of endurance depend more on the heart and lungs than on the limbs.

3. Functions of the forelimb

Whereas the hind limb of mammals is attached to the vertebral column by the ilio-sacral joint, the forelimb remains relatively free; the scapula is attached to the vertebral column only by muscles. There has been much discussion about the significance of this difference, since it is not immediately obvious that in a four-footed animal, such as a horse, there is less need for firm attachment of the forelimb than of the hind. The matter is especially puzzling because in the fishes that came on land to become the earliest tetrapods the pectoral girdle was jointed to the skull. Presumably this connexion was lost to allow greater mobility of the head. As the forelimbs were increasingly used to take weight they became connected with the main axis by a muscular sling rather than by bony or joint attachments.

The hind limbs usually play the main part in pushing the body forwards, but the front limbs carry more than half the weight in a quadruped in the standing position. Nevertheless, the forelimb retains greater freedom than the hind and has tended to acquire other functions such as digging, collecting food, or otherwise handling the

surrounding world. Moreover, the region to which the forelimb is attached (the thorax) must participate in respiration, necessitating independent movements. For all these reasons, therefore, power to move the limb in various directions is more important than firm attachment, and this power is conspicuously developed in man.

4. The shoulder girdle

The *shoulder girdle* in fishes was a point of attachment for the segmented body musculature, as well as a support for the limbs. In the amphibia and reptiles it has retained both functions and hence has a complex structure. With the raising of the body off the ground, however, it takes on the new function of transmitting the weight of the body to the limb, and for this purpose it becomes modified and simplified until it consists in mammals of two elements, the scapula and clavicle, or often of the former alone.

The *scapula* (Figs 8.6 and 9.3) is a flattened bony plate, carrying the attachment of many of the muscles that transfer the weight of the body to the limb and anchor the latter to the vertebral column. In four-footed mammals the direction of movement of the limb is mostly antero-posterior and all the parts of the shoulder girdle except the scapula are reduced. The clavicle disappears, or remains only as a very thin vestige, as, for example, in the rabbit. However, in those mammals, for example, man (Fig. 9.3), where the limb has acquired increased mobility and can be abducted away from the body, the scapula itself is able to rotate, allowing the glenoid facet to be turned in various directions. This rotation takes place about the axis provided by a well-developed *clavicle*, articulating with the sternum and with the scapula. Although this human condition involves retention of the primitive clavicle it must be regarded as a specialization; no such mobility of the shoulder girdle is possible in the primitive condition, seen still in the monotremes, where there is a plate-like pectoral girdle, with a well-developed ventral region, the dorsal region of the scapula being relatively small.

The girdle of placental mammals is relatively constant in structure (Figs 8.6 and 9.3) with a large plate-like scapula, carrying a ventrally directed *coracoid process* and bound to a clavicle if the latter is present. The coracoid ossifies in two parts but the significance of this is uncertain. The scapula bears on its outer side an *acromial spine*, characteristic of mammals, ending ventrally in an expanded knob, the *acromial process*, which is the point of articulation of the clavicle. A backward *metacromium* is developed at this point in four-footed animals that have reduced clavicles.

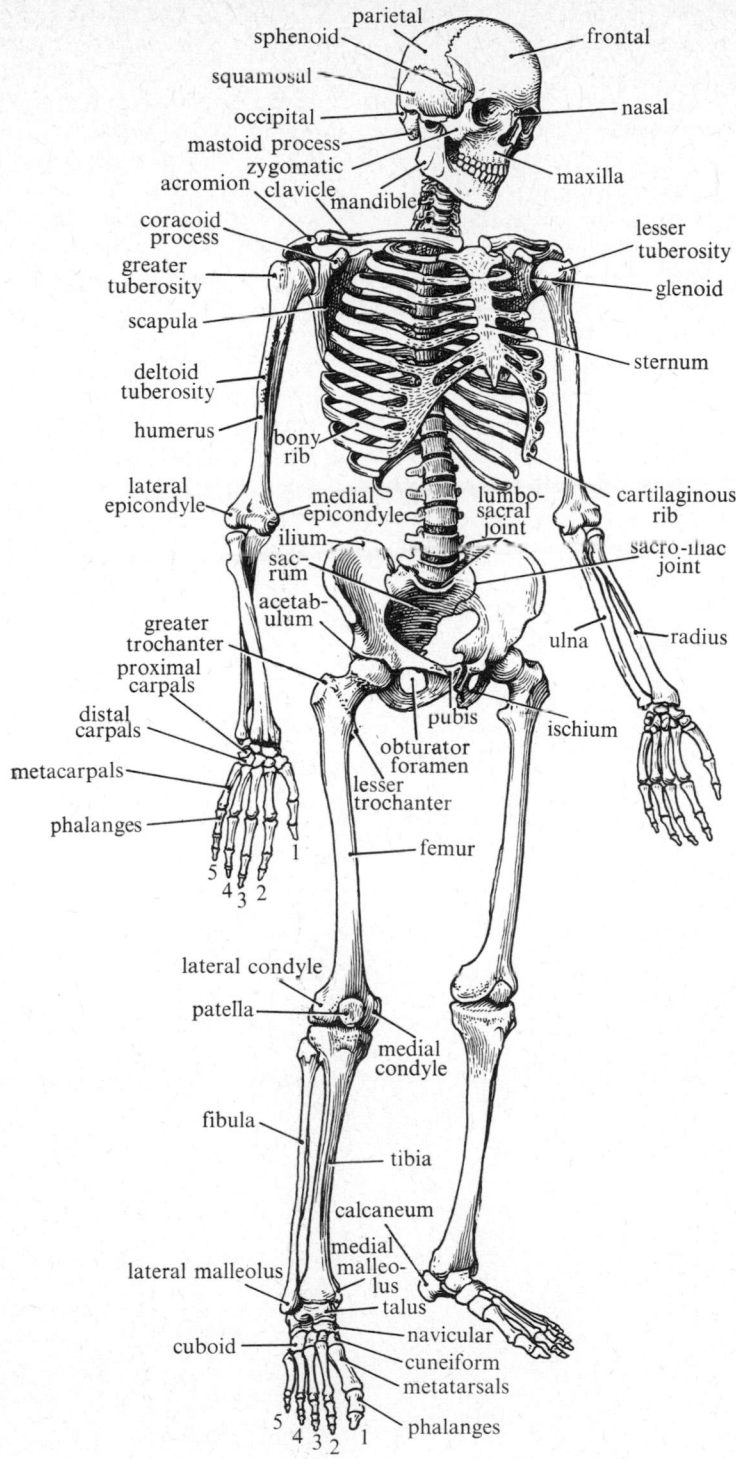

FIG. 9.3. Skeleton of man, approximately as it would be in the standing position.

5. The components of the forelimb

The glenoid cavity for articulation with the humerus is a relatively shallow cup compared with the deep acetabulum of the hip joint and this is so even in the four-footed mammals. In man, where mobility is the characteristic of the shoulder joint, the glenoid cavity is saucer-like and the head of the humerus is a large spherical structure, able to rotate in the saucer. The *humerus* itself is a typical long bone, carrying roughenings and protuberances where the muscles are attached (Fig. 9.4). At the upper end the *greater tuberosity* lies on the lateral side and is the point of attachment of muscles fixing the limb to the shoulder and abducting and externally rotating it. The *lesser tuberosity*, on the medial face, carries muscles that serve for adduction and internal rotation. The main muscles for moving the limb are attached farther down the shaft, where they obtain a greater leverage.

At the lower end the humerus is expanded into *lateral* and *medial epicondyles*, from which arise respectively extensor and flexor muscle masses for the wrist and hand. In man, where flexion is more powerful than extension, the medial (flexor) mass is the larger and the medial epicondyle is larger than the lateral one (Fig. 9.3). The humerus of the rabbit ends in a pulley-like *trochlea* for the ulna and the only movement possible is flexion and extension. In man there is also a second facet, the rounded *capitulum* for articulation with the radius.

The bones of the forearm, wrist, and hand vary greatly with the use to which the forelimb is put but in all mammals we can recognize a *radius* and *ulna*, though the latter may be reduced. These bones articulate with a proximal row of *carpals* at the wrist joint which allows a considerable range of movement. The proximal carpals articulate with a distal row and these

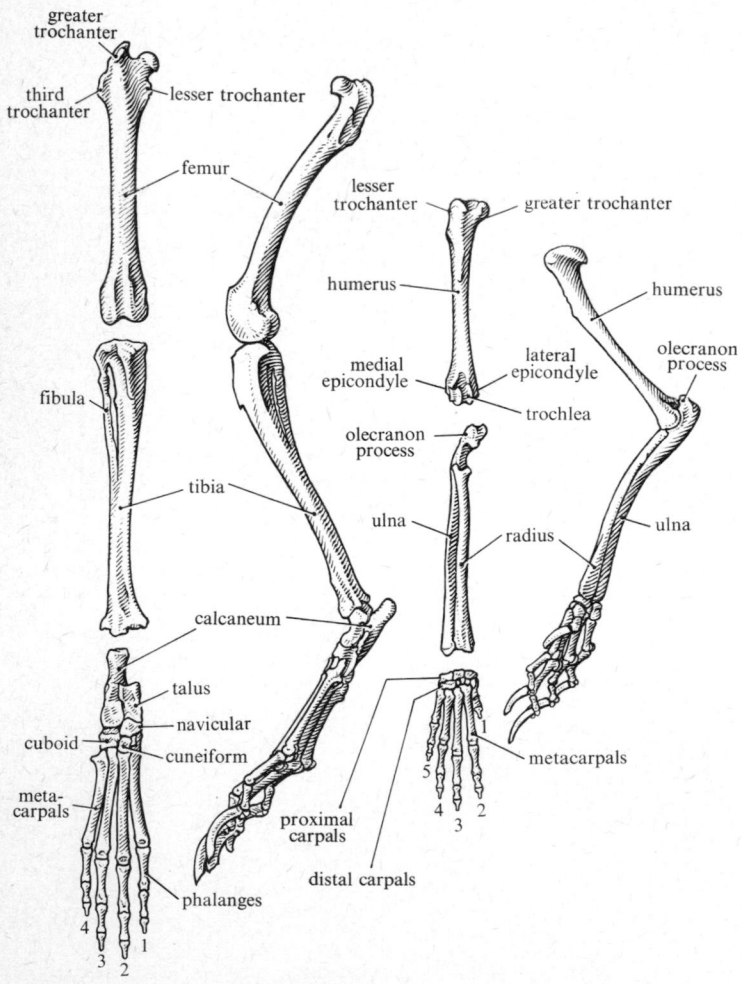

FIG. 9.4. Skeleton of the limbs of a rabbit.

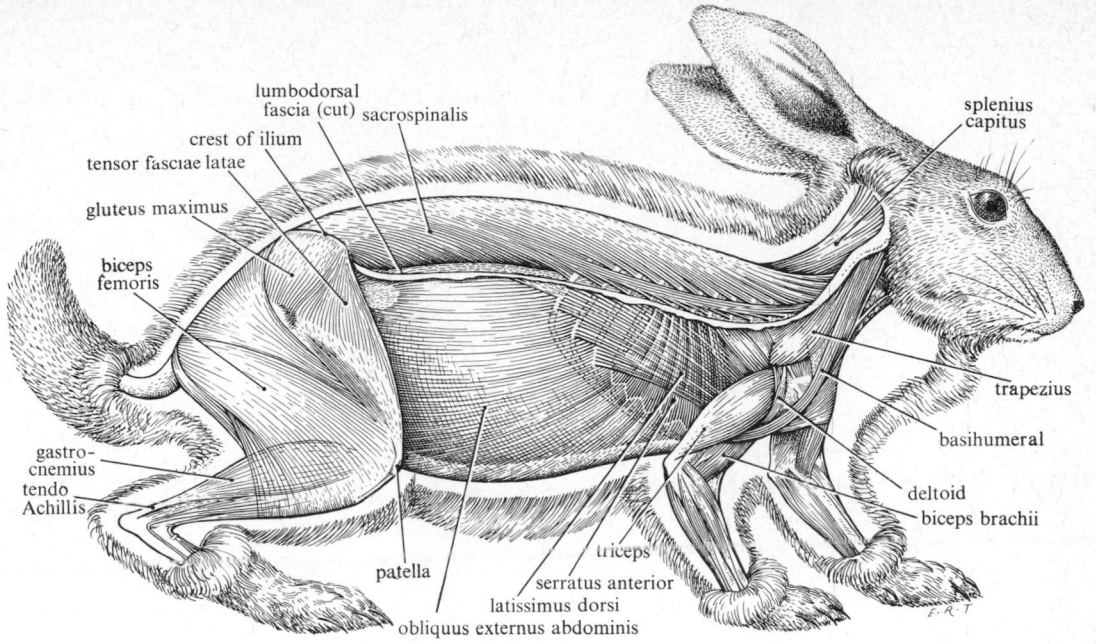

FIG. 9.5. Muscles of the limbs and back of the rabbit.

with *metacarpals*, but at the intercarpal and carpo-metacarpal joints there is usually little movement, except in animals where the thumb is free. The next important joint is therefore that between the meta-carpals and *phalanges*; here and at each of the two interphalangeal joints movement is possible, usually in one plane only.

The forelimb is thus a jointed strut consisting of four main segments: the scapula, humerus, radio-ulna, and hand, the last having three subsidiary terminal segments on each of the digits except the thumb. The most mobile joints are at the shoulder, elbow, wrist, and fingers.

6. Transmission of weight from forelimb to vertebral column

We shall consider the use and structure of the limb where it is an organ for weight-bearing, as in the rabbit; in the running digitigrade mammals, of course, it is still further developed for that purpose. The characteristics of the forelimb musculature in man, where they are suited mainly for manipulation of outside objects, will be referred to where necessary.

The forelimb is not jointed to the main vertebral axis, and the whole transmission of weight is accomplished by muscles. The muscle mainly concerned is the *serratus anterior* (Figs 9.5 and 9.6) which in the rabbit

forms a set of large bundles. The thoracic portion consists of slips running from the ventral part of the ribs to the upper border of the scapula and thus makes a sling by which the weight of the body, transferred through the vertebral column to the ribs, is hung on to the limb which acts as a pillar (p. 85). An anterior, cervical portion of the muscle helps to take the weight of the head and neck in the same way, running from the transverse processes of C3–7 and the first two ribs to the scapula (Fig. 9.6). In man, the weight-bearing function of the muscle is lost and the serratus anterior helps the trapezius to rotate the scapula upwards, thus suspending the weight of the arm; it also pulls the whole scapula forward round the chest and its serial slips can be seen contracting in actions involving pushing or punching.

The scapula is also attached to the body by muscles on its dorsal side. The most superficial and conspicuous of these is *trapezius*, a muscle mainly of lateral plate origin (p. 119), innervated by the spinal accessory nerve. This arrangement is presumably a relic of the time when the pectoral girdle was attached to the hind end of the gills. The trapezius muscle in all mammals is a broad band of fibres arising from the occiput and spines of the vertebrae all along the neck, thorax, and cephalic part of the lumbar region and converging to insertion on to the clavicle, if present, and along the

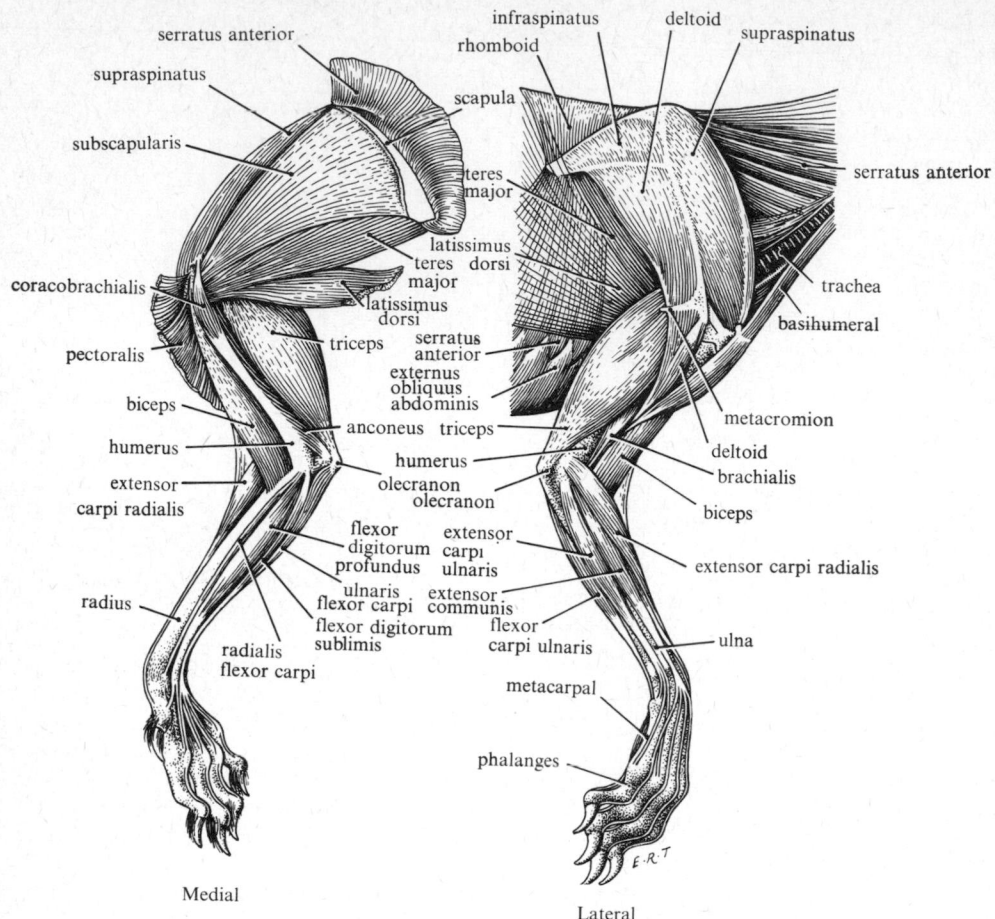

FIG. 9.6. Muscles of the foreleg of the rabbit. Lateral and medial views of dissection of right forelimb of the rabbit.

spine of the scapula (Fig. 9.5). The effect of the muscle is thus to hold the scapula against the body in quadrupeds and in man also to hold up and rotate the bone, so as to turn the glenoid upwards. The trapezius does not play a great part in weight-bearing in quadrupeds and is a less important muscle in such animals than in man, where it carries a large part of the weight of the arm and is active in the lifting of weights. This rotation of the scapula is characteristic of the freely moving scapula of man. It allows the glenoid cavity to be turned so as to face either upwards or downwards, thus greatly increasing the range of movement of the arm. The clavicle moves to some extent at the sterno-clavicular joint but the main movement is of the scapula on the clavicle at the acromio-clavicular joint. The rotation can be seen and felt every time that the arm is abducted.

Deep to the trapezius the *rhomboid* muscles (Fig. 9.6) pass from the ligamentum nuchae and spines of the anterior thoracic vertebrae to the dorsal border of the scapula. These are members of the set of muscles by which in quadrupeds the scapula is held against the chest wall. In man they work with the trapezius in holding up the scapula but since the fibres take a downward course their effect is to rotate the scapula downwards.

The cranial border of the scapula is held in to the body by the muscles known in the rabbit as *levator scapulae major* and *minor*, running from the skull to the metacromion and inferior angle of the scapula respectively (Fig. 8.3). The arrangement of these muscles is one of the chief differences between the forelimb of quadrupeds and man. In the latter the levator scapulae runs from the transverse processes of the upper cervical vertebrae to the medial border of the scapula; it thus holds the scapula up and also assists the rhomboids in rotating the glenoid downwards.

7. Ties around the shoulder joint

The scapula of the rabbit is firmly attached to the body, and we may consider it as the upper segment of the jointed strut on which the body is balanced and by whose lever action it is moved. Although mobility of the scapula is no doubt an important part of the locomotion of quadrupeds, the balancing of the body on the foreleg occurs mainly at the shoulder joint, and there are here a number of muscles that act as contractile braces to produce this balance (see p. 85). These braces occur all round the joint, preventing unbalance in any direction, but movement is much freer in the sagittal than in the transverse plane, and the braces are therefore strongest in front and behind, especially behind, since it is by drawing the whole limb backwards that the body is propelled forwards. Indeed, here, as throughout the limb, we can divide these muscles into two sets: (1) more cranial protractor, flexor-adductors, which are often more ventral, and (2) caudal retractor, extensor-abductors, lying more dorsally and innervated from more caudal segments. This division is not always sharp but it reminds us of the originally segmental arrangement of these muscles and that the movement of the limb, first forward and then backward, is related to the metachronal contractions of the segmented swimming muscles of a fish (Young 1962, p. 133).

Description of the braces around the shoulder joint is made difficult by the fact that whereas some of them run direct from the axial skeleton to the humerus others are shorter and arise from the scapula. We may consider first the longer and then the shorter ones. A conspicuously long anterior brace is the *basi-humeral* muscle of the rabbit (Figs 9.5 and 9.9), running from the base of the skull to the humerus and clavicle. The clavicular portion represents the cleido-mastoid part of the *sterno-cleido-mastoid*, the muscle that appears in the neck of man during the action of turning the head.

The chief long brace of the limb on its medial side is the *pectoralis*, which is ventral and cranial; it may also draw the limb forwards. The muscle is divided into several portions in the rabbit, some fibres running to the scapula and clavicle. The *latissimus dorsi* is a long brace behind the shoulder joint, running from the lumbo-dorsal fascia and ribs to the humerus. It plays an important part in locomotion by drawing the limb backwards.

The shorter braces around the shoulder joint are important not only for the movement they produce but also because they hold the head of the humerus into the shallow glenoid cup and make the joint stable and yet capable of a wide range of movement.

These muscles, whose tendons are inserted close to the upper end of the humerus, thus act as ties of adjustable length. In quadrupeds, where movement of the whole forelimb is an important part of locomotion, these shorter muscles become very large, particularly in the digitigrade and unguligrade types such as the horse, where they act as protractors and retractors and move the limb as a lever.

These muscles arise from the surfaces of the scapula. On the lateral side *supraspinatus* and *infraspinatus* are inserted on to the greater tuberosity and therefore act in quadrupeds as lateral and cranial braces of the humerus on the scapula (Fig. 9.6). In man they protect the upper part and back of the shoulder joint from dislocation. A further muscle that acts as a 'ligament muscle' at the shoulder is *subscapularis* (Fig. 9.6), running from the costal surface of the scapula to the lesser tuberosity. It acts as a medial brace in quadrupeds and in man protects the front of the joint. These muscles also act as rotators, the first two turning the humerus laterally, the subscapularis turning it medially. The action of infraspinatus during lateral rotation can easily be felt in man by a hand placed over the muscle.

Besides these muscles attached close to the head of the humerus there are also others attached lower down. *Coracobrachialis* passes from the coracoid process to the humeral shaft. The *deltoid* is one of the chief muscles that move the whole arm at the shoulder. In the rabbit (Figs 8.3 and 9.6) it arises from the spine of the scapula and the fascia covering the infraspinatus muscle and is attached to the conspicuous tuberosity on the humerus. The deltoid of man is the muscle that gives the rounded appearance to the shoulder. It arises from the clavicle, acromion, and spine of the scapula and is inserted on to the lateral side of the shaft of the humerus. The deltoid is important in all animals but is especially well developed in man since it lifts the arm sideways from its hanging position. In this action of abduction it is helped by supraspinatus, which holds the head of the humerus into the glenoid, making a fulcrum, as a man standing on the foot of a ladder helps another to push it up. Besides acting as an abductor the various parts of the deltoid of man can also act separately, the anterior fibres drawing the arm forwards and also producing medial rotation, the more posterior fibres having the opposite actions. Another muscle running from the scapula to the humerus is *teres major*, which draws the limb caudally and is an important retractor in quadrupeds (Fig. 9.6).

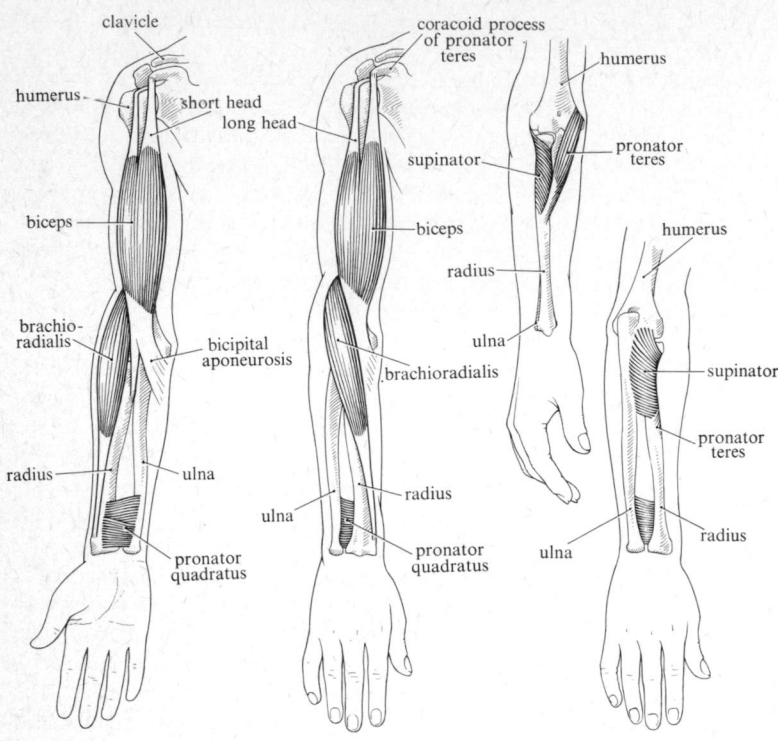

FIG. 9.7. Simplified diagram of some of the muscles producing pronation and supination of the hand.

Supine

Prone

8. Muscles of the upper arm

The limbs consist of articulated rods which, as Gray puts it, 'have little or no natural rigidity and whose ability to resist bending forces depends almost entirely on the activity of their associated muscles'. The internal musculature of the limbs is so arranged as to allow them to function as struts or levers, which in a quadruped are in equilibrium with three groups of external forces: (1) those exerted on the limb by the body, (2) the force exerted by the ground against the foot, and (3) the weights of the individual limb segments. In a plantigrade quadruped, such as the rabbit, it is not difficult to see how the strains resulting from these forces are met by muscles at the front and back of the joints. We shall consider this more in detail in connexion with the hind limbs (p. 97). In the forelimb the joints, besides that at the shoulder, are those (1) at the elbow, (2) at the wrist, (3) between the metacarpal bones and the phalanges, and (4) between the phalanges themselves (Fig. 9.6). There is also some movement between the carpal bones, allowing adjustment of the foot when moving over uneven surfaces. In plantigrade animals the whole sole may provisionally be considered as a single lever arm and the main joints to be considered are therefore those at elbow and wrist. The muscles of the foreleg may be divided into sets for stabilizing and moving these two joints. In the mobile hand of man movements at other joints will also have to be considered.

Movement at the elbow, between the humerus and ulna and/or radius is in many mammals restricted to simple flexion and extension about a single axis, but in man the movements of supination and pronation are also possible. The anterior, flexor muscle is the *biceps brachii,* so-called in man (Fig. 9.7) because it arises at the upper end by two heads, the short head from the coracoid process and the long head by a narrow tendon that runs through the shoulder joint to be attached to the scapula above the glenoid. This tendon serves as a strap helping to hold the head of the humerus into the glenoid. The two heads of biceps unite in the forearm of man to make a single muscle inserted below the elbow into the tuberosity that lies on the medial side of the radius and also by an aponeurosis to the deep fascia. By virtue of this insertion the biceps, besides acting as a powerful flexor of the elbow, also turns the palm forwards in the movement of supination (see below). In the rabbit and most quadrupeds only the short head of biceps is present (Fig. 9.6).

Biceps is a muscle crossing both shoulder and elbow

joints and serving to flex both. The second flexor of the elbow, *brachialis*, crosses the elbow joint only and is attached to the lower part of the humerus and to the ulna.

Extension at the elbow is a powerful movement, especially in digitigrade quadrupeds, where the raising up on the toes decreases the strain on the wrists and ankles but increases that at the elbows and knees. The extensors take this strain and serve to increase the length of the whole leg, which can thus act as a propellant lever. The action is performed by the *triceps* muscle, whose three upper origins include a long head from the scapula and lateral and medial heads from the humerus. At the lower end the tendon of the muscle is inserted into the *olecranon process*, which projects from the ulna. The olecranon is especially well developed in digitigrade animals, which have large extensor muscles. The *anconeus* is a shorter muscle extending the elbow, running from the humerus to the olecranon process (Fig. 9.6).

9. Forearm: pronation and supination

The structure of the lower part of the forelimb varies greatly according to whether the limb is used for walking, climbing, flying, swimming, or, as in man, as a sensory and manipulating organ. The two bones of this region, radius and ulna, originally lay one in front of the other and were unable to rotate. However, as we have seen, early in the history of the quadrupeds there was a rotation in order to allow the forefoot to point cranially when the whole foreleg was brought in beneath the body and the elbow was directed backwards. Thus the radius, primitively lying cranial and parallel to the ulna, comes to lie across in front of the latter, so that while its upper end remains lateral the lower lies medially. This prone position is, therefore, characteristic of the mammals, and the bones are fixed thus in monotremes and in many marsupials, insectivores, rodents, and other primitive mammals, including the rabbit (Fig. 9.4). In this condition both the forearm bones articulate with the humerus, and the only movement at the elbow joint is one of flexion-extension.

We can recognize two lines of development from this beginning. In the mammals that became highly specialized as quadrupeds, such as horses and ruminants, the bones of the lower end of the limbs are simplified, making stable pillars. The ulna becomes fused with the radius and reduced to little more than an olecranon process to serve as an attachment for the triceps muscle. All the weight is carried by the radius, which acquires large surfaces of articulation with the humerus and

with the carpus. A somewhat similar reduction of the ulna has occurred also in bats.

The other development increases the mobility of the hand by allowing a rotating movement of the radius on the ulna, and is especially characteristic of the arboreal primates and of the working hand of man. The mobility results from a change in the articular facets such that the main articulation at the elbow is that of the ulna with the trochlear facet of the humerus. The radius, however, carries the hand and is able to rotate with it around the ulna from the prone position by the act of supination, until the two forearm bones lie side by side and the palm of the hand faces forwards (Fig. 9.7). To allow this movement the radius of these animals has acquired a special circular facet at its upper end, by which it rotates on the rounded capitulum of the humerus; there are also joints between the radius and ulna at their upper and lower ends. The essence of the arrangement is that pressure from the hand is transmitted to the forearm mainly through the radius and then to the humerus by the ulna. The two bones are bound together firmly for this purpose by the *interosseus membrane*, running between the sharp ridges that mark their adjacent borders. It is easy to confirm the essentials of this arrangement on one's own arm. In the prone position, with the hand held facing backwards, the radius can be felt crossing in front of the ulna. As the hand is turned over to face forwards (supination), the radius moves but the ulna does not do so until the two bones lie parallel. Further, it can be verified that the head of the radius, which can be felt at the back of the elbow, lateral to the olecranon, rotates during his movement. The ulna does not articulate with the carpus but ends at a higher level than the radius above the wrist. This can be confirmed by feeling the styloid processes that mark the lower ends of the ulna and radius.

This movement of supination not only enables the individual to bring the hand into a variety of positions but is also in itself a powerful twisting action, which can be used to exert considerable force (Darcus 1951). It is effected mainly by the biceps muscle, whose attachment to the radius in man is such as to pull the latter around the ulna, an action easily verified on oneself during a screw-driving or cork-screwing action (Fig. 9.7). There is also a *supinator* muscle, running across the back of the elbow between the ulna and the upper part of the radius. The opposite movement, pronation, is performed by two bands of fibres running across the front of the forearm, the upper *pronator teres* and lower *pronator quadratus*. Where the position of the radius and ulna is fixed, as in the rabbit, these muscles are small.

10. The hand and digits

The terminal part of the forelimb of mammals varies profoundly with the use to which it is put. The fundamental plan is based on a row of 3 proximal carpals, 1 central, 5 distal carpals, and 5 digits, with 2 phalanges in the first and 3 in each of the others (2·3·3·3·3). Unfortunately, the nomenclature of the carpus has become involved by the use of various systems, especially in man. However, since the human carpus is of primitive type, retaining nearly all the bones, we may list first the bones of an ideal form and then give the names as commonly used in British human anatomy.

	Ideal carpus	*Human carpus*
Proximal row	Radiale	Scaphoid
	Intermedium	Lunate
	Ulnare	Triquetrum
	Centrale	(Fused with scaphoid)
Distal row	Carpal 1	Trapezium
	Carpal 2	Trapezoid
	Carpal 3	Capitate (Os magnum)
	Carpals 4 and 5	Hamate (Uncinate)

In the human hand (Fig. 9.3), which may serve as an introduction, there is movement only between some of the bones. The radio-carpal joint allows flexion and extension and some degree of the lateral movements of abduction and abduction, which are important for bringing the hand to a particular spot. Movements also occur between the carpal bones during flexion and extension, but there is little movement where the carpals articulate with the metacarpals. The next level of active movement is, therefore, the metacarpo-phalangeal junction, where there is flexion and extension, and also the power of drawing the fingers apart (abduction and adduction). Finally, the proximal and distal interphalangeal joints allow only flexion and extension. The thumb is peculiar in that a considerable range of rotation and flexion is possible at the carpo-metacarpal joint, allowing the digit to be brought across to meet the fingers in the movement of *opposition*.

The movements of the hand and digits are accomplished by a series of muscles lying partly outside and partly within the hand. They will be described mainly in man. The extrinsic flexors arise from the medial epicondyle of the humerus and from the front of the ulna, radius, and the interosseus membrane. Conversely, the extrinsic extensors arise from the lateral epicondyle and from the back of the two bones and interosseus membrane. Movement at the wrist or radio-carpal joint is controlled by two flexors and three extensors, some of whose tendons can be identified above the wrist. *Flexor carpi radialis* is attached mainly to the base of the second metacarpal. *Flexor carpi ulnaris* inserts on to a sesamoid bone, the *pisiform*, which is easily felt at the wrist, and from this the pull is transferred to the palm by ligaments running to the hamate and fifth metacarpal. Similarly, on the back of the hand, *extensors carpi radialis longus* and *brevis* pull mainly on the base of the second metacarpal and *extensor carpi ulnaris* on the fifth metacarpal. By various combinations these muscles can bring the hand into any of its many positions (p. 72). For example, the two flexor muscles work together as prime movers in flexion, the two muscles of the ulnar side in adduction. Flexor and extensor muscles of the wrist can also be found in the rabbit on both radial and ulnar sides, running from the lower end of the humerus to the carpus (Fig. 9.6).

The system for moving the fingers includes large extrinsic extensors and flexors for each finger and also small intrinsic muscles (Figs 8.3 and 9.6). The more superficial flexor mass, *flexor digitorum sublimis*, sends four tendons through the hand and each splits before it inserts on to the second phalanx of each digit, so that this muscle flexes the second phalanx on the first. Each tendon of *flexor digitorum profundus* runs between the two parts into which each sublimis tendon divides and is attached to the distal phalanx. *Extensor digitorum* similarly divides into one tendon for each of the four medial digits and pulls on the back of the phalanges.

Movements of the fingers are further controlled by muscles within the hand (intrinsic), which are well developed in the mobile hand of man but are minute in the rabbit. They are known as the *interossei* and *lumbricals* and are partly responsible for movements of abduction and adduction and also co-operate with the extrinsic muscles in the finer movements of flexion and extension.

The movements of the thumb are brought about by special muscles, both extrinsic and intrinsic, which correspond in the general plan to those of the other digits and have been derived by specialization of the latter. These muscles, of course, are well developed in man to enable the thumb to meet the other digits in its movements of opposition.

11. Various types of locomotion in mammals

The full plan of the bones and muscles of the wrist and hand is found in those mammals that walk on the whole sole of the fore and hind feet and are hence said to be *plantigrade* (Fig. 9.8); this is the primitive condition. In so far as locomotion is produced by the forelimb in plantigrade quadrupeds it is partly by movement of the whole limb by the upper extrinsic muscles

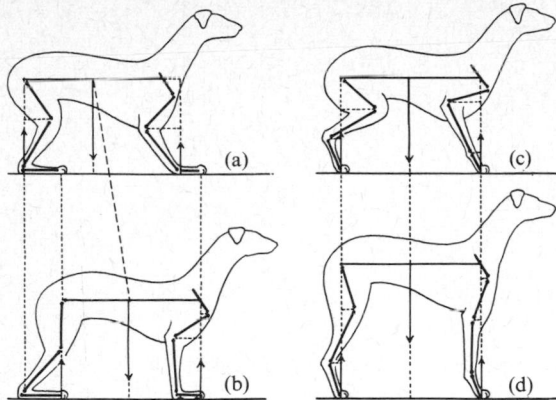

FIG. 9.8. Effect of change from plantigrade to digitigrade posture.
(a) Plantigrade habit, limbs advanced so that centres of pressure are at their posterior ends. (b) By retraction of the limbs the centres of pressure are moved to the toes, reducing the strain on extensors of elbows and knees, but increasing it at the wrists and heels. (c) By raising the knees and elbows the wrists and heels are brought nearer to the line of the reactions of the feet, reducing the strain on extensors of wrists and heels, but increasing it at elbows and knees. (d) By extending the knees, elbows, hip, and shoulder joints the strain on all extensor muscles is reduced. (After Gray 1944.)

and partly by lengthening of the lever by straightening at the elbow and wrist, brought about by the triceps and forearm extensors. With further development of quadrupedal locomotion the animal tends to rise up more and more on the limbs, which are straightened and converted into levers. In *digitigrade* locomotion, such as that of the dog, the metacarpals are raised off the ground and the weight rests on the underside of one or more phalanges of each digit (Fig. 9.8). With this raising on the toes the movement of locomotion becomes increasingly a movement of the whole limb, used as a lever and operated by the more proximal muscles. As already explained (p. 93) extra strain is thus put upon the extensors of the elbow but strain and movement at the wrist and the joints of the digits are reduced.

Finally, in the most rapidly moving quadrupeds the weight is carried on the tip of the distal phalanx. This is the *unguligrade* condition, associated with reduction in the number of digits either to two in artiodactyls, such as the cow and sheep, or one in the perissodactyls, such as the horse. In these animals there is further progress towards converting the limb into a single lever. Besides reduction of the ulnar bone (see p. 93) the carpals are arranged in an interlocking manner to give greater stability and the two metacarpals of Artiodactyla are fused together to give a 'cannon bone'. It is interesting to speculate why fusion of the carpals has not gone farther. Although the whole limb is held much straighter in unguligrade than in plantigrade forms yet joints still remain at radio-carpal, carpo-metacarpal, metacarpo-phalangeal, and interphalangeal levels. Probably no great thrust is developed by the muscles acting at these levels but the joints serve to give the limb the slight flexibility that allows it to carry the animal so fast over uneven ground. The radio-carpal joint of the horse allows considerable movement and is provided with elastic ligaments allowing it to act as a shock absorber and to conserve energy, the animal bouncing off the ground at each step. It must indeed be a wonderfully sprung system of levers and ties that will carry the whole weight of a horse without breakages, even allowing it to lift high off the ground and to land again securely. The possibility of carrying such a large load on relatively long and thin struts presumably depends partly on the flexible character given to the bone by its protein matrix (p. 37), but mainly on the continual adjustment of the many joints of the limb by variation in the tension of the muscles in such a way as to minimize strains.

10 The pelvic girdle and hind limb

1. Functioning of the hind limb

THE hind limb of mammals is usually involved mainly in propulsion and support and shows less tendency than does the forelimb to modification for special functions. The pelvic girdle has become articulated with one or more sacral vertebrae and forms functionally one piece with the vertebral column. The pelvic girdle and vertebral column, with their muscles, thus form a 'girder' for the support of the whole body (p. 74). The hind limb may therefore be thought of as an extensible, moveable pillar, on which the main girder is balanced and propelled at the hip joint.

When a quadrupedal animal is standing, the weight of the hinder part of the body is carried by the hind limb, the system approaching that of a balanced cantilever, especially in animals with a long tail. Gray's analogy of the table top (p. 85) may be applied; braces running from the vertebral column and pelvic girdle stabilize the limb at the hip joint. Movement is produced by retraction of the limb by shortening of the more posterior braces (Fig. 8.3, p. 76). With the foot fixed this has the effect of producing a horizontal force acting forward at the head of the femur. In this action the whole limb is used as a lever. Propulsion is also effected by lengthening of the limb by its intrinsic muscles, acting especially at the knee and heel in plantigrade and at the knee in digitigrade types when the hip has passed in front of the foot.

In considering the skeleton and muscles of the hind legs, therefore, we have to look at (a) the pelvic girdle as a means of balancing the weight of the body on the limb or vice versa transmitting the thrust of the latter to the body, (b) the muscle ties that balance the body on the limb and move the whole limb as a lever, and (c) the bones and muscles that make the limb a jointed extensible strut.

2. Sacrum and pelvic girdle

The modified sacral transverse processes and ribs articulate with the iliac bones by a joint that usually allows only little movement (p. 79). The attachment of the girdle to the vertebral axis is therefore performed by ligaments rather than, as at the pectoral girdle, by muscles. In man the weight of the body is transferred by the fifth lumbar vertebra to the sacrum and the latter partly rests on the ilia and is partly suspended from them (Fig. 9.3, p. 87). The sacral and iliac articulating surfaces are such that the sacrum acts partly as the key-stone of an arch formed by the ilia but the greater part of the weight is transferred by the ligaments, especially those at the back, by which the sacrum is slung between the iliac bones.

The mammalian *pelvic girdle* shows a characteristic development of its more dorsal portions and especially of the *ilium* and *ischium*. In quadrupeds the muscles acting from the surfaces of these bones are the braces that stabilize the balance of the body on the leg and propel it forwards or backwards (Gray 1968). The anterior ilia and posterior ischia are large dorsal expansions for the attachment of these muscles. The acetabulum lies between the ilium and the ischium, and the *pubis* forms a narrow tie member across the mid-ventral line.

In man the adoption of the bipedal posture has led to great modifications of the pelvis. Locomotion is no longer effected mainly by the upper muscles moving the the whole leg as a lever but by a complicated set of actions including those of the calf muscles, which raise the body on the toes. The ilium of man is large and broad and lies above the acetabulum. Attached to it are the gluteal and other muscles that are the posterior braces, extending the thigh, that is to say, keeping the body braced backwards and, in addition, the vertebral column in line with the legs (see Joseph 1960). The pubis and ischium are relatively small in man and form, with the ilium, a system of arches by which weight is transferred to the femora. From the sacro-iliac joint, which as we have seen carries the weight of the sacrum, a thickened strip of bone extends down to the acetabulum and constitutes the main portion of the arch. The pubis acts as a tie-beam, preventing spreading of the arch under the weight. The three bones that make up the pelvic girdle enclose an aperture, the *obturator foramen*, which in life is closed by the obturator membrane with muscles attached both to its internal and external surfaces.

3. The femur

The articulation of the head of the femur with the pelvis allows less freedom of movement than is found at the shoulder. The acetabulum is a deep cup, and the head fits far into it. This, together with the arrangement of the ligaments, limits the movements, which occur mainly in an antero-posterior plane both in the rabbit and in man.

The pillar supports of the ilio-sacral arch are continued downwards in the femur, the lines of the trabeculae in the head and neck being arranged to distribute the weight downwards on to the tubular shaft of the bone (see p. 51). The head of the femur is carried on a long neck that holds the bone out at a distance from the body and allows a swinging motion of the leg both in the rabbit and in man. At the upper end the femur bears large knobs (trochanters) for the attachment of muscles. The *great trochanter* is on the lateral surface and is divided in the rabbit into an upper *first* and lower *third trochanter* (Figs 8.6 and 9.4, pp. 78 and 88). On the medial side of the femur, below the head, is the lesser or *second trochanter*. The shaft of the femur is made of a tube of compact bone, thickest near the middle. The lower end is expanded into lateral and medial condyles, each bearing a facet for articulation with the tibia.

4. Braces around the hip joint

The pelvic girdle moves only slightly on the vertebral column and the hip joint is the point about which movements of the body on the hind legs take place. The muscles that act at this joint have been evolved, like those of the forelimb, within the dorsal (abductor) and ventral (adductor) sheets. The details of the stages of this transformation are obscure but, as in the forelimb, we may distinguish a cranio-ventral group that draws the limb forward and a caudo-dorsal group drawing the limb back. They have become arranged around the joint, however, to form the braces at front, sides, and back, which we have already seen serve to balance the body on the limb and move it as a lever.

In quadrupeds the more caudal braces are especially large and serve to draw the limb caudally, giving the main locomotor thrust. More cranial braces are also prominent, and lateral ones prevent the body falling medially. Medial braces prevent lateral falling, but this is less likely to happen. In man the braces all round the joint are important since the balance is likely to be

upset in any direction; however, the anterior (cranial) and posterior (caudal) braces are especially large, for obvious reasons.

The more cranial braces include the *ilio-psoas* muscles, which act to bend the body ventrally or to flex the leg on the body. They arise from the ventral surfaces of the ilium and of the lumbar vertebrae and are inserted on to the lesser trochanter of the femur (Fig. 8.3). This muscle mass forms part of the posterior (dorsal) wall of the abdominal cavity and is therefore conspicuous during dissection, though it is difficult to feel it in the living human body. It makes an anterior brace for the hip joint. Other muscles that stabilize the front of the joint and draw the leg forwards originate from the front of the ilium and from the pubis. Some of them cross the hip joint only and are inserted on to the femur; others run a long course to the tibia and therefore act as extensors of the knee as well as flexors of the hip. The most superficial of these muscles, *tensor fasciae latae* and *sartorius*, run this long course from ilium to tibia, the former attaching laterally, the other medially (Fig. 10.1). The fascia lata is thickened laterally to form a broad band of connective tissue on the side of the thigh, the ilio-tibial band, which is pulled upon by the gluteus maximus as well as by its own tensor (Fig. 10.1). At its lower end it makes a strong fibrous band, which can be easily felt in man at the lateral side of the knee. This band of fascia, kept tight by the muscles at its upper end, acts as a brace (see p. 48) that distributes the weight of the body symmetrically on the femur.

Deep to this band lie the four parts of the *quadriceps femoris*, the great muscle of the front of the thigh. This muscle assists the psoas by acting as an anterior brace at the hip. It also plays the main part in extending the knee joint. This action is obviously of great importance in the locomotion of quadrupeds, since it straightens the limb, lengthening it as a strut, and thrusting the body forwards. In man, the main thrust is delivered with the leg straight (see p. 101) and the quadriceps plays little part in it. It remains, however, an important muscle because of its action in balancing the body at the hip and in producing extension and balance at the knee. The great amount of use that we give to this muscle is shown by the rapidity with which it wastes if we do not stand for a week or two. During an illness the muscle may become so reduced that the femur can readily be felt along the front of the thigh.

Of the four parts of the quadriceps muscle one arises from the ilium, three from the femur. The *rectus femoris*, or straight portion of the muscle, runs from the ilium and is inserted with the others into the patellar tendon, a broad band reaching across the front

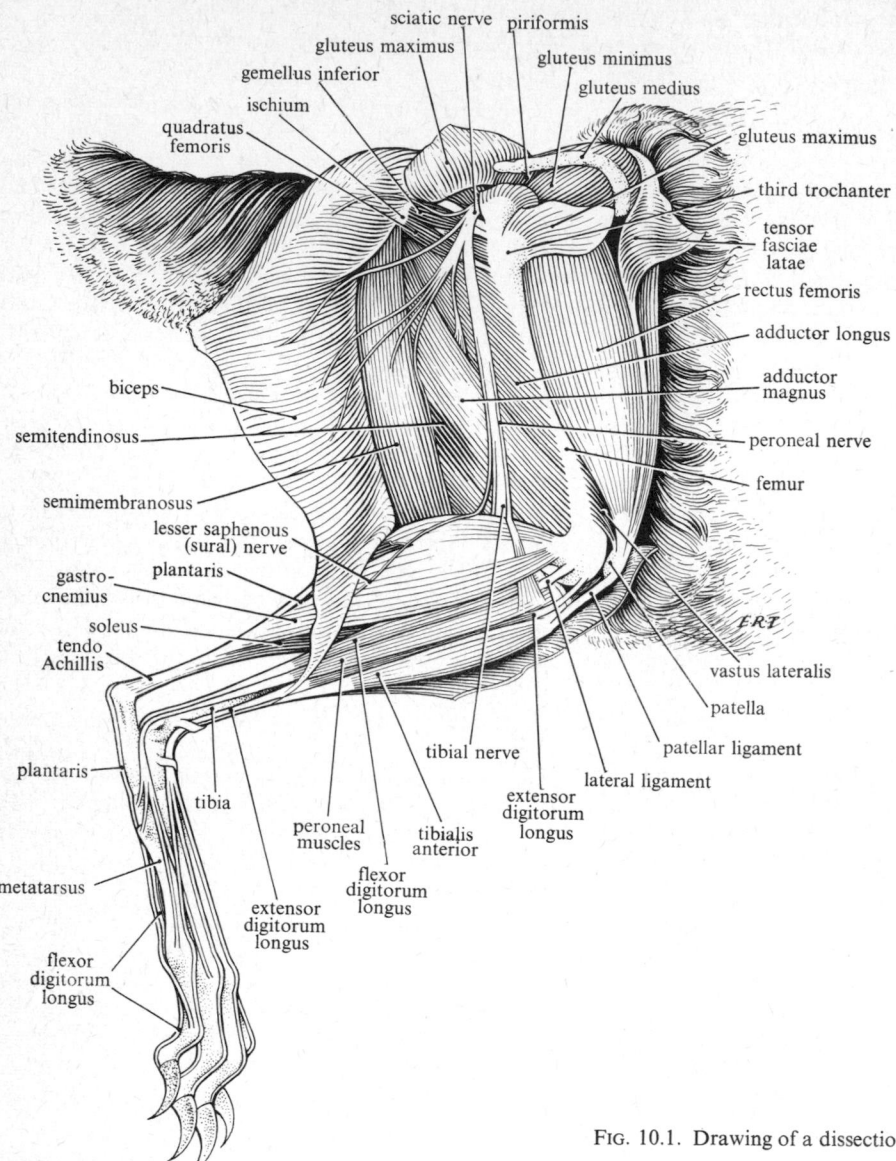

FIG. 10.1. Drawing of a dissection of a hind limb of a rabbit.

of the knee and containing the characteristic *patella*, or knee cap, perhaps the best known example of a *sesamoid* or tendon bone (Figs 10.1 and 10.2). Such bones are often found where the direction of action of a muscle changes as it crosses a joint and thus tends to exert a pressure on the joint and bones. The effect of the sesamoid is to 'increase the working distance of the muscle from the centre of rotation of the joint and provide a bearing surface for the main bones forming a joint' (Gray 1944). In this case the patellar ligament proceeds downwards to be inserted on to the front of the tibia.

The other three parts of the quadriceps are the *vastus medialis*, *lateralis*, and *intermedius*, all arising from the femur and inserted on to the patellar ligament, so that the whole muscle mass acts together for extension of the knee, such as is produced reflexly when the patellar tendon is struck. This tendon and the whole muscle are richly supplied with sensory nerve endings (proprioceptive), which are brought into action when the knee bends and pulls on the tendon. The resulting nerve impulses produce others that volley back from the spinal cord into the muscle and thus check the tendency for collapse of the knee. By this characteristic action

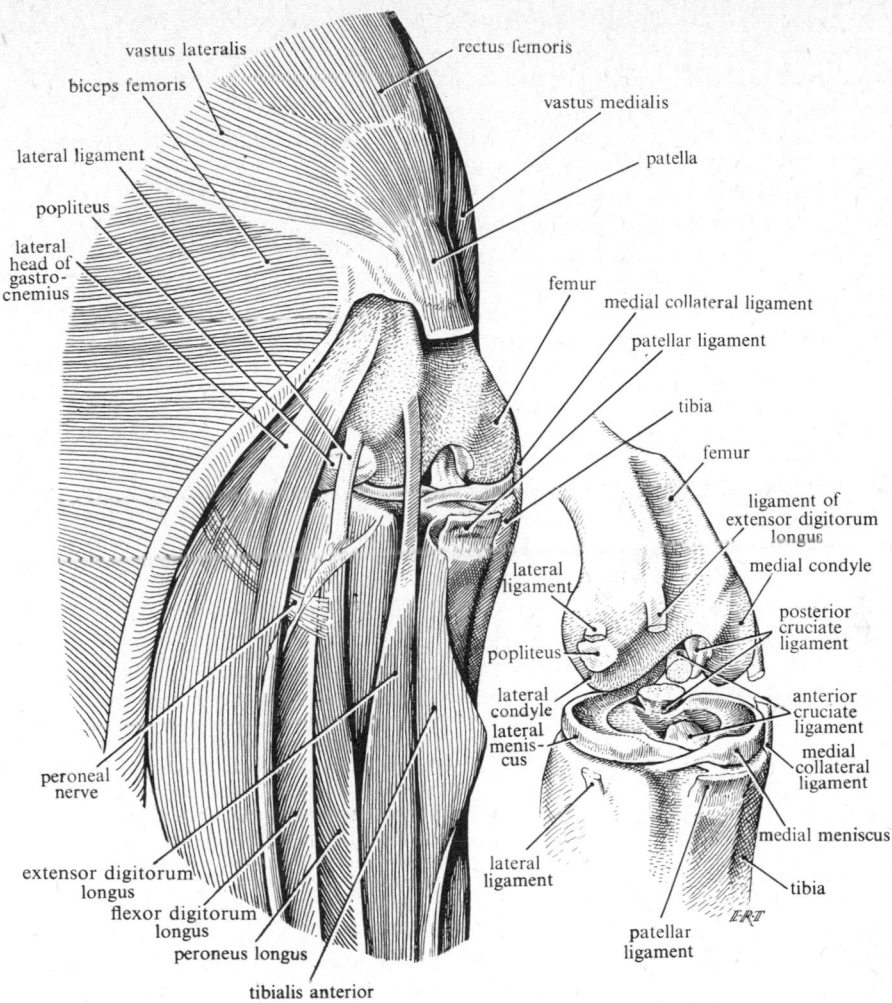

Fig. 10.2. Dissections of the knee joint of the rabbit.

the anti-gravity muscles maintain the body in a standing posture (p. 291). It is to be noted, however, that the quadriceps is only one of the muscles stabilizing the joints that it crosses. A balance is maintained by the proper co-operation of all the muscle braces, which must be tense when necessary. All act as anti-gravity muscles at some time and in some degree.

The remaining muscles of the flexor-adductor group of the thigh are the *adductor* muscles, *longus*, *brevis* and *magnus*, and *pectineus*. They all arise from the pubis and ischium and are inserted on the femur. They are the braces that in man stabilize the hip joint medially, preventing the body from falling laterally. In quadrupeds the true adductors also have this action and other adductors also assist in drawing the leg caudalwards.

The muscle known as *adductor magnus* of the rabbit is very large, running from the ischium to the femur and tibia (Figs 8.3 and 10.1); down its middle runs a red muscle, *semitendinosus*.

The posterior (extensor abductor) group of muscles includes those that stabilize the lateral side and back of the hip joint and also includes many that cross both hip and knee joints and thus serve at once for extension of the former and flexion of the latter. The uppermost members of the series are the *gluteal muscles*, which lie immediately above the hip joint and are well developed in man. These muscles arise from the sacrum and ilium; *gluteus maximus*, the most superficial, from the iliac crest, *gluteus medius* and *minimus* from the outer surface of the ilium (Figs 8.3 and 10.1). All are inserted on to the femur. Their action in quadrupeds is therefore to abduct the limb or, acting from the femur as a fixed point, to prevent the body falling medially, that is to say, to stabilize the lateral side of the joint.

In man gluteus medius and minimus retain this action as abductors or lateral braces, but gluteus maximus comes to act as the great posterior brace or extensor of the hip, preventing the body falling forwards. It thus acts to raise up the whole body and this action has led to the great development of gluteus maximus in the buttocks of man. This is the muscle by which we rise and stand upright, and in this sense it is responsible for the achievement of our characteristic posture, though, of course, many muscles act together to maintain the balance. In man gluteus maximus does not come into action strongly as a propellant during the movements of walking but it is at work almost constantly in balancing the body on the legs. Gluteus medius and minimus have a more strictly abducting action and are the braces of the lateral side of the hip joint and there-fore also very important for human posture. When one leg is off the ground there is obviously a tendency for the body to fall over to the unsupported side, that is to say medially. This is prevented by the action of these glutei on the supporting side, bracing the whole body on the leg that is on the ground.

Deep to these gluteal muscles lie various other short muscles running from the sacrum, walls of pelvis, and obturator membrane to the greater trochanter. These are the *piriformis*, an abductor, *gemelli*, *quadratus femoris* (Fig. 10.1), *obturator internus*, and *externus*, which all act as lateral rotators and adductors. Besides these actions these muscles help to hold in the head of the femur, rather as supraspinatus, infraspinatus, and subscapularis stabilize the shoulder joint. The move-ment of lateral rotation of the femur is important to keep the toes turned sufficiently outwards during locomotion.

The muscles of the more caudal group that cross both hip and knee joints are the *biceps*, *gracilis*, *semi-tendinosus*, and *semimembranosus*. These are at the back of the joint and their action is of the greatest importance in quadrupeds, since they change the angle at which the whole leg is held relative to the body and thus produce a main part of the forward thrust. To give some of these muscles good leverage the ischium of the rabbit and other quadrupeds is prolonged backwards. The tendons of these muscles are variously arranged in different animals as the *hamstrings* at the back of the knee.

In quadrupeds (rabbit) the muscles form long tendin-ous insertions extending from the knee all along the back of the leg, thus increasing the effectiveness of their

action (Figs 8.3, 9.5, and 10.1). In man these muscles are inserted by three compact tendons easily felt at the back of the knee and ending on the head of the fibula or the upper end of the tibia.

To summarize, the muscles acting at the hip joint are:

Protractor muscles	Ilio-psoas	L2 and 3
(anterior braces,	Rectus femoris	L3 and 4
flexors of the hip)	Sartorius	L3 and 4
	Tensor fasciae latae	L5
Retractor muscles	Biceps femoris	L5–S3
(posterior braces,	Semitendinosus	L5–S2
extensors of the hip)	Semimembranosus	L5–S2
	Gluteus maximus	L5–S2
Adductor muscles	Adductors	L2 and 3
(medial braces)	Gracilis	
	Pectineus	
Abductor muscles	Glutei medius and minimus	
(lateral braces)	Piriformis	
Rotators	Obturators	
	Gemelli	
	Quadratus femoris	

In the last column are shown the segmental nerves from which some of these muscles are supplied in man and it is clear that there is a distinct progression; the flexors (protractors) are innervated more cranially than the extensors (retractors). It is reasonable to see in this arrangement a survival of the action of the limbs in conjunction with the myotomal musculature in early land animals, swinging forwards and then back. Indeed action of these muscles, which is so important for the locomotion of quadrupeds, might still be said to be activated by passage of a wave of activity caudally through the lumbosacral segments, as in fishes.

For stabilization of the hip joint a variety of braces is provided. Besides strong ligaments there are muscles, some short and acting close to the joint, presumably mainly as balancers; others, much longer, exert greater turning moment at the joint (p. 71) and are used for locomotion. Some act only across the hip joint; others cross both hip and knee and serve as economical stabilizers and movers of them both. Although we still have not organized our knowledge of these muscles well enough to be able to consider them in a wholly simple and logical manner, an analysis such as that given by Gray has taken us a long way from the position where it was necessary to consider each of these muscles only as a separate morphological entity. There is room for the exercise of much ingenuity in further extending this treatment.

5. The knee

The bones and muscles of the lower leg and foot are arranged on a plan essentially similar to that of the forelimb but modified for the purpose of providing a locomotor thrust. In arboreal species, the limb has a grasping action. Of the two bones in the lower leg the pre-axial *tibia*, corresponding to the radius, is always the better developed and lies on the medial side of the leg (Figs 9.3 and 9.4, pp. 87–8). Usually it alone articulates with the femur at the knee. The *fibula* articulates with the tibia (Haines 1953), but little movement is possible between the two (Barnett and Napier 1953); the fibula is reduced in many mammals and fused with the tibia at its lower end. In man, the fibula is a separate bone along its whole length and there are distinct upper and lower tibio-fibular joints. In the rabbit the fibula is separate only at its upper end.

At the knee there is little rotation and movement is limited mainly to the transverse plane. The joint surfaces, however, are not closely congruous, as are those of the elbow joint. The lower end of the femur bears conspicuous lateral and medial condyles, easily felt in man, articulating with two slightly concave condyles on the upper end of the tibia. The large size and the shape of these facets allow full flexion at the knee, but would seem to leave the joint unstable and easily dislocated. To minimize this the depth of the tibial facets is somewhat increased by the two *menisci*, partial rings of cartilage, present in all mammals (Fig. 10.2). In man they are the 'cartilages' that may become·torn and displaced if gripped between the members of the joint during sudden movements. There are also two strong *cruciate ligaments* within the joint, holding the femur and tibia together (Fig. 10.2). Lateral and medial ligaments, such as are found at all hinge joints, also increase the stability, which is further assured by the muscles acting across it.

The tendons of the quadriceps, the patella and its ligament, support the joint in front and are the main extensors. The flexors are the biceps laterally and the semitendinosus, semimembranosus, and sartorius medially, assisted by the two heads of gastrocnemius and by plantaris (see below), which complete the stability by support at the back (Figs 8.3 and 10.1). The action of these muscles, especially the extensors, is an important item in maintaining posture, particularly in man, where the body is balanced so that the centre of gravity in the fully erect position often falls in front of the knee, the quadriceps being then relaxed and the patella movable.

The *popliteus* muscle, which runs from the lateral surface of the lateral femoral condyle through the capsule of the knee joint to the medial side of the tibia, flexes and medially rotates the latter on the femur (Fig. 10.2).

6. The foot and walking

The *foot* in both man and the rabbit is based on the plantigrade plan, the whole sole being applied to the ground. The tarsals are fundamentally arranged on the same plan as the carpals (p. 88), but only two members of the proximal row remain; the *talus*, also known as the *astragalus*, is the medial proximal tarsal (tibiale), fused with the intermedium, and it provides the main articulation with the tibia by a characteristic pulley-like facet. In man the talus is gripped between the lower ends of the tibia and fibula, the *medial and lateral malleoli*, which are the familiar ankle bones (Fig. 9.3, p. 87). The ankle joint thus allows movement only in one plane, the upward movement of the foot being known as dorsi-flexion, the downward as plantar-flexion (by comparison with the condition in more primitive vertebrates the upward movement, dorsi-flexion, would be considered one of extension, the plantar movement being one of flexion). The sides of the ankle joint are supported by strong lateral and medial ligaments, which may become strained, however, if the foot is turned over accidentally as in walking over rough ground (an ankle 'sprain').

The more lateral proximal tarsal (fibulare) forms the *calcaneum*, modified to support the talus and having a conspicuous and characteristic backward prolongation to receive the attachment of the great calf muscles *gastrocnemius*, *soleus*, and *plantaris* (Fig. 10.1). These are the muscles that produce a large part of the forward thrust, raising the body up on the toes. The action is essentially similar in man and the rabbit, which distribute the weight along the tarsals, metatarsals, and digits.

The essence of the action of *walking* may therefore be said to be a rhythmical change in the balance of tension in the braces about the hip, knee, and ankle. In both rabbit and man this produces a raising of the weight of the body on the foot acting as a second-order lever, the fulcrum being located at the front end of the foot. The details of the way in which the rhythm of movement is produced of course vary greatly with the build and balance of the animal (Gray 1968). Thus in the case of the rabbit the resting posture is squatting, with the knee and ankle joints flexed. This is presumably correlated with the burrowing habit and necessitates the hopping method of progression by lengthening and retracting the limb and extending the back.

In man, where the weight is balanced on the legs in standing, movement is produced by complicated rhythmical adjustments of the balance at all the joints so that the weight is first allowed to fall forwards. The fall is then stopped and the centre of gravity raised again, and so on. A considerable part of the thrust comes from the calf muscles, serving to raise the heel, with the heads of the metatarsals as a fulcrum. The bones of the foot are especially modified to make an arched structure for supporting and propelling the weight (p. 103).

The gastrocnemius muscle arises by lateral and medial heads from the back of the lower end of the femur and is attached by the *tendo Achillis* to the calcaneum (Fig. 10.1). It thus acts across both knee and ankle, whereas the deeper-lying muscle soleus runs from the back of the fibula and tibia to the heel. In the rabbit the more superficial muscle, acting across two joints, consists of white fibres, presumably quick-acting and used for movement, whereas the deeper muscle (soleus) is composed of red fibres and probably acts to fix the single joint in the flexed position of squatting.

The *plantaris* muscle, large in the rabbit (Fig. 10.1) but small in man, runs from the lateral femoral condyle to be inserted on to the second phalanges and is thus able to act upon the knee, heel, and metatarso-phalangeal and proximal inter-phalangeal joints. A muscle having a similar action is *flexor digitorum longus*, running from the back of the tibia and fibula to the distal phalanges. It will be seen that this muscle corresponds to flexor digitorum profundus in the arm, the plantaris being similar to flexor sublimis. In man a separate *flexor hallucis longus* acts on the great toe, serving to keep the under surface pressed against the ground. The *tibialis posterior* runs from the back of the tibia round the medial side of the ankle to be inserted on to the navicular and other tarsal bones. The muscle is highly developed in man as a support for the arch of the foot (p. 103).

The antagonists of these flexor muscles of the calf are weaker in action but have important functions to play, especially when acting from below to balance the leg on the foot. The more lateral members are the *peronei*, four muscles in the rabbit, three in man, arising from the fibula (*Lat. peroneus*: a skewer); their tendons run round the lateral border to be inserted on to the metatarsals. Peroneus longus runs round under the cuboid and across the foot to the first metatarsal; therefore, besides plantar-flexing the foot it raises the lateral border and supports the arch (p. 103). The more medial extensors of the ankle include *tibialis anterior* from the tibia to the first cuneiform and metatarsal and *extensor digitorum longus*, running in the rabbit from the femur through the knee joint to the phalanges; there is also a separate *extensor hallucis longus*.

The foot contains only two members of the proximal row of tarsals but retains the centrale, absent in the hand, as the *navicular* bone (Figs 9.3 and 9.4, pp. 87, 88). This articulates proximally with the talus and distally with the more medial members of the distal row of tarsals, the *cuneiform bones*, three in man, reduced to two in the rabbit, where the first digit

(hallux) is missing. The two lateral distal tarsals are fused, as are the medial in the hand, to a single bone, known here as the *cuboid*. The *metatarsals* provide long levers for the locomotor functions. The phalangeal formula is as in the hand 2·3·3·3·3, but the phalanges are short and provided in the rabbit with pads and claws by which they grip the ground and prevent the limb sliding forward as the heel is raised. The claws have also, of course, other functions—for example, digging, and cleaning the fur. The metatarsals and digits are all held parallel, but in man this is probably a secondary condition. The ape-like form from which man has descended may have possessed opposable great toes and a more freely movable first metatarsal.

The relative lengths of the toes, like that of the fingers, varies greatly in different mammals. Among primates the members usually regarded as more primitive (for example, lemurs) have the lateral toes relatively longer. In other primates the central digits are the longest, as in the hand, but in man the great toe is the strongest and, at least in modern man, often the longest. This is a modification connected with the assumption of the bipedal habit, the weight being carried on the medial border of the foot and all the metatarsals held together to make a compact structure.

7. Movements within the foot: inversion and eversion

The joint at which most of the movement in the distal part of the foot takes place is that between metatarsals and phalanges, the point at which the foot bends in order that the metatarsal heads shall act as a fulcrum. The interphalangeal joints also permit considerable flexion and extension but this mobility of the toes is reduced in man.

In order to serve as a lever and a support, the foot acts as a moderately rigid whole but some important movements take place in man between the tarsal bones, especially at the calcaneocuboid and talocalcaneonavicular joints, turning the sole of the foot inwards or outwards. To allow this the calcaneum and navicular are able to slide sideways on the talus, so that the whole foot turns either medially, *inversion*, or laterally, *eversion*. The former movement, which is of greater extent, corresponds to adduction, with a twist comparable to that of supination. Inversion is performed mainly by tibialis anterior and tibialis posterior acting together; eversion is performed by the peroneal muscles. Slight movements also occur between the proximal and distal tarsals and between the latter and the metatarsals, both during inversion and eversion and especially when weight falls on the foot. The movements are probably important in giving the suppleness that enables the foot

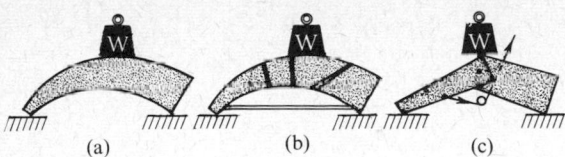

FIG. 10.3. Diagrams of the three components of the weight-bearing structures of the foot. (a) The beam. (b) The truss or arch. (c) The muscle mechanism. (Redrawn from Hicks, in Evans (ed.) (1961). *Biomechanical studies of the musculo-skeletal system*. Courtesy of Charles C. Thomas, Publisher, Springfield, Illinois.)

to take large loads suddenly without breaking, and especially in walking over rough ground.

8. The arches of the foot

A characteristic feature of the human foot is that the bones do not lie flat on the ground but are arranged to form a structure that may be described as consisting of longitudinal and transverse arches. The medial part of the longitudinal arch is the most marked and consists of the talus as apex, resting on the calcaneum posteriorly, and the navicular, cuneiforms, and medial metatarsals anteriorly. The arch serves as a means to distribute the weight of the body, which falls on the talus. The lateral part of the longitudinal arch is flatter and consists of the calcaneum, cuboid, and the two lateral metatarsals. The transverse arch is at the bases of the metatarsals and, with the other tarsal bones, has the effect of making the whole foot into half a dome, by which the weight is distributed all round from the talus to the ground. The weight-bearing system of the foot may be said to have three parts. It may act as (a) a rigid beam, (b) a true arch or truss by virtue of the ligaments, and (c) a jointed structure supported by the muscles (Fig. 10.3).

The arched form of the foot results partly from the shape of the bones but is to a considerable extent maintained by the the action of the ligaments that hold the bones together and by muscle action, so that when the muscles are weak the arches tend to collapse. The extra strain thus thrown on the ligaments may cause them to stretch and the condition of *flat foot* results.

The chief ligaments on the plantar surface are the *calcaneo-navicular*—the 'spring' ligament—which lies directly below the head of the talus, and the plantar ligaments running from the calcaneum to the cuboid and to the bases of the metatarsals. These ligaments support the longitudinal arches; deep transverse inter-metatarsal ligaments support the transverse arch and prevent the toes from spreading outwards.

The most important muscles supporting the arches are the tibialis posterior and the long flexors for the medial side of the arch and the peroneus longus for the lateral side. In addition there are short muscles within the foot, especially *flexor digitorum brevis*, which supports the longitudinal arches, and *adductor hallucis*, which supports the transverse arch and helps to hold the great toe parallel to the others. *Interossei* and *lumbricals* act upon the digits, as in the hand, but are of small importance either in man or the rabbit, since mobility of the toes is slight.

11 The head of mammals

1. Functions of the head

THE head has six main functions: (a) it contains the brain; (b) it carries the organs of special sense—nose, eyes, ears; (c) it carries the mouth and the jaws at the entrance of the alimentary canal; (d) the jaws may be used for offense or defence; (e) through the nasal apertures it provides passage for the oxygen needed for respiration; (f) in many mammals the head is used to produce characteristic expressions that perform an important part in communication and the social life of the species.

These functions are important in all animals, and the mammals show special developments of them, leading to great complexity of the organization of the head. The many small parts make this region difficult to understand; there are so many little bones with odd shapes, such a maze of passages, and so many nerves, blood-vessels, and irregularly arranged muscles that at first it is difficult to grasp the anatomy of the whole head. The significance of its parts will appear if we look at the head as a system for the performance of activities of the six types mentioned, that is to say, by consideration of the nervous, sensory, feeding, fighting, respiratory, and expressive activities that the mammals have developed from the organization they inherited from reptilian ancestors.

The mammalian head is sharply marked off from the body and carried on a long and mobile neck. It is therefore a single 'organ', whose structure has been moulded by its activities even more than has the head in fishes, amphibia, and reptiles, which could often be said to be merely the anterior part of the body. Moreover, the shape of the head in mammals is usually less affected by considerations of frictional resistance during motion than is the case in aquatic or aerial animals. Movement over the land surface is not often fast enough for the air resistance to be serious, and the head is therefore free to take on the different forms that we see in the horse, elephant, or man. In those mammals that have returned to aquatic life there is usually a reintroduction of the 'streamlined' form of the head and body.

Our thesis is, then, that the shape of the head of any particular type of mammal is determined by the way its parts are used for the purposes mentioned above, and we shall hope to be able to organize our knowledge about the various structures in the head in terms of these functions. We shall not recount lists of names or describe the shapes of numerous, unrelated structures but shall hope to find simple ways of describing the parts by considering how the special conditions in the mammals have been derived from the ancestral segmental type of organization.

First we shall deal with the container provided for the brain, showing how it develops a shape and size suitable for the sort of nervous structures that mammals use. This will be a way of describing the shape of much of the skull. Secondly, we shall consider the provision made for eyes, ears, and nose, the last-named being closely associated with the respiratory passages. Finally, the structure of the lower part of the head is determined by the activities of the apparatus for seizing the food— the jaws and teeth; while considering this we shall also deal with the muscles for moving the jaws and other muscles of the face, including those of expression. In this way, when we have described the activities that go on in the head, we shall have given a complete description of the whole organ and the various details will appear in their proper places.

2. The skull

The main outlines of the skull are determined by genetic factors, later bone tends to form in parts of the body that are liable to be stressed (p. 53). In the skull this stressing is mostly produced by the outward pressure during growth of the brain, sense organs, and nasal passages, but the form is also affected by the pull of muscles. The hereditary factors and these stresses produce bony capsules of shape corresponding to the organs they contain, perforated only where nerves or blood-vessels pass through them. The irregular shape of the whole skull is due partly to the multiplicity of the various capsules and passages and partly to the presence of special protuberances for muscle attachments (de Beer 1937).

The shape and appearance of the head as a whole are determined to a large extent by the structure of the box provided for the protection of the brain. The shape of the skull conforms with that of the brain; there is no wide space between the cranial wall and the contents. The factors determining brain shape are discussed on p. 320; they produce in lower mammals such as the rabbit an organ approximating to a tube of circular cross-section. The cranial cavity of these animals is therefore of approximately cylindrical form, with suitable enlargements where there are special dilations of the brain (Fig. 11.1).

The whole head has not a simple tubular form even in the lowest mammals because of the influence on the shape of the other systems (nose, jaws, etc.) that are associated with the cranium. In the rabbit the nasal apparatus is a large tubular structure placed in front of the brain case and internally divided to form a complicated set of passages (Figs 11.1 and 11.5). The teeth are carried on the premaxillary and maxillary bones above, mandible below, and are divided into the long incisors in front and a battery of grinding molars behind. The jaw muscles for moving the lower on the upper teeth are attached to the surface of the brain case and to the zygomatic arches that project on either side (Fig. 11.6).

In man (Figs 11.2 and 11.3) the cerebral hemispheres have become greatly enlarged relative to the rest of the brain and their nearly spherical form has imposed itself on the skull. The nasal apparatus is relatively much smaller than in the rabbit and causes only a slight deviation from the spherical in front. The jaws attached below also cause some deviation from the spherical form of the whole head.

We can thus consider the skull as a tube in a rabbit or a sphere in man, but as soon as we begin to look at the parts that make up its walls we see in the bones the marks of the past history of mammalian life. Bones that are separate in reptiles are often united in mammals. Sometimes they appear as separate bones during development but are fused in the adult. This obviously makes it difficult to provide suitable names for the bones in different animals, and thus it is that the names used by anatomists for the skull of man, where the bones show much fusion, are not always consistent with those used by zoologists.

In lower vertebrates there is a large number of small skull bones, but in mammals a relatively small number of large bones. This further justifies us in considering the shape of the skull as a whole as being more important than that of the constituent bones.

A large part of the skull is first developed in cartilage as the *chondrocranium*. This includes most of the skull base, parts of the side walls and capsules surrounding the nasal and auditory sacs. A series of rods forms the visceral (branchial) arches. In mammals nearly all the chondrocranium ossifies to make cartilage bones.

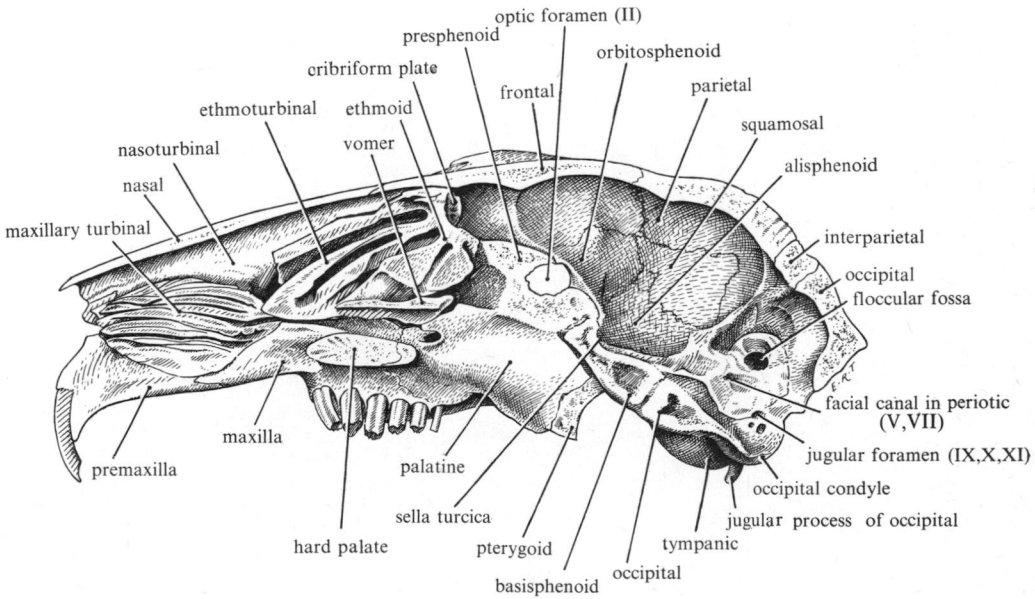

FIG. 11.1. Sagittal section through the skull of a rabbit.

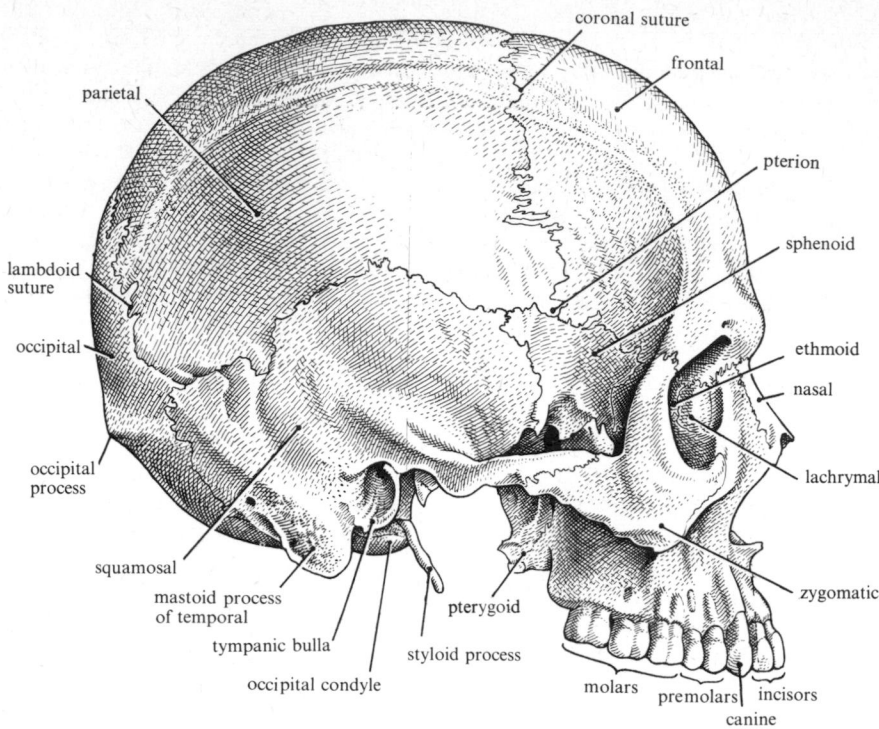

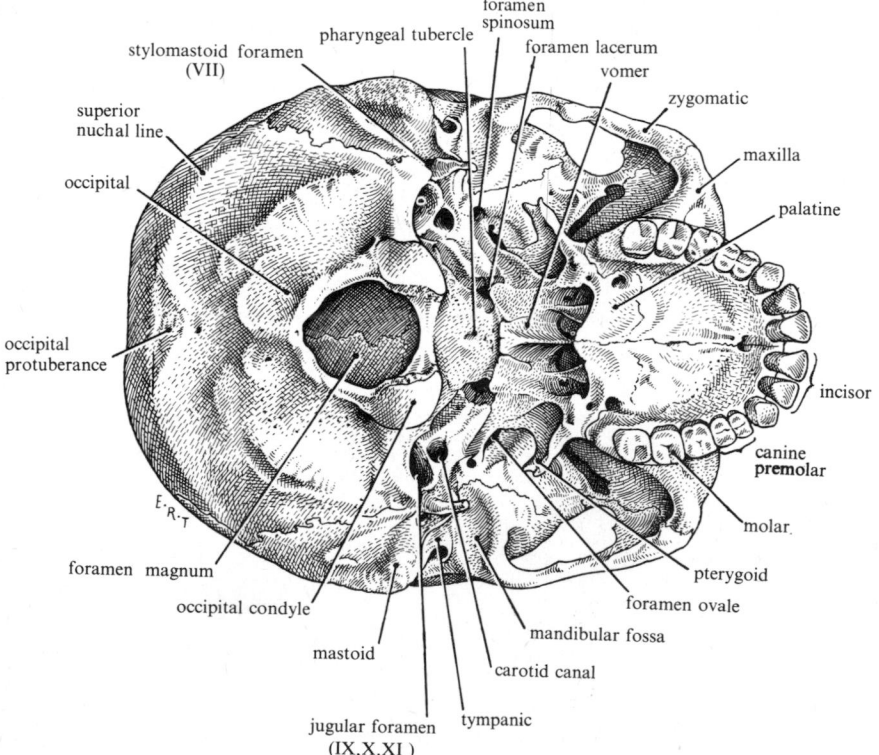

FIG. 11.2. Human skull seen from the side and from below.

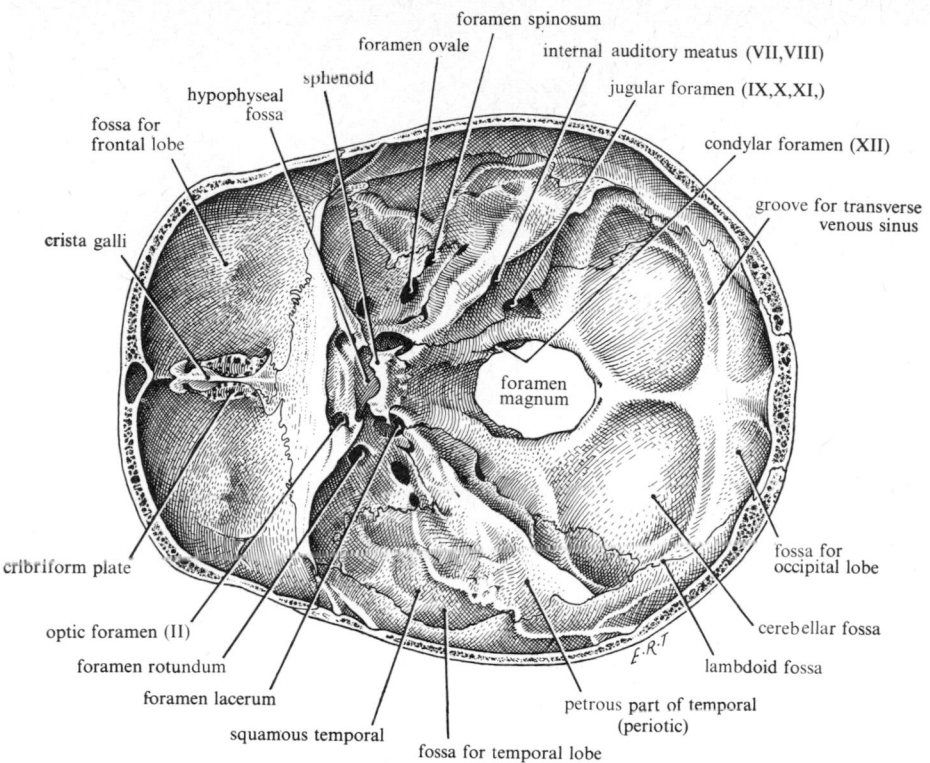

foramen spinosum

foramen ovale

internal auditory meatus (VII,VIII)

sphenoid

jugular foramen (IX,X,XI,)

hypophyseal
fossa

condylar foramen (XII)

fossa for
frontal lobe

groove for transverse
venous sinus

crista galli

foramen
magnum

cribriform plate

fossa for
occipital lobe

optic foramen (II)

cerebellar fossa

foramen rotundum

lambdoid fossa

foramen lacerum

petrous part of temporal
(periotic)

squamous temporal

fossa for temporal lobe

FIG. 11.3. Inside of human skull seen from above.

3. Bones surrounding the cranial cavity

Early in embryonic life dermal bones are laid down more superficially and they may become fused with the cartilage bones.

The bones that are present in a rabbit may be dealt with in three series, those of the roof, sides, and floor (Fig. 11.1), arbitrary divisions of the surface of a cylinder. In the roof (speaking all the time only of the brain box proper) are the paired *frontals*, *parietals*, and *occipitals* and a small median *interparietal*. The roof bones continue down over the sides and without abrupt changes of shape the surfaces pass into other bones. The *orbitosphenoid* forms the wall of the back of the orbit and behind this the *squamosal* and *alisphenoid* continue the tube back to the ear region, where the *periotic bone*, the ossification of the wall of the auditory capsule, makes part of the side wall of the brain box. The extreme posterior end of the lateral portion of the box is made by the *occipital* bones, and these carry the occipital condyles.

Proceeding forwards again from the foramen magnum the floor of the cranium is made by the occipital. In front of this the *basisphenoid* forms a considerable part of the skull floor and is hollowed out in the shape of a saddle, hence known as the *sella turcica*, in which lies the pituitary body. The remainder of the floor is made up by the *presphenoid* and *ethmoid*, and the cavity ends in front at the *cribriform plate*, through which pass the fibres of the olfactory nerve.

Parts of the surfaces of twelve bones, therefore, make up a continuous wall for the brain box. The inside of the box is approximately moulded to the form of the brain and externally there are departures from the smoothly cylindrical form where muscles are attached. Thus the back of the skull is flattened to form a nuchal surface, with a conspicuous *external occipital protuberance* above it. Over this area are attached the splenius, part of the trapezius, semispinalis capitis, obliquus capitis, and rectus capitis muscles. These are the muscles that hold the head up on the neck and tilt it backwards at the atlanto-occipital joint. In animals that have a heavy head conspicuous *occipital crests* are formed for the attachment of these muscles.

The *mastoid process* of the periotic bone marks the attachment of the sternomastoid muscle and of longissimus capitis—muscles that brace the head back on the neck but rock it forward on the atlas. A further irregular

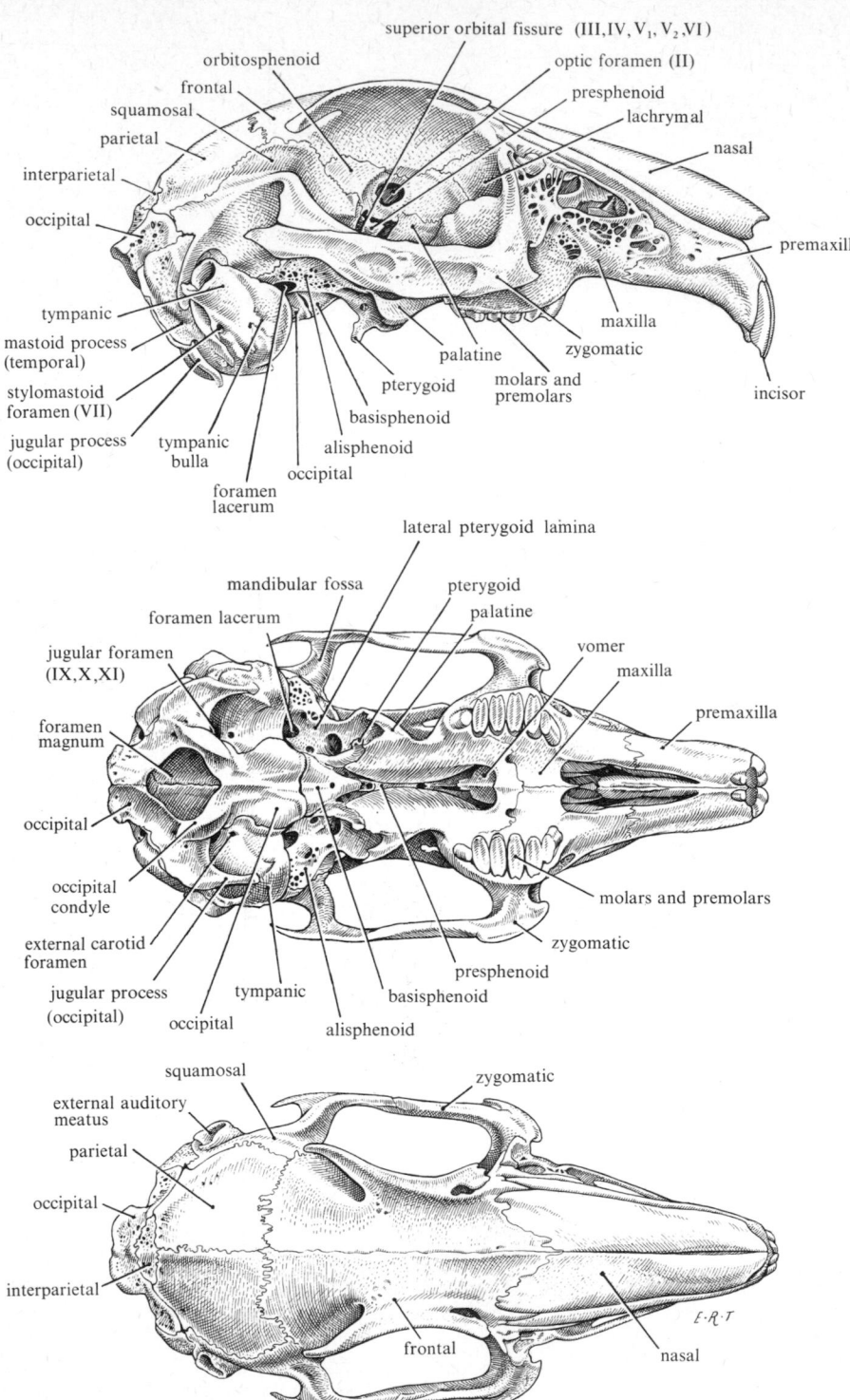

FIG. 11.4. Three views of the skull of a rabbit.

point is made by the *jugular process* of the occipital, where the muscles of the tongue and hyoid (p. 115) are attached. After removal of the temporal muscle of newborn rats it has been found that there is imperfect development of the crest on the skull and of the coronoid process of the mandible, to which the muscle is attached, showing that these ridges and processes on the skull are formed partly as a response to the stresses placed upon the bones.

4. Foramina of the cranial cavity

The cranium makes a complete box for the brain, pierced only by holes for the passage of nerves and blood-vessels (Figs 11.2, 11.3, and 11.4). The numerous foramina for the olfactory nerves pass through the ethmoid. The optic nerves pass through a large *optic foramen* in the orbitosphenoid bone, and behind this is the *superior orbital fissure*, lying between the orbitosphenoid and alisphenoid bones and providing the aperture for the exit of the eye-muscle nerves (III, IV, and VI) and the ophthalmic branch of the trigeminal (V_1). The remaining branches of the trigeminal leave by separate openings in most mammals, the maxillary by a *foramen rotundum*, passing into a canal in the alisphenoid bone, the mandibular through the *foramen ovale*, close to the lower jaw articulation. In the rabbit the skull is little ossified in this region and the maxillary branch of the trigeminal leaves with the ophthalmic, there being no foramen rotundum. The mandibular nerve of the rabbit leaves through an aperture that also transmits the internal carotid artery, which in other mammals passes in a separate *foramen lacerum*, just in front of the periotic bone.

The seventh and eighth nerves pass through the skull in the *auditory foramen* (facial canal), the eighth (auditory) passing directly to the inner ear, while the facial nerve leaves above the ear by the *stylomastoid foramen*. Between the periotic and exoccipital bones the *jugular foramen* gives exit to the ninth, tenth, and eleventh cranial nerves, and the internal jugular vein. The twelfth nerve, hypoglossal, leaves by the *condylar foramen*, through the occipital bone. Finally, there is the *foramen magnum*, by which the spinal cord leaves the cranium.

5. The nasal apparatus

The shape of the head and skull is much influenced in the rabbit, which is a macrosmatic animal (p. 317), by the very large nasal passages, set in the form of cylinders on the front of the brain box (Fig. 11.1). In some *microsmatic* animals, such as man, the head has a more nearly spherical form. The nasal tube is roofed above by the *nasal bones*, laterally by the *premaxillaries* and

maxillaries, tooth-bearing dermal bones. Medial palatine processes of these bones make the front part of the floor of the nasal tube (hard palate), and this bony palate is continued backwards by *palatine bones*. Behind these are conspicuous bony plates, the *pterygoid processes* of the basisphenoid, serving for the attachment of the pterygoid muscles (p. 112). The nasal tube opens in front by the large external nostrils and behind to the pharynx by the internal nostrils (Fig. 11.5). These lie much farther back than the hind end of the bony palate, the floor of the tube being continued behind the palatine bones as a *soft palate*. The walls here contain muscles that allow the tube to be closed to prevent the passage of food from the mouth to the nasal cavities. A pair of fine *nasopalatine ducts* connects the nasal cavity with the mouth by an opening behind the incisors. Jacobson's organ is a part of the olfactory organ opening into these ducts.

The nasal passages have two functions, first that of smell and secondly of moistening and warming the air that passes to the lungs and of filtering dust and other particles from it. This function of cleaning the air is especially well developed in rabbits, probably because they live much of their life underground and line their nests with fur. Anyone who has dissected a nursing rabbit will soon feel that the fine hairs are very irritating to the nose and will appreciate the value of the continual 'snuffling' movements, which the rabbit makes with the wings of skin covering the nostrils (Fig. 11.6). The *levator alae nasae* muscle and tendon responsible for these movements are well developed in the rabbit. Evidence of the dangers of nasal infection is shown by the fact that the nasal epithelium of nearly all wild rabbits shows signs of past disease.

The moistening and warming of the air is performed in all mammals by an extensive epithelium with rich blood supply, supported by thin bony plates, the *turbinals* or scroll bones, attached to the maxillae, nasals, and ethmoid bones. The lower part of the ethmoturbinals forms a long median bone, the *vomer* (Fig. 11.1). The surfaces of these turbinals are covered with a ciliated, mucus-secreting epithelium, which maintains a current of mucus towards the mouth. The olfactory epithelium itself (p. 410) occupies an area at the back of the nasal cavity.

The nasal cavity is, therefore, a complicated chamber, roughly cylindrical but with a form not easily referable to any simple mechanical principles. It serves the purpose of making the air pass over extensive moist, vascular surfaces, parts of which are olfactory. Communicating with the nasal cavity is a further system of spaces, the *maxillary* and *frontal sinuses*, lying within those bones. These spaces serve to lighten the bone; on

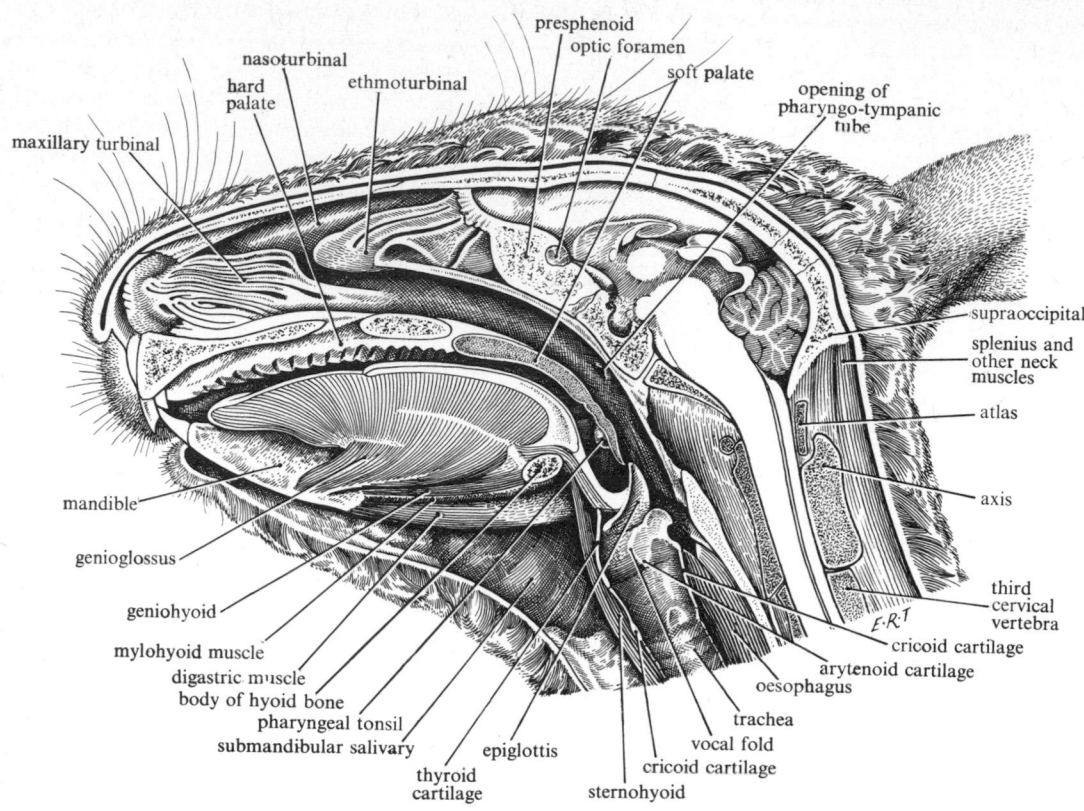

FIG. 11.5. Drawing of dissected sagittal section of the whole head of a rabbit.

account of their communication with the nasal passages they may become infected from the latter and if their openings into the nasal cavity are blocked, they may become painful in man.

6. The auditory capsule

The cartilaginous wall of the auditory capsule ossifies in mammals to form a single *periotic bone*, leaving three apertures, a foramen for the auditory and facial nerves medially and *fenestrae rotunda* and *ovalis* laterally, at which the membranous inner ear comes into contact with the wall of the middle ear. The *stapes* fits into the fenestra ovalis and carries sound waves to the perilymph (p. 400), whose vibrations are made possible by the presence of the other aperture, the fenestra rotunda. The periotic bone is a complicated sac enclosing the various parts of the membranous labyrinth (p. 398). Externally its *mastoid* region appears on the surface of the skull and is drawn out into the mastoid process for attachment of the sternomastoid muscle. The bone is not compact but contains many spaces (mastoid air

cells). The tympanic membrane, which serves to receive air-borne vibrations, is stretched across a tube communicating internally with the pharynx and externally with the outside air by the external auditory meatus. The tympanum develops approximately at the junction of the first pharyngeal (branchial) pouch and the first of the series of external grooves, which separates the mandibular and hyoid arches (p. 123). The whole passage thus retains the general character of the gill slits present in fishes.

The inner part of the passage, the *pharyngo-tympanic* (*Eustachian*) *tube*, leads from the pharynx (Fig. 11.5) to the cavity of the middle ear, medial to the tympanum. The outer canal, the external auditory meatus (Fig. 11.4), is surrounded by a bony tube, the *tympanic bone*, derived from the angular, a bone of the lower jaw of reptiles.

The whole arrangement thus serves to support the tympanic membrane as a receptor for sound waves, providing it with air on both sides so that it may vibrate freely. Blockage of the passage at either end in man, from within by the mucous secretions of a cold,

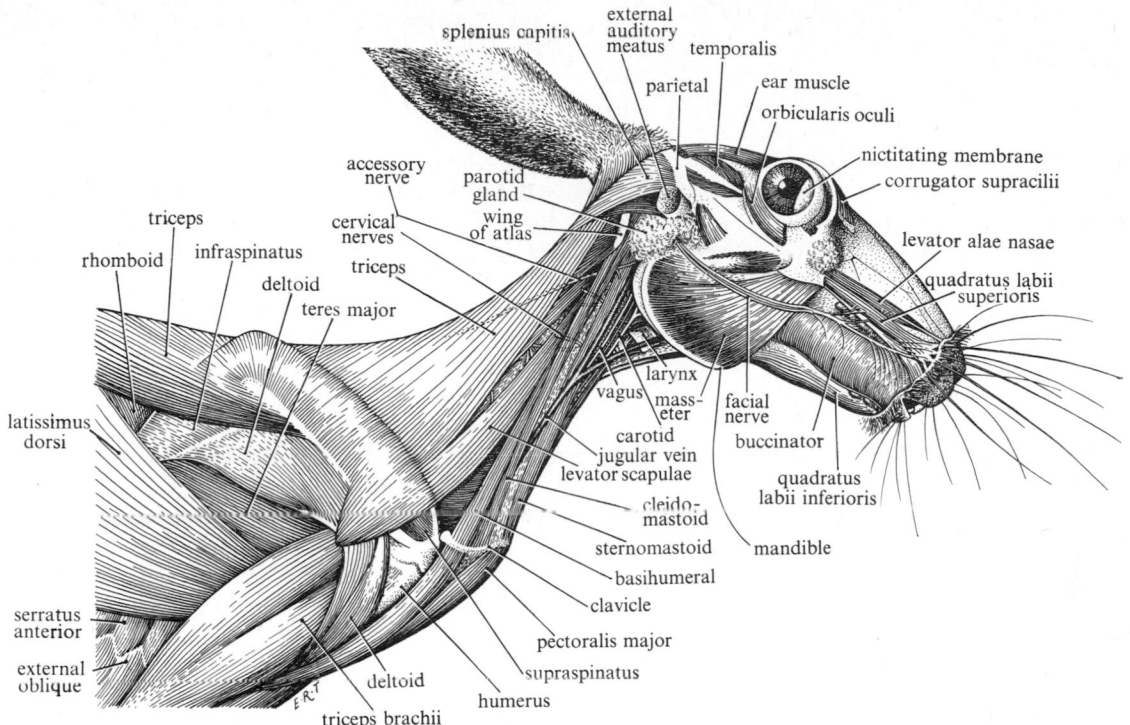

FIG. 11.6. Drawing of dissection of the head and shoulders of a rabbit.

or from without by water during bathing, produces a characteristic impairment of hearing.

The sound vibrations are carried from the tympanic membrane to the inner ear by the chain of auditory ossicles, *malleus, incus,* and *stapes* (Fig. 43.2, p. 398). These ossicles provide amplification of the force of the vibrations and their peculiar shape makes them vibrate aperiodically and therefore they transmit complex sound waves faithfully (p. 398). Small muscles, the *tensor tympani* and *stapedius,* are attached to the malleus and stapes respectively and serve to hold the ossicles and tympanum in a condition suitable for vibration (p. 399). The whole constitutes an elaborate arrangement for receiving sound waves, and is especially developed in mammals, for the detection of both the direction and form of the sound.

Study of mammal-like reptiles shows that the malleus and incus have been formed by modification of the bones that provided for the articulation of upper and lower jaws in early reptiles. Probably at first, sound waves reached the ear from the ground, through the jaw. When the first vertebrates came on land they possessed no tympanum and therefore little sensitivity to air-borne vibrations. When the tympanum first developed it probably served to increase the sensitivity

of the lower jaw to vibrations. The apparatus then became still more sensitive by freeing the bones at the hind end of the jaw, so that they could carry vibrations directly from the tympanum to the inner ear.

The large *external ears* of the rabbit serve to increase the time-difference of arrival of a sound wave at the two tympana and thus provide means for detecting sound direction, which is important to a defenceless herbivore. The *pinna* of the ear is strengthened by cartilage and is moved by a series of muscles attached to the back of the skull (Fig. 11.6). The large supply of sensory nerves to the pinna suggests that the 'ears' of rabbits also serve as tactile organs, perhaps responding to small air movements, a function that is still more highly developed in bats. The variation in size of rabbits' ears in different climates (p. 15) shows that they also influence the temperature of the animal. In the elephant, where the surface–volume ratio of the body is small, ear movements constitute an important part of the cooling mechanism (p. 17).

7. The orbit

The nearly circular eyeball is enclosed in a cavity lying in front of and below the brain case. The medial wall of the orbit is formed in man by the frontal, orbitosphenoid, ethmoid, and alisphenoid bones. The frontal forms a characteristic protecting supraorbital ridge above the eye, continued at its front end by a small bone, the *lachrymal* (Fig. 11.4). The side of the eye is protected by the *zygomatic arch*, made by the early fusion of a *zygomatic (jugal) bone* in front and the zygomatic process of the *squamosal* behind; this bar also serves for the attachment of the masseter muscle (p. 113). The lower side of the orbit is unossified in the rabbit, but is closed in some other mammals (and in man) by bony processes of the jugal and maxilla. In the rabbit, as in most mammals, the orbit is continuous behind with the *temporal fossa*, though a process of the zygoma further separates these spaces laterally.

The structures accessory to the eye are highly developed in mammals and provide protection from desiccation and irritation by dust and other particles (Fig. 11.6). The upper and lower lids are hairy folds, with long eyelashes and a row of modified sweat glands, the *Meibomian glands*. The lids are closed by a slight ring of muscle the *orbicularis oculi*, most conspicuous behind the eye and innervated from the facial nerve. The function of raising the upper lid is performed by the *levator palpebrae superioris* muscle, which is a derivative of the first myotome (p. 119) and therefore innervated from the third cranial nerve. The third eyelid (*nictitating membrane*) is well developed in the rabbit as a curved plate of cartilage covered with glandular epithelium and moulded to the shape of the front of the eye, across which it can be drawn by a sheet of smooth muscle, cleaning and moistening the cornea.

There are six extrinsic eye muscles in mammals, as in all other vertebrates, and they develop from the three myotomes at the front of the head, though the relations are less obviously segmental than in lower forms such as the dogfish. The oblique muscles are attached to the more anterior portion of the wall of the back of the orbit but in mammals the *superior oblique* takes a sharp turn through a pulley or *trochlea*, which gives its name to the fourth nerve. This change of direction is especially marked in the forwardly directed eyes of man, where the muscle originates with the recti at the back of the orbit. The *inferior oblique* arises from the lachrymal bone and is inserted on the postero-ventral portion of the eyeball. It is a large muscle in the rabbit, whose eyes are directed laterally and show much movement in an antero-posterior direction. The four *rectus muscles* arise from the orbito-sphenoid bone around the optic foramen and are inserted on the four quadrants of the eyeball. The six muscles together thus serve to turn the eye in any direction. The eye of the rabbit is also provided with a retractor muscle, which arises

with the recti and is inserted around the entrance of the optic nerve to the eyeball.

The glands of the orbit are highly developed in the rabbit, perhaps in order to keep the eye moist underground. *The Harderian gland* in front, between the oblique muscles, is composed of two parts, which have very different structures but open together inside the third lid. The *lachrymal gland* itself is smaller and lies in the temporal region of the orbit. The tears are carried away from the orbit by a *nasolachrymal duct*, opening into the nasal cavity. The cornea is thus well protected, both by mechanical means, through movement of the lids, and by abundant secretions. Stimulation of the sensory nerve fibres of the cornea brings both of these means of protection into action.

8. The jaws and muscles of mastication

The remaining features of the skull are largely concerned with the apparatus for taking the food by use of the teeth, which are highly differentiated in the rabbit. The anterior incisors serve for collecting materials by nibbling and gnawing, the premolars and molars for grinding the material obtained. The jaws and their muscles are highly specialized to make these movements possible. The teeth are carried in the premaxillary and maxillary bones of the upper jaw and in the single lower jaw bone, the *mandible (dentary)*, a membrane bone covering the embryonic Meckel's cartilage. The articulation of the dentary with the skull is a facet on the lower side of the squamosal bone (Fig. 11.4), and the articulation is such that it allows mainly opening and closing of the mouth but also some movement forwards and backwards and sideways. Muscles are present for closing the jaw and for drawing it forward and back, while combinations of these produce the rotation.

The main muscle raising the jaw is the *masseter* (Fig. 11.7), which is a prominent feature of the lateral side of the head, running from the zygomatic arch to the lateral surface and lower edge of the mandible. This muscle is large in the rabbit and its insertion gives the characteristic deep jaw. The *temporalis* muscle arises from the surface of the cranium in the temporal fossa and passes downwards medial to the zygomatic arch to be inserted behind the last tooth on the lateral surface of the *coronoid process*, a projection on the mandible in front of the articulation. The muscle thus not only closes the jaw but pulls the bone backwards. The temporal muscle and coronoid process are small in the rabbit. The *medial pterygoid* muscle runs from the pterygoid process of the basisphenoid bone to the medial surface of the mandible and therefore also closes the jaw. The *lateral pterygoid* has a similar origin

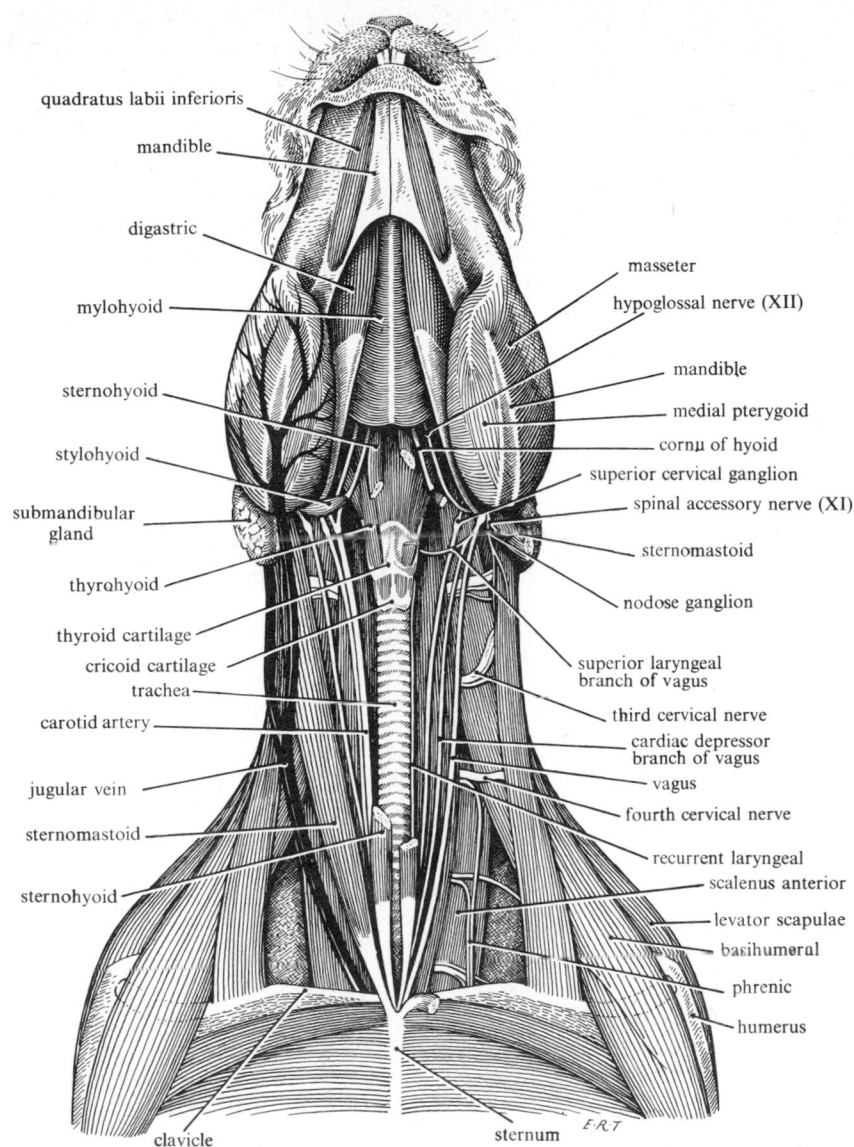

FIG. 11.7. Dissection of the neck of a rabbit seen from the ventral surface.

but is attached to the front of the neck of the mandible close to the joint and thus pulls the jaw forward. The movement of chewing in the rabbit and man is produced by alternate action of the elevators of the jaw of one side and the lateral pterygoid of the other, giving a slightly rotatory action between the surfaces of the teeth.

9. Comparison of the skull of man and the rabbit

The activities of the head are different in the rabbit from those in man and the shapes of the heads may be correlated with these differences in activity. Classification of the differences and interpretation of their significance serves to emphasize the characteristics of the two types. On account of the development of the cerebral cortex, the brain of man approaches a sphere in shape. The reduction of the olfactory organs leads to a reduction of the snout and the eyes have come to face forwards. The rounded head is carried balanced on top of the vertebral column, rather than on the front end of a vertebral girder (p. 77). The nasal passages are not essentially different in the two types but they are simpler in man, because of the reduction in the olfactory functions and perhaps also because of a lesser necessity for cleaning the air. The human dentition is suited for an omnivorous diet and lacks the specializations for gnawing and grinding that are found in the rabbit. Finally, in man the superficial musculature of the face, instead of controlling the vibrissae and snuffling movements, produces the complicated movements of expression, which form such an important feature of the communication system upon which man's social life is based.

These differences in function are accompanied by considerable differences in the bony structure of the skull, especially in shape and the development of prominences for attachment of muscles. The shortening of the snout has led to great reduction of the nasal bones in man and to the disappearance of the premaxillaries as distinct bones. The rounded shape and characteristic posture of the head lead to many differences. The foramen magnum lies underneath rather than at the back of the skull. The ridges for attachment of the muscles of the occiput are less prominent in man than in most mammals. The differences in method of mastication have led to a relative reduction in the size of the masseter and to some development of the temporal muscle in man, with corresponding differences in the shape of the mandible and skull. In carnivorous mammals the temporalis is typically the dominant muscle, in herbivorous mammals the masseter. Man is omnivorous, and both muscles are well developed.

Finally, certain fusions of bones take place in adult man. The two frontals come to form a single bone. The various sphenoids make a single *sphenoid bone*, which, however, has a complicated shape, with processes for the attachment of the pterygoid muscles. The squamosal fuses with the periotic and with the tympanic to give a single *temporal bone* of complicated form.

10. The lips and muscles of the face

The appearance of the head of a rabbit or other mammal is largely determined by the arrangement of the lips and their movements (Fig. 11.6). Around the nostrils the skin is highly specialized by the presence of the tactile *vibrissae*, long hairs with a special sensory innervation (p. 366). The skin is characteristically divided in the midline in the rabbit to give a cleft upper lip and the fold between this margin and the nostril, together with the skin above the nostril, in life perform continual rhythmical twitching movements, serving to keep the air passage clear (p. 109). These and other movements of the skin of the head are produced by special portions of the *platysma* muscle, a muscle of the skin (not myotomal, see p. 119), innervated by branches

of the facial nerve. The *facial muscles* are differently developed in various mammals according to the movements required of the face. They are large and highly differentiated in man as the *muscles of expression*, which have acquired great significance as a means of communication between individuals in a social species. Beginning perhaps from simple actions such as baring the teeth, which occurs preliminary to biting and hence is an indication of rage, the whole pantomime of emotional expression has developed into a means of communication second only to that afforded by speech.

In the rabbit the parts of this musculature that are well developed are concerned with moving the lips (Fig. 11.6). The *quadratus labii superioris* is the main muscle that produces retraction of the upper lip. It arises from the surface of the skull in front of the eyes and consists of several separate parts inserted forward on the lip. The skin around the mouth is also moved by a circular ring of muscle, the *orbicularis oris*; though large in man this is small in the rabbit. The *quadratus labii inferioris* is a large muscle occupying the side of the mandible and retracting the lower lip (Fig. 11.7).

The sides of the mouth are occupied by a large sheet of muscle, the *buccinator*, running from the upper jaw to the mandible and especially well developed in a chewing animal such as the rabbit, in which the food is passed backwards and forwards into cheek pouches through a gap in the tooth row, the *diastema*. The platysma includes several portions and its main sheet consists of longitudinal fibres running beneath the skin from the angle of the jaw down under the neck.

11. The mouth, tongue, pharynx, and larynx

The *tongue* projects as a conical lobe from the floor upwards into the mouth. It is covered with papillae containing the taste buds (p. 409) and is attached by a system of muscles whose action helps in the act of swallowing. The mouth continues backwards into a cavity under the back of the head, the *pharynx*, into which the nasal tube (nasopharynx) opens from above and from which the oesophagus dorsally and larynx ventrally open backwards (Fig. 11.5). The Eustachian tube (pharyngo-tympanic tube) opens into the back of the nasal tube. The pharyngeal cavity is thus an important meeting-point for several passages. Being a likely seat of infection it is guarded by collections of lymph glands (p. 184), especially around the *tonsils,* which are depressions in the side walls of the front of the pharynx (Fig. 11.5). In development the tonsil arises from the second pharyngeal pouch.

The ventral and side walls of the pharynx contain a series of bones and cartilages. These are derived from the branchial arches and serve to support the tongue and to give attachment to the muscles that control swallowing and the passage of air; similar functions have been performed by this apparatus from the fish stage of evolution onwards. The *hyoid bone* lies between the angles of the mandible and from it two cartilaginous processes, the *greater and lesser cornua*, pass upwards to be attached to the jugular process of the occipital bone by the *stylohyoid* muscles (Fig. 11.7). The anterior (lesser) cornu represents the lower portion of the hyoid arch, the more posterior (greater) cornu the first branchial arch; the body of the hyoid is the basibranchial cartilage.

The *larynx* (Fig. 11.5) is a cavity forming the first portion of the respiratory tract and with its front and sides strengthened by the *thyroid cartilage* (Adam's apple) above and *cricoid* cartilage below. These represent the lower portions of the second and third branchial arches. The *arytenoid cartilages* are a pair attached to the cricoids in the larynx, and the *vocal folds* are ridges projecting into the larynx between the arytenoids and the thyroids, in a position to be vibrated by the air stream. They are small in the rabbit. Muscles tighten the folds and regulate the note. The entrance to the larynx is protected by the *epiglottis*, a fold projecting upward from the floor of the pharynx and stiffened by elastic cartilage.

The tongue is moved by muscles attaching it to the inner surface of the mandible (the *genioglossus*), to the hyoid (*hyoglossus*) and to the skull (*styloglossus*) (Fig. 11.5), while its own internal fibres (*lingualis*) run in various directions and allow changes of shape and projection.

The act of swallowing consists first of a compression of the cavity of the mouth and raising of the hyoid. This is achieved by the action of the buccinator muscle and of the *mylohyoid*, a set of fibres running on each side between the mandible and the hyoid bone (Fig. 11.7). These are assisted by the hyoglossus, stylohyoids, and the *digastric*, a muscle running from the jugular process of the skull to the midpoint of the lower jaw, through a ligamentous loop attached to the hyoid. The resultant upward movement of the larynx raises the epiglottis to close the entrance, assisted by the action of sphincter muscles of the larynx itself. At the same time the opening to the nasal tube is closed by the contraction of the small muscles that raise the soft palate. The bolus of food is thus passed over the epiglottis into the oesophagus, which, provided with its own muscular wall, passes the food onwards by contraction behind the bolus and relaxation in front (peristalsis, p. 148). Long muscles, the *sternohyoid* and *sternothyroid*, then draw the whole apparatus downwards (Figs 11.5 and 11.7).

The larynx is provided with an internal musculature by which the air flow is regulated and the position of the vocal folds altered to modify the note of the voice. These muscles are, of course, much larger in man than in the rabbit, which makes only limited use of its voice.

12. The cranial nerves

The central nervous system of any vertebrate is connected with the periphery by one dorsal and one ventral root on each side in each segment (p. 257). In the spinal region these roots join but in the head they remain separate; hence in the cranial region the segmental plan is not at first sight obvious. Nevertheless we can recognize dorsal and ventral roots in the cranial as in the spinal segments. The ventral (anterior) roots innervate the myotomes and in the front part of the head these persist as the eye muscles. The first ventral root is thus the *oculomotor* (third cranial), innervating the superior, inferior, and anterior rectus and the inferior oblique, all of which are derivatives of the first myotomal segment. The second ventral root is the *trochlear nerve*, innervating the superior oblique, and the third is the *abducent nerve*, supplying the posterior rectus. All these nerves carry motor fibres to their muscles and also afferent fibres of proprioceptive function.

The dorsal (posterior) roots of the cranial region differ from those of the rest of the body in that they carry efferent (motor) as well as afferent fibres. This is explained by the fact that in the head much of the musculature is derived not from the myotomes but from the muscles of the branchial arches, the lateral plate musculature (p. 119). In the trunk all the striped muscles are derived from myotomes and are therefore innervated by ventral roots.

Thus the *trigeminal nerve*, which embryology shows to be derived from the dorsal roots of the first two head segments, contains, besides many afferent fibres, efferent fibres supplying the muscles of mastication associated with the first (mandibular) pharyngeal arch, namely, the temporalis, masseter, and pterygoids. The *facial nerve*, which is the nerve of the third head segment, whose ventral root is the abducent, carries motor fibres to the muscles derived from the hyoid arch, as well as many sensory fibres.

The segments behind these first three are obscured by the development of the auditory sac, which is formed as an inpushing from the surface of the head. The myotomes in this region are thus obliterated, but the series of dorsal roots continues as the ninth (*glossopharyngeal*), tenth (*vagus*), and eleventh (*spinal accessory*) cranial nerves. Each of these contains both sensory and motor fibres, and the last two are composed of the combined dorsal roots of several segments. The series of ventral roots begins again with the *hypoglossal* (twelfth cranial) nerve, which represents the ventral roots of the more posterior segments of the vagus-spinal accessory series and innervates muscles of the tongue, derived from myotomes.

Thus the third, fourth, sixth, and twelfth cranial nerves represent ventral (anterior) roots. They consist mainly of motor fibres (together with some proprioceptive ones) and they carry no ganglia. They arise from the ventral surface of the brain (Fig. 29.1), except the trochlear, whose cells of origin lie ventrally but the axons run round within the substance of the brain on their way to the superior oblique muscle. The fifth, seventh, ninth, tenth, and eleventh cranial nerves are dorsal roots with ganglia corresponding to those of the spinal dorsal roots; they carry many afferent and some efferent fibres. The arrangement of the cranial nerves thus shows that the head, in spite of its specialized structures, is produced by a modification of the fundamental segmental plan that can be traced throughout the vertebrates.

13. The trigeminal nerve

The *trigeminal nerve* retains the three main branches found in lower vertebrates. The *ophthalmic division* is purely sensory and runs from its exit at the superior orbital fissure forwards along the median wall of the orbit. It divides into branches supplying the skin of the upper eyelid and passing through a supra-orbital foramen reaches the skin of the front of the head and of the inside as well as the outside of the nose. The afferent fibres for the cornea also arise from the ophthalmic division within the orbit and run in the ciliary nerves. Some part of this division represents the dorsal root of the first somite (corresponding to the n. ophthalmicus profundus of fishes—see Young 1962, p. 152).

The second or *maxillary division* of the trigeminal is also purely afferent. As it leaves the skull it passes in most mammals into an *alisphenoid canal* in that bone, so that the foramen rotundum, through which the nerve leaves, does not appear on the surface; in the rabbit this canal is not ossified. The nerve then runs across the floor of the orbit, giving branches to the upper jaw, and finally passes through an infra-orbital foramen to reach the face. The *mandibular*, or third division of the trigeminal, leaves the skull by the foramen ovale and contains motor fibres for the muscles of the mandibular arch, namely, those of mastication, as well as sensory fibres for the lower jaw. It therefore gives off on its course branches to the temporal muscle, the pterygoids and masseter. A *buccal branch* passes through the buccinator muscle to the skin of the

mouth, and the *lingual nerve* continues down to supply receptors for touch and pain on the front of the tongue. The *inferior dental nerve* supplies the mylohyoid muscle and anterior belly of the digastric and then enters the mandible to supply the teeth. A large *auriculo-temporal* nerve turns backwards and upwards to supply the skin of the back of the head.

14. The facial nerve

The dorsal root of the third (hyoid) pro-otic somite includes the *auditory* (p. 400) and *facial* nerves (Fig. 11.6). In mammals the facial is largely a motor nerve supplying the lateral plate muscles of the head, face, and neck but also containing fibres for the sense of taste from the front part of the tongue. The nerve leaves by the stylomastoid foramen, near the ear, and passes through the parotid gland, branching where it emerges on the face to supply the platysma and its derivatives the muscles round the eye, nose, and mouth. In man, the facial nerve therefore controls the muscles that express the emotions.

The gustatory fibres of the facial run in the *chorda tympani* nerve, which crosses the tympanic cavity and is distributed with the lingual branch of the trigeminal nerve to the tongue. The facial nerve also carries preganglionic parasympathetic fibres for the control of the secretion of some salivary glands and the lachrymal gland. These run to the submaxillary, and spheno-palatine ganglia (p. 263).

15. Glossopharyngeus, vagus, and spinal accessory

In the region of the otic capsule and immediately behind it, no myotomal muscle remains and no ventral roots develop. The dorsal roots form a series, the first separate as the *glossopharyngeus* and the subsequent ones fused to form a complex set of trunks, which are ascribed by anatomists, rather arbitrarily, to two 'nerves', the *vagus* and *spinal accessory* (Figs 11.6 and 11.7). The *hypoglossus* represents the combined ventral roots of some of these hinder cranial segments.

The first of the post-otic dorsal cranial roots, the ninth nerve (*glossopharyngeus*), supplies gustatory fibres for the back of the tongue and motor fibres for some of the muscles of the pharynx wall. It has two branches, a lingual and a pharyngeal. The dorsal roots of several segments behind the glossopharyngeal unite to make the vagus and spinal accessory trunks. These two nerves are only conventionally separable; they leave the skull as independent trunks but then interchange fibres, so that some of the fibres in the 'branches' of the vagus actually leave the skull in the spinal accessory root. The series makes a system of fibres for innervating the striped muscles derived from the lateral plate in this region (laryngeal muscles, trapezius, sternomastoid) and also provides preganglionic parasympathetic as well as sensory fibres for many of the viscera. It is not clear how it has come about that the nerves of a few segments should fuse in this way and then spread so far outside their territory, but the arrangement is as old as the vertebrates.

The *vagus* or tenth cranial nerve (Fig. 11.7) is a large mixed nerve, providing sensory and motor fibres to a great part of the gut, respiratory system, and heart. The nerve emerges through the jugular foramen and the ganglionic swelling containing the cell bodies of its afferent fibres is divided into an upper *jugular ganglion* within the skull and a *ganglion nodosum* lower down. The cell bodies of the parasympathetic pathways of the vagus do not lie in these ganglia but more peripherally among the tissues (p. 279). Shortly below the ganglion nodosum the nerve gives off the *superior laryngeal nerve*, supplying muscles in the larynx, and the *cardiac depressor nerve*, which carries sensory fibres from the heart and aortic arches (p. 286). Other small cardiac branches leave the vagus in the neck and make a plexus of fibres in the sinu-auricular node of the heart (p. 198). These plexuses contain the fibres and cells of the vagal parasympathetic system, which slows the heartbeat (p. 286).

At the lower end of the neck the vagi give off *recurrent laryngeal nerves*, which loop round the subclavian artery on the right and ligamentum arteriosum on the left before running cranially again on either side of the trachea to reach the intrinsic muscles of the larynx. This peculiar course is a result of the elongation of the neck during development.

Below this point the vagi break up into a series of plexuses. The pulmonary plexuses carry sensory and motor fibres to the tissues of the lungs and below them the nerves continue as an oesophageal plexus so dense and tangled that separate right and left vagal trunks can hardly be recognized in man. From these plexuses fibres proceed on to the stomach and others join the sympathetic fibres of the solar plexus (p. 282) and reach back as far as the lower end of the small intestine. The motor (preganglionic) fibres of these vagal gastric and intestinal plexuses act upon a variety of functions, often in a direction 'antagonistic' to that of the sympathetic nerves to the same parts (p. 286). In addition to these motor functions the vagi also provide pathways for afferent impulses such as those of 'hunger', produced by contraction of the stomach (p. 151).

The *spinal accessory nerve* (Fig. 11.7) arises by rootlets not only in the skull but from the cervical cord as low as the fifth or sixth segment. These rootlets lie between the dorsal and ventral rootlets of the cervical nerves and their cell bodies lie in the grey matter dorsal to those of the ventral roots. More cranially this column of cells continues as the *nucleus ambiguus*, which in turn is continuous with the motor nuclei of the vagus and glossopharyngeus (p. 277). The morphology of the spinal accessory is therefore quite clear; it is a motor dorsal root, supplying muscle that is formed from the lateral plate and therefore innervated in a manner different from the muscles of the trunk, which are all myotomal. After emerging through the jugular foramen the accessory nerve makes anastomosis with the vagus, supplying it with the fibres that operate the striped musculature of the larynx (superior and recurrent laryngeal nerves); it then divides into branches to the sternomastoid and trapezius muscles.

16. The hypoglossal nerve

The twelfth cranial nerve, *hypoglossus*, is also a pure muscle nerve, but in this case representing the ventral roots of the more posterior of the vagal-accessory segments and supplying the derivatives of the ventral portion of the myotomes of this region, which are limited to the tongue. The so-called descending branch of the hypoglossal consists of fibres that have joined the trunk within the skull from the more cranial cervical nerves and this branch rejoins these nerves in the neck to form the *ansa hypoglossi*, from which fibres run to the sterno-hyoid and other muscles that draw the larynx caudally.

17. The cervical spinal nerves

The eight *cervical spinal nerves* emerge through foramina in front of the cervical vertebra and the first thoracic vertebra. They have the typical dorsal and ventral roots but they have no white rami communicantes, since no preganglionic autonomic fibres leave the spinal cord in the neck (p. 281). The first cervical nerve, leaving in front of the atlas, is small and lies at the back of the neck. The second nerve is also small but the third and subsequent nerves are conspicuous bundles. They innervate the muscles and skin of the neck and also provide the fibres of the *phrenic nerves* (Fig. 11.7). These arise mostly from the fourth cervical in the rabbit, with contributions from neighbouring nerves. The phrenic trunks run back over the pericardium and along the mediastinum to end in the striped musculature of the diaphragm. This course of the phrenic nerves is a result of the backward movement of the diaphragm, which begins its development in the septum transversum at the level of the heart and pushes backwards as the lungs develop (p. 120).

18. The cervical sympathetic trunks

The *cervical sympathetic trunks* (Fig. 11.7) are conspicuous nerves running along the neck. They consist of preganglionic fibres that leave the spinal cord in the white rami of the thoracic region and run forward to end in either the *inferior cervical ganglia*, at the level of the subclavian arteries, or the *superior cervical ganglia*, lying dorsal to the point of division of the carotid arteries. From the inferior ganglia, also known as *stellate ganglia*, postganglionic fibres run to a variety of organs, accelerator fibres to the heart, bronchomotor fibres to the lungs, and vasomotors and sudomotors to the forelimb. The superior cervical ganglion is the cell station for the sympathetic fibres to all the organs of the head, including vasomotors, pupillo-dilators, fibres for the nictitating membrane, and others that assist in control of salivation and sweating. The postganglionic fibres leave the superior cervical ganglion and pass with the branches of the carotid artery as the internal carotid plexus. They are distributed to the peripheral organs with the branches of the cranial nerves.

12 Development of skeleton, muscles, and coelom

1. Differentiation of the mesoderm

THE greater part of the mesoderm differentiates from tissue that has been invaginated over the lips of the blastopore or the primitive streak. It comes to lie between the ectoderm and the endoderm, and the more lateral part proceeds to differentiate forming layers applied to the ectoderm on the outside and endoderm within, leaving a cavity, the coelom, between (Fig. 12.1). The coelom has a complicated shape, changing greatly as development proceeds. The walls are not of the same thickness throughout. In the dorsal region of the embryo they are thick, forming the *paraxial mesoderm*, lying on either side of the spinal cord. By migration of the cells this part of the mesoderm becomes divided into a series of similar blocks, the somites, which are the fundamental basis of the segmentation of the embryo (Fig. 12.2). Each somite consists, in the early stages, of thick walls of cells and a small cavity, the *myocoel*. As development proceeds the cells of the walls differentiate to form the *dermatome* laterally and the *myotome* medially. The cells of the dermatome spread out beneath the ectoderm to form part of the dermis. The cells of the myotome elongate in the main axis of the body and become the myoblasts, from which the axial muscles differentiate. From the medial parts of the myotomes, cells separate off to form the *sclerotomes* around the neural tube and notochord. This tissue gives rise to the skeletal tissues of the vertebral column and some of those of the skull (p. 121).

The myocoels soon become obliterated and they leave no traces in any adult vertebrate. Below the somites the mesoderm narrows to form a strand, the *intermediate mesoderm*, and this is joined to the *lateral plate mesoderm* of the ventral part of the embryo (Fig. 12.3). The intermediate mesoderm gives rise later to the excretory system and gonads and its cavity becomes obliterated.

The somites extend from the level of the front end of the notochord all the way down the body. There are about 44 of them in the human embryo (Fig. 12.4). They appear from in front backwards and a count of the number of somites thus provides a useful measure of the age of an embryo.

The three somites in front of the otic capsule give rise to the extrinsic muscles of the eyeball (see Gilbert 1957). The first forms the superior, inferior, and

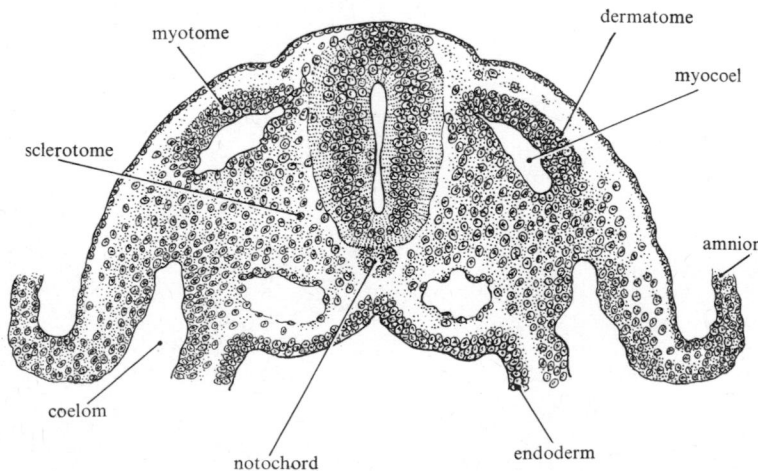

FIG. 12.1. Transverse section of a 16-somite pig embryo. (After Patten 1968.)

FIG. 12.2. Diagrams of stages in the differentiation of the somites in man. (a) Origin of sclerotomal cells from somite; (b) sclerotomal cells of neighbouring somites move together; (c) they form the rudiment of the centrum. The remainder of the somite differentiates into dermatome and myotome. (After Hamilton and Boyd 1972.)

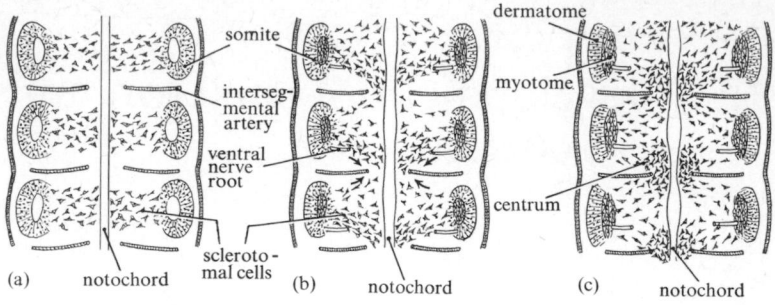

medial recti, and the inferior oblique; the second forms the superior oblique, and the third the lateral rectus. The occipital somites immediately caudal to the otic capsule regress and form no muscles, but those of the hinder head region produce the intrinsic muscles of the tongue (p. 114). The somites of the trunk region give rise to the axial musculature and downward extensions form the muscles of the chest and abdominal walls and of the diaphragm. The thoracic muscles retain their original segmentation. The muscles of the fins of fish-like vertebrates were formed from extensions of the myotomes but in a mammal the limb muscles develop independently (p. 125).

The lateral plate mesoderm differentiates to make the mesothelial lining of the coelomic cavities of the adult. It therefore comes to consist of an outer layer, in contact with the ectoderm, the somatopleure, and an inner splanchnopleure covering the alimentary canal and other viscera. The lateral plate mesoderm is not segmented and the coelom at first forms a continuous cavity ending at the pericardium in front. The caudal limit of the coelom, in front of the pericardial cavity, is called the *septum transversum*. As the head-fold develops and the embryo elongates, the septum transversum

comes to lie behind the heart instead of in front of it (Fig. 12.5). The connexion between the pericardium and the rest of the coelom remains as a pair of long *pericardio–peritoneal canals*. The developing lungs push into these canals, which thus become the *pleural cavities*. The communications between the pleural cavities and the pericardium and peritoneal cavities ultimately become closed.

The organs that lie in the coelom are supported above and below by folds of mesoderm (Fig. 12.6) and it is from these folds that the *mesenteries* of the adult are formed. They serve to attach the organs to the body wall and for the passage of blood vessels, lymphatics, and nerves. The lateral plate mesoderm of the somatopleur gives rise to little muscle in the trunk region but it has been held to contribute to the musculature of the abdominal wall. The muscles of the gut wall are derived from the splanchnic mesoderm.

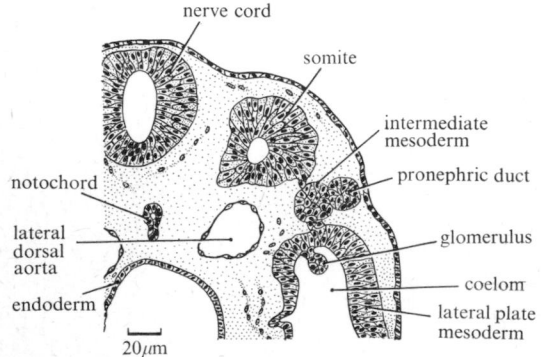

FIG. 12.3. Section through the tenth somite of a 14-somite human embryo.

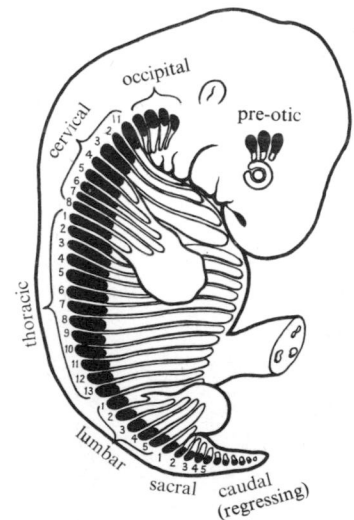

FIG. 12.4. Diagram of the myotomes of a human embryo. The blackened areas show the extent of the original meso-dermal somites. (After Patten, *loc. cit.*)

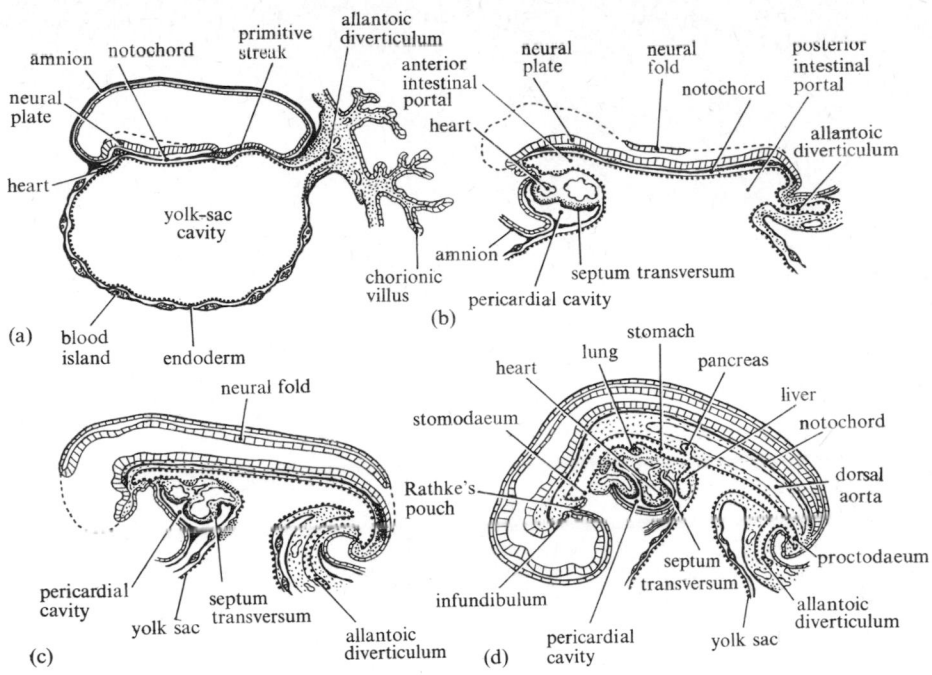

FIG. 12.5. Diagrams showing stages in the development of the neural tube and septum transversum in man. (After Patten, *loc.cit.*)

The lateral plate mesoderm of the head region produces the musculature of the branchial arches of fishes. Some of this persists in mammals as the muscles of mastication, tensor tympani, facial muscles, muscles of the pharynx and larynx, and parts of the sternomastoid and trapezius. These muscles show their origin in the fact that they are innervated by nerves that can be identified developmentally as dorsal roots (for example, the spinal accessory nerve). Derivatives of the myotomes are always innervated by ventral roots though afferent fibres may run in them.

2. Mesenchyme

The cells migrating from the medial aspect of the myotome to form the sclerotome are often said to consist of *mesenchyme*. This is a term used somewhat loosely to indicate the packing tissue of non-epithelial character that fills up the spaces between the developing organs. Mesenchyme is formed from the splanchnic and somatic mesoderm as well as from the somites, and in the head it may also arise from ectoderm (p. 347).

The cells of the mesenchyme of the young embryo are star-shaped and their processes make contact to give a loose network. Between the cells lies a jelly-like material, probably a muco-scleroprotein (p. 29). The mesenchymal cells have active powers of migration and the movements they make play a considerable part in determining the form of the organs. They may also become phagocytic. By differentiation, the mesenchyme cells give rise to the cells of the skeletal, connective, and blood vascular systems. Some of them also form fat cells, and others the macrophages of the reticulo-endothelial system. They include the wandering macrophages (macrocytes) and fixed macrophages (histiocytes).

3. Development of the axial skeleton

The mesenchyme of the sclerotomes differentiates to form the rudiments of the vertebrae. These first appear as condensations of mesenchyme around the notochord. The main aggregations of cells come to lie opposite the intervals between the myotomes, contributions being made to each group from two segments (Fig. 12.2). These aggregations make the centra of the vertebral bodies; the intervertebral discs develop from the somewhat less concentrated cells opposite to each myotome. Cartilage soon begins to form in the centra and the rudiment of the notochord is there obliterated. It persists, however, in the intervertebral regions to produce the nucleus pulposus of the discs (p. 77). Meanwhile further condensations of mesenchyme have arisen

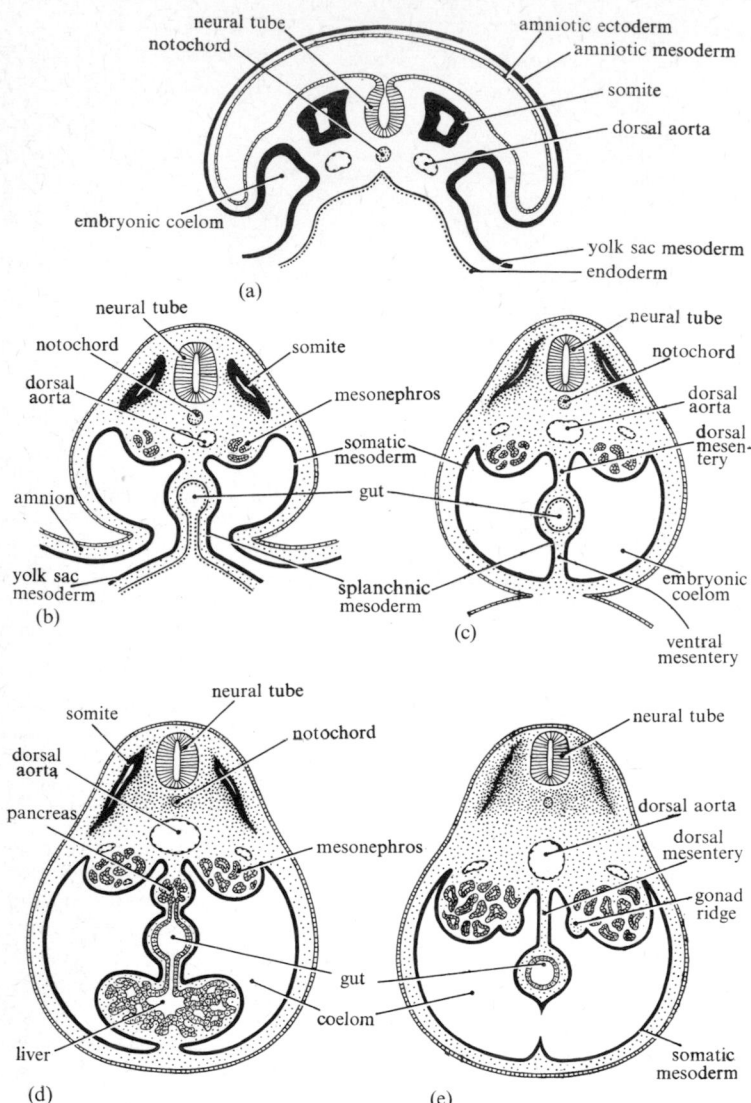

FIG. 12.6. Diagrammatic sections showing stages in closing off of the gut and formation of mesenteries in a mammal. (After Patten, *loc. cit.*)

dorsally and laterally to make the neural arches and transverse processes, which later join the centra.

The vertebrae form first as cartilaginous rudiments, which are later replaced by endochondral bone (p. 77). The other elements of the axial skeleton also develop as condensations in mesenchyme, forming the ribs and sternum.

4. Segmentation of the head

The development of the skeleton of the head has been much modified during evolution to produce the special systems present there for the support of the sense organs, brain, and feeding apparatus (see p. 104). Nevertheless, the developmental processes at work are

modifications of the fundamental segmental ones found in the rest of the body. Even in a mammal the head is essentially a segmented structure and this appears in its muscles, skeletal elements, and nerves. Added to these fundamental elements of the axial skeleton are the branchial or visceral arches, produced by modification of the processes that gave rise to the gill apparatus of earlier vertebrates.

5. Development of the skull

The first outline of the skull is formed of cartilage, which may be regarded as produced by three sets of morphogenetic processes: (a) those that form cartilaginous capsules around the special sense organs;

(b) the modified processes for the formation of the branchial arches; (c) those producing material corresponding to the sclerotomes (Fig. 12.7; see de Beer 1937).

The material corresponding to the sclerotomes lies on either side of the notochord. In the occipital region it shows signs of segmentation, but around the front of the notochord it forms a pair of unsegmented rods, the *parachordal cartilages*, which later join to make a parachordal plate. The segmented region of the head may be said to end at the front end of the notochord. The floor of the developing skull in front of this level is formed at first by a pair of rods known as the *trabeculae cranii*. These develop partly from the neural crest and probably represent the first of the series of visceral arches, lying in front of the mouth (p. 124). They become united with each other and with the parachordals, forming the front part of the floor of the neurocranium. The walls and roof appear somewhat later and the cartilaginous neurocranium is completed by the cartilage around the sense capsules (Figs 12.8 and 12.9). The mesoderm around the olfactory pit forms the olfactory capsule, which soon fuses with the trabeculae. The optic capsule is little developed in mammals. It becomes the sclera, which is not cartilaginous. The mesoderm around the otic sac forms the otic capsule. This fuses with the parachordal plate, leaving, however, a gap, which becomes the jugular foramen.

The cartilaginous neurocranium gradually becomes converted into the bony skull, partly by the laying down of bone within the cartilage already formed (*endochondral bone*) and partly by the addition of *membrane bones* outside the original neurocranium (Fig. 12.9).

The bone first appears at numerous discrete centres, which then join up to make the definite 'bones' of the adult skull. The membrane bones are the products of morphogenetic processes similar to those that form the scales of a fish. In the earliest land vertebrates the skull was covered, therefore, by a large number of small bones. As evolution has proceeded the number of dermal skull bones has become reduced, the rudiments fusing or being suppressed earlier and earlier in development. In mammals the adult skull consists of relatively few large bones; in man even these tend to fuse towards the end of life.

The skull grows in size mainly by the addition of new bone in the sutures but also by addition of material from the periosteum of the outer surface and its removal by osteoclastic action on the inside.

6. Branchial arches

The pharynx of all vertebrates is formed as a tube, whose walls show endodermal pouches and ectodermal

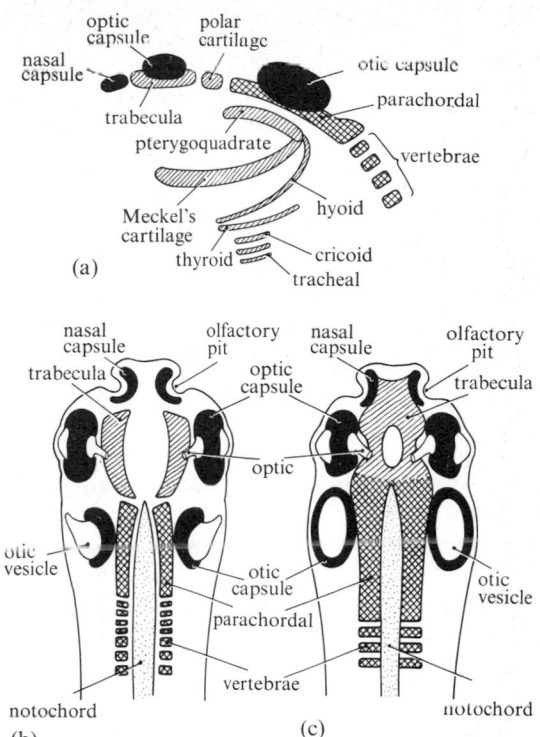

Fig. 12.7. The three sources of cartilage of the chondrocranium: (1) Neurocranial capsules of the sense organs (black); (2) Branchial arches (shaded); (3) sclerotomal material (cross-hatched). (a) Side view. (b) and (c) Dorsal views of an early and a later stage.

gill clefts. It is 200 million years since the ancestors of the mammals respired by gills, but embryological processes change only very slowly and the pharynx still develops on a plan based upon a series of branchial arches.

The *viscerocranium* consists of cartilaginous bars supporting the pharyngeal arches that separated the gill slits and developed from tissue of neural crest origin. These bars become partly ossified in the adult and are supplemented by membrane bones formed outside them (see Goodrich 1930). The fishes from which land vertebrates are derived probably had seven gill slits and traces of these can be found in mammals (Fig. 12.10). The first slit has joined the mouth; the second forms the hyoid gill cleft, and the remainder the branchial clefts. There are, therefore, six pharyngeal arches separating the pouches and clefts and the trabeculae cranii probably represent cartilaginous bars lying in front of the slit that has joined the mouth (p. 162).

The second cartilaginous bar of the whole series is

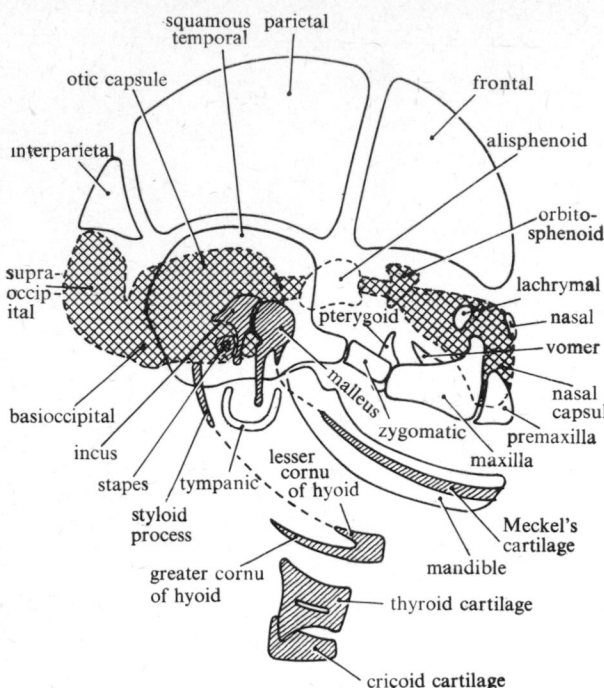

FIG. 12.8. Scheme of sources of the material of the cranium. Cross-hatched areas: cartilaginous neurocranium; shaded areas: visceral arches; white areas: membrane bone. (After Hamilton and Boyd 1972.)

thus that of the first or mandibular arch. Its upper portion grows forwards to form the *maxillary process*, the rudiment of the upper jaw. The arch forms a sharp angle behind the mouth and its ventral portion again runs forward as the *mandibular process*, the rudiment of the lower jaw. Each of these processes contains a cartilage, the pterygoquadrate bar in the upper and the Meckel's cartilage in the lower jaw. The jaws of the

adult are formed by membrane bones developing outside these cartilages, the premaxilla (where present), maxilla, zygomatic, and squamous temporal above, and mandible (dentary) in the lower jaw (Fig. 12.9). The hinder portions of the cartilages of the first arch become incorporated in the ear (Fig. 12.11). The end of the pterygoquadrate ossifies to form the *incus* and of Meckel's cartilage to form the *malleus*.

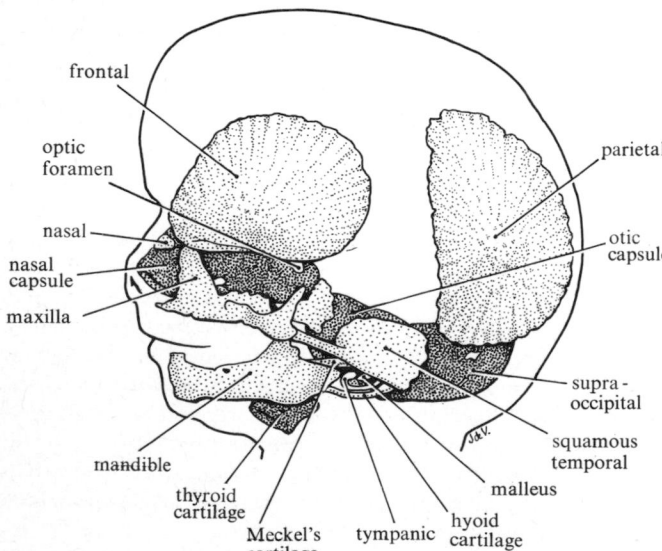

FIG. 12.9. Side view of skull of 80 mm human embryo showing chondrocranium (darkly stippled) and membrane bones (lightly stippled). (After Hertwig, in Hamilton and Boyd, *loc. cit.*)

The second of the original series of pharyngeal pouches, the spiracular or hyoidean, becomes the pharyngo-tympanic (Eustachian) tube, closed externally by the tympanic membrane (p. 398). It is usually (if confusingly) reckoned as the first of the series of apparent pharyngeal pouches, excluding the slit that has joined the mouth. The second (hyoid) pharyngeal arch, lying behind this pouch, contains *Reichert's cartilage*, the upper part of which forms the third of the chain of ear ossicles, the *stapes*. The middle portion of this arch ossifies as the *styloid process* and its ventral portion becomes the *lesser cornu* and part of the body of the hyoid bone.

The cartilages of the third branchial arches fuse with those of the ventral part of that of the second and become the *greater cornua* and part of the body of the hyoid bone. The cartilages of the fourth and fifth arches form the *thyroid cartilage* and that of the sixth the *cricoid*. The remaining cartilages of the larynx are probably also derived from these hinder arches. There is no sharp distinction between the cartilages of the larynx and of the trachea. The latter develops as an outpushing from the pharynx (p. 163), and the formation of cartilages in its walls is the result of an extension backwards of the morphogenetic process by which the branchial arch cartilages are produced.

7. Development of the limbs and their skeleton

The paired fins of fishes, from which the limbs have evolved, were folds of the body-wall, containing a segmental series of radial cartilages, muscles, and nerves

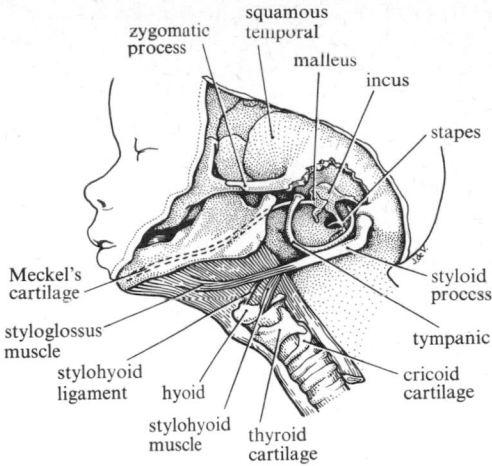

FIG. 12.11 The ear ossicles of a human foetus. (After Kollman.)

(p. 84). In the course of their change into limbs the base became narrowed and the segmental structure less obvious. The earliest limb-buds of a mammal are lateral folds with a broad base and they are not unlike fins (Fig. 12.10). The cranial or pre-axial borders become the pollex and hallux. The adult position of the limbs is reached by flexures at the joints (p. 84). The segmental nature of the limbs is shown by the fact that each is supplied by a series of nerves, the more cephalad of which innervate the pre-axial borders (p. 258).

The development of the skeleton of the limbs of a mammal shows no trace of segmental origin (unless it be in repetition of the digits), and the limb muscles do not develop in continuity with the myotomes. The bones of the girdles and limbs are all endochondral in origin, except the clavicle, which is a membrane bone.

8. Factors controlling development of the skeleton

The differentiation of the mesenchyme cells before they 'condense' to form the rudiments of a cartilage must be the result of some evocator action. In the chick the cells below the epithelium of the lower jaw are already determined at the third day and they will produce the cartilage if they are transplanted.

Soon after this evocation has occurred the tissues of each part become self-determining. This is well shown by transplantation of portions of limb-buds of the chick, which will produce whole isolated bones (p. 42). The general outline of the shape of the bones is therefore determined by the hereditary information, after appropriate evocator action. The detailed form depends upon the strains that are later placed upon the bone (p. 42)—'double dependence' (see p. 4). The muscles,

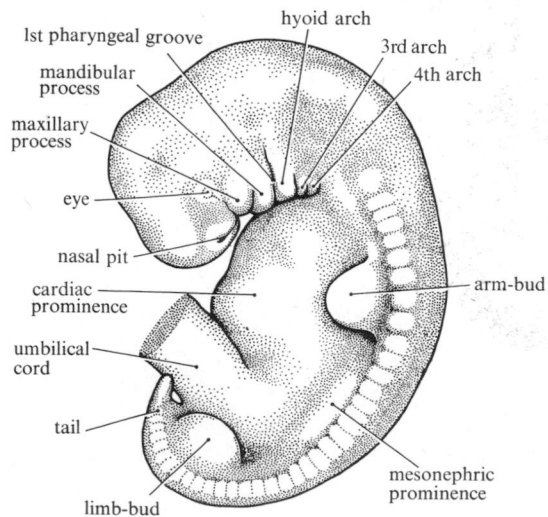

FIG. 12.10. Side view of human embryo 5 weeks after fertilization showing branchial arches.

bones, joints, and ligaments maintain an effective functional state only when they are provided with appropriate stimulation by the environment.

The differentiation of the *vertebrae* depends upon the neural tube, and they will form even around a piece of tube transplanted to the flank of a salamander embryo. The segmentation depends upon the sequence of spinal nerves and this in turn on the segmentation of the somites.

The determination of the limb-buds is a function of the mesoderm. The ectoderm can be removed and replaced from any part of the body. However, at a later stage the ectoderm forms a well-defined 'apical ridge', which is necessary for the differentiation of the distal part of the limb. Provided this ridge is present the apical mesenchyme of the bud gradually changes its capacities, as is shown by reciprocal grafting of the tips of buds of different ages (Fig. 12.12).

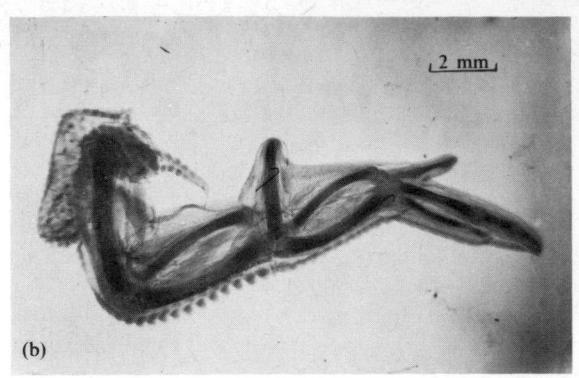

(b)

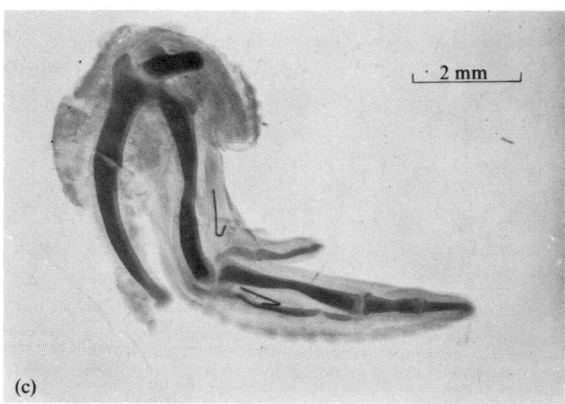

(c)

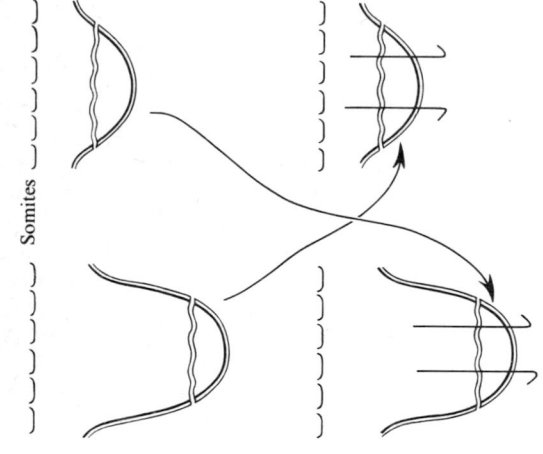

(a)

FIG. 12.12. (a) The whole of a stage 19 wing-bud is exchanged with the distal quarter of a later (stage 24) bud. (The grafts are held in place with pins.) Each part develops essentially as it would have done undisturbed. (b) The young bud on the older stump gives two humeruses and forearms, terminated by a hand. (c) The old tip on the young stump gives a humerus followed by a full length hand. In both cases the size of the host's contribution corresponds with that of the contralateral wing, and that of the donor with its control wing. (From Summerbell, Lewis, and Wolpert (1973), *Nature* **244**.)

13 The intake of food. Mouth and teeth

1. The supply of raw materials

THE matter within the system of any animal needs continual renewal; the body is in a steady state, not an equilibrium, and life depends upon the availability of fresh supplies. Moreover the homeostatic actions of the body depend upon the expenditure of energy for moving substances against concentration gradients and for many similar purposes. In a mammal, energy-giving foods are required to maintain the high temperature and to make possible all the many special actions that ensure homeostasis. The whole system of life of a mammal is thus based upon the expenditure of large amounts of energy, the information available through the nervous system ensuring that such expenditure brings the body into conditions in which life can be maintained.

A large part of the activity of every mammal, therefore, is devoted to obtaining supplies of material for building and rebuilding the body and for the provision of energy. Mammals have acquired the power to gain the necessary nourishment from all sorts of odd corners of the varied world of land, sea, and air. Yet the basic pattern of their digestive system shows few special features and is essentially that common to all vertebrates. It is surprising that such a wide variety of diets can be digested by means of the familiar apparatus of a stomach with its acid and pepsin, and an intestine with enzymes acting in alkaline solution.

The differences between alimentary systems are mainly in the apparatus for obtaining the food from the environment, the mouth and teeth, and in the presence of special chambers of the gut for the cultivation of symbiotic bacteria to assist in breaking down cellulose. The placental mammals of the Cretaceous Period, more than 70 million years ago, were insectivorous; from these all the specialized carnivores and herbivores have been evolved and many types, like man, have become omnivorous. Indeed most mammals show considerable flexibility in the way they obtain the materials necessary for their maintenance, as would be expected from animals with such an adaptable nervous organization.

Whatever their specializations, all mammals take some proteins, carbohydrates, and fats, together with water and certain special molecules, inorganic salts, and vitamins. Proteins, besides providing the necessary nitrogenous materials, can be used as fuel for the provision of energy and warmth, which are otherwise supplied by carbohydrates and fats.

2. Selection of food

The activities by which the animal seeks, selects, and eats its food are an important part of its behaviour and the types of food chosen depend largely on the nervous organization (p. 307). In this, as in other matters, the mammals show exceptional powers. They are provided even more fully than other animals with a system that provides a 'drive' to seek food, and they use great ingenuity until they find it. We know little of the source of this drive or how it is related to the needs of the organism. Whatever the mechanism may be it arranges that the precise amount required is taken. Thus an adult man eats about 12 000 kg of food in 25 years and may remain of the same weight to within 2 kg or less.

The basic regulation is almost certainly from the hypothalamus, where there are centres controlling both eating and drinking (Chapter 32). The feeding centres are partly controlled by the level of the sugar in the blood, but this is not the only factor, since diabetics, with a high blood-sugar level, may be often hungry. There is a ventro-medial hypothalamic satiety centre, stimulation of which, in a rat with implanted electrodes, produces reduced food intake, while destruction leads to obesity. A lateral feeding centre operates in the opposite direction.

There is some evidence that these centres operate by stabilizing the fat stores (Stevenson 1969; Kennedy 1953). When one of a parabiotic pair of rats was given a hypothalamic lesion it became fat and the other rat then reduced its food intake and lost weight (Hervey 1959).

It was shown long ago that gastric motility, measured by a swallowed balloon, is increased during hunger pains. Such contractions are often supposed to be the source of hunger, but they seem not always to be present, and hunger persists after denervation of the stomach. Its source is more complicated, but may also include the contractions.

In a young child it may be the hunger contractions that initiate the behaviour of crying, which ensures a supply of food from the mother. Probably as development proceeds, a pattern of food intake at various times throughout the day is set up, triggered partly by hunger contractions, partly by complicated memories within the central nervous system. The cerebral cortex controls the activities of digestion as it does nearly everything else in the body (Chapter 34). One proof of this control is the fact that after removal of part of the frontal lobe of the cortex the food intake may be greatly increased (p. 328).

It is a common belief that animals and men seek and find food that is 'good for them', for instance, game will move many miles to a 'salt lick' to obtain the small amounts of inorganic salts that they need. Little is known about the receptor and central nervous mechanisms related to these appetites. It is certain that they do not provide unfailing guidance for the creature, and yet it remains an astonishing thing that organisms with such complicated chemical constitutions continually renew themselves, seeking and eating the raw materials that they need. The chemo-receptors probably play a large part in finding the right food, those of the nose acting as distance receptors and those of the mouth serving to test the food at the point of intake (p. 409). Whether any given chemical stimulus leads to acceptance or rejection of the object as food no doubt depends largely on past experience as recorded in the cortex. In mammals, food choice is probably mainly determined by acquired rather than by hereditary information, but little evidence is available on this point.

An interesting example of the influences affecting food choice is the difference in behaviour between wild and laboratory rats when presented with food consisting of an assortment of purified substances in separate containers. Offered a protein, a fat, a carbohydrate, and twelve other dishes of pure minerals and vitamins, a laboratory rat makes a selection that allows normal growth. A wild rat eats only the fat and dies in a short time. These wild animals are so 'suspicious' that they will not taste new foods, and this, incidentally, makes it difficult to poison them. Richter, who conducted these experiments, found many other signs of the 'suspiciousness' of wild rats (p. 430).

The rabbit is a herbivore, able to eat not only relatively soft vegetable foods such as dandelion or lettuce leaves but also much harder roots and grasses. For this activity it possesses sharp, continually growing incisors to crop the grass, large grinding molars, and an enormous intestinal sac, the caecum, in which bacteria make the cellulose of the grass available to the rabbit. Such unpromising food is not easily digested and rabbits have the habit of passing it through the gut several times by eating their faecal pellets and thus ensuring maximum use of the material available. Many mammals have specializations allowing co-operative digestion by means of symbionts. In artiodactyls there are elaborate stomachs (Young 1962, p. 744). In perissodactyls (horses) there is a special fauna in the intestine. Leaf-eating monkeys have an enlarged stomach.

The dentition and gut of man, like his tastes, are those of an omnivore, able to obtain his nourishment from either plant or animal food or from a mixture of both. Neither the teeth nor gut of man are able to deal with the hardest plant foods.

3. Development and structure of the teeth

The features of the digestive system that are peculiar to mammals are found mainly at the point of intake, where the moist internal meets the dry external world, that is to say, in the mouth and teeth. Here the food is seized and the salivary glands are developed to moisten it for swallowing. The teeth have become differentiated from the original series of cone-like reptilian teeth, known as a *homodont dentition*. In mammals, the more anterior ones retain the single cusp pattern, but farther back along the jaws new developmental processes have led to the appearance of broad molariform teeth with uneven biting surfaces (Romer 1966). In this way the *heterodont dentition* has arisen, the incisors in front serving to seize the food, canines to pierce and tear it, and molars to cut and grind it. Moreover, the mammals have proved able to vary this dentition greatly according to their diet. In a few (for example, some toothed whales) there has been a reversion to the homodont condition, with a large number of single-cusped teeth, like those of fishes or reptiles.

Mammalian teeth are formed in two series, first a 'milk' or deciduous, then a permanent dentition. This condition is said to be diphyodont, in contrast to the polyphyodont dentition of lower vertebrates, in which several series of teeth appear. Each tooth is the product of the collaboration of a thickening of the ectoderm, the *enamel organ*, whose *ameloblasts* secrete the enamel, and a mesodermal *dental papilla*, whose *odontoblasts* secrete the dentine (ivory) (Fig. 13.1). The first sign of the appearance of the tooth rudiments is a line of thickenings of the epithelium of the mouth, the dental lamina. Along this line there form groups of cells, the tooth buds (Fig. 13.1). The cells of these buds proli-

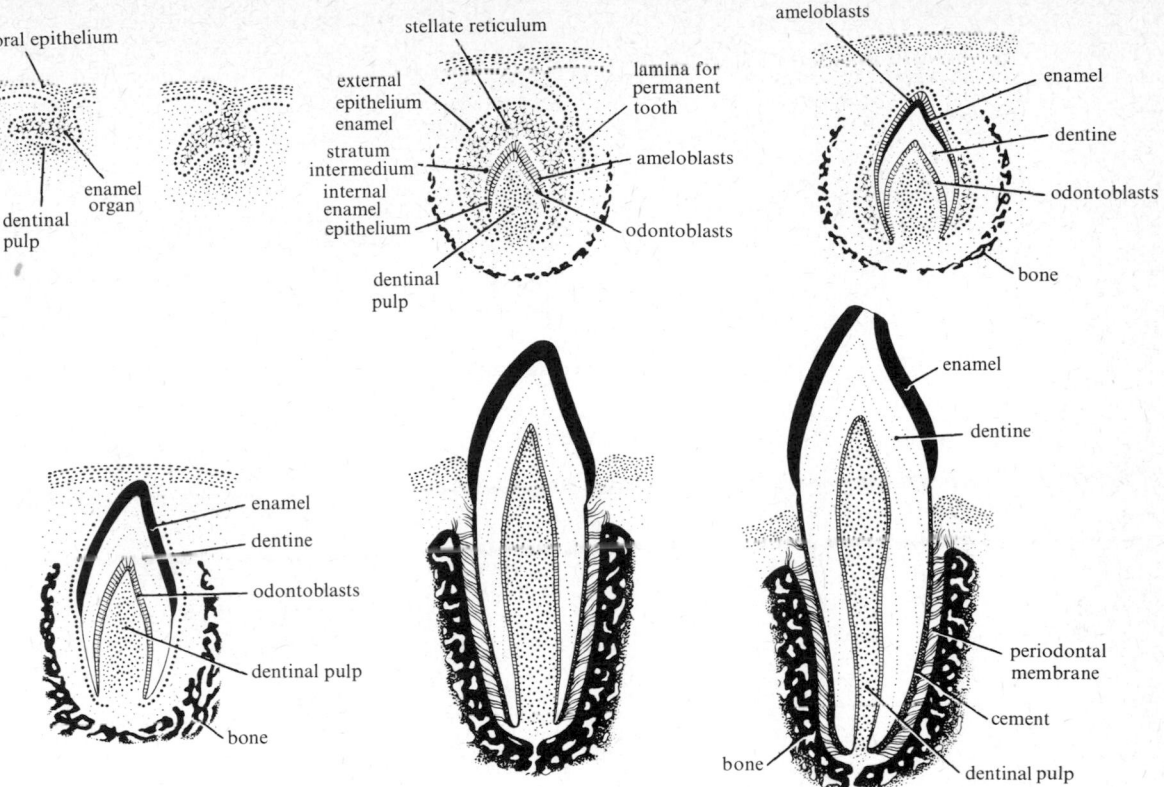

FIG. 13.1. Stages in the development of a tooth. (After Orban.)

ferate faster than their neighbours, so that each bud extends into the mesoderm. Here its tissues differentiate to form four layers, the inner and outer enamel epithelia, separated by a central mass of cells, the *stellate reticulum,* and a layer 3–5 cells thick, the *stratum intermedium*. The inner enamel epithelium differentiates into the ameloblasts. It soon invaginates to make a cup-like structure enclosing a mesodermal core, the dentinal papilla, within which the odontoblasts and pulp of the tooth develop.

The cells of the inner enamel epithelium differentiate to form a row of columnar cells whose first function is to induce the differentiation of the underlying mesodermal cells to form odontoblasts. As soon as the latter begin to produce dentine, inner enamel epithelium cells further differentiate into ameloblasts and begin to produce enamel (Figs 13.2 and 13.3). After a thin layer of dentine has been produced, the ameloblasts must rely on the blood supply to the enamel organ. It is believed that the intercellular substances of the stellate reticulum are utilized by the ameloblasts as the stellate reticulum collapses. Blood vessels are then in contact with the stratum intermedium layer which persists throughout amelogenesis. Enamel and dentine

secretion are interrupted at birth, and this causes discontinuities in their structure recognizable as 'neonatal' lines (Fig. 13.4).

The tooth is thus an elevation with a mesodermal core, covered by two sorts of hard material and set in a bony socket of the jaw known as the *alveolus*. A third material, the *cement*, is added to the surface to attach the collagen fibre bundles of the periodontal ligament to the surface of the tooth. Many herbivorous mammals have large, high-crowned (*hypsodont*) teeth which erupt and function before root formation commences—these have cement on the surface of the enamel of the tooth so that the tooth can be attached to the socket bone before it has a root proper. Cement also serves to bind together and pack the spaces between the high, flattened cusps or *ridges* of herbivorous molar teeth. The differing hardness of the three dental tissues ensures the maintenance of a rough grinding surface, which is especially important in the grazing herbivores for chewing the hard siliceous stems of grasses.

4. Enamel

Enamel is the hardest tissue in the body on account of its high content of mineral salts (96 per cent) and their

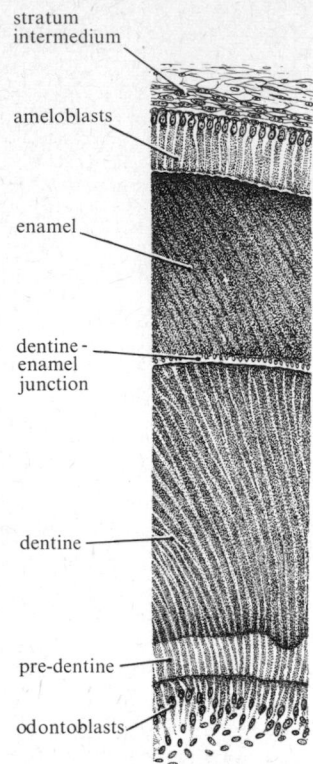

stratum
intermedium

ameloblasts

enamel

dentine-
enamel
junction

dentine

pre-dentine

odontoblasts

Fig. 13.2. Part of section of tooth of a kitten. (From a preparation kindly lent by Mr. K. C. Richardson.)

crystalline arrangement. It forms a layer up to 2·5 mm thick over the cusps of the teeth. The tissue is composed of a number of enamel prisms (rods). The prisms run in a general direction at right angles to the surface of the dentine, but they pass along oblique or wavy courses, forming patterns that minimize the likelihood of cleavage during mastication (Figs 13.5 and 13.6). Each rod is composed of a number of hexagonal submicroscopic crystals of an apatitic calcium phosphate, with axes arranged primarily in the direction of the long axis of the prism. The enamel crystals are only 30–60 nm in cross-sectional diameter, but very long— tens of micrometres at least and perhaps hundreds.

5. Dentine

The bulk of the tooth is made up of *dentine*, which is a material resembling bone but harder on account of the regular arrangement of its constituents and its higher level of mineralization. It is composed of 70 per cent of inorganic material and 30 per cent water and organic material, the latter being chiefly collagen. The tissue consists of a fibrillar calcified ground substance containing apatitic calcium phosphate crystals (size limits

2–>100 nm), through which run the protoplasmic processes of the odontoblasts (Tomes' fibres). These latter thus lie in canals, known as dentinal tubules, which may branch and anastomose (Fig. 13.7).

The collagenous fibrils run in a plane at right angles to these tubes (Fig. 13.8). Odontoblasts may remain vital throughout the life of the individual—this property is shared only with neurones. Odontoblasts, having completed the normal form of a tooth, may recommence dentine production at a later date, in response to mild injurious stimuli. Secondary, 'repair' dentine is produced, which prevents exposure of the dental pulp to infection from the oral cavity, which might result from a

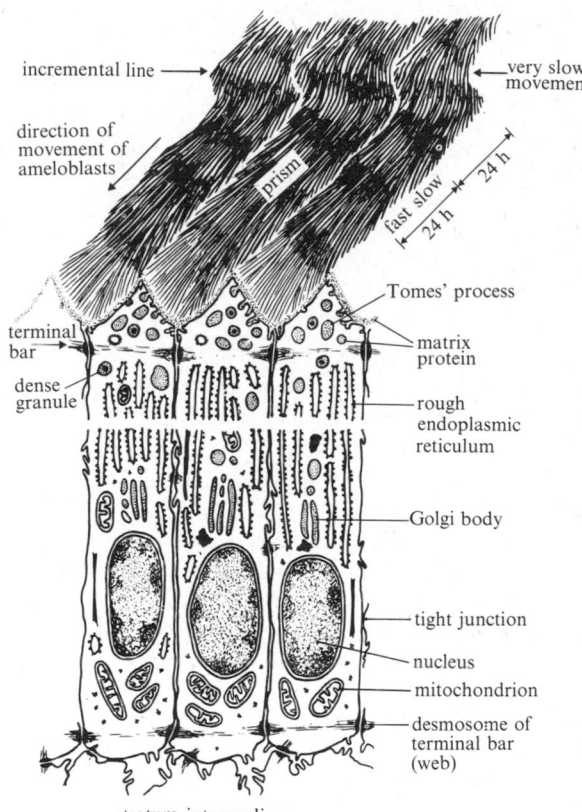

incremental line

very slow movement

direction of
movement of
ameloblasts

prism

fast slow   24 h
24 h   24 h

Tomes' process

terminal
bar

matrix
protein

dense
granule

rough
endoplasmic
reticulum

Golgi body

tight junction

nucleus

mitochondrion

desmosome of
terminal bar
(web)

stratum intermedium

Fig. 13.3. Diagram showing some features of amelogenesis. Tall columnar ameloblasts secrete the protein matrix in which extremely elongated crystals of hydroxyapatite grow. Bundles of *ca.* 10^4 crystals constitute 'prisms' which are demarcated from one another by a sharp change in crystal orientation which is generated by a sharp change in direction of the surface of the developing enamel. The cross striations of the enamel prisms are seen as undulations of the prism boundary—they correspond to 24-hour increments of formation. Growth disturbances cause more severe changes in prism direction (and perhaps disturbed mineralization) which are the incremental lines of *Retzius*. (Figure kindly supplied by Dr. A. Boyde.)

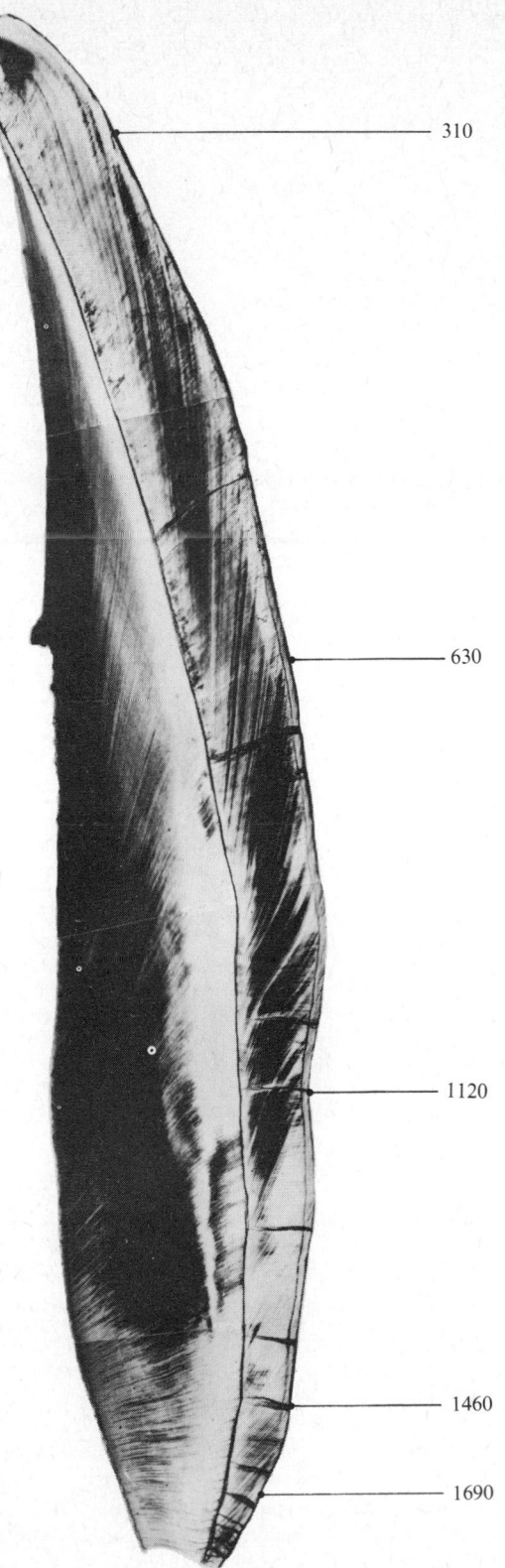

310

630

1120

1460

1690

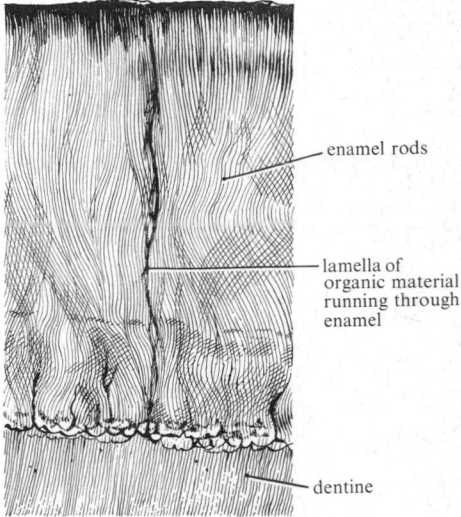

enamel rods

lamella of organic material running through enamel

dentine

FIG. 13.5. Section of enamel of a human tooth to show the direction of the enamel rods. (After Orban.)

tooth being worn heavily through use, fractured through the dentine, or attacked by dental decay. There are abundant nerve fibres around the odontoblasts and some of them extend into the dentinal tubules.

6. Cement

Cement is a modified type of bone which surrounds the roots of the teeth and serves to bind them to neighbouring structures. In some herbivores it extends between folds of the tooth so as to form part of the grinding surface. It is less hard than dentine. Between the cement and the surrounding bone of the jaw is a system of collagenous fibres, the *periodontal ligament* (or membrane) by which the tooth is attached (Fig. 13.1).

In most teeth the tip of the root narrows considerably, but remains open allowing blood vessels and nerves to enter the pulp cavity. In teeth that grow continually, such as the incisors of the rodents, the pulp is widely open below and these teeth are said to be 'without roots' and to have 'persistent pulps', they are worn away continually.

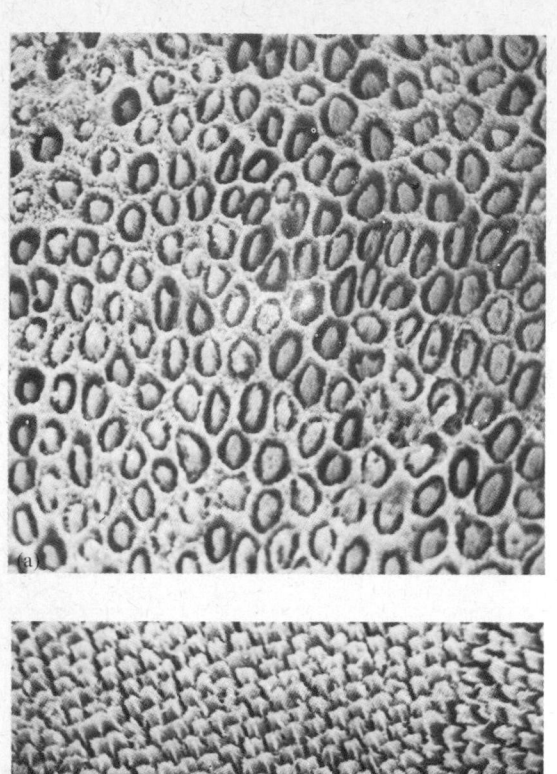

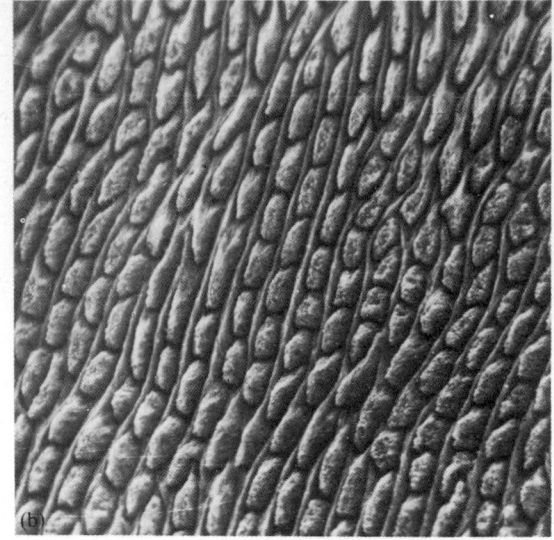

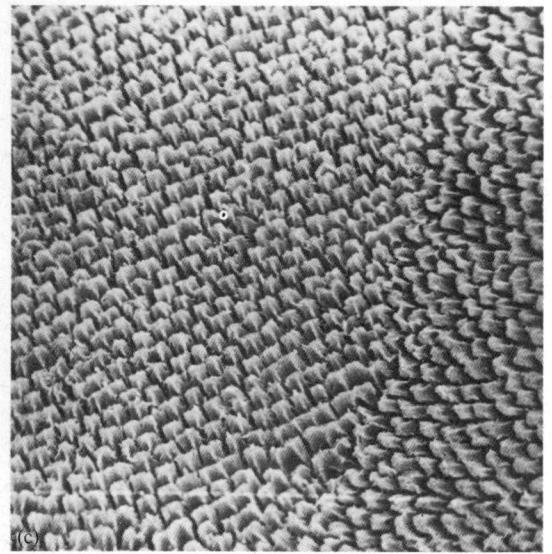

FIG. 13.6. (a) Scanning electron micrograph of EDTA etched African hedgehog enamel showing pattern 1 prisms with a number of circumscribed prism boundaries. Field width 64 μm. (b) Pig molar enamel, EDTA etched, showing longitudinal rows of pattern 2 prisms with longitudinal inter-row sheets of interprismatic regions. Field width 50 μm. (c) Scanning electron micrograph of elephant molar enamel prisms (pattern 3). This is very similar to the type found in man. All the enamel belongs to prisms which have 'tails' corresponding to the interprismatic regions in the other two patterns. Field width 165 μm. (Figures kindly supplied by Dr. A. Boyde.)

7. Innervation of the teeth

The periodontal ligament is well supplied with nerve fibres, which are probably mechanoreceptors. There are brush-like nerve endings in the dental pulp. These extend among the odontoblasts and into the inner dentine. Electron microscopy has shown that nerve fibres enter one in every few hundred dentinal tubules. They lose their Schwann cells and come into direct contact with peripheral fibrocytes and odontoblasts. This gives 50 nerve fibres per square millimetre, which is a high density of innervation. These provide the basis for the notorious pain sensitivity, which is the only

sensation elicited from a tooth. These nerve fibres do not reach to the amelo-dentinal junction which is nevertheless highly sensitive. Some effect therefore must be transmitted along the tubules (Frank, Sauvage, and Frank 1972).

8. The dentition

Rows of tooth germs arise along the lines of the developing upper and lower jaws and give rise to the *milk dentition*. Later in life each permanent tooth is formed by the development of a new rudiment from the stalk of the original ectodermal invagination (dental lamina), and this new bud acquires a dental papilla. The milk dentition contains fewer teeth than the permanent set. The incisors and canines are replaced, as are the anterior of the molariform teeth, the milk-molars. Behind these are the molars, which form only later in life and are not preceded by milk teeth. Thus in man there are 20 deciduous teeth, represented by the formula d.i.$\frac{2}{2}$.c.$\frac{1}{1}$ d.m.$\frac{2}{2}$ on each side. This set is complete by the second year. In the sixth year the first of the permanent teeth appears, a molar, behind the last of the milk-molars. Then, between 6 years and 12 years of age, the roots of all the milk teeth are resorbed and the crown portion is shed, and permanent teeth replace them. In the twelfth year a second molar is added, and finally, sometime after the seventeenth year, a third molar (wisdom tooth) appears. The adult set

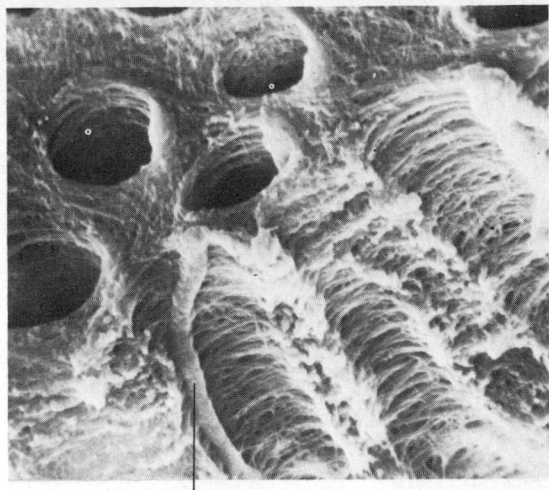

shrunken odontoblast process

FIG. 13.8. Scanning electron micrograph of forming dentine surface with fracture running 'at right angles to surface, showing orientation of collagen fibrils forming the felt-work matrix of the dentine. A shrunken odontoblast process adheres to the wall of one tubule. Field width 17·5 μm. (Figure kindly supplied by Dr. A. Boyde.)

therefore contains 32 teeth: i.$\frac{2}{2}$.c.$\frac{1}{1}$.p.m.$\frac{2}{2}$.m.$\frac{3}{3}$ on each side. This is a more complete set than is found in specialized mammals, such as the rabbit, but the primitive placental dentition probably contained as many as 44 teeth: $2 \times$ (i.$\frac{3}{3}$.c.$\frac{1}{1}$.p.m.$\frac{4}{4}$.m.$\frac{3}{3}$).

The shape of the teeth varies in a regular manner along the row, evidently by a gradation of the morphogenetic processes responsible for production of the hard matter. The shape of the surface of the tooth is determined by that of the inner enamel epithelium of the enamel organ, which falls into folds whose lines control the differentiation of the odontoblasts. Grafts of complete embryonic tooth germs to different parts of the body grow well and form complete teeth, but the dentinal papilla grafted alone produces only irregular bands of dentine. The isolated enamel organ cannot produce enamel; evidently therefore the two parts are mutually dependent. Isolated tooth germs can also differentiate in tissue culture. There is at present no information about the nature of the process that determines the folding of the enamel organ and hence the form of the cusps.

In many species there is no sharp boundary between the anterior, single-cusped incisors and canines and the series of molariform teeth. The more anterior premolars may be quite simple. There is usually an increase in complexity of cusp passing backwards, but there is no constant feature other than the history of the replace-

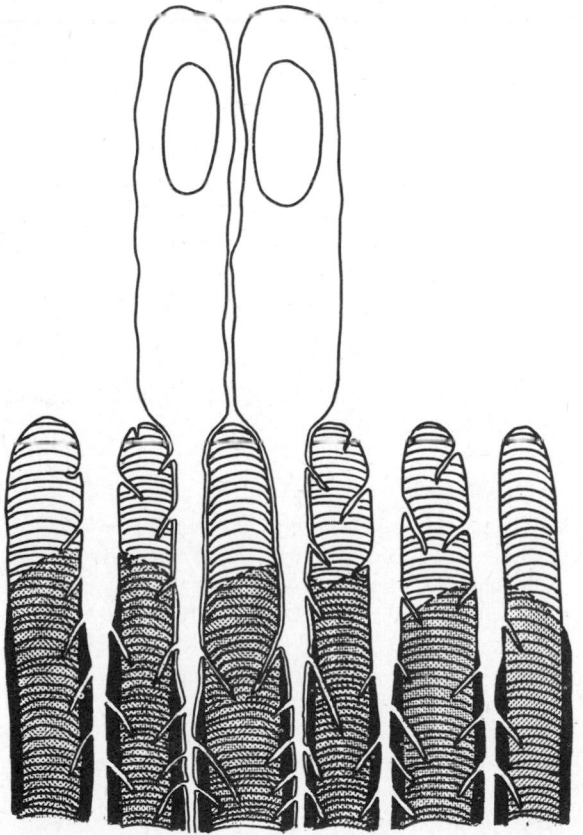

FIG. 13.7. Diagram showing relationship of odontoblasts in dentine formation. The odontoblasts leave a major cell process commonly called a 'Tomes' fibre in the dentine. Dentinal tubules may be partly or completely filled with a densely mineralized substance called peritubular dentine. Unmineralized predentine is hatched, mineralized dentine is hatched and stippled. (Diagram kindly supplied by Dr. A. Boyde.)

ment that distinguishes premolars from molars. The incisors are the teeth that in the upper jaw occupy the premaxillae, the canines being the first teeth of the maxilla. In the lower jaw the canine is the tooth that bites in front of the upper canine. Such definitions are necessary because in some animals these anterior single-cusped teeth are all closely similar, although typically the incisors have sharp cutting edges, whereas the canines form pointed cones for piercing.

9. The evolution of mammalian teeth

The nature of the changes in morphogenesis that have produced the mammalian molar are still obscure, in spite of long controversy. Two types of theory can be distinguished, according to whether the many-cusped molar is supposed to be produced by fusion of a number of reptilian teeth ('concrescence') or by formation of ridges and knobs on the sides of a reptilian cone-like tooth. Perhaps consideration of the processes of tooth development will show that there is less difference between these theories than their protagonists suggest.

In thinking about such questions it is important to bear in mind the elementary point that the teeth are produced by developmental processes! Evolutionary change consists in a change in these processes, so that some new form is produced. The morphogenetic processes that produce the reptilian tooth are extended both in space and in time. Change in the position in which the teeth rudiments form could produce 'concrescence'. Change in the time at which they appear might lead to the same result by forming one many-cusped instead of many single-cusped teeth. Thus the many single-cusped teeth could be replaced by fewer many-cusped teeth by reduction either of the distance between the tooth germs or between the times of their formation, or by reduction in both. In either sense the mammalian molar could be said to be formed by the 'fusion of reptilian teeth'. Some compound molars, those of the elephant, for instance, are formed by the union during development of many, at first separate, cones (Fig. 13.9).

10. The dimer theory

According to the 'dimer' theory of Bolk, the ancestral reptilian teeth each carried a row of three cusps, one behind the other. Such teeth are termed 'triconodont' teeth. Bolk suggested that each mammalian molar was formed by the fusion of two such teeth that came to lie side by side. He believed that there was evidence of this double or dimeric nature of the teeth in a septum sometimes seen in an early stage of the enamel organ. The fundamental pattern of the tooth would thus show

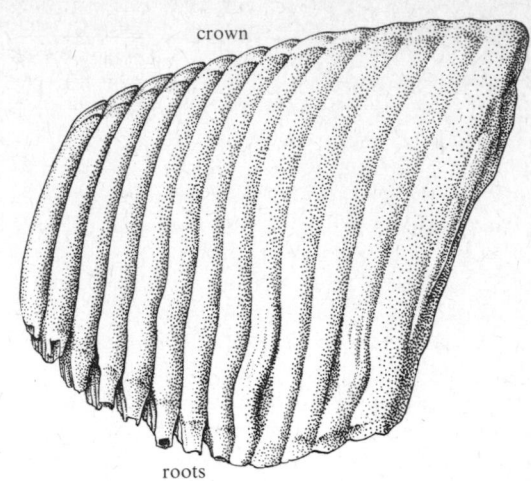

FIG. 13.9. Unerupted molar of an elephant to show formation by union of a number of separate units, each with its own pulp and root. The units are held together by cement. Those most recently formed are to the left.

six cusps, three on the labial and three on the buccal side.

Probably we shall never find out exactly how the embryological processes that made reptilian teeth became changed to give those of mammals. Our only evidence is the study of the embryology of modern forms and the teeth preserved as fossils. The remains of the molars of many of the earliest mammals show three cusps and therefore do not support Bolk's theory, which requires that the most primitive molars should have six cusps.

11. Trituberculy

An important generalization that makes possible an analysis of all cusp patterns was the recognition by the American palaeontologist Cope that the basic arrangement in all early mammals was a triangle of cusps in the upper molars (Young 1962, p. 548). In 1883, Cope pointed out that of 41 species known from the Puerco (Paleocene) period, 70 million years ago, all but four had three tubercles on the upper molars. From this observation the theory of *trituberculy* has been developed by Cope and his successors Osborn, Gregory, Scott, Simpson, and others (Gregory 1934). The theory has had some critics, mainly because of failure to distinguish between the use of morphological, embryological, and palaeontological criteria for its terms, but it constitutes the best means at present available for expressing the historical facts and for naming the cusps.

Observation of the jaws of the simpler types of mammals shows that the upper teeth bite outside the

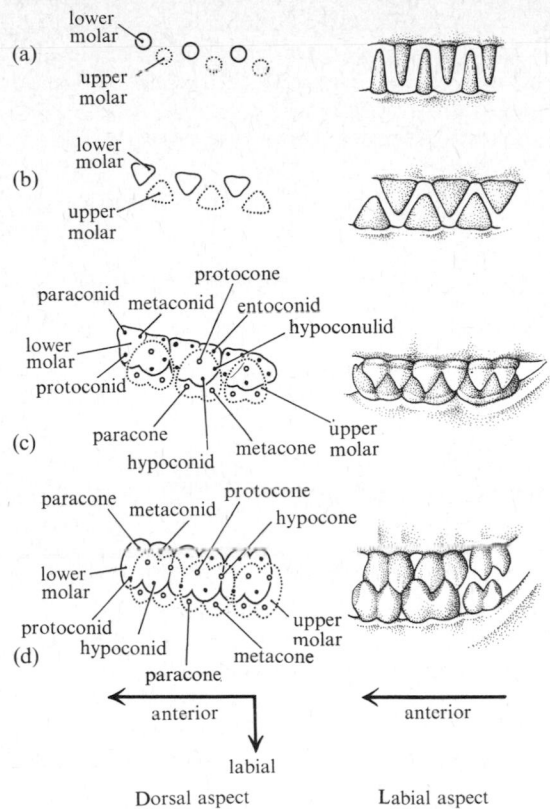

Fig. 13.10. Diagrams to show stages of arrangement of teeth and cusps.
(a) Reptilian pointed teeth with no occlusion. (b) Hypothetical stage in which the teeth have become triangular and rub against each other. (c) *Tarsius*. Upper molars tritubercular, lower tritubercular with a heel. The protocone occludes with the heel. (d) Old World monkey. Upper and lower molars quadritubercular, the hypocone of the upper molar biting between two lower teeth.

lower and are triangular and three-cusped in shape, leaving wedge-shaped spaces between them (Figs 13.10–13.13). This is true of the earliest fossil forms and of modern 'primitive' types such as the opossum and some Insectivora, and also of lower primates (*Tarsius*, Fig. 13.10). The lower teeth also carry triangles of cusps and their surfaces form pyramids fitting into the wedge-like gaps between the upper molars. Each lower molar has in addition a posterior extension or heel (talonid), which occludes (bites) with the apex of the triangle of the upper molar.

Both upper and lower molars therefore carry triangles of cusps, with the apical member of the triangle on the inside in the upper and outside in the lower jaw. Osborn proceeded to give names to the cusps by starting from the assumption that every tooth is based on the original reptilian cone, to which other cusps have been added. He supposed that in the upper molar the inner (palatal) cone, lying at the apex of the triangle, represents the reptilian cone, and accordingly he called it the *protocone*. The outer two cones were supposed to be secondary and the anterior was called *paracone* and the posterior *metacone*.

In the lower teeth the cone at the apex of the triangle, namely, this time, the *outer* (buccal) cone, was supposed to be the original one and was called *protoconid*, the other cusps being named consequently *paraconid* and *metaconid* and the heel called *talonid*. This system of names proved remarkably convenient and it has persisted and been extended. The above cusps can be recognized in the molars of most mammals and others may be added. An internal (buccal) posterior cusp on

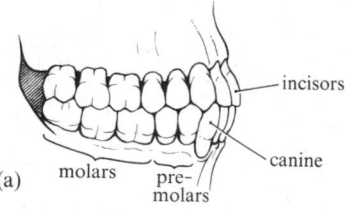

FIG. 13.11. Diagrams of the dentition of man. (a) In profile; (b) upper jaw; (c) lower jaw. In (b) and (c) the cusps are labelled with the nomenclature of human anatomists on the left, comparative anatomists on the right.

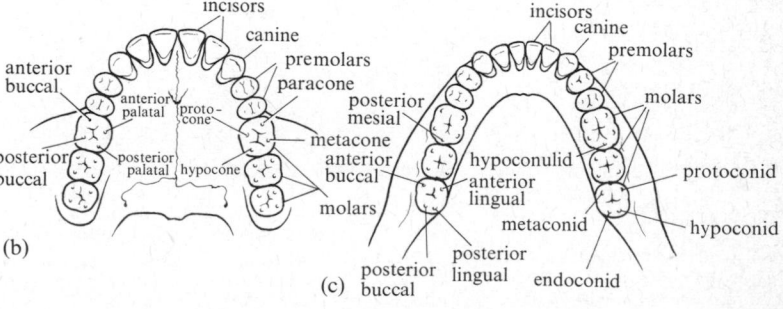

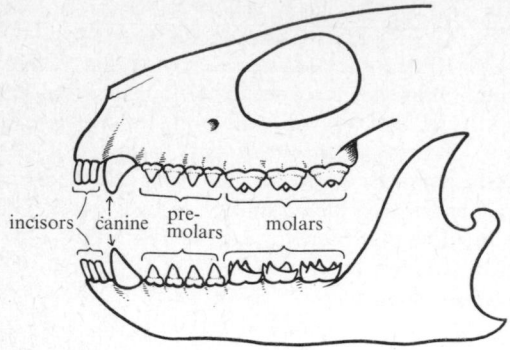

FIG. 13.12. Dentition of a hypothetical generalized eutherian mammal. (After Le Gros Clark (1934). *Early forerunners of man*. Baillière, Tindall, and Cox, London.)

the upper molar, called the *hypocone*, makes the tooth quadrilateral. *Protoconule* and *metaconule* added between the main upper cusps then produce a six-cusped molar, from which many modern types can be derived. Further ridges along the outer side of the tooth are known as styles. In the lower molar *entoconid* and *entoconulid* added to the heel also make a six-cusped tooth.

The tritubercular system of nomenclature lacks generality in that it assigns undue importance to the individuality of the 'cusps' and too little to the distribution of tissue-forming activities that produce them.

Those who support this theory have recognized the triangular pattern in the teeth of the early members of many of the orders of mammals. The relation of these names used by comparative anatomists to those given to the cusps in human anatomy shown in Fig. 13.11 is as follows:

Upper molars

Mesio-palatal	Protocone	
Mesio-buccal	Paracone	original triangle (trigon)
Disto-buccal	Metacone	
Disto-palatal	Hypocone	heel (talon)

Lower molars

Mesio-buccal	Protoconid	remains of original triangle
Mesio-lingual	Metaconid	(trigonid)
Disto-buccal	Hypoconid	
Disto-lingual	Entoconid	heel (talonid)
Distal	Hypoconulid	

12. Origin of the cusp pattern

It is almost certain that the identification of the 'protocone' as the original reptilian cone cannot stand. It was early noticed that if one follows along the tooth row, the cusps of the incisors and canines, which may reasonably be supposed to be the 'primitive' cones, lie in line with the outer cusps, paracone and metacone, of the upper molar (Figs 13.11–13.13). This 'premolar analogy' therefore suggests that these latter cusps represent the original centre of the tooth. This has been

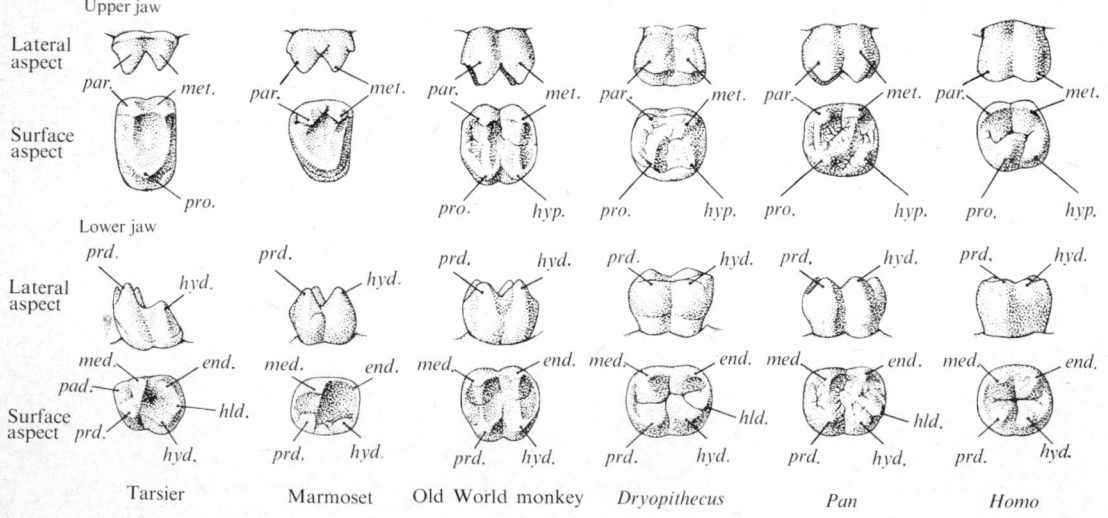

FIG. 13.13. Second molar teeth of various Primates.

end. entoconid; *hld.* hypoconulid; *hyd.* hypoconid; *hyp.* hypocone; *med.* metaconid; *met.* metacone; *pad.* paraconid; *par.* paracone; *prd.* protoconid; *pro.* protocone. (Drawn from skulls except *Dryopithecus*: upper molar from Gregory (1920). *J. dent. Res.* **2,** lower molar from a cast in the British Museum.)

confirmed by the discovery in the Cretaceous deposits of Mongolia of the remains of very early placental shrew-like animals (*Deltatheridium*) in which the centre of the tooth is occupied by a ridge (*amphicone*) that is hardly divided into paracone and metacone, while the 'protocone' is a less conspicuous tubercle on the inner face of the main ridge. In the lower teeth the protoconid may more reasonably be supposed to represent the original cusp.

The evolutionary changes in the arrangement of the cusps were of course connected with changes in the way the teeth were used. The original single-cusped reptilian teeth probably interlocked when the jaw was closed, and were used for piercing the hard cases of insects (Fig. 13.10(a)). The development of the triangular form of the teeth allowed their surfaces to become more closely related and thus to serve to some extent for grinding as well as for piercing (Fig. 13.10(b)). Only when the teeth began to overlap and meet each other in a bite could they serve for grinding. The stages by which this came about are not known with certainty but may have been somewhat as follows. The lower molar developed a heel carrying a hypoconid biting in the centre of the triangle formed by the three cusps of the upper molar (Fig. 13.10(c)). Then the upper molar developed a backward heel, which produced a grinding surface with the paraconid of the lower molar behind it (Fig. 13.10(d)). Thus the effect of the whole change was to convert the teeth from piercing styles into an apparatus capable of cutting and grinding. In herbivorous mammals further modification has been in the direction of producing extensive grinding surfaces, whereas in carnivores the cusps form blades for cutting. In an omnivorous type such as man the low cusps are preserved.

13. Various types of molar teeth

Terms have been devised for certain general types of cusp pattern. When there are separate low cusps (as in man) the tooth is said to be *bunodont*. If they are 'fused', that is to say, joined by intermediate masses of dentine to form loops or ridges, we speak of a *lophodont* dentition (as in the horse). When the cusps are crescentic the tooth is *selenodont* (sheep and cow). The molars of grazing animals often become very large and deep, allowing for grinding and continual replacement and are then said to be *hypsodont*, in distinction from the normal shorter or *brachydont* molars.

14. Teeth of rodents and lagomorphs

In rodents (for example, rats, squirrels, and guinea-pigs) and lagomorphs (for example, rabbits and hares) the teeth are highly specialized, the incisors being used for gnawing and the molars for grinding the food. The characteristic 'rodent' incisors are arranged to give continually sharp edges. Enamel is present only on the anterior (labial) and lateral aspects of the incisor, and cement only on the remaining lingual surfaces. Since enamel is harder than the dentine, a sharp cutting face is always available as the tooth wears away. To replace the wear the tooth grows continually at its base and the curved roots of these incisor teeth occupy a considerable part of the skull. Small second upper incisor teeth are present in the rabbits and hares. The Rodentia possess only a single pair of incisors and in other features shown an extreme specialization for gnawing and for allowing a grinding action of the molars. The division between the Lagomorpha and Rodentia occurred a very long time ago, and the similarities between them are probably due mainly to convergence, rather than common ancestry (Young 1962, p. 652).

The absence of canines and lower incisors in the rabbit makes a big gap or *diastema* in the tooth row, allowing the food to be passed into cheek pockets, which are chambers almost separated from the central cavity of the mouth.

The molariform teeth of the rabbit, six in the upper and five in the lower jaw, are arranged to give a large and continuous grinding surface. They no longer show a triangular pattern of cusps but are crossed by a series of transverse ridges, developed by fusion and extension of the cusps and deepening of the valleys between them, which become filled with cement. Continual chewing of hard grasses wears away the teeth. The three different materials composing them (dentine, enamel, and cement) wear at different rates, and a rough grinding surface is thus maintained. The material is replaced by persistent growth, made possible by the wide, open roots and good blood-supply.

15. The teeth of man

In the early primates the molar teeth resembled those of insectivores and this condition is retained by the living *Tarsius* (Figs 13.12 and 13.13) and in lemurs. Triangular molars are still found in some monkeys (marmosets, Fig. 13.13), but in most of the higher primates the upper molars are quadrangular and carry four cusps. The lower ones lose the paraconid and often also the hypoconulid; they thus also come to have four cusps (Fig. 13.13). In herbivorous monkeys the molars often develop transverse ridges, as in ungulates.

In the anthropoid apes the lower molars still carry a hypoconulid and are five-cusped. The human molars are derived from a similar type, perhaps close to that of *Dryopithecus* found in the Miocene 25 million years ago (Fig. 13.13). In the course of human evolution the hypoconulid has mostly become reduced, producing the four-cusped lower molar, but it usually remains in the lower first and third permanent molars (Fig. 13.11). A characteristic feature of primate dentition has been the development of spatulate cutting incisors, and in man there has also been reduction of the canines towards a similar form. The whole tooth row thus acquires a smooth unbroken outline.

16. The mouth

In mammals the food is more thoroughly treated in the mouth than it is in other vertebrates. Besides the tearing action of the anterior teeth the molars break up food by grinding, and this process is assisted by the action of the tongue, whose roughened upper surface rubs the food against the hard and ridged roof of the mouth.

17. The tongue

The *tongue* is a fleshy mass attached posteriorly to the pharyngeal floor. It contains bundles of striped muscle fibres running in various directions, and these give it the power to mix and manipulate the food. They are operated by the twelfth cranial nerve (hypoglossal). The front part of the tongue is covered by papillae carrying a stratified squamous epithelium from which flakes are continually falling to form the *saliva cells*. Some of the papillae carry the taste buds (p. 409) connected with the seventh and ninth cranial nerves, and the skin of the tongue also contains fibres for the senses of touch and pain. Like other parts of the wall of the mouth and

pharynx the tongue bears numerous mucous and serous glands and also follicles of lymphoid tissue. The mouth is a region that cannot be kept sterile, and it requires elaborate defence against infection. This is provided by the co-operation of the rapidly desquamating skin, secretions of the glands, actions of the lymphoid tissue, and not least by the movements of the lips, cheeks, and tongue, which keep the fluids in circulation, allowing no time for decay. The saliva has an antibacterial action owing to the enzyme *lysozyme* (which also occurs in tears). In spite of these defences, pockets of bacteria continually form in the mouth of man, especially at night, and may produce unpleasant-smelling products and substances with a deleterious action on the teeth causing dental caries.

18. Salivary glands

The rabbit shows a great development of the salivary glands that are characteristic of all mammals. These glands have evolved by increase in number of the gland cells found in the mouth epithelium of lower vertebrates, and they include two types of cell; *serous cells* secreting a mobile liquid and *mucous cells* a more viscous one. These two are present in varying proportions in different glands and animals (Fig. 13.14). Surrounding the secretory alveoli there are basket-like *myoepithelial cells*. These are innervated by parasympathetic fibres and contract when these are activated, squeezing out the secretion. In man it is possible to recognize three pairs of salivary glands, *parotid*, *submaxillary* (also called *submandibular*), and *sublingual*; in the rabbit

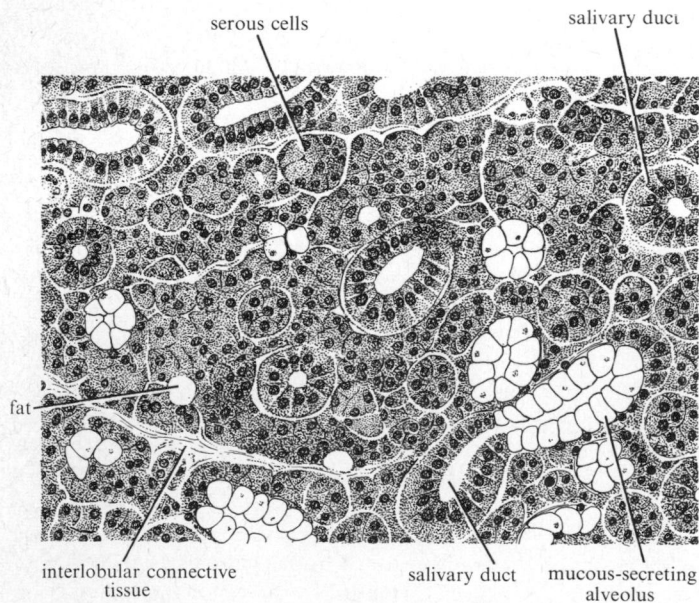

serous cells   salivary duct

fat

interlobular connective   salivary duct   mucous-secreting
tissue   alveolus

Fig. 13.14. Section of human submandibular gland, showing mixed serous and mucous cells. (From a preparation and photograph by Mr. K. C. Richardson.)

there are other large masses of cells in addition to these. All are caused to secrete by the action of a reflex originating from the presence of food in the mouth. The motor path to the serous cells runs through the preganglionic, parasympathetic fibres in the seventh and ninth cranial nerves and the postganglionic fibres pass to the glands from the otic and submandibular ganglia (p. 283).

Electrical stimulation of the chorda tympani nerve therefore produces a flow of watery saliva. Stimulation of the sympathetic pathways to the glands is followed by secretion of a concentrated solution of mucin; this is often held to show that the mucus-secreting cells receive motor nerves from the sympathetic system, but other interpretations are possible, since these sympathetic nerves control the blood-supply of the glands (p. 284).

Secretion of saliva is easily influenced by afferent stimuli from other sources besides the mouth and the famous experiments of Pavlov showed that impulses from almost any part of the body can be 'conditioned' to give salivary secretion if they are made to arrive shortly before, or together with, impulses in the 'unconditioned' afferents for the reflex (see p. 341).

The saliva serves to moisten the food and its mucin to lubricate it and make its passage easy. In addition it contains an amyloytic (starch-splitting) enzyme *ptyalin*, which, at the nearly neutral pH of the saliva, acts on starch. This can be followed by chewing potato and watching the changing action of the products on iodine. Instead of the blue colour produced by the starch itself there is first a red colour, due to the intermediate product erythro-dextrin, and then no colour when all the starch has been completely changed to maltose. The action of the ptyalin continues in the centre of the bolus of food after it reaches the stomach, until the acid there stops it. Evidently this mechanism for the breakdown of starch soon after the food has been eaten forms an important part of the digestive system of mammals.

14 Digestion

1. The dissolution of food materials

THE process of breaking up the food is begun in the mouth by the teeth, tongue, and by *ptyalin* in the saliva. The rest of the alimentary canal is concerned with completing the physical break-up of the tissues that have been eaten and the absorption from them of the molecules needed by the mammal.

The dissolution of the food is accomplished chemically by the combined effects of enzymes secreted by the gut and its glands and the action of symbiotic bacteria living within the gut. The high temperature of the body assists the action of the enzymes, and the gut becomes a great vat in which the macerated food is quickly acted upon by the enzymes and bacteria. The most striking special developments of the mammalian plan of digestion are indeed not regions of special enzyme production but large chambers harbouring symbionts, such as the elaborate 'stomach' of sheep, cows, and other ruminants, or the caeca of the large intestine of rabbits and horses. The movement of food along the gut and the secretion of the glands are so regulated as to keep the mass in the right physical and chemical conditions for these processes. The movement of *peristalsis* by which the food is propelled operates rhythmically and continually, rather as does the heart-beat (p. 198). It is regulated by the operation of receptors in the wall of the gut, which signal whether conditions are favourable for digestion, setting up reflexes that speed or slow peristalsis. Chemical signalling by hormones is also employed (p. 454).

Thus the regulation of digestion is ensured mainly by a series of mechanisms laid down by heredity. Like the food intake it is influenced also by the operation of higher nervous centres in the medulla oblongata, hypothalamus, and cerebral cortex. Electrical stimulation of the hypothalamus, or of parts of the frontal lobes of the cerebral cortex, produces changes in gastric and intestinal motility. Following removal of parts of the frontal lobes from monkeys it was found that the stomach becomes more active and excessively large amounts of food are consumed; similar symptoms may appear in man after injury or operations on the frontal lobes.

The processes of digestion thus also come under the influence of the higher cerebral centres, whose actions are influenced by memories set up during the lifetime of the individual. In man digestion may be profoundly influenced by learning processes. When habits of food consumption or digestion are interfered with, the balance of the organism may become severely disturbed.

2. Structure of the gut

The gut is a tube of varying width, extending from mouth to anus and including oesophagus, stomach, small intestine, caecum, large intestine, and rectum (Fig. 14.1). As the food mass passes along the gut it meets the characteristic secretions of each part and these, together with the fermenting action of the bacteria that live in some parts, change the food and make its constituents available for absorption. The *oesophagus* is a simple passage through which the food passes within a few seconds after eating. The *stomach* is a much wider sac where the food is received. Here it is sterilized by mixing with acid, a most important provision, especially for a warm-blooded animal. Here also the enzyme pepsin begins to act on the proteins. The *small intestine* is a long narrow tube where the food is mixed with enzymes and quickly broken down into simpler constituents, which are then absorbed into the blood-stream. The reaction of this region is only faintly acid, and its bacteria are 'fermentative', that is to say, they act upon carbohydrates and produce organic acids. The following *caecum* and *large intestine* have nearly neutral contents and are highly charged with bacteria, some of which are 'putrefactive' and act on proteins. In the rabbit there are other bacteria here that play an important part in breaking up refractory foodstuffs such as cellulose. An important function of the large intestine is the absorption of water from the gut contents. Residual matter is then extruded at the anus—in man, some 24 hours after eating.

The various portions of the canal therefore have

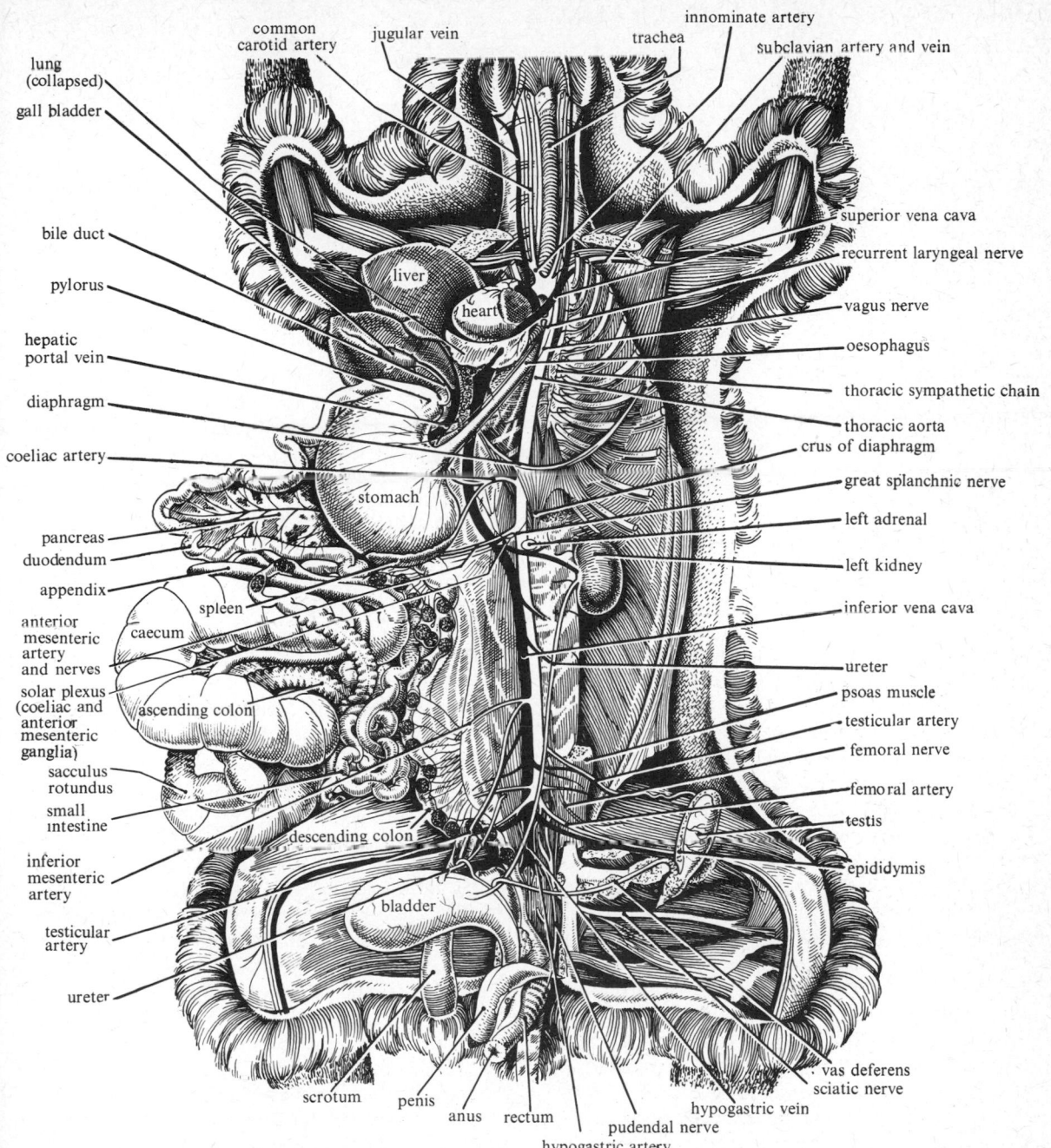

FIG. 14.1. Drawing of dissection of viscera of a rabbit.

different roles in the mixing, movement, and absorption of the actively fermenting mass that continually passes through the body. Nevertheless, all parts conform in structure to a common plan (Fig. 14.2). There is a central lining or *mucosa*, derived from the endoderm, and outside this is a thin layer of muscle, the *muscularis mucosae*. Below the mucous membrane the wall is composed of a layer of connective tissue, the *sub-*

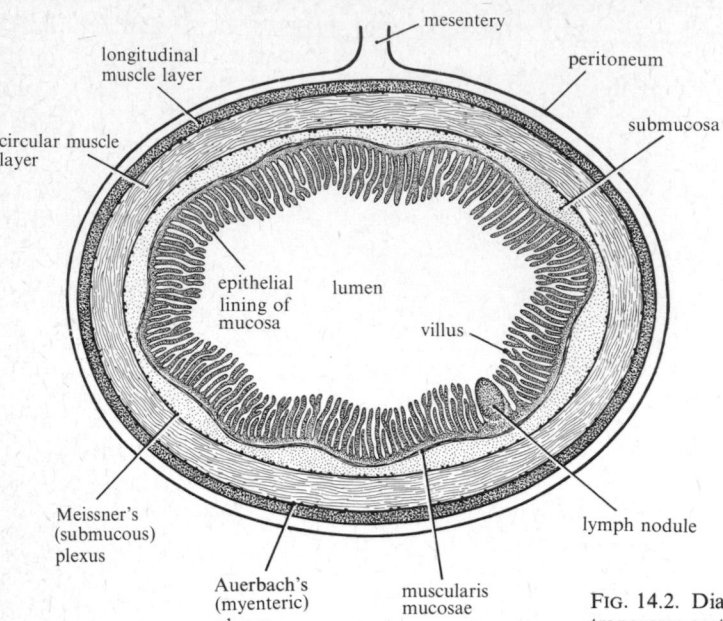

longitudinal muscle layer

circular muscle layer

mesentery

peritoneum

submucosa

epithelial lining of mucosa

lumen

villus

Meissner's (submucous) plexus

Auerbach's (myenteric) plexus

muscularis mucosae

lymph nodule

FIG. 14.2. Diagram of structure of intestine as seen in transverse section.

mucosa, and outside this are layers of *circular* and *longitudinal muscles*. The outer surface of the oesophagus is attached by connective tissue to other structures, but most of the remaining portions of the gut are covered by a layer of flattened mesothelial cells, making a smooth serous membrane, the *peritoneum*, which allows free movement within the peritoneal cavity. The gut is suspended from the dorsal wall of the coelom by a fold, the *mesentery*, which consists of connective tissue covered with mesothelium and carries the blood and lymph vessels and nerves to and from the gut.

The inner surface of the mucosa is folded in the small intestine into *villi*. Some of the lining cells secrete mucus, and there are other glands that have developed from the endodermal epithelium by sinking more or less deeply into the wall of the gut or even outside it like the pancreas. Secretion is controlled by the nerve fibres of the submucous or *Meissner's plexus*, lying between the muscular and mucous layers and connected with the sympathetic and parasympathetic systems (p. 283). Between the two main muscle layers is the *myenteric plexus of Auerbach*, which controls the main movements of the gut and is of course also part of the autonomic nervous system.

3. Movements of the gut: peristalsis

The movements of the gut show a certain similarity throughout, but characteristic features appear in each region. The food is moved along by the process of *peristalsis*, which consists essentially in a contraction of the circular muscles of the wall of the gut at and behind (oral to) any point distended by the presence of a bolus of food. The effect of this is to push the food onwards towards the anus. Bayliss and Starling described the movement by saying that the 'law of the intestine' is constriction on the oral side and relaxation on the anal side of any stimulated point. The contraction is usually more prominent than the relaxation. Peristalsis is well marked in the oesophagus and stomach. In the latter it occurs during digestion soon after a meal and probably begins again some hours later as the *hunger contractions*. These are accompanied by characteristic sensations, presumably owing to stimulation by the movement of receptor organs lying in the stomach wall. The afferent impulses from hunger contractions probably serve, at least early in life, simply to activate the higher centres much as do pain impulses. Thus they initiate the crying of an infant that has never yet been fed. The tendency to stomach peristalsis is therefore inborn and, far from being 'reflex', is itself a powerful initiator of action. It is, of course, possible that the actual contractions are initiated by the action of the gastric secretions, but either these or some previous stage in the chain of events must be inherent in the system rather than reflexly induced.

Peristaltic waves occur in the small intestine but here we find in addition the *segmenting movements*, which are simple non-travelling constrictions by which the food is churned and mixed. In the intestine there are also marked *pendular movements*, contractions primarily of the longitudinal muscle, shortening a loop of the intestine and throwing the chyme from one end to the other, mixing it as in a cocktail shaker. These movements occur rhythmically at a frequency as high as 20 per minute in the duodenum of the rabbit during certain phases of digestion.

The large intestine shows relatively infrequent movements, the contents being distributed and divided by a series of segmenting constrictions. Peristaltic waves occur and, in man, may take the form of mass peristalsis in which one large wave sweeps the contents forward into the rectum. The latter is normally empty, and its filling by mass peristalsis provides the stimulus for defaecation.

The movements of the gut are under the influence of the central nervous system, acting through the autonomic nervous system, but many of them are able to continue after isolation. The segmenting and pendular movements of the small intestine continue in strips of muscle from which all nerves have been removed. Peristalsis depends on the presence of the nerve plexuses and is usually increased by action of the vagus, decreased by the sympathetic nerves (see p. 284), though

it can continue in absence of connexion with both of these. Movements of the stomach have been shown to be influenced by stimulation of the hypothalamus and even of the frontal region of the cerebral cortex (p. 327).

4. The oesophagus

The bolus of food is passed from the mouth through the pharynx to the oesophagus by raising the tongue and contraction of the muscles of the palate, the soft palate being raised to close the nasopharynx and the hyoid bone to close the larynx as described on p. 115. The oesophagus is a straight tube with some striped and unstriped muscle in its wall and a lining of stratified squamous epithelium, with some mucous glands in man. Its muscles push the food along by an involuntary movement of peristalsis, by which the food is passed through the neck and thorax to the stomach.

5. The stomach

The stomach is a portion of the alimentary canal formed by special development of the lower end of the oesophagus; it is not found in the lowest vertebrates (lampreys). In mammals it shows its affinity with the oesophagus to various degrees in different species. In

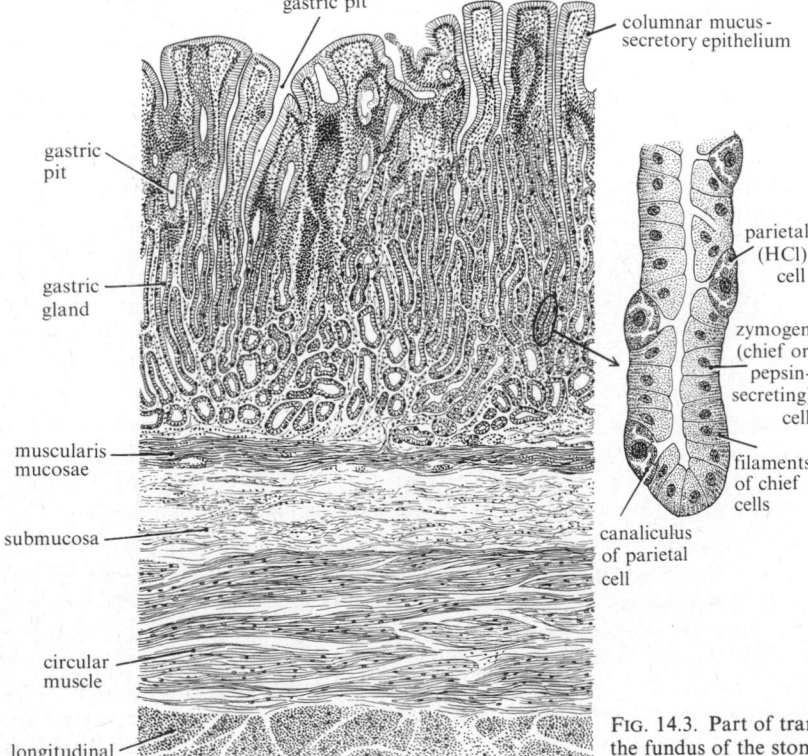

FIG. 14.3. Part of transverse section of gastric mucosa, from the fundus of the stomach of a cat. (From a preparation and photograph by Mr. K. C. Richardson.)

its most developed parts (Fig. 14.2) it has a lining of high columnar mucus-secreting cells, folded to form myriads of deep pits, into the bases of which open the ducts of long tubular *chief glands*. These glands are lined by two types of cell, the *pepsinogen-secreting* or *chief cells* and the larger *parietal* or *oxyntic* cells, which produce hydrochloric acid (Fig. 14.3). In fishes the pepsin and acid are both produced in a single type of cell, and the mammalian condition may be said to show a more differentiated or 'advanced' state in this respect. The oxyntic cells produce a fluid of pH far lower than anywhere else in the body by actively secreting H^+ and Cl^-. The OH^- left in the cells is neutralized by H^+ ions derived from carbonic acid, the HCO_3^- being passed into the blood-stream. There is much carbonic anhydrase in the oxyntic cells, producing the carbonic acid from CO_2 and water. The openings of the chief glands in mammals are guarded by special mucus-producing '*neck cells*' not present in fishes. They also produce a mucopolypeptide known as 'intrinsic factor', which is necessary for the absorption of vitamin B_{12} from the food.

On account of the presence of the gastric glands the mucosa of the stomach is very thick. It is renewed about every 3 days (in the rat) by new cells formed at the base from division of mucous neck cells. The division rate shows a daily periodicity (high in the morning) and is influenced by the hypophysis and adrenal and by the food intake (Gregory 1965).

The stomach is a bag whose shape varies in different mammals. It shows an upper *cardiac region*, close to the opening of the oesophagus, then a main portion, the *fundus*, and finally a *pyloric region*, close to the *pyloric sphincter*, which guards the opening to the duodenum. The cardiac region usually contains only mucous glands. The fundus has the typical stomach structure with 'chief glands'. In the pyloric region the glands have no parietal cells; the juice they produce contains no acid and is mostly mucus. In some mammals a large part of the stomach has the structure of the cardiac part, that is to say resembles the oesophagus. For instance, the rumen of sheep and cows is a large chamber of this type, in which the food ferments under the action of symbiotic bacteria (Young 1962, p. 744).

In man, the stomach varies in shape but is generally elongated, often turning up at its lower end, as a letter **J**. Food enters it almost immediately after swallowing, being only briefly retained by a weak *cardiac sphincter* at the entrance. Waves of peristalsis pass along the human stomach at the rate of about 3 every minute. With the arrival of each wave at the pylorus the latter opens and some of the food is passed into the duodenum. Some material therefore passes through the stomach within a few minutes after the beginning of a meal, especially if the food is liquid; solid food remains in the stomach for as long as 4 hours. While it is there it is acted upon by the *pepsin* of the gastric juice. This is formed from the precursor, pepsinogen, a protein of molecular weight *ca.* 42 000, which loses about 9 peptide bonds in the presence of acid to produce the active enzyme. This attacks nearly all proteins, at an optimum of pH 2, breaking them into polypeptides with about 7 amino-acid residues. These molecules are not absorbed, until they have been further broken up in the intestine. Little food is absorbed from the stomach, but under some conditions small molecules, for example, those of alcohol, may pass through its wall and quickly enter the blood-stream. Another action that goes on in the stomach is the clotting of milk, by means of an enzyme *rennin*, which is particularly active in new-born ruminant mammals.

The secretion of the gastric glands of the stomach is controlled partly by a reflex action, whose afferent pathway may be the sight or smell of food or the presence of food in the mouth or stomach itself, the efferent pathway being through the vagus nerve. In addition, the presence of food in the stomach causes liberation into the blood of a hormone that excites secretion of gastric juice, *gastrin*. This is a polypeptide with 17 amino acids whose COOH-terminal tetrapeptide sequence Tyr-Met-Asp-Phe-NH_2 causes gastric secretion, but only if the terminal acid is present.

6. The small intestine

In the stomach the food is churned up into a paste of mucus and enzyme known as *chyme*, ready to be passed in a liquid condition into the duodenum.

A sharp change takes place in the process of digestion as food passes the pyloric sphincter and enters the intestine. This corresponds to the point at which the mid-gut proper began in the lowest vertebrates, and here the breaking up of all the various types of foodstuff proceeds actively. The stomach, for all its importance, can be considered as a bag to contain the food that has been taken in. Until the food passes the pylorus, the only important enzymatic actions upon it are those of the ptyalin and pepsin, and little absorption has taken place. Beyond the pylorus the food is mixed with enzymes responsible for breaking down all the chief classes of foodstuff and absorption of the products into the blood-stream begins.

The gut distal to this point is essentially a tube with the basic plan already described, namely a mucous lining, connective tissue submucosa, two layers of muscles, and an outer covering of peritoneum. The intestine may be divided into two main sections: the

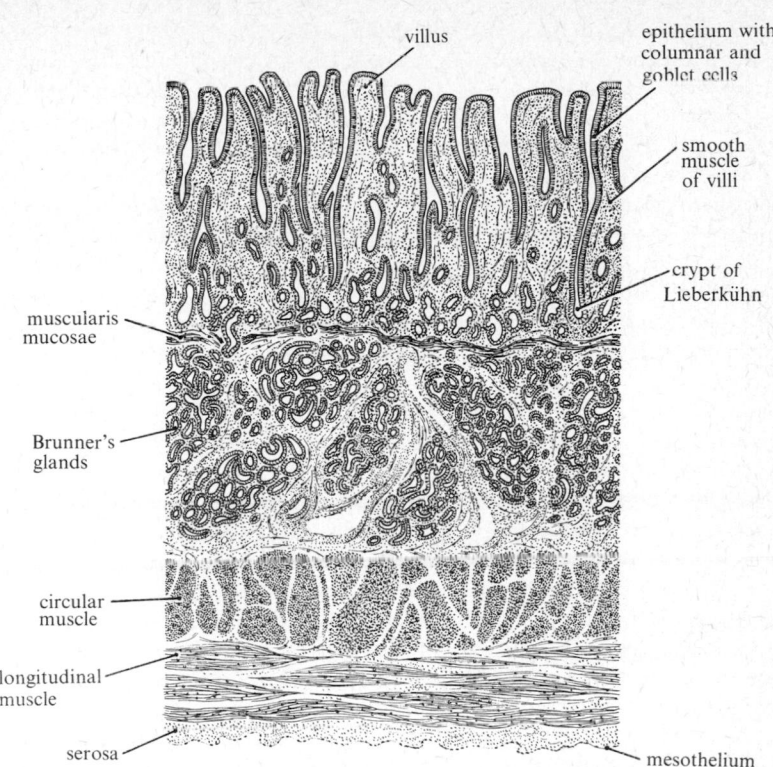

villus

epithelium with columnar and goblet cells

smooth muscle of villi

crypt of Lieberkühn

muscularis mucosae

Brunner's glands

circular muscle

longitudinal muscle

serosa

mesothelium

FIG. 14.4. Part of longitudinal section of human duodenum.

small intestine, in which the breakdown of foodstuffs is completed and absorption takes place, and the large intestine, which is a region concerned with the absorption of water from the food mass. Between these two great divisions of the intestine there lies a blind diverticulum, the *caecum*. This is small in man but very large in the rabbit and some other herbivorous mammals, serving as a receptacle in which the food stagnates while cellulose and other constituents refractory to the action of the animal's own juices are broken down by the enzymes of symbiotic bacteria.

The first part of the small intestine is the *duodenum* (Figs 14.1 and 14.4), which receives the juices of the liver and pancreas and has itself a very glandular wall. In the rabbit it makes a U-shaped loop within which lies the *pancreas*. The bile-duct enters at the upper end of the duodenum. The pancreatic duct in the rabbit enters at the distal loop, about 10 cm lower down. The chyme coming from the stomach is therefore mixed with bile and the secretions of the special duodenal glands (Brunners' glands) before it receives the pancreatic enzymes. In man, the bile-duct and pancreatic duct enter the duodenum together, shortly beyond the pylorus. Distally the duodenum passes without abrupt change into the second part of the small intestine the *ileum*. In some animals the small intestine below the

duodenum can be divided into *jejunum* and ileum, the former having a thicker, more vascular wall.

The mucosa of the whole small intestine is thrown into characteristic finger-like villi, which are continually contracting and re-expanding and serve to increase the absorptive surface and to move the liquid in contact with it. The surface of the extremities of the villi is covered by columnar cells, having on the face turned towards the intestinal contents a series of microvilli. These appear under the light microscope as lines running at right angles to the surface. The intake of food molecules is at least partly an active process. The outer region of the cell is rich in the enzyme phosphatase which is probably part of a system for making the necessary energy available. Mucus-secreting cells are also present on the villi and are known from their form as *goblet cells*.

The depressions between the villi are known as the *crypts of Lieberkühn*, and they are lined by columnar and goblet cells, together with other cells also apparently secretory, the *argentaffin* cells, rich in serotonin, and the *Paneth cells*, which are probably responsible for the production of some intestinal enzymes. The cells at the bases of the crypts show numerous mitoses and are continually replacing the lining of the intestine, the epithelial cells being continually shed. The entire epithelial lining of the human duodenum is renewed every 2 days. In the duodenum there are further numerous *pyloric* or *Brunner's glands*, lying deeper within the submucosa and having a structure and section similar to those of the glands at the pyloric end of the stomach. This secretion is alkaline and contains mucus but little enzyme. The mucous membrane of the intestine obviously requires protection against foreign invaders and is richly supplied with lymphoid tissues. In the ileum this forms large masses, the *Peyer's patches*. Lymphocytes are frequently seen migrating through the intestinal epithelium and apparently entering the lumen.

As the chyme enters the duodenum it is mixed with bile and the *succus entericus* from the intestinal wall and Brunner's glands. These render the chyme less acid and provide powerful enzymes, so that the whole becomes a copious liquid mass of rapidly changing composition. In spite of the alkaline juices added to it the reaction of the duodenal contents is usually acid, about pH 5·5 in man. The pancreatic enzymes therefore do not work at their optimum, which is alkaline. The lower regions of the small intestine are also acid, as a result of the presence of organic acids produced by bacteria. The contents of the large intestine are usually slightly acid or neutral.

7. The pancreas

The *pancreas*, occupying the space between the limbs of the duodenum, is a branching set of tubes ending in blind secretory sacs or *acini* (Fig. 14.5). The cells of these secrete the pancreatic enzymes. They contain much basophil material in the cytoplasm, which the electron microscope shows as rough endoplasmic reticulum. This becomes labelled within 5 minutes of an injection of radioactive leucine. Later the activity appears in the Golgi region (smooth reticulum) lying more peripherally, and in 12 minutes is found in the granules of zymogen, which contain the precursors of the pancreatic enzymes and are discharged into the ducts (Fig. 14.6). Among the alveoli lie a number of masses of cells, the *islets of Langerhans*, constituting an endocrine organ that is responsible for secreting into the blood the hormone *insulin*, whose presence is necessary to allow the tissues to utilize glucose (p. 432).

Control of secretion of the pancreatic juice is both hormonal and nervous. In 1902 Bayliss and Starling showed that the presence of food in the duodenum excites the production of the hormone *secretin*, which circulates to the pancreas and excites its secretion. This was the first clearly demonstrated case of chemical action at a distance in the body and its discovery led to the introduction of the word 'hormone' and to the development of the concept of endocrine secretion (p. 454). Secretin, like gastrin, is a polypeptide, with 28 amino-acid residues. Stimulation of the vagus also excites secretion of pancreatic juice in small amounts but especially viscous and rich in enzymes.

8. Digestion in the intestine

The pancreatic juice has a wide variety of actions, as its name implies. It is an alkaline liquid, containing the

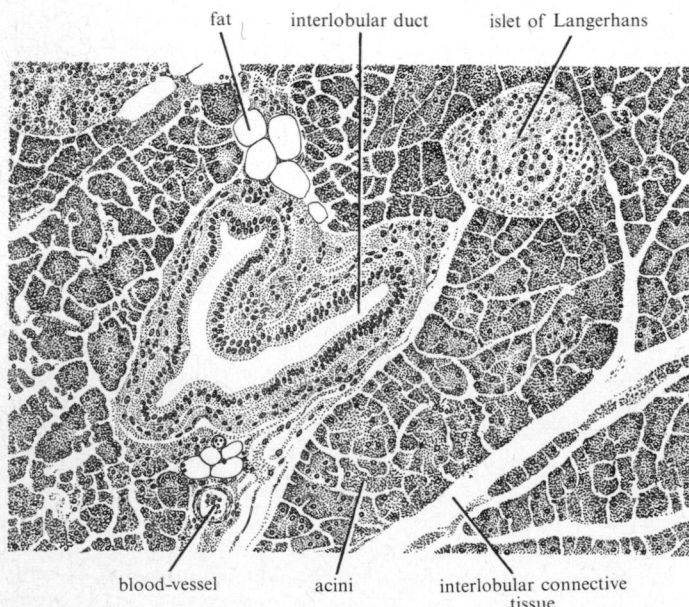

fat   interlobular duct   islet of Langerhans

blood-vessel   acini   interlobular connective tissue

FIG. 14.5. Section of the pancreas of the guinea-pig. (From a preparation and photograph by Mr. K. C. Richardson.)

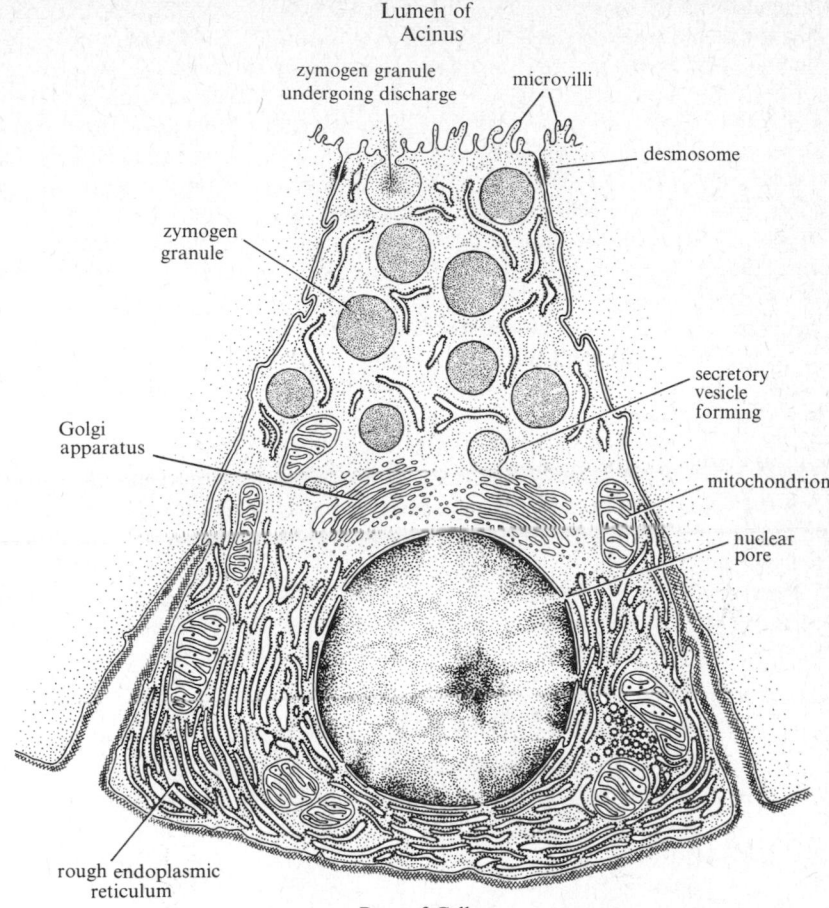

Lumen of
Acinus

zymogen granule
undergoing discharge   microvilli

desmosome

zymogen
granule

secretory
vesicle
forming

Golgi
apparatus

mitochondrion

nuclear
pore

Fig. 14.6. Diagram of pancreatic acinar cell. (From Ham
(1974). *Histology*. J. B. Lippincott, Philadelphia.)

rough endoplasmic
reticulum

Base of Cell

protease trypsin, also amylase, lipase, maltase, and the clotting enzyme rennin. The *trypsin* is secreted in an inactive form *trypsinogen*, which is readily activated in the intestine by an enzyme *enterokinase*, which removes a hexapeptide. Acting in alkaline solution trypsin splits proteins or polypeptides in the middle, being like pepsin an *endopeptidase*. The chymotrypsins are other endopeptidases, acting specifically on certain bonds. These actions produce fractions that contain two amino acids (dipeptides) or at most a few amino acids. These are then acted upon by a further set of enzymes the *exopeptidases*, which split off single amino acids, either that next to the COOH- (*carboxypeptidases*) or the NH$_2$-group (*aminopeptidases*). The cells shed by the walls of the intestine also contribute enzymes (cathepsins). The products of protein breakdown are absorbed by the intestinal cells, either as amino acids or small peptides, the latter being further broken down there by the contained peptidases, since only amino acids appear in the blood-stream. Absorption and transfer

are active processes in the course of which amino acids are moved against concentration gradients. The *pancreatic amylase* is more powerful than the amylase in the saliva and rapidly breaks down either boiled or unboiled starch. The digestion of starch and other carbohydrates is completed by enzymes of the succus entericus, such as *maltase* and *sucrase* (= invertase); the end-products that are chiefly absorbed are monosaccharides such as glucose.

For the digestion of fats, the pancreatic lipase is the most important agent; it splits them into glycerol and fatty acids, which are absorbed. Owing to the high surface tension, fat suspended in water tends to form large globules which would not be easily acted upon by the enzymes. The breaking up of these to make an emulsion of fine globules is largely the work of certain substances in the bile secreted by the liver, such as sodium glycocholate, whose parent acid is formed from the amino acid glycine, and cholic acid. These *bile salts* have great powers of reducing surface tension and thus of emulsifying the fats. In some forms of jaundice the bile-duct is blocked by infection or the liver otherwise disturbed; fat is not èmulsified, and foods containing it therefore remain undigested.

The fatty acids absorbed mostly do not reach the blood-stream directly but pass either through or between the intestinal cells into the lymph spaces lying behind. In the cells fats are resynthesized and discharged into special lymph vessels, the *lacteals*, hence ultimately reaching the thoracic duct (p. 184). If an animal is dissected soon after taking a fatty meal the mesentery will be seen crowded with rows of these beautifully white lacteal channels.

The secretions of the duodenal wall, pancreas, and liver therefore provide a number of enzymes able to split many types of ingested matter into products useful for the body. In addition to the enzymes mentioned there are also others in the succus entericus able to act on nucleotides and other materials, making them available for the body.

The food is passed from the stomach into the duodenum as fast as it can be acted upon in the latter. The pylorus is a thick ring of circular muscle that relaxes each time that a peristaltic wave arrives from the stomach, provided that the contents of the duodenum are not too acid. Probably there are receptors in the duodenum that respond to the presence of the acid, and their discharge reflexly inhibits the opening of the pylorus, thus ensuring sufficient neutralization of the stomach acid.

Food is passed along the intestine by waves of peristalsis, and it is mixed, shaken, and churned by the segmenting and pendular movements. During passage along the duodenum the processes of breakdown are completed and most of the absorption takes place in the jejunum and ileum.

9. The liver

To provide for the absorption of the food products, the intestine has a large blood supply from the branches of the *superior mesenteric artery*, which leaves the aorta close behind the coeliac artery and sends many branches spreading out through the mesentery. From the intestine the blood does not pass directly back to the heart

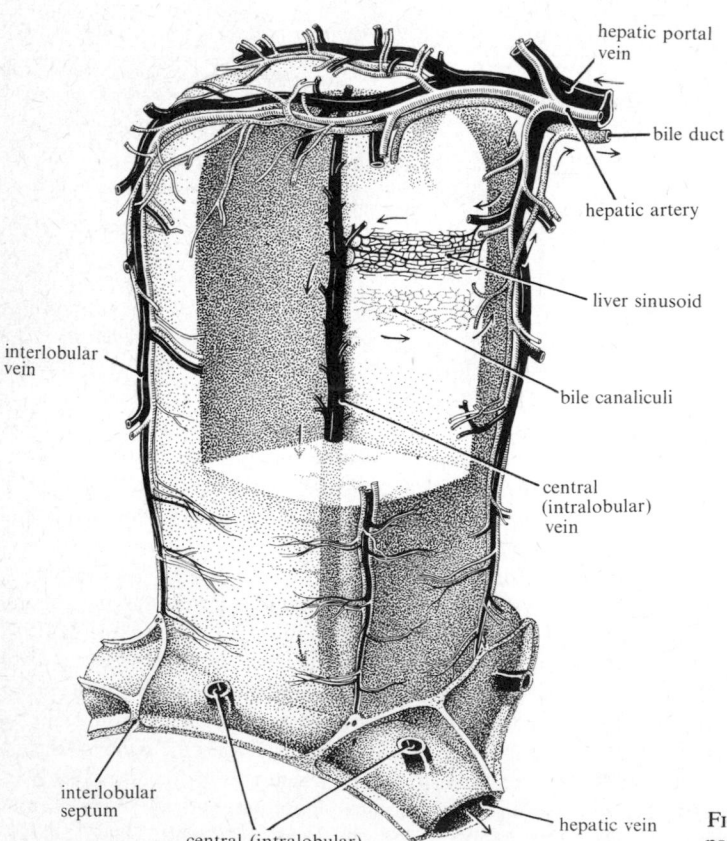

FIG. 14.7. Diagram of the arrangement of the parts of a lobule of the liver of a pig. (From a wax model, after Braus.)

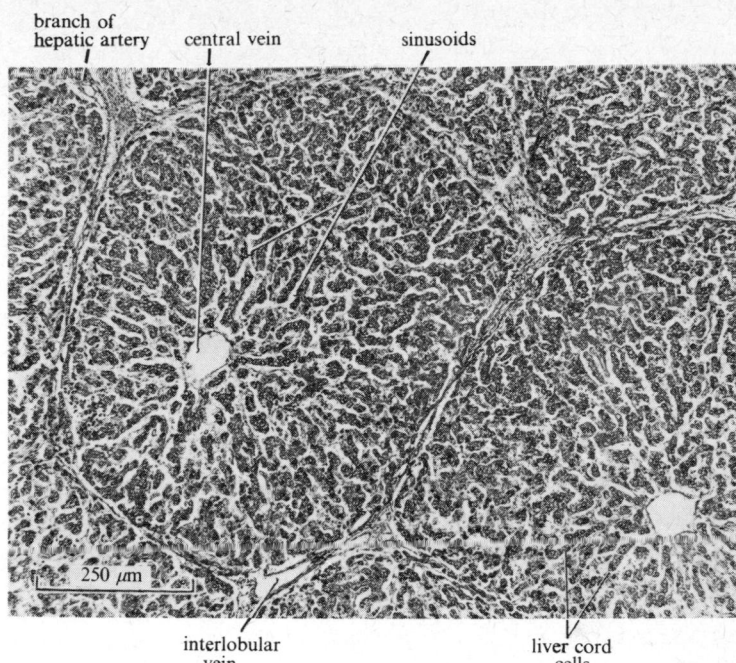

branch of hepatic artery   central vein   sinusoids

250 μm

interlobular vein   liver cord cells

FIG. 14.8. Light micrograph of pig liver lobules.

but to the liver, by the *hepatic portal vein* (Fig. 14.1). This enables the liver to perform those of its functions that are concerned with conversion and storage of the food taken in. Perhaps at an early stage of evolution the liver may have served directly as an organ for breaking down food. In *Amphioxus* the liver is represented by a hollow diverticulum, which secretes digestive enzymes (Young 1962, p. 32). This strictly digestive function of the liver is still represented in mammals by the action of the bile in emulsifying fats (p. 150). But the main action of the liver is to convert the foodstuffs into whatever products are needed by the body.

The liver develops as an outgrowth of the gut, but comes to form a compact mass of gland cells, *hepatocytes* (Figs 14.7 and 14.8). These are usually described as being arranged as plates, mostly one cell thick and organized into lobules (Fig. 14.8). Blood flows in sinusoids between the plates from branches of the portal vein around the outside (hence *interlobular veins*) in towards the *intralobular veins* at the centre, which are branches of the hepatic vein. The branches of the hepatic artery also run between the lobules and discharge into the sinusoids. The external secretion of the liver is discharged into minute *bile canaliculi* lying within the plates between the liver cells. These drain towards the outside of the lobule into ductules, which unite to form the bile duct. Thus blood and bile run in opposite directions in the lobule. A complication is that there are holes in the plates, so that the liver sinusoids

anastomose. Moreover, the vascular capillaries have a discontinuous endothelial lining, so that the blood plasma comes into direct contact with the cells. There is debate as to how this arrangement is arrived at from the original liver diverticulum.

The liver is an organ with many functions, which may be grouped under five headings: (a) those concerned with metabolism and storage of the food; (b) production of the bile; (c) activities concerned with the composition of the blood (such as production of plasma proteins including those needed for clotting), regulation of blood sugar and other substances; (d) protective activities and detoxification; (e) formation and destruction of red cells. Under several of these headings we could recognize subdivisions, and it is clear that the liver occupies a very important place in vertebrate life; its cells, primarily digestive in function, have come to constitute a chemical workshop for the body. It has been said that the liver has over 100 functions but only two types of cell, hepatocytes and macrophages. However, the former, although they all look alike, probably include many different clones, each specialized to produce certain enzymes.

With all these activities the turnover of proteins is very rapid in the liver. The cells themselves last for about five months (in a rat) but the half-life of total liver protein is only 2–4 days. The plasma proteins produced by the liver turn over with a half-life in man of about 20 days. The level is very precisely regulated;

greater protein degradation (outside the liver) is compensated by increased synthesis, but the mechanism of control is not well understood (see Tavill 1972).

The actions of the cells of the liver on the food are varied and depend upon the condition of the animal at the time (Rouiller 1963). Thus if amino acids are needed for building the body they are passed on in the blood-stream, but in an adult more amino acids are usually taken than are needed for building and the excess is *de-aminated* in the liver, the nitrogen being converted into urea which is passed to the kidney for removal, while the remaining parts of the molecules are available for combustion to provide energy. Similarly carbohydrate may be either passed on or converted into glycogen, in which form it is stored in the liver.

The liver plays an essential part in providing a continual supply of glucose for the support of the metabolism of the muscle, brain, and other organs. This glucose is formed by the process of *glycogenolysis*. But there is only enough glycogen in the liver to last for a few hours, and when it is exhausted the liver makes glucose from amino acids or other sources (*gluconeogenis*). The level of blood sugar is accurately regulated largely by the action of the hormones adrenaline and glucagon (from the pancreas), which promote glycogenolysis; and insulin, which increases the storage of glycogen from blood glucose. A third set of activities in the liver is concerned with fat metabolism and the liver cells are commonly filled with fatty granules. Many other substances are also dealt with by the organ, for instance vitamin A is synthesized from its carotene precursor, hence the richness of fish liver oils in this vitamin. Iron and copper are metabolized and stored in the liver and many other processes go on there. The bile is secreted continuously and stored in the *gall-bladder*, from which it is repeatedly discharged when food enters the intestine. If not used, its water is re-absorbed. The bile is alkaline in man but acid in the cat and dog. It contains, besides the bile-salts already mentioned, the *bile-pigments*, which are derived by breakdown of the haemoglobin of the blood, and also the polyhydric alcohol *cholesterol* and the phosphatide *lecithin*; the former of these is the main constituent of gall-stones. Several functions have been suggested for the bile; only those concerned with the digestion of fats are certain but the bile probably also has an action on the movements of the gut.

The detoxicating functions of the liver are perhaps connected with those that are more strictly metabolic. Many substances that have a pronounced action when introduced into the body, such as alcohol or anaesthetics, are broken down in the liver and rendered in-effective. Many of the pollutants now normal in our environment are dealt with in the liver including toxic gases, dyes, preservatives, and drugs. These are excreted only very slowly by the kidney, especially if they are lipid-soluble, but the liver cells have non-specific drug-hydroxylase enzymes on their endoplasmic reticulum able to oxidize lipid-soluble compounds. These require the co-operation of reducing vitamins such as NADPH, which may be absent in poorly nourished people who then show marked toxic effects of drugs tolerated by other people.

The liver also has an important function in regulating the amount of steroid hormones circulating in the blood. Thus, in man, the whole of the aldosterone is removed by a single passage through the liver.

The sinusoids of the liver contain in their walls many cells of the reticulo-endothelial system (histiocytes), here known as the *cells of Kupffer*. These are responsible for removing from the blood-stream particles of foreign matter, dead bacteria, etc. The liver is also concerned in regulating the amount and composition of the blood, destroying old red cells and, in the embryo, forming new ones, as well as producing some of the protein components of the blood, the fibrinogen and pro-thrombin (p. 171).

10. The large intestine

Nearly all the absorption from the gut takes place as the food is passing along the small intestine; the large intestine is a region in which the water secreted previously to make the fermenting mass is re-absorbed. This is evidently a very important function in mammals that live on land and are liable to be short of water. The large intestine may also have other functions. In the rabbit the small intestine is swollen at its lower end to form the *sacculus rotundus*, and just beyond this is attached an enormous blind sac, the *caecum*, where cellulose and other resistant materials in the food are acted upon by symbiotic bacteria. The caecum is a thin-walled sacculated structure whose surface is increased by a spiral valve. Presumably, materials liberated by the bacteria are absorbed through the wall of the caecum, as they are from the rumen of the sheep (Young 1962, p. 744). The caecum terminates in the *vermiform appendix*, which has thick walls and much lymphoid tissue, perhaps concerned here, as elsewhere in the gut, with the neutralization of bacterial toxins. In man and other mammals with a reduced caecum, the appendix perhaps retains some antibacterial functions, though it can be removed without any evident ill effects.

The *large intestine* proper (*colon*) proceeds from the sacculus rotundus and is characteristically constricted at regular intervals, especially in its lower part. The wall of the large intestine differs from that of the small in having no villi but there are glands of Lieberkühn and many mucus-secreting goblet cells. The action of the bacterial population on the contents of the colon, and especially on proteins, produces a variety of putrefactive products, some of which are toxic when injected into the blood-stream (histamine, skatole, indole, etc.). It has often been suggested that absorption of these products produces ill effects in man, especially in constipation, but probably the wall of the colon normally prevents the absorption of these substances almost completely. Like other parts of the intestine it contains large amounts of lymphoid tissue.

11. Digestion and homeostasis

The importance of intake of materials and digestion for maintenance of living organization needs no emphasis. Yet it is interesting to recall how little we understand about the system that controls the rate at which food is taken and hence growth and replacement proceed, although these processes limit the size and many other aspects of the organism. No doubt hereditary influences control the setting up of the peristaltic and reflex processes of the gut that are the basic triggers for the actions that lead to the intake of food. Nevertheless, the actual food-gathering behaviour of an adult mammal is largely controlled by the higher nervous centres (p. 365). Food is sought at certain times and in certain ways, according to a pattern that has been acquired during the lifetime of the individual.

We can say, therefore, that the basic pattern of feeding and digestion is controlled by the hereditary instructions, determining especially the structure of the teeth and of the alimentary canal. The regulation of the sequence of events during digestion involves many reflexes passing through the autonomic nervous system, and these are also laid down by heredity. Chemical signalling also plays a part. Information acquired during the lifetime greatly influences the whole process through the higher cerebral centres. There are also changes in the alimentary canal and its glands according to the type of food consumed. The gut of rats becomes longer on a vegetable than on a meat diet, and the enzymes of the pancreas becomes adjusted to suit the type of food that is habitually consumed.

So far as is known there is no detailed system of feed-back by which shortage of raw material in any single organ sets up processes that lead to intake of supplies of it. Shortages lead to hunger and hence to the search for food and consumption of it. The adequacy of these processes is ensured in the long run by the operations of natural selection. Those organisms whose methods of food selection and digestion are inadequate fail to grow or to survive.

15 Respiration

1. The significance of respiration

ALL living organisms are steady-state systems in which energy is continually expended to move materials against concentration gradients. Mammals depart widely in composition from the surroundings in which they live and this difference is maintained by the continual expenditure of large amounts of energy. We could express this by saying that mammals have specialized in methods by which energy is exchanged for information. Such a system is possible only if large supplies of energy are available, allowing reactions of the organism to changes in the environment. Regulation of respiration is thus an essential feature of mammalian organization.

The energy requirements of a mammal are all ultimately met by the oxidation of foodstuffs by oxygen taken in from the atmosphere. This process of oxidation takes place within the tissues by virtue of the enzymes found there (p. 158). This *tissue respiration* sets up a change in the haemoglobin of the blood and leads to the uptake of oxygen in the lungs and hence to the process of *external respiration*. The apparatus of the chest and lungs ensures that a supply of air is continually available, which is ensured by regulating the rate of breathing according to the condition of the blood.

The warm-blooded condition and high level of activity makes mammals, like the birds, acutely dependent on a good oxygen supply. An oxygen debt may be accumulated for a short time during powerful muscular effort (p. 67), but if the tissues are left without oxygen for more than a few minutes they become permanently impaired. In birds the large supply of oxygen is obtained by an arragement that ensures a complete and continuous change of the air by sweeping it through the lungs into air sacs. In mammals the air is not swept across the respiratory surface into air sacs but is drawn in and blown out of a closed 'respiratory tree', whose terminal portions constitute the region of gaseous exchange. This arrangement is made to function efficiently by the enclosure of

the lungs in a chamber, the thorax, whose walls can be expanded to allow air to rush in. The diaphragm closing off the thorax from the abdomen is a characteristic feature of mammals.

2. Trachea and lungs

The nasal passages with their vascular membranes for warming and cleaning the inspired air have already been described (p. 109). The respiratory system proper may be said to begin where the air enters the *larynx* (p. 115), whose inlet is guarded by the epiglottis.

The *trachea* is a tube strengthened by incomplete rings of cartilage which prevent its collapse during inspiration. At its lower end the trachea divides first into a pair and then into a number of secondary divisions or *bronchi*, leading in turn to smaller divisions, the *bronchioles* (Fig. 15.1). From each bronchiole a system of irregular alveolar ducts leads to the terminal *alveoli*, chambers wider than the bronchioles and each lobed to form a number of *air sacs*. The effect of this branching system of tubes is to allow air to penetrate to every portion of the lung, which is thus a spongy elastic tissue, honeycombed with passages (Fig. 15.2). In the embryo the branching system of tubes is sharply defined and the terminal branches are lined by a definite epithelium. In the adult the finer branches are less regular and the structure of the tree is not apparent in sections.

The walls of the finest chambers are excessively thin, allowing a very close relationship between the air and the blood. Outside the capillary wall there is a further thin membranous layer of flattened pulmonary epithelial cells (Fig. 15.3). Each sac is of the order of 100 μm across and there are probably more than 700 million of them in the two lungs of man, providing an immense surface (at least 50 m^2) for gaseous interchange. The presence of this great number of tiny sacs and of the elastic tissue between them make the lungs compact organs, very different from the flimsy sacs of the frog, which collapse to almost nothing when punctured. The capillaries of the alveoli arise from branches of the

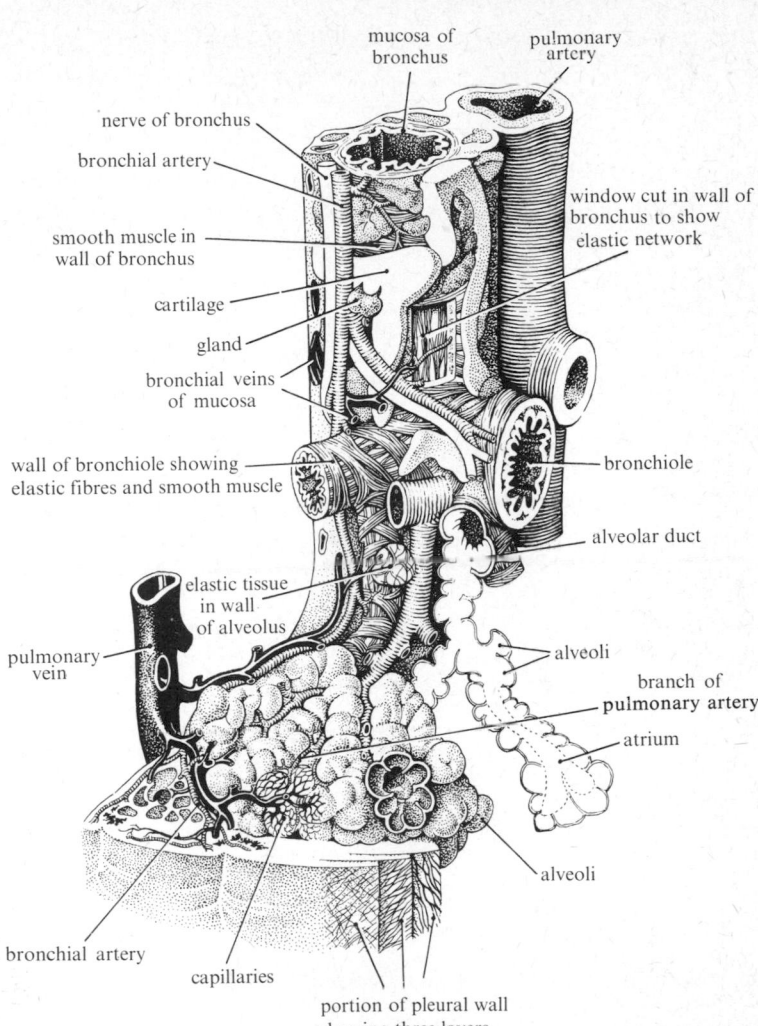

mucosa of bronchus

pulmonary artery

nerve of bronchus

bronchial artery

window cut in wall of bronchus to show elastic network

smooth muscle in wall of bronchus

cartilage

gland

bronchial veins of mucosa

wall of bronchiole showing elastic fibres and smooth muscle

bronchiole

alveolar duct

elastic tissue in wall of alveolus

pulmonary vein

alveoli

branch of **pulmonary artery**

atrium

alveoli

bronchial artery

capillaries

portion of pleural wall showing three layers

FIG. 15.1. Drawing of a reconstruction of part of a pulmonary lobule of a young man. (After Braus.)

pulmonary artery and discharge into the pulmonary veins (Fig. 15.1). In addition the whole tree is served by bronchial arteries, bringing oxygenated blood from the thoracic aorta. The blood from the tree as far as the second-order bronchi is returned by bronchial veins to the right atrium.

A striking characteristic of the lung is its elasticity; the walls of the alveoli and bronchioles are abundantly supplied with elastic fibres, giving them the power to contract after they have been dilated, which is important because expiration is otherwise largely a passive process, produced with only a slight muscular effort. The walls of the bronchioles also contain smooth muscles. These receive motor innervation from the vagus nerve and inhibitory sympathetic fibres. There are also many afferent fibres in the lungs.

The trachea and large bronchioles are lined by ciliated cells, which beat in such a way as to transfer substances upwards towards the pharynx, protecting the respiratory tree. Ciliated cells are highly specialized for their functions. Each cilium is a composite structure with an outer sheath and a central core of fibrils, which are attached to basal granules in the cell body.

When necessary the cleansing action of the cilia is reinforced by *coughing*. This action is initiated by the stimulus of particles in contact with receptors in the bronchi. Afferent impulses from these areas produce a sharp expiration, the glottis being at first closed and then suddenly opened to allow the particles to be swept out.

The lungs are covered by the mesothelium of the *visceral pleura,* the layer of splanchnic mesoderm covering the endodermal outgrowth that constitutes the embryonic lung (p. 163). The outer surface of the

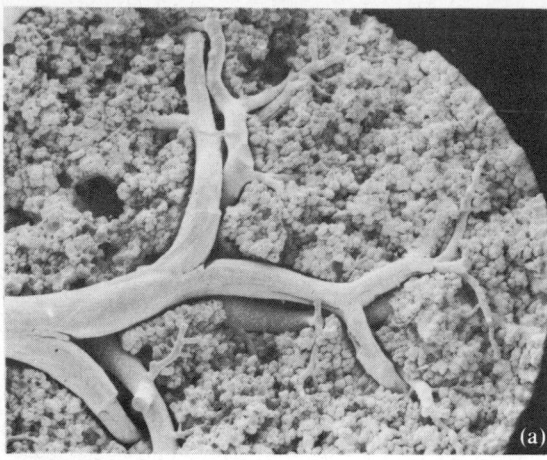

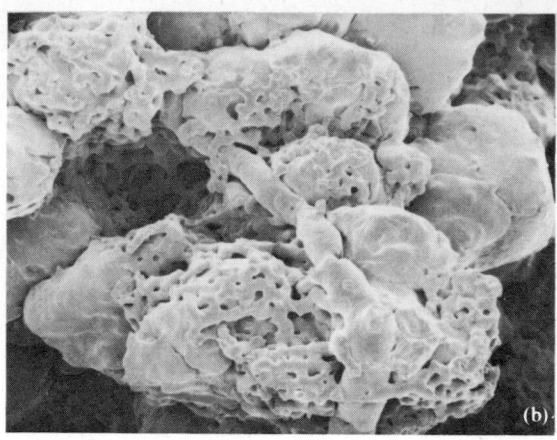

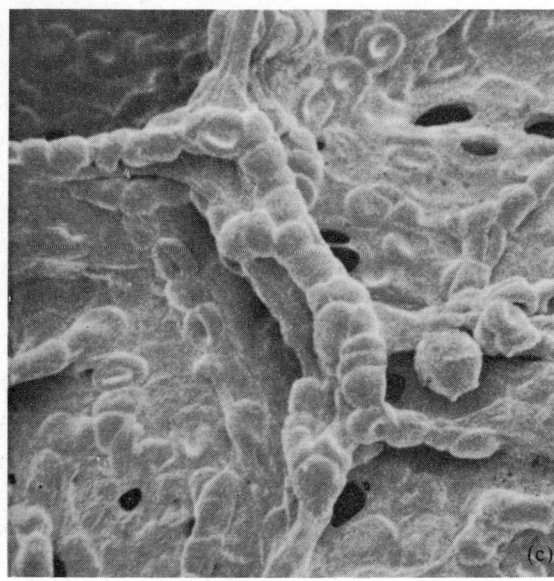

Fig. 15.2. Mammalian lung seen by scanning electron microscopy.
(a) Injection replica showing inter-relationship of blood-vessels and air passages in cat lung. A terminal bronchiole, overlying a branch of the pulmonary artery, gives rise to respiratory bronchioles, alveolar ducts, and alveoli. Field width 6.0 mm. (b) Injection replica of cat lung showing alveoli overlain by capillaries from a branch of pulmonary artery. Field width 250 μm. (c) Perfused hamster lung showing intimate relationship of alveoli and capillaries containing red blood cells. Field width 80 μm. (Figures kindly supplied by Miss J. A. Nowell.)

lung is thus shining and smooth and in life is closely apposed to the *parietal pleura*, the layer of somatic mesoderm that lines the inside of the thorax and has a similar smooth structure. In life these two layers are in contact, with no space between, and this close contact is essential to the working of the lung. If air is present between the pleura, the lung does not expand and contract with the movements of the thorax.

3. Respiratory movements

The tidal flow of air is produced by enlargement of the thoracic box, *inspiration*, allowing the air to rush in down the trachea under atmospheric pressure. The movement is the result of contraction of the respiratory muscles, chiefly the *diaphragm* and *intercostals*. When this contraction ceases, the movement of expiration follows, largely by elastic recoil, helped by the action of the transverse thoracic and abdominal muscles, tending to make the thorax smaller. These actions are characteristic of mammals, and they depend essentially on movements of the ribs. The fibres of the intercostal muscles run from each rib to the rib behind (Fig. 15.4). The external intercostals pass obliquely ventrally and caudally; the internal intercostals run in the opposite direction. There is also a thin layer of transverse thoracic muscles running from the sternum and to the ribs.

Attached to the upper ribs are the *scalene muscles*,

which tend to draw them cranially and thus to fix the whole sternum, so that contraction of the intercostals increases the size of the thoracic cavity. The fibres of the external intercostals are so directed that they have a mechanical advantage in drawing the more caudal rib of each pair cranially during inspiration. The action of the internal intercostals is probably mainly in expiration.

The most important factor in the inspiratory action is the *diaphragm*, a muscular partition separating the thoracic from the abdominal cavity. It arises during development from muscle fibres in the septum transversum in which the veins cross from the body wall to the heart (p. 214). This septum lies far forward and becomes invaded by striped muscle, innervated from the ventral roots of the cervical spinal nerves. As the lungs develop and the heart moves caudally the diaphragm also retreats and for this reason the *phrenic nerve* runs a very long course (see p. 257).

The diaphragm is a dome, concave on its abdominal side. It has a central tendinous portion, from which muscle fibres run to the sternum and ribs and by two cords, the crura, to the lumbar vertebrae. During contraction of these muscles the dome becomes flattened, thus enlarging the cavity of the thorax and pushing the abdominal viscera caudally. This process is limited by the action of the abdominal muscles. The main action of expiration is produced by the stored energy of the elastic tissues of the lungs. This is assisted by the muscles of the abdominal wall, pressing on the viscera and perhaps also by the internal intercostals.

Either the thoracic or the diaphragmatic movements alone are able to maintain respiratory exchange in man,

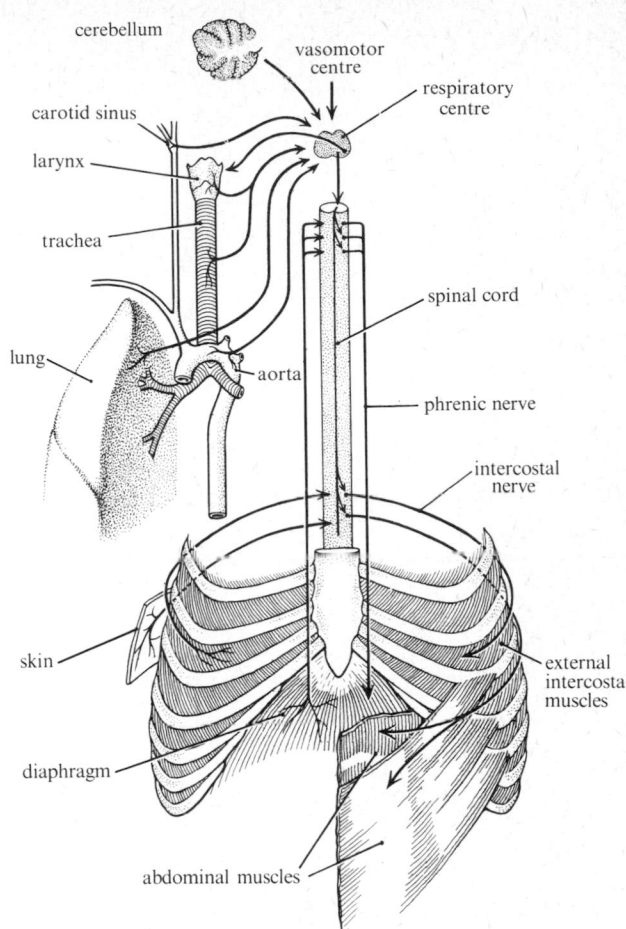

FIG. 15.4. Diagram of the respiratory muscles and the influences acting upon them.

but the diaphragm normally plays the greater part. In very deep inspiration the trapezius, serratus anterior, pectoralis, latissimus dorsi, and other muscles also come into play. The exact details of respiration vary greatly in different animals and even in individual men. In quadrupeds the weight-bearing functions of the thoracic wall (p. 77) tend to fix its parts and the respiration is therefore predominantly diaphragmatic. In these animals the thorax is usually narrow. In bipedal, arboreal, and flying mammals the thoracic muscles play a larger part in respiration and are especially conspicuous in the fully aquatic mammals, where all the weight is taken by the water.

4. Regulation of breathing

The sequence of movements in respiration is maintained throughout life by the rhythmical discharge of

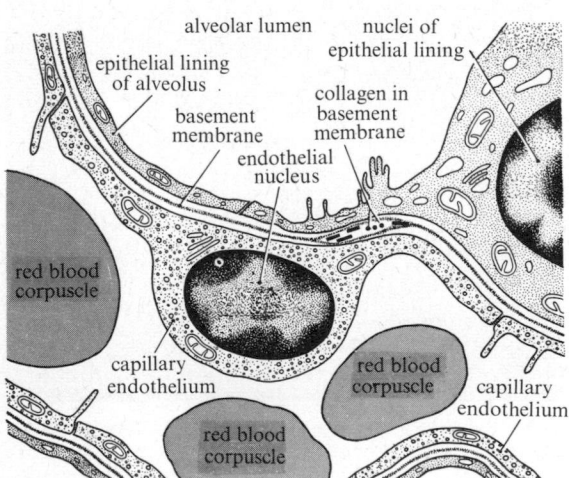

FIG. 15.3. Diagram of a section of a lung as seen by electron microscopy.

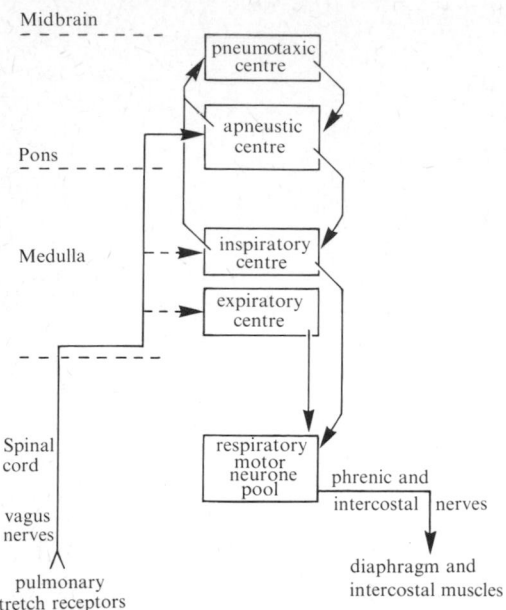

FIG. 15.5. Diagram of the organization of respiratory centres in brain stem of cat. (From Davson and Eggleton (1968), after Wang *et al.* (1957). *Am. J. Physiol.* **190**.)

impulses in the phrenic and intercostal nerves, producing a series of acts of inspiration. The rhythm is produced by the action of nerve cells in the medulla oblongata and pons. If the spinal cord is cut at the level of the first cervical nerve respiration stops. In the first part of this century a model of the respiratory system was developed out of experiments utilizing electrical stimulation and ablation. The model consists of a series of interacting *respiratory centres* (Fig. 15.5). The results of electrical stimulation suggested that the medulla contains two centres, one causing inspiration, the other expiration. These are connected so that local stimulation of part of the *inspiratory centre* excites the whole centre and inhibits the expiratory one and vice versa. These centres are able to maintain a rhythm even after cutting the vagi and sectioning the brain below the pons, presumably by a self-re-exciting mechanism. However, such breathing is of an abnormal gasping type with maximal action of all the inspiratory muscles. Normally inspiration is checked by the action of impulses in the vagus from pulmonary stretch receptors, acting through a pontine *apneustic centre*. If the vagi are cut and a section is made through the middle of the pons, this centre assumes an excessive control producing a cessation of respiration in the inspiratory position, except for occasional expiratory gasps (Fig. 15.6). Normal inhibition of inspiration, therefore, is produced by the

impulses in the vagus assisted by a still high *pneumotaxic centre* in the rostral portion of the pons (Fig. 15.5). This latter centre can produce a normal type of breathing even if the vagi have been cut; therefore it must receive information directly from the medullary centres.

The sequence of events in normal breathing is as follows.

(a) The inspiratory centre becomes increasingly active under the influence of the apneustic centre, discharging impulses both to the spinal centres and upwards to the pneumotaxic centre.

(b) As the lungs inflate impulses from them and from the pneumotaxic centre inhibit the apneustic centre and this in turn inhibits the inspiratory centre.

(c) The expiratory medullary centre completes the cycle. Breathing is also influenced by still higher centres in the hypothalamus and cortex.

There has been some difficulty in recent years in confirming the details of the respiratory centres model, described above, at the neuronal level. Microelectrode

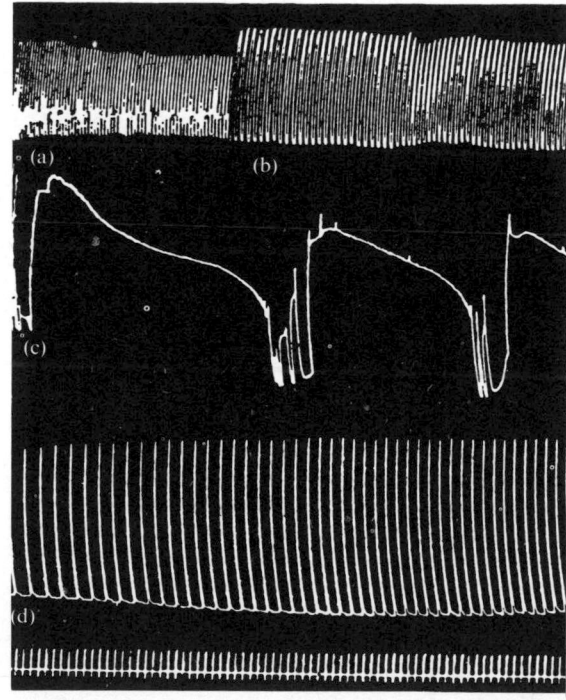

FIG. 15.6. Effect of brain-stem sections on respiration in cat. (a) Normal. (b) After vagotomy. (c) Prolonged inspiratory tonus (apneusis). (d) Gasping respiration. Time tracing (at bottom) at intervals of 5 s. Inspiration upwards (From Lumsden (1923). *J. Physiol.* **57**.)

recordings from the medulla oblongata show that large concentrations of cells with respiratory firing patterns are located in the extreme lateral parts of the medulla, not in the medial regions in which electrical stimulation and ablation are effective. Further, these cells do not appear to be connected to each other as in the model. The model, however, remains a useful summary of the effects of stimulation and ablation (see Merrill 1974).

The precise manner in which the respiratory pattern generated by the brain-stem is translated into movement is almost as important as the setting of the rhythm. Segmental reflexes involving intercostal muscle-spindles as well as vagal reflexes are essential. As in the muscles of the limbs (p. 264) the intercostal muscle-spindles operate to adjust the force of contraction to the demand in spite of variations in load. Thus if there is increased inspiratory resistance (as by occluding the trachea) the muscles shorten less, the spindles are less unloaded and their discharge therefore is greater, increasing the force of inspiration and maintaining the tidal volume (Corda, von Euler, and Lennerstrand 1965).

In quiet breathing, on the other hand, the intercostals contribute less to the inspiratory movement than the diaphragm, which has few muscle-spindles. Load-compensating mechanisms for the diaphragm are mediated through the spindle afferents of the lower intercostal muscles, via intrasegmental reflexes, as well as through descending vagal reflexes. The vagi and the lower intercostal nerves supply (in a sense) the proprioceptive input to the phrenic motoneurones, which is absent in the diaphragm itself (v. Euler 1973).

Regulation of breathing is effected by a complex of chemical and nervous factors besides the actual respiratory centres. These include the following.

(a) Areas on the ventrolateral surface of the medulla that respond to raised carbon dioxide (or acid) levels by increasing pulmonary ventilation (Schlaefke, See, and Loechcke, 1970). The receptors lie just beneath the surface of the brain, surrounded by extracellular fluid and perfused with blood. These receptors are not sensitive to oxygen lack but an increase of 1 per cent in concentration of carbon dioxide will cause a measurable increase in respiratory volume, though the subject is not conscious of this.

(b) Chemoreceptors in the carotid and aortic bodies are sensitive not only to increased carbon dioxide but also to a fall in oxygen tension. The frequency of impulses discharged at any given P_{O_2} will thus be higher the greater the P_{CO_2}, and vice versa.

(c) The discharge of the pulmonary stretch receptors inhibits inspiration (p. 156). There are spindles among the smooth muscle, especially at the points of bronchial branching (Hering–Breuer reflex).

(d) Other respiratory reflex effects come from changes in blood-pressure, stimulation of the mucous membranes of the nose or larynx, from impulses from muscle-spindles and tendon organs of the limbs, and from pain.

Some of these influences may be said to provide a positive feed-back, for example, the effect of the carbonic acid in the blood in stimulating the inspiratory centre. On the other hand, the flow of impulses from the alveolar receptors provides in the main a negative feed-back, stopping the action of inspiration and producing slowing of respiration and bronchoconstriction. Other receptors are present that respond to tactile and chemical influences. Impulses from them set in action the protective-cough reflex.

5. The respiratory exchange

The movements of respiration produce in each cycle only a partial change of the air in the lungs. During expiration the lungs are not completely emptied, for the alveoli are maintained distended by the small negative pressure in the thorax. Much of the alveolar air, however, is swept into the bronchioles and mixed with fresh air at the next inspiration. Atmospheric air contains about 21 per cent of oxygen, 79 per cent of nitrogen, and 0·04 per cent of carbon dioxide. By special methods alveolar air can be collected and shown to contain about 14 per cent of oxygen, and as much as 5·5 per cent of carbon dioxide, whereas expired air contains about 16 per cent of oxygen and 4 per cent of carbon dioxide, that is to say, it has a certain admixture of fresh air that has not reached the end of the respiratory tree.

The gases in the air sacs of the alveoli come into equilibrium with the blood by diffusion across the epithelium and capillary walls. On account of the thinness of these membranes the equilibrium is reached rapidly and the oxygen is taken up by the blood-plasma, and from this by the haemoglobin of the corpuscles, while the carbon dioxide is given up by the blood. The rates and extents to which these processes take place depend wholly on the partial pressures of the gases in the alveolar air and on the composition of the blood; it is not now thought that any secretion of oxygen across the lung surfaces is involved.

The solubility of oxygen in water is such that only about 0·3 ml or less is held in simple solution in 100 ml of blood. However, there are 15 g of haemoglobin in this volume, and this is able to hold 1·34 ml of oxygen per gram, that is to say, nearly 20 ml per 100 ml blood. The oxygen carried in simple solution is therefore only about 1 per cent of the whole and would not be adequate

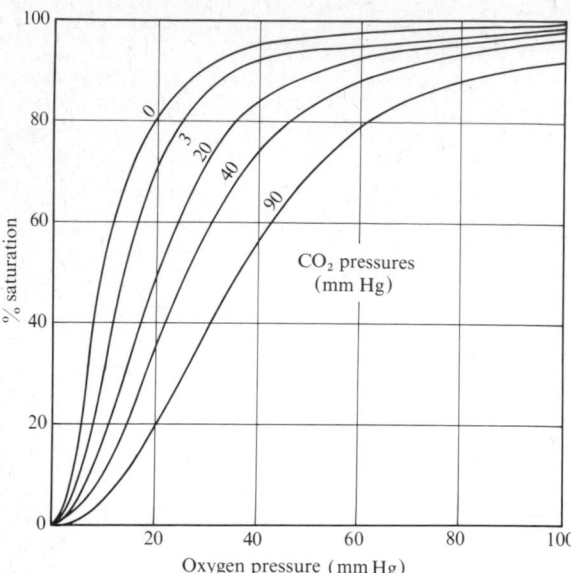

FIG. 15.7. Dissociation curves of human blood at different pressures of carbon dioxide.

The shape of the curves means that between oxygen pressures 100–50 mm Hg, little dissociation of oxyhaemoglobin occurs whilst in the range 50–0 mm Hg oxygen pressure, a small decrease in pressure produces a large fall in the percentage saturation, the oxyhaemoglobin breaking down readily. The effect of carbon dioxide pressure, which is similar to the effect of temperature and is called the Bohr effect, is to decrease the affinity of haemoglobin for oxygen, thus accelerating the dissociation of oxyhaemoglobin. (After Barcroft and Pulton (1913). *J. Physiol.* **46.**)

to maintain activity such as that of a mammal except by the very rapid transport of enormous quantities of liquid. The efficiency of the haemoglobin is therefore an essential feature of mammalian life, indeed for that of nearly all chordate animals.

The haemoglobin molecule is formed by the union of protein (globin) with an iron-containing *porphyrin* pigment or *haem* and it is this latter that gives it a power of loose union with oxygen to form oxyhaemoglobin. Under the conditions in which the haemoglobin occurs in the blood it becomes 95 per cent saturated with oxygen at the partial pressure of that gas in the alveolar air (100 mm Hg). There is therefore little to be gained in man by adding further oxygen to the air breathed under ordinary conditions, though this may be valuable when there is a low oxygen tension, for instance, at high altitudes. The haemoglobin remains saturated to as much as 80 per cent if the oxygen tension is reduced to 50 mm Hg, but below this tension it begins to give up oxygen (Fig. 15.7) and this normally occurs in the capillaries. Surrounded here by an

environment poor in oxygen the haemoglobin yields its oxygen store to the plasma, and through the capillary wall to the tissues. The shape of this *oxygen dissociation curve* of haemoglobin is obviously of great importance for the maintenance of a proper flow of oxygen to the tissues. The particular form of the curve depends on the composition of the blood; for instance, the presence of carbonic and lactic acids causes the oxygen to be given off more readily, shifting the curve to the right (Fig. 15.7). Haemoglobin alone, in simple aqueous solution, retains 80 per cent of its oxygen down to pressures as low as 20 mm Hg, and this would not provide an adequate supply (see Lloyd 1971). In worms that use haemoglobin as a means of storing oxygen the dissociation curve is flat, considerable quantities of oxygen being retained down to quite low tensions.

6. Tissue respiration

The process of taking oxygen into the lungs and its transport in the blood is only a preliminary to the essential feature of respiration, the using of oxygen by the tissues to provide energy by combustion. We can distinguish, therefore, between *external respiration* and the *internal respiration* of the tissues. The latter takes place by a system of enzymatic exchanges, associated with the mitochondria, making the oxygen able to undertake chemical transformations that would not otherwise occur under these conditions. Ultimately much of the energy of muscular contraction, for instance, comes from the combustion of glycogen to carbon dioxide and water, a reaction that would not take place to any extent at 37°C *in vitro*.

A central position in this sequence of changes is taken by *cytochrome*, a haem-containing substance (or set of substances) resembling haemoglobin, and present in all animal cells. Cytochrome has a characteristic absorption spectrum which, like that of haemoglobin, changes when oxidation occurs. This requires the action of an enzyme, *cytochrome oxidase*, which activates the oxygen brought to the tissues and transfers it to the cytochrome. The *dehydrogenases* are enzymes able to combine the oxidized cytochrome with the reducing substances produced by the metabolism of the tissues. Thus reactions such as the breakdown of pyruvic to lactic acid are made to proceed by removal of the hydrogen, and in this way the whole system of equilibria is shifted so that, for instance, glycogen can break down through the stages of the carbohydrate cycle (p. 433) through pyruvic to acetic acid and ultimately to carbon dioxide and water (see Lehninger 1970).

Various poisons act at particular points of these cycles. Thus cyanides inhibit respiration by preventing the action of cytochrome oxidase, leaving the cytochrome in the reduced state. Urethanes, on the other hand, by acting on the dehydrogenases, leave the cytochrome oxidized. Besides this cytochrome system there are other respiratory enzymes present, which are able to change the rate of the reactions in which oxygen is involved. Together these respiratory enzymes make possible the last stages of the flow of oxygen, which goes on throughout life from the air of the lungs to the blood and thence to the tissue fluids and into the cells, where it takes part in the reactions that are the life of the organism.

16 Development of the gut and respiratory system

1. Origin of the gut

THE tubular gut is first formed from the endoderm of the yolk sac when the embryonic disc is lifted up by head and tail folds (Fig. 12.5, p. 121). In this way head- and tail-guts become delimited, communicating by the anterior and posterior intestinal portals with the region that will form the rest of the gut. The latter is at first widely open to the yolk sac but the stalk gradually narrows and the embryonic and extra-embryonic regions become sharply separated. The gut is conventionally divided into fore-, mid-, and hind guts. The first includes the mouth, pharynx, oesophagus, and stomach. The mid-gut includes the whole small intestine, and the hind gut the large intestine and rectum.

The whole gut is lined by endoderm, covered with splanchnic mesoderm. By the formation of buds and pouches the endoderm gives rise to the various glands attached to the gut, and to the lungs. The splanchnic mesoderm produces the muscular coats of these organs and their mesothelial coverings (pleura and peritoneum). The gut and organs derived from it mostly lie in coelomic spaces, the pleural and peritoneal cavities. In the pharyngeal region no coelomic space is present (see Hamilton and Mossman 1972).

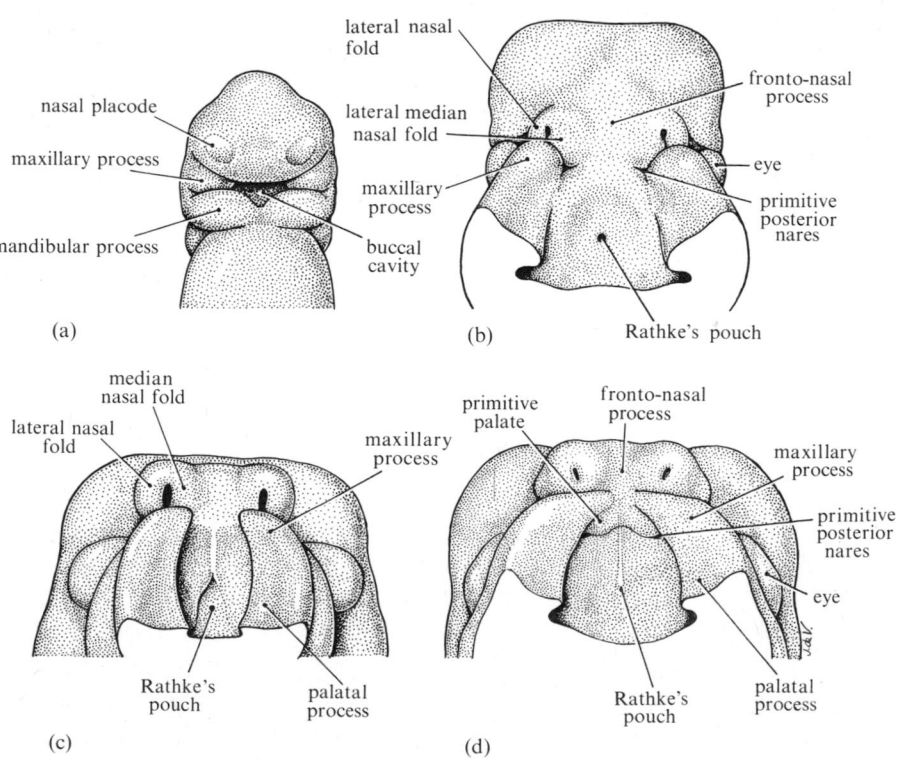

FIG. 16.1. Diagrams showing stages in the development of the mouth and nose.

2. Stomodaeum

The head-gut ends anteriorly in the buccopharyngeal membrane. Where this comes into contact with the ectoderm the latter forms a pit, the *stomodaeum* (Fig. 12.5), from the roof of which a diverticulum, *Rathke's pouch*, grows towards the floor of the brain. It has the effect of inducing a down-growth of the latter, the *infundibulum*, and from the two together the pituitary body develops (p. 436).

The buccopharyngeal membrane breaks down early in development, and thereafter there is no sharp dividing line between the ectoderm and endoderm of the mouth. This leads to uncertainty as to the layer from which structures in the mouth are derived.

3. Nasal cavities

These first appear as thickenings of the ectoderm, the *nasal placodes* (Fig. 16.1), which then sink in to form *olfactory pits*, with rather large slit-like openings, lying just above the mouth. The maxillary processes, developing from the first pharyngeal arches, grow forward and medially below the openings of the olfactory pits, which become the anterior (external) *nares*. The posterior (internal) nares form as passages between the olfactory sac and the oral cavity. The *palate* forms by a backward growth from the region between the olfactory pits, the fronto-nasal process, together with medial maxillary palatal processes. The posterior nares thus come to open farther and farther back and the hard and soft palates are formed.

The whole of the nasal cavity is produced by expansion of the olfactory pits, but the nasal placodes form only the olfactory epithelium of the upper regions (p. 410). The *paranasal sinuses* of the maxilla, frontal, ethmoid, and sphenoid bones develop as pouches of the wall of the nasal cavity.

4. Mouth and pharynx

The tissue around the pharynx soon shows condensations that are the rudiments of the branchial (pharyngeal) cartilaginous arches. This tissue is usually referred to as mesoderm, but it is formed by migration downwards of material of the neural crest (p. 345). The differentiation of this material to form cartilage presumably depends upon the position in which it finds itself rather than on its origin.

The pharyngeal arches are separated by depressions of the ectoderm and by grooves of the endoderm of the pharynx (Fig. 16.2), but in man these never meet to form open gill slits. Each pharyngeal arch contains, besides its cartilage, some lateral plate musculature, one of the aortic branchial arches, and branches of the corresponding segmental cranial nerve. Thus the trigeminal is the main nerve of the first arch, the facial of the second, glossopharyngeal of the third, and vagus branches of the remaining arches.

The first pharyngeal pouch of the whole series is probably represented by a gill slit in front of the first pharyngeal arch, and this becomes incorporated in the sides of the mouth (Fig. 16.3). The first of the actual series that can be seen in the human embryo lies between the maxillo-mandibular and hyoid arches (Fig. 16.2). It becomes the *pharyngo-tympanic (Eustachian) tube*, and the corresponding ectodermal cleft becomes the *external*

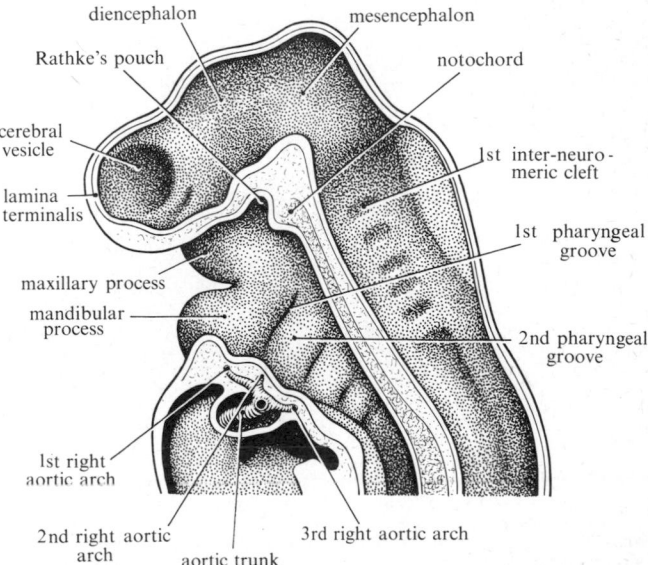

Fig. 16.2. Diagram of sagittal section of head and pharynx of human embryo.

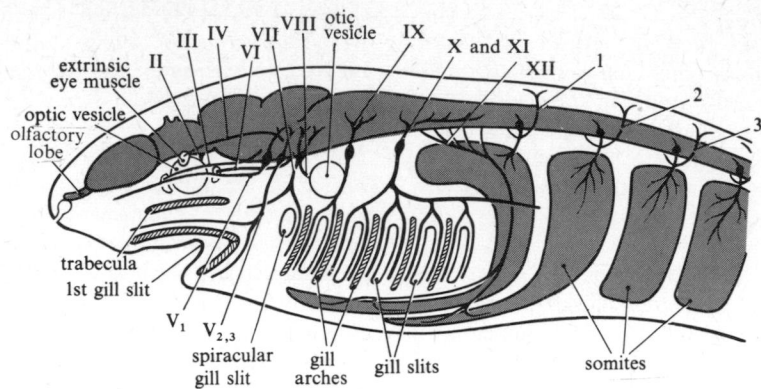

FIG. 16.3. Diagram of the anterior end of an early vertebrate showing the branchial arches and their nerves.
Cranial nerves: II, optic; III, oculomotor; IV, trochlear; V$_1$, trigeminal, ophthalmic; V$_{2,3}$, trigeminal, maxillo-mandibu-lar; VI, abducens; VII, facial; VIII, vestibulo-cochlear; IX, glossopharyngeal; X and XI, vagus and accessory; XII, hypoglossal; 1–3, spinal nerves.

auditory meatus. The *tympanic membrane* is thus covered on its outer side by ectoderm and inner side by endoderm, while its centre contains part of the mesoderm surrounding the pharynx. As development proceeds the tympanic cavity comes into relation with the structures of the inner ear that have differentiated from a separately invaginated ectodermal sac, the *otocyst.*

The second of the actual pharyngeal pouches becomes invaded by mesoderm, which differentiates into the lymphoid tissue of the *tonsil* (p. 114). The wall of the third pharyngeal pouch becomes thickened and part of it and of the fourth pouch differentiates to make the *thymus* (Fig. 16.4), the remainder of the third pouch forms the first of the pairs of *parathyroid glands.* The second pair is formed from the walls of the fourth pharyngeal pouch, with perhaps a contribution from the fifth and last pouch (ultimobranchial body) which is transient but may give rise to the calcitonin cells.

The floor of the mouth and pharynx gives rise to the tongue, the salivary glands, and thyroid gland. The *tongue* is formed in the central line of the mid-ventral portions of the mandibular, hyoid, and second pharyngeal arches. The greater part of its musculature is derived by ventral migration of three or more of the occipital myotomes and this explains the course of the

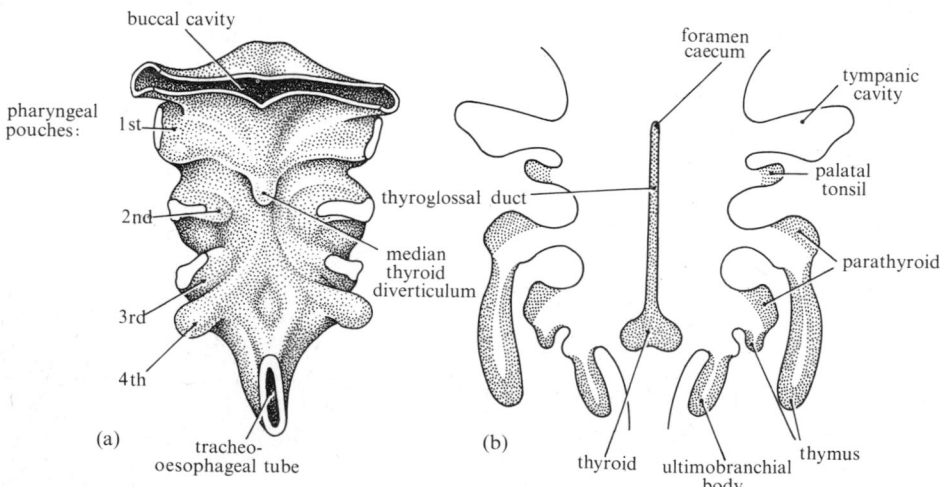

FIG. 16.4. (a) Pharyngeal endoderm of 4 mm human embryo from the ventral side (After Weller (1933). *Contrib. Embryol.* **24**.) (b) Diagram of pharynx of 10 mm embryo to show the fates of its parts.

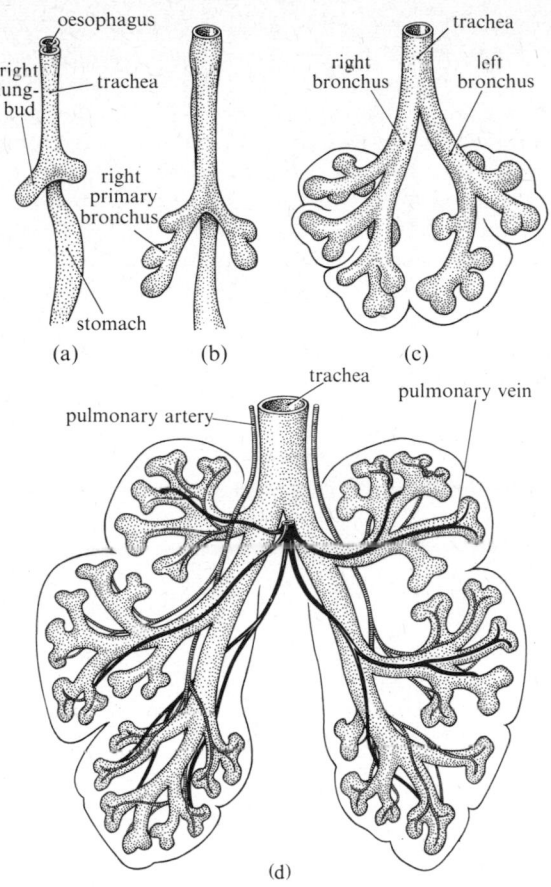

FIG. 16.5. Development of the lungs. (After Patten 1968.)

hypoglossal nerve, representing the ventral roots of that region. The tongue is therefore covered with pharyngeal endoderm and contains branchial mesoderm from these arches, as well as somitic mesoderm. Accordingly it receives nerves from several dorsal and ventral roots (V, VII, IX, X, and XII).

The *thyroid gland* develops as a thickening of the floor of the pharynx near its front end. This outgrowth elongates, making the thyroglossal duct, with the gland as a swelling on its lower end. There may possibly be a contribution to the gland from the wall of the fourth pharyngeal pouch (see Boyd 1950).

5. Development of the trachea and lungs

The first rudiment of the respiratory apparatus appears very early as a groove in the floor of the pharynx, which deepens and finally closes over to separate the larynx from the oesophagus. The caudal end of the groove then divides into two diverticula, the *lung-buds*, which grow backwards into the pericardio-peritoneal canals (Fig. 16.5). The resulting endodermal tubes, the bronchi, then divide as many as 18 times to form the adult bronchial tree. The splanchnic mesoderm covering the outgrowing lung-buds gives rise to the cartilage of the tracheal rings and to the musculature and connective tissue of the lungs. The epithelium lining the bronchial tree is cuboidal and ciliated during foetal life. At birth, when air enters the lungs, the terminal portions of the bronchi expand to form the alveoli (p. 152). The epithelium ceases to be cuboidal and becomes very thin in the adult lung (p. 154) (see Pattle 1969; Towers 1968).

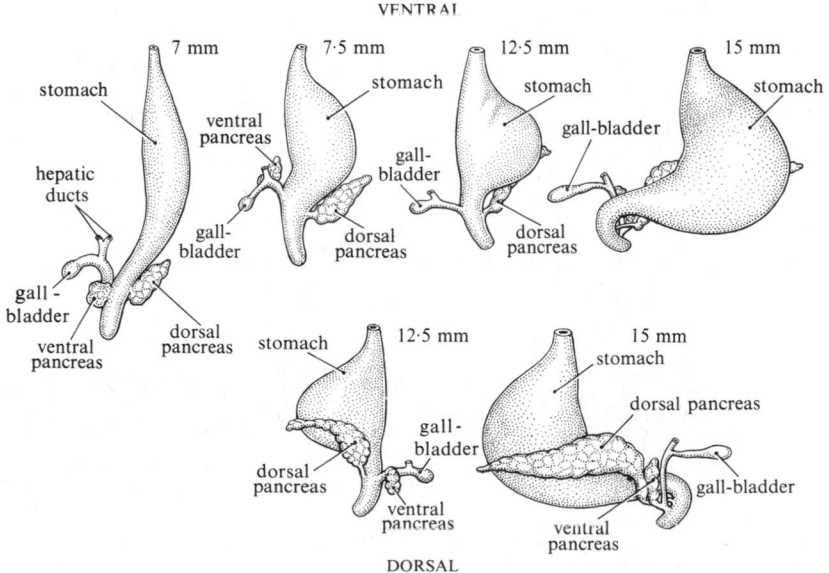

FIG. 16.6. Diagrams showing the development of the stomach of man. (After Pernkopf (1922). *Z. Anat. Entw. Gesch.* **64**.)

6. Development of the oesophagus, stomach, and duodenum

The oesophagus lies embedded in splanchnic mesoderm but is not surrounded by a coelomic space and therefore has no mesentery. Its endodermal lining is at first columnar and ciliated but later becomes stratified and squamous. It is not known whether this is a change in the oesophageal epithelium itself or whether squamous cells migrate from the epithelium of the mouth.

The stomach is at first a spindle-shaped tube, but its wall expands faster on the left than on the right, so that it soon comes to have greater and lesser curvatures (Fig. 16.6). The dorsal mesentery (dorsal mesogastrium)

is attached to the greater curvature and comes to form a sac-like fold, the *greater omentum*, allowing the stomach to expand (Fig. 16.7). This fold thus lies on the left side and contains a cavity the *lesser sac*, opening on the right side to the general peritoneal cavity, the *greater sac*. The ventral mesogastrium is attached to the transverse septum, which contains the developing liver. It becomes the *gastrohepatic (lesser) omentum* and the portion between the liver and the ventral body-wall forms the *falciform ligament*. The rest of the ventral mesentery disappears.

The duodenum elongates to make a loop, suspended in a mesentery so short that the caudal portion of the

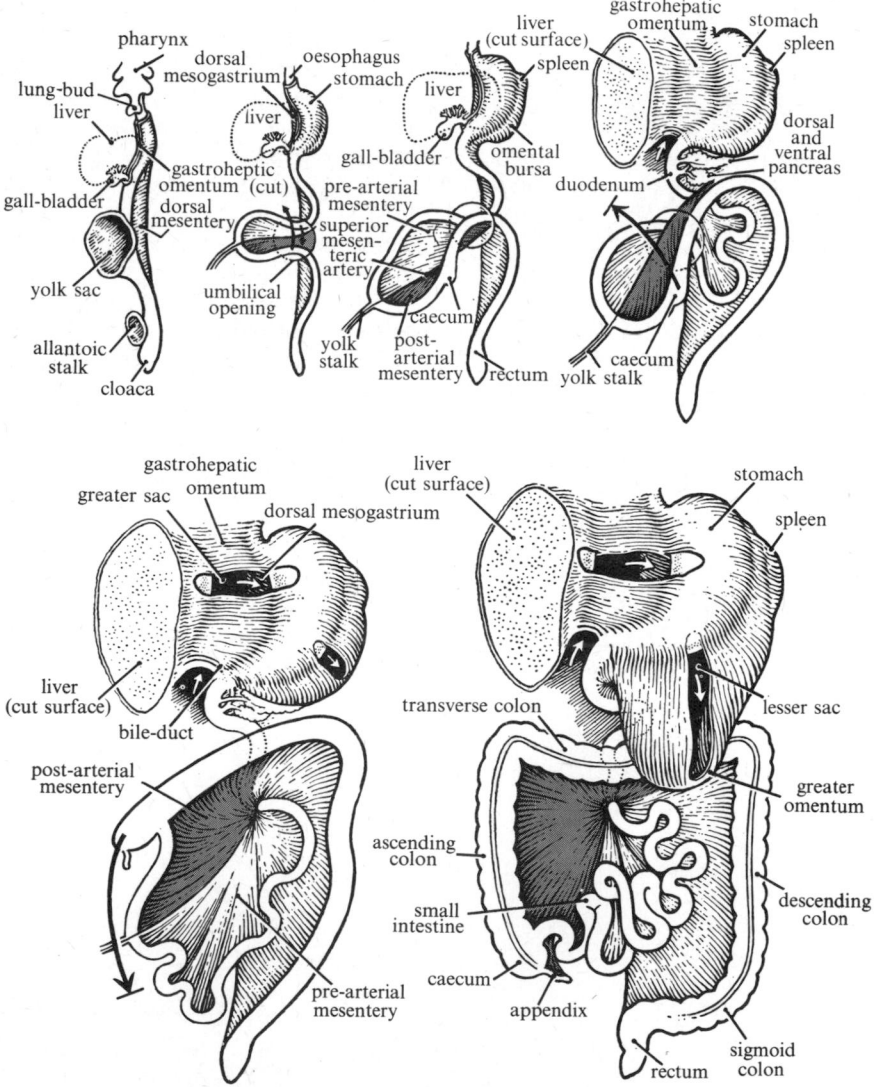

FIG. 16.7. Diagrams showing changes in position of the gut and mesenteries during development.

duodenum is fixed to the body-wall behind a sheet of mesothelium and is therefore said to be retroperitoneal.

The *liver* develops as a bud from the duodenum, extending into the ventral mesentery and septum transversum. The cavity of the original diverticulum remains as the *bile-duct*, which runs along the free caudal edge of the lesser omentum (Fig. 16.7) (see Du Bois 1963).

The *pancreas* in man arises from the duodenum as two buds, a dorsal and a ventral, the former lies cranial to the hepatic diverticulum, the latter opens with it. The two portions later fuse and their ducts usually join forming a single tube, opening with the bile-duct into the duodenum.

7. Derivation of mid- and hind gut

The intestine caudal to the duodenum develops from the mid-gut. At first it is a short tube but quickly grows in length and forms loops (Fig. 16.7). The dorsal mesentery also grows but remains fixed to the posterior abdominal wall by two strong ligaments (retention bands).

The hind gut gives rise to the lower part of the colon, the rectum, and upper part of the anal canal. The terminal portion at first consists of a single chamber, the *cloaca*, into which open the hind gut, the common excretory duct (p. 223), and the allantois (p. 489), whose proximal portion becomes expanded to make the bladder. The *urorectal septum* then divides this into a dorsal *rectum* and ventral *urogenital sinus*. The rectum is at first closed by an anal membrane and opposite this there develops an ectodermal depression, the *proctodaeum*. The anal membrane breaks down and the terminal portion of the anal canal is formed by the ectoderm of the proctodaeum.

8. Differentiation in the wall of the gut

The lining of the gut and its derivatives provide a set of cells specialized to respond to the presence of food substances in such a way that the latter are broken down and absorbed. These special developments of the digestive powers that are present in all living cells have become more and more marked in the course of evolution. In the lampreys, which are, perhaps, not very far from the ancestry of all vertebrates, there is no stomach or pancreas and the cells that produce trypsin and insulin are still partly incorporated in the wall of the intestine (p. 458). Nothing is known of the underlying developmental processes by which the parts of the gut and its glands acquire their particular characters. There must be a system of evocators and responses by which the stomach develops peptic and enzymic cells, the duodenum, Brunner's glands, the pancreas, its acini and islets, and so on. Probably these differentiations result from quantitative differences in a single set of processes. When we come to understand them further we should be better able to control digestive activities in health and disease. Comparative anatomy and embryology should provide clues as to the common factors that underlie the activity of the many different parts of the gut.

17 The internal environment. Blood-vessels and capillary circulation

1. The internal environment

MANY of the characteristic properties of cells depend on the difference between the composition of the cytoplasm and that of the surrounding fluids. One marked difference is that the substance of most cells is rich in potassium and poor in sodium, whereas in the blood and intercellular fluids the reverse is the case. The significance of this difference in the activities of the nervous system is discussed on p. 244. Similar differences maintained across the surface membranes are important in many other tissues. It was suggested by Macallum that the low sodium concentration within the cells represents the condition of the sea at the time when protoplasm first became separated from it, little salt having as yet been washed down from the land. Similarly, the concentration of the sodium chloride in the blood plasma (0·75 per cent in mammals), which is much lower than that in the sea today (3·5 per cent), has been held to represent the condition in the sea of the Cambrian Period 500 million years ago, when the vertebrate blood system became established (see Pantin 1964).

Whatever truth there may be in these historical speculations, the difference between cells and their surroundings is certainly fundamental to the life of mammals today. The system of spaces around and between the cells has become developed into an elaborate set of channels, sometimes called the *internal environment*. In some of these spaces the fluids circulate actively, notably in the vascular and lymphatic systems, but there is much in common between the cells that line the blood-vessels and the cells of the connective tissues that make up the packing between the special cells of each organ (see Cannon 1932).

Various systems are adopted to classify the fluids of the internal environment for purposes of description and analysis. These distinctions between blood, lymph, and connective tissues are important and valuable, but is must be remembered that even when a fluid is enclosed in a relatively complete set of vessels, as is the blood, it is still not by any means isolated from the rest of the body. Considerable amounts of fluid may leave and re-enter the blood-vessels through the walls of the capillaries, and we cannot regard 'the blood' as a distinct substance, constant in amount. Moreover, rapid interchanges take place between the cells and the 'spaces' in the body that contain fluid, including the cell contents.

We may list the intercellular spaces as: (1) the blood-system; (2) the lymphatic system; (3) the cerebrospinal spaces; (4) the coelomic spaces of the pleural, pericardial, and peritoneal cavities; (5) other tissue spaces. The liquids contained in these spaces, together with a few special liquids, such as the humours of the eye, provide the media within which the cells lie. Not all of the 'spaces' have equally definite boundaries. The coelomic cavities are limited by well-marked covering mesothelia. The blood and cerebrospinal fluids are enclosed in channels that are mostly circumscribed but are 'open' in some places, for example, the sinusoids of the liver. The lymphatic system includes some definite lymphatic vessels, which may communicate also with a system of tissue spaces between the cells. These spaces are not definitely demarcated nor lined by endothelial cells.

In view of this complexity and vagueness of the spaces within the body it is not easy to specify exactly the limits of the internal environment that is often supposed to be provided for the cells. Claude Bernard planted the conception of an internal environment in the brains of biologists with his dictum that '*la fixité du milieu intérieur est la condition de la vie libre*'. He was thinking mainly of the blood and in particular of the blood in the larger vessels of mammals, which is kept remarkably constant in its content of inorganic salts, oxygen, hydrogen ions, and materials necessary for the cells, such as glucose, protein, and fat. Even this fixity is only a steady state based on a continual change in the composition of the blood; as it passes through the capillaries, as we shall see, it loses water, salts, and other small molecules and becomes temporarily a more concentrated protein solution.

To regard the blood as a fluid of fixed composition is to ignore the first lesson that biology learned from the study of the circulation, namely that the apparent stability of living things is a steady state, achieved by intense underlying activity and change. The constancy of the composition of the arterial blood of mammals is a sign that these animals, like others, are systems that preserve a constant organization. As Barcroft has pointed out, mammals maintain a greater constancy in the blood than do lower animals and by this means they achieve the possibility of especially delicate adjustment of processes, particularly in the brain.

Investigation of the composition of the various body fluids has made it even more difficult to recognize any one 'internal medium', since these fluids differ among themselves in composition and also vary with the metabolic states of the tissues. It is necessary therefore to describe the various spaces, their linings, and the fluids they contain, considering each of them as part of the whole system of the body, neither more nor less 'living' than the cells, for which they provide an 'environment' only in the literal sense that they partly surround them. The interchange of ions across cell membranes is rapid and continuous. It is doubtful, therefore, whether we are even justified in regarding the cells as units, each with an 'environment'. Some of the most important interchanges go on at the surface of the mitochondria and other intracellular constituents. In describing living organization we have to take account of all these various barriers and try to discover how they play a part in providing a system that maintains a steady state in spite of fluctuations in the environment.

2. The blood-vessels

The blood-vessels are commonly divided into *arteries*, *capillaries*, and *veins*, through which the blood flows in that order. But distinctions are not sharp within the blood system; there are arteries of different sorts, grading down to arterioles and arterial capillaries, and similarly on the venous side. Moreover there is no sharp difference between the cells of the blood itself, of the walls of the vessels, and of the connective and other tissues. This is not a set of fixed tubes through which blood flows as does water in metallic pipes. The finer branches of the network are formed of cells continually growing and adapting themselves from moment to moment to the activities of the tissues. Even the larger vessels rapidly change their structure if they are differently used. Occlusion of the main artery supplying a tissue is followed by the opening up of a new circulation through roundabout channels. This regulation of the pattern of the blood-vessels proceeds by processes that are little understood. It is slower than the regulations produced by nervous or hormonal action, but it resembles these in ensuring that the body stores a representation of the events that have occurred in the past and are thus likely to occur in the future (p. 4).

The walls of arteries and veins (Fig. 17.1) have three layers: (a) the *tunica interna* or *intima*, (b) the *tunica media*, and (c) the *tunica externa* or *adventitia*. The intima consists of a layer of *endothelium*, composed of flat cells of polygonal outline which are similar to the fibroblasts of connective tissue and can probably become converted into these. Such cells line all the vessels and are the sole constituent of the walls of the capillaries. The intima of arteries is bounded externally

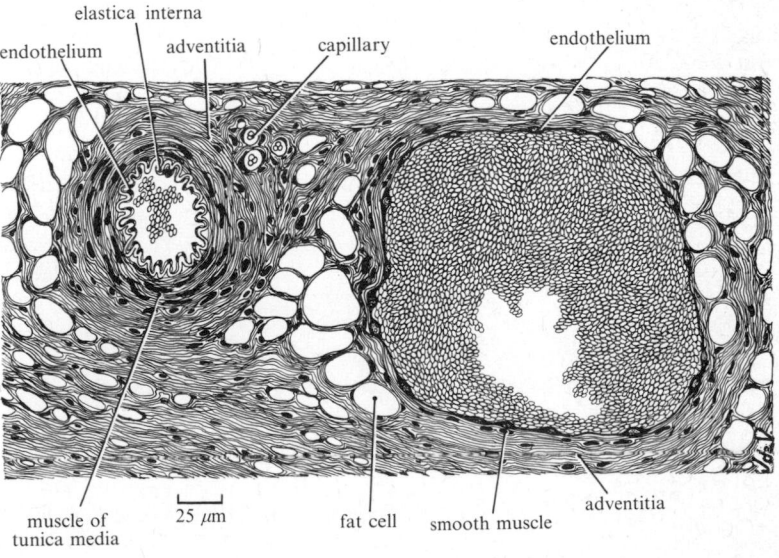

FIG. 17.1. Sections across a small artery and vein from the submucosa of the human intestine.

elastica interna
endothelium
adventitia
capillary
endothelium
muscle of tunica media
25 μm
fat cell
smooth muscle
adventitia

by a thin layer of elastic fibres (internal elastic membrane). Next to it lies the *tunica media*, which consists of smooth muscle fibres, running mostly round the vessel, together with varying amounts of collagen and much elastin in the large arteries. Outside this is the *tunica externa*, composed of collagen and elastic fibres, running mainly along the vessels. The adventitia also carries small blood vessels, the *vasa vasorum*. This is the only part of the wall that has its own blood and lymph vessels. The pressure in the inner layers would be too great for them, and these layers are nourished only by diffusion. Consequently they are especially liable to failure as in the various forms of arterial sclerotic disease.

The structure of the arteries varies continuously proceeding from the heart. The larger arteries contain little muscle, and their tunica media is almost wholly composed of elastic tissue. These vessels are therefore dilated by the systolic pulse wave, and they contract during relaxation of the ventricles (diastole), thus smoothing the flow of blood. The medium arteries have a well-developed muscular tunica media, and this, with its supply of nerves from the sympathetic system, enables regulation of their diameter so that the flow of the blood is directed to parts where it is needed (see p. 284).

The walls of the veins are essentially similar to those of arteries, but they contain less muscle and elastic

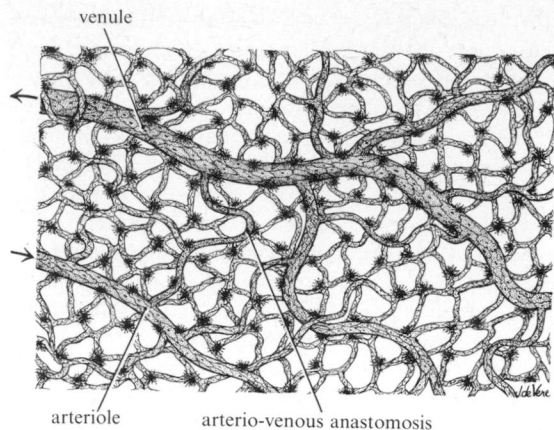

FIG. 17.3. Portion of capillary bed from web of frog's foot. Many star-shaped chromatophores are seen.

tissue and it is difficult to distinguish clearly between the various layers. Because of their thin, inelastic walls veins usually do not retain a rounded form when empty. In the medium-sized veins there are large valves to prevent the reflux of blood. These valves are covered by folds of the endothelium, and they contain a core of elastic fibres, continuous with those of the tunica interna.

Arterioles gradually become smaller and smaller until they become capillaries, the distinction being that the latter consist of an endothelial layer, without a muscular wall (Fig. 17.2). The final portion of the arteriole is sometimes called a metarteriole and its muscle fibres constitute a precapillary sphincter, whose contraction can stop the circulation through the capillary bed. The true capillaries are simple tubes, composed of a single layer of endothelial cells. They are about 8–10 μm in diameter (Figs 17.3 and 17.4). The capillary wall allows the passage not only of water and small molecules but of some larger ones; therefore it is postulated to have 'pores' of up to 6 nm diameter. The larger albumins and globulins do not usually pass through, but may sometimes do so. The 'pores' have never been seen, but may be somehow connected with the vesicles that are found in the capillary cytoplasm (Fig. 17.5). Where the endo-

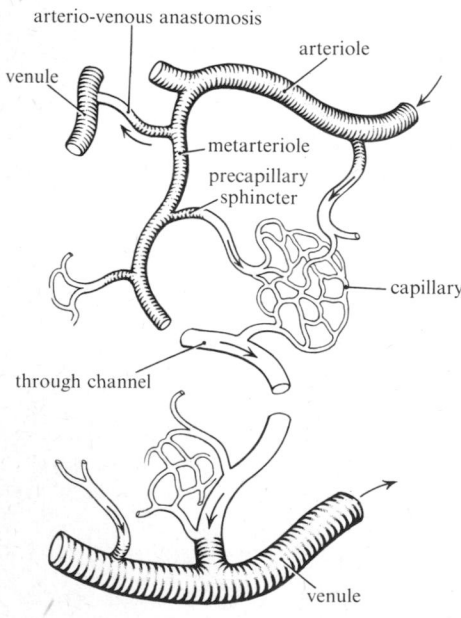

FIG. 17.2. Diagram of a portion of the capillary bed. The vessels with muscles in the wall are shown with cross shading

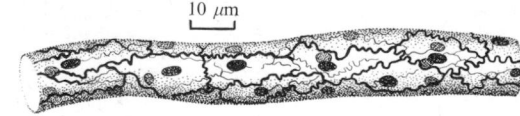

10 μm

FIG. 17.4. Portion of a capillary in the mesentery of a frog seen after staining with silver nitrate to show the boundaries between the endothelial cells. (After Ranvier.)

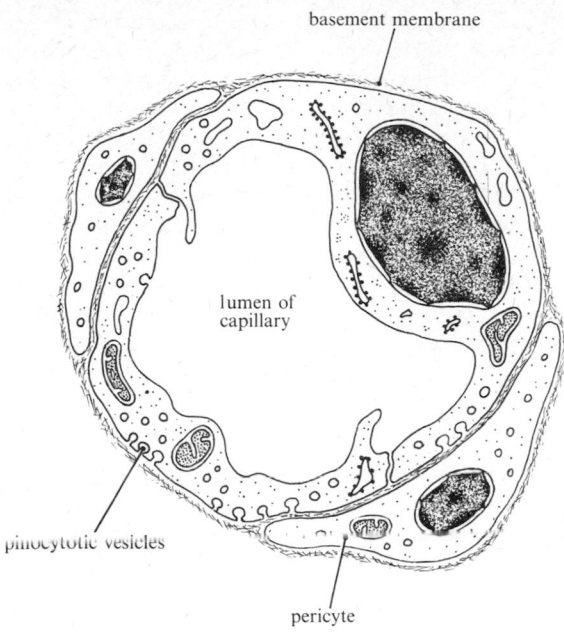

basement membrane

lumen of
capillary

pinocytotic vesicles

pericyte

Fig. 17.5. Diagram of skeletal muscle capillary in cross-section as seen by electron microscopy.

thelial cells join there is a space 10–20 nm wide. This is probably usually closed at some point by a *zonula occludens*, but these may perhaps open from time to time. The endothelial wall is covered by a basement membrane, within which is included a further type of cell the *pericyte* (Fig. 17.5). White cells can enter or leave between the cells. Various forms of supporting tissue have been described and probably some capillaries are surrounded by connective tissue sheaths composed of scleroprotein.

The true capillaries have no muscles in their walls (at least in mammals), but they may be able to change their diameter by changes of shape of the endothelial cells themselves. There are no nerves to the capillaries and any changes of diameter must be chemically controlled by metabolites and circulating hormones. These substances may also act upon the arterioles and pre-capillary sphincters, but these are provided with nerves.

Besides the channel from arterioles through capillaries to venules there are also direct *arterio-venous anastomoses* (Fig. 17.3). These are like capillaries in having no muscle in their walls, but they are larger. In tissues such as muscle, in which the amount of blood needed varies, there are many arterio-venous anastomoses acting as direct short circuits between large arterioles and venules.

The whole capillary system makes a fine network by means of which blood is brought to within a few micro-

metres of all parts of a tissue. Few cells are in direct contact with capillaries, however. Exchange must therefore take place through the medium of the fluid in the intercellular spaces, which are large channels running everywhere around the cells and among collagen fibres and other supporting tissues. The details of the arrangement of the finer vessels are able to change rapidly to provide for variations in demand. The vessels form a common anastomosing system covering a wide field, and if any arterial channel becomes severed or obstructed the tissues it formerly served soon come to be supplied by the opening of channels from neighbouring arterial fields (see Bloom and Fawcett 1968).

3. Composition of the blood plasma

The blood consists of a fluid, the *plasma*, in which float *blood cells* referable to three types: red corpuscles (*erythrocytes*), white corpuscles (*leucocytes*), and blood platelets (*thrombocytes*). The plasma of venous or arterial blood of man contains about 8 per cent of solid matter, distributed as follows:

(a) Protein (serum albumin, serum globulin, and fibrinogen)	7·0 per cent
(b) Other organic constituents (urea, amino acids, glucose, fats, etc.)	0·1 per cent
(c) Inorganic constituents (sodium, calcium, potassium, chloride, etc.)	0·9 per cent
(d) Hormones, enzymes, antibodies, etc.	traces

4. The circulation from capillaries to tissues

The composition of the blood is similar in arteries and veins but liquid does not pass unchanged through the capillaries. Starling first showed that considerable amounts of liquid may leave through the walls of the arterial ends of the capillaries and enter the tissue spaces or the tissue cells. Fluid is then re-absorbed into the venous portion of the capillary. This capillary circulation is an important part of the mechanism for supplying the cells. It depends upon the pressure provided by the heartbeat and on the difference in composition between the capillary contents and their surroundings, which is a result of the properties of the capillary wall.

In man, the hydrostatic pressure of the blood, derived from the heartbeat, is about 44 cm H_2O at the point where the capillary structure begins. This pressure tends to force water across the capillary wall, together with dissolved salts, oxygen, glucose, amino acids, and other small molecules that can pass the barrier. The back hydrostatic pressure in the tissue fluids or surrounding cells, due to the mechanical resistance of the tissues, varies from 2 cm H_2O upwards; it leaves an

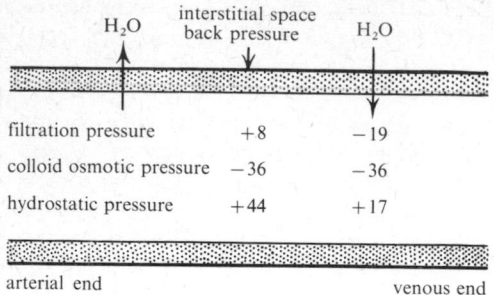

FIG. 17.6. Scheme of the fluid exchanges across the walls of a capillary. At the arterial end the hydrostatic pressure from the heartbeat exceeds the colloid osmotic pressure and the varying back pressure from the tissues. Water therefore leaves, but returns at the venous end of the channel. Pressure in cm H_2O. (After Davson 1970.)

outward pressure from the capillary of more than 40 cm H_2O.

In addition to these hydrostatic forces we must consider the osmotic pressures across the capillary wall (Fig. 17.6). The proteins of the plasma are molecules too large to pass out of the vessels and the tissue fluid is poor in protein. There is therefore a *colloid osmotic pressure*, of about 36 cm H_2O, tending to suck liquid from the tissue spaces into the capillaries. At the arteriolar end of the capillary there is thus a net outward pressure of the order of 4–8 cm water, driving fluid from the arterial end of the capillaries into the tissues. As the blood passes through the capillary towards the venous end the hydrostatic pressure within it becomes less and less, while the osmotic pressure of the proteins becomes greater. Eventually the combined hydrostatic pressure in the fluid around the capillary and the osmotic pressure of the plasma proteins produces suction of fluid back into the venous part of the capillary (see Davson and Eggleton 1968).

It is not known to what extent this circulation of fluid from the vessels to the tissue spaces takes place in normal life. Probably the proportion of the blood that passes through the capillary and extra-capillary circulations varies in different tissues and under different conditions. Solutes with small molecules pass through the walls of the capillaries by diffusion at least as fast as they would do if carried with the flow of fluid produced by the net filtration pressure. Any substance that is being used by the tissues (say glucose) will be present in lesser concentration in the intercellular fluids than in the blood-plasma. Diffusion down such concentration gradients is the main means of transport to the tissues. Nevertheless, the lesser circulation is important, and its existence can be shown in a variety of ways. The

most direct method is to compress a capillary and then to watch the movement of fluid along it proximal to the compression, owing to escape through the capillary wall (Fig. 17.7). In practice, the movement of blood cells along the capillary provides a satisfactory index of the rate of flow. Landis showed that this rate varies directly with the pressure in the capillary, measured by insertion of a micro-pipette used as a manometer.

The circulation between capillaries and tissue can also be shown to vary by altering the colloid osmotic pressure in the vessels. Oedema, that is to say swelling of the tissues, occurs if the blood colloid osmotic pressure is lowered, as by kidney disease or dietary deficiency, so that fluid accumulates around the cells in tissue spaces. It is uncertain whether such 'spaces'

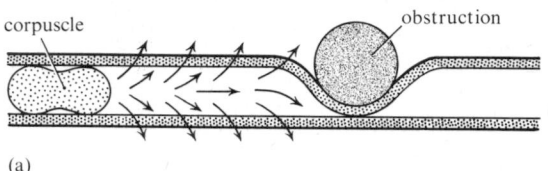

(a)

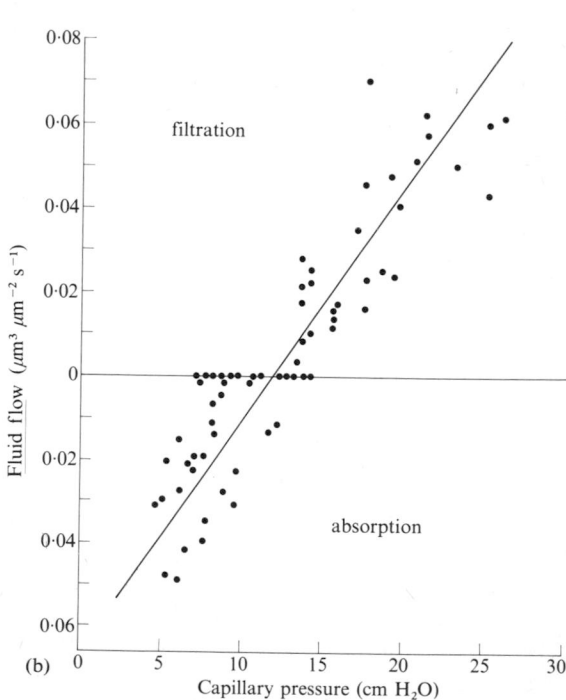

FIG. 17.7. Landis's experiment to determine the rate of flow across the wall of a capillary.

(a) The capillary is obstructed and the rate of movement of the corpuscle measures the rate of filtration or absorption of fluid. The results with various capillary pressures are shown in (b). (After Landis (1927). *Am. J. Physiol.* **82**.)

normally exist in the tissues. Oedema can occur also as a result of damage to the capillary endothelial cells or obstruction of the venous or lymphatic channels. Knowledge about this capillary circulation is obviously of first importance for understanding the condition of the tissues, their maintenance in good health, and the effects on them of trauma and disease.

5. Clotting of the blood

The power of the blood to become converted to a solid substance is one of the most familiar and important protective devices of the body. It depends on the presence in solution in the blood-plasma of a protein, *fibrinogen*, which under suitable conditions precipitates to form threads of *fibrin*. The change is brought about by the action of an enzyme *thrombin*, which is present in the circulating blood in the form of a precursor *prothrombin*. Damage to the tissues liberates a substance *thrombokinase* and this, in the presence of calcium, converts the prothrombin to thrombin and sets off the clotting. Circulating in normal blood there are numerous non-nucleated bodies, the *platelets* or *thrombocytes*, which also constitute a source of thrombokinase, breaking down when the blood is shed. The platelets are formed from the giant cells of the red bone marrow, *megakaryocytes* (Fig. 18.2, p. 173), and they serve further protective functions by sealing small gaps in blood-vessels (see Seegers 1969).

The mechanism of clotting can therefore be expressed as:

prothrombin + thrombokinase + calcium → thrombin;

thrombin + fibrinogen → fibrin.

The coagulation can be prevented by acting at various points in the chain. Perhaps the most familiar method is to remove all the free calcium. Blood treated by the addition of sodium oxalate will not clot, calcium oxalate being insoluble. *Heparin* is an anticoagulant substance, found in liver extracts, which prevents the action of thrombin on fibrinogen. The heparin probably arises from the *mast cells*, which are found in all connective tissues, and its presence normally serves to prevent clotting in the smaller vessels. The hereditary disease of *haemophilia*, transmitted as a sex-linked recessive gene, is a condition in which the activation of prothrombin is delayed, probably owing to the inability of the platelets to break down when the blood is shed. Men carrying the gene therefore bleed vigorously and continuously from the smallest wound.

A blood clot does not remain in the same condition after it has been formed. The particles of fibrin contract, probably by an intramolecular change somewhat similar to muscular contraction, fibrin being a protein related to myosin (p. 58). During this contraction the remains of the red cells become enmeshed in the clot and as a result a straw-coloured liquid, the *serum*, is pressed out. This liquid contains the salts and proteins present in the blood but, being devoid of fibrinogen, it remains liquid.

Blood clots play an important part in the *healing process* that follows an injury. Besides holding the damaged surfaces together the clot forms a matrix into which fibroblasts can move in order to lay down new collagen and make a firm union of divided surfaces. The ease with which this takes place depends on the direction of orientation of the fibrin fibrils, which is in turn influenced by the forces operating upon the clot. Within a day or so after injury the fibrin begins to be dissolved by enzymes. The first fibres to dissolve are those that do not lie in lines of stress and the clot is thus converted into a set of fibres orientated approximately along the stress lines. Fibroblasts and other cells grow along the surfaces thus provided and lay down collagen fibres in directions suited to prevent straining by the forces falling on the tissue.

18 The cells of the blood

1. The red blood corpuscles

THE red corpuscles are among the most specialized structures of the body. In adult mammals they are bags of haemoglobin, without nuclei, devoted solely to the transport of oxygen and surviving in man only for 100–120 days; then they are destroyed by the liver and spleen and replaced by others formed in the red bone marrow. There are about 5 million red cells per millilitre of blood in a normal man (less in woman, more in children), and therefore some 35×10^{12} of them in the body.

The shape of the corpuscles is characteristic and is usually described as showing a biconave or dumb-bell outline (Figs 18.1 and 19.2, p. 180). The effect of this shape is to give the corpuscle a 25 per cent larger surface area for its volume than would be presented by a sphere. It also allows rapid and equal diffusion from the surface to all inner parts. These are obviously efficient features for the purpose of oxygen transport. The surface is not one of minimum area, and it is difficult to see how the shape is produced; it must be maintained by some structure having considerable rigidity. This structure is presumably provided by the cell membrane. This is birefringent and contains phosphatide and protein components. The membrane can be pulled away as a distinct structure with micro-

dissection needles. The consistency of the contents of the corpuscle is that of a rather soft gel; the water content (60 per cent) is lower than that of many tissues. Yet the mechanical structure of the whole is such that the corpuscle can be deformed in shape, for instance, during passage through a capillary, and immediately afterwards resume its biconcave appearance.

The red corpuscles vary little in size, the mean longest diameter being $7 \cdot 2\,\mu$m and the thickness $2 \cdot 2$ μm as measured in fixed preparations. In the fresh state they are certainly over $8 \cdot 0\,\mu$m in diameter and may show fluctuations with the composition of the fluid in which they are placed.

The surface membrane is permeable to water but certainly not to all solutes, so that the cell shape is markedly affected by the tonicity of the surroundings. The 'normal' shape is preserved in $0 \cdot 9\%$ sodium chloride; various signs of plasmolysis ('crenation') occur in stronger solutions (Fig. 18.1), and with weaker ones the corpuscles become inflated and may allow escapes of the haemoglobin (haemolysis) leaving only a red-cell 'ghost'. The contents approach those of other cells in inorganic composition, that is to say, they contain more potassium and less sodium and chloride than does the blood-plasma.

The red corpuscle of an adult mammal thus retains many characteristic features of cells but has lost its whole apparatus for synthesis, including nucleus, endoplasmic reticulum, ribosomes, Golgi bodies, and mitochondria, leaving a homogeneous 'cytoplasmic' material composed almost wholly of haemoglobin, which indeed makes up 95 per cent of the dry weight of the corpuscles. The corpuscles of the foetus are, however, still nucleated.

2. Myeloid tissue

The life of red cells, lacking a nucleus, is limited to about 3 months in man. Production of these, as of the other components of the blood, is basically a function of the connective tissues of the body, which are indeed allied both to the endothelial linings of the vessels and

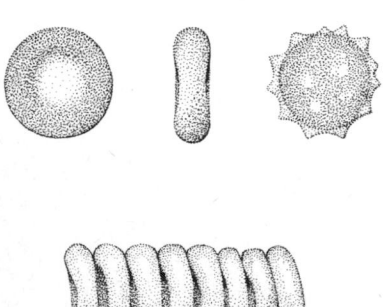

FIG. 18.1. Erythrocytes as seen under various conditions.

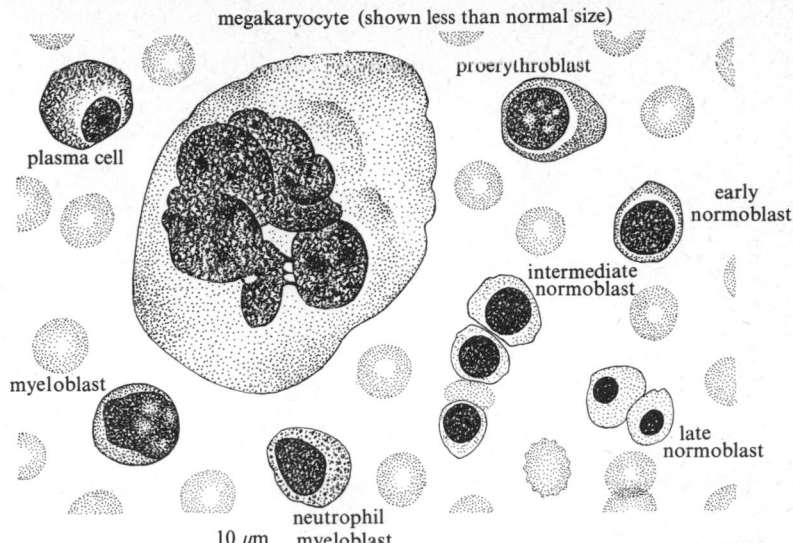

megakaryocyte (shown less than normal size)

proerythroblast

plasma cell

early
normoblast

intermediate
normoblast

myeloblast

late
normoblast

neutrophil
myeloblast

10 μm

FIG. 18.2. Some of the types of cell
seen in red bone marrow.

to the blood itself. In the earliest embryo the blood is
formed from the connective tissue of various parts of
the body, especially in the yolk sac, liver, spleen, lymph
nodes, and bone marrow (p. 488). The power becomes
gradually more restricted, until in adult life all the
haemopoiesis goes on in the *red marrow* of the ends of
the long bones and especially in the flatter bones such
as the bodies of the vertebrae, the ribs and sternum.
The shafts of the long bones are filled with a fatty
yellow marrow, which is not haemopoietic but may
become so when there is a demand for more red cells.

The red marrow or myeloid tissue (Fig. 18.2) is a soft
material having, like lymphoid tissue, a basic stroma of
reticular tissue. This consists of a fabric of fine scleroprotein fibrils among which lie 'primitive reticular
cells', which are presumed to give rise to all others in
the tissue. In addition, myeloid tissue contains fat cells,
many macrophages, and blood cells in all stages of
formation. The blood-supply is peculiar in that, instead
of capillaries, there are numerous large sinusoids, which
have a very thin wall and allow free passage of cells.

Although much remains to be discovered about blood
formation it is probable that, as the unitary theory
holds, all types of blood cell can be produced from a

TABLE 18.1

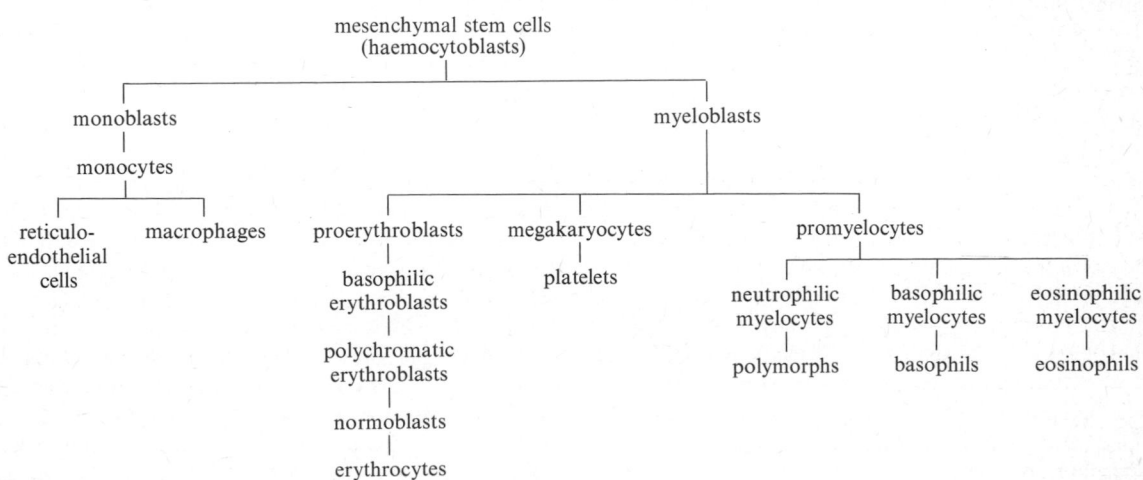

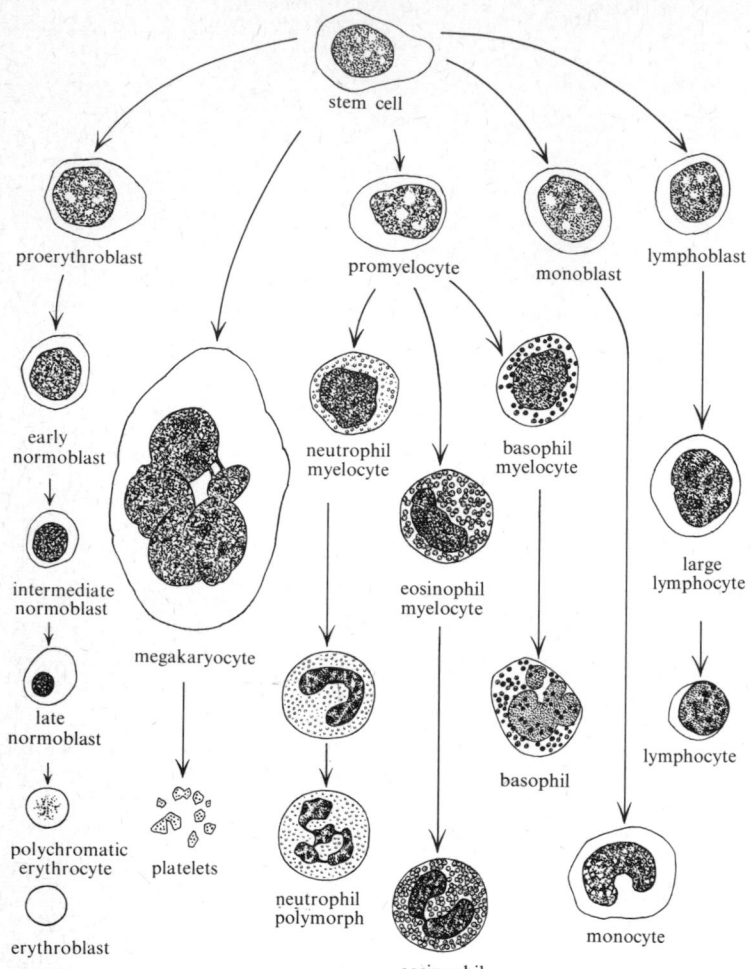

stem cell

proerythroblast

promyelocyte

monoblast

lymphoblast

early
normoblast

neutrophil
myelocyte

basophil
myelocyte

intermediate
normoblast

eosinophil
myelocyte

large
lymphocyte

megakaryocyte

late
normoblast

basophil

lymphocyte

polychromatic
erythrocyte   platelets

neutrophil
polymorph

monocyte

erythroblast

eosinophil

FIG. 18.3. Diagram of the processes of formation of red and white cells. The polychromatic erythrocyte is the reticulocyte of cresyl blue stained preparations.

single undifferentiated cell type of the mesenchyme (Fig. 18.3). These stem cells have not been identified for certain but they are probably those that are some-times called haemocytoblasts, which are rather like large lymphocytes, with a deeply staining nucleus and somewhat basophil, non-granular cytoplasm. The red marrow has the double function of producing new cells, red and white, and removing the old ones. The mesenchymal stem cells therefore must be able to produce the precursors of the two main lines, those leading to production of new cells (known as myelo-blasts) and those responsible for removing the old ones, the reticulo-endothelial cells of haemopoietic tissue or macrophages of loose connective tissue (Table 18.1). The myeloblasts in turn are the cells able to form the three main types: erythrocytes, granular leucocytes, and platelets.

3. Haemopoiesis: plasma proteins

The production of the cells and other materials of blood and lymph takes place in many parts of the body, but we can recognize two major types of haemopoietic (blood-forming) tissue: the lymphatic tissue of the lymph nodes, spleen, and thymus and the myeloid tissue of the bones. The functions of these overlap in that cells developing in one migrate to the other and to connective and other tissues elsewhere. The regulation of the numbers and proportions of these various types of cell in the body is a complex matter, fundamental for homeostasis, but little understood.

A large portion of the proteins of the blood is formed in the liver, including albumin, fibrinogen and α and β globulins. Thus when liver slices are incubated with labelled amino acids, albumin appears in the parenchymal cells in a few minutes. The γ globulins (antibodies) are formed by the plasma cells (p. 189).

Different plasma proteins vary in turnover rates from half-lives of hours to weeks. Some at least are degraded in the liver, and some probably in the intestine. After injury or stress there is usually a fall in plasma proteins, followed by increased synthesis in the liver. The stimulus for this is not known but may be partly from the histamine and other substances produced by tissue injury (see Owen 1966). Adrenal corticosteroids, besides increasing the amount of protein degraded and excreted, also stimulate its production.

The net effect of injury is to increase the rate of plasma protein turnover, perhaps with the effect of making amino acids more readily available for repair.

4. Stem cells of the myeloid tissue

Although the stem cells have not positively been seen, good evidence for their existence comes from lethally irradiated animals. In these the number of neutrophils falls and, rather later, that of megakaryotes and red cells. Life can be saved by injecting small amounts of isogenic marrow from a normal individual. The cells form clones which appear as little nodules on the spleen. There is evidence that each nodule comes from a single cell, for when the injected material was itself sub-lethally irradiated then, in some of the colonies, all the cells could be seen to have the same chromosome abnormality (Becker, McCulloch, and Till 1963).

Such splenic colonies are able to show differentiation, along all the three lines that derive from the myeloblasts. Probably the basic stem cells of the mesenchyme are similar. These cells are not yet fully undifferentiated since they retain the power of forming all the blood cells and perhaps others (such as fibroblasts) as well. We are beginning to understand how their multiplication is controlled and what causes the products to differentiate into the various cell lines as required (see below). Control of this would be extremely valuable in the management of many diseases. The differentiation into the red-cell line is promoted by a hormone, erythropoietin (p. 176). Presumably some such stimulus is responsible for initiating production of each of the lines deriving from the original stem cells. Probably the differentiation proceeds gradually, the first products being still capable of several further divisions to produce many cells of a given type.

5. The formation of erythrocytes

The myeloblasts are 12–20 μm in diameter, with a nearly round nucleus, not deeply basophilic. The cytoplasm is faintly basophilic and contains free ribosomes but little endoplasmic reticulum. In the transition to erythrocytes, the myeloblasts pass through the stages shown in Table 18.1. The cytoplasm becomes more basophilic in the stages known as proerythroblasts and basophilic erythroblasts, due to an increase in ribosomes. Then as haemoglobin begins to accumulate, the cytoplasm also takes up acidic dyes, hence these are called polychromatophilic erythroblasts. These may develop directly into erythrocytes, the nuclei degenerating (pyknosis). The cytoplasm then still contains some RNA, hence such cells show a basophilic network and are known as 'reticulocytes'. They occur in the blood when haemopoiesis is very active. More usually the polychromotophilic erythroblasts continue to divide and lose their cytoplasmic RNA. They are known as normoblasts and nucleus degenerates to leave a typical erythrocyte.

6. Control of red-cell production: erythropoietin

There are probably separate control systems for all the seven types of cell shown in Table 18.1 and indeed for subtypes as well. However, there are conditions that influence many of them together, where they share a common stem cell. There are various ways in which regulation could be achieved, and it is not certain which are used. It may be by single feed-back loops activated by increased demand and acting upon cell production. Alternatively, the situation may be more complex and demand may produce increased release from a storage compartment, whilst this might initiate maturation in a second compartment which in turn might activate a mitotic compartment. There is evidence that stem cells can divide only a limited number of times and there must therefore be some concatenation of compartments.

The control of the red-cell production is mediated by a hormone, erythropoietin, produced by the kidney in response to anoxia (Boggs 1966). Production is increased following cell-loss or decreased atmospheric oxygen and decreased by hypertransfusion of erythrocytes. The increase is faster than the decrease. Under normal conditions the feed-back is very precise, and no subtle waveform or other oscillations have been detected.

The evidence for the existence of the hormone is that plasma from anaemic animals injected to normal ones produces increased red-cell formation, but not after nephrectomy. If one member of a parabiotic pair of rats breathes an atmosphere low in oxygen then the red marrow of both shows hyperplasia. The juxtaglomerular cells of the kidney are the main source of the hormone (p. 222), but there may be others in the body. The hormone is a glycoprotein of molecular weight 40000–60000. It is not known how its basic level is set or related to erythrocyte removal from the circulation.

The attainment of a certain low oxygen level is presumably sensed by the juxtaglomerular cells in such a manner as to promote the synthesis of more erythropoietin. The setting of this level of sensitivity by appropriate synthesis during development is the prime determinant of the rate of haemopoiesis. The level itself, of course, may be capable of later adjustment by change in the amount of some receptor that is synthesized or indeed simply by production of more (or less) juxta-glomerular cells. In some such manner it is ensured that at all stages of growth, activity, and environmental oxygen level there is an adequate supply of red cells.

The erythropoietin itself probably acts on the bone marrow by stimulating the formation of proerythroblasts from the myeloblast stem cells. Evidence of this is obtained by stopping haemopoiesis in mice by hypertransfusion (polycythaemia) and then giving a single dose of erythropoietin. This produces a crop of new proerythrocytes whose progress can be followed. After 36 hours most of them were early normoblasts and after 72 hours they were reticulocytes ready for release. Single doses of actinomycin D, which prevents the formation of DNA-dependent RNA, if given 1 hour after the erythropoietin, prevented the process completely. Given at 16 hours or 24 hours it reduced the red-cell output, but less drastically. There is, therefore, certainly a marked influence of the hormone in inducing differentiation of stem cells. It may well be that this involves a de-repression of the genes appropriate to the programme of differentiation into an erythroblast. If prolonged actinomycin inhibition is stopped red-cell production begins within 18 hours. The stem cells have therefore not been permanently damaged.

There is evidence that the key enzyme in haem synthesis is γ-aminolevulinic acid synthetase (ALA). It appears in the spleen within 8 hours of a single dose of erythropoietin to polycythemic mice. This appearance is completely blocked by actinomycin given at the same time, but not if given 6 hours later. Therefore, this may well be the first enzyme induced, and its products may in turn induce others involved, such as ALA dehydrase and iron protoporphyrin chelating enzymes (see Weyer 1968).

The haemoglobin molecule is composed of four polypeptide chains, two known as α and two as β chains, together with four iron-containing haem groups. The molecular weight is 64 500 and the α chains have 141 amino acids and the β chains 146 (see Perutz 1969). The molecule has long helical portions and the remainder are bent or kinked, giving an elaborate tertiary structure, approximately globular over all.

During haemopoiesis each polypeptide unit is assembled, like other proteins, by stepwise polymerization on a polysome starting from the amino end of the chain, the sequence being determined by the mRNA. Each amino acid is activated by ATP and transferred to a molecule of tRNA, three nucleotides of which are attached by hydrogen bonding to the complementary sequence on the mRNA. A new peptide bond is then formed enzymatically between the carboxyl group of the previous amino acid and the amino group of the incoming one, and the tRNA is removed. The complete polypeptide finally must be released and associated with its fellows and with haem, but it is not known how this is achieved.

In agreement with this model the nucleated erythroid cells of a rabbit, which synthesize Hb rapidly, have many ribosomes, mostly in polysomes. As the nucleus is lost the rate of Hb synthesis decreases and the number of ribosomes falls (Wilt 1967).

Thus red blood corpuscles are produced continually in the red bone marrow, and after a life of a few months they are destroyed by the spleen. This continual replacement of the tissues of the body is another, clear example of the general rule that the living organism must be considered not as a particular set of matter carried from the cradle to the grave but as a way of so transforming energy that a particular plan of organization is preserved. It is calculated that, in man, *10 million red corpuscles are formed and destroyed every second* and this gives a clear idea of the great changes that are going on in spite of the constant appearance of the whole body.

7. The spleen

The *spleen* (Fig. 18.4) is a portion of the circulatory system having some characteristics of lymphoid tissue and others of bone marrow. It lies in the mesentery, close to the stomach, supplied with its own artery and vein. It assists in various ways in the adjustment of the blood. The spleen has a muscular contractile wall, but there is relatively less muscle in man than in many mammals. This capsule surrounds a pulpy substance in which the blood comes into contact with a network of macrophages that acts as a filter system. Around the branching arteries of the organ are little masses of lymphoid tissue ('white pulp'), and between these there lies the 'red pulp', which is a mass of connective tissue trabeculae and a mesh of reticular tissue. This is permeated by passages leading from the straight arteries called penicillary arteries to the venous sinusoids of the red pulp. Some of these channels have walls of reticuloendothelial cells. There has been much discussion as to the nature of the communication between arteries and veins in the spleen, and it is uncertain whether there is

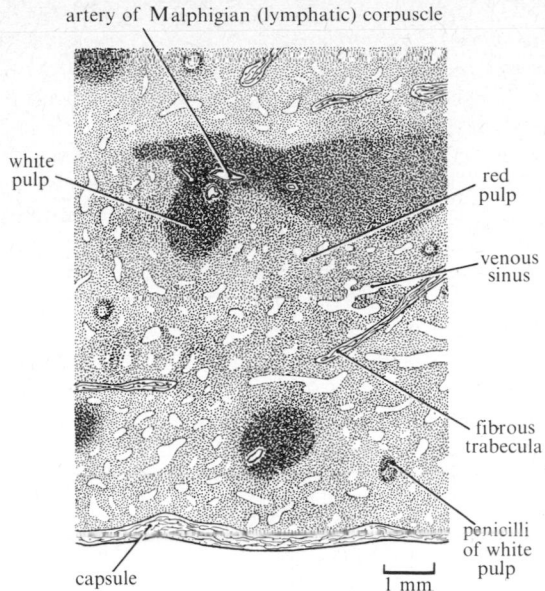

artery of Malphigian (lymphatic) corpuscle

white pulp

red pulp

venous sinus

fibrous trabecula

penicilli of white pulp

capsule

1 mm

FIG. 18.4. Low-power view of a part of a section of the spleen of a rabbit. (From a photograph lent by Mr. K. C. Richardson.)

a system of open sinuses. Certainly its substance provides, like the lymphoid tissue, a means by which the blood is brought into close contact with the elements of the reticular tissue. Red cells in process of destruction can be seen at all times in the macrophages of the spleen, and bacteria or coloured matter injected into the blood-stream soon appear in the red pulp. The organ thus functions as a filter for the blood rather as the lymph nodes do for the lymphatic system.

The spleen also produces lymphocytes, monocytes, and plasma cells. In many mammals the spleen stores some of the available red corpuscles, and these can be expressed into the circulation when exercise or reduced oxygen supply make extra calls on the oxygen transport system. This action is performed by contraction of the smooth muscle in the walls of the spleen, under the control of the sympathetic nervous system. A considerable proportion (one-fifth in the dog) of all the red corpuscles of the body are stored in the spleen when the body is at rest.

8. Regulation of the internal environment

Our analysis has shown that the conception of a single internal environment in which the tissues lie is only of limited value. Exchange between the cells and their surroundings is exceedingly rapid and the surfaces that are significant in regulating it may lie within the cells,

for instance, in the endoplasmic reticulum, Golgi bodies, or mitochondria (p. 2). Nevertheless, the composition of the fluids that circulate rapidly in the body has a special importance since it affects all cells alike and therefore provides a common factor producing a somewhat similar influence on all parts.

In order to be able to fulfil its functions it is not necessary that the blood should maintain an equally constant composition of all its components, and they are in fact regulated to different extents. The regulation is quickest and the constancy greatest in the oxygen content, which affects the performance of the tissues from moment to moment. The glucose that provides the material for oxidation to provide energy is also maintained nearly constant (Chapter 48). Materials for growth and replacement, although ultimately essential, are not constantly required and they fluctuate in amount in the blood. Changes in the composition of the blood may be actually used as a method of signalling, not only by the specific hormones (Chapter 45) but also by such effects as are produced by the accumulation of carbonic acid upon the respiratory centre of the brain or of amino acids on the liver. The presence of a common circulating fluid thus provides the possibility of an integration of activities in many parts of the body. Alterations in the amount of fluid in the body are regulated by intake and output rather than in the blood itself.

The oxygen-carrying power of the blood is one of its most important properties and is regulated both quickly by nervous action to anticipate rapid changes in demand and also by slower processes. The spleen provides a store of red corpuscles that can be discharged into the circulation during exercise. Moreover the speed of circulation of the blood can be increased (Chapter 21) and the direction of flow regulated by contraction of some of the arteries, so that blood is available where it is most needed. Slower adjustment of oxygen-carrying power is seen in the increase in number of red corpuscles produced by sojourn at high altitudes. No doubt there are still slower changes in the blood produced by the operation of the evolutionary processes of variation and natural selection. These have provided the characteristic features that give to mammalian blood its great capacity for transport of oxygen.

In its protective functions the blood is also regulated at various rates, but here the nervous system plays little or no part, since the regulation is slow. The capacity to form a clot is an inherited anticipation of the probability that vessels will at some time become damaged. Moreover the clot becomes invaded by fibroblasts and its fibrinogen molecules dissolved in a manner that provides for the laying down of collagen so that future stresses will be met (p. 34).

The capacity to resist invasion by foreign substances and organisms is conferred ultimately by heredity but in each case the antibodies are produced in the blood as a response to the presence of a particular type of extraneous invader. The hereditary system confers on the blood the power to carry a specific memory of past 'experience'. This is only one example of a whole complex of changes that take place in the capacity of the organism to react to invasion and to stress (p. 428). The entire white-cell system is continually changing, for example, in the number of macrophages or lymphocytes that are being produced according to the influences that have been falling upon the organism.

In many features therefore the internal environment provides a representation of the past and thus antici-pates events that are likely to occur to the organism in the future. This anticipation is secured by selection among various possible reactions of those that conform to the conditions that have occurred in the environment. Thus, of the fibrin fibres laid down randomly in a clot, those that are not pulled upon are removed first (see p. 171). This has been ensured during evolution by the process of natural selection. Organisms are produced with a variety of different sorts of fibrin and those whose clots do not show this property are eliminated. We can see vaguely how by these selective mechanisms it is ensured that the internal environment can receive 'information' and thus adequately represent conditions outside. Far more study is needed to complete our understanding of the relationship and to make the formulation of it precise.

19   Protection and defence of the body

1. Turnover, protection, and defence

ORGANISMS maintain their integrity by selecting certain environmental components which are then incorporated for varying periods of time. The instructions of the DNA ensure this continuity by organizing the formation of proteins of the numerous types needed. These include (a) those that prevent the entry of unwanted materials, for instance, the resistant keratins of the skin; (b) those that facilitate the entry of some molecules by providing the permeable surfaces of the lungs and gut and the appropriate enzymes inside and outside the cells; and (c) proteins that make possible the removal of unwanted material that has either entered in spite of the barriers or been produced within the body. This class will obviously include all the mechanisms usually called 'excretory', but we are concerned here with those that are considered as 'defensive' in that they deal with 'invasion' of bacteria, viruses, or other agents.

All of these living processes are interrelated, and it is really artificial to separate certain of them by making the military analogy with 'defence'. This is especially clear when we think that one of the chief agents is phagocytosis, by which the unwanted material is made part of the body itself, a process thus involving digestion and growth as well as defence.

In fact, since nearly all parts of the body are involved in turnover and replacement, the question of what is part of the body and what is 'foreign' to it is less clear than it seems. The enzyme systems must be so regulated as to produce at all times an appropriate balance between incorporation and rejection of the materials that are presented. It usually is held that this involves mechanisms for identification of the body's own macromolecules, so that they are not rejected by the defence mechanisms designed to destroy 'invaders'. This is an attractive view, but it presents difficulties, for in fact there is continual turnover with breakdown and replacement. The question of the maintenance of its 'identity' is too subtle to be properly handled by picturesque metaphors taken from warfare.

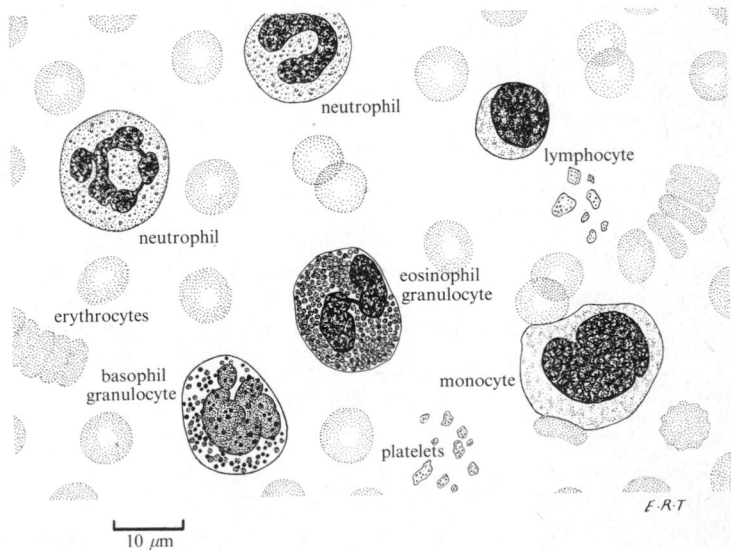

FIG. 19.1. The cellular constituents of normal human blood as seen after staining a smear preparation. (It would be very unusual to see all the types close together in one field in this way.)

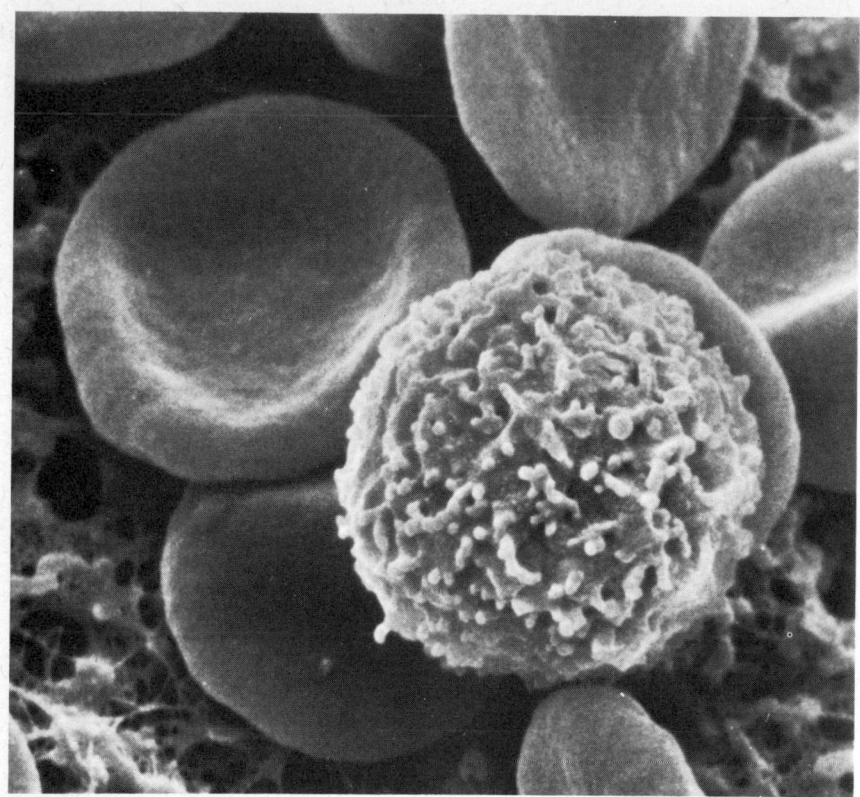

FIG. 19.2. Scanning electron micrograph of human red blood cells and one white cell (probably a neutrophil). This picture emphasizes the artefactual appearance of white cells in smear preparations. Field width 12·3 μm. (Figure kindly supplied by Elaine Bailey.)

2. The defensive tissues

Nevertheless, we can recognize in a mammal a distinct set of tissues that are concerned with the removal of unwanted substances and particles from the tissues. They include the white cells of the blood (*leucocytes*) and the related macrophages of the *reticulo-endothelial* systems of the tissues. Defence is ensured either by engulfing particles (*phagocytosis*) or by producing *antibodies* that unite with and destroy invading agents such as viruses. There are distinct cells for these two types of function, but the responses overlap and have much in common. Both depend upon the presence of some system of receptors for recognition of what shall be destroyed. Moreover, the entire 'defence' system is assisted by a series of systemic humoral agents including some steroids of the adrenal cortex. Indeed the response of the body to a massive infection may involve a great part of its resources so that many organs show signs of 'stress' (p. 429).

3. The white blood cells

There are about 7000 white corpuscles per millilitre in normal human blood (Figs 19.1 and 19.2), divided as follows:

agranulocytes	lymphocytes	20–30 per cent
	monocytes	3–8 per cent
	neutrophils (polymorphs)	
		60–70 per cent
granulocytes	eosinophils	1–3 per cent
	basophils	0·5 per cent.

Of these the monocytes and neutrophils become phagocytes, and the lymphocytes are concerned with antibody production (see Bloom and Fawcett 1968).

The neutrophils are so called because they contain numerous granules that are neither acidic nor basic. Their more usual name is polymorphs (polymorphonuclear leucocytes) because their nuclei show several lobes. The cells are 10–12 μm in diameter in dry smears but in fresh blood are spheres of 7–9 μm. They are formed in the myeloid tissue of the bone marrow. Their function is to collect around foreign bodies that have penetrated the tissues and to destroy them with enzymes. These enzymes are produced by the granules of the polymorphs, which are lysosomes (Fig. 19.3). These are the cells that collect around a splinter or in an infected wound. They are the first defensive cells to arrive, leaving the venous capillaries probably by passing between the endothelial cells. After they have

Fig. 19.3. Electron micrograph of a human polymorph (neutrophil). The plane of section is such that two separate lobes of the nucleus are seen. The most prominent cytoplasmic inclusions are the specific granules (lysosomes) which contain hydrolytic enzymes. (Figure kindly lent by Prof. H. Z. Movat.)

discharged their enzymes the whole cell often dies, and the aggregate of living and dying polymorphs constitutes pus.

The polymorphs destroy bacteria by surrounding them and taking them into a vacuole into which digestive enzymes are then passed from the lysosomes. It is not known how the neutrophil 'recognizes' the bacterium as a foreign body, nor how it avoids ingesting particles of the body itself. The number of polymorphs in the blood is increased during a heavy infection. Some signal, therefore, must be sent from the infected site to the bone marrow, but its nature is not known.

4. Formation of neutrophils (polymorphs)

These cells differentiate in the marrow from myeloblasts, first into large promyelocytes with a few granules and then into smaller myelocytes with an indented nucleus and numerous granules. The nucleus then becomes increasingly indented, forming first a horseshoe shape and then becoming constricted into five or more lobes (Fig. 19.1).

Little is known about the control of polymorph production. Half the total in the body are stored in the marrow so that a sudden appearance of more in the blood (as after infection) does not necessarily indicate increased production. Such shifts to the blood may occur from adrenal medullary secretion or exercise. However, it is probable that formation is increased by demand; for the number in the blood is chronically raised in severe infections or inflammations, where many are serving as phagocytes.

It is possible indeed that regulation of neutrophils is effected partly by stimulation of the marrow by the 'endotoxins' that are produced by the membranes of bacteria. Injection of any material containing endotoxin increases the number of circulating neutrophils. Indeed, the difficulty of avoiding endotoxin contamination makes it hard to decide whether there are any neutrophil-stimulating hormones. However, endotoxin stimulation cannot be the only method of regulation since animals kept free of bacteria still have polymorphs.

5. Eosinophils and basophils

The eosinophils (1–3 per cent of the leucocytes),

besides their acidophilic granules, contain large electron-dense granules. The cells are somehow concerned with immunological reactions and especially with allergy and with anaphylaxis. Hydrocortisone causes them to disappear from the blood and depresses allergic reactions. The eosinophils contain histamine, and either they produce it themselves or take it up from mast cells, (p. 29). They greatly increase during infection by parasitic worms.

Basophils are even less frequent in the blood (0·5 per cent). They resemble mast cells and like the eosinophils contain heparin and are concerned in allergic reactions.

6. Monocytes

These are the largest of the leucocytes, up to 20 μm in diameter when flattened in a blood smear. The nuclei are often kidney-shaped, and there is more abundant cytoplasm than in lymphocytes. There is a conspicuous Golgi body and some rough endoplasmic reticulum (Fig. 19.4). Living monocytes have characteristic ruffled membranes, in continuous motion. The monocytes are thus motile but not themselves active in defence; they are immature cells, ready to develop into phagocytes when they enter the tissues. They may also be able to give rise to fibroblasts. The monocytes probably develop from monoblasts in the bone marrow as a line of cells distinct from the lymphoblasts. It is not known how the rate of formation is controlled. If a rat is injected with tritiated thymidine, any newly formed monocytes become labelled. If the animal is first irradiated with 750 rad no labelled cells appear, because no new ones are formed. But if the tibial bone marrow is protected from the radiation new cells are formed (see Cohn 1968). Conversely, if labelled cells are infused into syngeneic recipients, only those of marrow (and to a lesser extent spleen) appear as monocytes.

Mononuclear cells appear in inflammatory exudates, where they differentiate into phagocytic macrophages. They appear later than the neutrophils (polymorphs), but probably partly in response to the same stimulating substances. There may be substances released by neutrophils that attract monocytes and others that are produced by the damaged tissues and act upon monocytes alone (Ward 1968).

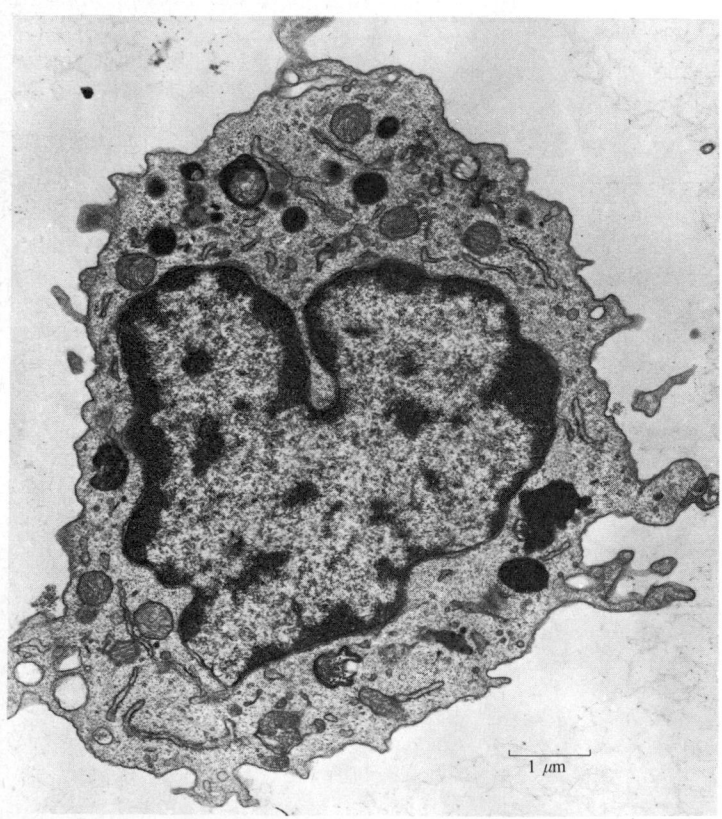

1 μm

FIG. 19.4. Electron micrograph of a human monocyte. The nucleus is horseshoe-shaped, and the cytoplasm contains mitochondria, distended Golgi vesicles, rough endoplasmic reticulum, free ribosomes, and lysosomes. The numerous pseudopodia of monocytes give the plasmalemma a characteristically ragged appearance in thin sections. (Figure kindly supplied by Prof. H. Z. Movat.)

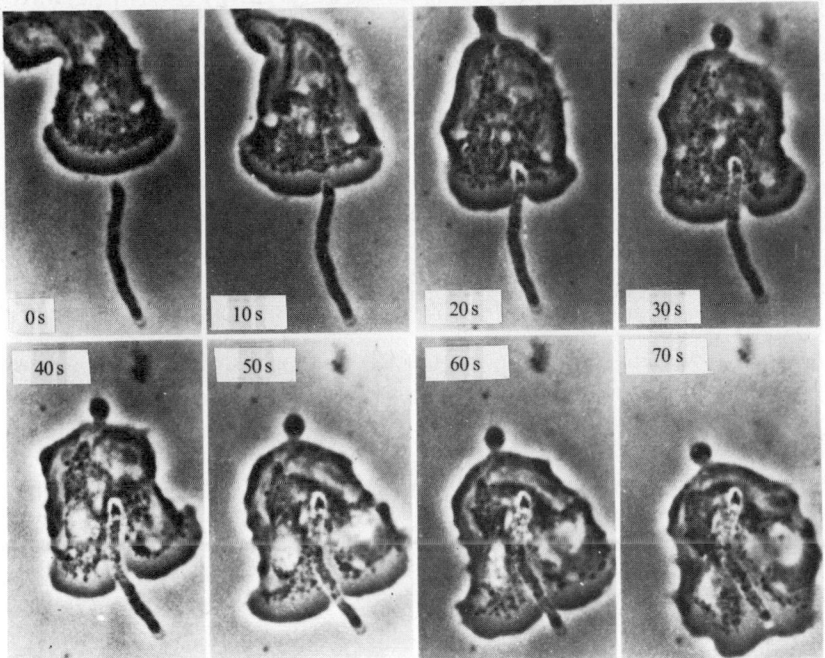

FIG. 19.5. Phagocytosis of *B. megaterium* by a human neutrophil in a sequence of phase-contrast pictures from a motion picture film. (Figure kindly supplied by Dr. J. G. Hirsch.)

7. Macrophages

Some 3.6×10^6 monocytes are produced daily in a normal rat and then disappear from circulation with a half-life of 3 days. The cells leave the vascular system at random in a wide variety of tissues, passing between the endothelial cells and intercellular space and basement membrane into the extravascular connective tissue. Here they become converted into the large phagocytic macrophages (histiocytes). They probably do not divide further and have been shown to remain in place for at least 50 days. These cells are important components of many organs, for instance, the Kupffer cells of the liver, sinusoidal macrophages of the spleen, microglia of the CNS, and 'dust cells' of the lung. Probably, but not certainly, all these are derived from monocytes but some may be formed locally. Little is known of their permanence or of the control of their formation. Tritiated thymidine studies show that 1 per cent of Kupffer cells are labelled in 24 hours.

Monocytes kept in culture develop into large macrophages with much rough endoplasmic reticulum and Golgi apparatus and many lysosomes. No doubt this conversion is the source of the macrophages of the tissues. These are the cells that give the reticulo-endothelial tissues their power to take up bacteria and other foreign particles. They have very varied shapes. The nuclei are smaller and darker than those of fibro-

blasts and are indented. In haemopoietic tissues the macrophages line the sinusoids as the 'reticulo-endothelial cells'. In loose connective tissue they are oval, but may be elongated or otherwise adjusted to their neighbours. The microglia cells of the CNS are highly branched macrophages, ready to round up and become phagocytes.

Macrophages protect the tissues by the processes of phagocytosis and pinocytosis, which essentially depend upon turning in some portion of the surface of the cell to enclose the foreign body in a vacuole. Here it is acted upon by enzymes, probably derived from the lysosomes of the cell. If it can be broken down the products are disposed of in various ways, mainly by liberation into the tissue fluids. It is not known how the macrophage 'recognizes' the material to be ingested. The first phase of phagocytosis is an attachment to the membrane of the cell, followed by formation of a vacuole. As has been remarked, 'the engulfment process *per se* is an astonishing event to behold under the microscope, the bacterium or other particle appearing to pass as if by magic through the cell membrane to reach a cytoplasmic site' (Hirsch 1965) (Fig. 19.5). A wide range of materials can be ingested, including bacteria, virus particles, antigen–antibody complexes, and inert materials, such as trypan blue or Indian ink. Besides these 'foreign' invaders macrophages will also consume 'effete' red cells, fragments of broken tissue cells, and perhaps quite a wide range of the body's own components. In metamorphosis of the frog the tissues of the tail are consumed by macrophages.

Macrophages do not make specific immunoglobulins but probably pass to lymphocytes the information as to which antibodies are needed (p. 190). Some macrophages are provided with characteristic 'dendritic processes' with which they embrace lymphocytes and then let them go. This may be the process by which information is transferred.

8. The lymphatic system

The lymph channels (Fig. 19.6) constitute an additional mechanism by which material is returned from the tissues to the blood-system. The lymph capillaries are vessels like the blood capillaries but they end blindly in contact with the cells or tissue spaces. These vessels therefore normally have no open ends, but their walls are permeable not only to water and crystalloids but also to colloids. Moreover, the finest branches of the lymph capillaries are modifiable structures, changing from moment to moment, so that gaps in their walls open and then close again. The lymph system is thus able to take up proteins and also any foreign particles

or bacteria from the tissues; one of its main functions is to filter off and neutralize such intruders.

The lymph capillaries unite to form larger vessels, essentially like veins but with even thinner walls. At intervals along the lymph vessels are the lymph nodes, which are the main protective filters of the body (p. 185). After passing through the nodes, the lymph is collected into still larger lymphatic vessels. The majority of these ultimately join a large sack, the *cisterna chyli*, lying on the aorta below the diaphragm. From this sack the *thoracic duct* proceeds headwards to open into the left innominate vein. The lymph vessels of the intestinal villi are known as *lacteals* (p. 148) and serve to carry away the fat, which is otherwise unable to enter the blood-stream.

There is a continual slow movement of fluid through the lymphatics back to the venous system, and by this means material such as protein that cannot re-enter the capillaries returns to the blood. The lymph is therefore a fluid of variable composition, containing less protein than the plasma and having a salt composition similar to that of the blood. It contains few cells at the periphery but some lymphocytes in the main channels. The flow of the lymph depends to only a small extent on the pressure within the tissues produced by the heart. The propulsion comes mainly from the massaging actions produced by the muscles when they contract. This pushes the lymph in one direction because the lymph vessels are provided at frequent intervals with valves similar to those of the veins. The lymph flow is increased in conditions that lead to extra permeability of the blood capillaries such as increased muscle movement and the presence of certain poisons and of oedemas of varied origin.

The network of lymphatics thus consists of a set of channels collecting material from the tissues. It may be considered to have four main functions: (a) to return excess fluid (in some conditions); (b) to carry fat via the lacteals and chyle; (c) to circulate lymphocytes and produce antibodies; (d) to provide cell-mediated reactions against foreign bodies such as grafts. The system is highly complex including a network everywhere in the body; certain regions have special lymphatic functions, especially the thymus, the gut-associated lymphatic tissues (GALT), the lymph nodes, and the spleen. Within this framework of channels and organs the lymphocytes proliferate, differentiate, migrate, and respond to antigenic stimuli (Gatti, Stutman, and Good 1970).

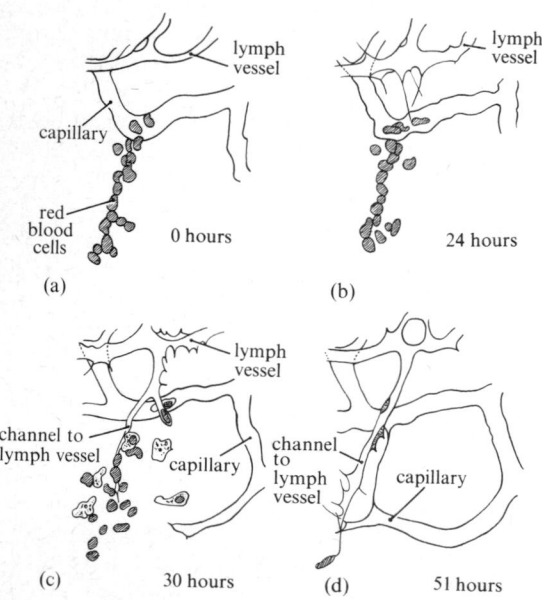

Fig. 19.6. Outgrowth of fine lymphatic vessels towards extravasated red cells.
(a) The cells have escaped from a capillary. (b) The lymph vessel puts out lateral processes. (c) Some of these approach the red cells and form a channel with a lumen. (d) The cells are absorbed. Note that the capillaries have also grown meanwhile. (From Clark and Clark (1926). *Amer. J. Anat.* **38**.)

9. The lymph nodes

The lymph nodes and lymph nodules are scattered along lymph drainage pathways throughout the body in

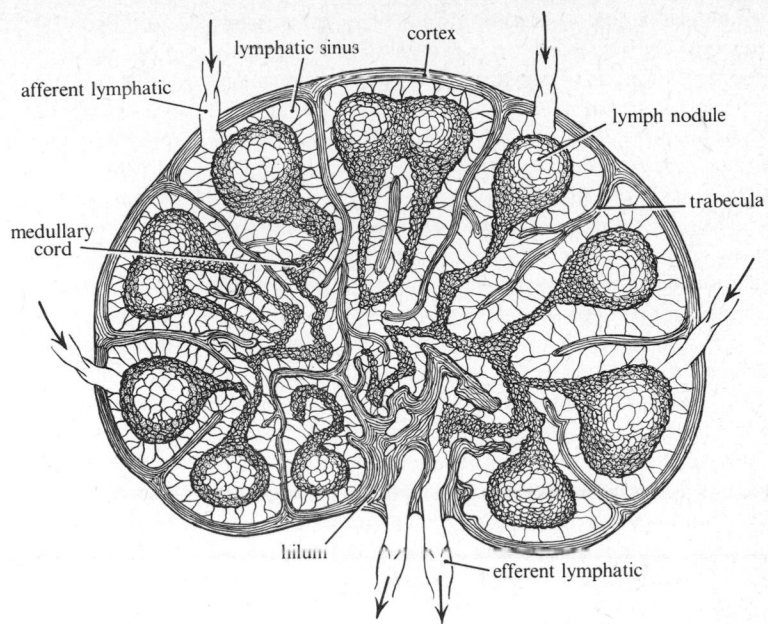

FIG. 19.7. Diagram of a lymph node.

strategic situations at which infection is likely, such as the tonsils, the walls of the intestine and appendix, and the lymph nodes of the groin and axilla where lymph vessels from the limbs converge. They are places at which the lymph vessels break up into fine channels so that the fluid is forced to pass through a fine system of reticular and collagenous connective tissue, where it comes into close contact with macrophages and lymphocytes (Figs 19.7 and 19.8). The lymph nodes vary in size from minute invisible collections to masses that are large and easily felt. The afferent lymphatic vessels enter the node round its outer surface and the lymph then filters through an ill-defined system of spaces. The liquid collects at a hilum and is carried away by large efferent lymph vessels.

In between the trabeculae that carry the lymph sinuses are solid masses of cells, the lymph nodules, each consisting of an outer mass of lymphocytes and a centre composed of lymph-producing tissue. This consists of a fine reticulum of supporting substance containing reticular cells, which, according to the unitary theory (p. 187), are able to develop into either macrophages, lymphocytes, or erythrocytes. The lymph nodules, therefore, are composed of these active reticular cells and masses of lymphocytes and macrophages, presumed to be their descendants. The lymphocytes are of various diameters between 8 μm and 12 μm. They have a very large nucleus, containing deeply basophilic nucleoli, and relatively little cytoplasm (Fig. 19.1). They are present in vast numbers in the

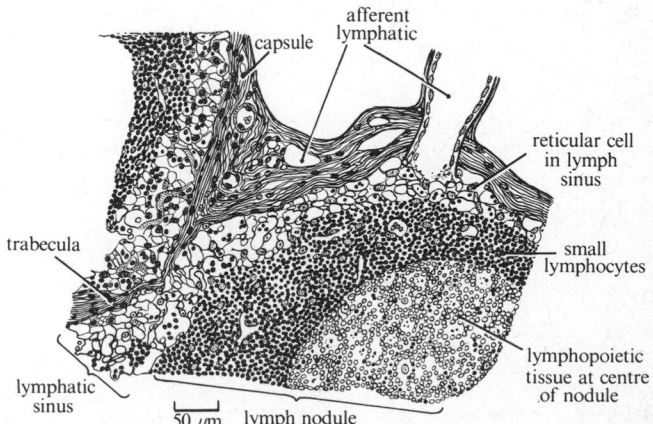

FIG. 19.8. Portion of a lymph node of a dog. (After Maximow, in Bloom and Fawcett 1968.)

lymph nodes and some pass into the lymph and blood-streams. In the adult they are mostly formed by mitotic division of existing lymphocytes, germinal centres within the nodes becoming active for a while and then subsiding. In the young, and to some extent in the adult, they are formed by rounding up and division of reticular cells.

The lymph nodes are characteristic of mammals and birds; in lower vertebrates, blood-cell producing (haemopoietic) and leucopoietic tissues are usually found together. The presence of well-developed discrete lymph nodes is perhaps a sign of a greater need for defence mechanisms by the warm-blooded animals, but there is still much uncertainty as to how the lymphoid tissue functions in this capacity. The nodes are certainly efficient filters; if fluids heavily infected with staphylococci are injected into the lymphatics of a dog's leg the lymph in the thoracic duct remains sterile. This filtering function is performed mainly by the macrophages of the nodes. Granules of dye injected into the lymph are taken up by the macrophages.

Lymph nodes are agents for collecting small lymphocytes from the blood and transmitting them to lymph vessels. Antigen enters the node in phagocytes (macrophages) which filter foreign material from the lymph stream. These then inform the visiting lymph cells, which transform into immunoblasts (= large lymphocytes = plasmoblasts). These turn into plasma cells (p. 189) in the medulla of the gland and are discharged into the efferent lymph vessels to disseminate the immune response and produce a systemic immunity. It is a main function of a lymphatic system to collect from the tissues the large molecules that cannot enter blood-vessels, and these include antigens. These large molecules are then filtered out by the phagocytic cells of the regional lymph node. Obviously, particulate matter such as bacteria or viruses is more likely to be retained and produce a reaction than are soluble antigens, which may pass through the lymph node.

20 Antibodies

1. The thymus

THE thymus is a large lymphoid organ of pink colour lying behind the sternum, around the great vessels. It arises as an outgrowth of the epithelium of the third branchial pouch, which forms cords of cells. Mesenchyme collects around these and forms the special lymphocytes known as thymocytes. The central groups of cells, derived from the epithelium, are known as Hassal's corpuscles. The mesenchymal cells are probably not locally produced but invade the thymus from the bone marrow and are then processed there. In post-natal life, if all the lymphocytes are destroyed by irradiation and replaced by those of the bone marrow of another animal then these latter re-populate the thymus. A similar invasion of bone-marrow cells probably also happens in development.

Each lobule of the thymus has an outer cortex with many lymphocytes and a central medulla, where there are fewer (Fig. 20.1). Mitosis is very active in the cortex, and it is here that lymphocytes are formed in a dense reticular tissue. The capillaries of this region are peculiar in that they are surrounded first by endothelial cells, sometimes containing pericytes, outside which there is a space separated in turn from the lymphocyte-forming reticular tissue by an epithelium whose cells are united by desmosomes (Fig. 20.2). The new lymphocytes are thus formed in an environment where they are protected from the larger molecules circulating in the blood. It is supposed that in this way cells are produced that are prepared to be antigen-reactive but are not yet committed by contact with antigens. A series of stages can be recognized by which original lymphoblasts divide probably 4 times and then form prolymphocytes, which after two further generations become lymphocytes. These are then supposed to leave the cortex for the medulla, where they may be stored for a while and then put into circulation.

It is still not certain exactly what part the thymus plays in the development of immunological responses. If it is removed at birth the animal becomes immunologically incompetent in the sense that it will not reject foreign grafts. Grafting of another thymus restores competence, but this is not due to a contribution of new lymphocytes that have already been processed, since

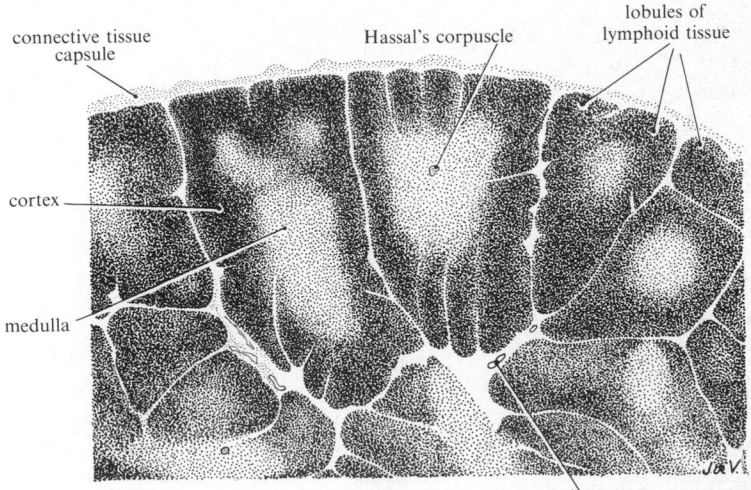

FIG. 20.1. Thymus gland of new-born human. (From a photograph by Mr. K. C. Richardson.)

connective tissue capsule

Hassal's corpuscle

lobules of lymphoid tissue

cortex

medulla

blood-vessel

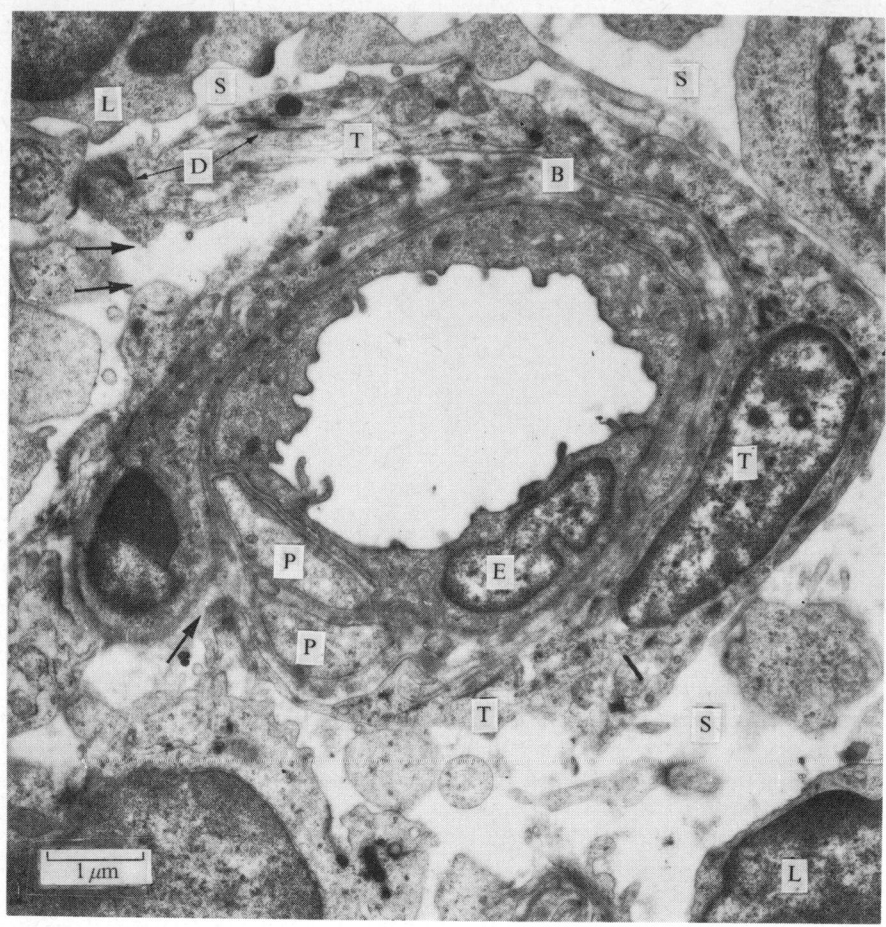

FIG. 20.2. Electron micrograph of capillary of mouse thymic cortex. The endothelial and perivascular cells (E and P) are surrounded by a basal lamina (B), epithelial cells (T), and a space (S). The thymic lymphocytes (L) are thus partly separated from the blood. The epithelial cells contain tono- fibrils and are linked by desmosomes (D), but there are gaps in the barrier (large arrows). Correspondingly antigens in the blood penetrate the parenchyma in much lower concentration than elsewhere (see Clark 1963 and 1964). (Figure supplied by the courtesy of Dr. S. L. Clark.)

those in the graft all die. There must, therefore, be some effect of the thymocytes upon the lymphocytes of the host, which cannot otherwise acquire competence. This function is probably related to the need for the lymphocytes produced after birth to begin to react to new macromolecules, though not to those already encountered, which are proper to the body. It is possible that this function includes the production of a lymphocytosis-stimulating factor (LSF). But thymic extracts have not been found to be able to prevent the effects of the thymectomy at birth.

Undoubtedly the thymus plays a central and special part in immune reactions. It is probably not unique in this, however, and perhaps we should recognize it as only one member of a set of central lymphoid tissues, in contrast to the peripheral tissue of the spleen and lymph nodes. The central tissue would then include much of that in relation to the gut (GALT), tonsils, thymus, Peyer's patches and appendix, and the bursa of Fabricius in the chick. Removal of the appendix as well as the thymus from rabbits at birth depressed antibody production more than the removal of the thymus alone (see Davies 1969).

2. Lymphocytes

Lymphocytes are concerned with maintaining the identity and integrity of the body. They and the macrophages help to recognize and destroy invaders or debris but must not harm the body's own tissues. Their basic capacities are derived from the inherited

DNA, but they must also learn to recognize and react appropriately to the body's own various cells and to an infinite range of possible invaders. The *small lymphocytes* have densely basophilic nuclei, of about 5 μm diameter, and a thin layer of cytoplasm. They have few mitochondria and little endoplasmic reticulum or Golgi material (Carr 1970). In fact, they are not actively producing cells. About 8 per cent of lymphocytes are larger (nuclei 7 μm, over-all 12 μm) and called *prolymphocytes* because they are supposed to divide and give rise to smaller ones.

Lymphocytes are actively migratory cells, and they are seen in large numbers passing through the lining of the intestine and other tissues. They often seem to be in a dying state, and until recently it was held that they were short-lived. They are poured out in great numbers from the lymph nodes and pass via the thoracic duct into the venous system (p. 184). But these are not all newly formed lymphocytes, for if they are artificially removed via the thoracic duct the number of lymphocytes falls but is restored if they are re-injected via the blood. Lymphocytes labelled, say, with radioactive adenine, can be recovered up to a year later. It is clear therefore that some lymphocytes continue to re-circulate from the blood to the lymph for a very long time.

Recent autoradiographic studies show that after injection with radioactive label, the small lymphocytes can be divided into a group that lives more than 2 weeks and one that lives less, and that most of the compartments of the lymphatic system contain a mixture of the two groups (Everett and Tyler 1967). Separate study of blood and thoracic duct lymph showed that many lymphocytes enter the blood directly and they are probably made in the bone marrow (hence B-lymphocytes) (Fig. 20.3). These are heavily labelled by the time they reach the blood. The cells that predominate in the thoracic duct are more weakly labelled. They are originally derived from the bone marrow and are then processed in the thymus where they acquire their specific power to produce antibodies (hence T-lymphocytes). This agrees with the evidence that cells of the thymus are partly isolated from the blood-stream (p. 188). The B- and T-lymphocytes play different parts in antibody production (p. 190) (see Greaves, Owen, and Raff 1973).

The long-lived lymphocytes recirculate repeatedly from blood to lymph by way of lymph nodes and Peyer's patches of the intestine. But they are not all continually on the move. Injection of labelled long-lived lymphocytes from thoracic-duct lymph shows that they 'home' to the lymph nodes, splenic white pulp, and Peyer's patches, but not to the thymus or bone marrow. In contrast, labelled small lymphocytes from

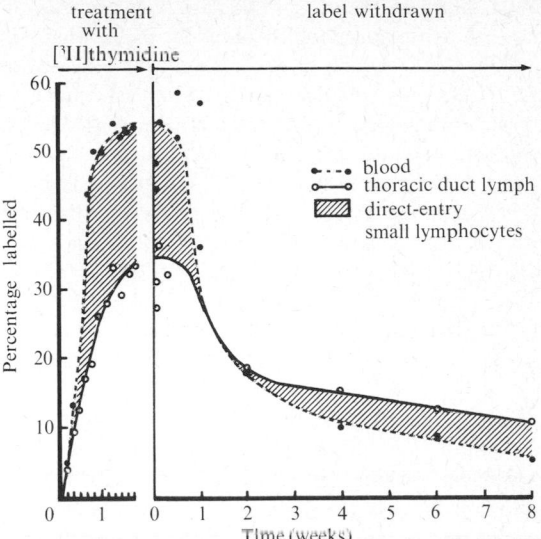

Fig. 20.3. Percentage of [³H]thymidine labelled small lymphocytes in blood and thoracic-duct lymph of same animals. (From Everett and Tyler 1967.)

bone marrow or thymus go to bone marrow or the red pulp of the spleen.

Further evidence of the importance of the thymus in lymphocyte production is that if it is removed at birth the lymphocyte population falls very low in the circulation and in the white pulp and other regions, but the other germinal centres remain normal. The depleted areas are in fact those that are also depleted by prolonged thoracic duct drainage.

Lymphocytes are among the most highly radio-sensitive cells of the body and following heavy irradiation of rats the long-lived cells are deficient for many months, whereas the short-lived ones recover in 1–3 weeks. If the thymus is removed before irradiation the long-lived population is never replaced.

The importance of this population of long-lived lymphocytes is that they are almost certainly the immunologically competent cells necessary to allow production of antibodies (p. 191). Animals deprived of this population by thoracic-duct drainage do not reject allografts or xenografts (p. 192). They are therefore immunologically incompetent. Injection of thoracic-duct lymph or of normal but unsensitized lymph-node cells of the same strain will cause destruction of the grafts (Billingham, Brent, and Medawar 1956).

3. Plasma cells and the production of antibodies

The circulating lymphocytes have none of the characteristics of actively producing cells, and there is now much evidence that their function is to be ready to produce antibodies, which they do by turning into immunoblasts, which circulate briefly and then enter the tissues and produce plasma cells, responsible for producing the antibody systemically, or graft rejection cells that produce it locally (p. 192). Plasma cells are found in the lymph nodes and in the loose connective tissue under the epithelial linings of the intestinal and respiratory tracts, in fact at strategic points where they are responsible for producing antibodies against invading materials. They have a nucleus with characteristic radiating 'cartwheel' appearance and a conspicuous basophilic cytoplasm. Electron microscopy shows that they have a very well-developed rough-surfaced endoplasmic reticulum and a large Golgi apparatus (Fig. 20.4). There is much RNA in the cytoplasm, which therefore stains strongly with pyronin. The plasma cells are obviously active cells, producing the antibodies, immunoglobulins (p. 191), which are either released into the circulation or applied locally, the cells being then called graft rejection cells. These are found around a graft from an individual of different genetic constitution from the host, and they resemble plasma cells but with numerous free ribosomes not attached to endoplasmic reticulum. They probably serve to produce antibodies only locally, for none are found in the circulation after such a graft.

Plasma cells only begin to appear after birth. Probably each produces a particular antibody. As life proceeds the individual acquires an increasing store of committed plasma cells (or their precursors). These provide a memory of the types of invasion that have been experienced and protection against their repetition.

There is evidence that the lymphocytes themselves are not directly sensitive to antigen. This primary response is a function of the macrophages, which then proceed to pass the information to small lymphocytes. Associations between the two types of cell can be seen. It is possible that the information is transferred in the form of a specific RNA passed by the macrophage. The specificity of the antibody produced is probably not due to the actual presence of antigen in the plasma cells (see Lennox and Cohn 1967). The effect of the information passed by the macrophage to the virgin lymphocyte is to allow it, in co-operation with a T-lymphocyte, to develop into a plasmablast (immuno-

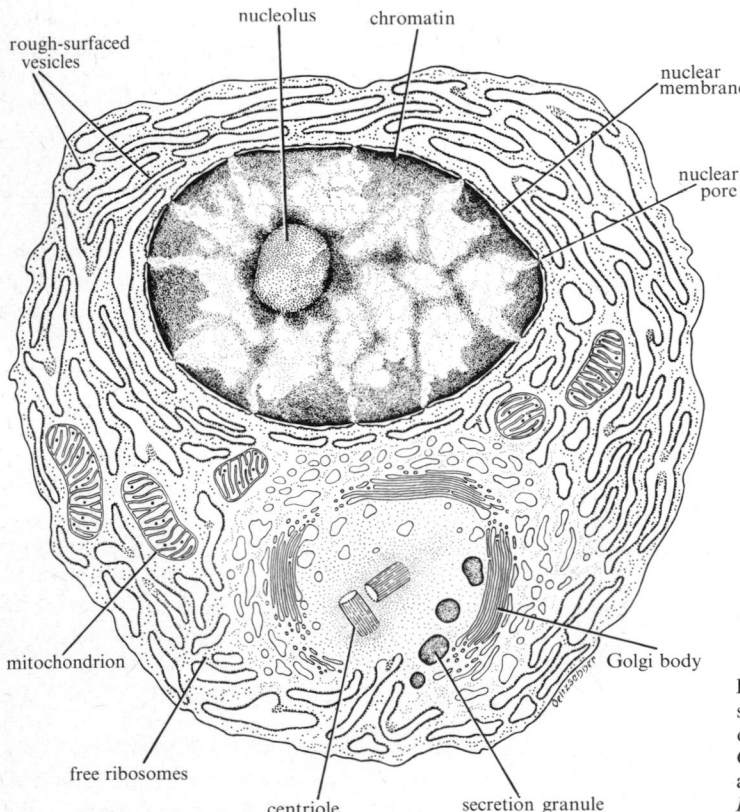

nucleolus chromatin

rough-surfaced vesicles

nuclear membrane

nuclear pore

mitochondrion

Golgi body

free ribosomes

centriole secretion granule

FIG. 20.4. Drawing of mature plasma cell as seen by electron microscopy. The abundance of rough surfaced vesicles and prominent Golgi bodies indicate intense secretory activity. (From Leblond, in Ham (1974). *Histology.* J. B. Lippincott, Philadelphia.)

blast), which migrates to the tissues and there makes plasma cells, which produce the antibody.

As a result of the introduction of various antigens the body thus comes to contain a 'library' of T-lymphocytes, allowing antibodies to be made to any antigen that has already been encountered. Subsequent low doses of antigen trigger antibody formation by the action of those experienced cells on other lymphocytes possibly without obligate intervention of macrophages. High doses of antigen may kill the experienced lymphocytes and lead to immunological paralysis.

The plasma cells (and graft rejection cells) contrast with the lymphocytes in showing all the signs of active secretion, and it is probable that they produce the immunoglobulins that are the specific antibodies. The evidence now suggests that the circulating lymphocytes provide a system of cells competent to develop into plasmablasts and then plasma cells able to produce the antibody appropriate to an invading antigen.

4. Antibodies

One of the body's most powerful methods of defence is to produce these proteins, which have the property of combining with the soluble antigen that elicits them (Roitt 1971). Antibodies can be produced to most but not all molecules, but there is still uncertainty as to what it is that makes a molecule antigenic. Most proteins are antigens, and so are many bacterial polysaccharides, but few nucleic acids. There seems to be a need for a certain characteristic shape of the molecule; proteins substituted by fatty acids with long flexible chains cease to be antigenic. Conversely, certain non-specific protein and other colloidal molecules can act as 'adjuvants', increasing the tendency of other molecules to elicit antibodies. Many smaller molecules can combine with antibodies to make soluble complexes, and such molecules are known as haptens. They cannot themselves elicit antibody formation.

There are several different classes of immunoglobulin molecule, distinguished by letters. Thus IgG is the most abundant of the five classes in man. Purified IgG antibodies are seen in the electron microscope as **Y**-shaped molecules, which swing open after combination with antigen (Fig. 20.5). Each consists of two heavy

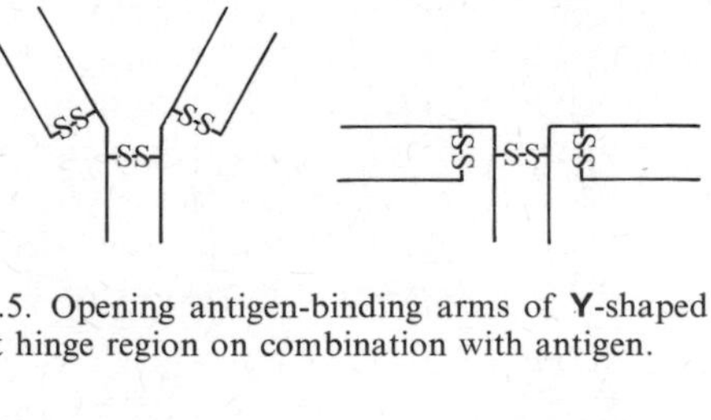

FIG. 20.5. Opening antigen-binding arms of **Y**-shaped antibody at hinge region on combination with antigen.

and two light chains linked by disulphide bonds. The amino-acid compositions are partly known. Each sub-unit has at the amino-terminal end a sequence of amino acids that varies and provides the specificity, and at the other end a sequence that is the same for all members of the species and similar for all animals of a class. Higher vertebrates have 2 or 3 types of light chains and 10 types of heavy chains, whereas in lower vertebrates there is only one of each (Lennox and Cohn 1967).

The recognition of the antigen by the antibody probably depends, as does enzyme specificity, upon a cleft-like site on the latter into which the outer electron shell of the antigen fits. The recognition is thus more by its configuration than its chemical nature.

When the lymph nodes of a rabbit are excited to antibody synthesis by injection of foreign protein into a foot pad, two species of rapidly labelled RNA appear in considerable quantity. These have sedimentation constants of about 10 S and 13 S and molecular weights of 220000–370000, which are close to those expected for mono-cistronic RNAs coding for the light and heavy antibody chains, using a triplet code. These RNAs are found only in extracts of polysomes and only after stimulation with antigens. It is likely that they are the messenger RNAs responsible for the chains. Such messenger RNA molecules have rarely been observed in mammals. They stand out here because these experiments were so designed that the lymphs nodes were producing much more of the globulins than of other proteins (Kuechler and Rich 1969).

5. Immunological memory

After a first injection of an antigen there is a lag period of 3–5 days before measurable amounts of antibody can be detected. The total amount of antibody produced is very much greater than that of the antigen injected. Isotopic labelling shows that the antibody molecules turn over at the same rate as other serum globulins (half-life of 13 days in man). Nevertheless, the animal or man retains a 'memory' of the first injection for much longer, and this is shown by the fact that if the same antigen is injected at a later date the antibody appears in the blood sooner than at the first injection, reaches a higher level and remains in evidence for much longer. This secondary or 'anamnestic' reaction has been explained in two ways. Some hold that small quantities of the antigen have persisted. In particular if it is a virus it may have continued to multiply in the body. More widely held is some version of the *clonal selection theory* of Burnet and others, according to which certain cells, originally of the T-lymphocyte series, becoming differentiated during development, perhaps by somatic mutation in such a way as to become capable of producing each a particular kind of antibody. If the corresponding antigen arrives, the cell proliferates and produces a clone of antibody-forming plasma cells. Some of the committed cells remain and provide the memory.

In support of this theory it has been shown that (in general) each cell forms antibody of only one specificity (see Nossal 1969), moreoever the progeny of antigen-initiated mitotic divisions have very variable lifetimes—some last only for 48 hours, others for many months. The problem then is to show how the somatic mutants could be selected in this way.

The information responsible for producing a large range of antibodies is, therefore, according to this theory, already present in the organism before exposure to the relevant antigens. Particular antigens serve to stimulate those antibody-forming cells that synthesize antibodies having appropriate complementary binding sites. It is not clear precisely how this is ensured but clearly the number of sites must greatly exceed the number of different antigens. A very efficient system is provided for amplifying the production of a specific antibody.

This theory is very attractive because it can be expanded to explain the facts that (a) macromolecules present in the body before birth are not thereafter antigenic and (b) plasma cells appear only after birth and only then do antibodies begin to form. There is evidence discussed later that the thymus gland is the place where cells are 'taught' before birth which macromolecules belong to the body. This organ then constitutes a reservoir of 'trained' but uncommitted cells, ready to produce antibodies to substances that enter after birth.

6. Mutant-breeding organs

In the development of the lymphoid system, multi-potential undifferentiated stem cells from the foetal liver or bone marrow migrate and differentiate according to the environments through which they pass. Those that pass through the thymus become the T-cells, responsible for cellular immunity and allograft rejection and for collaboration with the B-cells, which somehow acquire the power to develop into cells that produce immunoglobulins. A good classification is thus into thymus-derived and thymus-independent populations.

If a given antigen-sensitive lymphocyte is committed to produce only antibody molecules of one specificity, we have to explain how so many types arise (perhaps 10^4–10^5). It can hardly be that the inherited genes provide for them all. One theory is that the variety is produced by somatic mutation and that the thymus and other central lymphoid organs are *mutant-breeding organs*. Stem cells both enter and leave the thymus at relatively slow rates but there are estimated in a mouse to be 10^8 new cells produced there every day. Most of these cells must die. The suggestion is that there is a selective process by which all except mutant cells are killed. This would be achieved if the body, during development, produces by virtue of its genes, self-recognition histocompatability antigens and complementary 'antibody' molecules at cell surfaces. These would allow the self-recognition of the various specialized tissues. In the thymus all cells that produce these antibodies, determined by the individual's own genes, are killed. The variety of mutant clones arising by somatic mutation are the only ones to survive. Such a mechanism would produce both self-tolerance and a library of potential antibody producers. The known rates of somatic mutation are said to be adequate to support this view.

7. Tissue grafts

When living tissue is transplanted from one species of animal to another the *xenograft* becomes surrounded by a mass of lymphocytes and is then destroyed. *Allografts* from one individual to another of the same species also usually produce this reaction, unless they are *isografts*, that is, between syngeneic individuals. *Autografts* of tissue from one part of the body to another are not destroyed. For example, in the rabbit allografts of skin never survive. The host at first tolerates the graft but then after a few weeks comes to contain a substance (antibody) that reacts against the graft and causes it to fall off. Fig. 20.6 shows the condition 8 days after grafting a piece of rabbit's thyroid under its own skin. In this autograft the outer follicles have survived, the centre of the piece becoming necrotic because of lack of blood-supply. Fig. 20.6(c) and (d) show that if the rabbit's own thyroid has been removed these grafts may go on to develop into large masses of tissue. (b) shows that in an allograft of a piece of thyroid from another rabbit there is survival of some follicles at 8 days, but numerous lymphocytes are present and, as (e) and (f) show, these penetrate the cells of the graft and apparently produce enzymes that lead to the break-up of the grafted tissue. The details of the process are not known, but the clear zone around each lymphocyte is characteristic. At the time when the graft is being most actively dissolved, immature plasma cells (p. 189) appear, apparently by conversion from lymphocytes. After the reaction has subsided many mature plasma cells remain in the tissue as *graft-rejection cells*.

8. Immunological tolerance

Much of our understanding of immunity comes from study of the situations in which antibodies are or are not produced to grafts. Allografts and xenografts are usually rejected, and a second graft is rejected more quickly than the first. There are, however, certain

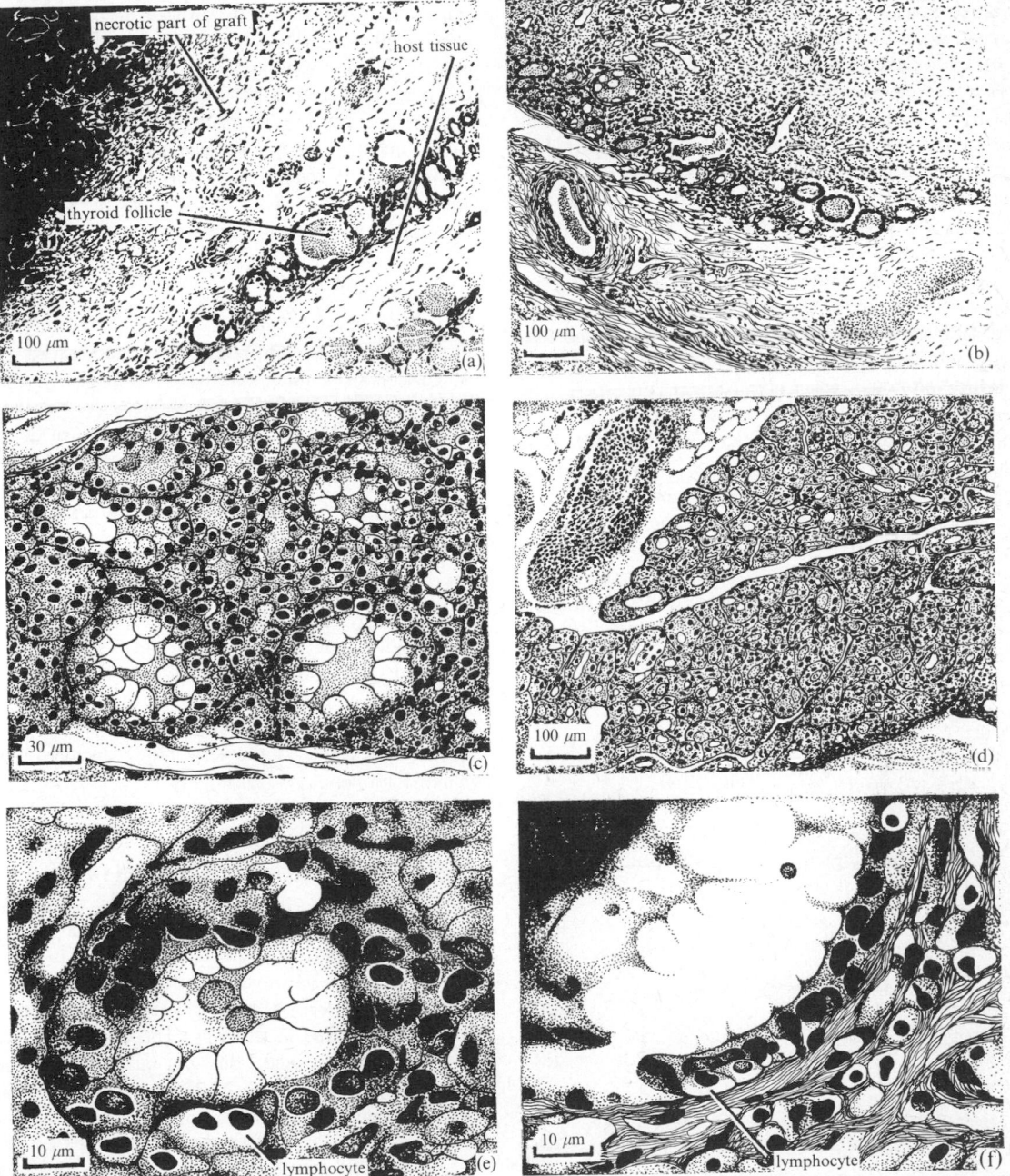

FIG. 20.6. Pieces of thyroid tissue of the rabbit grafted beneath the skin.

(a) Autograft 8 days. (b) Allograft 8 days. (c) Autograft 1 year, from partially thyroidectomized rabbit. (d) Autograft 6 months from completely thyroidectomized rabbit. (e) and (f) Allograft 8 days; lymphocytes (surrounded by haloes) are invading the follicles, and in (f) are actually within the epithelial cells, which are caused to come away from the basement membrane. (Drawn from photographs in thesis by Dr. D. A. D. D'Arcy by permission of the author.)

tissues that can be transplanted from one individual to another of different genetic make-up, notably the cornea of the eye or valves of the heart. These are organs that obtain nourishment by diffusion, without vascularization. The rejection of an allograft must depend upon some circulating factor.

Rejection does not necessarily follow even if donor and host differ genetically. Thus the offspring of a cross between two inbred lines of mice will accept a graft from either parent. Evidently the organism learns to recognize the macromolecules that have been built under the instructions of each set of genes. Further evidence about this tolerance comes from non-identical twins of cattle, in which the foetal circulations are partly joined ('free-martens'). Skin grafts from one such twin to another are not rejected (Billingham, Brent, and Medawar 1956). Following this clue it was found that antigens injected before birth (or shortly after) do not produce antibodies when injected later (Fig. 20.7). One theory is that the cells that are the precursors of the plasma cells (or graft rejection cells) become 'conditioned' before birth. The part of the DNA responsible for producing the antibodies to macromolecules encountered before birth is blocked. The alternative is that there is clonal selection, according to Burnet's theory, in the thymus, which at birth thus comes to contain a reservoir of prolymphocytes, whose progeny are emitted into the circulation ready to produce antibody to any newly entering macromolecules but not to those already encountered, thus avoiding the danger of production of antibodies against the body's own tissues.

Tolerance to particular antigens can be produced in adults by injection of certain 'immunosuppressive' drugs at the same time as the antigen. These presumably have the effect of killing off all the cells that have been induced to proliferate at that time and thus removing the clones that are competent to react to the particular antigen injected. Dosage with X-rays at the time of antigen injection has the same effect.

Tolerance in general acts only for a limited time unless it is stimulated by further injection of antigen. Its duration is increased by thymectomy.

9. Autoimmunity

This theory also explains why it is that the body does sometimes produce antibodies to certain of its own proteins, namely, those that have not gained access to the circulation. Such secluded antigens are the thyroglobulin of the thyroid follicles, the proteins of the lens of the eye, and the spermatozoa (since these appear only post-natally). All of these tissues, and some others, can produce antibodies if introduced into the circulation of their own parent body.

There is evidence that this may occur spontaneously in certain conditions and produce *autoimmune disease*. Thus the thyroid occasionally becomes invaded by lymphocytes and plasma cells, and its follicles are destroyed. Antibodies to thyroid protein have been found in this condition. It is possible that some diseases of the nervous system are due to production of antibodies to the myelin of the nerve fibre, which is another secluded antigen.

10. Antilymphocytic serum

If animals of one species are immunized by injection of lymphocytes from another (for example, rabbits and horses) a serum specific against lymphocytes is produced (ALS or the corresponding globulin ALG). This appears to act by selectively depleting the pool of long-lived lymphocytes. Allografts are not rejected by animals immunized in this way. The capacity to reject is gradually recovered, but not after thymectomy. Lymphocytes coated with ALG may remain immunologically unresponsive for several cellular generations (Levey and Medawar 1966).

ALG may be useful, in man, in conjunction with the immunosuppressive drugs azothioprine and prednisone to prevent rejection of allografts of kidney, liver, or heart. If the host could be made permanently unreactive to the tissue of the donor further drug suppression would be unnecessary (see Woodruff 1969).

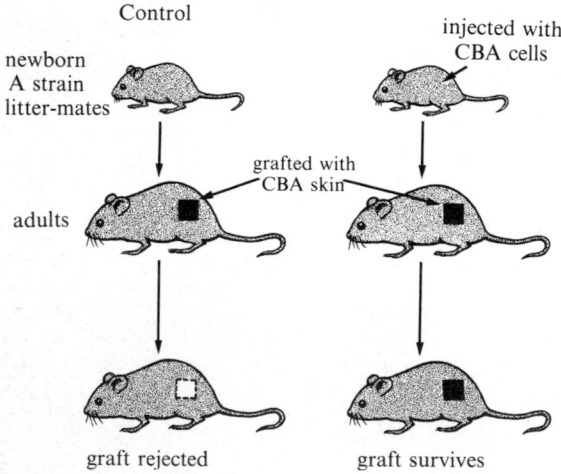

FIG. 20.7. Diagram illustrating induction of tolerance to foreign CBA graft in A strain mice by neonatal injection of antigen. (After Billingham, Brent, and Medawar, from Roitt 1971.)

21 The heart and circulation

1. The functions of the blood

THE circulation of the blood is one of the most striking signs of the continuous activity of life. The body appears outwardly a relatively fixed or stable thing, but there is continual change going on in almost every part within. Much of this change is chemical and cannot be seen, but the circulation provides a direct sign of it. At a short distance from the skin surface millions of capillaries carry enormous volumes of fluid continually round the body. Here indeed is evidence of the active state of the living organism. The blood is the supply system, providing materials for the ceaseless changes by which the steady state of life is maintained.

The modern period in biology might be said to begin with the discovery of the circulation of the blood by William Harvey in 1616. Once it had been demonstrated that a flow of blood goes continually round the body it became possible to realize the activity underlying all life; 300 years later we are still in the process of adjusting ourselves to this view. As perhaps the most obvious and important of the inner animal activities, the circulation has been intensively studied since Harvey's time, and we possess a better knowledge of it than of almost any other part of the body. Moreover, we can speak exactly about some aspects of the circulation by using the language that physicists and engineers have developed for description of the flow of liquid in tubes. Harvey's original discovery was largely based upon comparison of the heart and its valves with a man-made pump. Since that time hydrodynamic analogies have been the basis of our knowledge of the circulation, though it is now clear that some of the liquid leaves through the walls of the finer branches of the vessels (p. 169).

The maintenance of a rapid circulation is especially important for mammals, whose life is more intense than that of any other animals, except the birds. To maintain a high and constant temperature they need to transport large quantities of fuel for combustion and oxygen with which to burn it. With the high temperature all reactions can be accelerated and all movements made faster, with consequent further increase in demands. Moreover, the nervous organization of mammals rarely leaves them passive, but drives them on to activity, making them seek persistently for their requirements even if these can be obtained only by indirect means, involving prodigal expenditure of energy.

For all these purposes mammals require a circulation that is both abundant and adjustable; it would be inefficient for them to maintain a maximum circulation to all parts simultaneously. The whole life and activity of the animals therefore depend in an especially intimate manner on the performance and regulation of the circulation, and this has found recognition in the detailed observations of its functions by physiologists and physicians (see Greenfield 1965; Ross 1971).

It is perhaps ridiculous to ask what are the 'functions' of so important a part of the body as the blood; it is an indispensable part of the life of the animal. However, it is convenient to make a list of the more important activities in which the action of the blood plays a part.

(a) Transport of water, the universal solvent in the living tissues.

(b) Respiration, by carrying oxygen to the tissues and carbon dioxide back to the lungs.

(c) Nutrition, by carrying raw materials for growth and replacement and fuel for combustion in the tissues.

(d) Excretion, by transport of waste matter to the kidneys, lungs, or elsewhere for removal.

(e) Temperature regulation; by its high specific heat the water of the blood is able to carry heat from regions where it is produced to other parts of the body. Because so much of the body is made of water temperature variations between different regions are slight.

(f) Transport of hormones and other chemical stimulants produces a considerable part of the regulation and integration of the activities of the various parts of the body.

(g) Protection against infection is largely a function

of the blood. Its white cells are able to take up foreign particles and it carries the antitoxins and antibodies from their places of formation to other organs.

No one of these functions is peculiar to the blood of mammals; for example, the properties of water provide some degree of temperature regulation even in amphibia, but in mammals all the activities become intensified. To take one example, protection against infection is more difficult and more necessary in a warm- than a cold-blooded animal.

2. The heart

A great part of the success of the mammals depends on the efficiency of their circulatory apparatus, achieved by complete separation of the main or *systemic vessels*, as a greater circulation, from the *pulmonary* or lesser circulation, supplying the lungs. By this means well-oxygenated blood is delivered to the capillaries at a far higher pressure than is achieved by a system such as that of a fish with a single circulation, in which the blood passes through two sets of capillaries under the influence of a single pressure source in the heart. The advantage of the separate circulations is probably especially marked in the case of the muscles, which are regions of high oxygen consumption. By means of this arrangement, quick and sustained movements become possible and the whole mammalian organization can

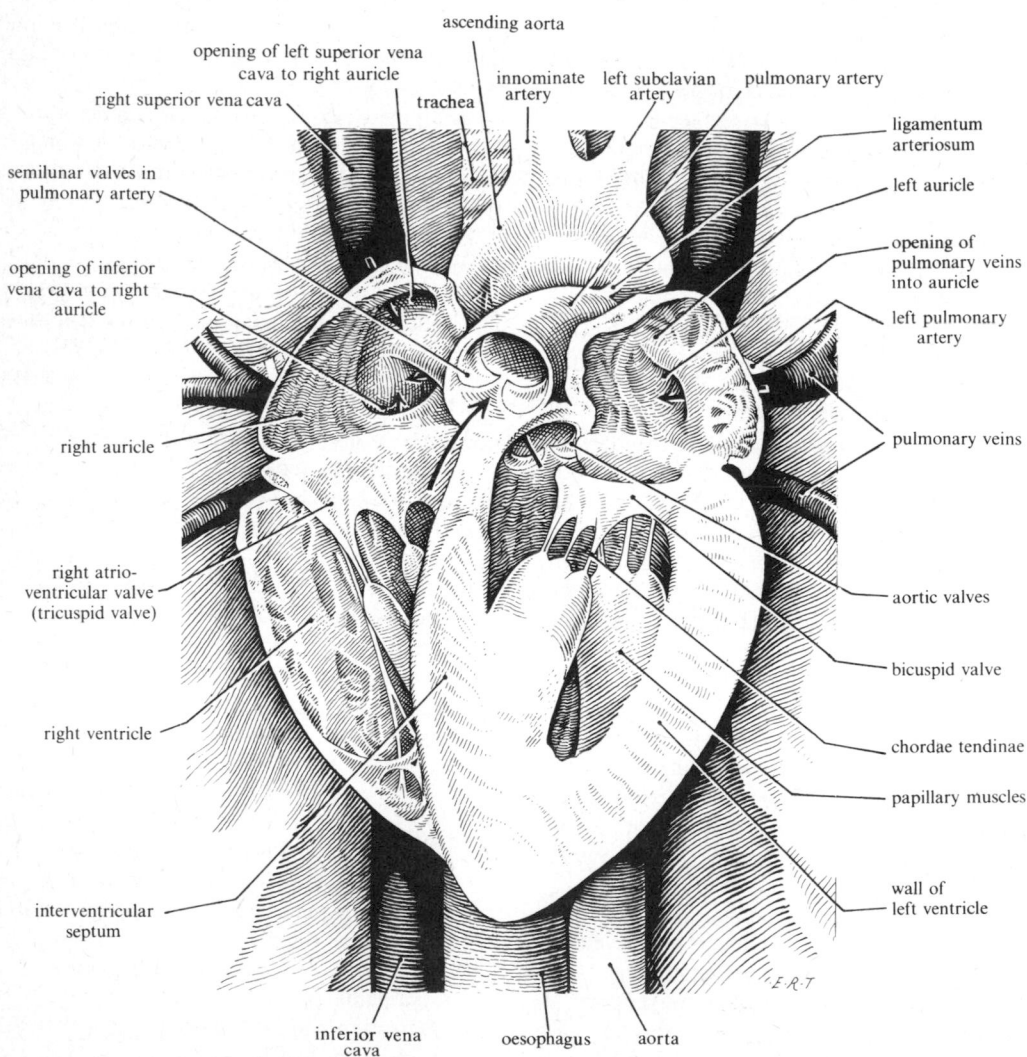

FIG. 21.1. Drawing of heart of the rabbit, dissected from the ventral surface.

reap the advantages of the high body temperature and active chemical processes. It is less important that the interchange between the blood and the internal organs should be rapid and in the liver a *portal system* still remains, in which blood passes successively through two sets of capillaries. Blood from the arteries to the gut is passed first through capillaries in its walls from which it is collected into the *hepatic portal vein*. This proceeds to the liver where the blood passes through the sinusoids around the cells and is then returned by the hepatic vein to the heart.

The separation of the two circulations has been achieved in the mammals by a method quite different from that adopted in the existing reptiles and birds. The original ventral aorta of the fish-like plan of the circulation becomes completely divided into two. The pulmonary artery arises as a single trunk from the right ventricle and supplies the lungs. The aortic arch is also single; arising from the left ventricle it curves over on the left side of the body.

With these modifications the general arrangement of the heart remains in mammals as in lower vertebrates. The organ is formed in the mesoderm below the gut by folding (p. 208) in such a way that it lies in a *pericardium,* whose inner wall (splanchnopleure) adheres closely to the heart surface, while the outer (somatopleure) makes a sac (the fibrous pericardium in the anatomical sense) separated from the inner layer by a small space containing coelomic fluid. The heart in its sac thus lies in a septum, the *mediastinum,* which runs down the centre of the thorax and separates the two lungs. The pericardium is attached to the diaphragm caudally. The surface of the heart and the inner surfaces of the pericardium are covered with smooth layers of mesothelium, allowing freedom of movement. The outer surface of the pericardium is loosely attached to the ventral thoracic wall and to the mediastinum.

The great veins (superior and inferior venae cavae) return their blood to the right atrium, a thin-walled chamber, which leads into the right ventricle by an opening guarded by the flap-like right *atrio-ventricular valve* (Fig. 21.1). This valve (the 'tricuspid' valve of man) allows forward passage of blood but closes during contraction of the ventricle, being prevented from eversion into the auricle by the pull of a number of *papillary muscles,* attached to the valve by fibrous *chordae tendinae.* The opening from the right ventricle to the pulmonary artery is guarded by three *semilunar valves,* arranged to prevent reflux into the ventricle.

Blood returns from the lungs to the left atrium and from here passes through the *left atrio-ventricular* (*bicuspid* or *mitral*) *valve* to the left ventricle, the valve being checked by a series of papillary muscles even

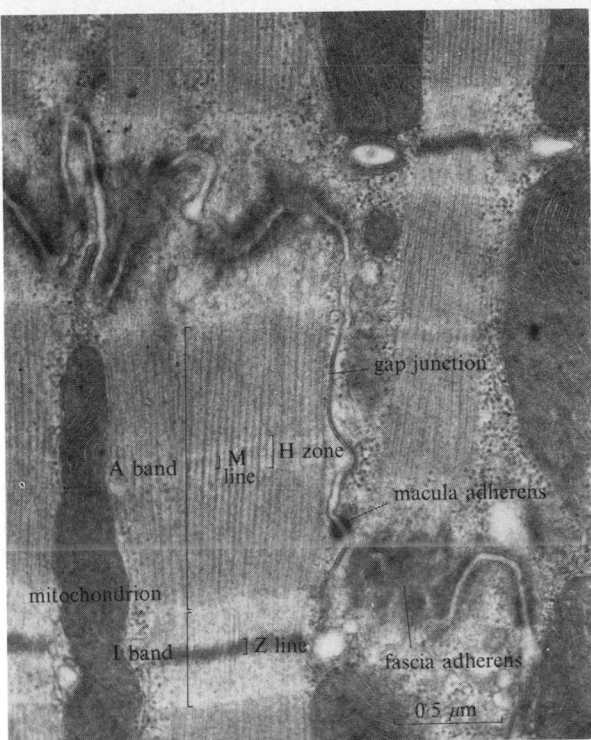

FIG. 21.2. Electron micrograph of rat cardiac muscle in longitudinal section showing part of an intercalated disc. Two elaborately specialized transverse segments at different levels are connected by a relatively unspecialized longitudinal segment. (From Muir (1965.) *J. Anat.* **99.**)

stronger than those on the right. The left ventricle is by far the most muscular of the four chambers and pumps the blood into the aorta, whose entrance is guarded by semilunar valves.

It is essential to the proper functioning of this double circulation that there should be proper co-ordination of the working of the two sides of the heart. The heart musculature consists of a form of muscle in which the myofibrils are striated, but there is also much sarcoplasm and many mitochondria (Fig. 21.2); the nuclei lie at the centre of the fibres. Frequent partitions, the *intercalated discs,* cross the fibres, which are thus divided into a series of cells. These discs are supported by desmosomes, but over part of their length they form tight junctions (*zonulae occludentes*). Another characteristic feature is that the fibres branch and anastomose, so that the whole forms an elaborate net-like arrangement (Fig. 21.3). This allows for much endomysium between the fibres, carrying an abundant blood-supply. The contraction of heart muscle fibres occurs rhythmically and continues without external stimulation from nerves. Characteristic of the action of heart muscle is

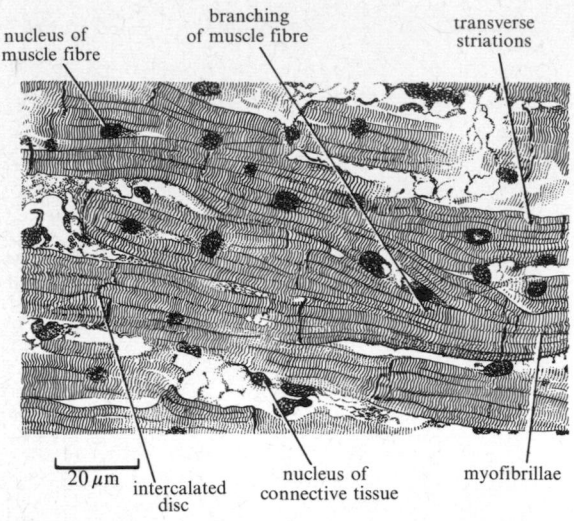

nucleus of
muscle fibre

branching
of muscle fibre

transverse
striations

20 μm

intercalated
disc

nucleus of
connective tissue

myofibrillae

FIG. 21.3. Cardiac muscle of a horse. (From photograph lent by Mr. K. C. Richardson.)

its very long refractory period. Following one contraction no other will take place during the period of relaxation. This enables proper timing of the beat of the various parts. Contraction originates in one centre, the *sinu-auricular node*, a small patch of tissue in the wall of the right atrium, close to the point of entry of the great veins. This corresponds to the position of the sinus venosus of the fish heart, that is to say the posterior end of the region of the sub-intestinal vessel from which the heart has evolved.

In mammals the sinu-auricular node acts as the *pacemaker* of the heart. It consists of a specially modified form of muscle fibre faintly striated, intermixed with nerve cells and fibres from the vagal and sympathetic branches, both of which regulate the heart. The beat originates 'spontaneously' in the node, but its frequency is controlled by the nerves, the vagal fibres decreasing and the sympathetic increasing the rate (p. 284). The contraction wave spreads away from the sinu-auricular node at about 1 m s^{-1} by conduction along the muscle fibres that fan out from this region. The whole musculature of each chamber of the heart thus acts as if it were a single muscle cell. It is probable that electrical propagation between the individual cells occurs by means of the tight junctions (p. 197). The atria thus contract, approximately together. Transmission to the ventricles does not take place in the same way, there being no muscular continuity, and the atria are separated from the ventricles by a ring of connective tissue, the *annulus fibrosus*. Control of the beat, therefore, is taken up by another special region,

the *atrio-ventricular node*, from which impulses are transmitted to the ventricles by a strand of tissue, the *atrio-ventricular bundle of His*. This is formed of modified muscle fibres, small near the node but larger distally and known as *Purkinje tissue,* which conduct impulses at 5 m s^{-1}. This bundle spreads out to join the musculature of both ventricles, whose beat is thus initiated nearly simultaneously at all points. After destruction of the bundle of His the ventricular muscle is able to initiate its own beats, but these are not co-ordinated with those of the atria.

3. Regulation of the blood-pressure

The work involved in the circulation is immense. A man's heart beats over 2600 million times in an ordinary life, pumping a total of 150000 tonne of blood from each ventricle. All this is done without any serious irregularity and with no rest longer than 0·75 s.

The rate of the heartbeat varies greatly in different animals, being lower in larger than in smaller animals (roughly 25 per minute in the elephant, 70 per minute in man, 300–500 per minute in the mouse). It also varies with activity, with 'emotional' and other factors, and with 'training', being lower in athletes than in sedentary individuals. The means by which the 'basic' rate is determined are not known but the nervous control is reflex, there being receptor systems in the heart itself, in the main blood-vessels and in the lungs, whose discharge produces alteration of the rate of the heartbeat.

The vagus carries afferent fibres from receptors in the arch of the aorta. In the rabbit they run in a separate *depressor nerve*. They discharge when the pressure is raised. The effect of these impulses is then to lower the blood pressure by action through a centre in the medulla oblongata on the efferent sympathetic and vagal nerves, the latter reaching the sinu-auricular node through the cardiac plexus. If the depressor nerve is cut stimulation of its central end produces slowing of the heart; but stimulation of the peripheral end is without effect. Other receptors are those of the *carotid body* and *carotid sinus*, which lie close to the bifurcation of the carotid arteries and contain cells sensitive to changes of oxygen and carbon-dioxide content of the blood (chemoreceptors) as well as to pressure changes (baroreceptors). By the discharge of this complex set of receptors, as well as of others in the lungs, the heart-rate is adjusted to suit the requirements of the moment. The vagal efferent fibres slow the heart by inhibiting the atria (cholinergic). The sympathetic efferents accelerate it by direct action upon the pacemaker and ventricular muscle (adrenergic). When a man rises from the lying to the standing position there is an increase in systolic pressure. While this adjustment is being

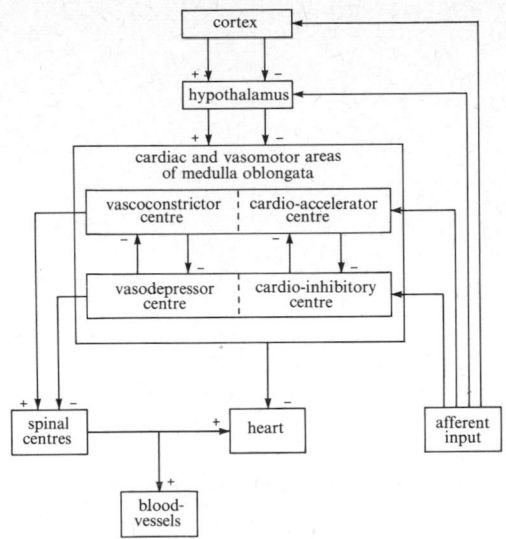

FIG. 21.4. Diagrammatic scheme for central control of the cardiovascular system. (After Peiss (1965). In *Nervous control of the heart* (ed. Randall). Baltimore.)

made there may be a temporary reduction of the supply of blood to the brain, which is the reason for the familiar feeling of giddiness.

The regulation of the blood pressure in response to changing conditions operates through the action of neurons of the medulla which can be considered as four 'centres' (Fig. 21.4). The vagal afferents operate a cardio-inhibitory centre, with efferents in the vagus. Paired with this are, probably, connexions depressing the sympathetic accelerator effect. Close to this is a vasomotor centre whose output is through the accelerator sympathetic pathway.

The neurons of the cardiovascular centres in the medulla thus receive information with each heartbeat about mean pressure and rate of change of pressure from the aorta and other strategic regions in the circulation. The chemoreceptors and receptors signalling inflation in the lungs give information about respiratory performance. These signals together with less specific ones from somatic and visceral sensory inputs all influence not only the medullary neurons but also higher cardiovascular centres in the diencephalon and cortex (Fig. 21.4). These can produce conditioned vascular responses, including those associated with emotion. However, the main homeostatic adjustments are reflex and ensure return of the blood pressure towards a set-point determined by the medullary centre.

Like other physical control systems the blood-pressure regulation depends upon the relationship between the output M and the error signal given by the feed-back transducer e, in this case mainly the baro-receptors. In proportional control $M = k_1 e$, where k is constant; and in proportional plus derivative control $M = k_1 e + k_2 (\mathrm{d}e/\mathrm{d}t)$, which diminishes hunting behaviour. Proportional and integral control, $M = k_1 e + k_2 \int e \, \mathrm{d}t$ tends to eliminate steady-state errors. The circulatory control system may show each of these properties at different times (Korner 1971).

The arterial baro-receptors adapt very slowly and thus provide a continuous information about the blood pressure. Following a sudden rise, the firing rate increases greatly but drops after about $1\frac{1}{2}$ min to a steady level, which is maintained for up to an hour. Over the important part of the range the relation of firing rate to pressure is nearly linear (Fig. 21.5). The chemo-receptors signal changes in arterial $[O_2]$, $[CO_2]$, or pH, but large changes in impulse frequency occur only when there are large changes in the blood.

The effect of increasing carotid sinus pressure at lower pressures is to reduce sympathetic impulses to the heart (Fig. 21.6). As the pressure increases the vagal efferents are brought into play to lower the blood-pressure. This can be considered as a relatively crude regulator type of action. It has a 'gain' of only 1–2 and compensates only incompletely for blood-pressure disturbance over long periods, with a large error signal provided by the difference between the actual blood-pressure and the previous returning value. When there is persistent change in blood-pressure there is re-setting of the operating range to the new elevated or lowered pressure. Such changes occur regularly during sleep or in chronic conditions such as renal hypertension. There

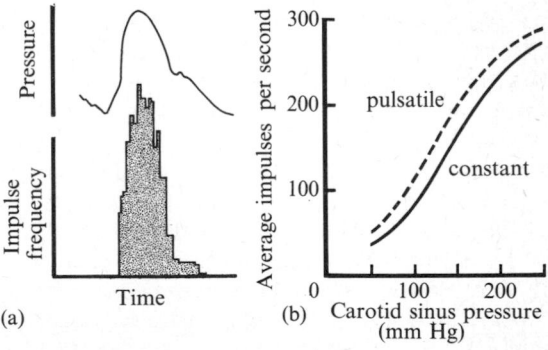

FIG. 21.5. (a) Frequency of impulse firing in a carotid baroceptor and the pressure pulse at normal arterial pressure. (From Korner 1971, after Christensen *et al.* 1967.) (b) Average impulse frequency in carotid sinus nerve relative to mean sinus pressure whilst pressure in sinus is either pulsatile or constant. (From Korner 1971, after Spickler *et al.* 1967.)

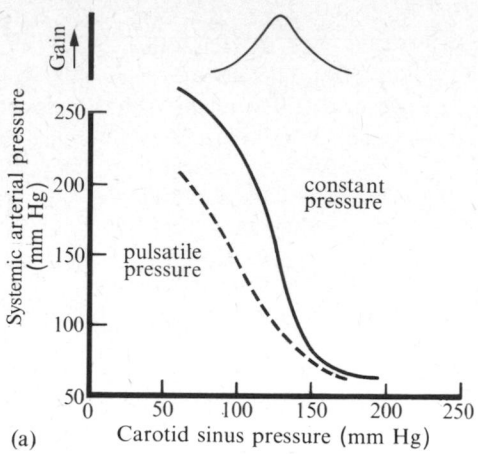

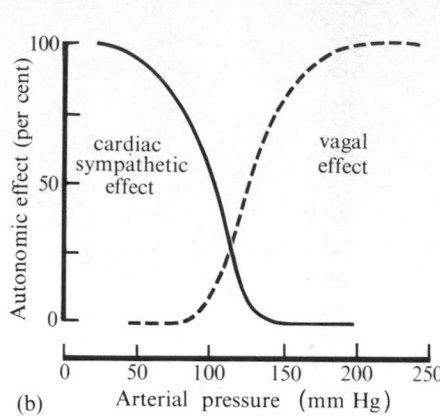

FIG. 21.6. (a) Relationship of mean carotid sinus pressure and mean systemic arterial pressure when pressure in isolated sinus is either constant or pulsatile. Uppermost curve shews relationship of steady sinus pressure to gain of reflex. (b) Diagrammatic relationship of cardiac sympathetic and vagal autonomic effects at different arterial pressure levels. (From Korner (1971). *Physiol. Rev.* **51**.)

is an inhibitory influence of the cerebral hemispheres whose effect is to limit excessive autonomic action on the heart so that it acts as a relatively low-power regulator. The need of the various organs for blood can thus be ensured by the operation of local auto-regulatory mechanisms.

These local factors are so important that it is often considered that local control rather than control of the heart is the main regulator of cardiac output. Diminution of local oxygen supply is followed by immediate vasodilatation, due either to the lower P_{O_2} itself or to raised P_{CO_2}. Changes in body-fluid volumes and electrolytes play a large part in regulating the blood-pressure. When the latter is raised, the kidneys immediately excrete excess water and salt until the pressure is back to normal. This is an integral control system with infinite gain and can over-ride all the others.

The kidney plays a special part in blood-pressure regulation by the production of the enzyme renin, which acts on a plasma globulin to produce the pressor substance angiotensin. This raises blood-pressure by causing vasoconstriction and causing retention of water and salt, also by stimulation of the production of aldosterone (p. 428), which enhances these effects. Renin is produced by the juxtaglomerular cells of the kidney (p. 222), production increasing as pressure falls.

4. Measurement of blood-pressure

Blood is driven along the main arteries by contraction of the left ventricle at a pressure that is much higher in mammals and birds than in cold-blooded animals. The pressure may be measured using a sphygmomanometer by connecting an artery with a column of mercury (manometer) so that the pressure can be recorded with only small changes in the volume of the whole system. The pressure can be measured also without opening any vessel by wrapping round a limb a cuff that can be inflated with air in such a way as to press upon the tissues and stop the blood-flow. The restarting of the flow as the pressure is lowered can be heard with a stethoscope and a reading of the pressure in the cuff at that moment provides an estimate of the blood-pressure. The pressure recorded during the contraction of the ventricle is said to be *systolic*, whereas the lesser pressure during relaxation is *diastolic*.

In man, the systolic blood-pressure of the young, healthy adult is about 120 mm Hg, dropping to about 80 mm at each diastole (Fig. 21.7). In the dog and horse, the pressures are rather higher but the variations are not as extreme, and for nearly all mammals, large or small, the systolic pressure lies between 100 mm Hg and 200 mm Hg, whereas in reptiles, amphibia, and fishes it is about 40 mm Hg.

The pressure falls only slightly as the blood passes along the great and medium arteries, much more rapidly along the arterioles (Fig. 21.7), until it reaches about 22 mm Hg, which is the pressure available for driving liquid through the capillary wall (p. 169). The difference between systolic and diastolic pressures gradually disappears along the arteries and in capillaries and veins the flow is uniform. In the vessels immediately beyond the capillary bed there is a pressure of about 10 mm Hg, and there is a further fall along the veins, so that only a tiny pressure of about 5 mm

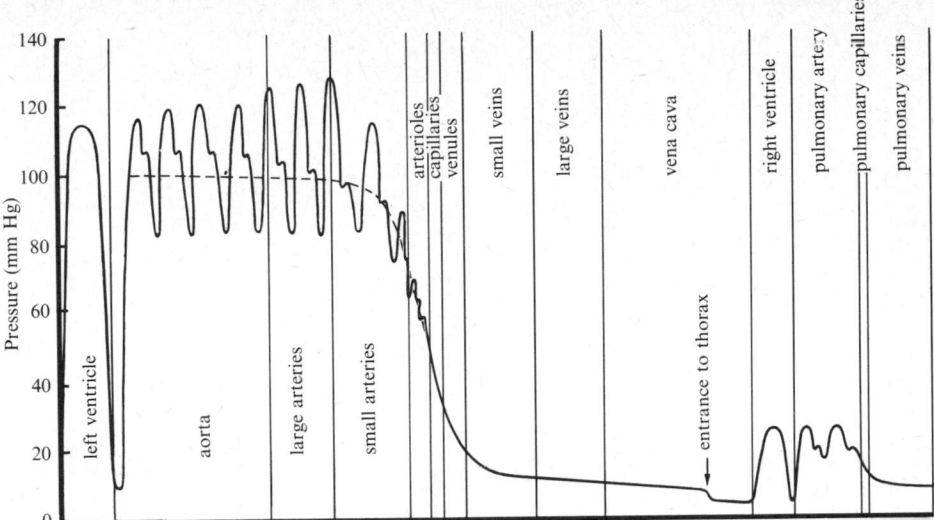

FIG. 21.7. Blood-pressure in man in the systemic circulation. The abscissa represents the relative length of vessels except between the arterioles and venules where the scale is en-larged. (After Selkurt (ed.), (1971). *Physiology*. Little, Brown, Boston.)

water remains to drive the blood along the great veins to the heart. Muscle contraction assists the venous return flow but the last portion of the return journey of the blood is effected by a suction pressure exerted on the veins and on the right atrium during inspiration. Enlargement of the whole thoracic cavity in which the heart is enclosed reduces the pressure to some 6 mm Hg falling to 2·5 mm Hg during expiration. The mainten-ance of an adequate venous return pressure is a matter of special importance to man because of his upright position, and various means are adopted to prevent a fall of this pressure on changing from the horizontal to the vertical position. Quadrupedal animals, which lack means for such adjustment, show greater changes if they are made to stand upon their hind legs.

5. Arteries and veins of the rabbit

The course of the arterial and venous trunks of mam-mals follows closely the pattern seen in lower verte-brates, with such changes as are made necessary by the presence of special features (single aortic arch and a long thorax). The first vessels to leave the aorta are the right and left *coronary arteries,* supplying blood to the wall of the heart itself. They lie at the level of the union of atria and ventricles. The first large vessel is the *innominate* artery, the root of the right common carotid and subclavian arteries; it represents a portion of the original right aortic arch (Fig. 21.8). The left carotid arises just beyond this and except for their origins the two *carotid arteries* are similar. Each

divides at the top of the neck into a small internal carotid, which enters the skull, and an external carotid. The latter divides into five branches, an occipital artery to the back of the head, lingual to the tongue, external maxillary to the submaxillary gland and masticatory muscles, internal maxillary to the orbit, and superficial temporal to the cheek.

The right and left *subclavian arteries* are also similar except for their origins. Each gives rise to a *vertebral artery* running forward in the foramina of the cervical vertebrae to the base of the brain, where the two fuse to form the *basilar artery*, from which vessels pass to the medulla oblongata and cerebellum. In front, the basilar artery divides again to form two *posterior cerebral arteries*. These join the internal carotids and at the level where they do so they give off *anterior cerebral arteries*, which join in the midline and thus complete the *circulus arteriosus* or circle of Willis. There are considerable differences among mammals in the arrangement of the arteries of the head. In the cat, sheep, and many others the internal carotid artery is very small or absent and the circle of Willis is then mainly supplied with blood from the branches of the external carotid through an abundant plexus of arteries known as the carotid rete. This network lies in a venous lake and is presumed to have some special haemodynamic significance. It is absent in mammals such as the rabbit and man where there is a well-developed internal carotid.

From the subclavian arteries (Fig. 21.8) there also

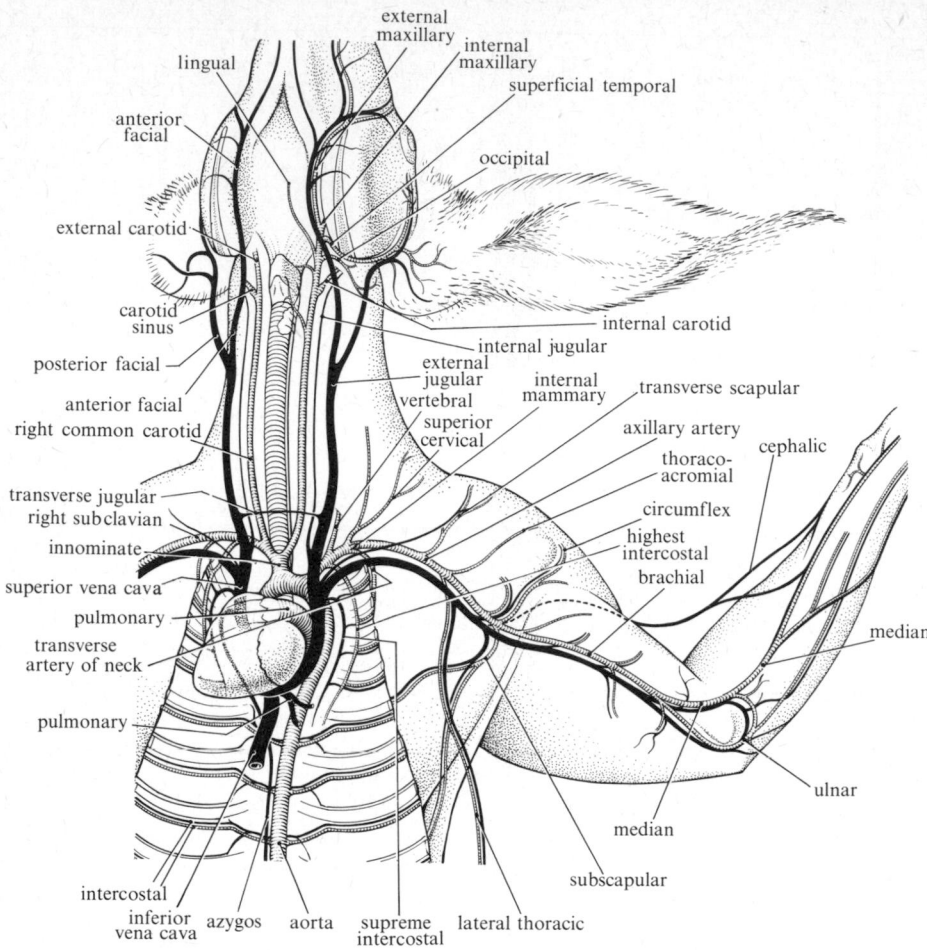

FIG. 21.8. Arteries and veins of the head, neck, and forelimb.

arise backwardly directed vessels, the *internal mammary arteries*, which run along inside the ventral chest wall and supply its muscles. The main trunk of the subclavian continues as the *axillary* and then *brachial artery*, supplying numerous branches to the muscles of the arm and dividing at the elbow into the *median* and *ulnar* arteries, which pass down the forearm to the hand.

The aorta gives no large branches as it passes through the thorax but a series of small though important *intercostal arteries* (Fig. 21.9) supplies the muscles of the back and thorax. At the lower end small *phrenic arteries* supply the diaphragm. The first large branch of the aorta is the *coeliac artery*, given off shortly below the diaphragm. This divides almost immediately into (a) the *splenic artery*, with branches to the pancreas, stomach, and spleen; (b) the *left gastric artery* to the stomach; and (c) the *hepatic artery* to the liver,

stomach, and duodenum. The stomach, like other organs, thus receives several arteries, whose terminal branches make anastomosis. Blood from one of these may flow into the capillary bed previously supplied by another if the latter has been damaged.

Shortly caudal to the coeliac artery arises the *anterior mesenteric artery*, another large vessel, supplying the small intestine and upper part of the colon. The abdominal aorta continues to give off segmental arteries, similar to the intercostals and here known as the *lumbar arteries*. It also provides paired *suprarenal*, *renal*, and *internal spermatic* arteries, in that order. The internal spermatic differs in its course in the two sexes. In the male it runs caudally and supplies branches to the epididymis and vas deferens and then runs as a much twisted vessel to the testis. In the female it runs more laterally to supply the ovary and uterine tube. A

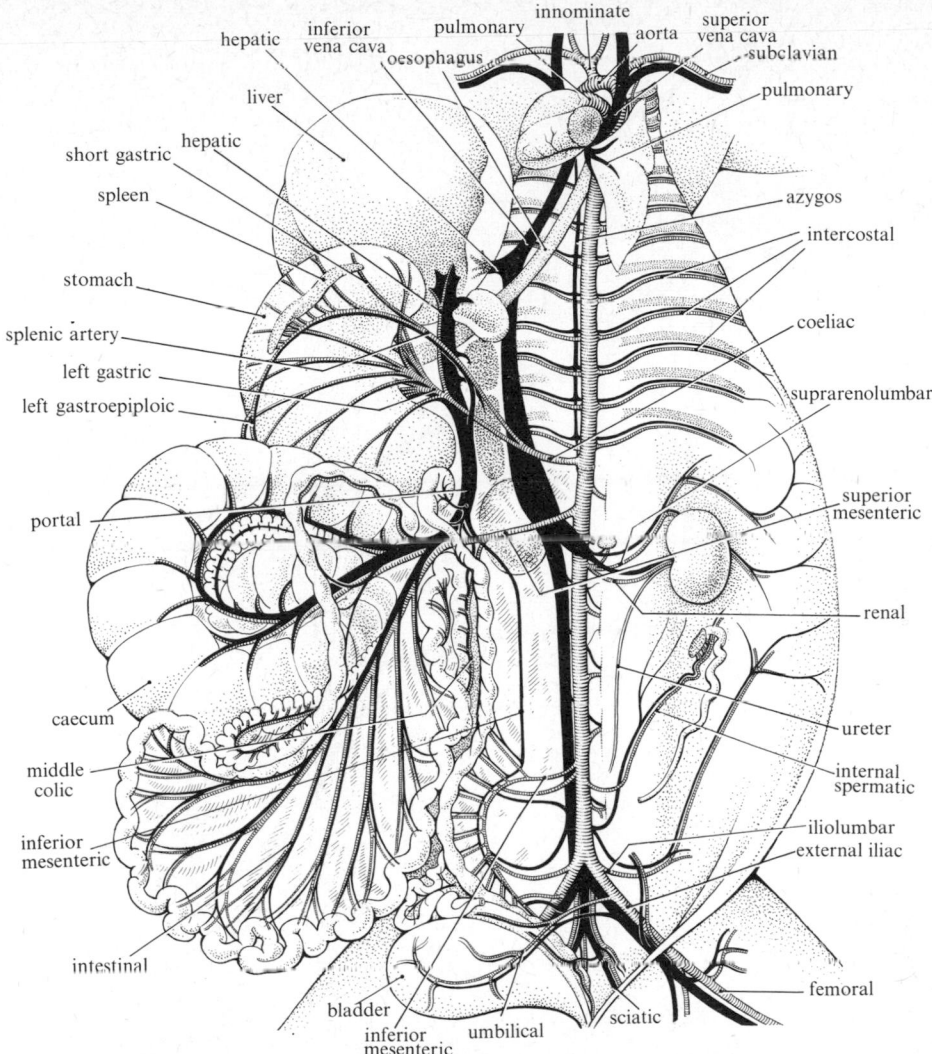

FIG. 21.9. Arteries and veins of the thorax and abdomen.

median *inferior mesenteric artery* supplies the lower part of the colon and the rectum.

Within the pelvic cavity the aorta divides into two *common iliac arteries*. There is a small continuation of the aorta, the median sacral artery, which becomes the caudal artery. Near the origin of the common iliac arteries arise the *ilio-lumbar arteries*, but they may originate from the aorta itself. Each common iliac artery divides into an *external iliac artery* supplying the hind leg, and a *hypogastric (internal iliac) artery*, passing caudally to supply the organs of the pelvis, including the uterus in the female. At this division the *umbilical artery* arises from the external iliac. Passing under the inguinal ligament the external iliac emerges as the *femoral artery*, which has three branches supplying the thigh (Fig. 21.10).

Above the knee the femoral artery supplies the *great saphenous artery*, which continues as the *posterior tibial artery* to the heel. The *genu suprema artery* to the knee arises from the femoral at this level. The femoral artery then continues behind the knee as the *popliteal*, which branches to form the *small saphenous artery* and continues down to the foot as the *anterior tibial artery*.

6. Venous system of the rabbit

The arrangement of veins is more variable than that of arteries and there are considerable differences between the condition in the rabbit and man. Only the veins of

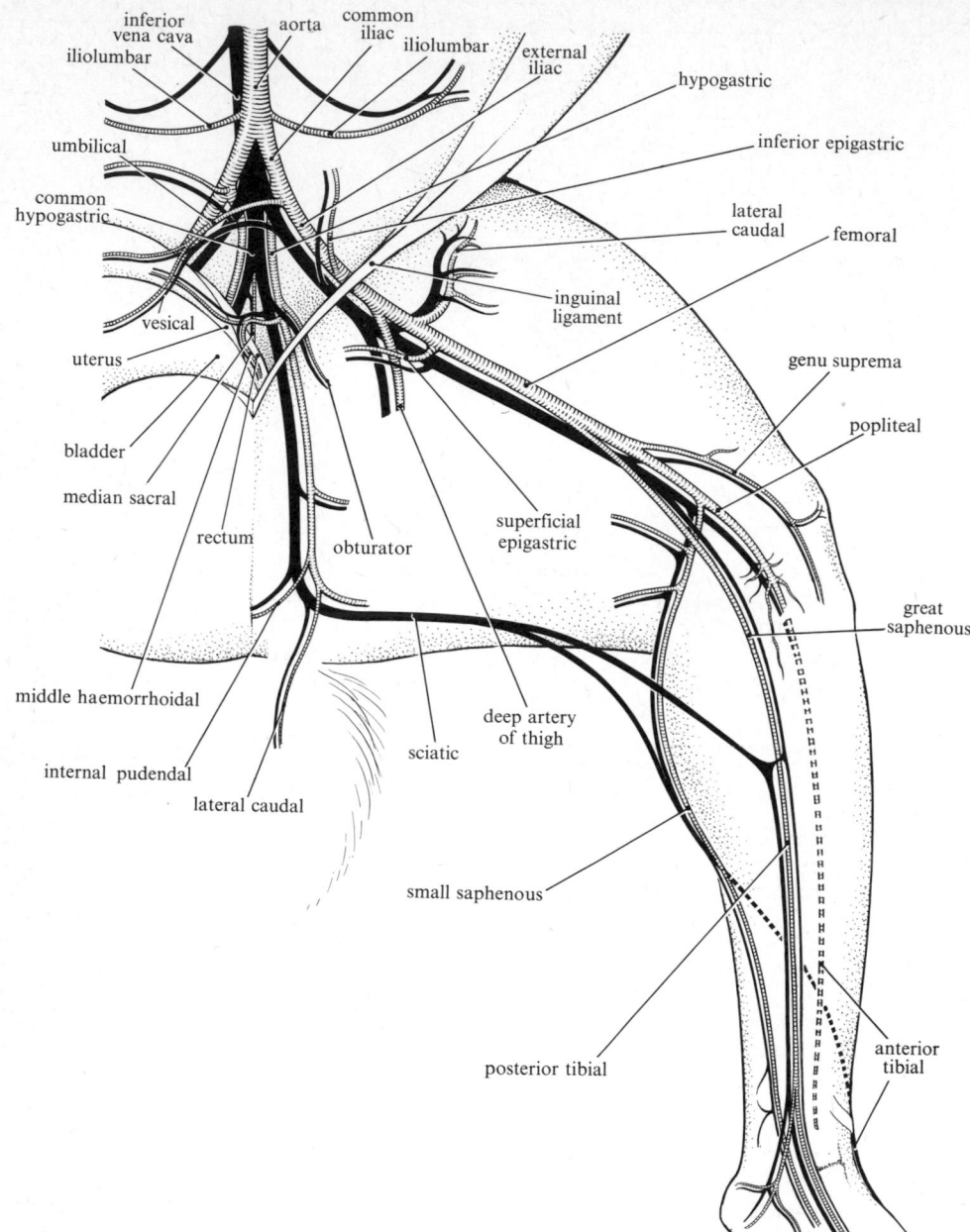

FIG. 21.10. Arteries and veins of the hind limb.

the rabbit are described here, except where specific reference is made to man.

The finer divisions of the deeper veins run mostly with the arteries; often there is a pair of vessels one on each side of an artery, hence *venae comitantes*. There is also a system of superficial veins below the skin; these are often quite large and not accompanied by arteries. Such veins are visible on the human fore-arm, and by emptying them with a stroking movement the position of the valves can be made out (Fig. 21.11). There are communicating vessels between the super-ficial and deep veins and since the latter accompany their arteries cooling of the skin may produce cooling of the arterial as well as the venous blood.

Veins communicate freely with each other, making a system of channels less regular than that of the main

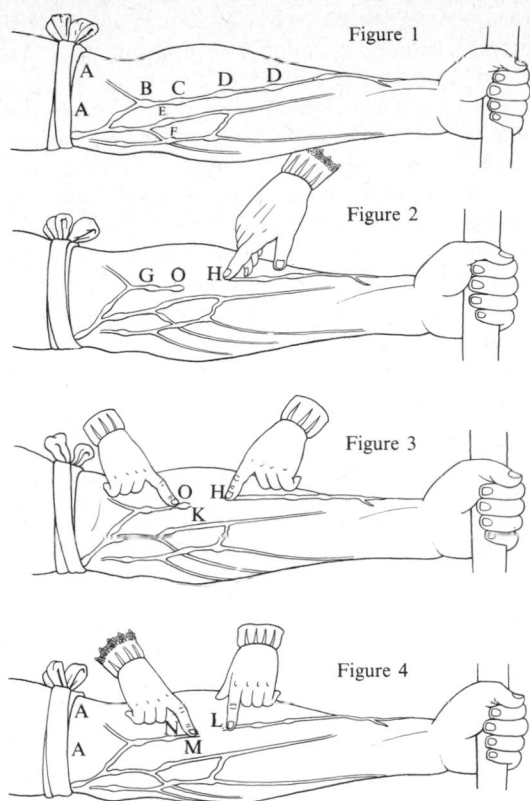

Figure 1

Figure 2

Figure 3

Figure 4

FIG. 21.11. Harvey's experiment to show the action of the valves in the veins of the forearm.

'Let an arm be tied up above the elbow at A, A (Fig. 1). In the course of the veins, especially in labouring men and those whose veins are large, are certain knots or elevations as at B, C, D, E, and F, which will now be seen. These knots are not only at the places where the veins branch, as at E and F, but also where they do not, as at C and D. These knots are formed by valves, which thus show themselves externally.

'If you now press blood from the space above one of the valves, as from H to O (Fig. 2), and keep the point of the finger upon the vein below, you will see no influx of blood from above. The portion of the vein between the point of the finger and the valve O will remain empty. Yet the vessel will continue sufficiently distended above that valve, as at O, G . . . If you now apply a finger of the other hand upon the distended part of the vein above the valve O (Fig. 3), and press downwards, you will find that you cannot force the blood through or beyond the valve . . . You will only see the portion of vein between the finger and the valve become more distended, while the portion of the vein below the valve (H, O, Fig. 3) still remains empty.

'Further, the arm being bound at A, A as before, and the veins full and distended, compress a vein with one finger L (Fig. 4), at a point below a valve or knot. Then with another finger stroke the blood upwards beyond the next valve N. You will now perceive that this portion of the vein L, N still continues empty . . . But if the finger first applied (H, Fig. 2; L, Fig. 4) is removed, the vein is immediately filled from below, and the arm becomes again as at D, C (Fig. 1).' (Copied from the figure in *De motu cordis,* 1628. Translation from Singer, C. (1922). *The discovery of the circulation of the blood.* Bell.)

arteries. On account of their thin walls they easily become dilated with blood, especially after death, and it is often difficult to proceed with dissection without breaking them, for instance, in the rabbit's neck.

The drainage from the front part of the body is based on the embryonic anterior cardinal veins (p. 214) lying in the dorsal wall of the coelom and represented in the adult chiefly by the large *external jugular veins.* These are connected in the rabbit by a transverse jugular in the neck and each receives smaller *internal jugular veins* from the brain and the large *subclavian veins* from the foreleg. In the rabbit, the two vessels enter the right atrium separately as the two *superior venae cavae,* the last portion before the heart representing the *ductus Cuvieri* of the embryo. In man, the left ductus disappears (except for some rudiments in the coronary system) and all the blood is passed to the right side by a left *innominate vein.* This joins the right innominate to form the single superior vena cava. The venous drainage of the head and neck is thus mostly into the external jugular veins, though blood from the brain returns largely by the internal jugular.

The venous system of the hind part of the body is based partly on the posterior cardinal (dorsal) vessels found in the embryo (p. 214) but this channel is largely replaced by the great development of the *inferior vena cava,* which proceeds ventrally through the mesentery and past the liver to the heart (p. 215). This very large vessel runs through the whole length of the abdomen and thorax, receiving on its way a number of *lumbar veins* from the muscles of the back and body wall, *testicular* or *ovarian veins,* *renal veins,* *suprarenal veins,* and finally three large *hepatic veins* from the liver.

In the groin the *external iliac vein* (Fig. 21.10) joins the *common hypogastric vein* to form the inferior vena cava. The external iliac vein receives blood from the abdominal wall and the bladder. On the distal side of the inguinal ligament the external iliac vein is called the *femoral vein.* This receives tributaries from the thigh and the knee. In the region of the knee it is called the *popliteal vein* and receives the *great* and *small saphenous veins* from the foot. In the distal part of the leg the great saphenous becomes part of the posterior tibial vein, a paired vessel lying on either side of the posterior tibial artery.

The common *hypogastric vein* is formed by the union of paired hypogastric veins. Each hypogastric vein receives blood from the uterus, rectum, and abdominal wall via small tributaries and blood from distal parts of the limb by the large *sciatic vein.* Each sciatic vein is joined by one of the posterior tibial branches, anastomosing with the other, and by the anterior tibial vein.

The drainage from the dorsal wall of the thorax and

to some extent of the lumbar region still takes place in the adult by dorsal channels directly derived from the embryonic posterior cardinal (p. 214). Only that on the right is continuous, however, forming the *azygos vein*, opening to the right superior vena cava. The left-hand channel is broken into several sections, which cross to join the azygos as the *hemiazygos vein*. These vessels receive the intercostal veins on each side.

7. Regulation of the peripheral circulation of the blood

The life of every tissue of a mammal depends upon the maintenance of its blood-supply. Ischaemia lasting for a few minutes leads to paralysis of brain tissue or muscle, and if prolonged it is followed by irretrievable damage and necrosis. A most elaborate system of hereditary instructions has been developed to ensure that so far as possible every part shall be provided with an adequate supply of blood. This is possible because the blood-vessels are not a fixed set of tubes but a system of membranes and spaces delicately responsive to changes around them.

These adjustments take place at all levels of time scale from momentary reflex changes of blood-flow occurring within a few milliseconds of the contraction of a muscle to evolutionary changes of arterial pattern requiring thousands or millions of years. The quicker changes depend upon reflex actions that are the result of signals set up by the receptors provided by the hereditary instructions. They lead to changes of heart-rate in response to pressure changes or chemical changes in the main arteries and carotid body (p. 199).

Regulation of the blood-flow to a tissue is produced largely by the influence of the sympathetic nerve fibres. All the arteries and many veins receive sympathetic vasoconstrictor nerve fibres (p. 284) which normally show a low tonic discharge rate of 1–3 impulses per second. The flow is varied by changes in this rate. The capillaries are not innervated but change their diameter in response to the chemical conditions produced by local activity. There are also specific dilator nerve fibres in some parts of the sympathetic and para-sympathetic system. There are neurons in the medulla oblongata constituting pressor and depressor centres. Electrical stimulation of them causes rise or fall of the vasoconstrictor tone. One part of this system is the heat-regulating centre causing vasodilation of the skin, under control from the hypothalamus (p. 308). Proper training of the response of these higher centres is no doubt part of the education that produces a good athlete.

The muscles of the walls of the heart are subject to the effects of use and disuse as are other muscles. After training, it is possible to obtain a much greater output than can be given by the untrained heart.

The degree of vascularity of any organ is also influenced by growth processes, operating with a time-course of hours or days. If the channels that have been providing blood to a tissue become blocked, or inadequate because of increased demand, others will rapidly be opened from neighbouring vessels. Arterioles may enlarge and new capillaries become established. Little is known about the mechanisms by which these very sensitive adjustments are made. Ultimately they depend upon the growth properties of the tissues, conferred by the hereditary instructions. The vascular system possesses in a high degree the power to acquire in this way a memory or record of the demands that have been made upon it in the past and therefore are likely to be made again in the future. Major changes in the pattern of arteries and veins, of course, are made by the operation of the processes of mutation, recombination, and natural selection, acting here as in other parts of the body.

Variation in the pattern of the blood-vessels between individuals is familiar in the dissecting room, and the surgeon is continually aware of it. Little is known of the underlying genetic basis, but there are major differences between the arterial systems of different mammals, for instance, in the internal carotid system (p. 201). The supply of blood is an essential feature of every organ, and the genetic mechanism is no doubt very sensitive to the advantage conferred by each particular arrangement. On the other hand, the capacity for adjustment during the lifetime may to some extent protect the organisms against the effect of genes that produce an unsatisfactory original pattern of blood-vessels. Here as elsewhere we find it difficult to disentangle the effect of genes that produce a particular structural pattern and those that confer the power of reacting to the conditions that the organism experiences.

Thus numerous factors enter into the regulation of the circulation. The adjustments that take place keep the system extremely stable in spite of disturbance at any one point. It is now possible to submit the circulation to a full systems analysis (Guyton, Coleman, and Granger 1972). This consisted of 354 blocks each representing equations for some facet of the circulation. It was then possible to use simulation to predict the effect, for example, of removal of three-quarters of the mass of the kidneys, which causes less than 1 per cent change in body-fluid volume and 7 mm Hg rise in arterial pressure. This is an area where physiology 'is on the verge of changing from the realm of a speculative science to that of an engineering science' (*loc. cit.*).

22 Development of the blood vascular system

1. Differentiation of blood islands

THE heart- and blood-vessels develop from the mesoderm. In mammals the first vessels appear as isolated
blood islands in the splanchnopleur of the yolk sac.
Some of the cells of these islands become flattened to
form endothelial cells, which arrange themselves to
make tubes. Other cells, at the centre of the tubes,
become the first haemocytoblasts, from which the
cellular elements of the blood are derived (p. 172). At
first all the corpuscles are nucleated but as development
proceeds an increasing proportion lose their nuclei.

Further blood islands form in the mesenchyme
throughout the developing embryo, partly from new
centres and partly by spreading of vessels from those
already formed. The main vessels are thus formed
separately and later become linked up and extended to
provide a circulation for each part as it develops. The
later vessels arise by outgrowth from the endothelium
of existing vessels.

The differentiation of mesenchymal cells to haemocytoblasts is presumably an evoked response, but it is
not known what stimulus is responsible. The course of
the main vessels is determined by heredity but the
stimulus of the demand made by developing tissues has
a powerful influence on the blood-system. Actively
growing centres come to have a large blood-supply and
the vessels leading to them become enlarged. The
responsiveness of these vessels to the amount of blood
flowing through them is thus a major factor in determining the vascular pattern (double dependence).

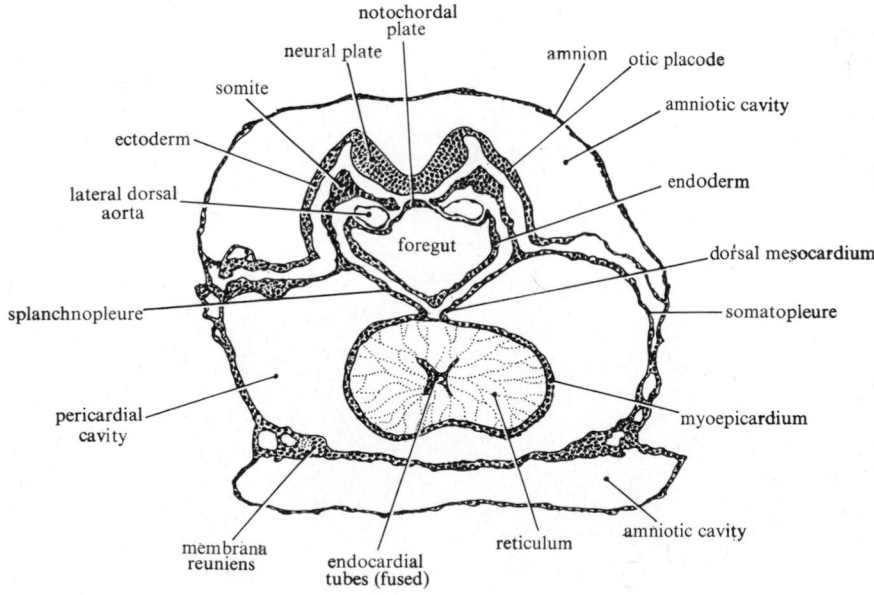

FIG. 22.1. Transverse section of a seven-somite human embryo showing the developing heart. (After Payne (1925). *Contr.
Embryol.* **16**.)

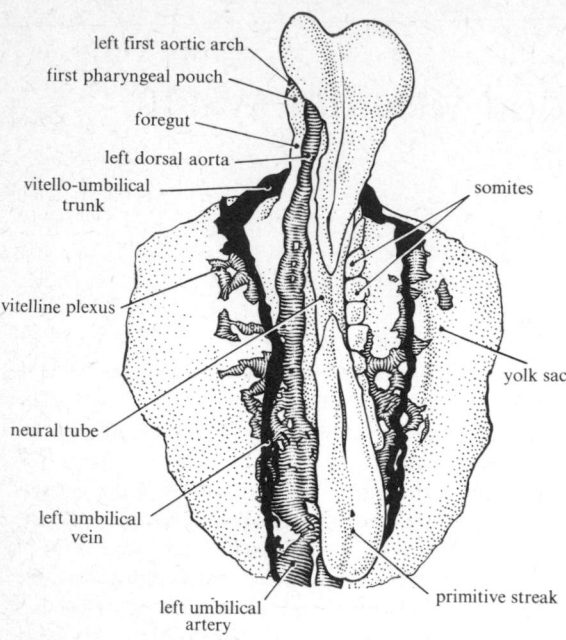

FIG. 22.2. Diagram of dorsal view of seven-somite human embryo showing the early blood-vessels of the yolk sac circulation. The somites have been removed from the left side. (After Payne, *loc. cit.*)

2. The development of the heart

The heart of the earliest chordates was a simple tube below the gut, pumping blood forwards from the intestine to the gills. In the development of a mammal this same plan can be seen. The heart is the first of all the organs to begin functioning. The first signs of it are collections of mesenchymal cells between endoderm and splanchnic mesoderm below the foregut (Fig. 22.1). These cells associate to form a pair of endocardial tubes, which soon approach the midline and join to make a single tube. This union is the result of a folding up of the splanchnic mesoderm, which forms the myocardium and epicardium. The embryonic heart thus projects into the pericardium, attached to the foregut by a dorsal mesocardium. The endocardial tubes lead in front to a short *ventral aorta* and this by the *first aortic arches* to *dorsal aortae*. From the dorsal aorta branches communicate with the networks on the yolk sac and in the chorio-allantoic placenta, becoming the *vitelline* and *umbilical arteries* (Fig. 22.2). The corresponding veins lead to the hind end of the heart (Fig. 22.7, p. 212). The vitelline arteries become the superior mesenteric arteries and the umbilical arteries represent the vessels of the precociously developed bladder (allantois).

The heart thus arises as a continuous tube, which

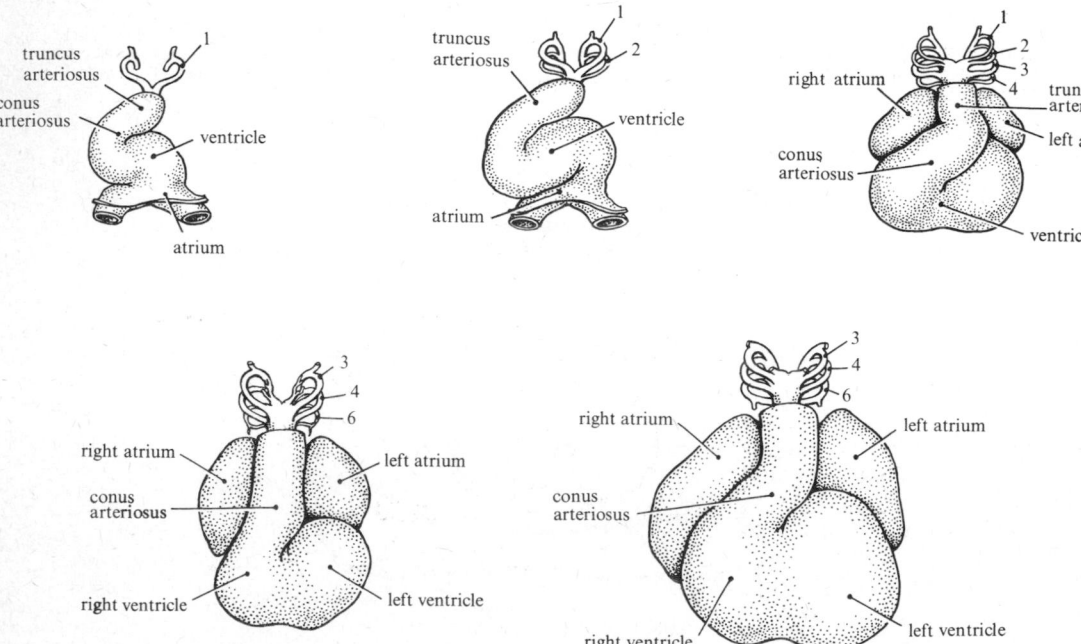

FIG. 22.3. External ventral views of the heart showing its development by bending of the cardiac tube. 1–6, aortic arches. (After Kramer (1942). *Am. J. Anat.* **71**.)

then increases in length and is first bent to the right and then twisted to form a loop (the dorsal meso-cardium disappearing) (Fig. 22.3). The region where the veins enter at the hind end becomes the *sinus venosus*. The part in front of this forms the first un-divided *atrium*, with a ventricle cranial to it. The right-hand loop of the heart forms the conus and truncus arteriosus. By a further bend the originally caudal end of the heart (the sinus and atrium) comes to its adult position cranial to the ventricle. The determination of tissue to form cardiac muscle occurs in the early gastrula stage of amphibians or early blastoderm of the chick, probably through an inductive action of the endoderm. Pieces of the presumptive heart tissue taken at the end of the neurula stage cultivated *in vitro* undergo self-differentiation to form a tube of rhythmic-ally pulsating cells, with characteristic coiling, whose contractions proceed in an orderly way moving it about in its dish like a worm. The presence of blood is not therefore necessary for early differentiation, but the streams of blood determine the details of later develop-ment. If half the rudiment is explanted it regenerates to form a whole one, conversely two rudiments can be fused to make one (see Hamilton and Mossman 1972; Langman 1969).

3. Development of the atria and ventricles

The sinus venosus shifts to the right and the division of the atrium begins. A ridge, the *septum primum*, grows down from the dorsal wall of the atrium towards the ventricle (Fig. 22.4). The atrio-ventricular canal becomes divided into two by the outgrowth of *endo-cardial cushions* of connective tissue. The septum primum does not at first reach as far as the cushions but leaves an opening, the *ostium primum*, between the two atria. Later, as this ostium closes, a second aper-ture, the ostium secundum, appears in the septum primum, maintaining the continuity between the atria (*foramen ovale*). A fold of tissue, the *septum secundum*, grows from the atrial wall to the right of the septum primum and thus lies across the opening of the ostium secundum. At birth the apposition of the septa closes the foramen (p. 215).

Meanwhile the ventricle also becomes divided, by an

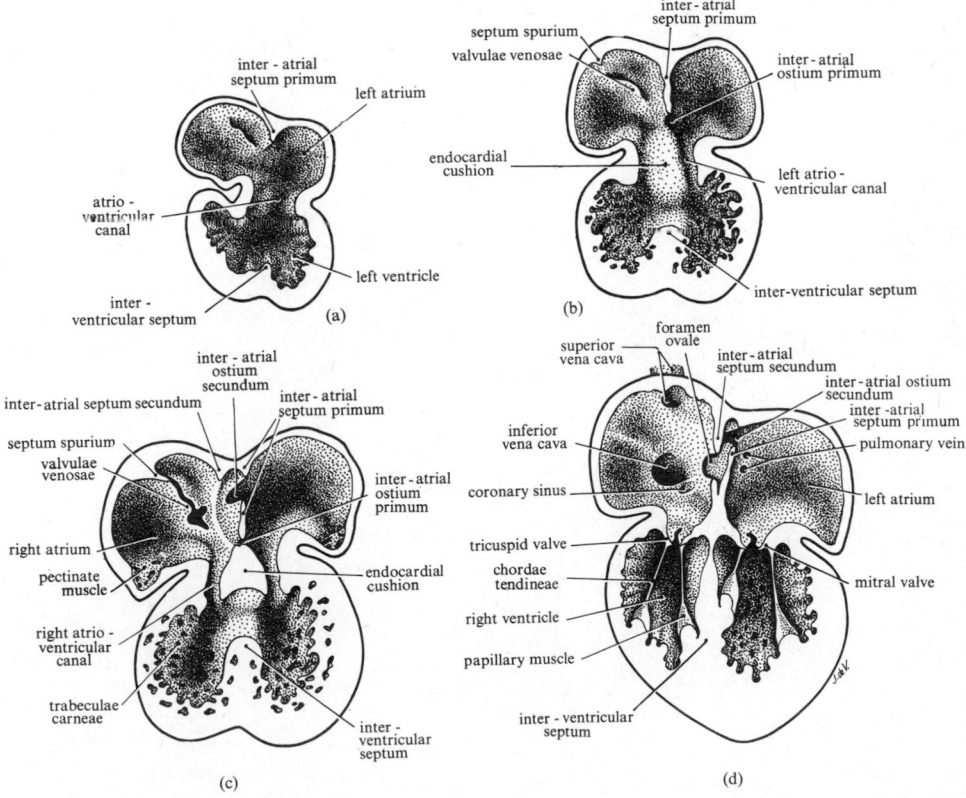

FIG. 22.4. Diagrams showing the development of the septa of the heart. (After Patten 1968.)

inter-ventricular septum growing from the base towards the atrio-ventricular cushion, which it reaches at the end of the second month in man. The truncus arteriosus becomes divided by a spiral fold in such a manner that the right ventricle communicates with the pulmonary, the left with the systemic arch.

4. The aortic arches and arterial system

The arterial system of a mammal is based upon a plan fundamentally similar to that of a fish in that it consists of a ventral aorta, a series of six pharyngeal (branchial) arches, and paired dorsal aortae (see Goodrich 1930, 1958). The morphogenetic processes that gave rise to the arterial system in the fish-like ancestors of the mammals are still retained, producing the appearance

of 'recapitulation', which has so often puzzled zoologists. Evidently this formation of branchial arches is a fundamental feature of the process of vertebrate embryogenesis; its occurrence in all members of the group leads to the apparent 'recapitulation'. In mammals there are considerable modifications from the arrangement found in fishes. The aortic arches appear in a cranio-caudal series, and the anterior ones quickly disappear. The whole series of six cannot therefore be seen at any one time. Moreover, the fifth is at best a transient and incomplete vessel and hardly appears in the figures given by even the most careful investigators.

In a human embryo with 22 somites we can recognize on each side a short ventral aorta, a first (mandibular) arch, and a dorsal aorta (Fig. 22.5). The latter is not at

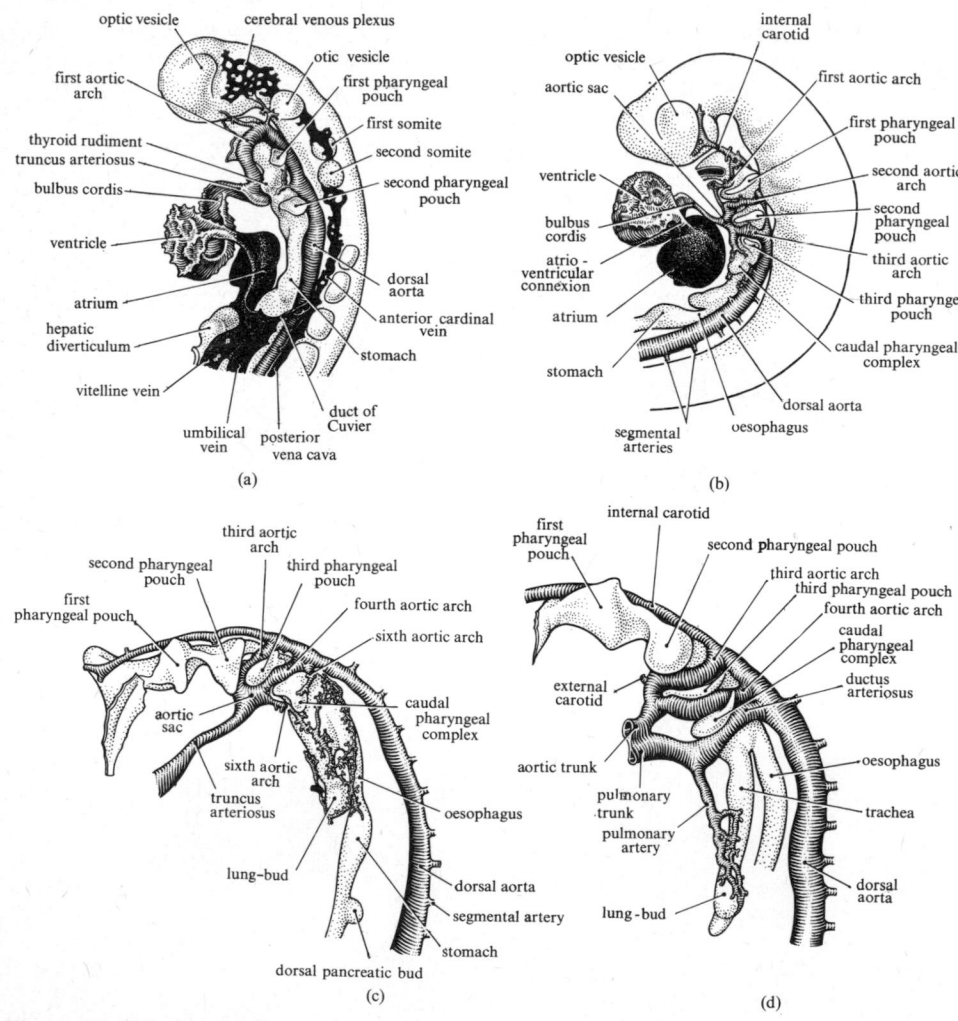

FIG. 22.5. Views of the left side of (a) 22-somite; (b) 4 mm human embryo; (c) 5 mm human embryo; and (d) 11 mm human embryo showing the principal blood-vessels. (After Congdon (1922). *Contr. Embryol.* **14**.)

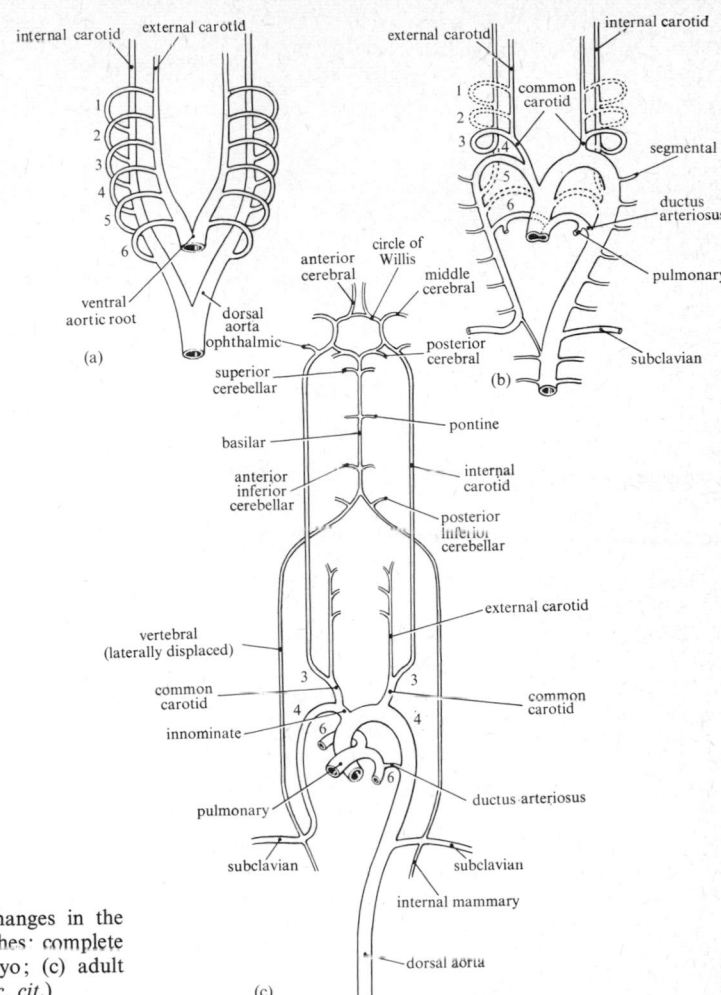

FIG. 22.6. Diagrams of the developmental changes in the mammalian aortic-arch system. (a) Aortic arches: complete set as in a fish; (b) early changes in embryo; (c) adult condition. 1–6, aortic arches. (After Patten, *loc. cit.*)

first a continuous vessel but a series of intercommunicating channels, from which vessels proceed to the somites and yolk sac and to a large umbilical artery to the placenta.

The ventral aortae proceed to fuse to make the short truncus arteriosus (aortic sac). The dorsal aortae fuse in their caudal portions, but remain separate above the pharynx. Meanwhile arterial arches differentiate and disappear in series. As the third becomes complete the first disappears and so on. Yet considerable traces of the original branchial arches remain even in the adult. The first and second probably disappear completely but have been held to contribute to the maxillary and stapedial arteries. The front parts of the ventral aortae become the external carotids and of the dorsal aortae the internal carotids. The dorsal aortae disappear between the third and fourth arches, the internal

carotids being thus fed through the remains of the third arches (Fig. 22.5). The left fourth arch becomes the aorta, while the right fourth remains only as the base of the right subclavian artery (Fig. 22.6).

The fifth and sixth pharyngeal arches appear when the embryo is about 5 mm long, the fifth being incomplete and very short lived. From the sixth arches vessels accompany the developing lung-buds. The ventral portions of the sixth arches thus become the pulmonary arteries. The dorsal portion of the left sixth remains connected with the dorsal aorta as the *ductus arteriosus*, which provides a channel short circuiting the lungs during intra-uterine life. Blood reaching the right ventricle is passed through the ductus to the aorta. This channel closes at birth, when the full circulation of the lungs becomes established (p. 215).

The branches given off from the dorsal aorta are at

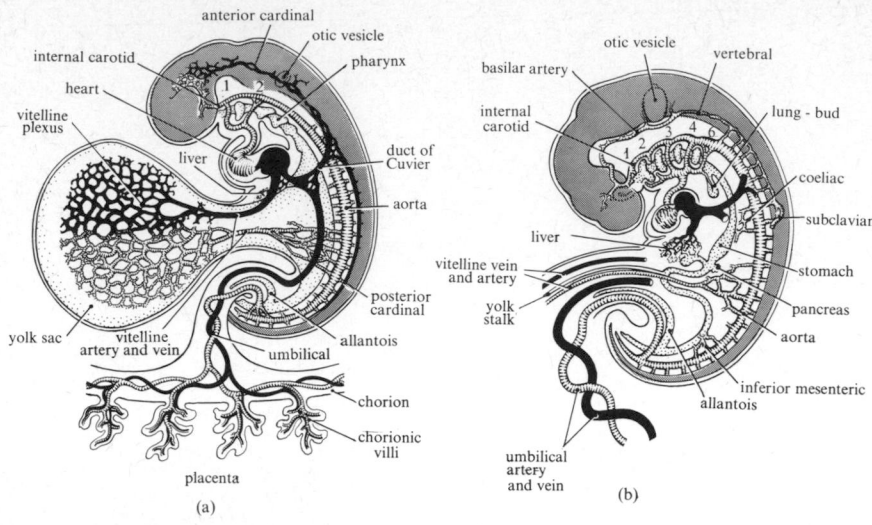

FIG. 22.7. The aortic-arch system and other blood-vessels of the human embryo seen from the left side. (a) 3–4 weeks old; (b) 4–5 weeks old. 1–6, aortic arches. (After Congdon, *loc. cit.*)

first segmental vessels passing to the somites, viscera, and limb-buds. As development proceeds only the somitic intersegmental arteries retain the segmental plan. The coeliac and superior and inferior mesenteric arteries are splanchnic vessels that were originally connected with the seventh cervical and third and fifth thoracic segments; later they make anastomosis more posteriorly. The umbilical arteries can be regarded as the arteries of the allantois (p. 489) and they open from the common iliac arteries (Fig. 22.7).

The arteries of the limbs develop as four or five segmental vessels, which are later reduced to single arteries arising at about the seventh cervical and fifth lumbar levels.

5. Development of the venous system

The veins, like the arteries, are formed upon the plan of development found in fishes, but with great modification of the time of appearance of the various vessels. The heart is essentially a subintestinal vessel receiving

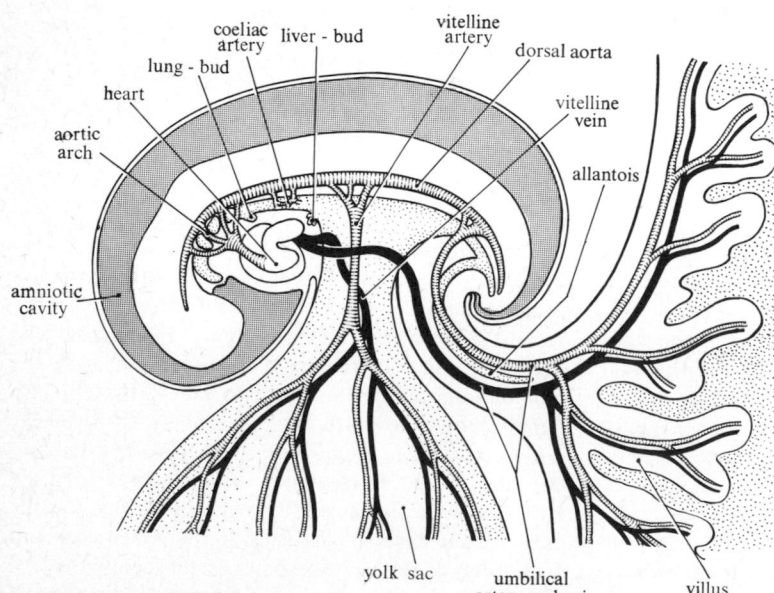

FIG. 22.8. Diagram showing the early embryonic blood-vessels and their relations with the maternal tissues.

blood from the gut. The *vitelline veins* form very early in the splanchnopleure of the yolk sac and join the sinus venosus (Figs 22.7 and 22.8). Later, as the liver develops in the septum transversum, they contribute to make parts of the portal vein and inferior vena cava.

The circulation in the yolk sac is of far less importance in most mammals than that of the chorio-allantoic placenta, which can be regarded fundamentally as derived from the vessels of the allantois (p. 489). The

umbilical veins are therefore developed very early, passing around the yolk-sac stalk to the sinus venosus. As the liver develops, the blood from the umbilical veins passes to it, joining that from the vitelline veins. The right umbilical vein disappears at about 7 mm, all the placental blood then returning in the left umbilical vein. As development proceeds a direct channel, the *ductus venosus*, appears so that much of the blood from the umbilical and vitelline veins proceeds direct to the

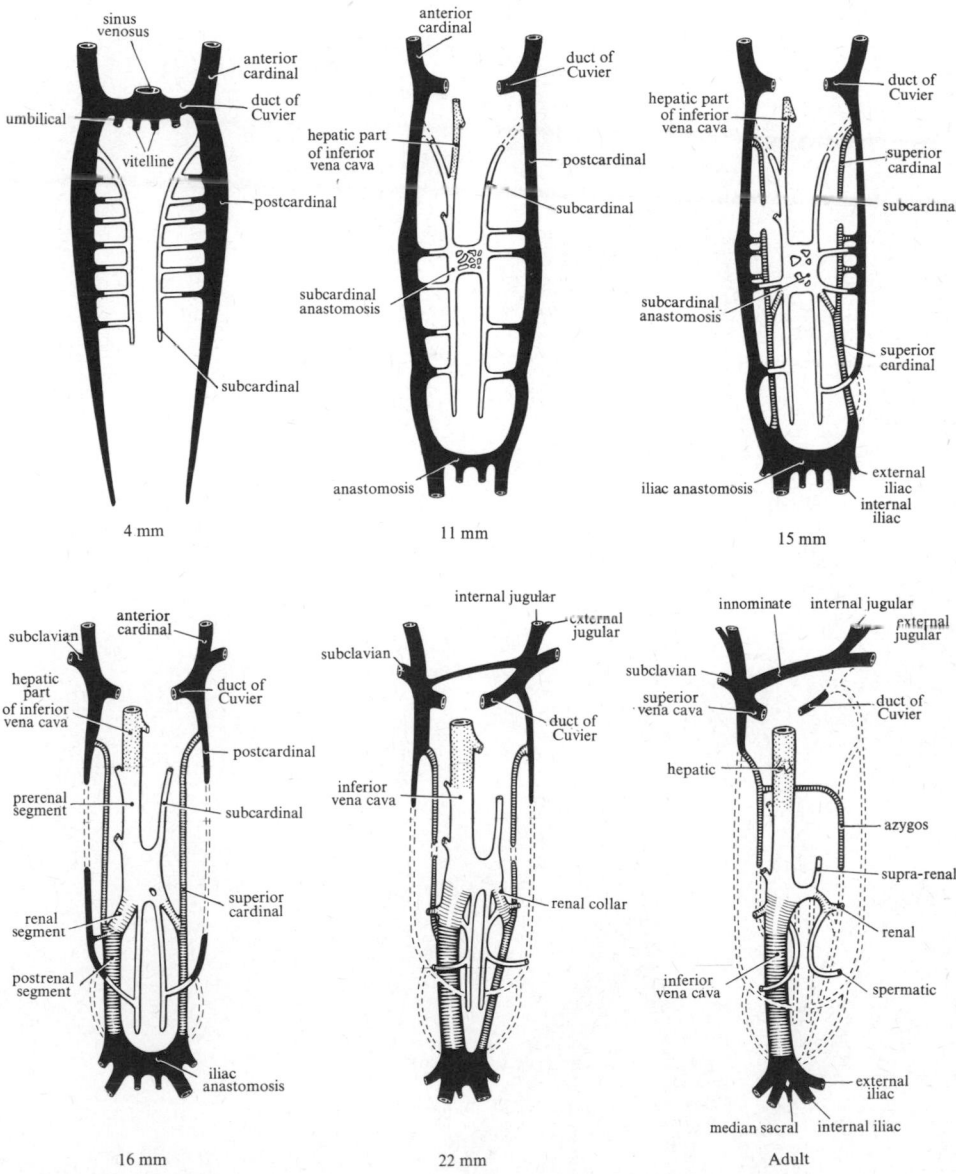

FIG. 22.9. Diagrams of the venous system of man at various stages of development showing the origin of the adult veins. (After McClure and Butler (1925). *Am. J. Anat.* **35**.)

right atrium. However, the liver performs important functions in the foetus and receives a good blood supply via *venae advehentes*, and the blood is returned by *venae revehentes*.

The main venous drainage channels of the body of the embryo are at first the anterior and posterior cardinal veins, lying in the dorsal body-wall, lateral to the aortae (Fig. 22.7). These return blood to the heart by paired *ducts of Cuvier*, lying in the septum trans-

versum and entering the sinus venosus.

The venous drainage of the front part of the body comes to be concentrated on the right anterior cardinal and duct of Cuvier, which becomes the *superior vena cava*. In the hind part of the body, the original dorsal venous channels provided by the posterior cardinals are replaced by a series of more ventral vessels, leading directly to the heart through the inferior vena cava. A series of *subcardinal* and *supracardinal* vessels is formed

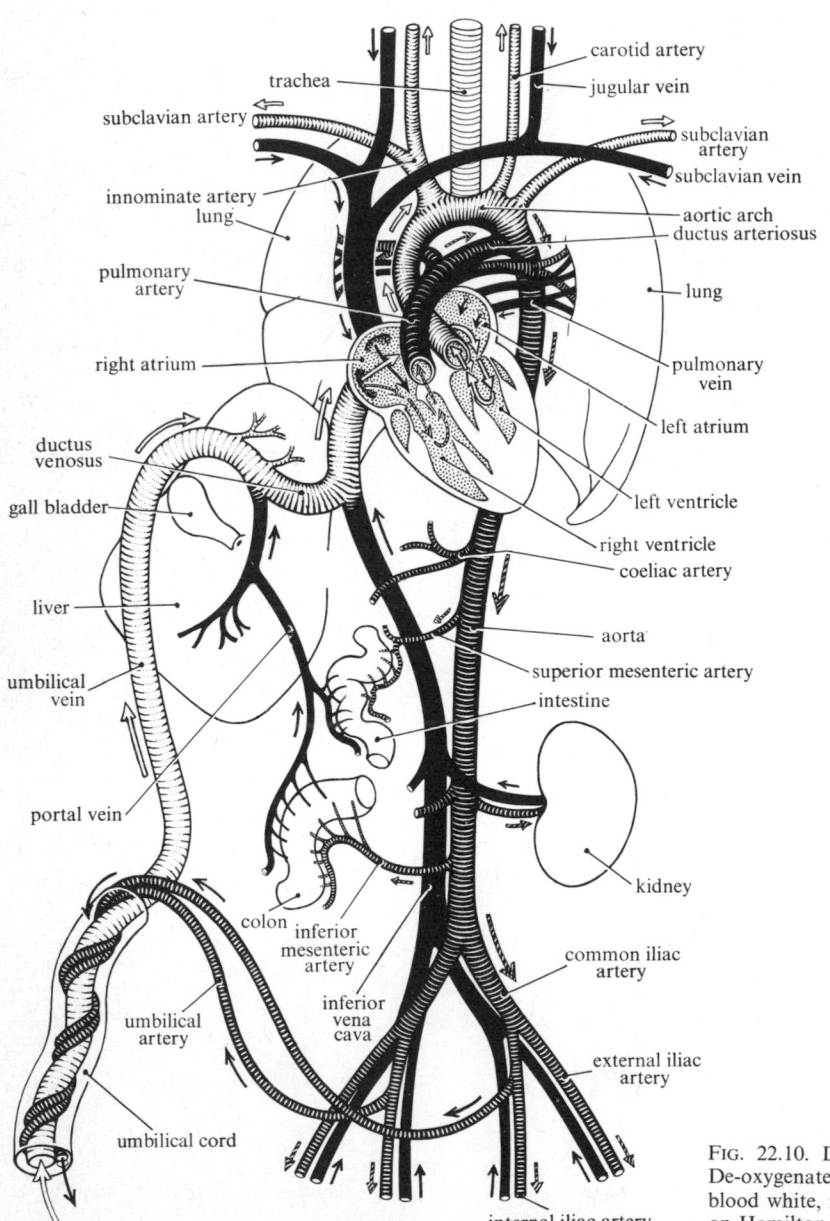

FIG. 22.10. Diagram of the foetal circulation. De-oxygenated blood shown black, oxygenated blood white, mixed blood cross-hatched. (Based on Hamilton and Boyd 1972.)

draining the meso- and metanephros and the thorax (Fig. 22.9). From these there is developed a median vessel, the inferior vena cava, which comes to receive the blood from the hind limbs. It passes down through the mesentery, receiving the hepatic veins and proceeding to the right atrium. This is thus a more direct pathway from the hind part of the body to the heart that was provided by the posterior cardinals. The dorsal portion of the thorax is drained by the *azygos* and *hemiazygos* veins, derived from the supracardinals (Fig. 22.9).

6. Course of circulation in the foetus

The placenta provides the foetal blood with oxygen and other raw materials. A special set of foetal conditions ensures that the blood returning in the umbilical vein to the right atrium is sent mainly direct to the developing tissues of the body, only a small part of it passing through the lungs (Fig. 22.10). Conversely, the blood returning in the venae cavae, being poor in oxygen and rich in excretory products, requires to be sent as directly as possible to the placenta.

The positions of the openings of the superior vena cava, ductus venosus, and inferior vena cava suggest the 'figure-of-eight hypothesis', that the mainly oxygenated blood from the inferior vena cava passes directly through the foramen ovale to the left atrium and ventricle, heart wall, and head. The blood from the superior vena cava passes directly down through the right atrium to the right ventricle and so by the ductus arteriosus to the aorta, umbilical artery, and placenta (see Assali, Bekey, and Morrison 1968).

It is still not certain to what extent separation of the streams is achieved. It is not complete because venous blood in the caudal part of the inferior vena cava is mixed with oxygenated blood returning in the umbilical vein. No doubt there is also some mixing of the streams within the right atrium, but X-ray studies of foetal sheep after suitable injections of opaque material show that a considerable degree of separation is achieved.

7. Changes in the circulation at birth

At birth a series of changes in the walls of the vessels leads to a rapid alteration of the circulation to its adult condition with closure of the three shunts provided by the ductus venosus, foramen ovale, and ductus arteriosus. The umbilical arteries contract so that no further blood leaves the body. The umbilical vein closes some minutes later than the arteries and therefore blood actually may be drawn back from the placenta. As the child becomes anoxic by deprivation of maternal oxygen, the receptors controlling its own respiratory system begin to operate and respiratory movements begin. Air enters the lungs and the circulation of blood through them increases. Large quantities of blood begin to return to the left atrium and the pressure there becomes greater than in the right atrium. The flow through the foramen ovale stops and the septum primum is forced against the septum secundum to close the aperture. The closure of the ductus arteriosus is produced by a mechanism depending upon the change in oxygen concentration of the blood. The foetal P_{O_2} is low and this has the effect of constricting the vessels of the lungs. It is not known what stimulus starts breathing. The high concentration of oxygen from the air dilates the pulmonary vascular bed but causes contraction of the muscles of the wall of the ductus arteriosus. Blood from the right ventricle is thus directed through the lungs and the adult circulation is established. Later the foramen ovale and lumen of the ductus arteriosus become obliterated by the growth of fibrous tissue.

These changes at birth are possible because of anticipatory growth processes, preparing the body for actions that occur only on the single occasion of birth. By selection of suitable genetic mechanisms modifying previously existing morphogenetic processes, the system has been brought to operate with a high degree of precision. Presumably in the early ancestors of mammals, where much yolk was present and the placenta was less developed, the changes at birth were less abrupt.

23 Excretion and the control of water balance

1. Excretion

THE foodstuffs taken into the body are not used with complete efficiency. Some products of metabolism, for instance, lactic acid, can be used again but many others, such as urea, have to be removed from the tissues and ultimately from the body by the process of excretion. Most animals, therefore, are provided with some region in which materials superfluous or harmful to the internal environment can be allowed to escape to the outside. The mammalian kidney not only gets rid of unwanted substances, such as urea, but also serves to regulate the water and salt contents and acid–base balance.

This delicate regulation has developed from a much simpler system in earlier vertebrates, consisting of a series of funnels leading from the coelom to the exterior. Primitively these served to transport genital as well as excretory products, and the urinary and genital systems remain related to each other even in mammals. During development the channels are formed in a series, the pro-, meso-, and metanephros (p. 223), but in a mammal they are never open funnels. Instead each tube is closed at its inner end, where there is a tangle of capillaries, the glomerulus. This arrangement is probably related to the need for a land animal to conserve water (see Windhager 1969). Water and solutes filter out from the blood in the glomerulus and then such part of the filtrate as is needed by the body is reabsorbed during passage down the tube. Thus only unwanted water and solutes are excreted (Morel and de Rouffignac 1973; Orloff and Burg 1971). The tubules of this type develop as a metanephros behind and quite separate from the mesonephros, which remains to perform genital functions (p. 227). The urinary system of the adult, therefore, is entirely separate from the genital system, except where the two open to the exterior.

2. The kidney

The kidney of a mammal consists of a mass of tubules, the nephrons (Fig. 23.1), each about 35 mm long, end-ing blindly internally in sacs known as Bowman's capsules, whose walls are invaginated by special capillary loops, the *glomeruli*. The capsule and glomerulus together are known as a *Malphigian corpuscle*. These closed ends of the tubes lie in the cortical region of the kidney and from each proceeds a tube divided into three regions: (a) the proximal convoluted tubule; (b) the loop of Henle or 'thin segment'; and (c) the distal convoluted tubule.

Finally the lower ends of the distal convoluted tubules unite and thus several nephrons discharge into a common collecting tubule.

The proximal and distal convoluted tubules lie in the cortex and the loops of Henle proceed towards the centre of the kidney and then back again to the surface, each having thus a descending and an ascending portion.

3. Filtration in the glomeruli

It is now generally accepted that the kidney functions by a system of filtration in the glomeruli and re-absorption in the tubules. Under the pressure of the left ventricle of the heart there is forced into the Bowman's capsule (the cavity of the corpuscle) a solution containing all the non-colloidal constituents of the blood. As the liquid passes along the proximal convoluted tubule, salts, sugars, and other substances not normally found in the urine are re-absorbed, and much of the water is taken back in the loop of Henle, which is specially developed in connexion with life on land. Many features of kidney structure agree with this cycle of filtration and re-absorption and earlier theories ascribing the production of the urine to an active secretion into the tubule are not generally applicable, although it may be that some dyes and other substances can pass from the cells into the lumen of the tube.

The lining of Bowman's capsule is a simple epithelium of flattened cells, invaginated at one point by the glomerulus (Fig. 23.2). The endothelium of the glomeru-lar capillaries is pierced by numerous pores (Fig. 23.3).

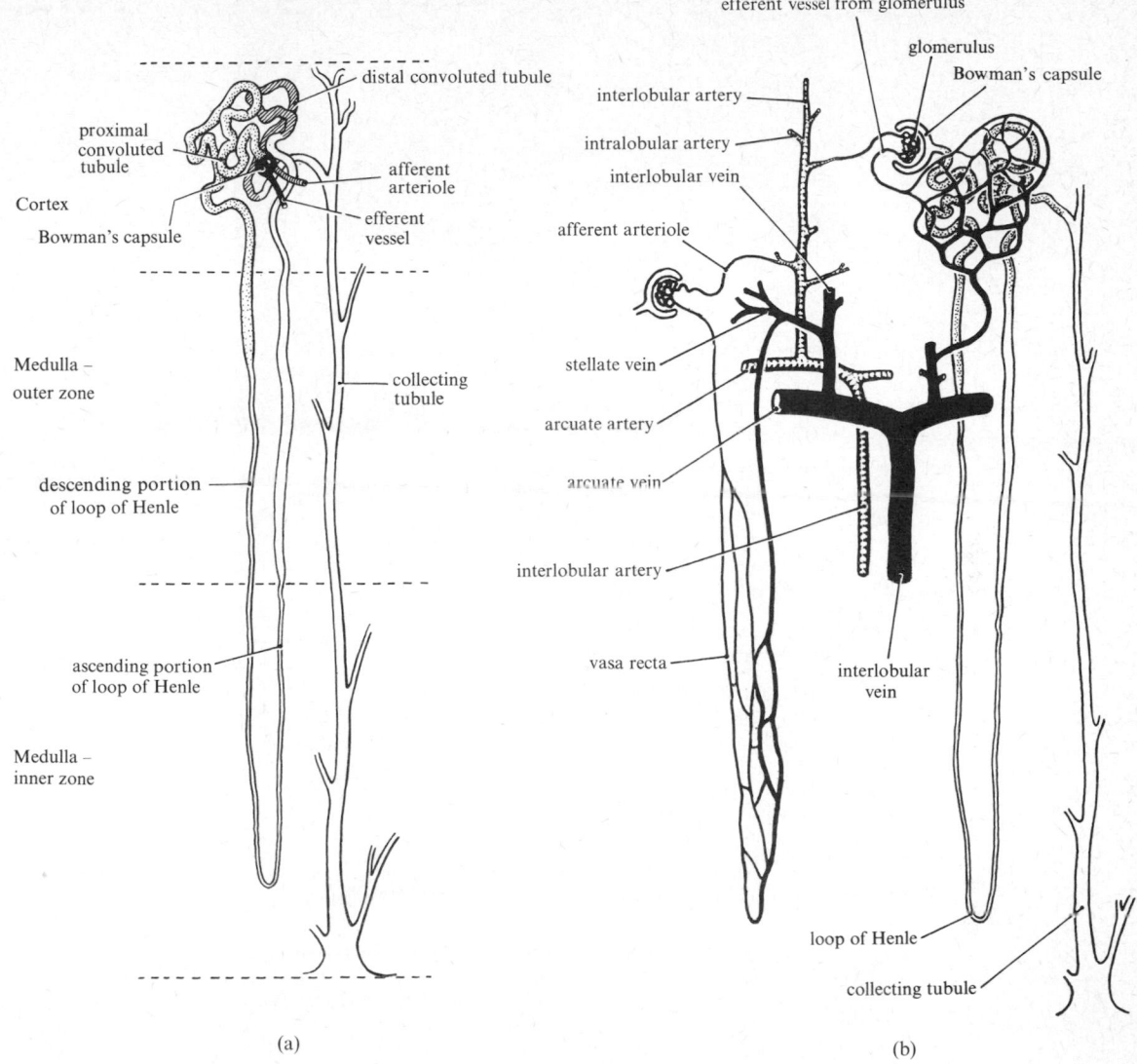

(a)   (b)

FIG. 23.1. (a) Diagram of a human nephron. (b) Diagram of blood-supply to capillary beds of renal cortex and medulla.

The outside of the vessel is covered by the processes of a special set of cells, the *podocytes*. Between the surfaces of the endothelial cells and podocytes lies a continuous basement membrane, which is the effective filter. The renal blood-supply is so arranged that nearly all the arterial blood passes first through the glomerular capillaries and then on to the tubules (Fig. 23.4). The vessels provide for a high filtration pressure; each renal artery arises directely from the aorta and divides rapidly into arterioles, which divide again each into some 50 loops, running with little anastomosis through the glomerulus to join again to form an efferent arteriole, which then proceeds to distribute blood to the tubule. The glomerulus thus provides a region where a solution can be forced through the capillary wall into Bowman's capsule, just as liquid normally passes into the tissues (p. 169). The colloid osmotic pressure tends to act in the opposite direction but is lower than the hydrostatic pressure. As a result a concentrated solution is carried away in the efferent blood-vessel of the glomerulus, and this aids in re-absorption from the tubules. The tubules of Henle's loop also receive blood from glomeruli as straight vessels (*vasa recta*) running close to the tubules and looping back as veins (Fig. 23.1(b)).

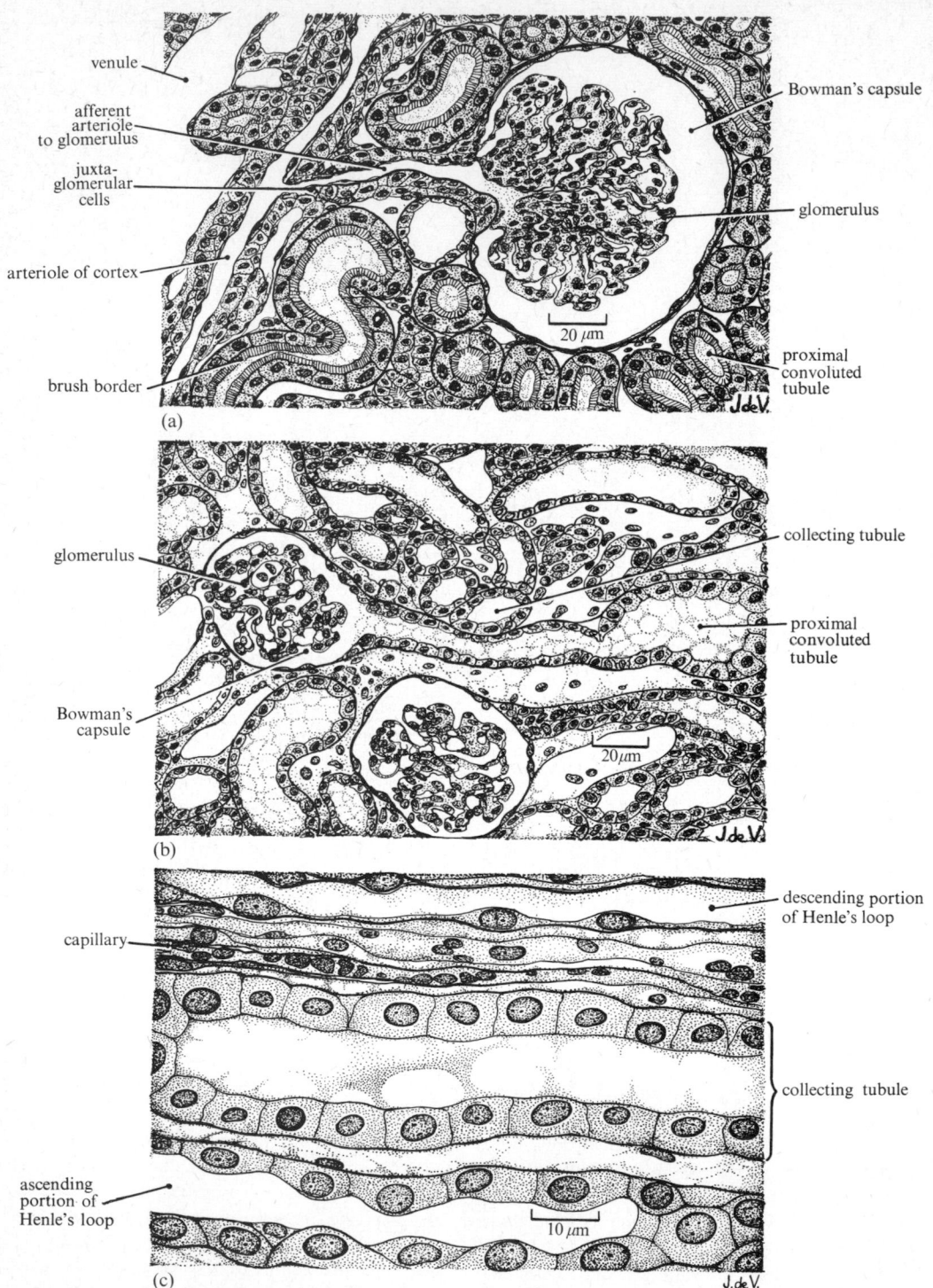

venule

Bowman's capsule

afferent
arteriole
to glomerulus

juxta-
glomerular
cells

glomerulus

arteriole of cortex

20 μm

proximal
convoluted
tubule

brush border

(a)

collecting tubule

glomerulus

proximal
convoluted
tubule

20μm

Bowman's
capsule

(b)

descending portion
of Henle's loop

capillary

collecting tubule

ascending
portion of
Henle's loop

10 μm

(c)

FIG. 23.2. Sections of the kidney of (a) dog, showing the entrance of an afferent arteriole to a glomerulus; (b) sala- mander, showing the proximal convoluted tubule arising from Bowman's capsule; and (c) cat, in a section passing longitudinally through part of the outer zone of the medulla. (From photographs and preparations by Mr. K. C. Richardson.)

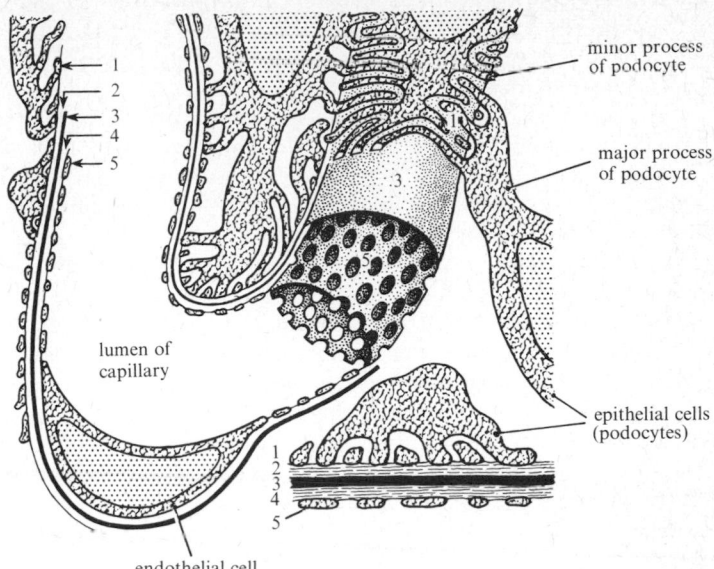

FIG. 23.3. Diagrammatic reconstruction of part of a capillary loop in the glomerulus.

The endothelial cells (5) are perforated by holes about 0.1 μm in diameter. Outside these cells is a complex basement membrane consisting of inner and outer cement layers (4 and 2) with a 'structural portion' (3) between them. Embedded in the outer cement layer are the end-feet (1) of a set of epithelial cells (podocytes) lying in the urinary space. These feet interdigitate in a complicated manner, providing a covering for the capillary, but leaving spaces. The only continuous barrier between blood and urine is thus the basement membrane (3). It remains to be discovered whether the epithelial cells have contractile or elastic properties and thus help to regulate the flow. (Modified after Pease (1955). *J. Histochem. Cytochem.* **3.**)

Many experiments have confirmed the occurrence of filtration through the glomerular wall, the most direct being those in which fluid is withdrawn by a pipette from the Bowman's capsule of a frog and found to have the composition of blood-plasma free of protein and fat, that is to say it contains substances such as sugar and chloride that are absent from the normal urine. The capsular membrane acts as a filter whose pore size allows the passage of molecules up to about a molecular weight of 50000. Thus the large serum

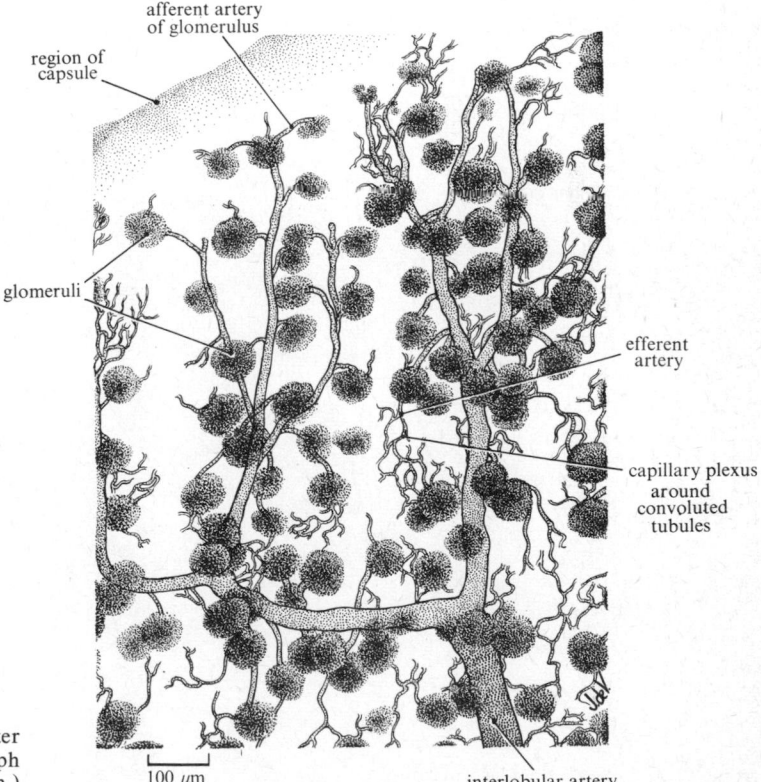

FIG. 23.4. Portion of the renal cortex after injection of the arteries. (From a photograph and preparation by Mr. K. C. Richardson.)

proteins do not normally pass through it. According to the filtration theory a rise in the tubular pressure should prevent urine formation, and it can easily be shown that raising the pressure in the ureter will do this, as will also a fall of blood-pressure. Formation of fluid by active secretion, for instance in a salivary gland, does not depend on the maintenance of this difference between blood and duct pressures.

A large portion of the circulating blood flows through the kidney, perhaps a fifth of the output of each discharge from the left ventricle, and as much as 25 per cent of the liquid entering through the renal artery leaves the glomerular vessels and enters the lumen of the capsules. It is calculated that 170 l a day filter across the 2 million glomeruli of the two kidneys. This tremendous flow continues throughout life and nearly all of it, 168·5 l, are absorbed back into the blood.

4. Reabsorption in the tubules

Each *proximal convoluted tubule* is a much twisted tube 15 mm long, formed of cells whose inner margin has a characteristic striated structure, the so-called *brush border* (Fig. 23.5). Electron microscopy shows that this region contains a number of tubes or ducts

extending towards the interior of the cell. These are probably not hollow tubes but special compartments of the cytoplasm, lined with a denser surface structure. The base of the cell is also divided into compartments by folds of the surface membrane (Fig. 23.5), and in this region lie many mitochondria (Bulger 1965). A somewhat similar 'brush border' is seen in the cells lining the small intestine (p. 145) and in other cells where absorption occurs, and it is presumably related to the transport across the membrane of material from the lumen. Phosphatase is abundant in these kidney cells and is presumably related to the energy used in the transfer. It is supposed that organic constituents (sugar, etc.), water, and salts are removed from the filtrate in the proximal convoluted tubule by means of the secretory activity of the tubule cells. This transfers sodium ions to the blood and chloride follows. Glucose is also secreted and some water is removed by osmosis. Urea and potassium are partly re-absorbed by diffusion. Some substances that cannot pass the glomerular membrane are secreted into the tubule. These include penicillin and dyes (for example, phenol red) that are bound to plasma proteins.

At the end of the proximal tubule the fluid is about isosmotic with blood. In the rest of its course it is

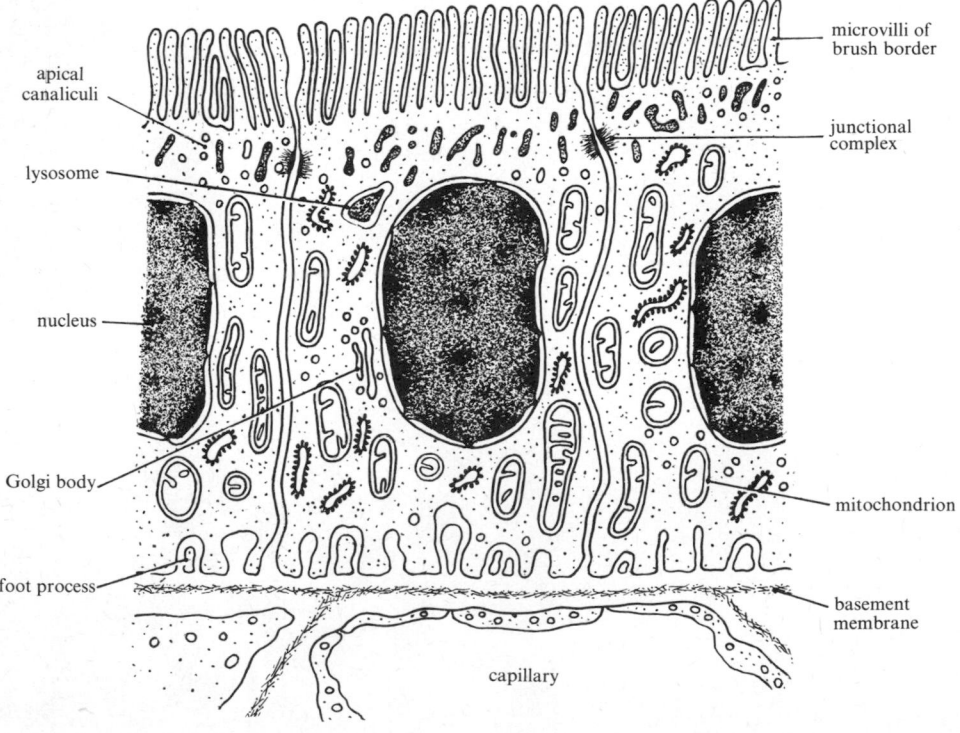

FIG. 23.5. Diagram of part of the wall of a proximal tubule, as seen by electron microscopy.

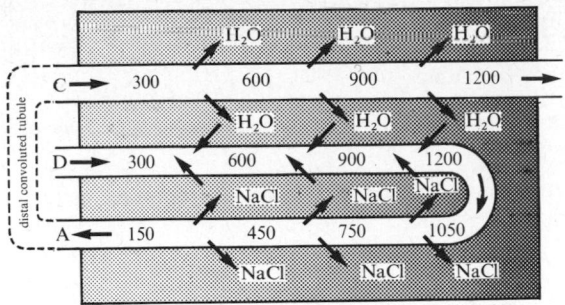

FIG. 23.6. Schematized hairpin countercurrent system in the kidney. There is a steady rise in the concentration of fluid in the descending loop of Henle (D) and conversely a steady fall in the ascending loop (A). The distal convoluted tubule connecting the ascending loop with the collecting tubule (C) actively reabsorbs NaCl and allows water to pass to the blood from the tubular fluid so that they become isotonic. Stippling indicates the increased concentration of interstitial fluid towards the tip of the loop of Henle corresponding to the milli-osmolar concentrations of the fluid indicated at various points in the loop and collecting tubule. (After Lippold and Winton (1968). *Human physiology*.)

converted into urine of the appropriate concentration, more water being absorbed if the animal is over-hydrated. In the succeeding *descending loop of Henle*, about 15 mm long, the walls have a thin flattened epithelium. In the recurrent ('ascending') loop, how-ever, the cells are of deeper cuboidal form. The passage of water from the lumen back into the blood-stream is operated partly by the osmotic suction exerted by the concentrated proteins of the blood, as is the return from tissue spaces to venous capillaries (p. 169). The transport of sodium and glucose back into the blood has a similar effect.

The concentration of the urine is assisted by a counter-current mechanism by which the osmolar con-centration of the tissue fluids is increased in the medullary region (Fig. 23.6). The ascending loop of Henle secretes sodium into the tissue fluids and this draws water from the descending loop. The effect is that although the concentration is much the same at the two ends of the loop, the salt concentration of the tissue fluid is much greater in the medullary region. As a result the fluid in the final collecting tubule is subjected to powerful osmosis leading to the final concentration and production of hypertonic urine. In animals that have a need to conserve water the loops of Henle are especially long (Davson and Eggleton 1968).

The 170 l of glomerular filtrate that are formed each day in man carry as much as 1100 g sodium chloride, 425 g sodium bicarbonate, and 145 g glucose; 95–9 per cent of these are re-absorbed, for example, only 5–10 g

sodium chloride appear in the daily output of 1·5 l urine (Smith 1953).

These figures give an idea of the magnitude of the changes that take place along the tubules, whose total length if placed end to end would be nearly 50 miles!

5. Composition of the urine

The main constituents of the urine of man are urea and the chlorides, sulphates, and phosphates of sodium and potassium. Ammonium salts formed by the kidney are found in small amounts and there is some uric acid formed by the endogenous metabolism of nuclei throughout the body. Protein is normally absent (but there may be some mucus from the walls of the bladder and urinary passages).

6. Ureter and bladder

The ureter is a tube of mesodermal origin, whereas the bladder of mammals is formed as an endodermal pouch of the embryonic cloaca. The two organs are, however, of similar structure, being tubes with characteristic lining epithelium and muscular wall. The so-called *transitional epithelium* consists of several layers of cells, not regularly arranged and presumably allowing for con-siderable increase in area when stretched, especially in the wall of the bladder. The ureters show continual rhythmical peristaltic contractions, by which fluid is passed along into the bladder, the terminal portion acting as a sphincter to prevent reflux.

The smooth muscle fibres of the wall of the bladder run in various directions and they are able, like other smooth muscle fibres, to adjust their length so that with distention of the organ the pressure increases only slightly. As fluid accumulates, however, receptors in the wall of the bladder are stimulated and ultimately by reflex action they produce the rhythmical contractions of the bladder muscle which, together with relaxation of the sphincters of the exit tube from the bladder, the urethra, lead to micturition.

The bladder muscles are controlled by nerves from the hypogastric (sympathetic) and pelvic (parasym-pathetic) systems (p. 284). The fibres of the pelvic nerve produce contraction of the bladder muscle and relaxation of the sphincter, whereas the hypogastric fibres have the opposite effects. The pressure at which the micturition reflex is initiated is about 15 cm water in man, but during micturition the contraction of the bladder wall produces much greater pressures (up to 130 cm water).

The two ureters enter the bladder near its lower end at some distance from each other. The line between them forms the base of a triangle, the *trigone*, at whose apex is the exit from the bladder to the *urethra* (Fig.

24.4). The area of the trigone is recognizable by the smooth stretched state of its mucous membrane. The urethra is a tube leading to the exterior either alone, in the female, or after union with the vas deferens in the male. The lining of the urethra resembles that of the bladder in the upper part ('transitional epithelium') but lower down becomes of the stratified squamous type. Mucous glands occur at intervals and serve to keep the wall of the tube moist. There are also muscles in the walls of the urethra to serve as sphincters for the retention of the urine; the more distal portion of this musculature consists of striated fibres under voluntary control.

The bladder is the most ventral of the structures lying in the pelvis and the urethra passes through the pelvic ring immediately dorsal to the pubic symphysis. The male duct (vas deferens) opens into the urethra on its dorsal side; in the female the vagina runs as a separate passage above the urethra (Fig. 24.1). The rectum is the most dorsal of the structures running through the pelvic ring and lies above the urethra in the male or vagina in the female.

7. Renin

The kidney produces a substance *renin* that raises arterial blood-pressure. The cells responsible for this are probably certain *juxtaglomerular* cells, which surround the afferent artery to each glomerulus, occupying a region between the latter and the distal convoluted tubule nearby. After partial closure of a renal artery there is a marked rise in blood pressure accompanying a hypertrophy of the juxtaglomerular cells. Renin is not itself a pressor substance but is an enzyme serving to release the active pressor *angiotensin* from a plasma globulin *angiotensinogen*. The active material acts upon peripheral blood vessels and also stimulates the secretion of aldosterone. It is 40 times more active than noradrenaline.

8. Regulation of excretion and of composition of the body fluids

The amount of water re-absorbed from the glomerular filtrate depends upon the quantity of antidiuretic hormone (ADH) reaching the kidney. Increased concentration of the blood (as after sweating) stimulates osmoreceptor cells of the hypothalamus, whose axons produce a neurosecretion that regulates the posterior lobe of the pituitary (p. 445). This lobe produces ADH, which stimulates the tubules to absorb more water. Conversely more urine is produced after drinking, after a delay of 30 minutes or more, which is the time in which circulating ADH is removed. The effect of the ADH is to make the distal convoluted and collecting tubules more permeable to water. This is probably done by depolymerizing the hyaluronic acid that acts as a cement substance between the cells (p. 29).

The amount of sodium in the body (and the sodium–potassium ratio) is regulated by the amount of aldosterone produced by the adrenal (p. 428). This in turn is probably controlled by the hypothalamus. After increased sodium intake the aldosterone output falls and more sodium is excreted.

The regulation of the acid–base level of the blood is also assisted by the kidney when necessary, for example, by excreting phosphate or sulphate and retaining sodium. These various regulations each requires some instructional system of control. This is seldom to be found in any simple form but the setting of the hypothalamic osmoreceptors presumably determines the level of water excretion.

24 Development of the urogenital system

1. The coelom and genital and excretory ducts

THE urinary and genital systems are derived from the intermediate mesoderm and adjacent coelomic wall, and they show considerable traces of segmental origin. The excretory system in particular develops as a series of organs, the pro-, meso-, and metanephros, arising in that order along the length of the body.

The intermediate mesoderm forms a solid cord of cells, joining the somites to the dorsal wall of the coelom (Fig. 12.3, p. 120). In many vertebrates it contains cavities opening by a series of funnels to the coelom. The fundamental method of excretion was thus to allow escape of coelomic fluid, and this or a similar channel also provided for escape of the genital products. In mammals this situation no longer exists; no open nephric funnels remain, and the urinary and genital systems are mainly separate in the adult. Nevertheless we see in this, as in other systems, signs of the conservatism of organisms. In mammals the kidney develops as it did in earlier vertebrates by the formation of pro-, meso-, and metanephros (Fraser 1950; Gruenwald 1952).

In lampreys the genital products of both sexes are shed into the main coelomic cavity and this is still the condition in the female mammal. The coelomic cavity can indeed be regarded as the much expanded genital sac (Goodrich 1935). The system of nephric funnels and ducts may have served to carry away genital products before it became excretory. In modern mammals derivatives of the mesonephros, the Wolffian ducts, provide for the transmission of the genital products in the male and in the female this function is performed by a funnel and tube, the Mullerian ducts, which develop nearby.

Both the Wolffian and the Mullerian systems of ducts make their appearance during development in every individual irrespective of sex, and it is only at a relatively late stage that one or the other becomes definitively developed as a result of the influence of the hormones proceeding from the developing gonad. The gonad itself is at first of bisexual nature but the particular combination of X and Y chromosomes then ensures that either an ovary or a testis develops (p. 226). The secretions of the gonad in turn produce further development of either the Wolffian or the Mullerian duct. Castration can be performed *in utero*, and it is then found that male embryos become feminized; the Wolffian duct system fails to develop but the Mullerian system continues. Female castrates, however, develop nearly normally. The hereditary make-up of the tissues thus ensures development of female characteristics in the ducts and derivatives of the urinogenital sinus. The effect of male hormones is to suppress the female and promote the male tendencies, but probably female hormones play little part at this stage of development. In birds the situation is reversed and this is perhaps correlated with the fact that whereas it is the female that is heterogametic (XY) in birds, it is the male in mammals.

2. Pronephros and mesonephros

The *pronephros* of man consists of a series of cell masses between the levels of the fourth and fourteenth somites, joined by a backwardly growing cord, the rudiment of the pronephric duct (Fig. 24.1). The only sign of excretory capacity in the pronephros is the formation of glomeruli protruding into the coelom, but it is not known that any excretory products are formed. The pronephric duct grows back to the cloaca, and as it passes through the more posterior regions it provides the stimulus for the differentiation of the meso- and metanephros. Waddington (1938) has shown that if the pronephros is removed from chicks the remaining portions of the urogenital system do not develop.

The *mesonephros* (Figs 24.1 and 24.2) forms a prominent ridge behind the pronephros, containing a number of tubules with walls invaginated to form glomeruli. The tubules open into the pronephric duct, now known as the Wolffian duct, and they probably secrete urine, especially in those mammals where there are several layers between foetal and maternal blood and there is a large allantoic bladder (p. 489).

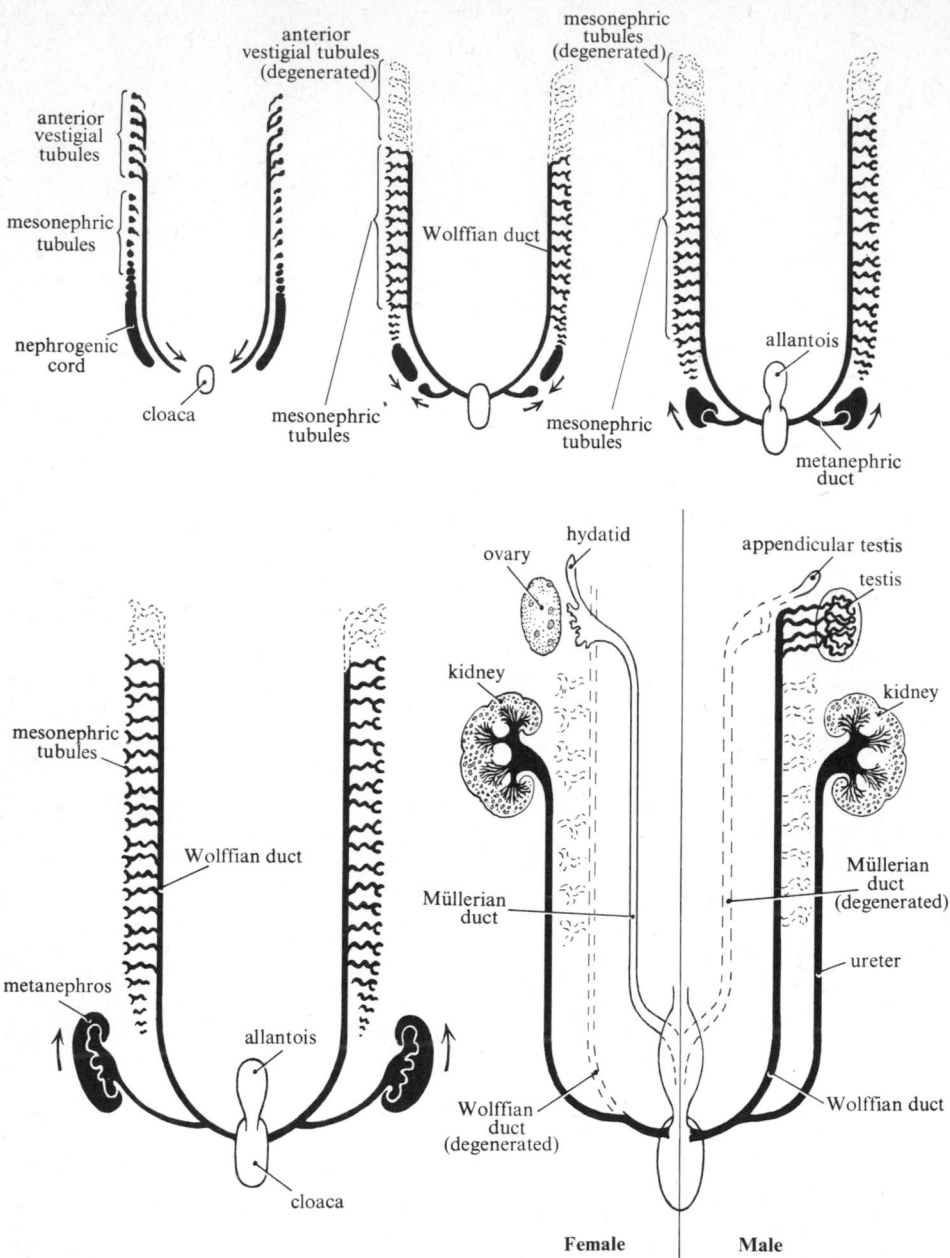

Fig. 24.1. Diagrams showing stages in the development of the kidney and ducts of the urogenital system in a male and female mammal. (From data supplied by Dr. E. A. Fraser.)

The pro- and mesonephros begin to degenerate early, so that the whole series is not to be found at any one time. The pronephros disappears completely, the mesonephros gives rise to the vasa efferentia of the testis and the paradidymis in the male, and to the epoöphoron in the female (see Du Bois 1969).

3. The metanephros

The adult kidney or *metanephros* develops from the intermediate mesoderm behind the mesonephros. Tubules and glomeruli differentiate within a solid nephrogenic ridge. The *ureter* forms as an outgrowth from the lower end of the Wolffian duct, which grows

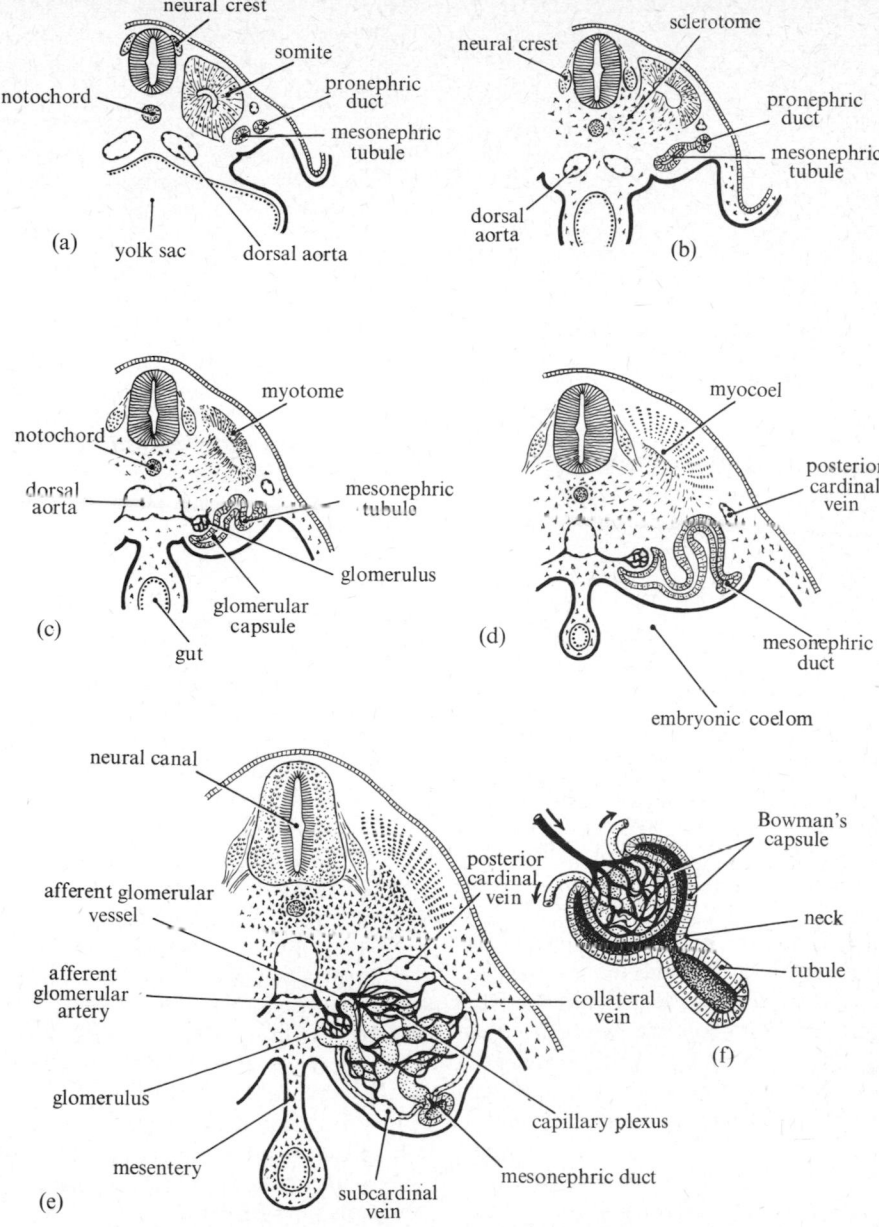

FIG. 24.2. Development of mammalian mesonephric tubules and their vascular relations.
(a) Tubule primordium still independent of duct; (b) union of tubule with primary nephric duct; (c) early stage in development of glomerulus and capsule; (d) further development of capsule and lengthening of tubule; (e) relations of blood-vessels to well-developed mesonephric tubule; (f) glomerulus and capsule, enlarged. (After Patten 1968.)

forward into the metanephros and divides to form the collecting tubules (Fig. 24.1) (see Vernier and Smith 1968).

4. The gonads

The gonads develop from two distinct sources, which are at first widely separate. The genital ridges of the coelomic epithelium, lying medial to the mesonephros, provide what may be called the *structural elements* of the gonad. The *primordial germ cells*, which ultimately give rise to the ova or spermatozoa, arise in the endoderm of the yolk sac and later migrate by active

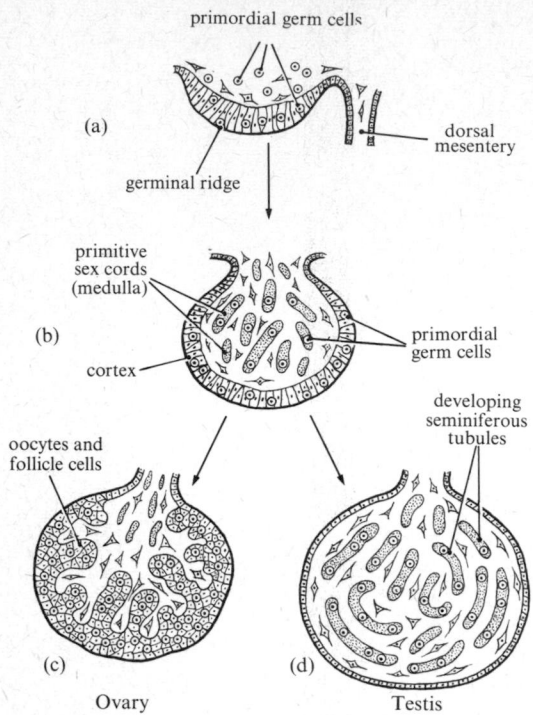

FIG. 24.3. Development of gonad in higher vertebrates. (a) Primordial germ cells partly embedded in epithelium of genital ridge and in mesenchyme. (b) Indifferent gonad with germ cells in cortex and in primary sex cords. (c) Differentiation into ovary: primary sex cords reduced and cortex containing germ cells proliferating. (d) Differentiation into testis: cortex reduced and germ cells in sex cords.

amoeboid movement to the genital ridges. They are round and much larger than the epithelial mesodermal cells of the gonad and have large vesicular nuclei. The germinal ridge forms a hollow sac with an outer epithelial layer, the cortex, and a central medulla of mesenchyme. Within this strands of cells develop (probably from the epithelium), the *primitive sex cords* (Fig 24.3). In the male the primordial germ cells all migrate into these cords, which then become hollowed out as the *seminiferous tubules*. In the female the primitive sex cords are absorbed and all the primordial germ cells migrate into the cortex. Here they become surrounded by epithelial cells and undergo the first steps of oogenesis to become primary oocytes.

The developing gonad thus has a bisexual composition, the outer cortex (germinal epithelium) being the female part, the central medulla (sex cords) the male part. In fishes and amphibians both parts may remain well-developed to quite a late stage, so that the gonad is visibly bisexual, but in amniotes the bisexual stage is passed through rapidly. In a normal male the medulla

suppresses the cortex and in the female vice versa. Sex is primarily determined by a balance between the actions of the sex chromosomes and autosomes within the nucleus. Every cell thus contains factors tending to development both of maleness and femaleness. Presumably the conditions in the cortical region of the gonad favour the enzymatic processes leading to femaleness, those in the medullary region to maleness (see Mintz 1960).

The sex of the gonad is determined by the structural elements. In the frog the material that will form the germ cells is already localized near the vegetal pole of the egg and can be destroyed by irradiation. A sterile gonad then develops, yet acquires the characteristic structure of ovary or testis. Moreover, by suitable transplantation, germ cells of one genotypic sex can be made to grow in a gonad with structural elements of the opposite sex; the latter will then determine the nature of the gametes that result (Willier 1955).

There is evidence that during normal development the cortex and medulla produce specific sex hormones with mutually inhibitory effects (Burns 1961). These hormones are probably at first effective by diffusion through the tissues and they may also reach the nearby ducts in this way. Later, these or related substances escape into the blood-stream as the specific sex hormones that control the development of the genital ducts and the secondary sexual characters. It has long been known that in cattle, if there are twins of opposite sex, the female member develops into an intersexual animal known as a *freemartin*. This confirms that the male element of the gonad inhibits the female.

Administration of steroid sex hormones to the mother during pregnancy in a mammal causes signs of sex reversal or hermaphroditism both in the gonad itself and in the ducts. Thus injection of male hormones causes development of the medulla of the gonad and of the Wolffian duct system in genetic females. There is therefore good reason to suppose that the differentiation of the gonad in normal development is controlled by sex-differentiating substances that are related to the sex hormones of the adult.

Reversal of the genetically determined sex can be produced in various ways. Thus in male toads a portion of the cortex of the gonad persists attached to the testis and is known as Bidder's organ. If the testis is removed from an adult toad the Bidder's organ develops into an ovary. Sex reversal as a result of conditions of nutrition, parasitism, disease, and even temperature has been reported from many vertebrates. Gonads containing both eggs and sperms are occasionally reported in adult mammals and their occurrence is to be expected as a deviation, given a mechanism that depends upon a balance of male and female tendencies.

5. The testis and male ducts

The sex cords of the testis are separated from the germinal epithelium by a fibrous layer, the *tunica albuginea*. Since there is reason from experiments on other animals to think that the centre (medulla) of the gonad has male tendencies, it is interesting that the definitive male cells of mammals are thus formed at the centre, whereas in the female the cortex remains active throughout life (p. 471). The sex cords of the testis form a network of cellular strands, the *rete testis*, which become hollowed out to form the *seminiferous tubules*. Here the primordial germ cells form the spermatocytes and the walls form the supporting Sertoli cells (p. 463). The interstitial cells are formed from the mesenchyme between the sex cords.

The developing testis, meanwhile, has become attached to the mesonephros, which lies lateral to it. The seminiferous tubules make connexion with some of the mesonephric tubules to form the *vasa efferentia* (Fig. 24.4). The portion of the mesonephric duct into which these tubules lead becomes greatly coiled as the *epididymis*. The lower portion of the duct forms the muscular *vas deferens* and a diverticulum from the lower end of this makes the *seminal vesicle*. The vas opens to the urethra by a short ejaculatory duct. The male embryo shows at early stages a Mullerian duct similar to that of the female but this retrogresses in later stages. Rudiments of it persist as the *hydatid* attached to the testis, and the prostatic utricle. The *prostate gland* is formed as numerous buds from the endodermic wall of the urethra. The prostate produces substances with very powerful actions on smooth muscle, the *prostaglandins*. They are based upon a C_{20}

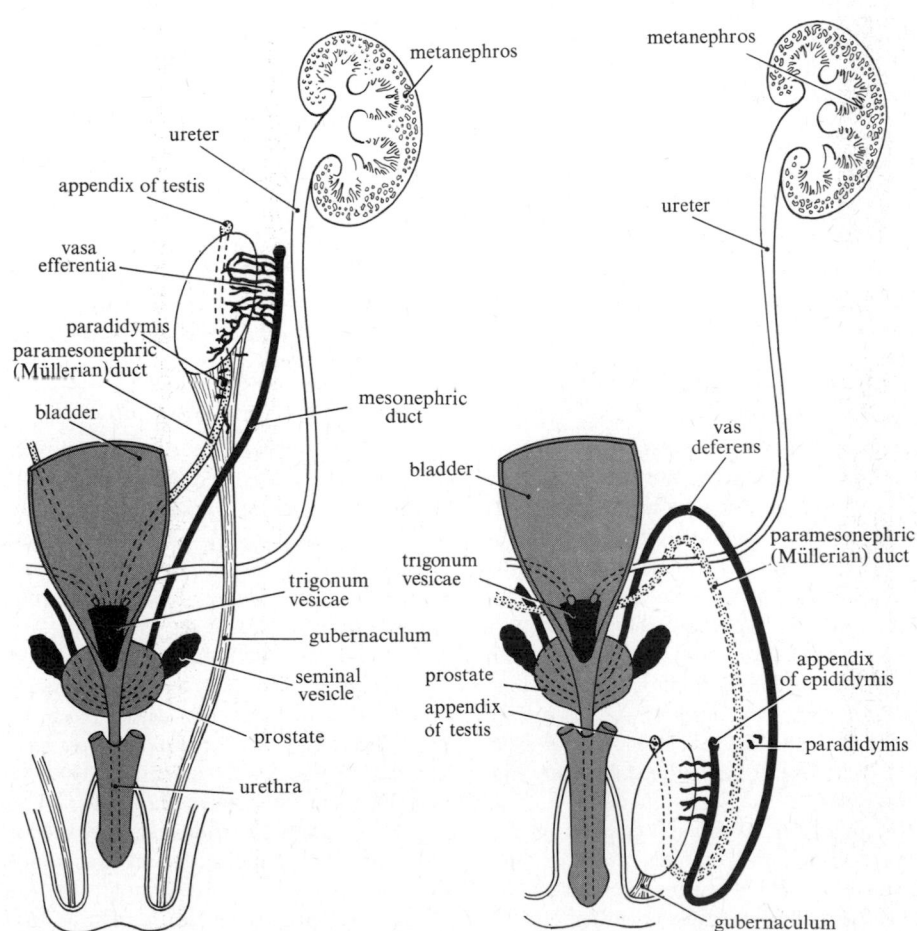

Fig. 24.4. Diagrams showing the development of the male genital ducts. (After Hamilton and Boyd 1972.)

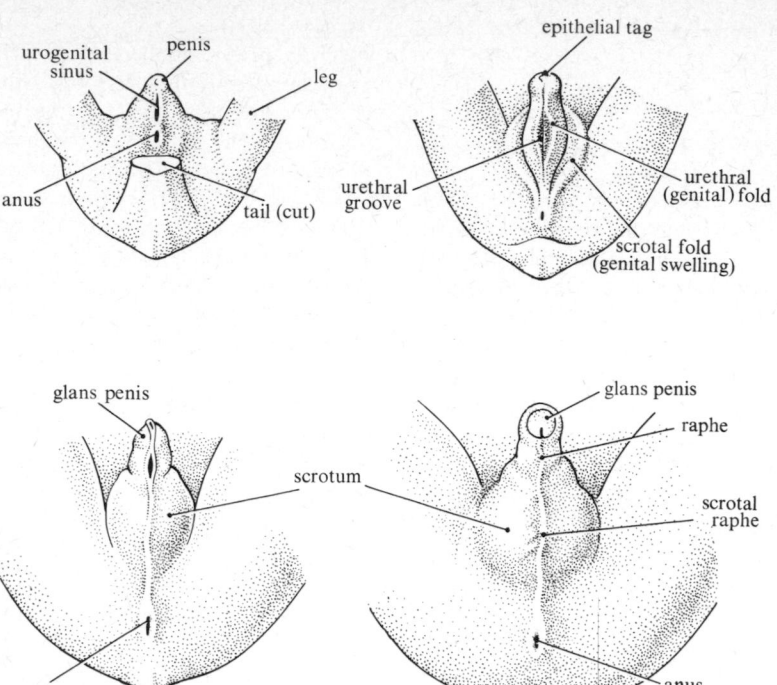

FIG. 24.5. Stages in the development of the male external genitalia. (After Spaulding (1921). *Contr. Embryol.* **13**.)

acid, prostanoic acid, containing a five-membered ring. They have various actions on uterine muscle according to the stage of the menstrual cycle or pregnancy. The high concentration in the semen may produce uterine contractions and play a part in fertilization. The action is said to be particularly marked at the time of ovulation (see Pharriss and Shaw 1974).

Prostaglandins are produced at many other places in the body and have marked effects on the contraction of blood vessels and other smooth muscles. They are also produced in the skin, where they control the rate of keratinization. The action of aspirin (salicylic acid) is said to be to inhibit the production of prostaglandins.

The *penis* has evolved in phylogeny from the vascular lips of the cloaca. In development it forms from an elevation, the genital tubercle, in front of the urogenital aperture (Fig. 24.5). This becomes elongated and a urethral groove forms and finally closes over in such a manner as to provide a continuation of the urethra to the tip of the penis.

The testis descends from its original position on the posterior wall of the abdomen into the scrotum, preceded by the *processus vaginalis*, a diverticulum of the coelom, surrounded by muscle layers corresponding to those of the body-wall. The testis is attached to the scrotal wall by a cord, the *gubernaculum*; when elonga-

tion of the whole body takes place this cord fails to lengthen and the testis is thus drawn down into the scrotum. The *inguinal canal* through which it passes becomes constricted in man, but remains open to the coelom in the rabbit.

6. The ovary and female ducts

The problem of whether ova are continually formed throughout life is considered on p. 472. The germinal epithelium certainly remains conspicuous and there is no outer fibrous coat corresponding to the tunica albuginea of the testis. The interstitial cells that form the walls of the ovarian follicles are produced by the mesenchymal stroma of the ovary.

The female genital duct system develops not from the mesonephric duct but from an independent *paramesonephric* or *Mullerian duct*, which forms in the mesoderm alongside the Wolffian duct and grows back to reach the urogenital sinus (Fig. 24.6). This forms at first in males but later disappears. In females its anterior (cephalic) portions remain paired and differentiate into the *uterine tubes* and their funnels, opening to the coelom near to the ovaries. The lower sections of the Mullerian ducts fuse in man to form the *uterus* and *vagina*. In most other mammals they remain separate, providing paired uteri. The uterine epithelium

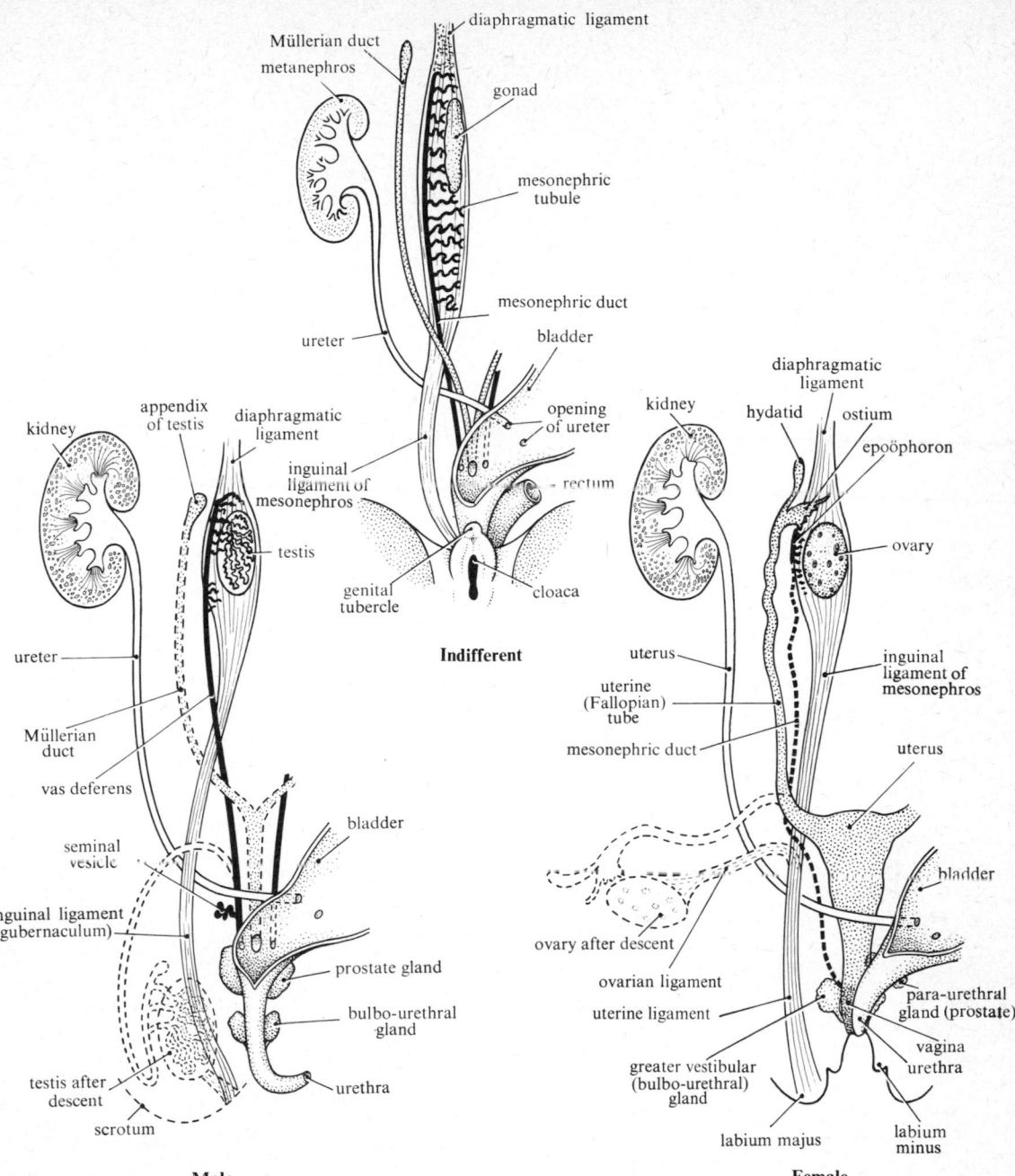

FIG. 24.6. Diagrams showing the development of the male and female genital ducts from the early indifferent stage. The testis and ovary are outlined by broken lines in their final positions. (Partly after Patten 1968.)

of the foetus undergoes considerable differentiation towards the end of pregnancy, presumably stimulated by maternal hormones passing through the placenta.

The caudal end of the utero-vaginal canal forms a solid cord of cells, meeting a cord growing from the posterior wall of the urogenital sinus. The vagina is formed by hollowing of these strands of cells, but it is disputed how much of its wall is lined by mesodermal (Mullerian) and how much by endodermal (urogenital sinus) epithelium.

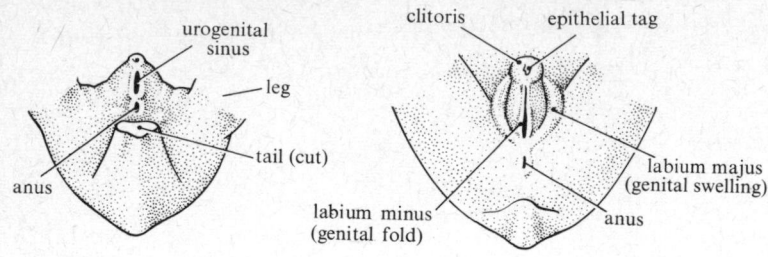

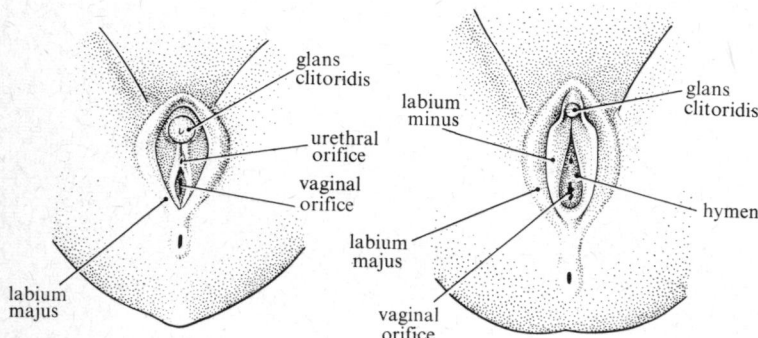

Fig. 24.7. Stages in the development of the female external genitalia. (After Spaudling, *loc. cit.*)

The genital tubercle is formed in the same position in the female as in the male but fails to become grooved and joined to the urethra (Fig. 24.7). It develops, however, into the clitoris, with sensitive skin and erectile tissue corresponding to those of the penis.

The mesonephric duct, well-developed in the early female embryo, later retrogresses, but a part remains in the adult as the epoöphoron, a body of unknown function attached to the ovary.

25   The nervous system

1. Nervous activity

IN micro-organisms and plants, changes in the surrounding conditions influence the metabolism of the cells directly and locally. Higher animals, on the other hand, using the information stored during their long history, operate as wholes in the anticipation of the course of events. They go out and make great efforts to reach the conditions where they can obtain the materials for their life. The various tissues are not controlled directly by the environment but by the nervous system, which regulates the performance of the whole animal. There is an elaborate system of receptors to signal information about events in the world in a code. In the central nervous system this is combined with the information stored there, and then translated into action by effector organs, especially the muscles. A nervous system thus enables the animal to provide a more perfect and quickly available representation of conditions in the environment than can be produced by changing the hereditary instructions by the slow processes of variation and natural selection. The receptors provide a flow of information from the surroundings, which ensures the production of appropriate immediate actions and also in many nervous systems leaves an altered state constituting a memory store, so that future prediction is improved.

The nervous system of mammals thus becomes the main agent regulating activity. In describing its properties we may concentrate from the start on the conception that its characteristics are to *act* and to *regulate*. It has long been usual to lay emphasis on the fact that nerves *conduct signals* (known as *nerve impulses*) from one part of the body to another and thus provide transmission lines that produce the familiar *reflex actions*. In higher animals, such as mammals, however, the nervous system is far more than a transmission system or even than a co-ordinating system between the parts; it has become the controlling agent, whose actions regulate almost every aspect of behaviour.

The nerve cells (*neurons*) are able to act in this way because they have developed the capacity to give discharges that serve as signals. This capacity is a special elaboration of the responsiveness to change that is characteristic of all living systems. It depends on the fact that the surfaces of the cells are electrically charged systems, ready to discharge when there is a small change in their neighbourhood. The setting up of the charge by the cell depends upon chemical processes involving the expenditure of energy, and the discharge is probably initiated by molecular changes at the surface, followed by ionic movements (p. 244). These movements set up electrical currents that promote discharge of neighbouring parts of the nerve fibre and thus make possible the transmission of signals. Nervous action therefore requires investigation by both bio-chemical and biophysical methods.

2. Nerve cells

Neurons are cells elongated to form nerve fibres, so that the effect of the discharge of part of the surface can spread away rapidly to a distance. This elongated form is typical of nervous tissue, but many nerve cells have only short processes and therefore their discharges produce their effect, whatever it may be, in the immediate neighbourhood and not at a distance, for example, the stellate cells of Fig. 25.1.

It is convenient, however, to describe first a neuron of classical form (Fig. 25.2), which possesses (a) a *cell body* with nucleus, (b) one or more *dendrites* that receive the signals from the environment or from other neurons, and (c) an *axon* that carries signals away. The whole constitutes a single trophic entity in the sense that any part becoming disconnected from the region containing the nucleus rapidly breaks up and is said to 'degenerate' (p. 240).

The nerve cells differ greatly in shape, especially in the length of the dendrites and their method of branching. Little is known about the properties of dendrites, which must be of great importance in determining the conditions under which nerve impulses are set up in each type of cell (p. 253).

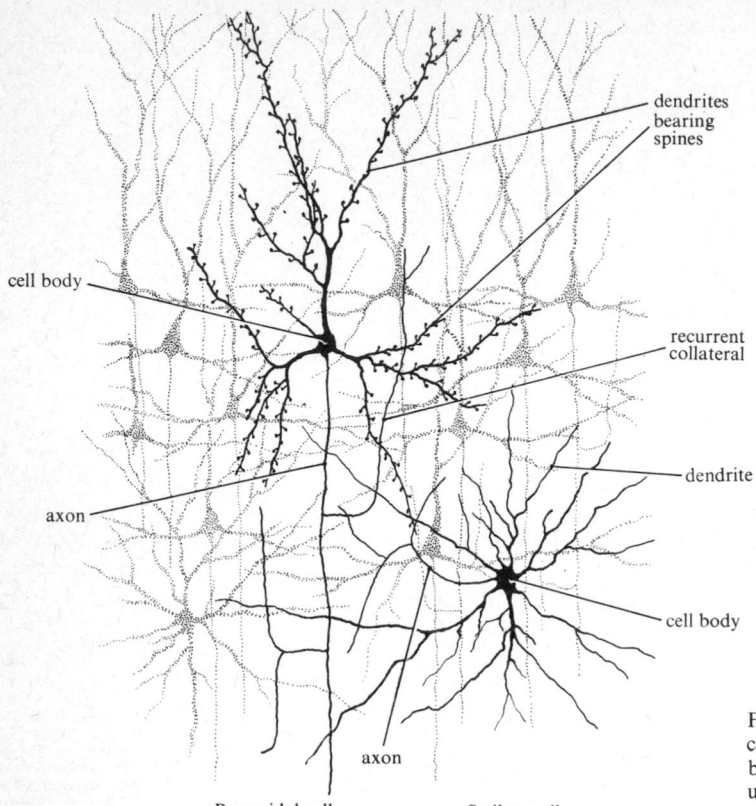

dendrites
bearing
spines

cell body

recurrent
collateral

dendrite

axon

cell body

axon

Pyramidal cell   Stellate cell

FIG. 25.1. Cells of the cerebral cortex of the cat stained by Golgi's method, which blackens a few cells but leaves the others unstained.

The cell body of nerve cells contains the *Nissl substance* (Fig. 25.3), composed of material that stains with the basic dyes that also colour nuclei. This material consists of granular endoplasmic reticulum with ribosomes, whose RNA is produced by or under the influence of the nucleus. The fact that nerve cells, whose cytoplasm is so very large, contain so much RNA outside the nucleus is presumably related to the part that this substance plays in the synthesis of the characteristic materials of the cell. Although adult nerve cells never divide, the turnover of the materials within them is perhaps more active than in any other part of the body.

3. Axoplasmic transport

There is evidence that materials are transported from the cell body along axons. After injection of various materials (for example, radioactive leucine) either in the region of, or actually into, nerve-cell bodies, the incorporated material can be detected further and further along the axons, finally appearing in the nerve terminals. This has been shown in, for example, the optic system (Grafstein 1967) and in sensory and motor spinal nerves (Ochs 1972).

The rates at which materials are transported cover a wide range, but most appear to be either 'slow' (a few millimetres per 24 hours) or 'fast' (100–400 mm per 24 hours). Differences in rate probably reflect the different materials being transported and to some extent the type of nerve cell involved and also the experimental conditions used. The most dramatic illustration of axonal transport comes from phase-contrast time-lapse film of nerve cells in culture, for example, Pomerat's which Kasten (1966) has described. These show movements of particles within growing axons both away from and towards the cell body. Another experimental approach has been to constrict axons locally and to look for signs of redistribution of materials within the axons (Fig. 25.11, p. 240). Such constrictions probably produce some local effects, but there is now evidence that accumulation of materials such as labelled protein, neurotransmitters, and their synthesizing enzymes proximal to a constriction is the result of damming up of proximo-distal axonal transport.

The movement of at least some materials can continue unabated for several hours after nerve section in the segment of nerve cut off from the cell body; so it is not likely that static pressure from the cell body is

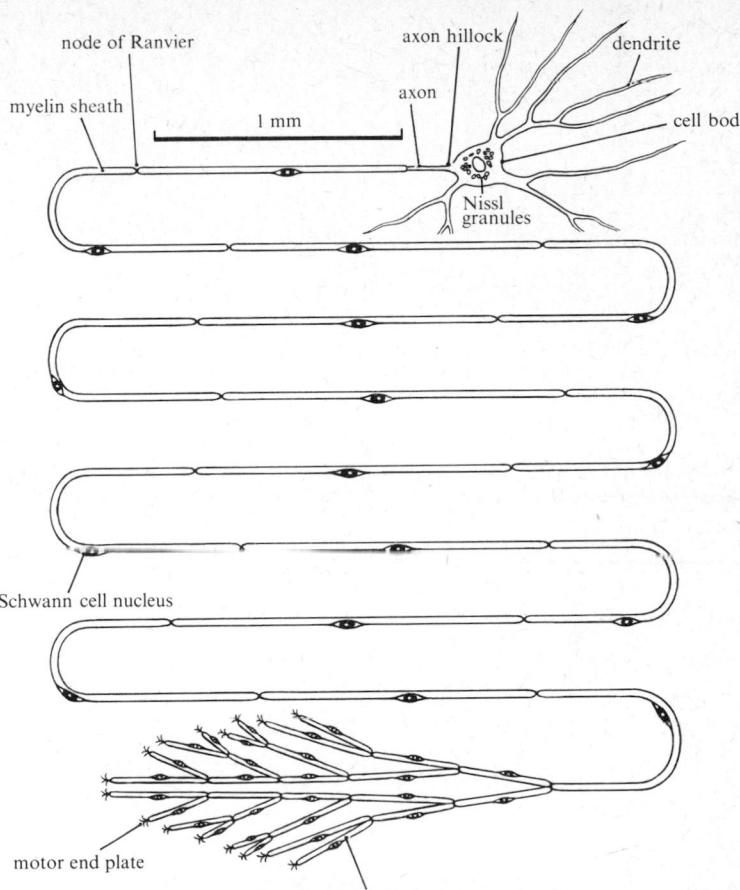

FIG. 25.2. Diagram of a motor nerve cell of a rabbit.

The cell is one of the shortest motor neurons in the body, reaching from the spinal cord to the nearby psoas muscle of the back. The cell body lies in the ventral horn of the spinal cord. Its dendrites conduct towards the cell body. The axon leaves the spinal cord in a ventral root and finally breaks up into a number of branches reaching end-plates in the muscle. The axon is about 17 mm long and 0·05 mm in diameter. It is divided into 16 internodal segments (excluding those at the end). One Schwann cell is attached to the fibre at the middle of each internode. Drawn to scale (except that the final branches are too short).

the mechanism of transport. Some dynamic process such as peristaltic movement either of the axon membrane itself or imposed upon it by the surrounding Schwann cells has been suggested. At present, it seems more likely that subcellular axonal structures, notably the neurotubules, are involved, perhaps by some actomyosin-like mechanism.

Functions suggested for this axonal transport include
(a) replacement of components concerned in turnover;
(b) secretion of chemical transmitters—the fibres of the hypothalamic tract are known to carry material to the pituitary (p. 442), and peripheral nerve transmitters noradrenaline and acetylcholine also appear to be transported from cell body to periphery;
(c) axonal growth in regeneration and maturation (p. 241).
(d) transport of trophic substances to innervated cells—for example, to prevent muscle wasting and to control the receptive fields of sensory

nerves. Such trophic substances have not been identified nor have their transfer from neuron to innervated cell been proved to occur.

Axonal transport is a very promising field of cell biology; the great lengths of the axons provide excellent material for experiment and there is still much to be discovered (Barondes 1967; Schmitt 1970).

4. Nerve fibres

The axons differ in size and in the presence and thickness of an outer *myelin sheath* (p. 235). The velocity of impulse conduction is different in the various types of axon, but so far as is known the impulses are all fundamentally alike (p. 244). The nervous system, therefore, may be regarded as a transmission network composed of a large number of units all transmitting the same type of signal (but with frequency differences) and with differences in shape, connexions, excitability, and perhaps other features of the units.

Axons in mammals range in diameter from less than 1 μm up to 20 μm (including the myelin sheath). They

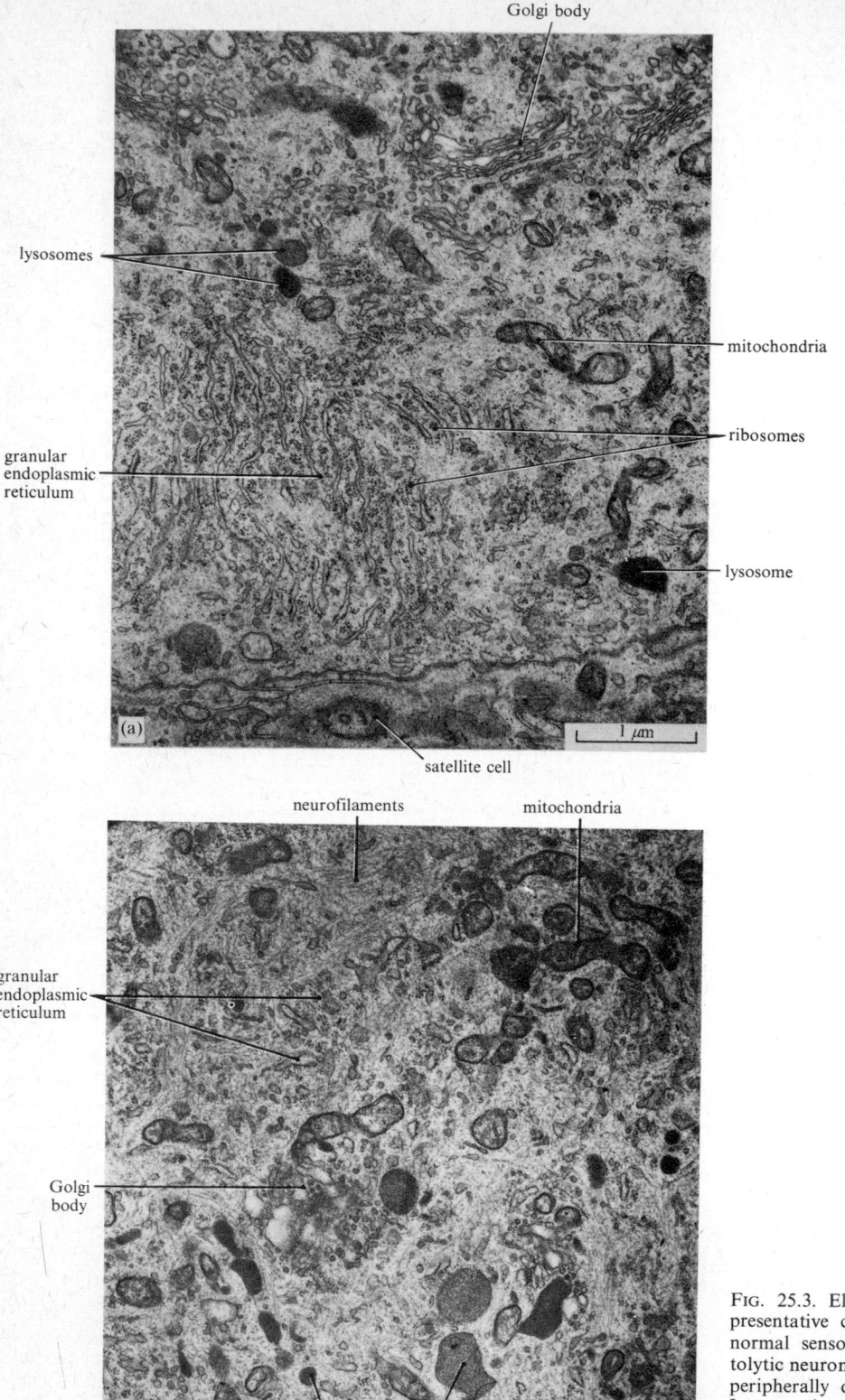

Golgi body

lysosomes

mitochondria

ribosomes

granular
endoplasmic
reticulum

lysosome

(a)

1 μm

satellite cell

neurofilaments

mitochondria

granular
endoplasmic
reticulum

Golgi
body

(b)

lysosomes

FIG. 25.3. Electron micrographs of re-
presentative cytoplasmic fields of (a) a
normal sensory neuron, (b) a chroma-
tolytic neuron (28 days after section of its
peripherally directed axon (see p. 242).
In normal neurons, the granular endoplasmic reticulum is studded with ribosomes. In (b) the granular endoplasmic
reticulum is dramatically reduced and
lysosomes are increased, with increased
acid phosphatase activity. (Figure kindly
supplied by Dr. A. R. Lieberman.)

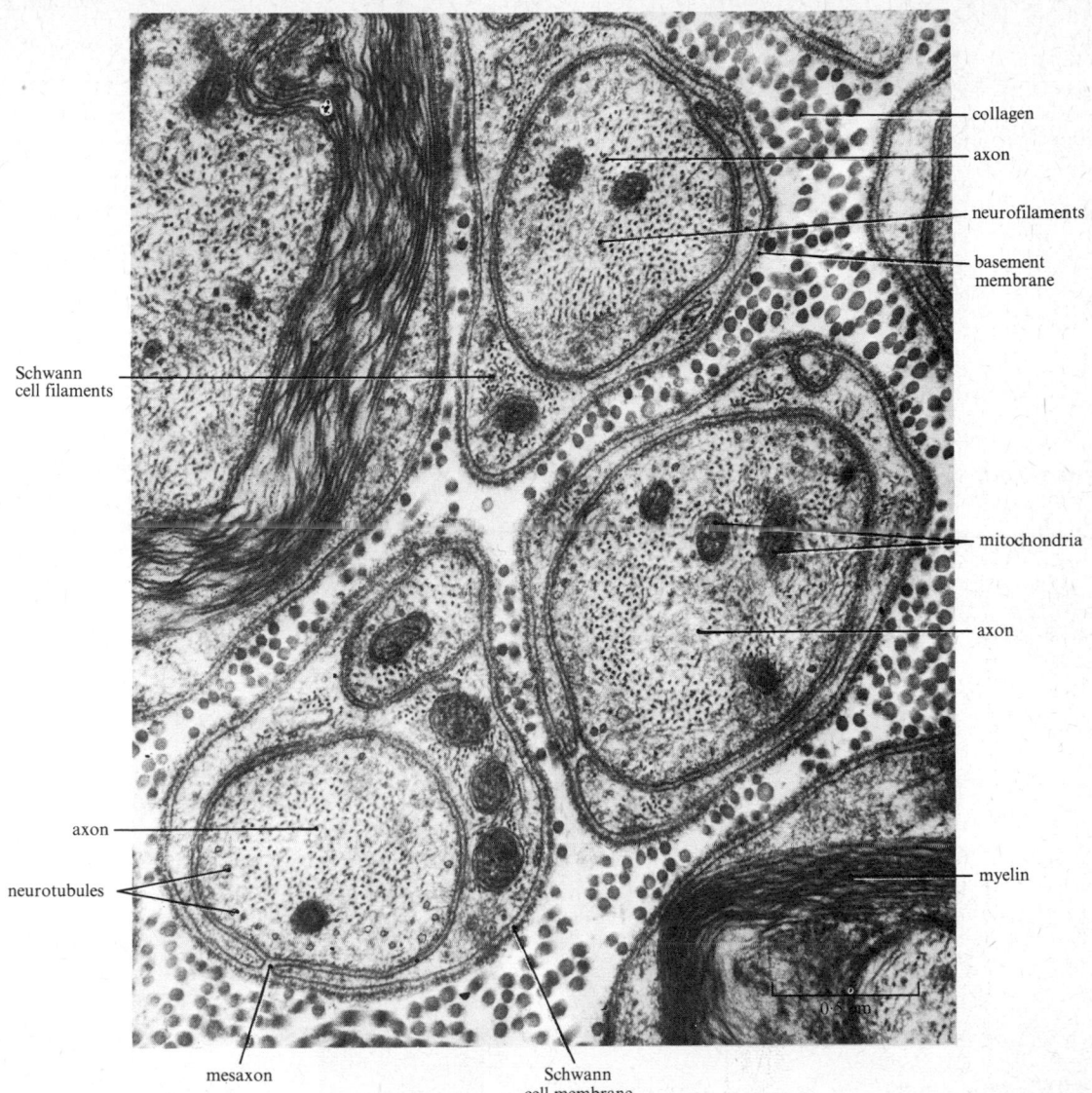

Schwann
cell filaments

collagen

axon

neurofilaments

basement
membrane

mitochondria

axon

myelin

axon

neurotubules

mesaxon

Schwann
cell membrane

FIG. 25.4. Electron micrograph of rabbit peripheral nerve in cross-section showing three unmyelinated axons and part of two myelinated fibres. (Figure kindly supplied by Professor E. G. Gray.)

vary in length from a few tenths of a millimetre to more than a metre. The *axoplasm* of which the fibre is composed behaves sometimes as a gel, sometimes as a viscous liquid. This material possesses a considerable degree of organization, and in polarized light it shows a weak birefringence. It contains elongated neurofibrils and neurotubules, which can be seen in some fresh nerve fibres and more clearly by electron microscopy (Figs 25.3(b) and 25.4).

The functioning of a nerve fibre depends largely on the difference of ionic composition between the inside and outside (see p. 244). The membrane at the surface of the axon ('axolemma') is so thin that it cannot be seen by the light microscope, but electron microscopy shows a distinct membrane around all nerve fibres (Fig. 25.8, p. 238).

In mammals every peripheral fibre greater than about 1 μm in diameter is covered by a myelin (medullary) sheath (Fig. 25.5), up to 2·5 μm thick on the larger fibres. In the central nervous system there may be myelinated fibres as small as 0·1 μm. The myelin consists of layers of material of fatty nature (largely

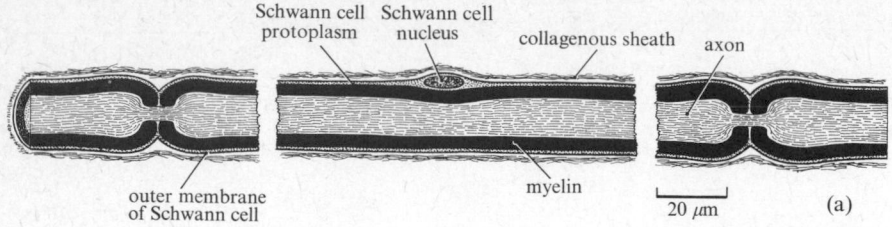

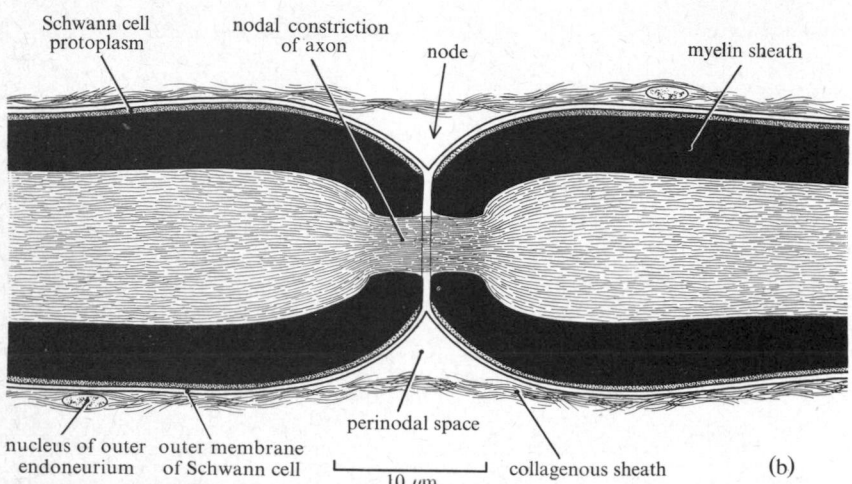

FIG. 25.5. Diagram of the structure of a medullated nerve fibre as seen by light microscopy: (a) in longitudinal section; (b) enlarged view of a node.

phosphatides), whose molecules are regularly arranged, with their long axes radially. Between these are thinner layers of protein (neuro-keratin) (Fig. 25.8). Each lamella (phosphatide + protein) is about 16·5 nm thick seen by X-ray diffraction of fresh material, but in fixed material seen by electron microscopy varies from 11 nm thick, where the sheath has more than 100 lamella, to 12 nm thick, where there are fewer. This regular arrangement of the molecules gives to the myelin a strong birefringence in polarized light (Fig. 25.6).

The myelin of the fibres in peripheral nerve trunks is produced at the surface between the axon and a series of satellite *Schwann cells*, which arise from the neural

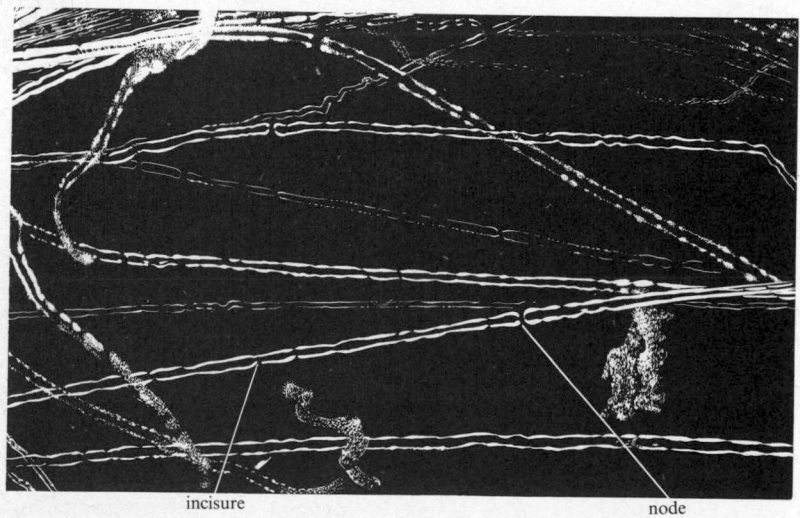

FIG. 25.6. Nerve fibres of a peripheral nerve of a rabbit, teased in Ringer's fluid and seen in polarized light. The fibres have adopted an unduloid outline.

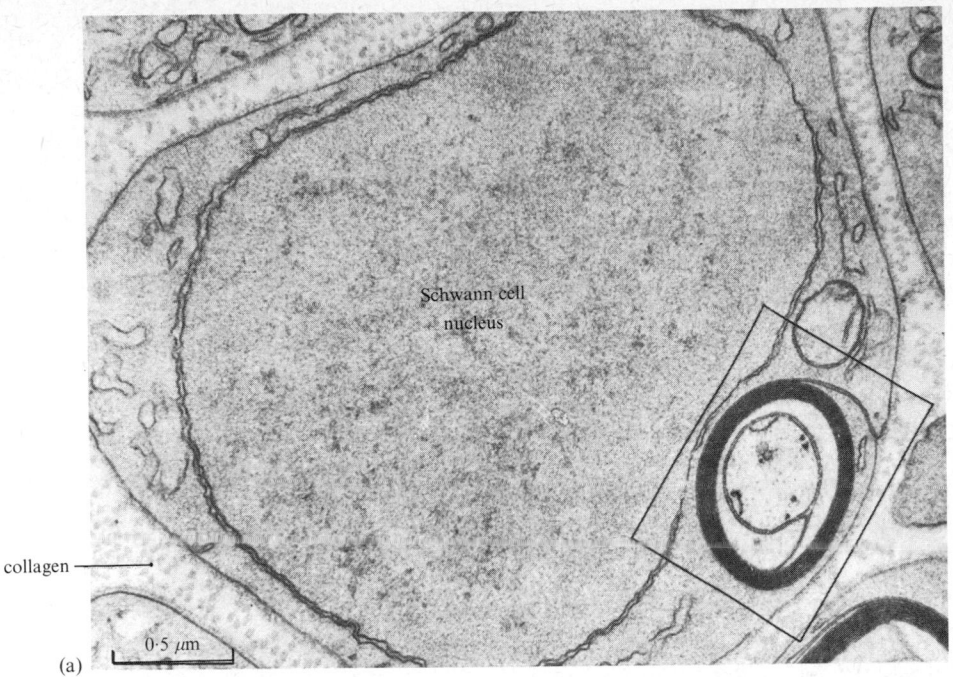

(a)

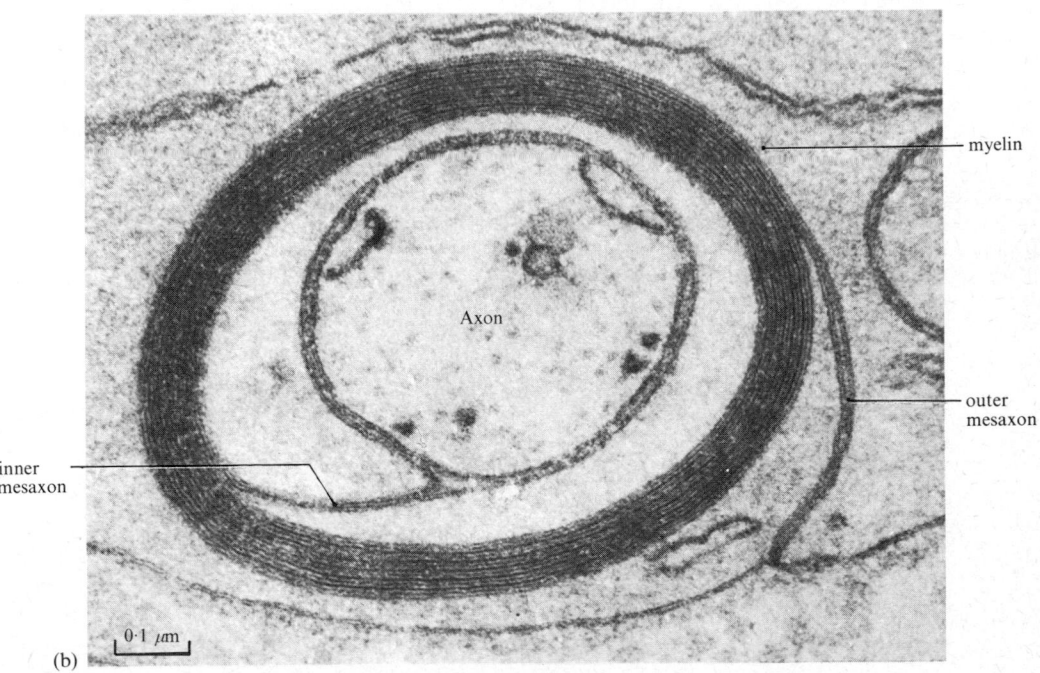

(b)

FIG. 25.7. Electron micrographs showing (a) general relation between Schwann cell and axon, and (b) detail of myelin sheath. (Figure kindly supplied by Dr. J. D. Robertson.)

crest during development (p. 345) and apply themselves to the axons. The axon therefore lies within the protoplasm of the Schwann cell and the myelin is formed from layers of the membrane of the satellite cell wound around the axon (Fig. 25.7) (Davison and Peters 1970). The presence of the myelin greatly increases the velocity of the conduction of the signals (p. 246).

The myelin sheath is easily recognized by the fact that it stains black with osmium tetroxide. Very small nerve fibres, such as those of the postganglionic sympathetic nerves (p. 279), do not show this sheath and are therefore said to be unmedullated (or non-myelinated) (Fig. 25.4). The myelin sheath of the larger fibres is interrupted at intervals, forming the *nodes of Ranvier* (Fig. 25.8), one Schwann nucleus being found between each pair of nodes. The internodal lengths vary from 0·3 mm to 1·5 mm in peripheral nerves of mammals and are greater on the larger nerve fibres. When the nodes are first laid down the distance between them is short, about 200 μm in the peripheral nerves of mammals. As growth proceeds the internodes become stretched, and the final length reached is proportional to the amount of growth that takes place after the time of first myelinization. The number of nodes thus remains constant throughout life. After regeneration of nerve fibres in an adult all the internodes remain short. The internodal lengths are much less in the central nervous system, down to 40 μm in mammals and as little as 5 μm in fishes. In addition to this long-period nodal segmentation, nerve fibres examined after removal from the body sometimes also show oblique cracks, the *incisures of Schmidt–Lanterman* (Fig. 25.6)

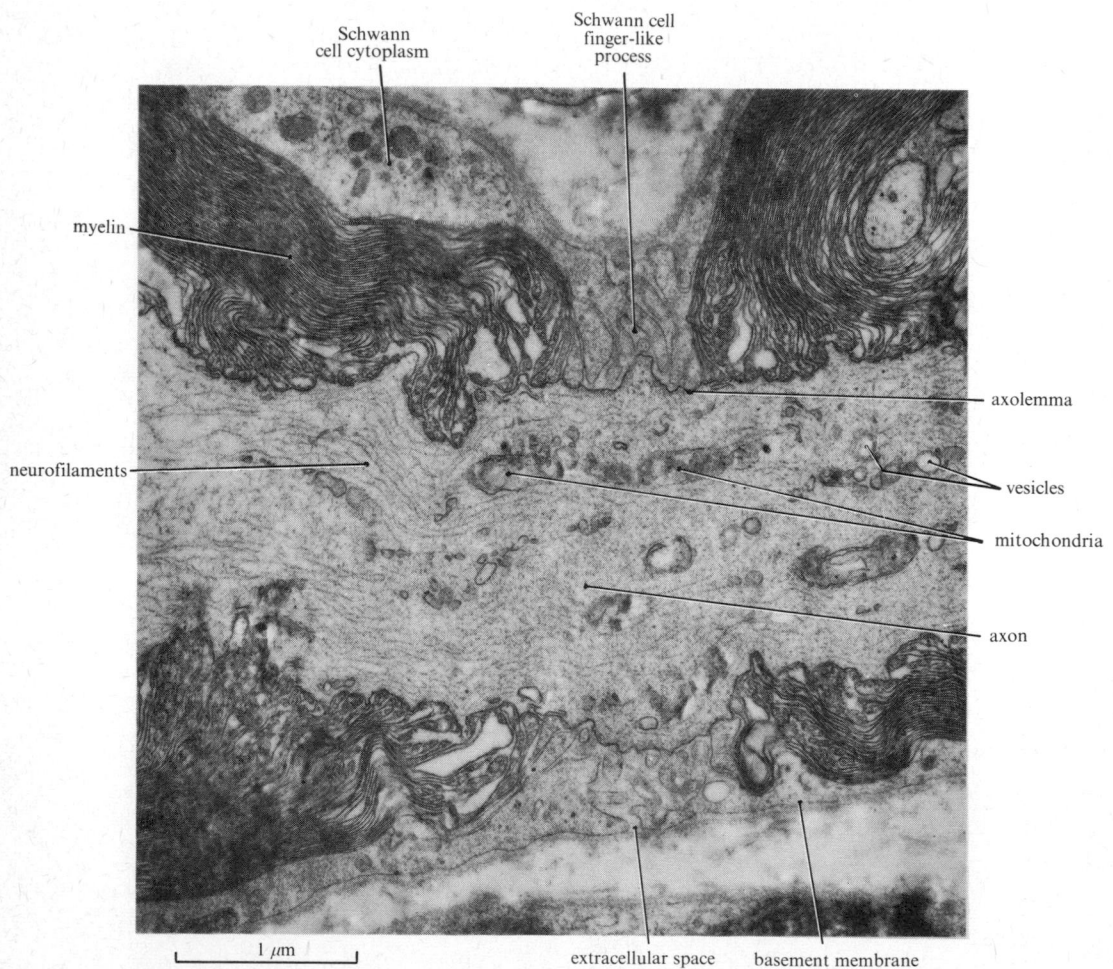

FIG. 25.8. Longitudinal section of a node of Ranvier from rat sciatic nerve. (Figure kindly supplied by Dr. G. Allt.)

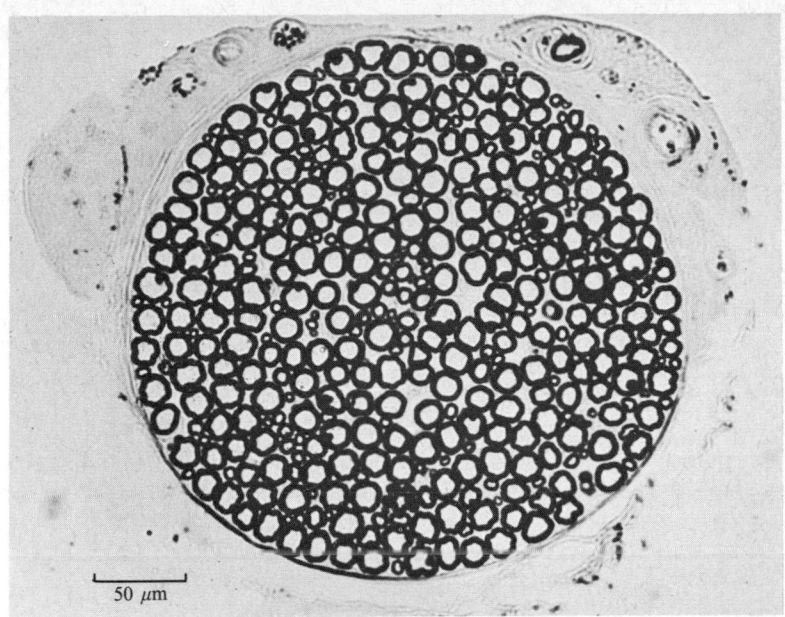

FIG. 25.9. Transverse section of the nerve innervating the gastrocnemius muscle of the rabbit, stained to show the myelin sheaths. Note the presence of fibres of various sizes. The membrane around the nerve is the epineurium.

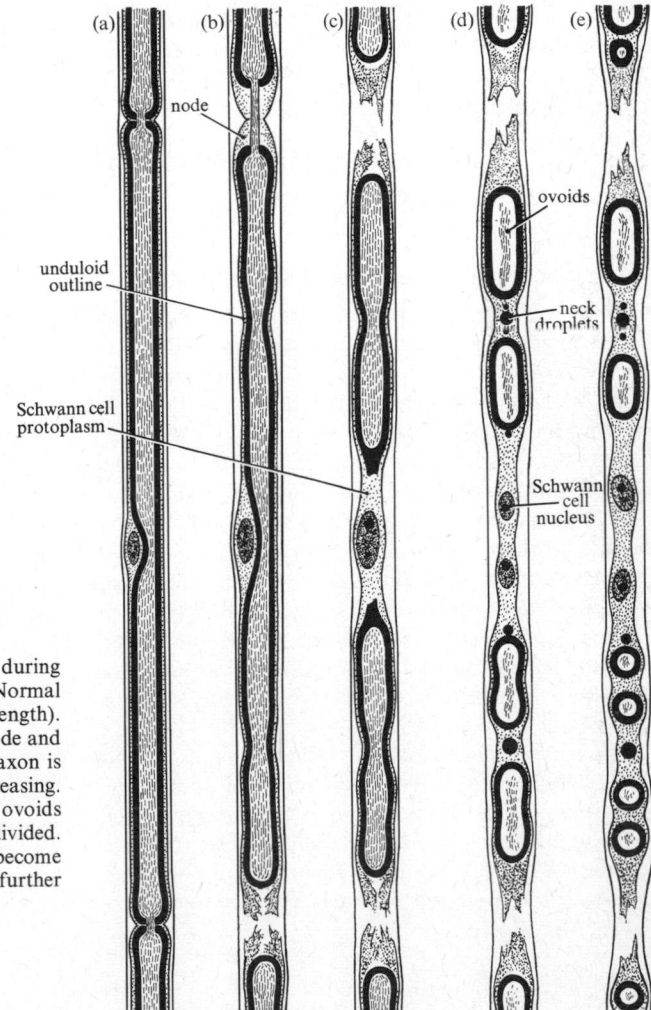

FIG. 25.10. Stages in the breaking up of a nerve fibre during the first week after severance from its cell body. (a) Normal internode (shown disproportionately wide for its length). (b) About 48 hours; the myelin has retracted at the node and shows an unduloid outline. (c) About 60 hours. The axon is interrupted and the Schwann cell protoplasm is increasing. (d) About 4 days. The myelin has segmented into ovoids leaving neck droplets. The Schwann cell nucleus has divided. (e) About 7 days. Some of the ovoids have now become spherical droplets, the Schwann protoplasm has further increased. (After Young (1949). *Adv. Surg.* **1.**)

spaced at intervals of about 50 μm. The state of these cracks in the intact fibre is not certain.

The outer membrane of the Schwann cell and the basement membrane external to it were not separately resolvable and were considered to be a single sheath, by light microscopy and given the name *neurilemma*. This term has also been used to include the sheath of delicate reticular fibrils round each nerve fibre, which is also known as the *sheath of Key and Retzius* (or *sheath of Henle*). It is continuous, with the fine longitudinal collagenous fibres (the *endoneurium*) running in the spaces between individual nerve fibres, which layer in turn connects with the *perineurium*, a dense concentrically arranged layer of connective tissue around bundles or *fascicles* of nerve fibres in peripheral nerves. The outermost sheath in nerve trunks enclosing groups of fascicles is the *epineurium* (Fig. 25.9), composed mainly of longitudinally arranged collagenous fibres.

Nerve fibres within the central nervous system differ from those in peripheral nerves in that they do not run in tubes of connective tissue. The white matter is therefore a characteristically soft, almost fluid material. The myelin sheaths of fibres in the central nervous system resemble those of peripheral nerve and are interrupted at regular short intervals by nodes of Ranvier (Waxman 1972). The oligodendroglia cells between the fibres (p. 277) correspond to the Schwann cells of peripheral nerves and collaborate in the formation of the myelin, but each cell may provide part of the sheaths of several fibres.

5. Degeneration and regeneration of nerves

Even the most remote parts of a nerve fibre are maintained intact by the activities of the nerve-cell body; when any part of a nerve fibre is cut off it undergoes degeneration. The isolated portion breaks up, the axon and myelin first adopting an unduloid outline (Fig. 25.10) and then falling into droplets. This is the behaviour that would be expected in a long cylinder of liquid under surface tension. It is suggested that the trophic influence that maintains the fibres includes an internal pressure, which prevents the breaking up under surface tension, somewhat as a sufficient flow of water from a tap prevents breaking up of the outflowing stream (Young 1945). This is certainly too simple a view, but the nature of the trophic influence remains unknown.

The nerve fibre consists of a cylinder, the axon, and the myelin is pressed against the inside of an inelastic neurilemmal tube. There are no exact estimates of the pressure and flow along the nerve fibres. The axoplasm is not a simple Newtonian liquid and indeed may approach the condition of a gel. Yet it certainly has some liquid properties; when a nerve trunk is constricted the axoplasm becomes dammed up within the fibres behind the obstruction (Fig. 25.11).

The breaking-up of the part of a nerve fibre that has become isolated from the nerve-cell body is known after its discoverer as *Wallerian degeneration*. The Schwann cells, however, do not disappear but undergo multiplication. After the remains of the axon and myelin

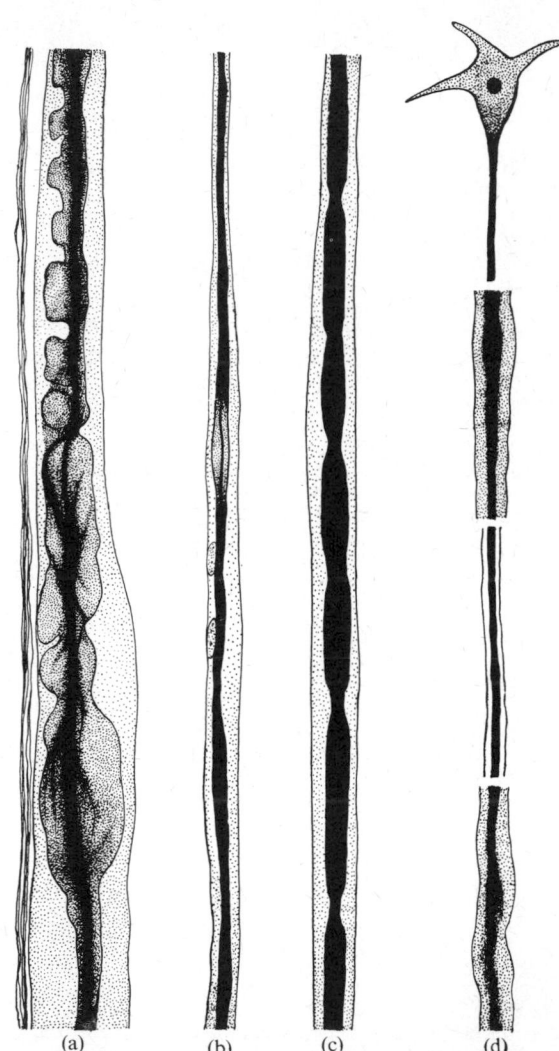

(a)	(b)	(c)	(d)

Fig. 25.11. Effects of chronic constriction on a mature uninterrupted nerve fibre. (a), (b), and (c) are three sections of the same fibre.

(a) Proximal to constriction, showing telescoping and bending of axon. (b) Inside constricted zone: axon narrow. (c) Distal to constricted zone: axon wider. (d) Axon after release of a constriction that had been applied for 25 days, showing cell body and three sections equivalent to (a), (b), and (c). (Modified after Weiss and Hiscoe (1948). *J. exp. Zool.* **107**.)

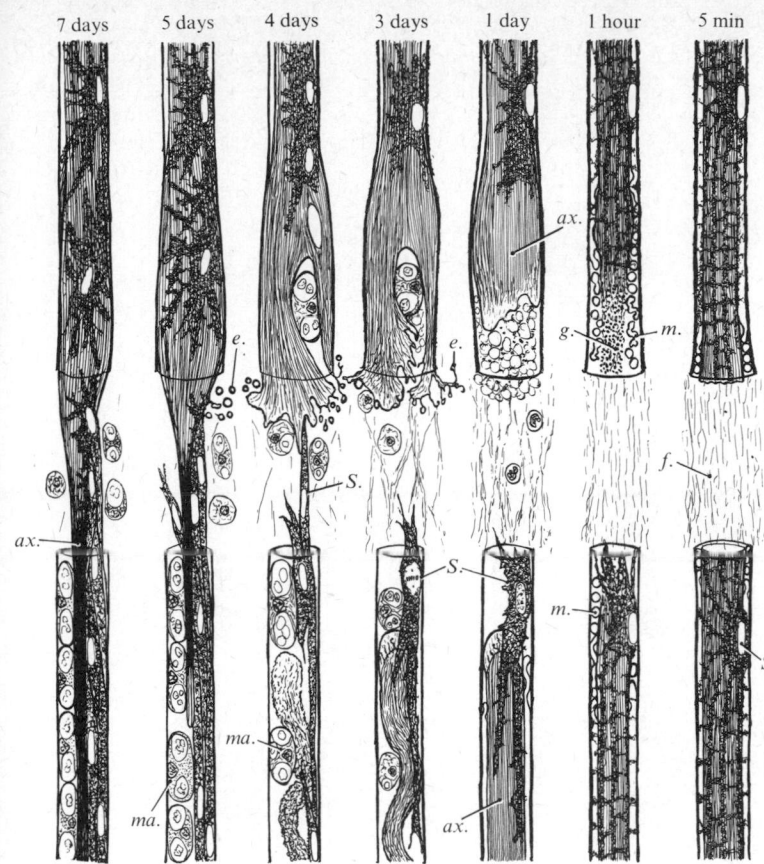

7 days   5 days   4 days   3 days   1 day   1 hour   5 min

FIG. 25.12. Changes seen in a nerve fibre that has been severed. Central stump above, peripheral stump below.
ax. axon; *e.* ends of new fibres formed on central stump; *f.* fibrin clot; *g.* granules formed by degeneration in central axon; *ma.* macrophage; *m.* degenerating myelin; *S.* Schwann cell. (From Young (1949). *Adv. Surg.* **1.**)

sheaths have been removed the severed part of a peripheral nerve therefore consists of the connective tissue tubes ('Schwann tubes'), now filled with the protoplasm of the multiplied Schwann cells (Fig. 25.12).

Meanwhile the nerve fibres of the *central* stump, which, of course, are still connected with the cell body, begin a process of *regeneration*. The severed end of the fibre swells and streams of axoplasm and filopodia emerge from the cut ends of the tubes and push out in all directions through the new tissue formed at the site of injury (Fig. 25.11). The elongated Schwann cells emerge mainly from the peripheral stump and grow in all directions in the scar between the severed stumps. Some of them form bridges between the two cut ends, along which the outgrowing nerve fibres proceed and are thus led to enter the tubes of the distal stump. There the Schwann cells provide the fibres with favourable conditions for growth and they proceed along the tubes to the original end-organs. The tips advance at a rate of as much as 5 mm per day. At first the new fibres are very small but when they reach the end-organ they receive a stimulus to grow in diameter (Fig.

25.13). By 3 months after the injury they have returned nearly to their original size, with well-formed myelin sheaths. Only after they have grown considerably do the fibres function with the conduction velocities and frequencies of those in a normal body needed to produce effective sensory and motor discharges. There is therefore a delay between the arrival of the tips of new nerve fibres at a muscle and the recovery of its normal activity.

Most nerves contain fibres of different types, connected with various end-organs, some sensory and some motor. During regeneration only some of the fibres reach the appropriate endings and many false connexions are produced (but see p. 349). The functions of the limb or other part are therefore often far from perfect after nerve regeneration. However during regeneration of the optic nerves of amphibians or fishes the new nerve fibres grow back rather precisely to remake their old connexions. Some directing agent (*neurotropism*) has been postulated, but its possible nature and effectiveness are still debated.

During the process of degeneration and regeneration

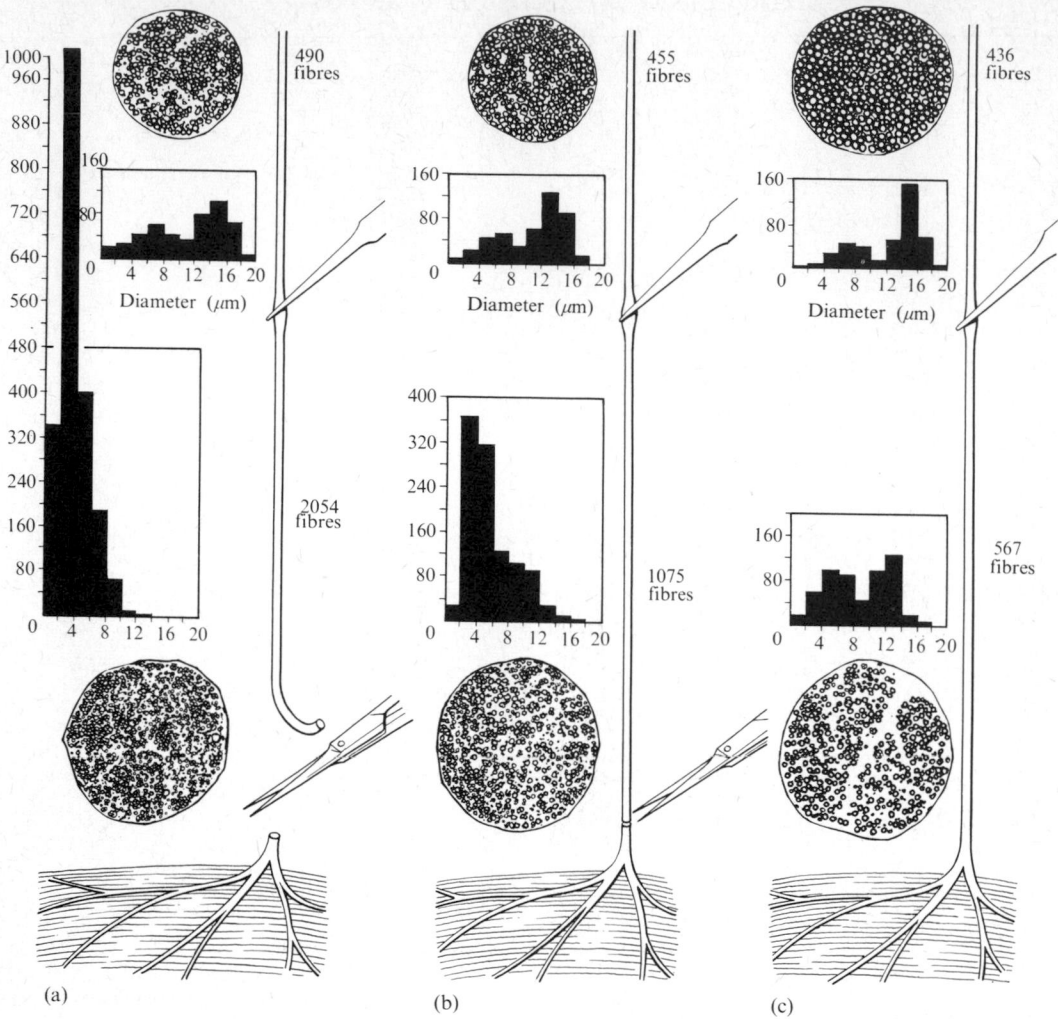

FIG. 25.13. Effect of peripheral connexion on nerve regeneration.
All three nerves were crushed with forceps above but treated differently below: (a) was cut and left without union; (b) was cut and sutured; (c) was untouched. All three were then left for 100 days. At the top the sections and histograms of the central stump show the shrinkage of fibres in (b) and especially in (a). Below are shown the conditions of the nerves 1 cm below the crush; there are very many small fibres in (a), a unimodal distribution including some larger ones in (b) and almost normal nerve in (c). (From Aitken, Sharman, and Young (1947). *J. Anat.* **81.**)

changes also take place in the nerve-cell body. The Nissl substance (endoplasmic reticulum) at first breaks up, a process known as *chromatolysis* or retrograde degeneration (Fig. 25.3). Later the full complement is restored. These changes are presumably connected with the changes in the synthetic activities of the cell consequent on alteration in the length of axon that has to be maintained. They are most marked when the axon is severed close to the cell body (see Singer and Schadé 1964, 1965).

If the new processes growing out from the cut end fail to make new peripheral connexions, retrograde degeneration may continue until the whole nerve cell atrophies and disappears. Conversely, nerve cells that are cut off from all their sources of normal afferent stimulation may also disappear, a process known as *transneuronal* degeneration. The nerve cells, therefore, are said to show double dependence (p. 4); they remain intact only when they receive proper stimulation at the dendrites and when the axons make proper connexion at the periphery.

26 Signalling in the nervous system

1. Plan of organization of the nervous system

THE nervous system consists of a vast set of material, each part triggered to respond to a suitable change in its environment. The special receptor surfaces found in eye, ear, or skin are points where cells are arranged to respond to small changes in the environment by producing generator potentials able to set up signals in the sensory or afferent nerve fibres (Chapter 38). Within any one receptor field, say the retina, cochlea, or skin, each individual sensory cell records a change slightly different—at least in position—from that recorded by its neighbours. All the signals carried in any one fibre are alike in amplitude, though they vary in frequency.

The nervous system therefore does not use amplitude modulation for signalling in the manner of most human communication systems. Nor is there any elaborate code of frequency modulation; in most nerve fibres the variation in frequency records only the magnitude of the change at the transmitting end. The great amounts of information received, therefore, can be transmitted only by having very numerous channels, each carrying a single type of information. Such a multi-channel system is very different from transmission of information along a single channel as in a telephone or television system. Therefore first we have to investigate the nature of the signals in the nerve fibres. To understand the system we also have to know how the signals in different fibres interact in the central nervous system.

2. The nerve impulse

Much is known about the changes that serve for the transmission of signals along nerve fibres (Katz 1966). Studies of the electrical changes in nerve have provided a great part of this information, although propagation of the nerve impulse is not a matter of simple electrical conduction such as occurs in metallic or electrolytic conductors. The essential feature of all nerve activity is

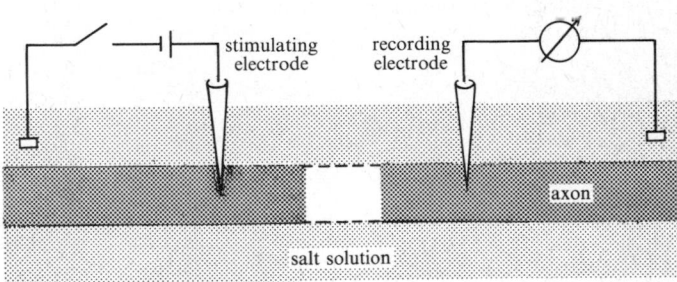

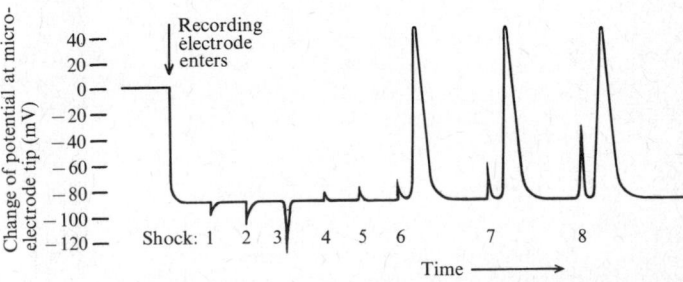

FIG. 26.1. Stimulation of a single axon.
Two intracellular micro-electrodes are placed far apart. After insertion of the recording electrode, eight brief shocks (1–8) of varying polarity and magnitude are applied. A shock of the right polarity exceeding a critical threshold strength (6–8) produces an action potential of fixed magnitude (all-or-none response). Note the interval between the record of the delivery of the shock and the appearance of the action potential, due to the relatively slow conduction of the nerve impulse. (From Katz (1966). *Nerve, muscle, and synapse.* © McGraw-Hill Book Company.)

discharge of a part of the charged surface of the cell. In a long nerve fibre, propagation is produced because the discharge of each region provokes that of its neighbours and is thus self-regenerative.

3. Nerve potentials

When a nerve fibre is at rest there is a potential difference of up to 90 mV across its surface, the inside being negative to the outside (Fig. 26.1). This *resting potential* can be recorded by placing an electrode on the outside of a nerve and another on its severed end, the latter being then effectively in connexion with the insides of the nerve fibres. A more direct demonstration can be obtained in the giant nerve fibres of the squid. These are as much as 1 mm in diameter, and an electrode can be inserted inside one of them so that the potential difference across the surface is measured directly (Fig. 26.2). There is a large difference in ionic composition between the contents of a nerve fibre and its surroundings. The inside is rich in potassium and poor in sodium and chloride, the last two being abundant in the tissue fluids around (Table 26.1). The differences between inside and outside are almost certainly mainly due to the properties of the axolemma, the membrane at the surface of the axon. This membrane appears under the electron microscope as two dark lines and a lighter one between. This is probably the image of a lipoprotein system where the hydrophilic ends of the phosphatide molecules are directed outwards and bound to protein and their long fatty radicles directed towards each other. Such 'unit membranes' are common around and within cells (Robertson 1964).

The significance of the axon for nervous activity is that it (a) alters the diffusion of various ions in such a

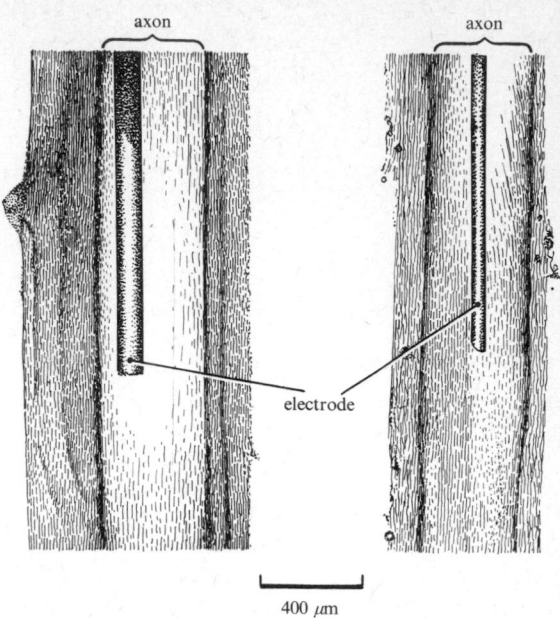

FIG. 26.2. Giant axon of the squid with capillary electrode inside it. The right-hand image is formed directly by the microscope, and the left-hand one is a side view obtained by a mirror. The electrode does not touch the side of the fibre at any point. (After Hodgkin and Huxley (1945). *J. Physiol.* **104.**)

TABLE 26.1

Squid axon and external electrolyte concentrations in mM 1^{-1}

	Internal	External
Na$^+$	50	460
K$^+$	400	10
Cl$^-$	40–100	540
Isethionate$^-$	270	0
Aspartate$^-$	75	0

The potential difference across the cell membrane is −60 mV.

way as to set up potential differences; (b) is involved in actual pumping operations by which ions are transferred against concentration gradients; and (c) is so constructed that when electric currents pass through it its properties change (it 'breaks down'), setting up greater potential differences, which in turn cause breakdown of further sections of membrane and hence propagation of a signal along the nerve fibre. In fact the nerve membrane, with the contained axoplasm, serves to set up a store of potential energy, which can be called upon to transmit signals and then later be replaced. Moreover, elaborations of the system by which the signals are set up (synapses) serve to perform the computational functions by which the nervous system makes decisions as to the actions that are appropriate to each environmental situation.

The nerve membrane is permeable to sodium and potassium and chloride ions. The concentrations of the last two in a resting nerve fibre are approximately what would be expected given their relative mobilities and concentrations, inside and outside the fibre, assuming that the resting potential assists the inward movement of potassium. The concentration of sodium, however, is very much less than it would be under simple physical forces, and the essential properties of the fibre depend upon the fact that sodium is actively pumped outwards. This movement continues while the fibre is provided with oxygen and is prevented if the energy-yielding mechanism is poisoned, for instance, by cyanide. It is this pump that charges up the fibre, providing the main part of the potential difference upon which the whole action of the fibre depends. The movements of the other ions, however, are not entirely determined by a simple diffusion but are complicated by being linked with the movement of sodium by a coupled exchange mechanism. Thus potassium mainly moves through the membrane by diffusion, assisted by the potential difference, but its movement is slightly influenced by a metabolic pump. Part of the potassium may also be bound to large negative ions inside the cell, though most of it is certainly free.

The way in which the potential across the membrane is used for signalling can be shown by inserting two fine capillary electrodes filled with strong KCl close to each other into a large squid's nerve fibre (Fig. 26.3).

Rectangular pulses of current (say 10^{-7} A for 2 ms) are passed through one and the other is appropriately connected to an oscilloscope. This latter will record the potentials applied through the other electrode, but with a characteristic rounded time course—this is known as the 'electrotonic' potential (Fig. 26.4). The distortion of the applied current is due, of course, to the fact that it must flow through the axolemma, which has not only a very high resistance ($10^3 \ \Omega$ cm^{-2}) but acts as a leaky condenser (capacity 1 mF cm^{-2}) with a characteristic charging time. By moving the recording electrode away this electrotonic potential can be seen to be purely local, it attentuates by some 50 per cent along each 1–2 mm.

If the current direction is such as to pass inwards through the membrane it simply *hyperpolarizes* the fibre locally (Fig. 26.4). With the local electrode as a cathode, however, the fibre is *depolarized* by the outward-flowing current reducing the internal negativity. When the resting potential drops to about 50 mV the membrane changes and admits sodium. After the

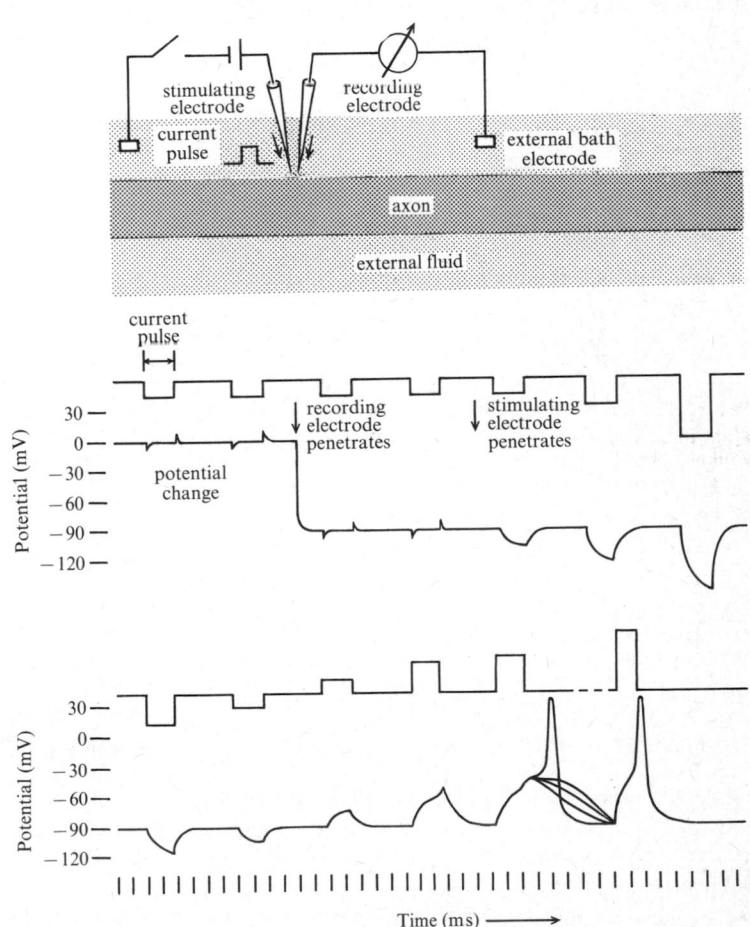

FIG. 26.3. Membrane potentials at site of stimulation. Stimulating and recording microelectrodes are inserted into an axon side by side. Square current pulses are applied producing 'electrotonic' potentials. If the current is strong enough and flows outward through the axon membrane, an action potential it initiated. (From Katz *loc. cit.*)

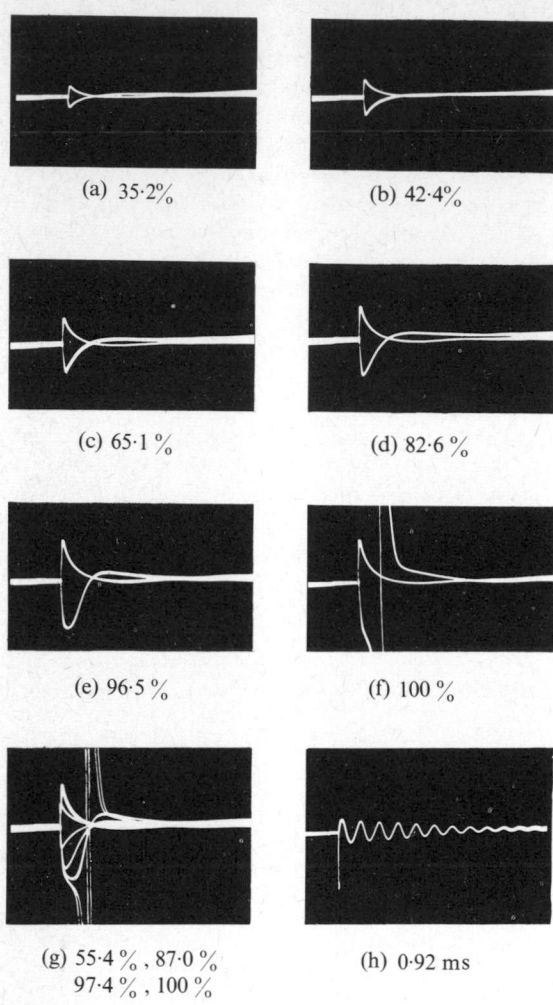

(a) 35·2% (b) 42·4%

(c) 65·1% (d) 82·6%

(e) 96·5% (f) 100%

(g) 55·4%, 87·0% (h) 0·92 ms
97·4%, 100%

FIG. 26.4. Records of response of a giant nerve fibre of the squid to cathodal and anodal polarization. Each record is a double exposure showing the response of the fibre at the stimulating electrode to an anodal stimulus (upwards) and a cathodal stimulus (downwards).
The percentages show the strength of the stimulus in terms of the threshold, which is reached in (f). With stimuli of less than half threshold (a) and (b) there is no marked difference between the passive anodal and cathodal responses. With stronger stimuli the cathodal response rises rapidly and assumes the characteristic form of the local action potential. In (g) the responses to four strengths of anodal and cathodal stimuli are superimposed. The double cathodal responses at 100 per cent are due to minute variations in threshold; each cathodal and anodal record has a double trace to check constancy. (From Pumphrey, Schmitt, and Young (1940). *J. Physiol.* **98.**)

applied current is removed either a small *local response* remains (Fig. 26.4) and then decays, or with slightly greater depolarization, the system flares up into a much larger potential change, the *action potential or spike*. This is a process no longer controlled by the

initially applied current pulse; it is a transient self-amplifying potential change that exceeds the threshold displacement by a factor of 4–10 times and crosses the zero line. At its peak it reaches a level of 40–50 mV *positive inside*, from which it rapidly swings back to the resting level. Once this spike potential has been elicited, it propagates along the whole length of the fibre at constant velocity and without attenuation of signal strength. It leaves behind a short refractory period— a silent interval of one or a few milliseconds during which the fibre is unable to carry a second signal. Thereafter the system is ready to be re-excited and to fire another propagated 'impulse'.

This regenerative response is of course the result of the effect of the current released by the initial membrane change in producing a similar breakdown at a distance away from the point of origin. The breakdown is the essential explosive process, but its initiation and speed of propagation depend upon the cable characteristics of the fibre. The current is carried by the electrolytic gel of the axoplasmic core and it must pass through the resistance of the membrane. It can be shown that it will fade exponentially with length constant Vr_m/r_i, where r_m and r_i are the membrane and internal resistances. This suggests that increased spread along the cable and hence increased velocity will be obtained either by increasing the diameter or the membrane resistance. In fact both factors are used by animals. The velocity of conduction is one of the chief variables among the nerve fibres. Those that are concerned with actions in relation to the environment must conduct fast and they are large, reducing the internal resistance and are often surrounded by a thick myelin sheath, increasing the membrane resistance (p. 249). Some groups of inverte-brates seem not able to produce thick sheaths and conduction velocity is increased only by having larger fibres. Since the velocity only increases with the square root of the diameter 'giant fibres' are needed to obtain even moderate velocities (Fig. 26.5). Obviously such very large fibres cannot be numerous. The rapid trans-mission of large amounts of information was only made possible by the evolution of thick myelin sheaths.

The action potential spreading over the surface of the nerve fibre is thus the agent of signalling making possible all the activities of the nervous system. It is a process dependent upon the properties of the nerve membrane, which when depolarized changes its charac-ter and suddenly allows the entry of the sodium which it had previously extruded. The positive charge carried by the sodium reinforces the lowering of the resting potential, which in turn increases the sodium per-meability and the process becomes explosive. This happens if the rate of sodium entry becomes high

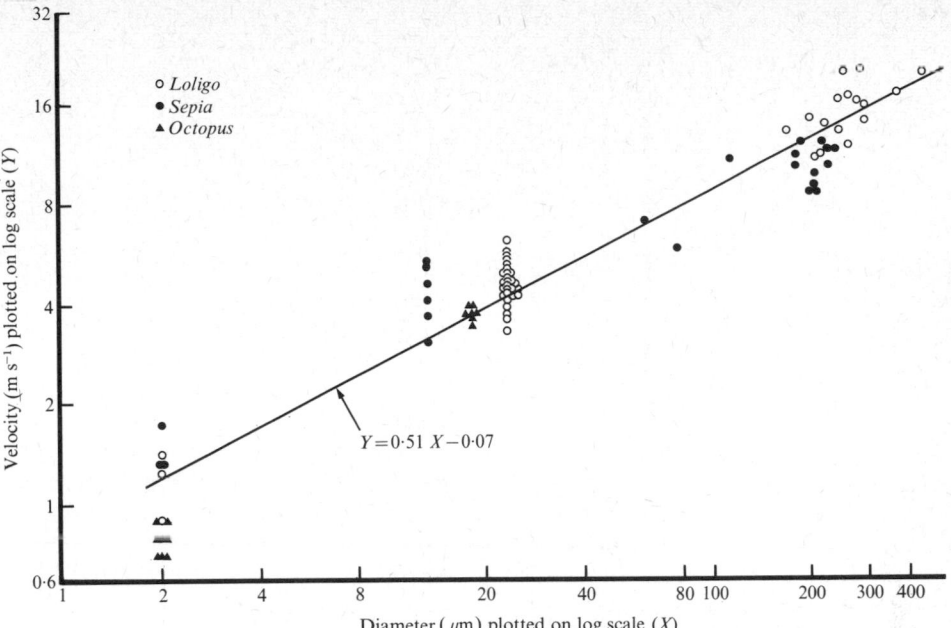

FIG. 26.5. The relationship between conduction velocity and nerve fibre diameter in three species of cephalopod. $V \propto \sqrt{D}$ ($Y = 0.51X - 0.07$). The points are plotted on a log scale, but the axes are labelled in ordinary units. (Figure kindly supplied by E. A. Bradley.)

enough to balance the simultaneous efflux of potassium and chloride, which will tend to return the fibre to its original level. Once the explosion has begun it will continue, until the axoplasm becomes sufficiently positive to balance the gradient of sodium ions ($\sim 50\text{--}60$ mV). Actually this level is not quite reached because the potassium gate also opens. Moreover, the sodium gate remains open only for a short time. Unfortunately we do not know what changes in the membrane open and close the sodium gate. During passage of a nerve-impulse there is a large fall in the resistance of the membrane, from 1000 Ω cm^{-2} to 20 Ω cm^{-2}, with no change in capacity. However, the resistance is still some 10^6 times greater than that of the surrounding fluids, which suggests that the increased conductivity is a matter of opening only pores that represent perhaps less than 1 per cent of the membrane surface.

As a result of the passage of the action potential the fibre has gained sodium and lost a quantity of potassium, estimated at one millionth of its store in a squid's giant axon. Even a poisoned fibre can therefore conduct hundreds of thousands of impulses. Smaller fibres with their greater surface–volume ratio lose a higher proportion. Normally the fibre is recharged by its pump extruding sodium and allowing entry of potassium. These events are accompanied by various small slow changes of potential level (afterpotentials) and an increased oxygen uptake and heat production. A very small amount of heat is also produced during or shortly after the impulse passage, but this initial heat change is so slight that it reinforces the evidence that the change occurs only in a small part of the membrane.

The effect of the explosive action potential is to amplify, by some 5 times, the local depolarizing charge of 20 mV applied to a fibre. The resulting charge of nearly 0.1 V is sufficient to produce a current flow causing a similar depolarizing action on a stretch of membrane one millimetre, or more, away. In spite of its poor properties as a cable conductor the fibre is thus able to allow propagation across a short paralysed area. Indeed the fact that the propagation is fundamentally electrical can be shown by assisting it to cross a block (produced say by cold) by boosting it with a sub-threshold current beyond the block. Thus a failing nerve impulse can be augmented electrically (Fig. 26.6).

4. The code of the nervous system

The characteristics of conduction determine the method of signalling that is adopted by the nervous system. An essential feature is that activity is broken up into a series of discontinuous signals, the *nerve impulses*. The impulses in any one fibre are all alike, the amplitude of

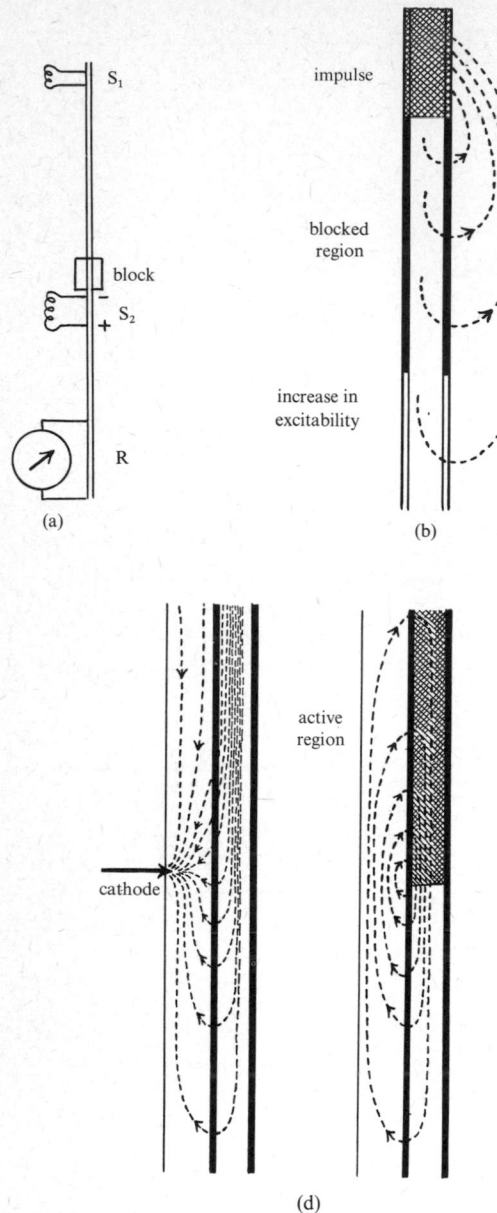

10 mV

above

below

1 mV

1·4 mm

2·5 mm

4·1 mm

5·5 mm

8·3 mm

Time (ms)

(c)

FIG. 26.6. Hodgkin's experiment to show evidence for electrical transmission in nerve.

(a) The sciatic nerve of a frog is laid on two stimulating electrodes S_1 and S_2, and recording electrodes R. Just above S_2 is an apparatus for blocking the nerve by cold or pressure. A sub-threshold stimulus applied at S_2 will only set up an action potential if it is delivered at such a time as to summate with the electrical effects of an impulse arriving at the block. The current responsible is shown in (b). (c) Potentials recorded above the block and at various distances below it, the latter with 5 times greater amplification. (d) Comparison of the stimulating effects of an applied current with those of the local circuits produced by the action potential. (From Hodgkin (1937). *J. Physiol.* **90**.)

the potential generated at each point depending only on the state of the membrane at that point. Once the membrane has discharged it cannot immediately do so again, it passes through a *refractory period*. There is therefore a maximum frequency of conduction of about 500–1000 impulses per second in a mammal. In normal functioning the fibres mostly discharge at much lower frequencies, for example a motor nerve fibre will carry about 20 impulses a second when it is producing the contraction involved in a defensive action such as with-

drawal of a limb from a painful stimulus (p. 256). Probably motor nerves seldom normally conduct above 100 impulses per second. This discontinuity of impulses means that all information by nerve signals is carried in a code of similar pulses.

The intensity of the impulse arriving at the end of a nerve fibre therefore bears no relationship to the intensity of the stimulus by which it was set up. In this respect the nerve fibre is said to follow the '*all-or-nothing law*'. Transmission in a nerve may be compared with that along a train of gunpowder, where each point fires the next. Variation in the intensity of stimulation may produce variation in frequency of impulses but the intensity of each impulse is determined only by the conditions at the point that is discharging.

The fundamental property upon which nervous activity depends is the possibility of changing the permeability of the surface, so that the energy stored as a concentration difference between the inside and outside of the cell becomes available to produce activity at some distant place. In a peripheral nerve fibre a relatively small depolarization (15 mV) triggers off a response yielding a large potential difference (120 mV) and the resulting local currents produce propagation over the whole fibre, with a high safety factor. This is the condition required for reliable transmission of a signal. There is a high probability that events at one end of the fibre will produce similar events at the other end.

Since the nerve impulses are identical pulses, which can vary only in frequency, the amount of information transmitted by any one nerve fibre is small. The transmission of large amounts of information is achieved by having large numbers of nerve fibres and providing arrangements by which the impulses that they carry are allowed to interact within the central nervous system (p. 251).

5. Conduction velocity and the myelin sheath

The velocity of propagation of nerve impulses is not the same in all axons. The largest mammalian fibres, about 22 μm in total diameter, conduct at about 120 m s^{-1} in man, rather more slowly in the rabbit. Rates grade down to less than 1 m s^{-1} for the smallest (non-medullated) fibres. The *myelin sheath* functions as an insulator, interrupted at intervals by the nodes of Ranvier (p. 238). The tiny cylinder of axon exposed at the node is the only active portion of the nerve membrane. The depolarization of this cylinder causes a flow of current along the inside of the fibre and out through the next nodal membrane, which is then depolarized. The impulse jumps from node to node. The effect of the insulating sheath is to ensure that the local circuits that produce conduction are spread out along the fibre. Each such unit propagates at high velocity and the whole process therefore proceeds faster than in non-medullated fibres, where the current produced at each point is allowed to leak through the membrane nearby. For the currents to be effective they must be conducted along a large tube and then canalized through a small area of membrane; correspondingly we find that the fastest nerve fibres are of large diameter, but they become narrow at each node (Fig. 25.5, p. 236).

On either side of the node the axon is enlarged, more above (that is, towards the cell body) than below. The Schwann protoplasm is also thickened here, to such an extent that it grooves the cytoplasm (Fig. 25.8, p. 238). Fingers of Schwann-cell cytoplasm extend into the node and make contact with the axon surface. This cytoplasm is packed with mitochondria and probably has some special function concerned with the ionic interchanges at the node. Between the fingers of satellite protoplasm the axon surface is in contact with a 'gap substance', an extension of the basement membrane and rich in sulphated mucopolysaccharide. Presumably the currents responsible for the propagation of the action potential pass through this substance, whose properties may help to provide the maximum electrochemical gradient (Langley and Landon 1967).

The velocity of conduction is an important feature of nerve action since it affects not only the time of arrival of impulses but also the dispersal of volleys in neighbouring fibres. Nerve fibres for each function possess characteristic diameters (Fig. 25.9, p. 239). The motor nerve fibres to striated muscles are large and medullated, as are the proprioceptor fibres that report back the tension on the muscle and play an essential part in producing a steady contraction (p. 65). Preganglionic autonomic fibres are small and medullated, while sympathetic postganglionics and some fibres for pain conduction are non-medullated. Slow conduction is more efficient in making the slow changes necessary in the viscera (p. 290).

6. The synapse

Nerve fibres thus transmit information in the form of a code of nerve impulses. Each impulse propagates over the whole fibre with a high safety factor. At the end of each axon the 'message' has to be 'decoded' and put to use by the receiving system. The study of the way the nerve fibres transmit their effects to other nerve cells is therefore of special importance (Eccles 1964; Iverson 1974; Weight 1974).

Single nerve fibres conduct in an all-or-nothing manner but, in the intact organism, responses are usually produced by the effectors only when groups of nerve fibres are active in particular patterns. This is ensured by the provision of regions, the *synapses*, at which the combined action of several nerve fibres sets up impulses in other nerve cells. We can speak therefore of *presynaptic fibres* conducting towards the synapse and *postsynaptic dendrites*, which, when appropriately activated, initiate nerve impulses in the axon leading to some other nerve cell or to an effector. The question of the relation of pre- and postsynaptic fibres is obviously of crucial importance for neurology; the processes that go on in this region and lead to excitation of the post-synaptic cell make possible the interaction by addition or subtraction between the signals in different nerve fibres. This interaction is the basis of all the 'computations' by which the nervous system ensures appropriate responses to external and internal conditions.

Impulses arriving in a single presynaptic fibre seldom cross a synapse; they produce in the next cell of the chain local responses (*synaptic potentials,* p. 251), which can summate with those set up by impulses in other fibres that converge on the same cell. An impulse set up anywhere on a nerve fibre normally spreads over the whole surface of that fibre. It follows that if excitation is not to spread indiscriminately throughout the nervous system there must be points of discontinuity, and these are provided where each cell process makes synaptic contact with some part of another cell. After much controversy it is now agreed that the contents of the individual nerve cells do not unite completely at these synaptic points, but remain separated by the membrane barriers of their surfaces, so that the potassium spaces of the two are not continuous. The effect of an impulse arriving in the presynaptic fibre is therefore not the same upon the postsynaptic cell as it would be upon a further extension of its own axon. The synapses thus provide regions at which the probability of transmission is low; impulses arriving in any one fibre are not always passed further. The next cell in the chain is usually excited only when impulses arrive from a number of sources, with suitable timing. In other words, the existence of synaptic regions allows for the interaction of signals coming from different sources, so that a properly co-ordinated response is produced.

The reduction in effect of the impulse produced by the break in continuity at the synapse is very large. An impulse passing along an ummyelinated fibre of 5 μm diameter uses a leaking cable with an input impedance of 20 MΩ and must lower the resting potential ahead of it by 20 mV in order to activate the membrane and produce the amplification of 5–10 times that is provided by the electrochemical relay mechanism. A membrane across the fibre with the same specific resistance as the surface membrane would place 3000 MΩ across the line, reducing the effectiveness of the current to less than 1 per cent. The capacity for transmission would be improved if the attenuation between the fibres were reduced. This is ensured in various ways, for example, by increased area of contact, or perhaps by alteration of the properties of the membranes when they are brought into close contact. There are some 'tight

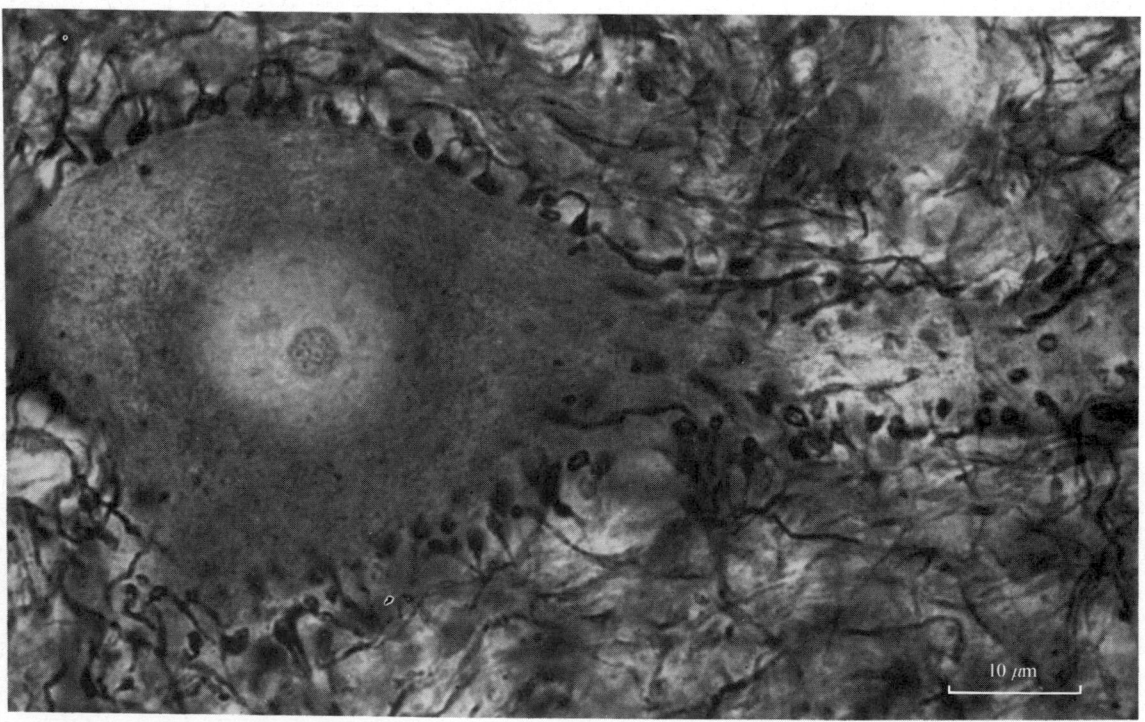

Fɪɢ. 26.7. Motor neuron from cat spinal cord showing boutons over cell body.

junctions' at which electrical synaptic transmission occurs (Katz 1966). However, at the great majority of synapses transmission requires a special chemical amplification system, so arranged as to allow the addition and subtraction of converging influences, by means of which 'decisions' are made in the nervous system.

There is great variation in the number of fibres that are involved in the synapses of nerve cells in different parts of the nervous system. In some parts impulses in one or a few presynaptic fibres are sufficient to fire the postsynaptic cell; other cells require an input from many sources. The regions of synaptic contact differ in appearance in different parts of the nervous system. Attached to the cell bodies and dendrites of the ventral horn cells of the spinal cord of mammals there are numerous small knobs, the *'boutons terminaux'* or *end-feet* (Fig. 26.7), which are the endings of the presynaptic fibres. There are 20000 or more of these endings in contact with a large motor cell.

Where the surface of the bouton makes contact with the surface of the nerve-cell body or dendrite the two membranes show various forms of thickening (Fig. 26.8). The gap between the two opposed membranes is usually at least 15 nm, often more. There is sometimes evidence of electron-dense material within it.

The boutons contain numerous mitochondria, which is evidence of the active metabolic processes that are involved in the production of materials able to produce the necessary amplification of the signals. These substances are the *synaptic transmitters*, and they may be either excitatory or inhibitory. The evidence for chemical transmission is complicated and comes largely from studies of neuro-muscular junctions (p. 62) or simpler parts of the nervous system such as autonomic ganglia (p. 279).

The transmitters are released on arrival of impulses at the presynaptic terminals, and they serve to depolarize the postsynaptic membrane. This action has been studied chiefly at nerve endings in muscle. Here the transmitter (acetylcholine) is being released even at rest in the form of minute packets (quanta) each producing a small depolarization of about 0·5 mV, the *miniature end-plate potential*, about once a second (Fatt and Katz 1952). Upon arrival of a nerve impulse, hundreds of these quantal events occur within less than a millisecond. Together these produce sufficient depolarization of the muscle membrane to initiate a propagated impulse.

There is sufficient evidence to show that chemical transmission in the central nervous system is similar. Among known transmitters in the CNS are acetylcholine, noradrenaline, dopamine, and 5-hydroxytryptamine, whose action appears to be predominantly excitatory, and glycine and γ-aminobutyric acid, whose actions at least in the spinal cord and cerebellum respectively are inhibitory. Other possible transmitters are glutamic acid, which exerts an excitatory effect on neurons in the cortex, and substance P, a polypeptide, which may be the transmitter at sensory input terminals in the CNS. A purine-like substance, probably adenosine triphosphate (ATP), acts as one of the transmitters in the autonomic nervous system.

The quantal packets of transmitter released are probably related to the *synaptic vesicles* (Fig. 26.8), which are membrane-bound droplets abundant in all synaptic endings and often crowded close to the presynaptic membrane. There is some evidence that they discharge into the synaptic cleft.

The precise means of action of the transmitters on the postsynaptic membrane is unknown, but there is evidence that it produces a local change, the *postsynaptic potential*, which may be excitatory or inhibitory (EPSPs and IPSPs). The interaction of the effects of impulses arriving at the various parts of the dendritic tree determine the firing of impulses by the postsynaptic cell, but it is still uncertain how this interaction is achieved. It may differ considerably according to the function of the neuron. The impulses probably arise in the postsynaptic cell at the *axon hillock* (Fig. 25.2, p. 233), and this presumably requires a depolarization of 20 mV there. On the other hand the synapses are scattered all over the cell body and dendrites—some may be over 1 mm away. We do not know what potentials are produced by their depolarization, but presumably they are so attenuated as to have little effect on the axon hillock. There is evidence however that in some neurons the dendrites produce not merely local changes but also propagate disturbances. Single impulses in fine dendritic twigs are probably blocked at the branch points, but if impulses arrive together they then propagate (Wall 1965). It may be that the dendritic tree, though mainly electrically inexcitable, provides a complex computing organ with several booster (trigger) zones. This would allow for the setting up of impulses upon the arrival of presynaptic impulses in particular spatial and temporal patterns. If the neuron operates like this it would act as a cascaded array of filters allowing for considerable computation within each neuron (Fig. 26.9). It has often been assumed in the study of cybernetics that each neuron functions as a simple on–off element and that computation is the result of the interaction of neurons. Evidently if computations can occur within neurons the situation is more complicated than this.

The excitatory and inhibitory synapses are not distributed at random. It is probable that those on the cell

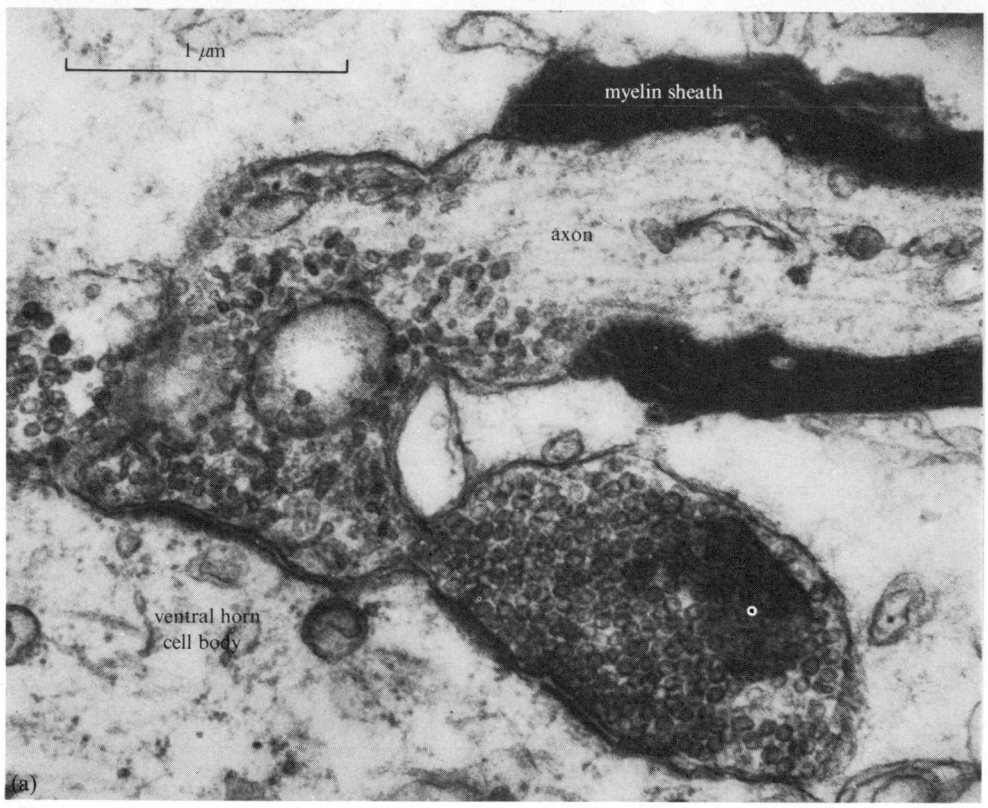

(a)

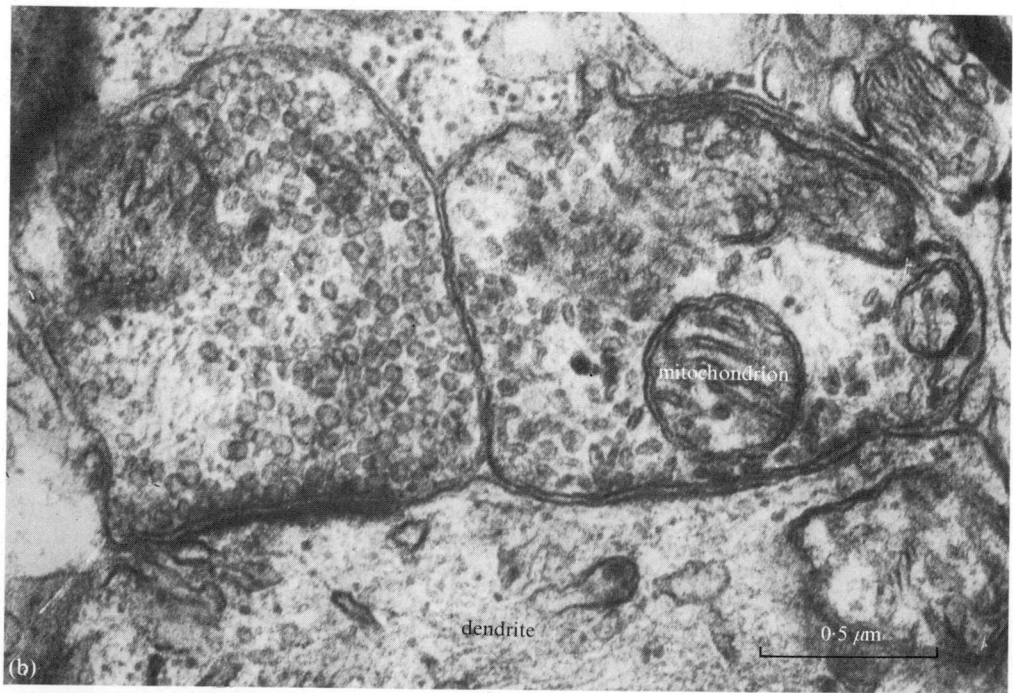

(b)

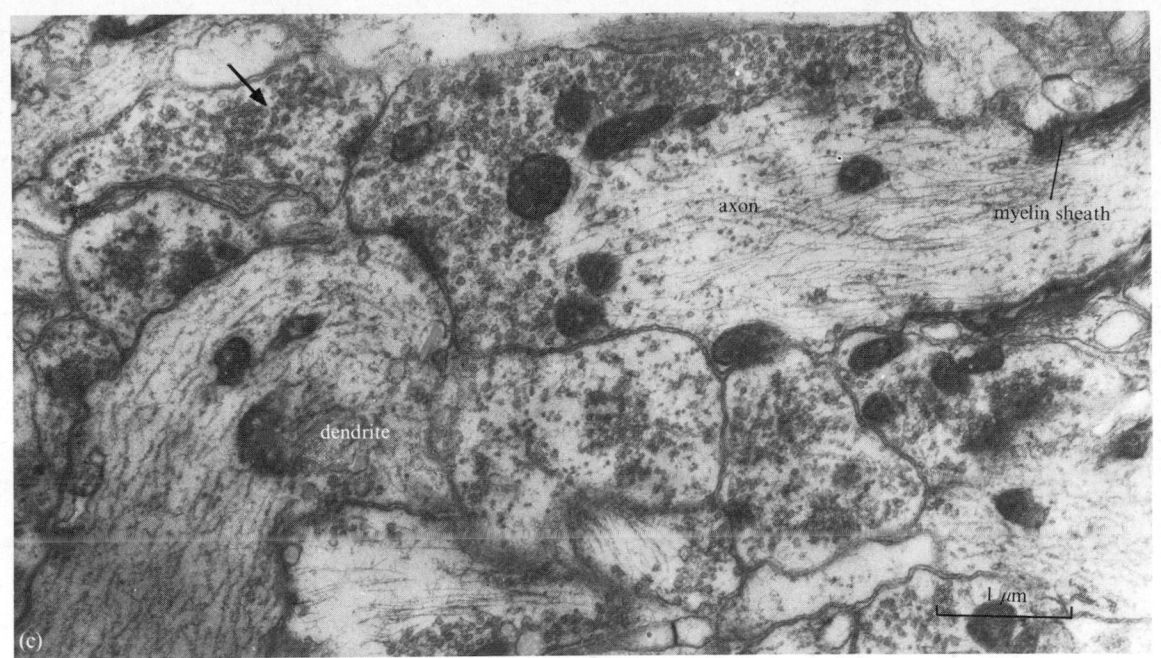

FIG. 26.8. Electron micrographs of synapses in goldfish spinal cord.

(a) Axosomatic synapse on ventral horn cell; (b) type 1 synapse (presynaptic excitatory), with round vesicles on the left, and type 2 synapse (postsynaptic inhibitory), with flat vesicles on the right, contacting a dendrite; (c) serial synapse; a presynaptic terminal (arrow) (inhibitory) contacts an axon itself synapsing with a dendrite. (Figures kindly supplied by Prof. E. G. Gray.)

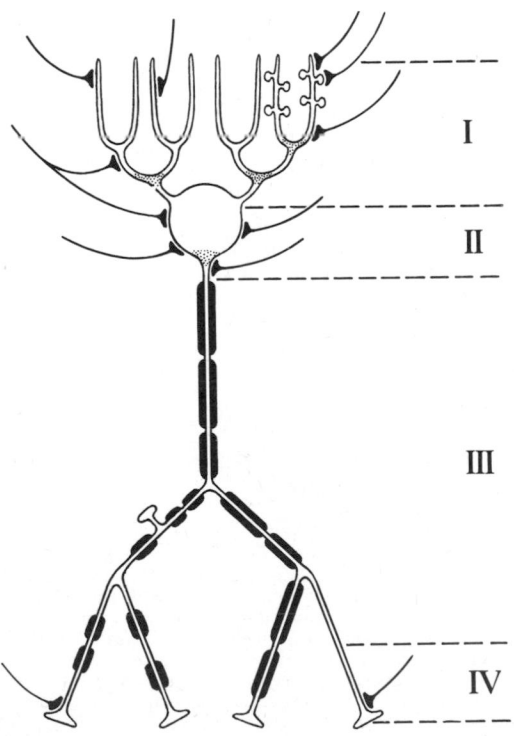

body are largely inhibitory, producing hyperpolarizing IPSPs. They show some peculiarities of structure (Fig. 26.8) and a particular form of oval vesicles, presumably carrying a distinct inhibitory transmitter. By studies of reflex excitation it can be shown that there can be summation of inhibitory as well as of excitatory effects. Sherrington as a result of such experiments spoke of *central excitatory* and *inhibitory states*, (c.e.s. and c.i.s.) without deciding on the physical basis for these (Creed, Denny-Brown, Eccles, Liddell, and Sherrington 1932). The effect of inhibition is sometimes produced in a different way by depolarizing the incoming nerve endings (presynaptic inhibition). Where this occurs the inhibitory fibres make contact with the excitatory ones

FIG. 26.9. Diagram of possible computational functions of what has been called a 'multiplex neuron'. The information is transferred sequentially, first in dendrites (with or without spines to isolate the effects of the boutons) (Phase 1). Booster zones may occur at branching points. The cell body and axon hillock receive further inputs, including inhibitory (Phase II). Series of action potentials are initiated in the axon and may be transferred in the axonal tree (Phase III). Finally the axon terminals may be modulated by presynaptic inhibition (Phase IV). (After Waxman (1972). *Brain Res.* **47**.)

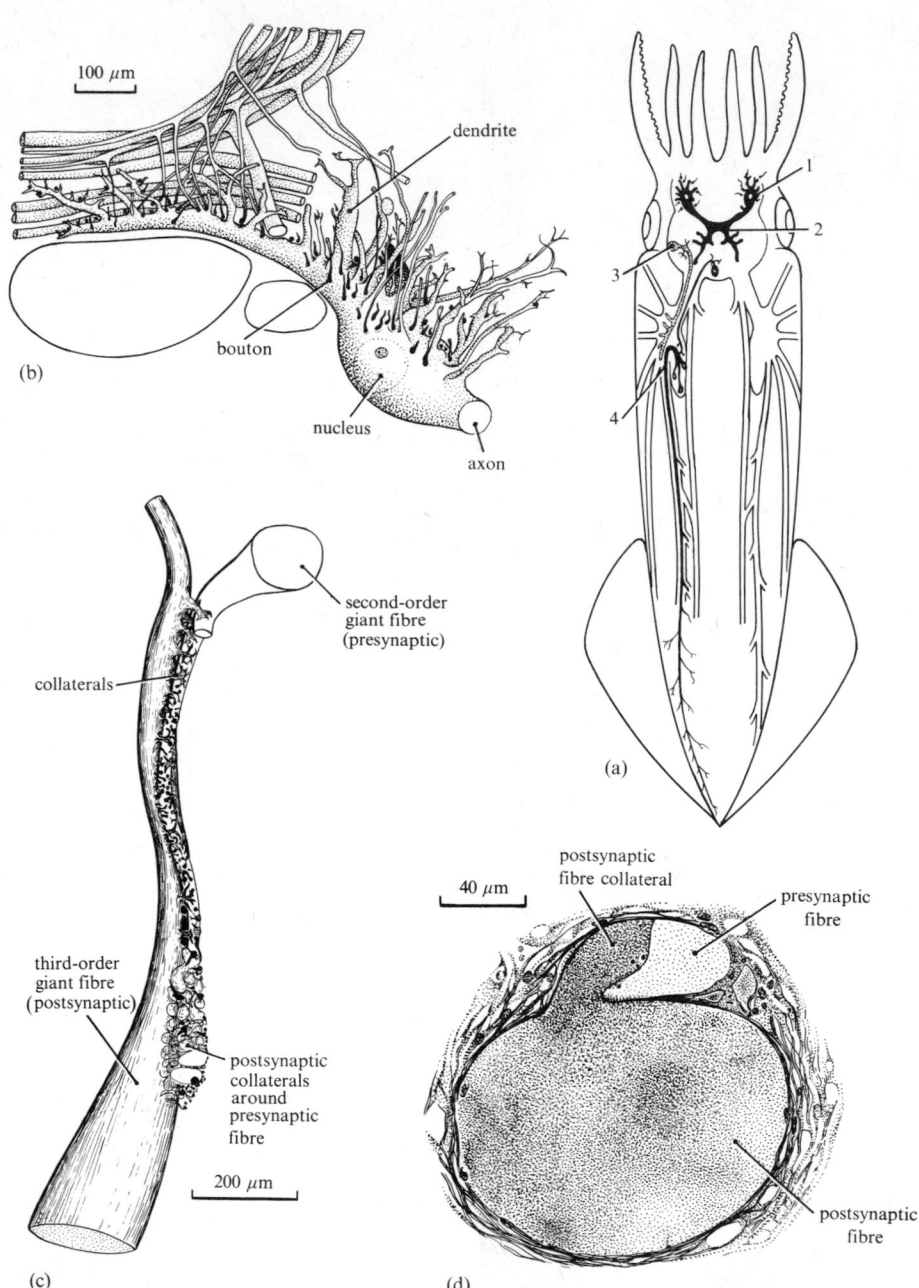

FIG. 26.10. Synapses in the squid.
(a) Diagram of the system of giant nerve fibres. 1. The two first-order giant cells are covered with boutons bringing excitation from optic and other stimuli. 2. The two giant axons fuse across the midline. 3. They make connexion with second order cells. 4. These make connexion with third-order giant fibres, which are formed by the fusion of the processes of many cells and innervate the muscles of the mantle by which the animal is squirted along. (b) Drawing of the first-order giant cell. (c) Low-power view of synapse between second-order giant fibre (presynaptic) and third-order fibre (postsynaptic). Collaterals from the postsynaptic fibre wrap round the presynaptic fibre. (d) Transverse section across this synapse showing the contact between a collateral and the presynaptic fibre. The latter is the more faintly stained with the haematoxylin dye used. (After Young (1939). *Phil. Trans. R. Soc. B229.*)

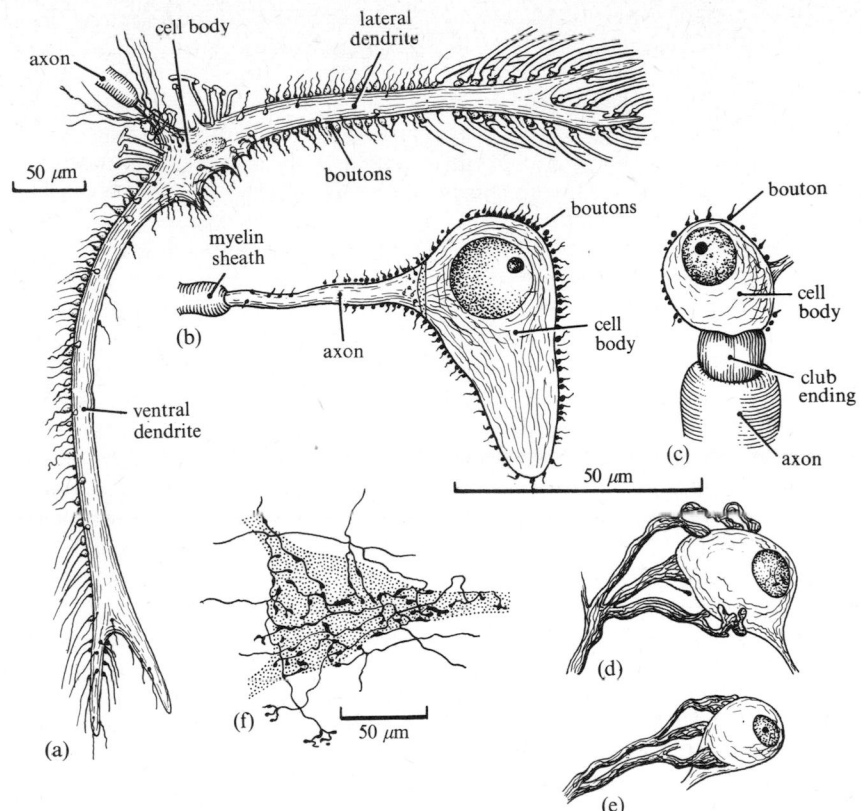

FIG. 26.11. Various types of synaptic system.
(a) A semi-schematic representation of the synaptic apparatus found on Mauthner's cell in the goldfish. The endings on the lateral dendrite are all of vestibular origin while those on the ventral dendrite and cell body come from various other sources. (b) A large motor type cell from the reticular formation of the goldfish showing relatively uniform distribution of homogeneous boutons on the cell body and proximal part of the axon. Mallory–Azan stain. (c) A cell from the reticular formation of the goldfish showing a single large club ending as well as small boutons. Mallory–Azan stain. (d) and (e) Two cells from the oculomotor nucleus of the goldfish showing a basket-like system of club endings derived from a single large branching axon. Bodian stain. (f) A large interneuron from the spinal cord of a 15- to 16-day-old cat. Golgi method. (From Bodian (1942). *Physiol. Rev.* **22.**)

and the situation can be recognized since there are vesicles on both sides of the synapse (Fig. 26.8). Suitably timed impulses in the inhibitory fibres will discharge the excitatory ending and reduce the EPSP that a subsequent excitatory impulse can produce.

At those synaptic junctions where there are few convergent sources of input the area of contact between pre- and postsynaptic units is large. In the case of the stellate ganglion of the squid, shown in Fig. 26.10, two input fibres co-operate in control of the giant postsynaptic fibres and each makes a large area of contact. Probably much will be learned about the nervous system of vertebrates when we know more about the extent of the areas of influence at synapses of various types (Fig. 26.11).

7. The reflex arc

Under artificial conditions impulses can pass in either direction along a fibre but, on account of the refractory period, impulses proceeding in opposite directions are unable to pass. A wholly separate set of *efferent nerve fibres* is therefore used to carry signals back to the muscles and other effector organs. Throughout life, series of impulses pass up the afferent fibres and down the efferent ones. Each act involves the co-operation of tens, hundreds, thousands, or millions of afferent fibres alone, whose impulses provide volleys bombarding the neurons of the central nervous system.

Setting our imagination to work we can now attempt to follow the internal sequence of events that occurs while the body produces a simple *reflex action*, such as the *flexor reflex*, by which the hand or foot is withdrawn on touching a hot object. The change in conditions (*stimulus*—see p. 362) activates a number of receptor organs in the skin, in this case the fine terminal branches of the nerve fibres for the 'sense' of pain (p. 364). Nerve impulses travel in these fibres to the spinal cord and there initiate activity in a number of *motor neurons*, either directly or after passing through *inernuncial neurons*. The motor neurons discharge impulses down to the flexor muscles, by which the limb is withdrawn. This simple movement, one of the most primitive protective or 'nociceptive' responses, thus depends upon connexions within the spinal cord that are determined by heredity. The afferent fibres entering the cord in any one spinal dorsal root may produce flexion in many muscles by the use of intersegmental spinal pathways.

This arrangement, in principle rather simple, is often used as the prototype of all nervous activity. It provides a model and allows us to introduce as it were a causal principle, by which we can seek the antecedents of all actions of the body. It is important to emphasize the limitations of the picture and the need to use it with caution and with imagination. Sometimes the body may act in this way reflexly, like a simple man-made machine, such as an electric front-door bell. Such simple machines are so designed that they respond only to one aspect of change in the world around and produce only one action. Even the lower parts of the nervous system, however, allow of variation of response as a result of the interaction of input from various sources. Each nerve cell is acted upon not by one or a few agents, as in a simple machine, but by many. Experiment shows that the pool of motor neurons that controls any muscle can be excited from various afferent sources. Moreover, not all of these sources produce excitatory effects. When we tread on a hot surface with both feet we do not flex both legs. Impulses that excite flexion in one limb *inhibit* it in the other. Indeed these same nerve impulses initiate contraction of the extensor muscles of the opposite leg, producing the *crossed extension reflex* by which the body is held up. The system is able to work therefore not like a simple bell system but as a computer, like that of a guided missile. Its instructions are to keep the body as nearly as possible upright in the light of the information

received from various sources, including feed-back arrangements from the proprioceptors of the muscles (p. 265).

In the higher parts of the nervous system this computing function becomes even more evident, so that only in a general sense is it possible any longer to use comparison with a simple reflex machine. Signals from large numbers of receptors are brought together in the brain centres and allowed to interact to produce elaborate and varied results. Moreover, the control at any moment depends not only on the instructions that are embodied in the hereditary design of the system but on the residual information about past events that has been stored (p. 341). The input interacts with this information store to produce a statistical estimate of the actions best calculated to ensure homeostasis.

These higher parts of the nervous system are composed of an immense number of separate units. There are more than 15×10^9 nerve cells in the human brain. Simple electrical machines contain a relatively small number of wires and other parts and we can therefore forecast exactly what will happen when current flows in any one part of the machine. In order to speak about the performance of a system with the immense number of parts found in the brain the best we can do is to use the methods that enable us to forecast the *probable* course of events. The application of such statistical methods to study of the nervous system is beginning to give us means for speaking about the behaviour of the vast number of cells that is involved.

Finally we must not imagine the nervous system to be in any fixed state of rest before the arrival of afferent impulses. Each part, such as a motor neuron, continually receives impulses from other parts of the nervous system, and these impulses affect its excitability and hence its reflex actions. After severance of the spinal cord, cutting off the normal flow of impulses from the brain, a state of *spinal shock* follows and few or no reflexes can be obtained from the motor neurons for some days. In other parts of the nervous system, for instance, in the brain, many of the neurons probably do not come to rest at all, but pass through continual cycles of discharge and recharge. We understand only little of these central activities, but must continually think of their existence as a background upon which the streams of impulses from the receptor systems produce their effects.

27 The spinal cord

1. Plan of the spinal cord

HAVING made a preliminary survey of ideas about the parts of the nervous system we can now begin to try to visualize how it works. The nervous system of simpler animals serves mainly for the performance of local reflex actions. In the starfish and other echinoderms there is no central nervous system but a network of nerve fibres all over the body beneath the skin, thickened in certain regions. The vertebrate nervous system has probably been derived from such a condition, by collection of the nerve cells and regions of synaptic contact to form the brain and spinal cord. Even in mammals the central nervous system shows by its development that it is essentially a sub-epidermal plexus rolled up into a tube (p. 345). The advantage of concentration of the nervous system is that it allows the animal to act more efficiently as a whole, bringing together all the information received by the various receptors. With such centralization it is possible for a store or memory of past information to be usefully put into relation with the information arriving at each moment.

The process of centralization has been going on throughout the evolution of vertebrates, with progressive increase in the number of routes that carry impulses up from the spinal cord to the brain and down again from the latter. This process culminates in man, where the main computations that determine the course of action are performed by the operations of the cerebral hemispheres. These are provided with information by all the main receptor systems and are the main stores of past experience. The 'lower' centres of the spinal cord remain of great importance in regulating the performance of each action.

The nervous system is often considered as consisting of a hierarchy of levels (following Hughlings Jackson). The spinal cord constitutes the lowest level, containing the motor neurons, whose processes pass out to innervate the muscle fibres, and thus provide the *final motor path*, through which all actions are produced. The motor neurons are partly controlled by impulses that reach them from the higher levels, but the details of performance are regulated at the spinal level. The degree of contraction of each muscle is determined by the information sent from the receptor organs in the muscles, the proprioceptors (p. 265). This information is used at the spinal level, though it is also transmitted to the brain.

The spinal cord therefore contains the motor neurons, the pathways by which they are controlled, mainly from the proprioceptors, and the pathways for certain quick defensive actions, such as the reflex withdrawal from a painful stimulus. The cord also contains numerous fibres ascending to the brain and descending from it. In all its actions the cord is profoundly influenced by the operations that go on at higher levels (Eccles and Schadé 1964a, 1964b).

2. The spinal nerves

The spinal cord of mammals is built on the same plan as in other vertebrates, giving off a dorsal and a ventral root on each side in each segment (Fig. 27.4). The spinal *dorsal roots* are wholly afferent, and each carries a ganglion containing the cell bodies of the nerve fibres. The *ventral roots* contain the efferent fibres, whose cell bodies lie within the spinal cord. Passing outwards the two roots join and then immediately divide into dorsal (posterior) and ventral (anterior) primary rami, each carrying motor and sensory fibres to the periphery. In many segments one or more *rami communicantes* join the sympathetic nervous system (p. 279).

In the cervical region the spinal nerves serve the muscles and skin and the more caudal ones (4, 5, and 6 in the rabbit) give rise to the *phrenic nerve*. The forelimb is served by parts of the ventral primary rami of a number of segments, usually C4–T1, which anastomose in a complex *brachial plexus* (Fig. 27.1) and distribute their fibres to form the nerves of the arm. The chief of these are the *radial nerve*, running mainly to the flexor muscles of the arm and forearm, and the *ulnar* and *median nerves*, to the muscles and skin of the forearm and hand.

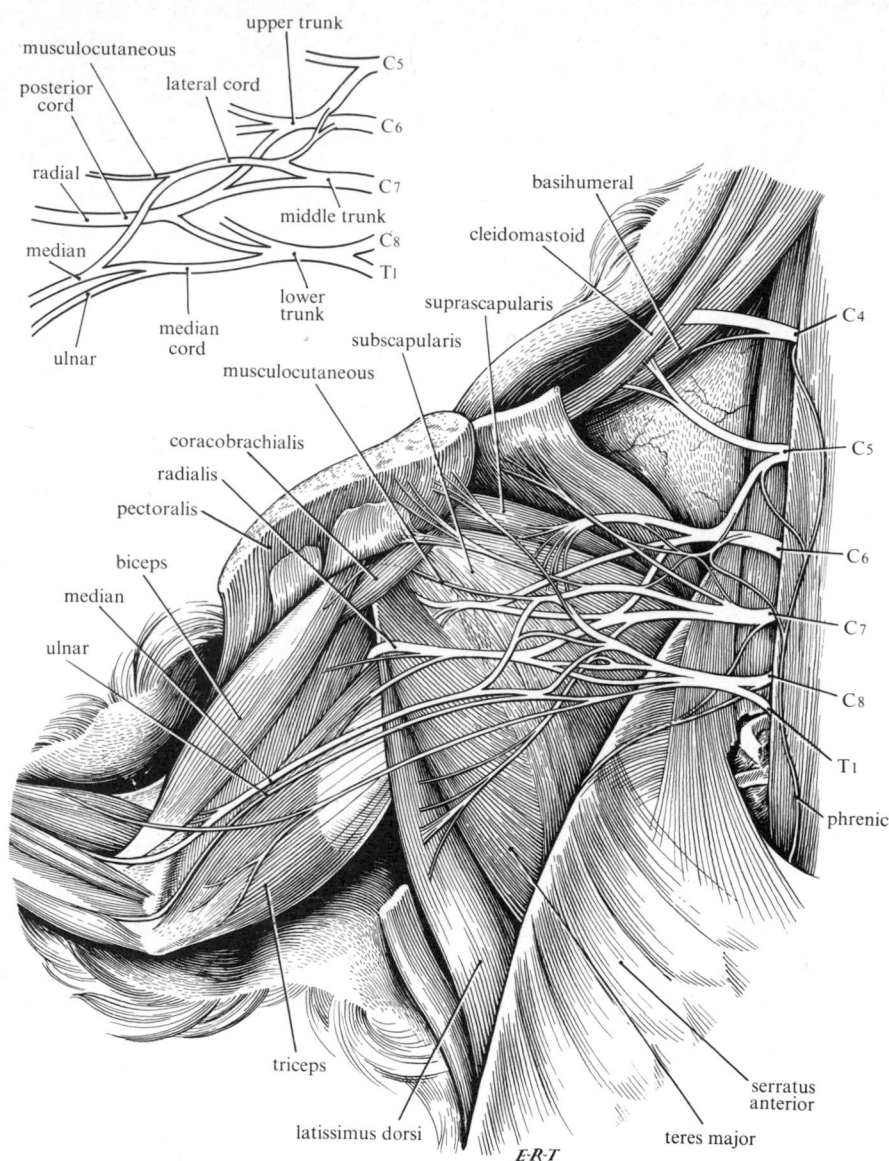

FIG. 27.1. Dissection of the brachial plexus of the rabbit. The simplified diagram at the top left shows the arrangement of the trunks and cords. C4–T1, spinal nerves.

In the thoracic region the ventral primary rami form the intercostal nerves, which supply the muscles and skin of the chest. Caudal to the thorax are the lumbar nerves, whose more cranial members innervate the abdominal muscles and skin. The more caudal lumbar nerves unite with those emerging in the sacral segments (usually L4–S3) to form the *sciatic* or *lumbo-sacral plexus*, which gives rise to the *femoral*, *obturator*, and *sciatic nerves*, innervating the hind limb (Fig. 27.2).

Caudal nerves are well developed where there is a tail but are reduced to small coccygeal nerves in man.

The fibres in the spinal roots and spinal nerves and their branches vary greatly in diameter and therefore in velocity of conduction (p. 249). The nerve to each muscle or part of the skin carries fibres showing a characteristic 'spectrum' of diameter. Thus the nerves innervating the main muscles of the body contain numerous large and small fibres but few of intermediate

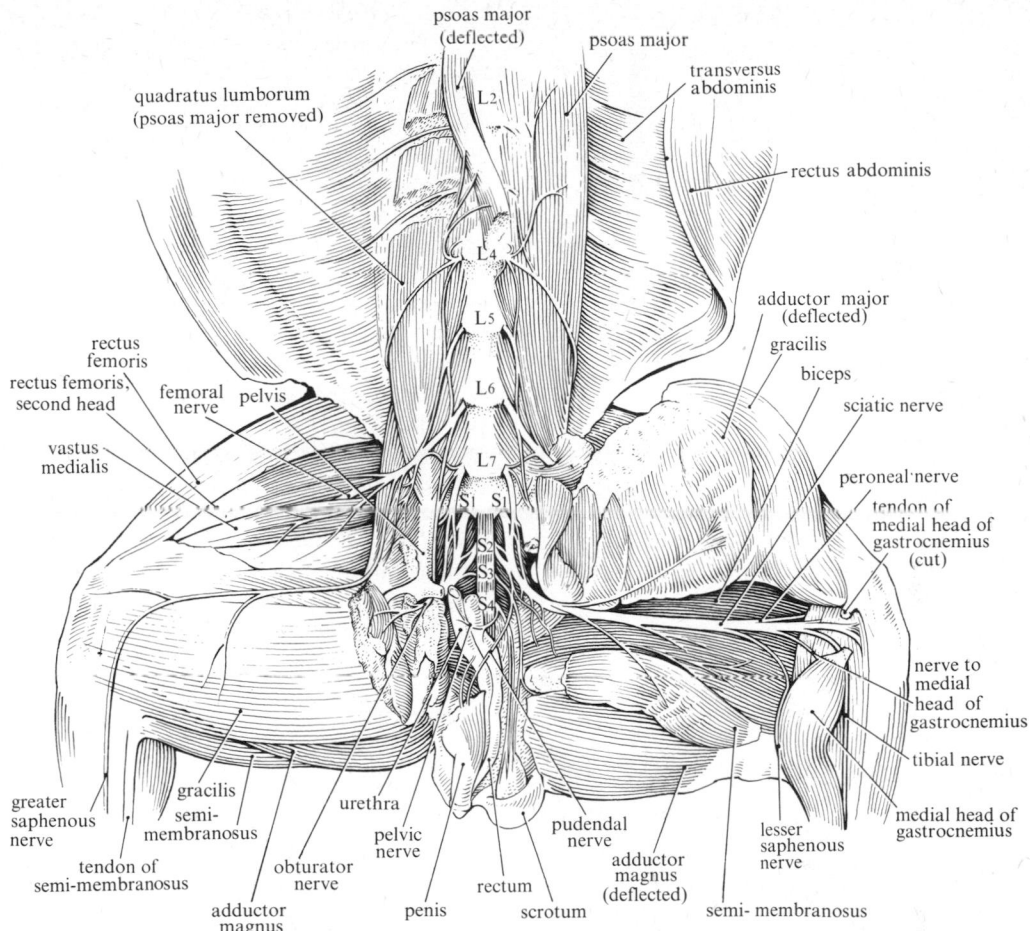

FIG. 27.2. The lumbo-sacral plexus of the rabbit. L2, L4, L5, L6, L7 lumbar vertebrae with corresponding ventral rami of lumbar nerves. S1, S2, S3, S4, sacral nerves.

diameter and are hence said to be 'bimodal' (Fig. 27.3). Some of the large fibres of a muscle-nerve are afferent (proprioceptor), others are motor. The small group contains the motor fibres that regulate the muscle-spindles and also some afferent fibres. The diameters are important because the proper functioning of the whole system depends upon the correct timing of arrival of impulses at the spinal cord or muscles. The complicated organization of the nerve to a single muscle shows the various types of fibre needed in a multi-channel system, where each fibre carries a different item of information (p. 249). The afferent nerves of the skin are equally complex but even less is known about them.

3. Structure of the spinal cord

The spinal cord terminates in mammals in the lumbar region, even if there is a tail. The roots for the hinder segments therefore run back for a considerable distance within the vertebral canal and form a bundle, the *cauda equina*, in the middle of which is a narrow *filum terminale*, the non-nervous continuation of the cord.

The spinal cord itself is composed of central *grey matter*, in which lie the nerve-cell bodies, and an outer *white matter*, composed of ascending and descending tracts, whose medullated fibres give the white colour (Figs 27.4 and 27.5).

The cells of the grey matter form nine rather distinct layers, each with cells of characteristic structure and connexions (Fig. 27.6) (Rexed 1964, Wall 1967). The layers lie as sheets throughout the length of the cord, but some layers are specially developed in certain regions. We can also recognize a functional division into four columns: somatic sensory, visceral sensory, visceral motor, and somatic motor. The *somatic sensory*

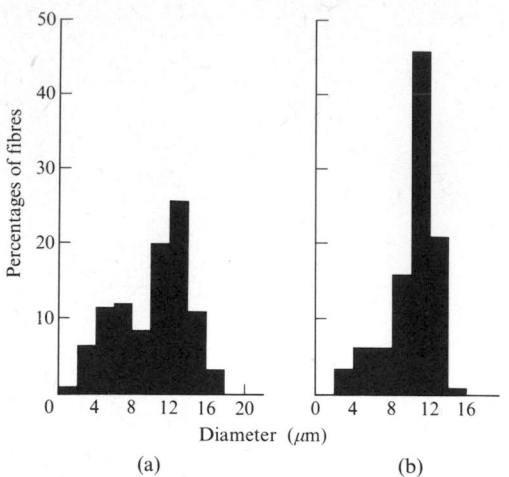

(a)   (b)

FIG. 27.3. Histograms showing the sizes of the fibres of two muscle-nerves in the rabbit. (a) In an antigravity muscle (gastrocnemius) with many spindles and a bimodal set of nerve fibres; (b) in the sterno-thyroid with few or no proprioceptors and a unimodal distribution of fibres in the nerve. (From Fernand and Young (1951). *Proc. R. Soc. B.* **139.**)

column, dorsally, includes laminae 1–4 and 6 and parts of 5. The more dorsal of these layers are mainly concerned with skin sensation and are present throughout the cord. Lamina 6 receives afferents from the proprioceptors of the muscles and is therefore only well developed at the level of the brachial and lumbar enlargements.

The *visceral sensory column* is included in lamina 5 and thus lies mainly ventral to the somatic sensory. It is well developed in the regions that receive afferents from the viscera, through the splanchnic nerves, that is, the thoracic and upper lumbar segments. We have also to consider in this group what may be called tissue afferents from the blood vessels, including those of the skin and other fibres indicating trauma to muscles and other deep structures. These also end in lamina 5. The system for signalling pain in fact involves the interaction of several of the dorsal horn laminae (p. 262).

The *visceral motor column* contains the motor neurons of the preganglionic fibres of the sympathetic system, whose axons leave in the ventral roots to form the white rami communicantes (Chapter 30). Therefore

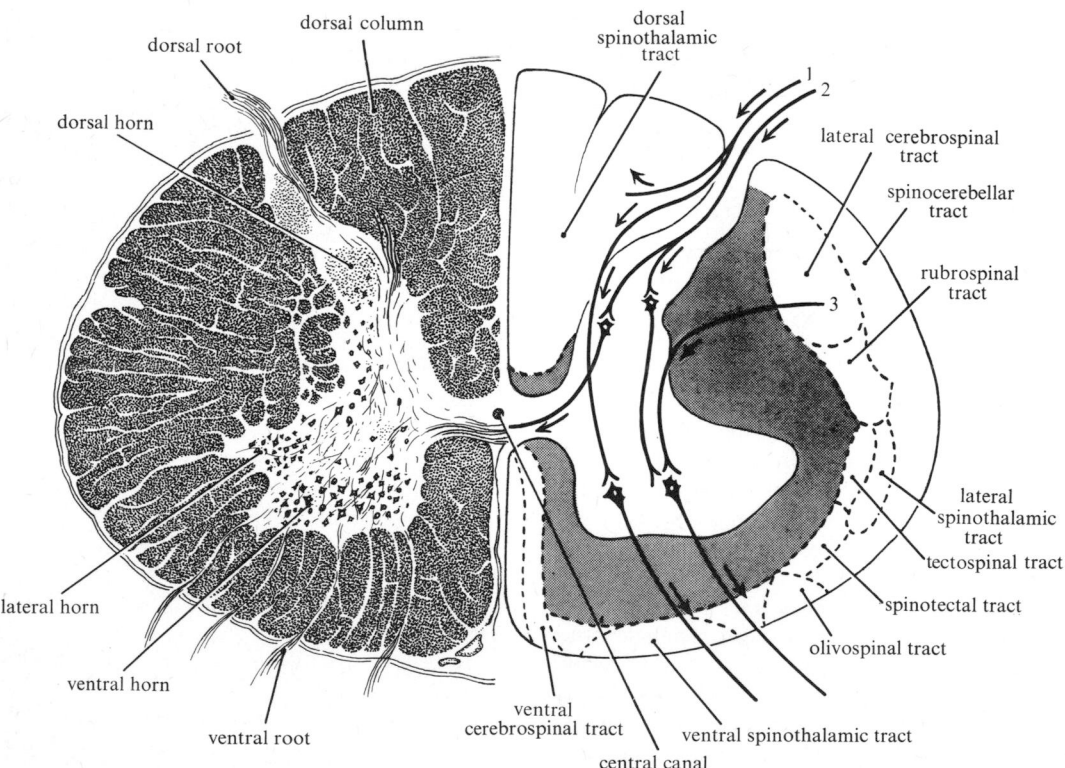

FIG. 27.4. Transverse section of the human spinal cord in the cervical region showing on the left the histological appearance with the medullated nerve fibres and the nerve cells stained. On the right a diagrammatic representation of some fibre connexions and pathways. 1. Pathway of the monosynaptic proprioceptor reflex. 2. Pathway of the flexor and crossed extensor reflexes. 3. Corticospinal pathway.

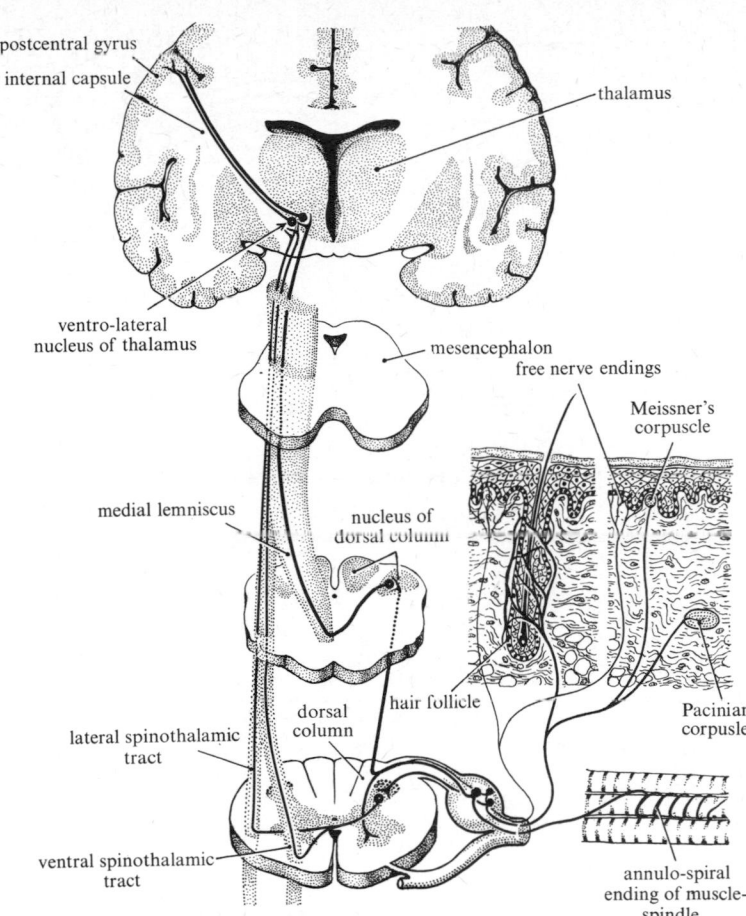

FIG. 27.5. Diagram of ascending pathways from the spinal cord to the cortex. (Modified from original illustration by F. Netter, from Ciba Collection.)

it is well developed in the thoracic and upper lumbar regions, where it forms a distinct part of lamina 7, called the lateral horn of the grey matter. However, lamina 7 is not purely visceral motor. It also plays an important part in the control of somatic motor activities, both by the receipt of impulses from the brain (rubro-spinal and tectospinal tracts, p. 300) and by the colla-terals of motor neuron axons, which here stimulate cells that in turn inhibit the motor neurons (Renshaw cells, p. 269).

The *somatic motor column* contains the cells that directly control the striated muscles, forming lamina 9 in the ventro-lateral part of the grey matter. Here are the large motor neurons whose fibres pass in the ventral roots to activate the striped muscles. Mixed with them are the smaller γ-motor neurons sending motor fibres to the muscle-spindles (p. 267). The motor neurons of each muscle are sometimes arranged as a separate group, and the neurons that control the muscles moving

a single joint tend to be grouped together (Romanes 1966). The groups innervating extensor muscles lie in the lateral part of the ventral horn; those for the flexors lie more medially. The cells of lamina 8, occupying the more medio-ventral part of the gray send axons mainly to the opposite ventral horn. The vestibulospinal, reticulospinal, and other descending tracts end in this region.

This is of course only a very crude representation of the detailed organization of the cord. There are said to be about 5×10^5 neurons in a single L5 segment of man, therefore perhaps 15×10^6 in the whole cord (Gelfan 1964). The number of synaptic knobs per cell varies from 800 for an 'average-size cell' to 10 000 for a large motor neuron. Only about 0.5 per cent of these are 'monosynaptic' endings of incoming dorsal root fibres, reaching down into the ventral horn (Chapter 28). The rest come either from the axons of other cells within the cord or from fibres descending from the brain.

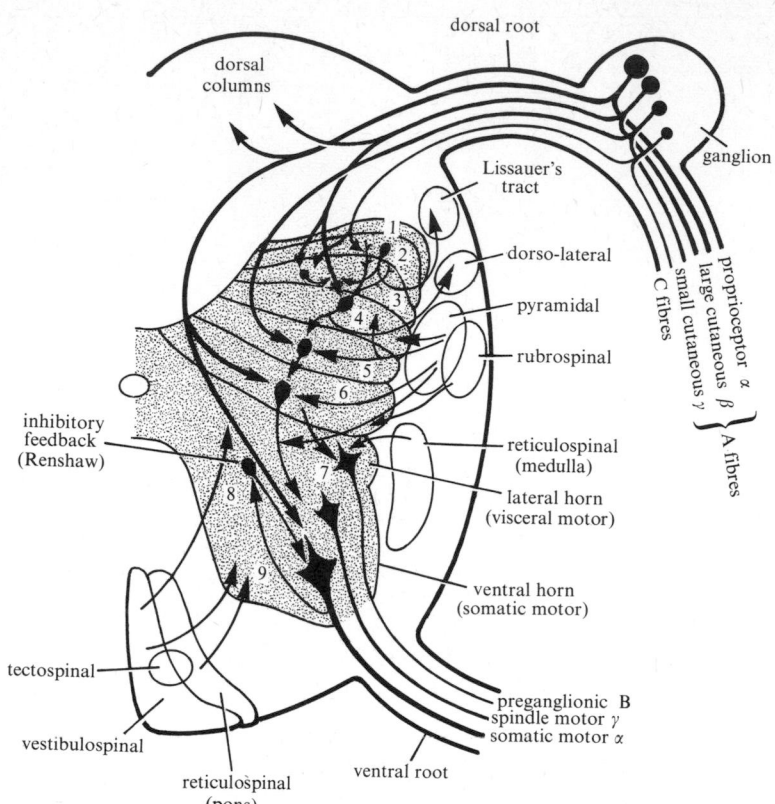

Fig. 27.6. Diagram of some of the fibres and tracts of the spinal cord of the cat (lumbar region). The grey matter is divided into nine laminae (1–9).

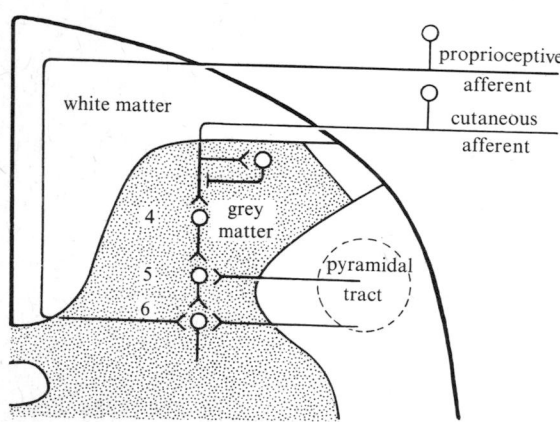

Fig. 27.7. Diagram of connexions in the dorsal horn of the spinal cord.
The cutaneous afferents are shown ending on cells of lamina 4 with a presynaptic control exerted on them by small dorsal cells. Lamina 5 cells are affected by lamina 4 cells. Incoming fibres that respond to passive movement of the limb (proprioceptive afferents) end on lamina 6 cells, which also receive from lamina 5. All laminae are affected directly or indirectly by descending impulses from the brain stem and pyramidal tract. (After Wall 1967.)

4. The receptor laminae of the dorsal horn

Lamina 1 is a thin sheet of rather large cells covering the surface of the dorsal horn. They may be concerned with relaying impulses from fibres signalling trauma (pain fibres). Lamina 2 consists of tightly packed small cells, a tissue that was early recognized by its curious microscopic appearance as the 'substantia gelatinosa Rolandi'. The cells probably receive the endings of small diameter fibres from the skin. Their axons end within this same lamina or in lamina 3, and their action serves to modulate transmission from the same afferent fibres that activate them (Fig. 27.7). Some of the axons run to neighbouring segments as the bundle of fine axons known as Lissauer's tract (Fig. 27.6). Lamina 3 also consists of small cells, and contains the terminals of large diameter cutaneous afferents.

The cells of laminae 4–6 are larger and their axons run either to the more ventral parts of the cord or else turn cranially and run to the brain in white matter other than the dorsal columns. The dendrites of the large cells of lamina 4 reach up through the upper laminae and receive the endings of the incoming dorsal root fibres. Those from the specialized endings in the skin (for example, Pacinian corpuscles, hair follicles) terminate more dorsally on cells of lamina 4. Endings from the tissue afferents (vascular and pain afferents) end in lamina 5. The proprioceptors from the muscles end more ventrally in lamina 6.

Recordings can be made from these large cells while the skin is being stimulated, say by stroking or pressure in a decerebrate or anaesthetized animal. Each cell of lamina 4 will respond to gentle pressure given within one small area of skin. They are regularly arranged, with those whose areas are more distal on the limbs or more ventral on the body placed medially. The cells of lamina 5 have larger fields and are probably stimulated by impulses from lamina 4 cells, as well as by tissue afferents. If the stimulus is repeated the response of lamina 5 cells rapidly declines, therefore they are said to be 'novelty detectors'.

Although all of these cells respond to all types of cutaneous stimuli, the separate qualities such as touch, pressure, and pain are probably discriminated by a 'gate control'. The axons of the small cells of the substantia gelatinosa serve to inhibit the effects of the incoming axons on the large cells, perhaps by presynaptic inhibition (p. 253). These incoming fibres are of various sizes, probably connected with different receptors (p. 259 and Fig. 27.7). The large afferent fibres, besides sending endings to the large cells, also send endings to the small ones, activating them and thus stimulating them to suppress excitation of the large cells. Conversely, the small afferent fibres inhibit the small cells and thus open the gate to the large cells. The output of the latter thus varies from short and sharp to prolonged discharge according to the combination of input, and the impulse patterns ascending to the brain vary accordingly and are 'recognized' as, say, touch or pain (p. 368).

28 Servo-control of movement

1. The stretch reflex

THROUGHOUT life a continual stream of nerve impulses proceeds from the spinal cord to the muscles, serving for the regulation of posture and the initiation of movement. Many influences play upon the ventral horn cells, increasing and decreasing the frequency of the impulses that are sent to particular muscles. The pattern of these discharges is set by the afferent impulses arriving from the many receptors. For example, maintenance of the standing posture depends upon impulses from the labyrinth of the ear, eyes, proprioceptors of the neck and limbs, and tactile organs in the feet (Chapter 31). Again, the initiation of walking movements may depend upon change affecting any of the afferent fields (say in the nose and eyes) and the combination of this information with that stored in the cerebral cortex.

All muscular contraction, whatever its ultimate source of initiation, requires to be performed in an orderly manner, and this is ensured by mechanisms situated primarily at the spinal level. The muscles are capable of producing very great forces. A maximal discharge of nerve impulses reaching the muscles of the human arm as a synchronous volley is capable of producing a tension approaching 1000 kg in 20 ms on the short arm of the lever. This would produce a very jerky start and an enormous kinetic energy in the arm, followed by an overfling. Effective movements depend upon proper limitation and control of these great forces that the muscles can produce. Actions must start and stop gradually or be steadily maintained. All of this is ensured by a control system based fundamentally on feed-back from the proprioceptor organs within the muscles, which record the length and/or tension within the muscles. The impulses they discharge pass to the spinal cord and brain, allowing comparison to be made between the state of the muscles and the instruction provided by the brain, which is itself dependent on the impulses received from the outside world by the receptors.

The system for making this comparison is elaborate and is imperfectly understood; at the spinal level it constitutes the *stretch reflex* (myotactic reflex), which ensures that a muscle contracts when the tension upon it is increased. This tension can be shown to depend upon receptors in the muscle by an experiment in which the effect of the brain on the spinal cord is first removed by transection to eliminate the influence of the higher centres (decerebration). The quadriceps muscle is then attached to a myograph to record its tension (Fig. 28.1). The table on which the cat lies is quickly lowered, as indicated by the line T, and the tension in the muscle increases markedly (line M). If now the nerve to the muscle is cut the increase of tension is only about one-fifth of that with the nerve intact (line P). The remaining four-fifths of the tension must be the result of nerve impulses set up in the muscle by the stretch, producing increased discharge of the motor neurons. That this is so is confirmed by cutting the dorsal roots, which has the same effect as severing the

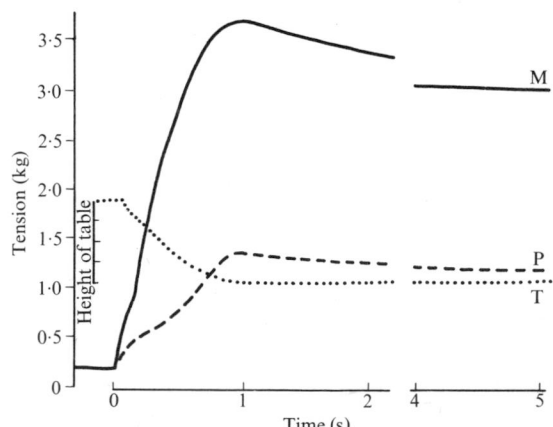

FIG. 28.1. Recording of stretch reflexes in a decerebrate cat. The quadriceps muscle is attached to a lever whose point moves upwards when the muscle contracts. The table on which the animal lies is dropped as shown by the line T, stretching the muscle. M shows the tension change with the nerve intact, P when it has been cut. (After Liddell and Sherrington (1924). *Proc. R. Soc. B.* **96.**)

whole nerve supply to the muscle. Therefore there must be proprioceptor organs within the muscle that are able to provide impulses when the muscle is stretched.

2. Sensory innervation of muscle

About 40 per cent of the myelinated nerve fibres innervating skeletal muscle are sensory, and histograms of their diameters generally show three peaks consisting of large, medium, and small fibres designated as groups I, II, and III. Unmyelinated sensory fibres outnumber

the myelinated fibres, sensory and motor, by about 2 to 1 and are recognized as group IV. A majority of the smaller sensory fibres (some group II and III, and all of group IV) terminate as free endings on blood-vessels and in fat and connective tissue. These are nociceptors (pain fibres) responding to deep pressure or squeezing the muscle, and compare with similar receptors found in skin. Exclusive to muscle, however, are two types of mechano-receptors responding to stretch: namely, *tendon organs*, in series with the contracting muscle

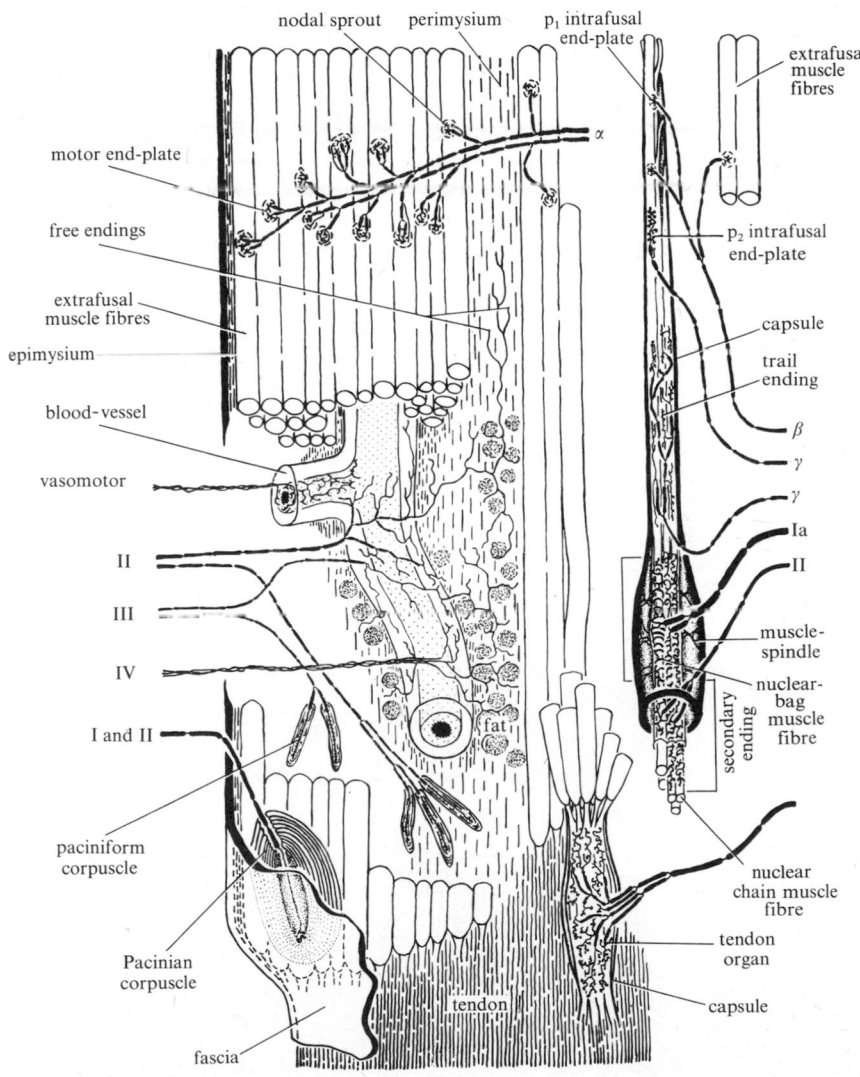

FIG. 28.2. Schema of the innervation of mammalian skeletal muscle based on a study of cat hind-limb muscles. Those nerve fibres shown on the right of the diagram are exclusively concerned with muscle innervation; those on the left also take part in the innervation of other tissues. Roman numerals refer to the groups of myelinated (I, II, III) and unmyelinated (IV) sensory fibres; Greek letters refer to motor fibres. The spindle pole is cut short to about half its length, the extra-capsular portion being omitted. (Figure kindly supplied by Prof. D. Barker.)

fibres, and *muscle-spindles* that are in parallel with them (Fig. 28.2). These receptors receive most of the group I and II fibres. Muscles also contain a few paciniform corpuscles, such as occur in joint capsules and periarticular tissue, while their much larger cousins, the Pacinian corpuscles, may occasionally occur underneath muscle fascia (see Barker 1974*b*).

Muscle-spindles are particularly conspicuous in small muscles concerned with the regulation of fine movements, such as the muscles of the hands. They are fewer in the large muscles of the more proximal joints. Thus there are 29 spindles per gram of muscle in the human abductor pollicis but only 1·4 spindles per gram in latissimus dorsi.

Tendon organs each consist of a small encapsulated bundle of tendon fibres innervated by a group Ib sensory nerve fibre. A muscle-spindle is composed of a bundle of small striated muscle fibres which for part of their length are enclosed within a fluid-filled capsule. Where they pass through this encapsulated 'equatorial'

region each muscle fibre is full of nuclei but with very few myofibrils, so that this short length, is virtually non-contractile. A group Ia sensory nerve fibre (up to 22 μm diameter) forms a series of rings and spirals (primary ending) around these nucleated portions, whilst on either side, both within and outside the capsule, the 'poles' of the spindle receive a complex motor innervation. The receptor nerve fibre is therefore wound around a region that can be passively stretched by pulling on the spindle, or actively stretched as a result of the contraction of its two poles.

Most spindles also receive another type of sensory ending, less regular in form that the primary ending and supplied with a group II nerve fibre (4–12 μm). These secondary endings lie along the poles mainly of the smallest muscle fibres in the spindle, in which the equatorial nuclei are arranged in a single row (hence nuclear-chain fibres). There are other larger muscle fibres in which the equatorial nuclei aggregate together to form a bag (nuclear-bag fibres). The primary ending

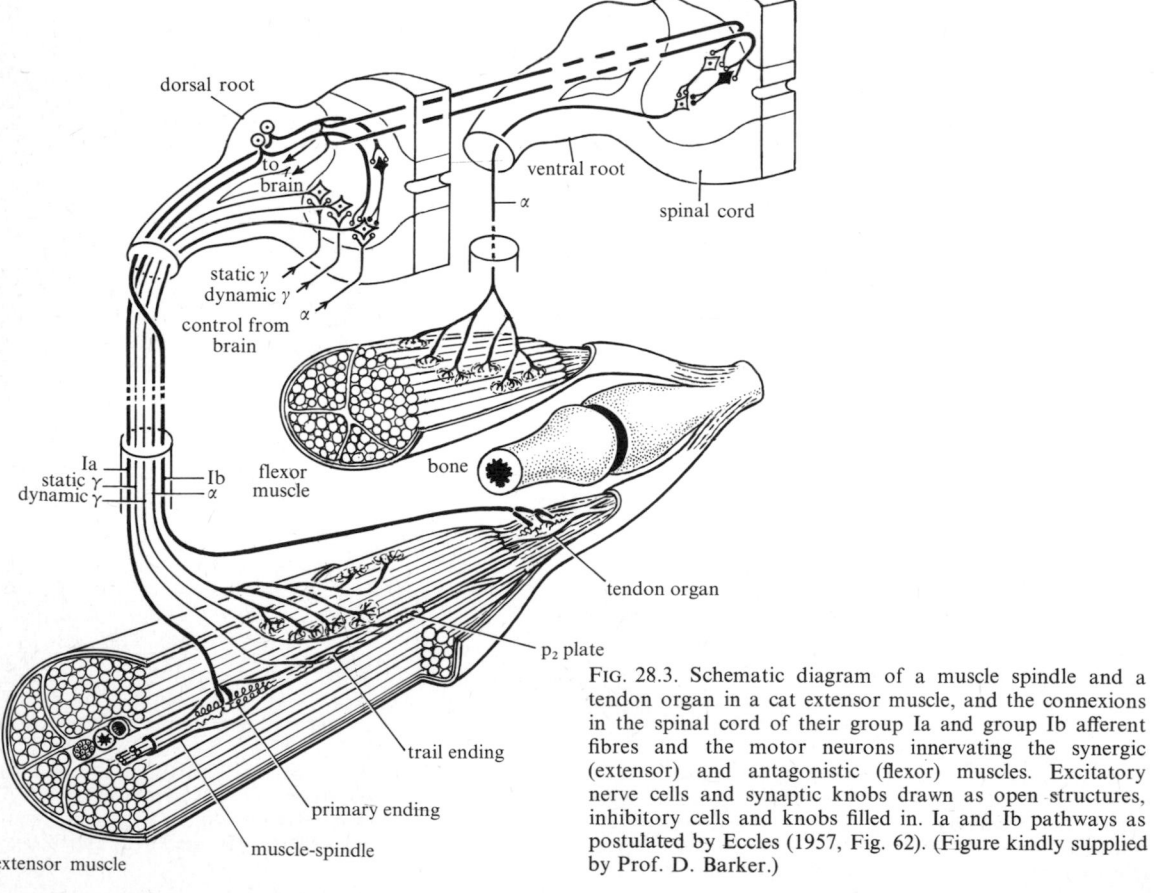

FIG. 28.3. Schematic diagram of a muscle spindle and a tendon organ in a cat extensor muscle, and the connexions in the spinal cord of their group Ia and group Ib afferent fibres and the motor neurons innervating the synergic (extensor) and antagonistic (flexor) muscles. Excitatory nerve cells and synaptic knobs drawn as open structures, inhibitory cells and knobs filled in. Ia and Ib pathways as postulated by Eccles (1957, Fig. 62). (Figure kindly supplied by Prof. D. Barker.)

is distributed to both the bag and the chain fibres.

Recordings made from single group Ia fibres show that a volley of impulses is discharged when the muscle is stretched, and this is responsible for the stretch reflex, ensuring that a muscle contracts when it is pulled upon. The large group Ia afferents enter the dorsal horns and send branches that reach directly to the ventral horn cells (Fig. 28.3). By this *monosynaptic reflex*, impulses are set up in the ventral horn cells with a delay shorter than is found in any other reflexes. Ascending branches of the group Ia afferents pass in the dorsal columns to the cerebellum, which also plays a large part in the regulation of muscular action (p. 298). The secondary endings have a higher threshold of response to muscle stretch than the primaries, and the reflex pathway of their afferents includes an intermediate neuron. There is evidence which suggests that they also play an excitatory role in the stretch reflex.

The discharge of impulses over the monosynaptic reflex pathway is a factor of first importance in regulating the discharge of the motor neurons. If the motor nerve fibres to the ordinary muscle fibres are stimulated electrically, the discharge of impulses from a stretched spindle is interrupted while the muscle is shortening (Fig. 28.4). The effect of this 'silent period' is to reduce the flow of excitation reaching the muscle over the monosynaptic pathway during any period in which it is shortening. Suppression of this excitation, as well as that produced by the secondary afferents, is further ensured by impulses discharged by the tendon organs, since the reflex effect of their activity is to inhibit the contraction of the muscle in which they are located (Fig. 28.3). This is an important part of the mechanism by which the tension in the muscle is adjusted to the task it is called upon to perform.

The motor nerve fibres supplying the spindle ('intrafusal') muscle fibres emerge in the ventral roots with the motor fibres that innervate the ordinary ('extrafusal') muscle fibres, but whereas the latter are large 'alpha' fibres (up to 20 μm) and conduct rapidly (50–120 ms^{-1} in the cat), those that control the spindle are smaller (2–8 μm) and conduct between 10 ms^{-1} and 40 ms^{-1}. These small 'gamma' fibres can be stimulated electrically under experimental conditions without stimulation of the larger alpha fibres, and the result is found to be a discharge of the afferent nerve impulses from the spindles (Fig. 28.4(b)). If alpha and gamma fibres are stimulated together there is a continuous discharge of afferent impulses (Fig. 28.4(c)). This is presumably the condition that occurs in life when a muscle contracts against resistance and the flow of afferent impulses serves to maintain the muscle in a state of tetanic contraction.

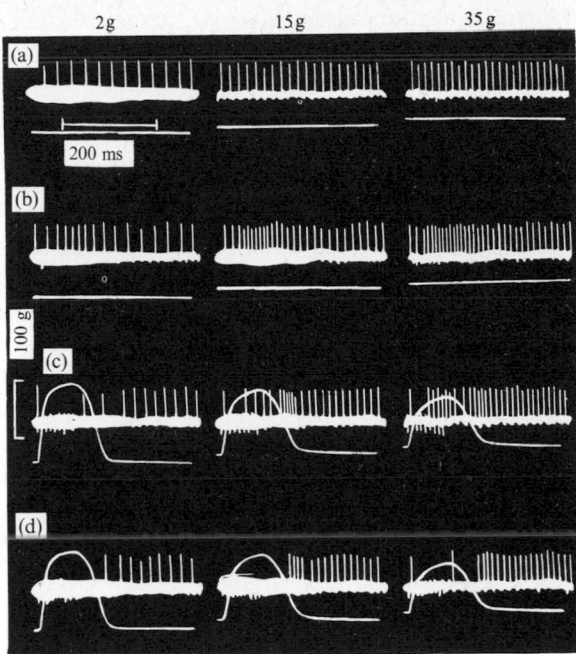

FIG. 28.4. Discharge of nerve impulses in a single nerve fibre from a muscle-spindle of the cat. The motor nerve has been cut so that there is no reflex discharge through the spinal cord. The white line below records the pull of the muscle upon a strain gauge.

Row (a) shows the increasing frequency of discharge under three levels of tension. In row (b) is shown the effect of electrical stimulation of a single small motor nerve fibre; no tension is produced but the afferent discharge is increased. In row (c) large and small motor nerve fibres are stimulated together and the muscle contracts. Under 2 g load there is a pause in the discharge but under the higher loadings the spindle discharge is maintained throughout. In row (d) only the large motor nerve fibres are stimulated: during the contraction there is a pause in the afferent discharge. (From Hunt and Kuffler (1951). *J. Physiol.* **113.**)

The muscle-spindle is sensitive to muscle stretch at constant length (static stretch) as well as to the rate of change of length during stretch (dynamic stretch). Primary afferent endings respond to each component of the stretch stimulus so as to give both a static and a dynamic response, the latter being characterized by a marked increase in the frequency of discharge (Fig. 28.4(b)). Secondary endings give a static response only, being essentially length indicators. The behaviour of these two types of response has been studied in experiments in which gamma fibres are stimulated so as to bring about contraction of the intrafusal muscle fibres at the same time that the muscle is stretched. Recordings of the discharge of the primary ending under these conditions show that stimulation of some of the gamma fibres greatly increases the dynamic response while stimulation of others decreases it. The gamma fibres are thus of two functionally distinct kinds and are designated as dynamic and static (Matthews 1972). At constant lengths of muscle stretch, both kinds of fibre increase the static response of the primary ending, though the rate of discharge is increased more by the static fibres (Fig. 29.1, p. 272). Secondary endings are activated by the static fibres only. In the cat, static gamma fibres are about 3 times more numerous than dynamic gamma fibres.

Static and dynamic control of the afferent response of the spindle in the frog and other lower vertebrates is exercised by branches of some of the same motor fibres that innervate extrafusal muscle fibres. Independent control by gamma motor fibres is an exclusively mammalian feature, as is the presence of secondary endings. The independence is not complete, however, for among the motor fibres that innervate mammalian muscle there are some that branch to supply both intrafusal and extrafusal muscle fibres. In the cat these 'beta' fibres conduct slowly ($31-61$ ms^{-1}), and the effect of the contractions produced by their intrafusal branches is to increase the dynamic response of the primary ending. They terminate in the spindles as motor endplates ('p$_1$ plates'), while the dynamic gamma fibres form larger and more elaborate plates ('p$_2$ plates') and the static gamma fibres form diffuse, multi-terminal 'trail' endings (Barker, Stacey, and Adal 1970). Every spindle receives trail endings and either one or both types of plates (Fig. 28.2). A low frequency of p$_1$ plates among the spindles in a muscle is offset by a high frequency of p$_2$ plates, and vice versa, dynamic control being provided mainly by gamma fibres in some muscles and mainly by beta fibres in others.

The three types of motor ending in spindles are not selectively distributed to the nuclear-bag and nuclear-chain muscle fibres. Bag fibres, for example, may receive both types of plate ending as well as trail endings. This implies either that the different types of motor ending supplied to the same intrafusal muscle fibre can cause it to contract in different ways (recalling the situation that obtains in arthropod muscle); or that each type of intrafusal muscle fibre gives only one kind of contraction regardless of the type of ending initiating it. The evidence for these alternatives is at present equivocal (see Barker 1974*a*).

3. Control of muscles through their afferents

The muscle-spindles are thus receptors in a feed-back system. In the absence of stimulation from their motor fibres the spindles discharge when the muscle is stretched, this discharge ceasing when the muscle itself contracts and they are unloaded. The brain, by activating the spindle motor fibres, not only causes the spindle to be active during muscle contraction but also increases the response of its sensory endings which relay the information about the length of the muscle to the central nervous system. The spindle thus signals the difference between the muscle shortening and the intrafusal shortening. If the spindle shortens more quickly than the muscle its discharge increases. Observation in man shows that the spindle discharge always increases during a 'voluntary' contraction. Since the effect of the afferent impulses is to produce a degree of reflex contraction, this means that the spindle motor fibres are agents that drive the muscle in addition to the alpha fibres.

The influence of the brain on the muscular system is exerted therefore not only by fibres that control the muscle directly but also by the spindle motor fibres. Evidence for this can be obtained by inserting stimulating electrodes into different regions of the cat's brainstem and recording their effects on the afferent discharge of spindles. Stimulation of some points produces increase of frequency, others decrease, at all levels and rates of muscle extension. Static and dynamic gamma fibres are controlled by different centres whose activities are capable of varying the bias of the spindles over their whole physiological range. In this way it is ensured that the instruction proceeding from the higher levels of the brain is translated into movements that are effectively adjusted to the resistance that is met with at each phase of the action. The system involves many further components at both spinal and cerebral levels, ensuring minimum transient errors and absence of oscillation.

The functioning of a feed-back system of this sort must depend upon the proper timing of the arrival of the signals. If the information about the tension in the muscle is to be effective it must be quickly available and the group Ia afferents of the spindles and group Ib afferents of the tendon organs are among the largest in the body and therefore the fastest conducting. There are also two fast-conducting pathways leading from the brain-stem down the spinal cord that relay the descending control of static and dynamic gamma fibres (Fig. 28.4).

4. Reciprocal innervation

Nearly all muscles are arranged in pairs that work in opposite directions. The classic examples are the flexors and extensors of the limbs. Obviously, when one of these antagonists contracts the other must relax and this achieved by reciprocal inhibition. In the example shown, stimulation of a nociceptive fibre in the skin of the foot produces contraction of the flexor of the knee on that side and relaxation of the extensor. On the other side the effects are reversed, thus stiffening that leg. Reciprocal innervation involves a complicated pattern of action of the proprioceptors (Fig. 28.5). The tendon organs (Ib) discharge when a muscle contracts against resistance, and this produces disynaptic inhibition (through an interneuron) of the muscles' own motor neurons (shown on the right side only). When a muscle is lengthened by gravity or the action of the opposing muscles the group Ia axons of the primary endings elicit the monosynaptic contraction of the same muscle and inhibition of its antagonist. These supply information not only about the length of the muscle but also of the velocity with which it is being extended. The group II axons from the secondary endings of the spindle provide information only about the length; their central connexions are not known and they, and those of the afferents from the joint therefore are not shown in the diagram. In the absence of fusimotor discharge active shortening of a muscle unloads the spindles and silences their discharge. But if the fusimotor neurons (*d* and *s* in Fig. 28.5) are activated, either reflexly or from the brain, then the group Ia input can continue (Phillips 1970).

The static receptors (*s*) discharge under brain influences when the muscle is stationary. The dynamic receptors (*d*) are activated when the muscle is lengthened. If the fusimotor cells are silent, contraction of the muscle unloads the spindles and their discharge stops.

But a 'command' to the fusimotor neurons, either reflex or from the brain, will allow the group Ia input to increase the effect of commands to the motor neuron and inhibit its antagonists. If the movement is resisted, the spindle discharge will increase and support any command to the motor neuron. Fig. 28.5 also shows how collaterals of the motor neuron axons, acting through smaller interneurons known as Renshaw cells (R), inhibit the motor neuron itself and also inhibit the inhibitor of its antagonist. Moreover, if the muscle cannot shorten the spindles are not unloaded, and there is thus a servo-mechanism that adjusts the contraction to the external load. The whole gamut of posture and movement is regulated in this way by the built-in connexions in the spinal cord and the influences from the brain upon them. The muscles are, of course, differentiated into those mainly concerned with movement (fast or 'white' muscles, p. 68) and those for posture (slow or 'red' muscles). The stretch reflex can hold a limb in a constant position for a long time by the red fibres, whose motor neurons discharge at a low rate (5–10 per second). The muscle fibres relax slowly and the tension is maintained at low energy cost.

The basic reflex actions of a cat are organized within the spinal cord and can be performed when it is isolated from the brain. This includes actions of the somatic musculature, such as standing or walking, and many visceral activities, such as micturition or sexual acts. The brain normally regulates behaviour by excitation or inhibition of these spinal or 'segmental' reflexes. In man the spinal cord has less independence.

5. The servo-control of movement

The accuracy with which movements are performed is influenced by many factors. Every movement follows some change inside or outside the organism resulting in a 'command' to the muscle either through a spinal reflex or from the brain. To be effective the movement must start at the right times and proceed with the correct speed and force. The efforts of engineers to produce controlled movements that have these properties have found that this can be achieved by a 'servo-system', in which the output effect is measured and the result used to regulate the input. The proprioceptors are obviously such servo-systems and there has been much discussion of how they influence the commands to muscles. Total de-afferentation of a limb does not completely prevent its use, but fine movements are grossly impaired. It cannot be that the command from the brain to the muscle is exercised only by increasing the bias on the muscle-spindles and hence a discharge over the monosynaptic group Ia pathway. Nevertheless, this is a very effective servo-system for regulating the

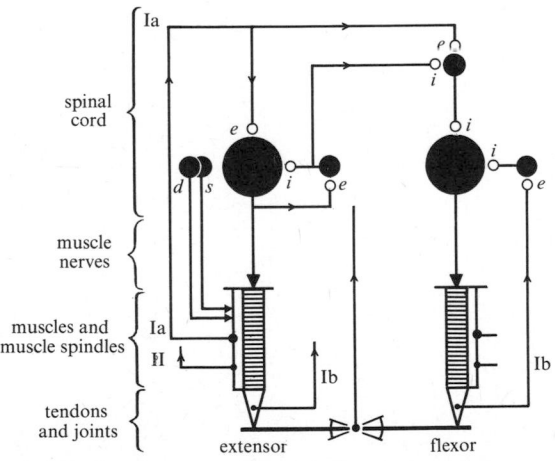

FIG. 28.5. Diagram of reciprocal innervation. Flexor and extensor muscles are shown as if attached to a seesaw, whose bearing represents a joint. The neurons innervating them are shown as black circles. Excitatory synapses are marked *e*, inhibitory ones *i*. See text for further explanation. (After Phillips (1970). *Paraplegia* **8**.) Reproduced by permission of E. & S. Livingstone.

extent of the contraction. The muscle shortens until the group Ia discharge returns to its original value. Probably in life the commands from the brain are partly direct to the motor neurons and partly by adjusting the bias on the servo-loop through the motor neurons. In a few situations the detailed timing of the various elements involved has been studied. In man, the consciously produced movements of the fingers have been monitored experimentally, including recording of the discharge within their nerves by electrodes (Merton 1951). The main contraction generally begins before the acceleration in spindle firing and is therefore a direct effect of the cortex on the motor neurons (p. 324). Yet a servo-system is undoubtedly also involved, though there is not yet enough evidence to show how the information from the muscles is used. In engineering practice the gain of a servo is generally made large, so that commands are followed in spite of any obstruction. High gain, however, brings increasing liability to overshoot and oscillation, and the chief agent producing this is the delay between the setting up of the error signal and the appropriate correcting action. In living servo-loops the rapid conduction in the large fibres of the monosynaptic loop obviously helps to reduce this danger, but the slow development of tension in the muscles is a major factor. In artificial servos, the danger of oscillation is reduced by introducing an element giving a response proportioned to the rate of change of the signal. The primary spindle endings do this by responding both to length and velocity of stretch. The velocity response thus 'predicts' the length of the muscle after delay time of the reflex and so ensures that it will be appropriate (Matthews 1972).

The precise way in which the complicated apparatus of primary and secondary endings of the muscle-spindles operate remains to be discovered. It is evident that the body has developed a system more subtle than a simple servo with a high gain. The commands issued are an anticipation of the force needed (p. 324) (Evarts 1967) and are then modulated by servos with gains also adjustable according to experience. This produces springy gaits and delicate movements rather than a series of jerky actions following stereotyped patterns.

29 The organization of the brain

1. The functional columns in the brain

THE presence of the four functional columns provides a convenient frame of reference around which to organize knowledge of the brain. The brain is the part of the neural tube that is modified by development of the various columns at places where there is an inflow of fibres connected with special receptor systems or where there are special motor centres. The general arrangement into the four functional columns seen in the spinal cord is preserved but the higher ('supra-segmental') centres, such as the cerebral cortex and cerebellum, cause much distortion of the simple tubular plan (Nauta and Karten 1970).

In the *medulla oblongata* (Fig. 29.1) the spinal plan can be clearly seen but the nervous tissue is spread out laterally, leaving an extensive roof of thin non-nervous tissue, the *choroid plexus*. In front of it the *cerebellum* represents a great development of the somatic sensory column, connected with the inflow of nerve fibres of the eighth cranial nerve concerned with balance. The medulla and cerebellum constitute the *hind brain* (metencephalon).

In front of this lies the *midbrain* (mesencephalon). Here the somatic sensory columns form the *corpora quadrigemina*, which are higher centres concerned with impulses from the optic and auditory nerves. They are relatively smaller in mammals than in lower vertebrates. The somatic motor column is especially developed in the floor of the midbrain as the *tegmentum*, a region that has important influences in controlling the pattern of motor behaviour.

The *forebrain* is greatly developed in all mammals and includes the unpaired between-brain (*diencephalon*) and the cerebral hemispheres (*telencephalon*). The more dorsal part of the diencephalon includes the *thalamus*, where fibres from the various receptor systems make synapse on their way to the cerebral cortex. The ventral part of the diencephalon (*hypothalamus*) represents the front end of the visceral sensory and visceral motor columns and serves to regulate many of the internal (visceral) activities. The *pituitary body* (hypophysis) is

attached to the lower side of the hypothalamus by a stalk (infundibulum).

Besides the regions of the brain that are conspicuous externally the central core is also specially developed as the *reticular system* (Ramón-Moliner and Nauta 1966). This by its nature is difficult to define anatomically. It consists of a set of mainly large neurons in the central grey matter reaching from the medulla oblongata to the hypothalamus and forebrain. The neurons have long, relatively straight dendrites (isodendrites), and the larger ones may reach for as much as 300 μm up and down the length of the brain, overlapping with those of other cells and with fibre tracts. They may have very large receptive fields. The processes of motor, sensory, and internuncial neurons are all mingled in this formation. Its functions are obscure, but it exercises some sort of general control over many brain activities. Stimulation by electrodes implanted in this region may produce activity of the animal or conversely sleep, hence it is called the *reticular activating system*. It has been suggested that this system is responsible for the maintenance of conscious awareness in man. Certainly it plays an important part in integrating all the activities of the brain to ensure homeostasis. In these generalized functions it may be distinguished from the specialized regions of the brain where the neurons have dendrites with special shapes, that is, curved, tufted, or pyramidal, and these are called allodendrites. The classification is rather coarse, but it calls attention to the presence of this isodendritic core and to the fact that the neurons of each part of the brain have their own peculiar features, such as those concerned with input discrimination or detailed output control (Pompeiano 1973).

The *cerebral hemispheres* represent an immense development of the front end of the somatic sensory column. This region was originally concerned with olfactory functions (as it still is today in fishes and amphibians). In mammals the olfactory centres are limited to the more ventral parts of the cerebral hemispheres. The remainder, the cerebral cortex proper, is

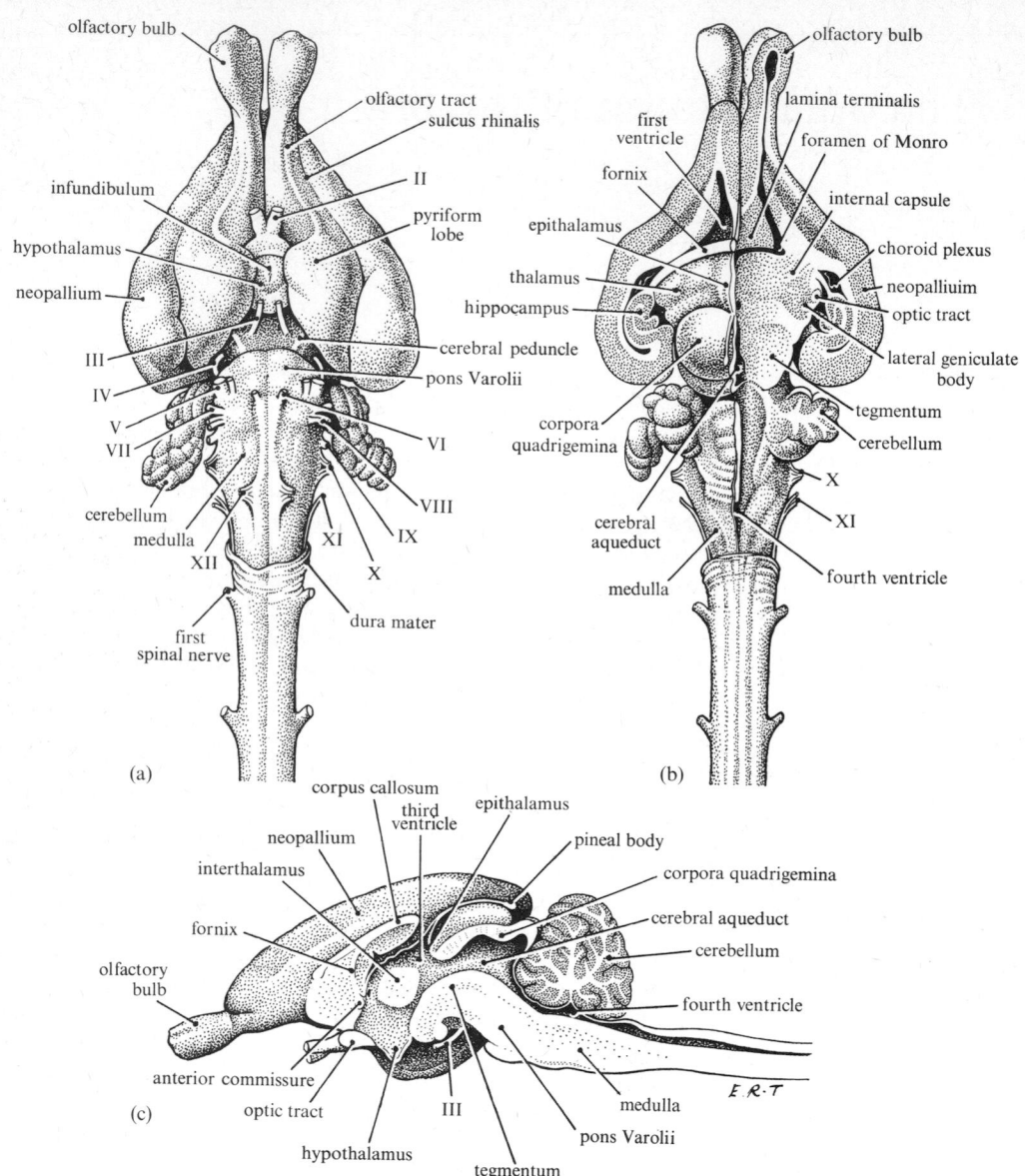

FIG. 29.1. Views of rabbit's brain.
(a) Ventral; (b) partially dissected from dorsal surface; (c) sagittal section.
II, optic nerve; III, oculomotor nerve; IV, trochlear nerve; V, trigeminal nerve; VI, abducent nerve; VII, facial nerve; VIII, vestibulo-cochlear nerve; IX, glosso-pharyngeal nerve; X, vagus nerve; XI, accessory nerve; XII, hypoglossal nerve.

made up of a vast number of neurons receiving impulses from the various receptor fields (eye, ear, skin, etc.) and serving to analyse the information that arrives from these receptors and to compute appropriate responses in the light of the memories of past events that are recorded there. The efferent pathways from the cerebral hemispheres reach to many parts of the brain (hypothalamus, tegmentum, cerebellum, spinal cord) and they dominate the whole action system of the

animal. The characteristic of the mammalian brain is thus that, instead of local centres, each concerned mainly with one receptor system, there is developed a very large cerebral cortex, which deals with afferent impulses of many sorts and has direct control of the muscles and other effectors.

The general plan of the brain is similar in all mammals but the cerebral hemispheres are relatively larger in the 'higher' forms. This development must have occurred independently in several distinct lines of mammals. In the living mammals that remain close to the earliest mammalian type, for example, shrews, the cerebral cortex is small, especially its more anterior (frontal) part, but there is a relatively large development of the olfactory parts of the forebrain. The brain of the rabbit (Fig. 29.1) shows some further development of the cortex but this remains relatively much smaller than in man (Fig. 29.2).

2. Ascending and descending pathways

Throughout the brain and spinal cord of mammals there is a great development of long pathways leading upwards from the primary sensory centres towards the cerebral hemispheres and downwards from the latter towards the spinal cord. Some of the lower centres become reduced as their functions are taken over by the forebrain. For example, the optic lobes of the midbrain, which are large and important in pre-mammalian forms such as the frog, are relatively small in mammals. On the other hand, some stations along both the ascending and descending pathways become greatly developed in mammals, for instance, the walls of the diencephalon (thalamus) are thickened because all the ascending pathways make synapse there. Similarly the red nucleus of the tegmentum and the inferior olive in the medulla, which are parts of the brain concerned with motor functions, become increased rather than reduced as the cortical control develops, and all of these reach an exceptional size in the higher primates.

The long pathways of the brain run in the white matter, which in the lower parts of the brain lies, as in the cord, around the grey matter. In certain parts these long tracts make thick strands that influence the external form of the brain. Thus the descending corticospinal fibres form a marked pair of columns, the *pyramids* (Fig. 29.3) along the lower side of the medulla oblongata. Beneath the cerebellum these join the tracts of the *pons Varolii*, which unite the cerebellum with the cerebrum. Together they form, beneath the midbrain and between-brain, an even larger pair of bundles, the *basis pedunculi*. The tracts ascending from the cord are externally visible where they make synapse in *nuclei of the dorsal columns* above the hind end of the medulla, but cranial to this level they run within the medulla and midbrain as large bundles close to the midline, known as the *medial lemniscus* (p. 278), eventually reaching the thalamus.

3. The cerebral ventricles, cerebrospinal fluid, and meninges

The development of special thickenings gives to the brain its curiously irregular shape. The central canal,

also, is not simple. In the medulla oblongata the canal forms the large *fourth ventricle*, then narrows again as the *cerebral aqueduct*, passing through the midbrain to the *third ventricle*, which occupies the centre of the diencephalon. From this cavity paired *interventricular foramina of Monro* lead to the cavities of the lateral ventricles. The ventricles are occupied by the *cerebrospinal fluid*, produced by the *choroid plexuses*. These latter are thin, vascular, folded membranes, formed from the roof of the ventricles. The anterior choroid plexus is an extensive membrane pushed through the foramina of Monro into the lateral ventricles. Here most of the cerebrospinal fluid is produced and passes through the foramina and backwards to the third and fourth ventricles. It leaves the fourth ventricle through three foramina beneath the cerebellum, which lead to *arachnoid spaces* between the inner membrane or *pia mater* that closely covers the brain and spinal cord and an outer, more loosely attached *dura mater*. The arachnoid space is crossed by a fine web of fibres, from which it gets its name, and here there are special regions (arachnoid villi) in which the cerebrospinal fluid is absorbed back into the venous system.

4. Neuroglia

The tissue of the brain is not, by any means, wholly made up of neurons. It contains many blood-vessels and also a set of cells, the *neuroglia*, even more numerous than the neurons, which are presumed to act in a supporting and nutritive capacity. These cells are of three types: *astrocytes*, *oligodendrocytes*, and *microcytes* (Figs 29.4 and 29.5). Together they make a packing tissue that fills up all the spaces between the nerve cells and fibres in the central nervous system. Their protoplasm is in close relation to the surfaces of the neural elements and may play a part in nourishing the latter and perhaps also in the specific functions of conduction and synaptic excitation. The neuroglia cells have a membrane potential of $+80$ mV, and this may change with activity of neighbouring neurons, perhaps related to the release of potassium ions. The glial protoplasm contains numerous mitochondria and is evidently an actively metabolizing tissue (see Glees and Meller 1968).

Astrocytes are classified as protoplasmic or fibrous, but these two are not sharply distinct. The protoplasmic type are found in grey matter and have thick, branched, spiny processes. The fibrous sort have longer, thin processes with little branching. They contain characteristic bundles of very fine fibrils. Both sorts send many end-feet to the blood-vessels, forming a covering layer over the basement membrane (Fig. 29.5). Nutrient substances do not need to pass through neuroglial

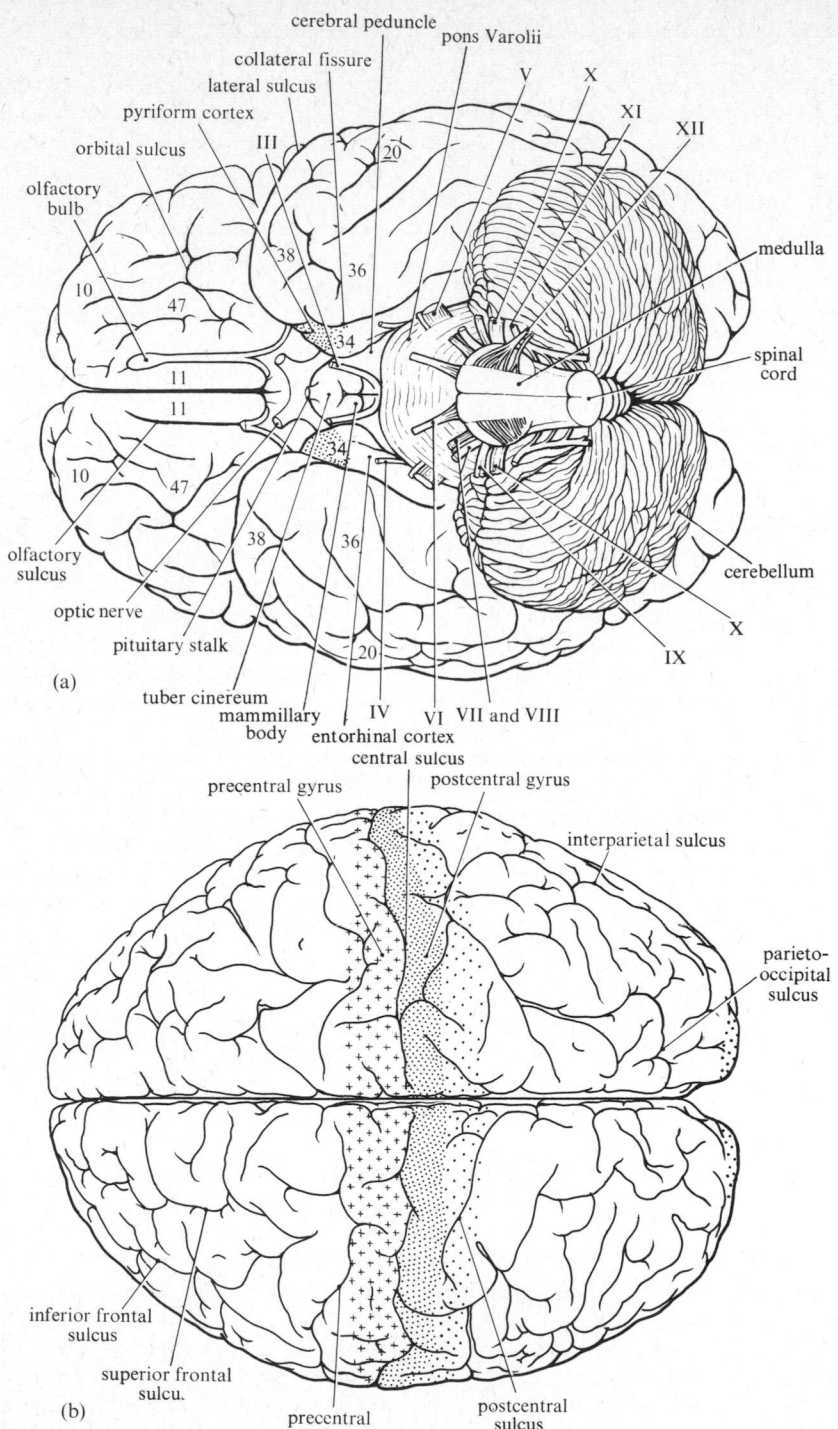

FIG. 29.2. The human brain seen (a) from below, (b) from above, (c) in median sagittal section, and (d) from the side. The numbers of the areas are those used by Brodmann. Primary afferent areas close stipple, secondary areas dotted. Crosses indicate motor cortex.

III, oculomotor nerve; IV, trochlear nerve; V, trigeminal nerve; VI, abducent nerve; VII and VIII, facial and vestibulo-cochlear nerves; IX, glosso-pharyngeal; X, vagus; XI, spinal accessory; XII, hypoglossal.

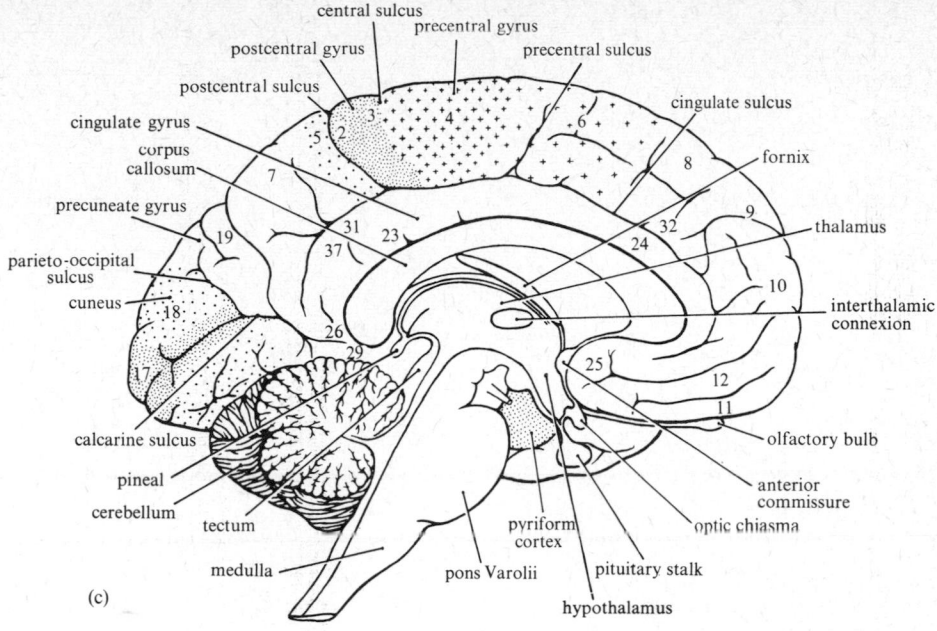

(c)

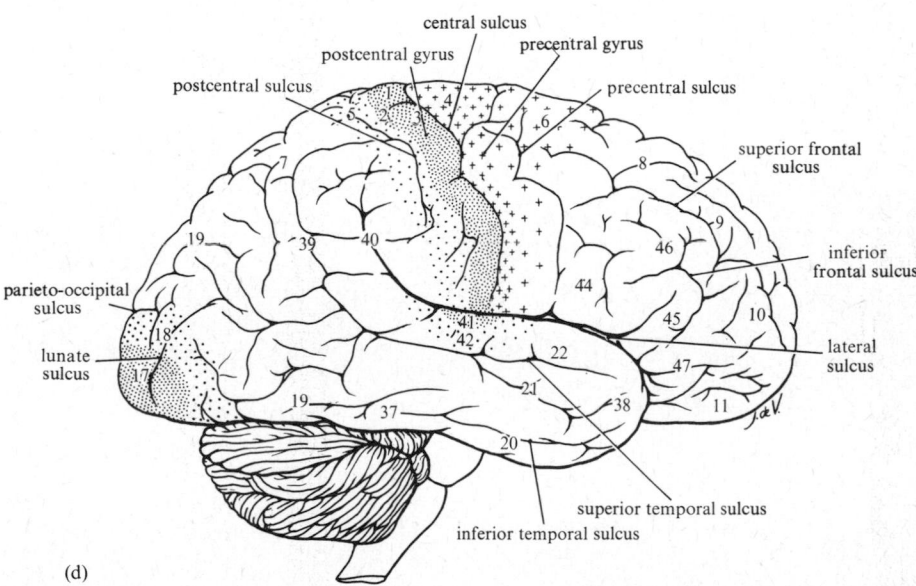

(d)

processes however, since there are open gaps between them at least 3–7 nm wide. The glia therefore do not function as a blood–brain barrier. Large particles, such as ferritin, however, do not pass either through or between the endothelial cells of the capillaries of the brain as they do in muscle (p. 169). This particularity of the cerebral vessels protects the nerve cells from the

action of many large molecules in the blood. The function of the glial cells is thus still uncertain. It may be that by restricting the distribution of potassium ions extruded during activity they allow complicated interactions between the neurons wrapped in the same glial compartment.

The oligodendroglia (= oligocytes) are especially

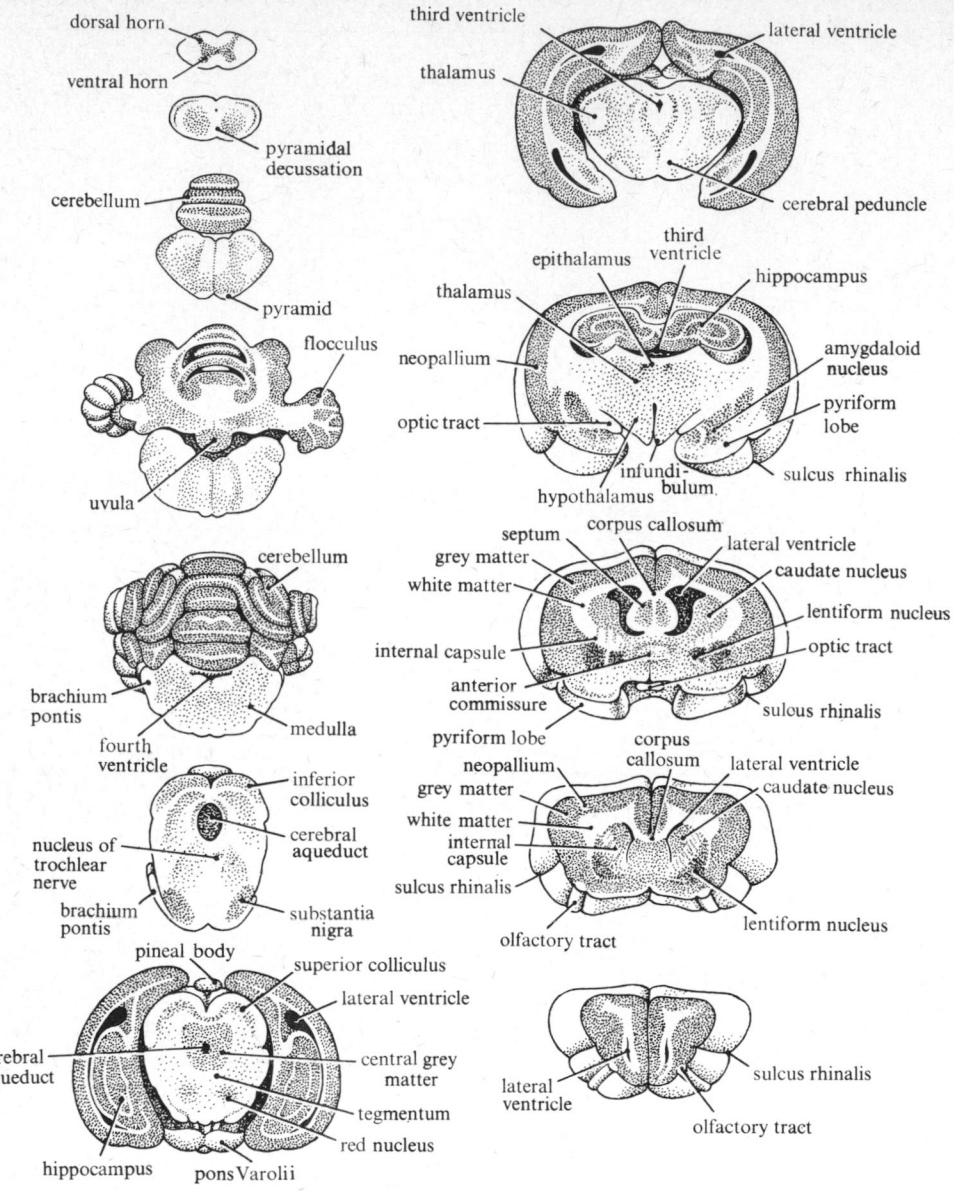

FIG. 29.3. Transverse hand sections of the rabbit's brain proceeding from back to front.

associated with myelinated fibres, but are found in both grey and white matter. Their protoplasm is more electron-dense than that of astrocytes. The oligocytes produce the myelin by wrapping round the fibres in a manner similar to the Schwann cells (p. 236).

Astrocytes and oligocytes are ectodermal cells but the microglia are mesodermal and enter the brain from outside. Their protoplasm is even more electron-dense

than the oligocytes. They correspond to the histiocytes found in other tissues (p. 183). If the brain is damaged their processes become phagocytic.

5. The medulla oblongata. Regulation of visceral activities
The medulla oblongata is the region of origin of the cranial nerves that innervate the muscles of the head and the viscera (p. 116). These nerves bring information

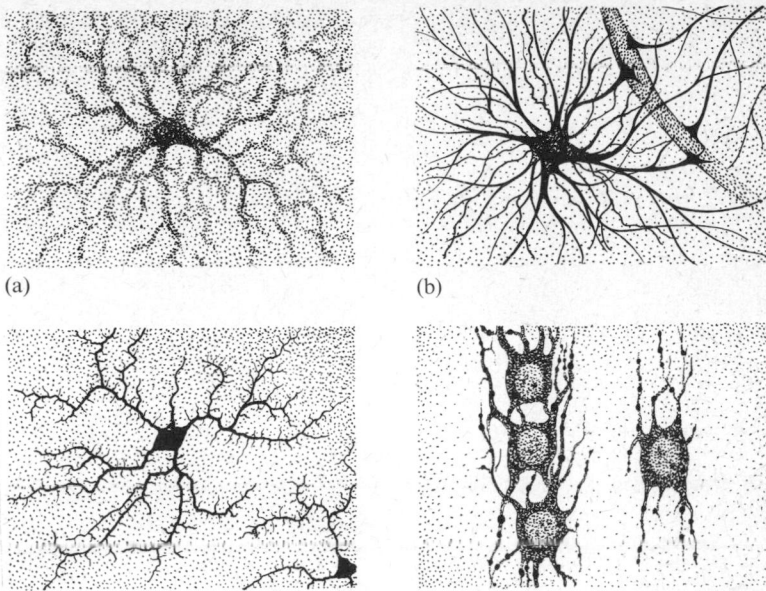

(a)

(b)

(c)

(d)

FIG. 29.4. Neuroglia, as seen by light microscopy after silver staining. (a) Protoplasmic astrocyte. (b) Fibrous astrocyte with vascular feet. (c) Microcytes (microglia). (d) Oligocytes (oligodendroglia).

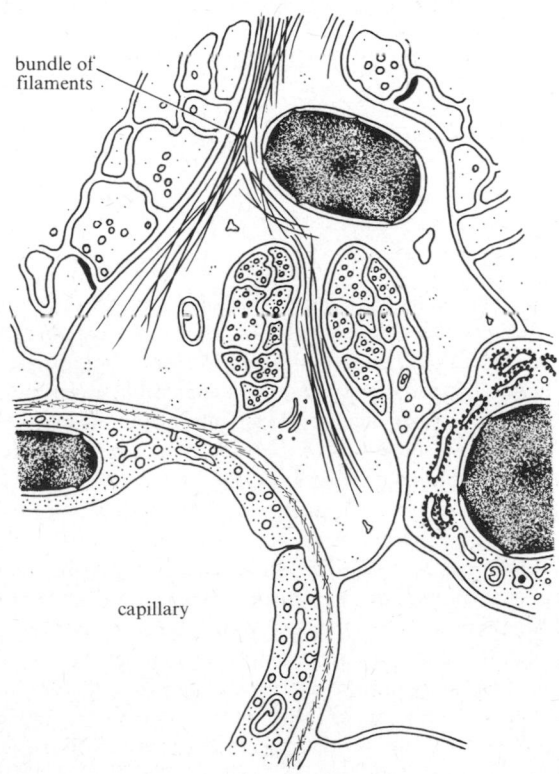

bundle of filaments

capillary

FIG. 29.5. Drawing of electron-microscope picture of fibrous astrocyte, characterized by sparse contents, with end feet on capillary. (Figure kindly supplied by Prof. E. G. Gray.)

from the viscera and send back motor impulses to them after appropriate synaptic action by the nerve cells aggregated as 'nuclei' within the medulla. This region thus has important receptor and effector functions and it shows the four main functional columns clearly. Indeed it retains more nearly the primitive organization than does any other part of the brain. Its roof is thin and non-nervous, but the functional columns can be recognized in the sides and floor (Fig. 29.6). The somatic sensory column is represented at the caudal end of the medulla by the large *nuclei of the dorsal columns* and more cranially by the nuclei connected with the somatic sensory systems of the cranial nerves, especially the trigeminal. The visceral sensory system is largely developed where the gustatory and other afferent fibres of the vagus and glossopharyngeus enter.

The motor nuclei of the medulla are complicated by the fact that in this region the dorsal and ventral roots do not join, and, moreover, the dorsal roots besides their afferents also carry efferent fibres to the musculature that is derived from the branchial arches (see p. 162). The somatic motor column is present in the hind and midbrain, lying medially as the sets of cells giving rise to the fibres of the ventral cranial roots, III, IV, VI, and XII. The cells of the fibres to the lateral plate musculature of the visceral arches arise in a special column somewhat lateral (dorsal) to the somatic motor nuclei, the axons running in the dorsal roots V, VII, IX, X, and XI. This column may be regarded as a special part of the somatic motor column and is some-

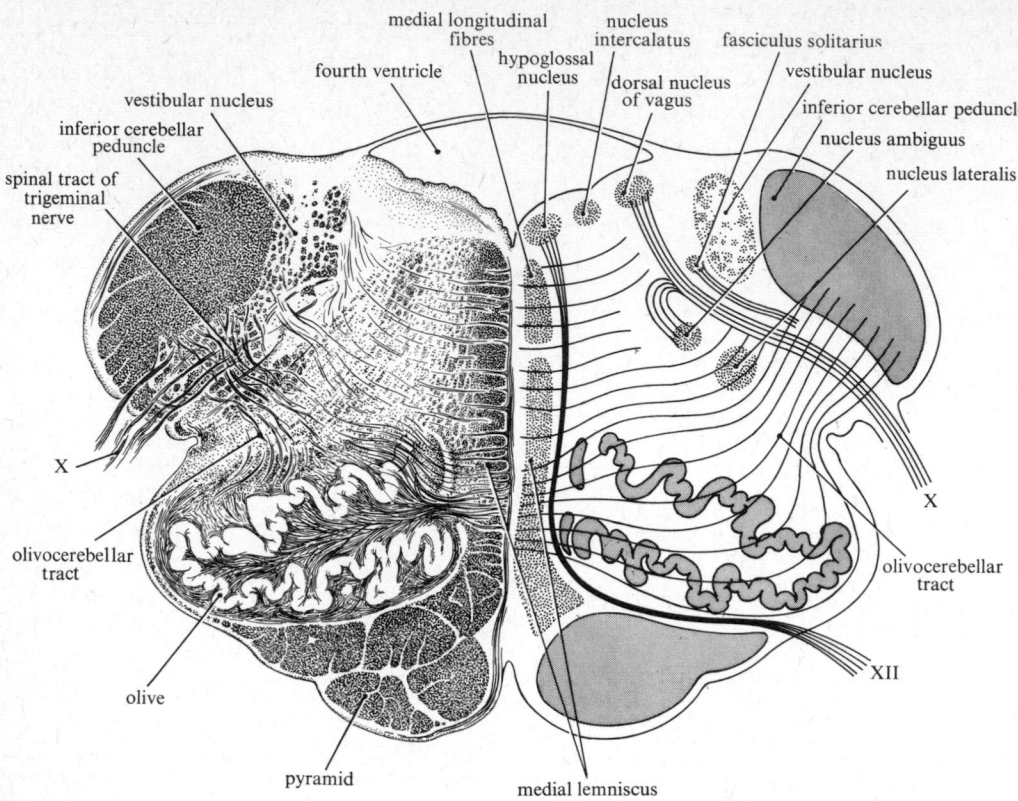

vestibular nucleus

inferior cerebellar
peduncle

spinal tract of
trigeminal
nerve

X

olivocerebellar
tract

olive

fourth ventricle

medial longitudinal
fibres

hypoglossal
nucleus

nucleus
intercalatus

dorsal nucleus
of vagus

fasciculus solitarius

vestibular nucleus

inferior cerebellar peduncle

nucleus ambiguus

nucleus lateralis

X

olivocerebellar
tract

XII

pyramid

medial lemniscus

FIG. 29.6. Transverse section through the medulla oblongata of man below the middle of the olive. X, vagus nerve; XII, hypoglossal nerve.

times called the *special visceral efferent* or branchial motor column. Its cells are large, like those of the somatic motor column in the cord. The similarity of function is emphasized by the collaboration of some of these muscles (sternomastoid and trapezius) with true somatic muscles in producing movement of the shoulder in man (p. 83).

The (general) visceral efferent column is represented in the medulla and the midbrain by the cells of origin of the preganglionic autonomic fibres that run in nerves III, VII, IX, and X (p. 162).

The detail of the organization of the medulla shows a bewildering variety of nuclei concerned with various functions. The appropriate level of activity of the heart, lungs, alimentary canal, and other viscera is computed by these nuclei in the light of signals received in afferents from these organs and from the hypothalamus and other parts of the brain. Local injuries in the medulla produce irregularity of many activities. There are groups of cells concerned with control of respiration, heart-rate, diuresis, sugar metabolism, and other functions. The

medulla, being the centre of origin of the fibres of the cranial nerves, has developed into a most elaborate system for the regulation of the life of the whole animal. No doubt it also provides, by interaction between these various centres, for the proper integration of many of the internal functions of the body. It is a region of the brain absolutely indispensable for the life of a mammal, and its cells are deeply involved in the production of the rhythm of action that dominates the life of every animal (p. 364). We can only dimly imagine the complicated activities that produce these various drives. Afferent impulses from the stomach, heart, and lungs, and elsewhere produce reflex effects in the medulla, but this region cannot be adequately described as a 'reflex centre'. Many of its cells do not come to rest, in particular those of the respiratory centres discharge rhythmically from birth to death (p. 156), though their periodicity is influenced by afferent discharges. Such rhythmical changes going on in the medulla and in other parts of the brain are as much a part of the 'life' of the animal as is the heartbeat.

30　The autonomic nervous system

1. The control of internal activities

THE somatic part of the nervous system controls those organs of the body that are in contact with the outside world. It serves to bring the body to places where food and drink can be obtained and enemies avoided; the striped musculature is the agent through which it works. For the continued maintenance of the body it is also necessary to have a mechanism by which the actions of the internal parts can be controlled and co-ordinated. In the simpler Metazoa the co-operation of the cells is mainly ensured by the effects that the working of each has upon others, but as the specialization of tissues proceeded during evolution an internal co-ordinating mechanism became developed, serving to ensure that there is proper adjustment of the functions of the organs to each other, the heart being made to beat faster when the muscles require more oxygen, the movements of the gut being accelerated during digestion and so on.

The visceral nervous system that accomplishes this co-ordination is already present in *Amphioxus*; throughout the chordate series it has become progressively elaborated, allowing more and more detailed control of the workings of the special parts that became developed as the animals colonized more and more 'difficult' habitats. This visceral nervous system works with an endocrine co-ordinating system of increasing complexity (Chapter 45). Study of the evolution of these two parts of the body therefore gives some indication of the nature of the change by which animals have become 'higher' or more complex.

The separation between outer (somatic) and inner (visceral) nervous systems is obviously largely artificial; they work together for the maintenance of the integrity of the body and can do so effectively only if their actions are co-ordinated. When the somatic nervous system sets the muscles into action, the heart and blood-vessels must adjust their output to provide a suitable flow of blood; moreover this blood must be made to carry enough sugar for fuel and enough oxygen to burn it. Nevertheless the visceral motor section of the nervous system forms a rather distinct anatomical unit, and this separation has been recognized by calling it the *autonomic nervous system*, although, in spite of its name, it is not independent in its working. According to the classical view it does not even allow the performance of reflex actions but is simply an arbitrarily separated portion, the visceral motor system (see p. 260). It receives its activation, as does the somatic motor division, from the central nervous system. The term 'involuntary nervous system' is also often used for this set of visceral nerves. It is true that we do not usually speak of exerting conscious control of the activities that it regulates, but by suitable training some people can obtain 'voluntary' control of these powers, for instance, they can narrow the pupil or quicken the heartbeat 'at will'. Conversely, many somatic actions are 'involuntary' (e.g. the knee jerk).

2. Preganglionic and postganglionic cells

The characteristic feature of the arrangement of the visceral motor nervous system is that the cell bodies of its final motor neurons are outside the central nervous system not within it as are the ventral horn cells of the somatic system (p. 261) (Kuntz 1953). The cells are aggregated into ganglia, which lie either along the course of the nerves or actually within the organs that they control. These peripheral autonomic motor neurons are controlled by nerve cells lying within the central nervous system. On every autonomic pathway, therefore, we can distinguish *preganglionic neurons* in the central nervous system, which send axons to make connexion with the *postganglionic neurons* of the peripheral ganglia (Fig. 30.1). In mammals the preganglionic axons are usually small medullated nerve fibres, whereas most of the postganglionic fibres are unmedullated.

The preganglionic synapses of both the sympathetic and parasympathetic fibres use the transmitter acetylcholine, whose action is paralysed by the alkaloids nicotine or curare or an analogue such as *d*-tubocurarine. This provides a convenient means of locating the

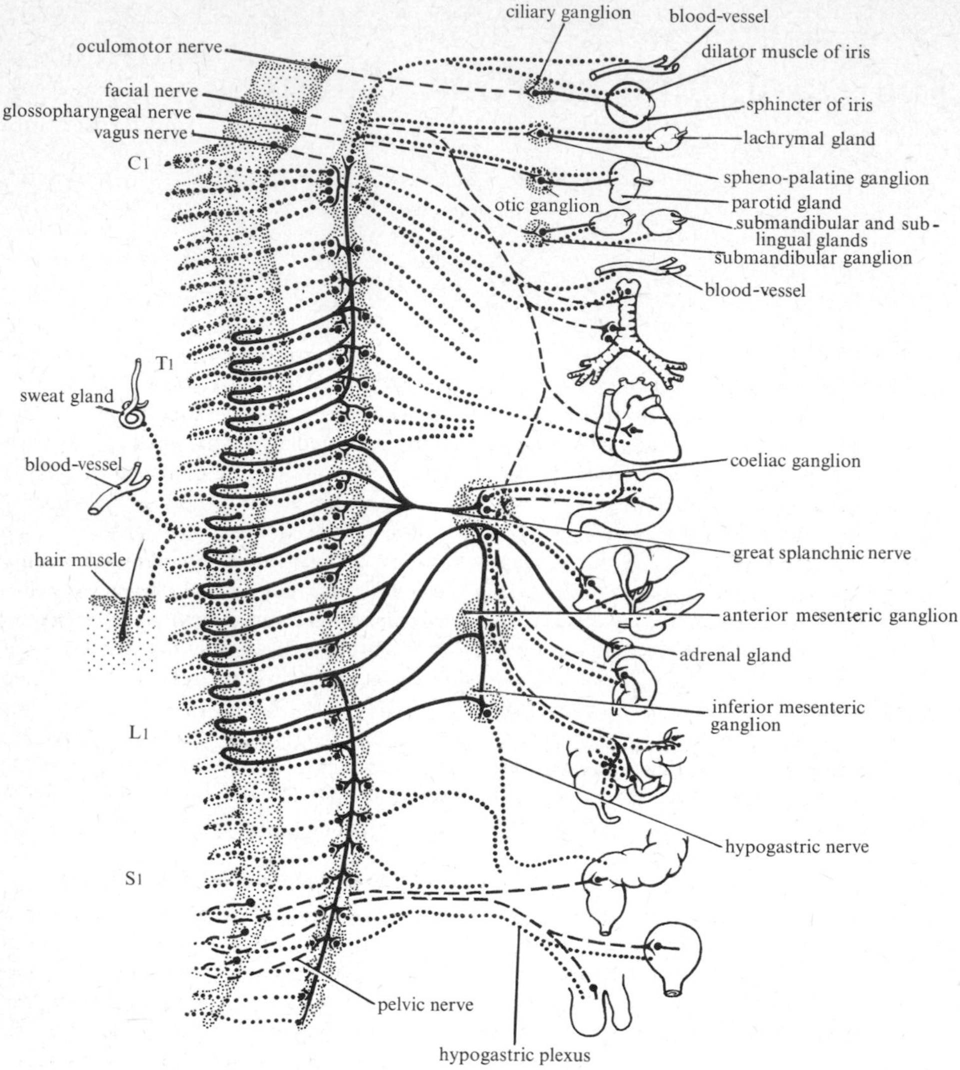

FIG. 30.1. Diagram of the autonomic nervous system. Preganglionic sympathetic pathways are shown as solid lines, postganglionic sympathetic pathways as dotted lines. Preganglionic parasympathetic pathways are shown as broken lines, postganglionic parasympathetic pathways are solid lines. The parasympathetic pathway to the gut passes mainly in the gut wall, partly through the coeliac ganglion as illustrated.

point at which the break (synapse) occurs. If applied to an autonomic ganglion these drugs first excite the cells to activity and then paralyse all the synaptic junctions. We can discover therefore whether a given set of fibres ends in a ganglion by seeing whether the effect produced by stimulating them is abolished by painting the ganglion with a 1 per cent solution of nicotine. For example, preganglionic fibres that produce contraction of the dilator muscle of the iris of the eye of a mammal leave the spinal cord in the ventral roots of the upper thoracic region, run forwards in the cervical sym-

pathetic trunk, and make synapse in the superior cervical ganglion with postganglionic nerve cells whose fibres ultimately reach the eye. Nicotine interrupts this pathway only when it is applied to the superior cervical ganglion (Fig. 30.1).

The functional significance of the fact that the visceral motor neurons lie outside the central nervous system is not clear. In spite of orthodox beliefs the visceral ganglia may be to some extent reflex centres (see p. 286). However, visceral afferent nerve fibres do not end in the autonomic ganglia but proceed through the dorsal roots (where their cell bodies lie) to make connexions in the brain or spinal cord. All activation of the visceral motor system is on the classical view through the preganglionic fibres, unless, of course, the postganglionic neurons show rhythmic spontaneous activity of their own or are directly stimulated by local influences. Each preganglionic fibre is connected with many postganglionic neurons, and this diffusion of the effects may be the reason for the presence of a cell on the motor pathway, but there may well be some further recoding or other change that at present escapes us.

3. Divisions of the visceral motor system

The classical analysis of the visceral motor system, developed by Gaskell and Langley (Langley 1921), recognizes two major divisions—*sympathetic* and *para-sympathetic* (Fig. 30.1). These are distinguished by the facts that they are (a) anatomically distinct, (b) physiologically opposite or 'antagonistic' in their actions, and (c) that they use different sympathetic transmitters and therefore their actions are mimicked by two distinct sets of drugs. The parasympathetic postganglionic endings secrete acetylcholine (*cholinergic*), whereas those of the sympathetic often liberate monoamines (*aminergic*). We have therefore three separate criteria for decision as to whether any given visceral nerve or ganglion is to be called sympathetic or parasympathetic. We may decide by its anatomical position and connexions, by its physiological effects, or by pharmacological study of the substances that stimulate it or imitate its action. There is sufficient agreement between these criteria to encourage us to believe that there are two different sets of nerve fibres, but there are divergences, suggesting that the nerves responsible for the control of the viscera are not simply divisible into two sharply distinct systems. Moreover, the two sets of nerves not only act (usually antagonistically) on the same effector organs, but also can act directly upon each other.

4. The sympathetic system

The chain of sympathetic ganglia extends in mammals from the base of the skull to the sacrum (Fig. 30.1). The head and tail are supplied by fibres originating in ganglia of the neck and trunk. The preganglionic neurons that control this system leave the cord as medullated fibres in the ventral roots of about segments T_1 to L_2 forming the *white rami communicantes*, which run from the spinal nerves to the ganglia. The non-medullated (grey) axons of the postganglionic cells pass from the ganglia to the muscles of the blood vessels and of the viscera and to various glands. Some of these fibres reach their end organs by passing back from the

ganglia to the spinal nerves as the *grey rami communicantes*. The muscles and glands of the abdominal viscera receive their nerves mainly not from the ganglia of the sympathetic chain itself but from certain splanchnic or collateral ganglia, including the coeliac and anterior mesenteric ganglia of the *solar plexus* and the inferior mesenteric ganglia. These collateral ganglia are connected with the sympathetic chain by a series of *splanchnic nerves* (Fig. 30.2), the most important being the greater splanchnic, which arises in the thorax and runs to the solar plexus.

The most cranial member of the sympathetic chain is the *superior cervical ganglion*, lying dorsal to the carotid artery, approximately at the level of its division into internal and external branches. This ganglion supplies postganglionic fibres to numerous structures in the head. It receives its preganglionic fibres from the white rami communicantes of the upper thoracic segments, and for this reason interruption of these spinal nerves by injury or operation produces many signs in the organs of the head, such as narrowing of the pupil and drooping of the eyelid.

The sympathetic ganglia of the neck do not show a regular segmental arrangement. Besides the superior cervical ganglion there is also a large *inferior cervical (stellate) ganglion* and often also a small middle cervical. From the three cervical ganglia grey rami communicantes pass to the cranial nerves, to the cervical spinal nerves and to the heart, lungs, and other organs of this region.

In the thoracic and lumbar regions the sympathetic ganglia are more regularly arranged, one to each spinal nerve. There are white rami communicantes in the thoracic and upper lumbar segments and these usually run obliquely, especially lower down, so that each ganglion receives a white ramus from the spinal nerve of a different (higher) segment (Fig. 30.2). The grey rami on the other hand run transversely.

The splanchnic nerves arise from the sympathetic chain mainly in the thorax. They consist chiefly of small medullated preganglionic fibres, which pass from the white rami through the ganglia of the sympathetic chain to end in the solar plexus, where they make synapse. Non-medullated postganglionic axons proceed from the coeliac and anterior mesenteric ganglia to the stomach, intestines, and other viscera. They reach the gut wall (p. 286) to end mainly on the neurons of the intrinsic plexuses, and partly on the smooth muscles, glands, and blood-vessels.

The viscera of the lower abdomen (bladder, rectum) receive their sympathetic fibres from the *inferior mesenteric ganglion*, whose preganglionic fibres leave the chain in several lesser splanchnic nerves in the

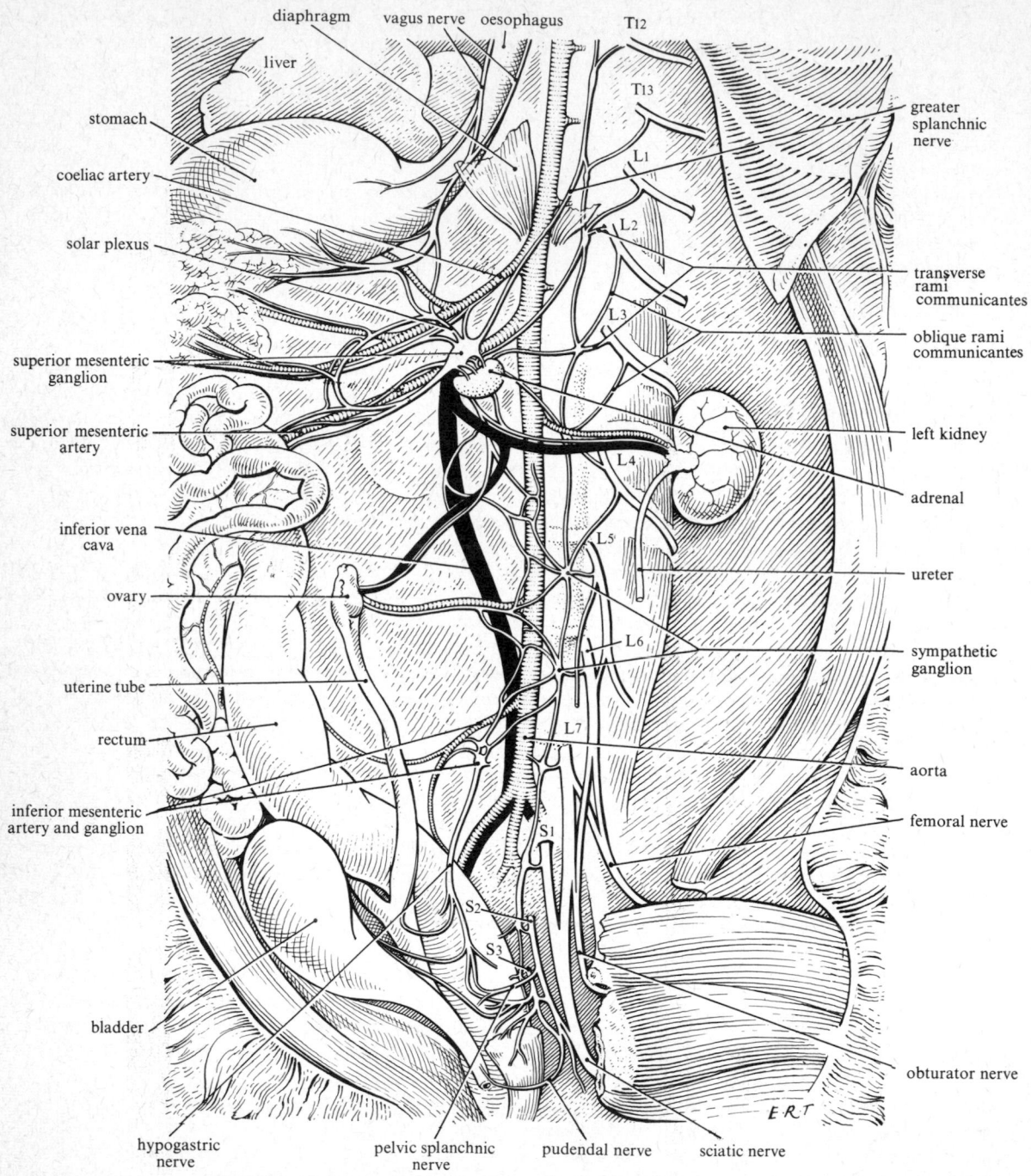

FIG. 30.2. Dissection of the autonomic nervous system of the cat.

lumbar region. The system in this region forms a complicated network of fibres on either side of the aorta, which is known as the *hypogastric plexus* and continues downwards at the side of the rectum as the

pelvic plexus; this is also connected with the sacral parasympathetic system (p. 284).

The sympathetic chains continue through the lower lumbar and sacral regions as thin strands with segmental ganglia, sending grey rami to the nerves. The two chains lie close together on either side of the aorta and are joined by transverse commissures.

The sympathetic system can be described as consisting of a set of ganglia, but the ganglia are only denser aggregations of cells within a rather irregular network of cells and fibres. Postganglionic cells are found at numerous places along the sympathetic chains and the rami communicantes, splanchnic nerves, and plexuses; the whole set constitutes a network of cells and fibres lying on either side of the aorta and extending among the viscera.

5. The parasympathetic system

The fibres whose actions are in general opposite to those of the sympathetic system run in nerves of the head and sacral region and are less easily characterized as an anatomical entity than are the sympathetic fibres. The sympathetic preganglionics all leave the central nervous system in ventral roots, but the parasympathetic fibres leave in both ventral and dorsal roots. Fibres regarded as parasympathetic are found in the oculomotor, facial, glossopharyngeal, and vagus nerves and in the ventral roots of the sacral segments at about the level of the third sacral spinal nerves.

6. Cranial parasympathetic system

The preganglionic fibres in the oculomotor nerve run to the ciliary ganglion, in the orbit. They make synapse with postganglionic cells there, and these in turn send fibres in the short ciliary nerves to produce contraction of the muscles of accommodation and the sphincter muscle of the iris (Fig. 30.1). This system is connected therefore with a ventral root (oculomotor) and it carries a ganglion (the ciliary). Thus it resembles the sympathetic nerves, and the only clear reason for placing it in a distinct system is that the action of the sphincter iridis muscle is 'antagonistic' to that of the dilator muscle, which receives its motor fibres from the superior cervical ganglion. Yet we do not place flexor and extensor muscles of the elbow in opposite 'systems' and it is difficult to see the justification for doing so in this case. The pharmacological criterion, however, provides a further test (p. 287).

The *ciliary ganglion* is a minute body lying in the orbit. It may be said to have three roots: short, long, and sympathetic. The short root runs from the oculomotor nerve and carries the preganglionic parasympathetic fibres that end in the ganglion. The postganglionic fibres then pass in the short ciliary nerves to the eyeball. They are unusual in being myelinated, since the actions they produce are fast. The long root of the ciliary ganglion contains sensory fibres from the trige-

minal nerve. They pass through or near the ciliary ganglion into the long ciliary nerves running to the eyeball.

The sympathetic root of the ciliary ganglion consists of postganglionic fibres arising in the superior cervical ganglion and passing along the carotid plexus (p. 280). They pass with branches of the trigeminal nerve through the ciliary ganglion without synapse and are distributed in the long ciliary nerves to the dilatator iridis muscle.

The main bulk of the parasympathetic system consists of fibres in the seventh, ninth, and tenth cranial nerves, controlling the secretion of the salivary glands and the actions of many of the viscera. The preganglionics in the seventh nerve run a complicated course to reach the submandibular and sphenopalatine ganglia. The *sphenopalatine ganglion*, which controls the lacrymal gland, lies just below the exit of the maxillary branch of the trigeminal nerve but fibres reach it from the facial nerve. It also receives sympathetic fibres from the plexus around the internal carotid artery and ultimately from the superior cervical ganglion (p. 281). The sphenopalatine like the other parasympathetic ganglia of the head is, therefore, said to have two motor roots, but of course the sympathetic fibres, being postganglionic, run through the ganglion without synapse. These ganglia also have sensory 'roots', in the case of the sphenopalatine from the maxillary nerve. The *otic ganglion* receives its preganglionic fibres from the glossopharyngeal nerve; it controls the parotid gland. The facial fibres for the *submandibular ganglion* pass through the chorda tympani and lingual nerves and control the submandibular and sublingual salivary glands.

The *vagus* is a large nerve and is seen to contain numerous small medullated preganglionic fibres in addition to many sensory ones. The preganglionic fibres run to the heart, lungs, walls of the stomach, small intestine, and colon, and the alimentary glands, such as the pancreas. The fibres end in contact with postganglionic cells lying close to the muscles or glands of these organs. In the gut wall there are two plexuses of cells and fibres, an outer *myenteric* or *Auerbach's plexus*, between the circular and longitudinal muscle layers, and an inner *submucous* or *Meissner's plexus*, beneath the mucosa (Fig. 14.2, p. 142). The cells of these plexuses are usually presumed to be the postganglionic parasympathetic neurons that have migrated out during development along the vagus, but the details of their development, arrangement, and functioning are still obscure, and Langley suggests that it would be wise to regard them as belonging to an enteric nervous system, distinct from the remainder of the autonomic. They receive many synapses from postganglionic adrenergic fibres whose significance is not known. Their axons run partly to the smooth muscles and glands of the gut, but the neurons of the intrinsic plexuses are also connected with each other synaptically.

7. Sacral parasympathetic system

The ventral roots of certain sacral segments contain preganglionic fibres that run to the viscera in the pelvic splanchnic nerves. These *pelvic nerves* join the hypo-gastric plexus to form the pelvic plexus, from which branches pass to the neighbouring organs. The pelvic plexus therefore contains both sympathetic fibres through the hypogastrics) and parasympathetic fibres (through the pelvic splanchnic nerves). The parasym-pathetic fibres are preganglionic and make synapse with cells lying either in the pelvic plexus or in the walls of the uterus, bladder, external genitalia, lower colon, and rectum. This sacral parasympathetic system has effects opposite to those of the sympathetic fibres that reach these viscera.

8. Functions of the two parts of the autonomic nervous system

It is sometimes said that the function of the autonomic nervous system is the regulation of the *milieu intérieure*, or internal environment, which is maintained constant in the higher animals. Cannon expresses a similar idea in his conception that the autonomic nervous system maintains the homeostasis by which the internal environment is kept constant (p. 8). It is true that the autonomic system is involved in many regulatory actions. For instance, the sympathetic system, by its control of the blood-vessels of the skin, regulates temperature. But the maintenance of the integrity of the organism is the 'function' of *all* its parts. The characteristic by which we recognize a creature as alive is that its actions tend to promote its own continuance (p. 1). It is necessary, therefore, to keep a sense of proportion and not ascribe to the autonomic system any particular mysterious influence on homeostasis not possessed by other organs, say the brain, the bladder, or even the less active parts like the hairs or the bones.

The autonomic nerves have indeed actions that regulate the activities of many internal organs and the sympathetic and parasympathetic parts of the system often act in opposite directions, setting up a balance. It is convenient to generalize these activities by saying that the sympathetic nerves make the body ready for action in an emergency whereas the parasympathetic ones promote the restorative activities, which go on while the animal is at rest.

9. Effects of sympathetic nerves

The effects observed on stimulating sympathetic nerves electrically are found to be such as would be expected for the purpose of preparing the body for action in attack or defence. Thus the sympathetic fibres running to the lungs produce dilatation of the bronchi, allowing increased ventilation. The nerves from the cervical sympathetic ganglia to the heart are known as the *nervi accelerantes*, because they increase the frequency and strength of the heartbeat. The action on the coronary arteries of the heart is, however, to produce relaxation, providing an increased blood-flow. In other parts of the body the sympathetic nerve fibres to the blood-vessels usually produce constriction (Chien 1967). The diameter of each artery or arteriole can be in-creased, when necessary, by inhibition of the sym-pathetic tonus, allowing a greater blood-flow. The state of the vessels thus comes under a complicated set of reflex and hormonal influences, which ensures that the blood-supply to any organ is made large when the organ becomes active. When the whole body is in action the sympathetic (constrictor) tone of the vessels to the viscera is increased, depriving them of blood, which is transferred to the muscles, whose arteries dilate. Conversely, during digestion the blood-supply of the gut is increased. In some active animals (dog, hare, and perhaps man) there are sympathetic nerve fibres that actively dilate the blood-vessels in the muscles, but usually the blood-flow is controlled by variation in sympathetic constrictor tone. This is exercised in such a way as to make the most efficient use of the blood by transferring it to whatever part of the body needs it at the time. This mechanism for effective use of the blood is probably a major factor in allowing the success of the higher vertebrates.

The list of sympathetic controls that prepare the body for action is very long. Many details of metabolism are controlled by monoamines produced either by sym-pathetic nerves or from the adrenal medulla. The sympathetic fibres cause the spleen to contract and pass out its store of red corpuscles into the circulation (p. 176). Glucose is mobilized from glycogen stores by the action of monoamines upon the liver. Activity of the viscera uses a large supply of blood and is reduced in an emergency; the sympathetic nerve fibres passing from the coeliac and anterior mesenteric ganglia to the gut therefore mostly inhibit the movements of peristal-sis, but they cause contraction of the pyloric and ileo-colic sphincters, preventing movement of food along the canal. Similarly there is constriction of the blood-vessels of the pelvic viscera and external genitalia and relaxation of the wall of the urinary bladder.

The muscles in the walls of the male reproductive

tract (vas deferens, seminal vesicles, and prostate) are caused to contract by their sympathetic nerves during ejaculation. The sympathetic nerves to the uterus vary in effect according to the state of the organ, producing relaxation when the uterus is empty, or contraction at the end of pregnancy. There is probably no antagonistic parasympathetic innervation of the internal reproductive organs.

Finally, in addition to these internal regulatory activities the sympathetic also has effects on organs at the periphery of the body that make preparation for action. The pupil of the eye dilates because of sympathetic action on the radial dilatator muscle, perhaps to allow more light to enter (though this would not necessarily always increase the usefulness of the eyes!). The dilatation of the pupil may serve to produce a sudden presentation of dark, round eye-spots eliciting a flight reaction in the opponent.

The sudden bristling of the hairs (associated with arching of the back) is another action controlled by sympathetic nerves, whose effect is apparent size-increase can be seen in any cat and dog fight. The control of the hair position also has the effect of regulating the heat loss. The flow of sweat, which is produced by impulses in sympathetic fibres, is another action that may be necessary during the stress of battle.

10. The sympathetic system and the adrenal glands

Many of these actions of the sympathetic nervous system are also performed by adrenaline (= epinephrine) (p. 425). The association between the sympathetic and the adrenal medulla is certainly close and there is evidence that sympathetic postganglionic nerve fibres produce their excitatory and inhibitory effects by secretion of a substance like adrenaline, probably noradrenaline, close to the muscles or glands that they innervate (p. 57). The adrenal medullary cells develop, like sympathetic cells, from the neural crest, and they are to be regarded as postganglionic sympathetic cells that pour adrenaline into the blood stream instead of producing it in the peripheral tissues.

l-noradrenaline and some *l*-adrenaline occur in post-ganglionic sympathetic neurons and nerve trunks and in larger amounts in the organs that these nerves innervate. These substances disappear from an organ after degeneration of its nerves but reappear as the nerves regenerate. Amine oxidases present in the tissue are able rapidly to inactivate these mediators. Evidently the sympathetic system and the adrenal medulla act together in the regulation of many parts of the body. The sympathetic nerves act by carrying signals to particular tissues, where they cause liberation of adrenaline-like substances. The adrenal medulla liberates similar substances into the blood so that they influence tissues throughout the body.

Groups of chromaffin cells, similar to those in the adrenal medulla, occur within many sympathetic ganglia. Presumably their secretion assists in the regulation of the activities of the postganglionic cells. This is a further sign of the close relationship of the two systems, which should repay further investigation.

11. Effect of removal of sympathetic chains

The part played by the sympathetic system in the life of the animal can be investigated experimentally by removing the entire sympathetic chain. The arduous operations involved were performed by Cannon and his colleagues at Harvard, where cats and dogs were kept alive for many months without any sympathetic nerves. Absence of the sympathetic can also be produced by injecting new-born mice with an antiserum to the nerve growth factor (p. 351). Providing that the surrounding conditions were not severe the animals without the sympathetic system showed no gross abnormalities. When they were exposed to conditions of stress, such as extremes of temperature or lack of oxygen, they died sooner than did normal animals. The cats were found to be considerably less active after the operation but this was less evident in the dogs.

In summary then we may say that the sympathetic division of the autonomic nervous system produces effects throughout the body that tend to make preparation for strenuous action.

12. The functions of the parasympathetic system

The parasympathetic nerves form a less compact unit anatomically than the sympathetic and functionally they control a number of separate activities; the system seldom acts as a whole. Nevertheless one can say that in general its actions are such as to assist in restoring the energies of the body after action, in preparation for further efforts. Thus parasympathetic nerves play a large part in promoting digestion. Fibres in the seventh and ninth cranial nerves control the secretion of the salivary glands, including vasodilator fibres that cause the blood-vessels of the glands to relax and produce a flow adequate to allow a copious secretion of saliva. Throughout the gut the rule is that the parasympathetic nerves increase peristalsis and promote secretion of the glands, whereas the sympathetic fibres inhibit these actions but cause contraction of the muscles of the sphincters. Thus the flow of gastric and pancreatic juice, succus entericus, and bile is increased by stimulation of the vagus. The nerve cells in the myenteric and submucous plexuses of the gut wall include many post-ganglionic parasympathetic cells, which have moved out during development along the vagus nerves. Their function is presumed to be simply the motor one of distributing the impulses for reflexes whose pathway is from afferent fibres of the gut or elsewhere through the spinal cord and brain and the preganglionic parasympathetic fibres.

It has often been suggested that these plexuses can perform reflex actions when isolated from the central nervous system. Co-ordinated peristaltic movements continue in isolated pieces of intestine suspended in a suitable liquid, and stimulation of one point in such a preparation may produce effects at a distance. Some of these phenomena may be due to 'axon reflexes'; the cells of these plexuses have many long and branching processes and impulses set up anywhere in one cell will thus become widely distributed. Further work is needed on the action of this rather isolated portion of the nervous system, to determine whether its cells are capable of initiating action either by their own rhythmic properties or reflexly.

There may well be shorter autonomic reflex chains through very small fibres which leave the gut and make synapse in the prevertebral ganglia. In addition there is an extensive system of afferent fibres from the gut and other viscera which project to the spinal cord and to the brain (Neil 1972).

The parasympathetic system has less general influence on the vascular system than has the sympathetic but produces effects in certain parts. The nerves from the vagus to the heart slow the rate of beat and there is usually a tonic discharge in the vagus restraining the heart. Doses of atropine, which paralyses this vagus action (p. 288), may produce a doubling of the rate of heartbeat in man. The dilatation of vessels in the salivary glands has already been mentioned, and the sacral parasympathetic fibres passing in the pelvic nerves cause relaxation of the vessels of the external genitalia, producing erection of the penis or clitoris.

Many other more or less isolated functions throughout the body are controlled by nerves classed as parasympathetic. Thus the muscles of accommodation of the eye and the sphincter muscle of the iris are caused to contract by impulses passing through the third nerve and ciliary ganglion. The lachrymal gland is made to secrete by fibres of the facial nerve, which synapse in the sphenopalatine ganglion. The effect of the vagal nerve fibres to the lungs is to cause constriction of the bronchi, an action clearly opposite to that of the sympathetic nerves.

The sacral parasympathetic nerves produce the reflex contraction of the wall of the urinary bladder in micturition, the internal sphincter muscle being caused to relax.

13. Antagonism of sympathetic and parasympathetic systems

Evidently the two sets of autonomic nerves often work in opposite directions on an organ and this is clearly brought out in Table 30.1. The sympathetic and parasympathetic nerves to the heart, blood-vessels of the salivary glands, and external genitalia and to the musculature of the gut clearly work in such a way that when the one excites muscular contraction the other inhibits it. In other organs, especially in the iris (p. 285), the antagonism is between muscles that pull in opposite directions and does not involve the presence of nerve fibres that actively promote relaxation. Moreover, it is important to recognize the large number of cases in which there is no clear 'antagonism' between the two systems. These exceptions are of two sorts. Some organs are innervated by one system only, thus the sweat glands, hair muscles, and many blood-vessels receive only sympathetic, while muscles of accommodation, salivary glands, and all the glands of the gut only parasympathetic fibres. Conversely, the nerves from one system may in some organs produce both excitatory and inhibitory effects. Thus the sympathetic nerves to the blood-vessels of the limbs contain in some animals vasodilatator, as well as vasoconstrictor fibres (p. 284). Again, electrical stimulation of the splanchnic nerves of the cat may produce contraction of the stomach musculature instead of the more usual relaxation, depending on the duration and frequency of the shocks used.

It is clear that the visceral nervous system cannot be divided into two separate and antagonistic divisions, and it is possible that this attempt at a classification has obscured the outstanding fact that *visceral nerves sometimes act by inhibiting muscular contraction, whereas the somatic nerves always promote it*. Thus in the heart the effects of the sympathetic are abolished by the action of the vagus. There is a similar antagonism in the nerves to the blood-vessels of the salivary glands, external genitalia, and the coronaries of the heart, as well as of the muscles of the lungs and the gut wall.

Consideration of these interesting peripheral inhibitory effects is made difficult because some of the organs contain nerve cells, and we do not know how the two sets of nerve fibres end in relation to them. For instance, the effect of the sympathetic nerves in causing inhibition of the movements of the gut may possibly be produced by action upon the nerve cells in the myenteric plexus, rather than directly upon the muscle of the gut wall (Paton and Zar 1968). In the heart the problem is even more obscure since the inhibitory (vagal) pathway makes synapse there, whereas the excitatory (sympathetic) one does not.

TABLE 30.1

Table of actions of autonomic nerves in man

(Partly after Best and Taylor (1955). *Physiological basis of medical practice* (6th edn). Baillière, Tindall, and Cox, London.)

Organ	Sympathetic	Parasympathetic	Organ	Sympathetic	Parasympathetic
HEART			**GLANDS**		
Rate	Increased	Reduced	Salivary	Some mucous secretion	Secretion
Auricular muscle	Stronger contraction	Weaker contraction	Gastric	No innervation	Secretion
Ventricular muscle	Stronger contraction	No innervation	Pancreas	No innervation	Secretion
			Intestinal	No innervation	Secretion
BLOOD-VESSELS			**LIVER**	Glycogenolysis	Flow of bile
Cutaneous	Constriction	No innervation	**UROGENITAL SYSTEM**		
Muscular	Constriction (usually, dilation also in some)	No innervation	Bladder muscle	Relaxation	Contraction
			Bladder sphincter	Contraction	Relaxation
Visceral	Constriction	No innervation	Vas deferens, seminal vesicles, and prostate	Contraction	No innervation
Coronary	Dilation	Constriction			
Cerebral	Constriction	Dilation	Uterus (non-pregnant)	Relaxation	No innervation
Pulmonary	Constriction	Dilation	(pregnant)	Contraction	
Salivary glands	Constriction	Dilation			
External genital	Constriction	Dilation	**EYE**		
LUNGS			Iris sphincter	No innervation	Contraction
Bronchial muscles	Relaxation	Contraction	Iris dilator	Contraction	No innervation
			Ciliary muscle	Relaxation	Contraction
ALIMENTARY CANAL			Muscles of eyelids	Contraction	No innervation
Muscles			Lachrymal gland	No innervation	Secretion
Oesophagus	Relaxation	Contraction			
Stomach	Relaxation (sometimes contraction)	Contraction	**SKIN**		
			Hair muscles	Contraction	No innervation
Intestine	Relaxation	Contraction	Sweat glands	Secretion	No innervation
Pyloric sphincter	Contraction	Relaxation			
Ileo-colic sphincter	Contraction	Relaxation	**ADRENAL MEDULLA**	Secretion	No innervation
Internal anal sphincter	Contraction	Relaxation			

14. Pharmacology of the autonomic nervous system

The whole subject is made especially interesting by the fact that the nerves produce their opposite actions by secreting different active chemical mediators. The study of these 'neurohumors' and of drugs that mimic or inhibit their actions has already shed much light on the nature of the action of this part of the nervous system (and indeed of other parts too).

The transmission of the nerve impulse depends on a series of electrical discharges (p. 244) but the effect of nerve fibres on their muscles or glands is produced by chemical mediators liberated by the nerve fibres close to these effector cells (p. 251). It has been known for some time that the ester acetylcholine is able to produce contraction of some smooth muscles (for instance, those of the gut or bladder) at extremely great dilutions (1 part in 10^8). It has also been satisfactorily demonstrated that this substance is produced at the endings of many autonomic nerve fibres. All preganglionic fibres (whether sympathetic or parasympathetic) produce it where they connect with their postganglionic cells. Further, parasympathetic postganglionic fibres produce acetylcholine at their nerve endings in muscles and glands and the sympathetic postganglionics to the sweat glands also operate by its action in man, though in the horse they liberate an adrenaline-like substance, as do other sympathetic postganglionic fibres (p. 285). Dale first suggested that synapses and motor endings where acetylcholine is produced shall be called *cholinergic*. Wherever they are found there is also present the enzyme, *cholinesterase*, which rapidly removes the acetylcholine and prevents the spread of its action.

Certain alkaloid drugs, such as *pilocarpine* and *arecoline*, imitate the action of acetylcholine and they are said to be parasympathomimetic in action. Other substances, notably the alkaloid *atropine* or the synthetic *d*-tubocurarine, inhibit the action of acetylcholine and of parasympathomimetic drugs. A third interesting set of substances inhibits the action of cholinesterase and thus makes the effects of acetylcholine more pronounced; these include the alkaloid eserine and the synthetic substance diisopropyl fluorophosphate (DFP).

The interactions of these various substances, especially those produced naturally in the body, has provided a fascinating study, which is also of great practical importance because of the valuable medicinal effects of the drugs. Thus atropine (belladonna) has long been known to quicken the heartbeat and dilate the pupil.

The inhibitory drugs probably act by competing with the acetylcholine for places on the sensitive or 'receptive' parts of the effector cells. The situation is complicated by the fact that not all of the actions of acetylcholine are inhibited by atropine. Thus, although atropine inhibits the effects of the parasympathetic postganglionic fibres on the heart, pupil, urinary bladder, and muscles and glands of the gut, it has no action on the preganglionic fibres (including those to the adrenal medulla) or on motor nerve fibres to striped muscle, which are also cholinergic (see p. 64).

At the synapse between sympathetic pre- and postganglionic neurons, acetylcholine is the transmitter. Indeed this was the classical site at which chemical synaptic transmission was first proved by Dale and his colleagues. The endings of the preganglionic nerve fibres in contact with the postsynaptic neurons contain clear vesicles (Fig. 30.3). However, there are some curious small neurons in the ganglia which besides receiving synapses of this type themselves make synapses at which the vesicles are of the dense-core type, usually associated with catecholamines (Matthews and Raisman 1969). These puzzling cells are presumably interneurons, and their presence introduces a possible complication to

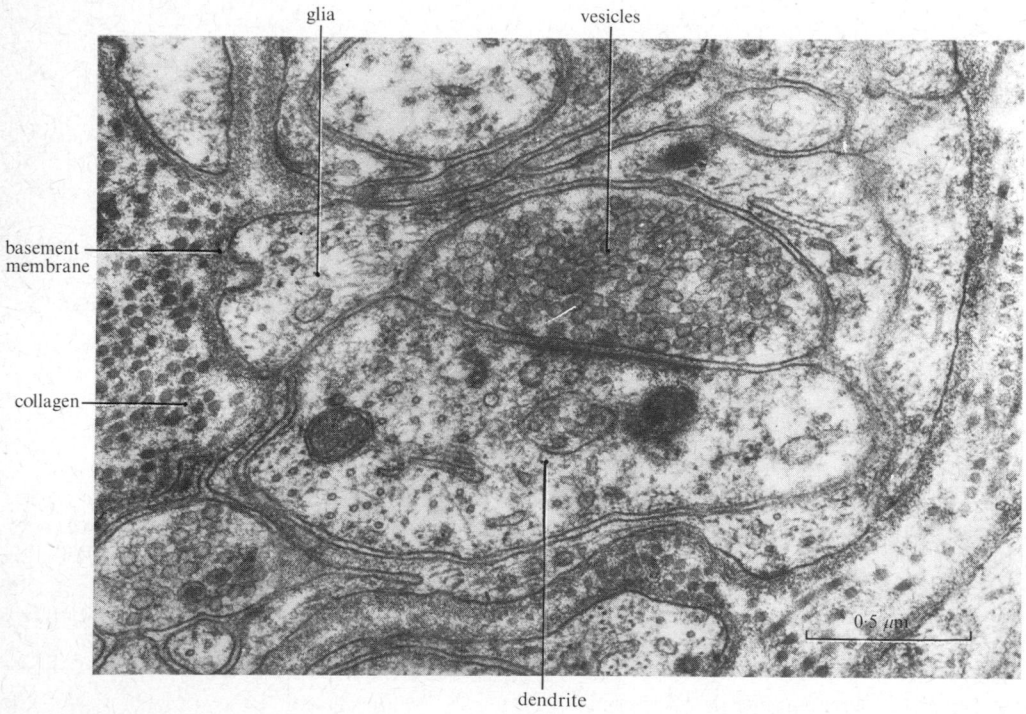

FIG. 30.3. Electron micrograph from superior cervical ganglion of rat. A nerve bouton, from a preganglionic fibre, filled with vesicles makes synapse with a dendrite. (Figure kindly supplied by Dr. G. Gabella.)

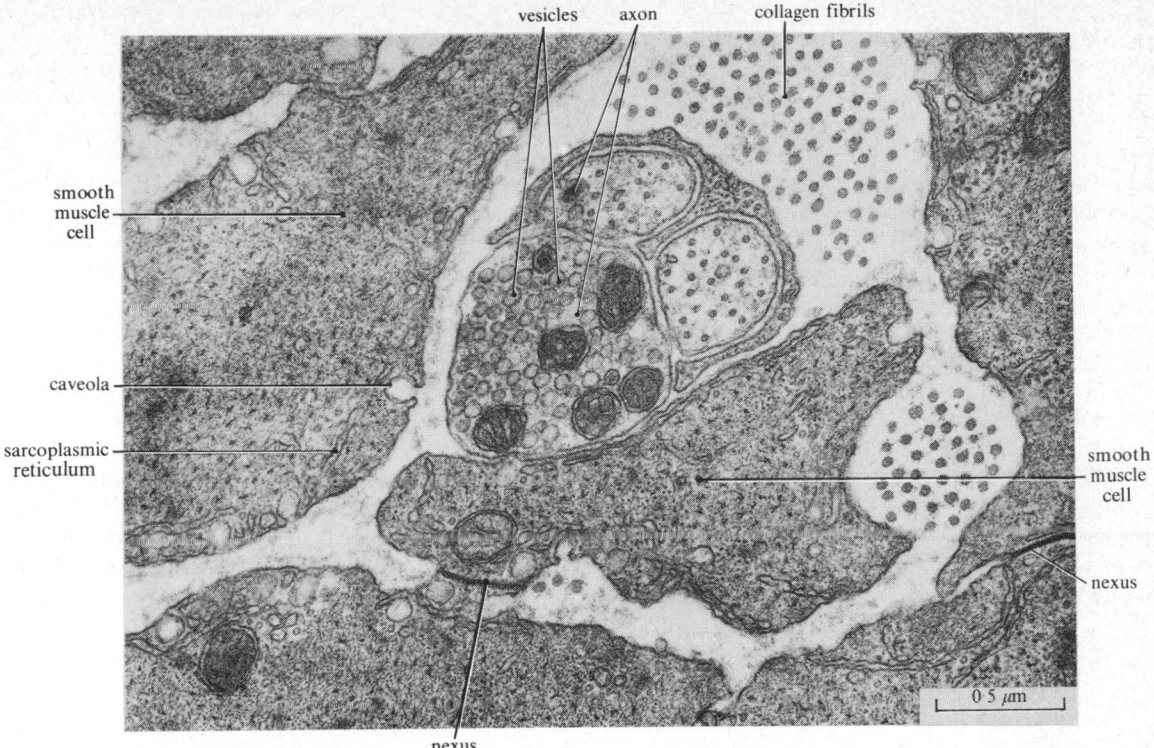

FIG. 30.4. Electron micrograph showing small nerve bundle amid smooth muscle cells of the sphincter pupillae from a guinea-pig iris. Two of the axons show regularly spaced microtubules, while the third one is filled with electron-lucent ('agranular') vesicles and mitochondria. The smooth muscle cells show filaments, mitochondria, and abundant sarcoplasmic reticulum. Their cell membrane has characteristic invaginations (caveolae) and areas of close contact (gap junctions) with other muscle cells (nexus). (Figure kindly supplied by Dr. G. Gabella.)

the classical scheme of organization of the sympathetic.

The sympathetic postganglionic fibres usually exert their actions by the liberation of catecholamines, often *l*-noradrenaline; in Dale's terminology they are said to be adrenergic. The terminal fibres ramify among the smooth muscle or gland cells. Usually they form no specialized endings, but the Schwann cell covering is absent for a stretch (Fig. 30.4). The separation between nerve and muscle fibre varies here from 20 nm to 400 nm, that is, it is considerably greater than at somatic motor endings. The catecholamines in the terminals can be identified by the specific fluorescence that they give after treatment with formaldehyde (Eränkö 1955; Falck and Owman 1965). Studies by this and other methods show that the events at the nerve endings are complicated. The release of the amine following the passage of a nerve impulse probably allows the entry of calcium. There is some evidence that acetylcholine may be released first as an intermediary (Burn and Rand 1965). Much of the amine released is quickly reabsorbed into the ending, but some is destroyed by monoamino oxidase (MAO) or catechol *O*-methyl transferase (OMT) and some overflows into the blood. Small amounts of transmitter are probably released continually even at rest and there is evidence that they produce miniature end-potentials. On the arrival of impulses a large proportion of all the transmitter is released at once (as much as $\frac{1}{3000}$ of the total store in the muscles of the spleen) and then rapidly reabsorbed.

Finally there is evidence that at least in the gut, there are also fibres that operate by release of another transmitter, possibly ATP or a similar purine derivative (purinergic fibres) (Burnstock 1972).

15. The autonomic nervous system and homeostasis

The visceral motor nerves thus play a part in a variety of actions by which the steady state of the body is maintained. These actions are mostly slower adjustments than those mediated by the somatic nervous system, but they are faster than the chemical changes and growth processes that are regulated by endocrine systems (p. 455). Correspondingly autonomic nerve fibres are small, often non-medullated and they conduct slowly. Such fibres may be more effective than larger more rapidly conducting ones for producing long-sustained influences. It may also be significant that some autonomic effects are produced by liberation in the tissues of mediators that are destroyed relatively slowly.

The activities of the autonomic nervous system show clearly the predictive character common to many living activities. Upon receipt of suitable signals from the central nervous system the sympathetic nerves prepare the body for an impending emergency. The parasympathetic nerves on the other hand operate when a period of recuperation is available, during which digestion and other restorative processes can proceed.

By arbitrary definition the autonomic nerves constitute only an effector system; they include no afferent or computing mechanisms. Therefore they are activated only by signals from the central nervous system and are not truly autonomous. But of course both parts of the autonomic system are under central control. Indeed there are regions in the hypothalamus said to be predominantly concerned with the central regulation of each of the divisions (Chapter 32). However, it is convenient to restrict the term autonomic system to the peripheral parts. They serve to distribute throughout the body the signals that ensure that the behaviour of the internal organs is consonant with the condition of the environment and the needs of the rest of the body. In this sense they assist the other parts in ensuring that the organism adequately represents the changing conditions of its surroundings. Their capacity to do this is largely dependent upon the hereditary instructions, which control the arrangement and functioning of the nervous pathways and allow them to receive information through the preganglionic fibres from the central nervous system. There is no evidence that the autonomic nerves themselves carry memories of their individual past history, but it may be that, like other nerve fibres, they are influenced by use.

31 The control of posture and movement. The cerebellum and midbrain

1. Representation of movement patterns in the brain

THE basic mechanism for the control of posture and movement lies in the spinal cord. Each muscle is controlled by its pool of motor neurons and the firing of these is under the influence of local input, especially the impulses arriving from its own proprioceptors (p. 265). Yet in normal life the patterns of behaviour of a mammal are not produced only by these spinal responses of the muscles but are called into action in a well-organized time-series by the brain.

The mechanism by which posture and movement are produced is only partly understood. Many parts of the brain are not at rest even when the individual is still, and we have to discover how these internal rhythms are affected by the pattern of stimulation that falls upon the receptors, especially on the main exteroceptors, eye, ear, and nose. Impulses from these organs, after suitable combination with memories stored in the cerebral cortex, are able to produce action by stimulation or inhibition of groups of neurons in the basal portions of the brain. Electrical stimulation has shown that the neurons of these basal regions play upon the spinal centres to produce patterns of movement. For example, activity of a centre in the hypothalamus of a cat will produce all the actions appropriate to rage (p. 308).

Each muscle participates in many different actions and in each action it requires to be co-ordinated in a different way with other muscles. For example, flexor carpi ulnaris and radialis muscles work together in some actions but antagonistically in others (p. 72). Such co-ordination is the result of the action of cells in the basal centres of the brain, which may be called the *representatives* of the spinal motor neurons. The whole basal region, from the corpus striatum to the medulla, is involved in this production of patterns of action, but there is little knowledge available as to how the co-ordination is achieved, and of course both cerebral and cerebellar cortices are involved.

2. Control of segmental reflexes by the brain

The brain exerts a complicated set of excitatory and inhibitory effects on the spinal cord. Of course its various parts interact, and it is not really possible to speak of the particular effect of any one part as if it acted in isolation. Nevertheless, we can recognize different effects that follow from removal or stimulation of different regions and so achieve some understanding of 'how the brain works'.

Some of the levels that can be recognized are shown in Fig. 31.1. One method of investigating their effects

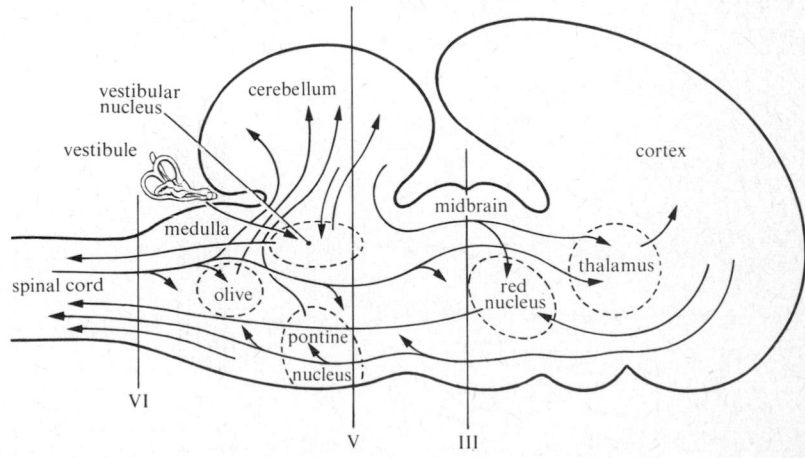

FIG. 31.1. Diagram showing some of the conduction paths in the central nervous system. Note the existence of alternative paths between primary sensory and motor pathways via cortical association areas. The section levels III, V, and VI correspond with those in Fig. 31.14.

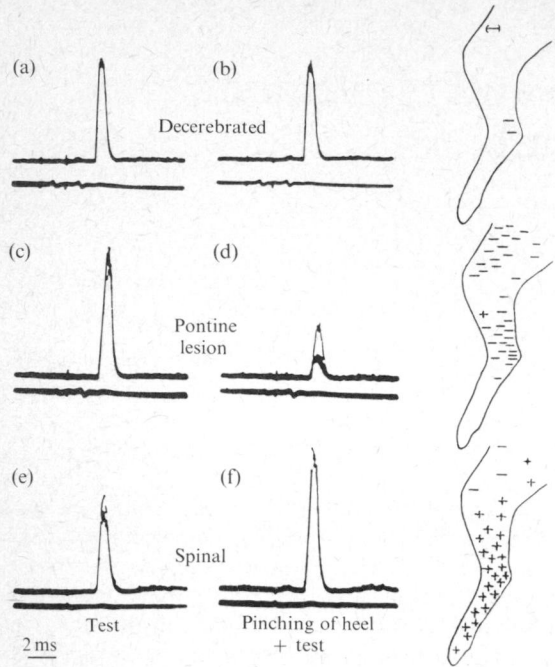

FIG. 31.2. Changes in response of flexor motor neuron pool on pinching cat's leg after successively caudal transections of the C.N.S. After the response to a standard test (controls: (a), (c), and (e)) comparison is made with the test plus an unknown input (in this instance, pinching). If the latter is inhibitory fewer neurons fire (d), if excitatory more neurons fire (f). There is no difference observable in the case of the decerebrate. (From Holmquist and Lundberg (1961). *Acta. physiol. scand.* **54**, Suppl. 186.)

is to transect the brain at various levels and to study the response of the resulting preparation. Transection of the brain behind the level of the red nucleus is known as decerebration (Fig. 31.14, p. 301). A cat so operated shows exaggerated extensor reflexes, the condition of

decerebrate rigidity. This can be shown to be due to an actual inhibition of flexor reflex activity (Fig. 31.2). If a section is made at the level of the pons this inhibition is even greater; indeed stimulation of the skin now actually decreases a monosynaptic flexor reflex (p. 267). However, when the spinal cord was isolated completely from the brain the flexor reflex reappeared in exaggerated form. Evidently the inhibitory effect arises from neurons both in and behind the pons. Perhaps those behind were themselves inhibited by others further forward. In a normal animal the cerebral hemispheres would ensure a proper balance between them all. This simple example will warn against expecting to find any easy description of the actions of cerebral centres. The brain contains a vast number of cells each with specific excitatory and/or inhibitory effects and all connected into complicated circuits. We have not yet developed methods suitable for description of such a system. It cannot be treated either as a homogeneous whole or as a set of discrete parts.

3. Input and output pathways of the cerebellum

The cerebellum is a part of the neuraxis concerned originally with responses to gravity and thence to movements mediated by muscle and skin senses and finally with those controlled by the cerebral cortex. It consists of a huge extension of the somatic sensory column at the front end of the medulla oblongata between the entrance of the eighth nerves and optic nerves, which in lower vertebrates mostly end in the midbrain (Bell and Dow 1967; Eccles, Ito, and Szentágothai 1967; Jansen and Brodal 1954; Llinás 1969).

The cerebellum is divided externally into a great number of lobes. The vestibular fibres all end in the posterior part of the cerebellum, the flocculonodular lobe (Fig. 31.3). The proprioceptor (spinocerebellar)

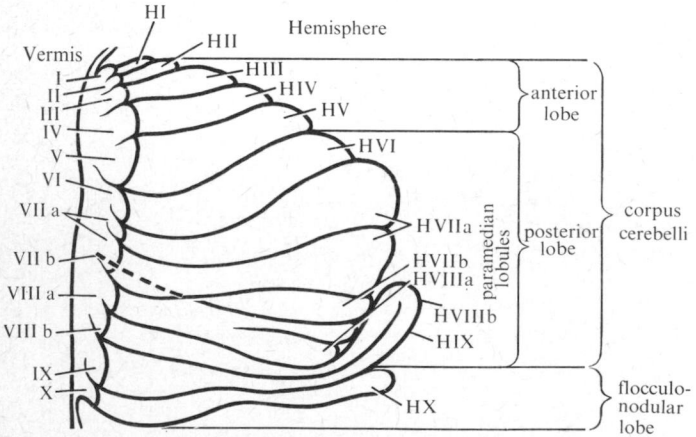

FIG. 31.3. Generalized structure of mammalian cerebellum. Numerical designations follow Larsell (1952). (From Bell and Dow 1967.)

fibres end here and also in the four most anterior lobes. These front and back parts are present in all vertebrates and may be called the *paleocerebellum*. They have become divided into anterior and posterior sections by the development at the centre of a region with pontine (cerebral) connexions, the *neocerebellum*. The cerebellum

is connected with the rest of the brain by three *cerebellar peduncles* or brachia: the inferior peduncle (restiform body), middle peduncle (brachium pontis), and superior peduncle (brachium conjunctivum) (Fig. 31.4). Fibres of the vestibular nerve run in the inferior peduncle, either direct or after synapse in the vestibular

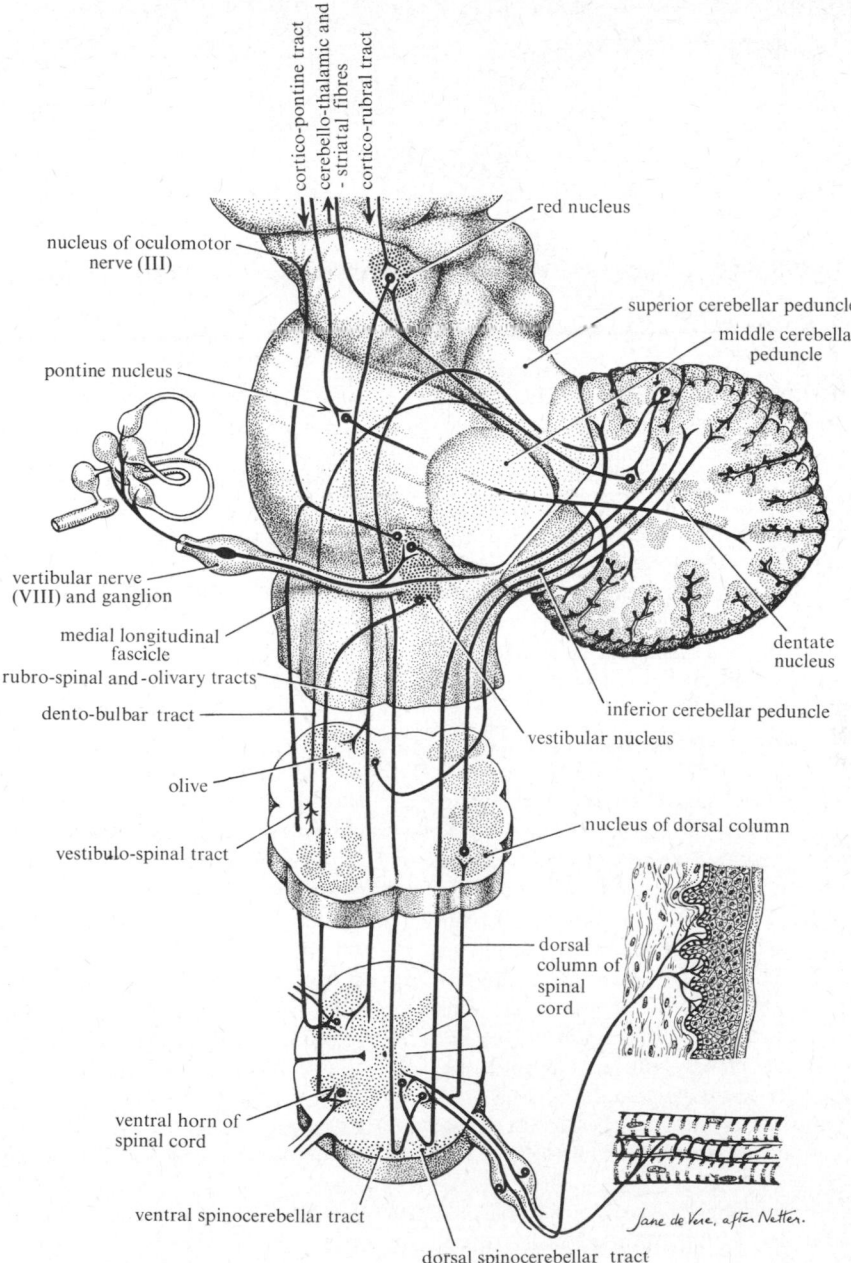

FIG. 31.4. The main afferent and efferent connexions of the cerebellum of man, seen from the left side. (Modified from original illustration by F. Netter, from Ciba collection.)

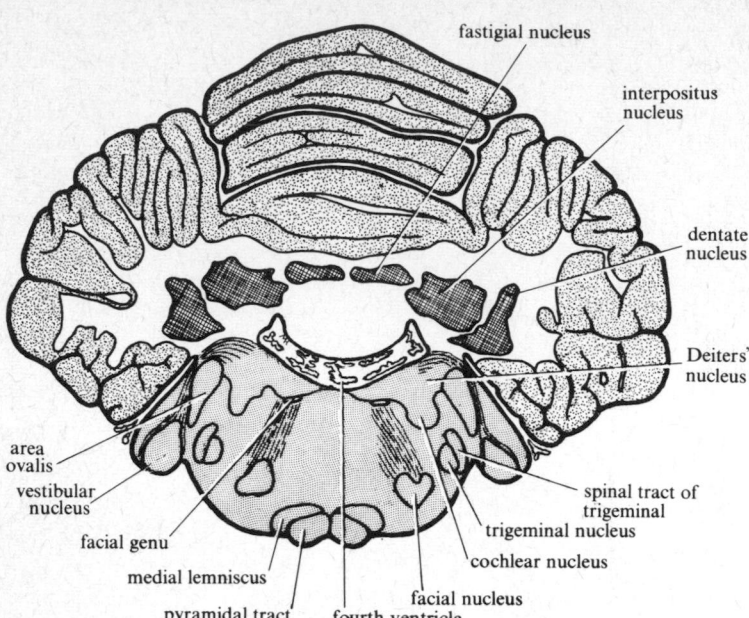

FIG. 31.5. Cross-section of cat cerebellum and brain-stem.

nucleus. Fibres from the proprioceptors and skin reach the cerebellum direct via the spinocerebellar tract and the inferior and superior peduncles. A second and important indirect path is from the spinal cord to the inferior olive and thence in the inferior peduncle to the cerebellar cortex. Fibres from the cerebral cortex pass to the nuclei of the pons, a large mass below the cerebellum, from whose cells fibres pass up in the middle peduncle to the cortex.

Fibres also reach the cerebellum from the midbrain, a pathway that was probably very important in lower vertebrates. A set of fibres has recently been shown to pass to the cerebellum from the locus coeruleus of the medulla, a centre perhaps concerned with passing signals of reward to higher parts of the brain (p. 343).

Fibres leaving the cerebellum arise in the three pairs of *cerebellar nuclei*, lying embedded in the white matter above the ventricle (Fig. 31.5). These nuclei are the medial (fastigial), intermediate (interpositus), and lateral (dentate). Their cells send tonic influences to the motor nuclei. Thus the medial nucleus activates the Deiters' nucleus of the vestibular complex and through it the extensor muscles of the limbs, while the intermediate nucleus produces flexor tonus through the red nucleus. Actually the effects are upon precise muscle groups, and the cerebellum controls movements by delicately varying the action of these nuclei which control the 'tone' of each muscle. The interpositus and lateral nuclei both project to the ventro-lateral nucleus of the thalamus, which in turn projects somatotopically

to the motor cortex. There is thus reciprocal action between cerebral cortex and cerebellum, each influencing the other.

4. The cerebellar cortex

Control is effected by the action upon the nuclei of three regions of the main cortex of the cerebellum: the median (vermis), intermediate (paravermis), and lateral (hemisphere) zones, corresponding to the three pairs of nuclei (Fig. 31.3). These are functionally significant divisions, but there are still more conspicuous divisions of the cerebellum into transverse bands or folia, numbered from 1 to 10, with some subdivisions (Fig. 31.3). The cerebellar cortical neurons are strictly orientated in relation to these folia (Fig. 31.6). The central units are the *Purkinje cells* with large cell bodies in the outer part of the granular layer and huge dendritic trees spreading strictly in the sagittal plane, that is to say, at right angles to the axes of the folia. The Purkinje cells send their axons to the cerebellar nuclei, whose cells they inhibit. Regulation of the action of the muscles is achieved by varying this inhibitory output of the Purkinje cells.

The efferent fibres that provide the necessary information form two groups (*a*) *climbing fibres* from the cells of the inferior olive, which reach the Purkinje cell dendrites directly, and (*b*) *mossy fibres* from vestibular spinal and cerebral sources, which take a longer course (Fig. 31.6). Both sorts send collaterals to the cerebellar nuclei, which they excite. The control is thus a balance

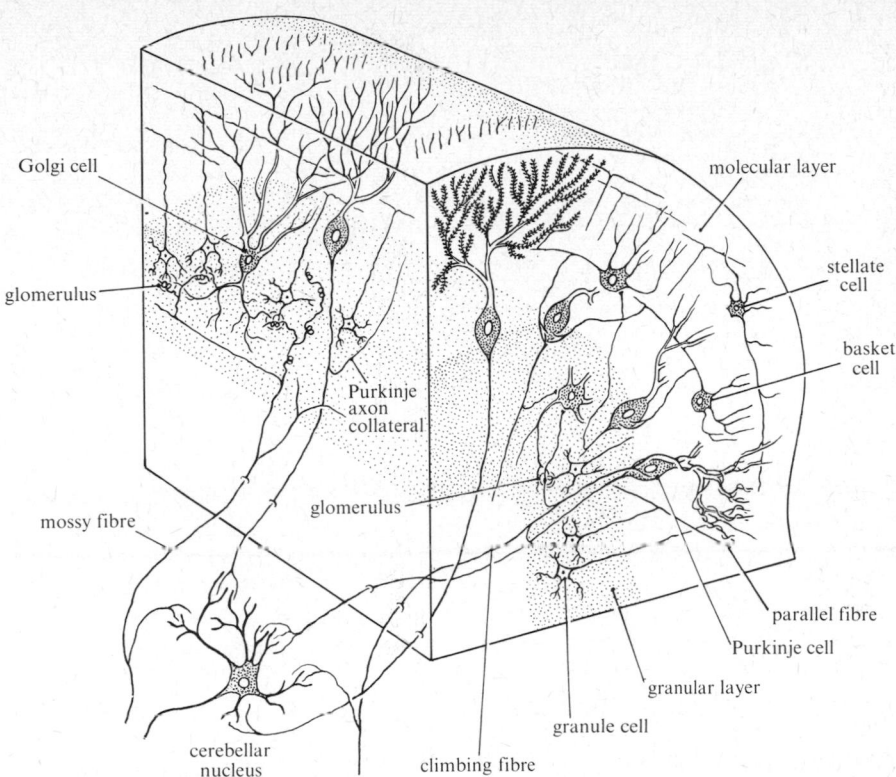

FIG. 31.6. Diagram of the principal cell types and connexions in one folium of a typical mammalian cerebellar cortex. Mossy and climbing fibres send branches to granule cell dendrites, Golgi cell bodies, and Purkinje cells. In the granular layer (shaded) mossy fibres and climbing fibres form associations in glomeruli with granule cell dendrites and Golgi cell axons. Granule cell axons pass to molecular layer and divide to form parallel fibres which synapse with Purkinje cell dendritic spines (indicated on only one cell). Purkinje cell dendrites are restricted to planes perpendicular to the parallel fibres. Collaterals from the Purkinje axons connect with Golgi, basket, and stellate cells, and these two latter inhibit the Purkinje cells. Purkinje axons constitute the only output from the cortex.

between this excitation and the inhibition continuously imposed by the cerebellar cortex and increased or decreased by it (Ito 1974; Palay and Chan-Palay 1973).

5. The two inputs to the cerebellar cortex

The input to the inferior olive from the spinal cord is somatotopically organized, and the cells send climbing fibres to particular areas of the cerebellar cortex (Fig. 31.7). Electrical stimulation of particular points on the cerebellar cortex will facilitate or inhibit the movements of individual muscles that are elicited by stimulation of the cerebral cortex.

Each climbing fibre coming from the inferior olive divides, following exactly the branching of the Purkinje cell dendrites and making synaptic contact only with smooth portions of its branches, not with spines. These contacts are powerfully excitatory, so that any impulse arriving in the climbing fibre excites one in the Purkinje

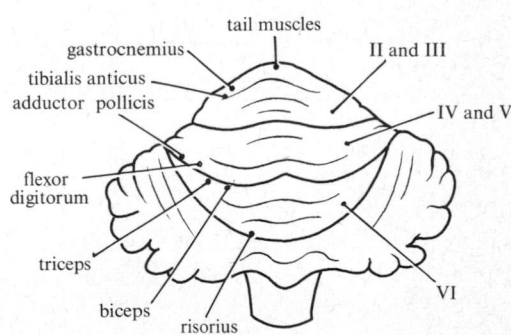

FIG. 31.7. Functional localization in the anterior lobe of the cerebellum of the macaque. Electrical stimulation of the points indicated inhibits or facilitates cortically induced movements of the muscles shown. (From Fulton, after Nulsen, Black, and Drake (1948). *Fedn. Proc. Fedn. Am. Socs. exp. Biol.* **7**.)

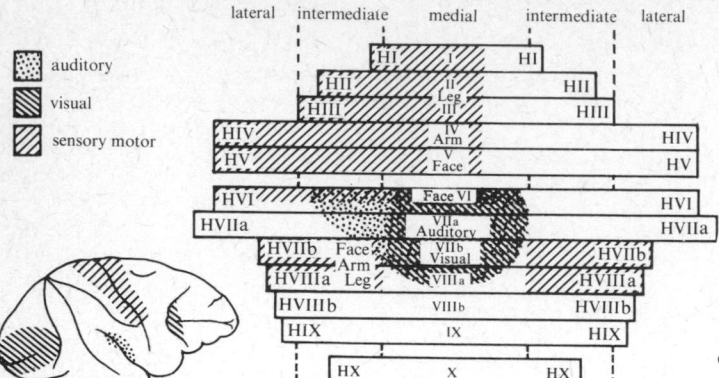

auditory
visual
sensory motor

lateral   intermediate   medial   intermediate   lateral

FIG. 31.8. Schematic representation of some cortical projection areas on a monkey cerebellar cortex. (From Bell and Dow 1967.)

cell, that is to say, increases its inhibitory effect on the nuclei. Since there are several times more Purkinje than olivary cells this represents a powerful amplification of inhibition, presumably by branching of the climbing fibres in the white matter.

The input from the cerebral cortex consists of mossy fibres from the pons. There is a general correspondence in the distribution from the spino-olivocerebellar and cerebro-cerebellar projections, but the exact relations have not been determined (Fig. 31.8). The mossy fibres make contact with very numerous small *granule cells* lying below the Purkinje cell bodies. The axons of these are very fine ($0\cdot1$–$0\cdot3$ μm) and run up into the molecular layer and then *along* the length of each folium, that is, across the Purkinje cell dendrites at right angles (Fig. 31.6). Each mossy fibre ends in a series of rosettes in

glomeruli, where it meets the dendrites of 20 or more granule cells. It excites these and they send impulses along their axons which make contacts with the spines of each Purkinje cell as they cross it. These contacts *inhibit* the Purkinje cell discharge, that is, release the tonic influence of the cells of the cerebellar nuclei from inhibition by the Purkinje cells.

The spinal mossy fibre input is also highly organized somatotopically. Single fibres of the dorsal spino-cerebellar tract (DSCT) have highly restricted fields of skin or come from a single muscle. Those of the ventral tract (VSCT) have wider fields and there are also reticulocerebellar fibres (RCT) with very wide fields indeed. Each mossy fibre projects to a region of cerebellar cortex similar to that receiving climbing fibres from the same region via the olive.

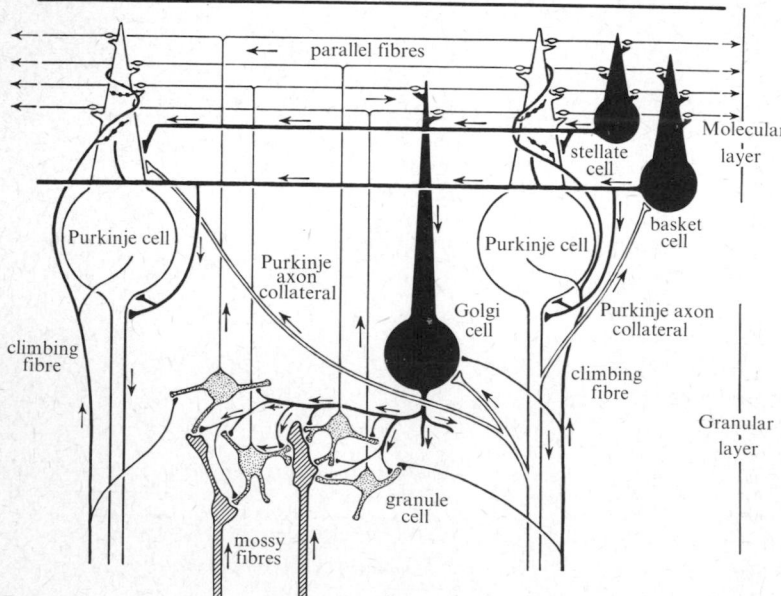

FIG. 31.9. Diagram showing principal neuronal connexions in the cerebellar cortex. Arrows indicate the directions of impulse propagation. (After Eccles, Ito, and Szentágothai 1967.)

6. Inhibition within the cerebellar cortex

The response of the parallel fibres is probably further focused by other neurons, the *Golgi cells*. These have cell bodies just below the Purkinje cell layer and huge dendritic trees in the molecular layer. Unlike the trees of the Purkinje cells, these are not restricted to a single plane but are cylindrical, extending for 300 μm in both directions (Fig. 31.6). These dendrites carry spines that make contacts with the parallel fibres. The Golgi-cell body also receives contacts from mossy fibres and other sources.

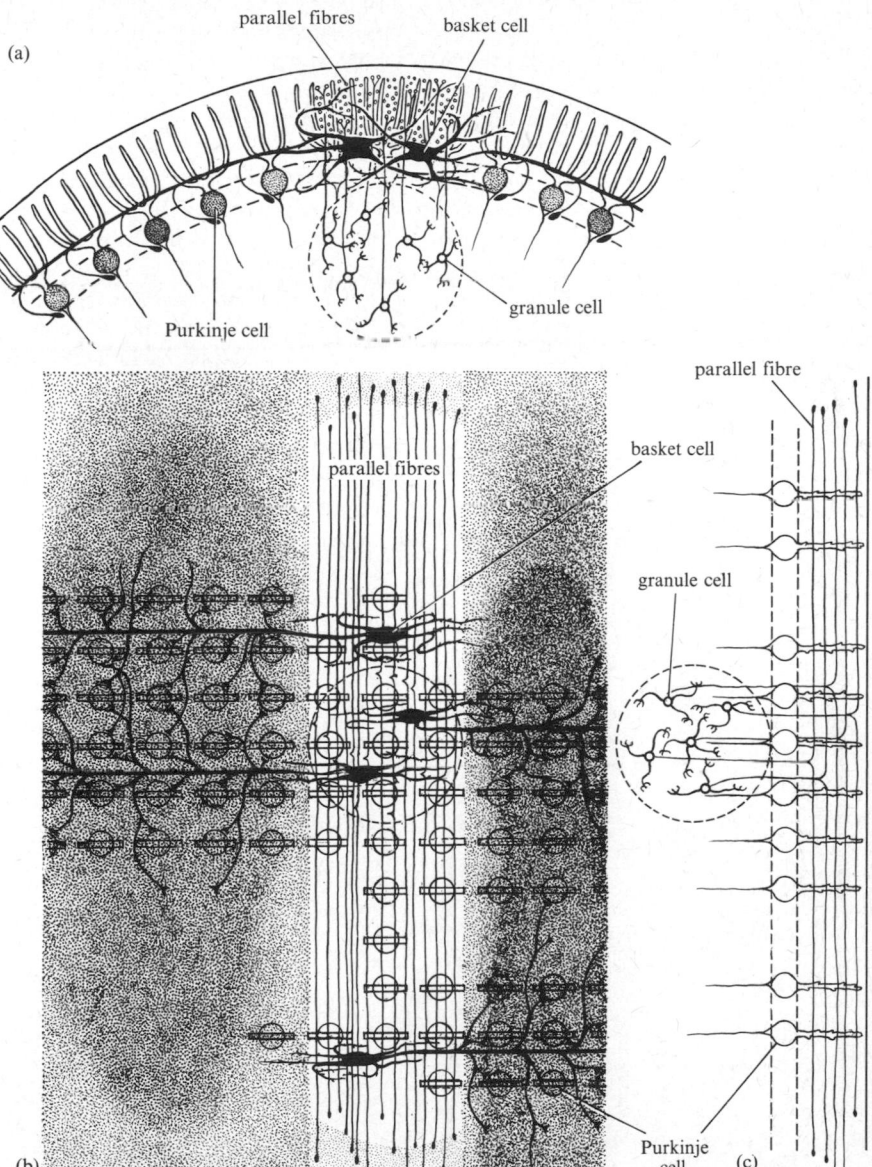

FIG. 31.10. Diagram illustrating the concept of the higher-order intergrative unit of mossy afferent-parallel fibre neuronal chain.

Part of a folium is represented (a) in transverse section, (c) in longitudinal section, and (b) in surface view. It is assumed that all granule cells in the broken circles are excited simultaneously resulting in the stimulation of all the Purkinje cells (depicted conventionally) along a longitudinal strip of length 3–4 mm (b). Basket-cell neurons in the strip are also excited and exert an inhibitory effect on either side of it. The degree of inhibition is indicated in (b) by general shading and in (a) by hatching of the Purkinje cell bodies, and depends on the number and magnitude of the connexions. Normally it extends to 10 rows of Purkinje cells. (From Szentágothai 1965.)

The Golgi cell axons branch very extensively in the granular layer, where they enter the glomeruli and inhibit the granule cells. This pathway thus provides feed-back that serves to focus the effect of mossy fibre input from a single muscle or other source. The parallel fibres excite the Golgi cells, which in turn prevent the discharge of all but the most strongly excited granule cells.

A further inhibitory pathway to the Purkinje cells is provided by the *basket cells*. The cell bodies of these lie superficial to the Purkinje cells, and their dendrites ramify in the molecular layer in the same plane as the Purkinje dendrites (Fig. 31.6). The axons give the basket cells their name by their terminations around the Purkinje cells, which they inhibit directly. The *stellate cells* are yet another set with essentially the same action. There are thus complicated feed-back and feed-forward loops regulating the activity of each Purkinje cell (Fig. 31.9).

The effect of mossy fibre excitation is, therefore, to produce a beam of excitation (Fig. 31.10). How can such beams serve for co-ordination of muscular action?

7. The cerebellum and the control of movement

The effect of the discharge of proprioceptors from a muscle is first to inhibit the Purkinje cell output over the mossy fibre–granule cell pathway, with a delay of 10–20 ms. This inhibition is then turned off by the powerful excitatory action of the climbing fibres, which also operates after a delay of about 20 ms, imposed by the synapses on the way up. By the interaction of these inputs with those from vestibular, cerebral, and other sources the cerebellum with its nuclei and the upper motor nuclei carries a pattern of activity representing the state of the whole muscular system from moment to moment and relates this to the environment and to the commands issuing from the higher centres. We have at present no way of picturing this complex pattern of action either in words or any other symbols. Physiological investigation has shown much of how the various pathways operate, but little of the total pattern.

There is certainly a background of activity in many parts of the system. The cerebellar nuclei show sustained activity that is excitatory to the higher motor nuclei and produce a pattern of posture. This pattern

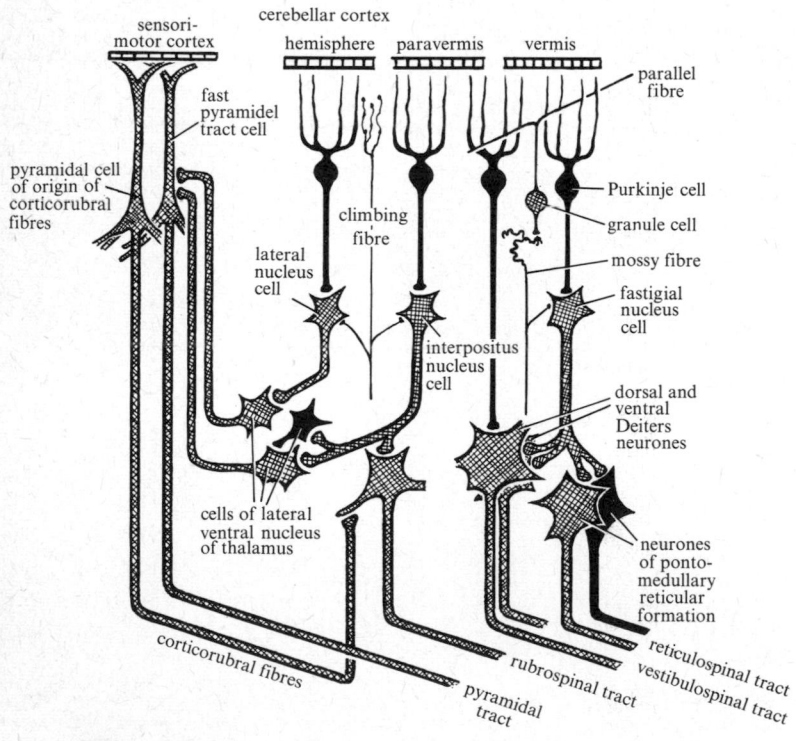

Inibitory

Excitatory

FIG. 31.11. Connexions in the cerebellar efferent system. (After Eccles, Ito, and Szentágothai 1967.)

is partly the result of 'spontaneous' activity of the neurons concerned and also of circuits sustained by gravitational or other stimuli. There is also a background of low-level sustained activity in the Purkinje cells. The pattern of action produced by any external event or internal cerebral command plays upon this complicated background activity of the motor nuclei, cerebellar nuclei, and cerebellar cortex. The actions of the various muscles produce shifting patterns of bands of inhibition of Purkinje cells passing across the cerebellum. Single narrow bands originating from small muscle groups (via the DSCT) are presumably integrated and controlled by the larger elements produced from whole limbs (by the VSCT) and indeed from the whole body (through the RCT).

The cerebral cortex is the prime generator of patterns of movement in mammals and sends its instruction to muscle groups partly direct (p. 323), but also through the pontine nuclei, to the cerebellar regions, which also receive information from these same muscle groups by the spinocerebellar and spino-olivocerebellar pathway. The pattern of input and output of the cerebellum is shown very diagramatically in Fig. 31.11. This of course includes only a few of the pathways—it is hard to imagine the ceaseless flow of activity in such circuits, remembering that there are 10^7 Purkinje cells and no less than 10^{10} granule cells.

8. The cerebellum—a timing device?

For efficient action it is essential that movements should be properly timed, and it may well be that one particular function of the cerebellum is to ensure correct sequencing and timing. The information from the direct spinocerebellar path arrives ahead of that from climbing fibres. We do not know how the cerebropontine information relates to these other two. Between them they allow the possibility to programme movements so that they start and stop at the right times. The very small diameter of the parallel fibres allows them to conduct at the low rate of $0 \cdot 3$ m s^{-1}, and thus perhaps to introduce delays in the sequences of inhibition and disinhibition via the Purkinje cells. It is uncertain exactly how this mechanism operates but most actions are pre-programmed (ballistic), and the cerebellum may provide the means for such actions (Braitenberg 1967). The commands for many activities include instructions not only when to start but also when to stop. This shows well, for example, in playing a musical instrument. A 'cellist or pianist must move his fingers over long distances and stop at exactly the right point *without any further information or feed-back to guide him*. From its initiation the movement is timed when to stop as well as when to start. The cerebellum

has possible equipment for such timing, since a command from the cortex can both begin the movement and set in action a delay line to stop it after a selected period.

The cerebellum is especially large in those animals that depend upon accurate timing in the use of their sense organs, especially those that emit a beam for detection by reflection as in radar. Thus the Mormyrid fishes, which live in muddy river water, detect objects by emitting electric pulses which are reflected back and may then be detected by special receptors. The cerebellum of these animals is enormous and its numerous narrow folia cover the whole of the rest of the brain (Fig. 31.12). Similarly in bats and in whales, which use sonar, the cerebellum is very large. In fact it is well developed in all vertebrates that make rapid, well-timed movements (for example, fishes and birds) but quite small in lampreys, amphibians, and many reptiles. The macroscopic shape of the cerebellum—its division into transverse folia—indicates how it is organized as a system of delay lines. The parallel fibres run for great distances, perhaps as much as 10 mm (Braitenberg 1967), presumably allowing for the projection of actions for an appreciable time ahead. The full length of a folium is 10 cm or more in man, allowing for timed delays of

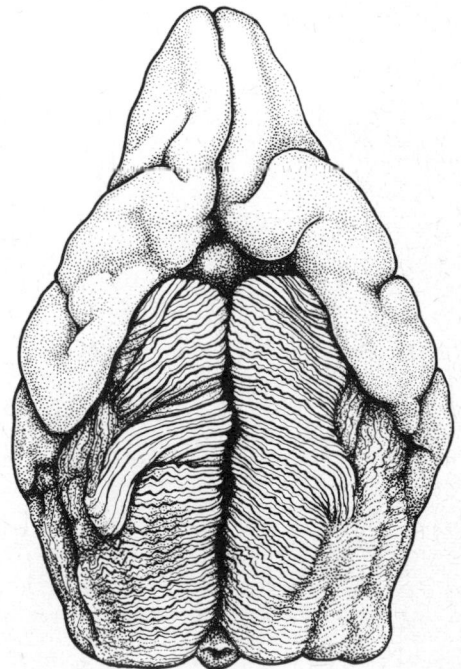

FIG. 31.12. Drawing of the dorsal aspect of the brain of *Mormyrus*. The enormously enlarged valvula cerebelli cover all other parts. (From a photograph in Nieuwenhuys (1967).)

up to 200 ms, which is a long time in a fast-moving creature. However, this is only one theory of the functioning of the cerebellum; quite a different one is that it serves to contain memories of patterns of movement. These could be encoded by the activation of particular sets of parallel fibres (Marr 1969).

9. Effects of removal or injury to the cerebellum

After removal or disease of the cerebellum a man or animal is unable to perform properly the voluntary movements initiated by the cerebral cortex. The actions are jerky, performed with too much or too little force and accompanied by oscillations—the classical 'intention tremor' of cerebellar disease. Movement may be in leaps, and progress once started cannot be stopped.

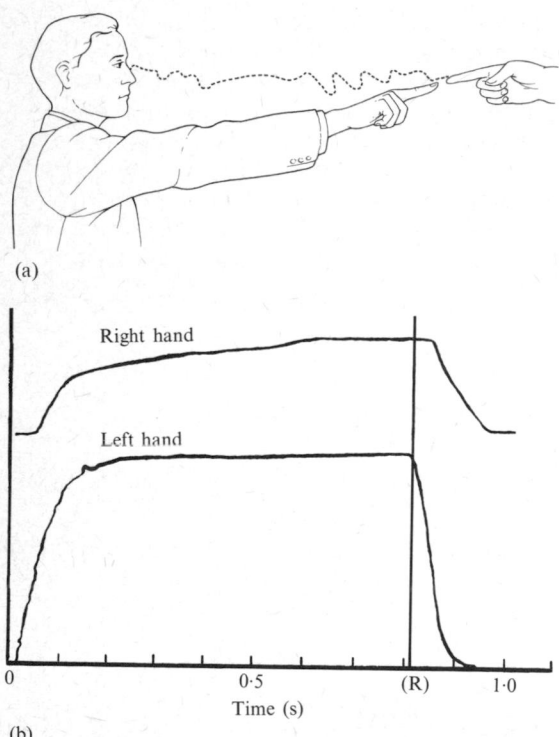

(a)

(b)

FIG. 31.13. Effect of cerebellar lesion on voluntary movement.
(a) Diagram showing how a man with a cerebellar lesion moves his hand. The patient is asked to move an index finger from his nose to the examiner's finger held at arm's length from the patient. The dotted line represents the course along which his hand moves and shows the typical tremor, most marked at the beginning and end of movement. (After Lyman (1950). *Medical diagnosis*. Saunders, Philadelphia.) (b) Voluntary muscle contraction in normal (lower curve) and affected arms (upper curve) of a man with unilateral cerebellar lesion. At the origin the signal to contract (grasp) was given and at R the signal to relax. In the affected arm the latency is longer, the contraction slower and weaker, and relaxation is delayed. (After Holmes (1917). *Brain* **40**.)

The course of the defect shown in Fig. 31.13 is often said to be that the movement is not guided by feedback. On the contrary, it may be that in the absence of correctly aimed and timed ballistic movement the action is performed as well as is possible under the guidance of visual feed-back. Speech depends very much on proper timing and after a cerebellar lesion words are pronounced only separately and deliberately. It may be that the exceptionally large development of the cerebellum of man is related to speech as well as to precise manipulation by the hands.

10. The midbrain

The dorsal and ventral parts of the midbrain are large and important in lower vertebrates but in mammals many of the visual and auditory computational functions of the dorsal parts have been transferred to the cerebral cortex. The roof (tectum) is divided into four hillocks, the *superior* and *inferior corpora quadrigemina* (Figs 29.1 and 29.2, pp. 272, 274). The superior are optic centres and receive some direct fibres from the optic tract; injury to them in man causes disturbance of eye movements. The inferior corpora quadrigemina receive, through the lateral lemniscus, fibres ascending from the cochlear nucleus of the medulla and they are intimately concerned with hearing (p. 333). Some of their neurons send axons forward to the medial geniculate body of the thalamus (p. 314).

The floor of the midbrain contains the nuclei of origin of the third and fourth cranial nerves, which are portions of the somatic motor cell column (Fig. 31.14). The greater part of this region is made up of a mass of grey matter, the *tegmentum*, including the *red nucleus* and neighbouring cell masses, which are part of the *reticular formation* (p. 271). This area receives fibres from the forebrain and cerebellum and sends fibres downwards to the spinal cord. It is an important higher motor centre, and because it is additional to the direct or pyramidal motor pathway (p. 323) this region, together with the corpus striatum, inferior olive, and parts working with them are known as the *extrapyramidal pathway*.

As already mentioned, the midbrain base constitutes morphologically the front end of the somatic motor column and contains the representational neurons whose activities produce movement by large groups of muscles. Stimulation or lesions of the tegmentum produce curious aberrations of locomotion. For instance, there has been described in the cat a 'syndrome of obstinate progression'. Following an injury to the tegmentum the animal persistently walked forward until it damaged itself by battering against the wall. The effect of the lesion was evidently to release one of

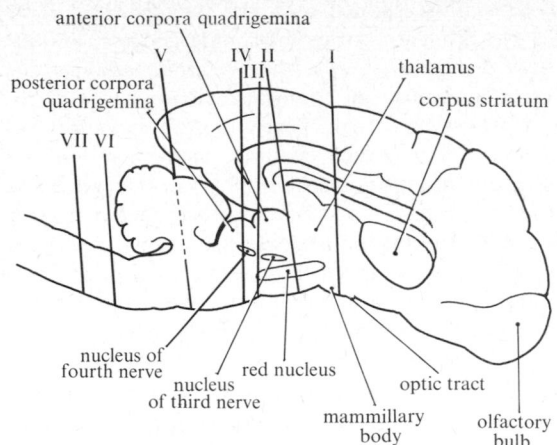

FIG. 31.14. Effect of section of the brain of the cat at various levels (I–VII), see text.

the motor patterns that are represented here. We can analyse these further by investigating the patterns of 'reflex' nature that are seen after isolation of this region from the influence of the higher nervous centres.

11. Postural reflexes

The basic mechanism for the control of the posture of a mammal is the stretch reflex, which has already been described (p. 264). If a cat whose spinal cord has been cut caudal to the medulla (Fig. 31.14, VII) is suspended off the ground the limbs hang limply. If one of them is now pressed against the ground it will extend, as if to bear the weight. This response is due to the setting up of nerve impulses by the proprioceptors of the extensor muscles. Such a *spinal animal* is not able to stand; the antigravity reflexes of the spinal cord operating independently cannot produce a pattern of contraction that ensures the balance of the animal (Creed, Denny-Brown, Eccles, Liddell, and Sherrington 1932).

If a cut is made across the midbrain behind the level of the red nucleus (Fig. 31.14, III) an exaggerated form of standing known as *decerebrate rigidity* is seen. When such an animal is hung with a sling round the belly the limbs are held stiffly, thrusting downwards and backwards. The tail and head are held up in spite of their weight. The animal maintains this rigidity for days, but it disappears if a second cut is made at a level caudal to the vestibular nucleus (Fig. 31.14, V); the limbs then hang flaccid, as in a spinal animal. Moreover, if the dorsal roots to a limb are severed the decerebrate rigidity disappears from that limb, showing that a supply of proprioceptor impulses is necessary for the phenomenon.

In decerebrate rigidity we see an exaggerated form

of part of the organized set of nervous patterns by which posture is maintained. Freed from the inhibitory influence of higher cerebral centres the vestibular nucleus uses the signals provided by the proprioceptors of the antigravity muscles to produce a combined action of all these muscles. This is a manifestation of the pattern of interactions in the nucleus that represents the posture of the animal. The rigid position adopted is a sort of exaggerated standing but it does not serve to support the animal properly. With care a cat showing decerebrate rigidity may be propped up on its limbs, but it is readily pushed over.

The proper maintenance of position depends upon the discharge of receptors that provide the information that ensures adjustment of the tension of the muscles to meet the changing environmental conditions. Four main sets of receptors are involved. (a) Those in the skin of the feet and the proprioceptors of the small muscles of the toes. (b) The proprioceptors of the antigravity muscles of the neck. (c) The gravity receptors in the labyrinths. (d) The eyes. This information is assembled in the centres in the midbrain and medulla oblongata and a dog or cat with the brain-stem severed in front of the midbrain (Fig. 31.14, I or II) stands normally, indeed it can walk. In primates such a lesion produces abnormal posture and no walking is possible; in these animals, therefore, the representation that controls posture lies mainly in the forebrain.

The skin of the feet provides part of the information for the postural reflexes. Gentle pressure with a finger on the skin of the foot of a dog after decortication is followed by extension of the limb to meet the finger. This *magnet reaction* or supporting response is due primarily to stimulation of the skin but slightly stronger pressure causes separation of the toes and stimulation of the proprioceptors of their small muscles, which also play a part in the phenomenon.

The *neck reflexes* can be investigated if the labyrinths are first removed. Passive turning of the head to one side then produces a thrust of the limbs of that side. In a mammal the head carries most of the main exteroceptors and slight changes in its position occur continually as eyes, ears, or nose are directed towards some change in the surroundings. All of these movements produce corresponding changes in the posture of the limbs, preparing for subsequent advance or retreat. Similarly, if the head of an animal without labyrinths is turned dorsally, the forelimbs extend and the hind ones become flexed. The converse movements are seen if the head is turned downwards.

In order to study the effects of the labyrinths on posture the animal is decerebrated and the neck reflexes also eliminated by cutting the dorsal roots of the three upper pairs of cervical nerves. It is now found that, for every position of the head relative to gravity, contractions appear in the muscles of the back and limbs, these contractions being such as would be appropriate to maintain the head and back in a horizontal position. The contractions are elicited by the two sets of receptors in the labyrinths, those of the semicircular canals, which discharge when there is angular acceleration of the head in any direction and those of the maculae of the utricle and saccule, which by the pull of the otoliths upon the hairs provide positional information (Chapter 43).

Every movement of the head relative to gravity thus sets in action a series of *righting reflexes*, which tend to keep the head the right way up. In life these are combined with further righting reflexes initiated by the pull upon the muscles that follows movement of one part of the body upon another. These ensure that the rest of the body follows the head into the correct relation to gravity. This relation will, of course, differ according to whether the animal walks upon four legs or two and, as mentioned already, in the sloth, which hangs from branches, the 'correct' position is 'upside down'.

Posture and righting thus depends upon the interplay of a large set of afferent impulses from the skin, muscles of the legs and neck, and the labyrinths. In life the eyes provide a further most important contribution. All of this information is assembled in the nuclei of the vestibular region and brain-stem, whose neurons compute at every instant the output to the muscles that will ensure correct posture. In lower mammals this can be assured by these brain-stem centres working alone, but in primates the appropriate representation also involves action by the cerebral hemispheres.

This whole complicated system of receptors, nerve impulses, and centres is obviously a predictor that is of fundamental importance to the maintenance of the life of the animal. It is mainly the product of heredity, at least in non-primate forms, though in the present state of knowledge it would be difficult to deny that learning plays a part in the establishment of the representation.

During the normal life of a mammal this whole apparatus is continually at the service of the associational systems of the forebrain. These compute in any situation the output that is likely to be effective in the light of past experience. This output passes over the fibres of the cerebrospinal, cerebrorubral, and other motor pathways, calling into action appropriate parts of the motor systems that they contain.

32   The hypothalamus

1. General plan of the forebrain

THE region of the third nerve marks the anterior end of the somatic motor column, and the whole forebrain consists of an expansion forwards of the dorsal portion of the neural tube. In the main this represents a vast extension of the somatic sensory region, developed originally as a computer for the signals from the olfactory receptors, but coming later to dominate all the rest of the brain.

In mammals the olfactory system is still connected with the forebrain separately from all the other receptor systems and by a simpler pathway (p. 410). It occupies only the basal region of the anterior portion of the hemispheres. The greater part of their bulk is made up of the tissue of the cerebral cortex, which receives signals from nearly all the receptor systems of the body. It records the patterns in which these signals have been associated and so stores a representation of the external environment. This information store is then used to compute outputs that produce anticipatory actions adequate to maintain the life of the animal. In other words, the cerebral cortex is the part of the animal particularly devoted to the establishment of stores of information received during the life of the animal, rather than from heredity. But of course its activities are profoundly influenced by genetic information and, on the other hand, it is not the only part of the brain concerned with memory.

In order to understand how the pathways and centres are arranged we may consider that the forebrain develops from a simple tube (Fig. 37.13, p. 354). Near the front end a pair of lateral pouches, the *cerebral hemispheres* (*telencephalon*), develops and their cavities are the two lateral (first and second) ventricles, communicating by the *interventricular foramina* with the median third ventricle, whose walls form the unpaired between-brain or *diencephalon*, connecting the hemispheres with the rest of the brain. The pathways from the receptor organs (other than olfactory) to the cortex make synapse on the way in the dorsal part of the diencephalon, the *thalamus*. The cortex exerts its effects on the rest of the brain by a number of pathways lying more ventrally. Some of these proceed directly to the spinal cord (the *pyramidal tract*), but others make synapse in centres in the base of the forebrain, including the *corpus striatum* and other parts of the *extrapyramidal motor system*. The *hypothalamus*, the part of the diencephalon ventral to the thalamus, is concerned with the production of co-ordinated visceral and other activities and is also connected with the cortex by fibres passing in both directions.

These *basal ganglia* of the forebrain constitute a series of groups of cells that set into operation elaborate patterns of motor activities. There is little understanding of how they do this, but we may consider that there is a hierarchy of levels from the spinal cord up through the medulla, midbrain, hypothalamus, and thalamus to the cortex. Information is passed upwards and commands downwards, but there is also very much reciprocal action along these pathways and the analogy of a social hierarchy is in fact inadequate. The system monitors its own input, and so produces motor actions adequate to the needs of the organism. The ventral parts of the cerebral hemispheres (basal ganglia) have a special rôle in this system of control. The corpus striatum acts with the cortex in the control of movements. The amygdala and hippocampus may be said very generally to be involved in the regulation of the exploratory and emotional responses (p. 310). There is evidence that in animals and men damage to these parts leads to disturbances that may in part be called emotional, whereas injury to the corpus striatum leads to errors of movement.

So there is an elaborate series of sets of cells regulating behaviour by reciprocal interactions upwards and downwards. Only in a general way should we say that the cells at each level *control* groups of cells at lower levels. All levels are involved in the more elaborate and highly co-ordinated movements. The appropriate predicting and sequencing of movements probably depend upon feed-back action in which 'lower' centres send 'efference copies' of the commands they issue back to the 'higher' centres.

The whole forebrain forms an immensely complicated apparatus, and it is absurd to expect to be able to understand it in terms of simple analogies. All that can be done here is to give an account of the main masses of tissue and their connexions, with some indications of the results of removing them or of activating them by electrical stimulation. No language is available with which to speak adequately of the functioning of the forebrain; suitable terms will probably only evolve gradually, perhaps helped if men make machines that imitate these complicated actions more closely.

2. Transverse connexions in the forebrain

The front wall of the tube, between the hemispheres, is the *lamina terminalis*, through which fibres run between the two sides to form the *anterior* and *hippocampal commissures*. In the human brain the enormous growth of the cerebral hemispheres results in a backward displacement of the hippocampal commissure and in the adult only the anterior commissure can be seen in its original position in close relationship to the lamina terminalis at the front end of the third ventricle.

As the hemispheres grow large they bulge forwards and backwards and come to lie over the diencephalon. The floor of the evaginated hemisphere, the corpus striatum, thus lies immediately lateral to and in contact with the thalamus (Fig. 37.15, p. 355). New commissural fibres come to join the hemispheres, forming the *corpus callosum*.

3. Hypothalamus

The thick walls of the diencephalon form the thalamus above and hypothalamus below, separated by a sulcus which is probably a continuation of the sulcus limitans, separating the alar and basal plates (p. 352). The hypothalamus, therefore, can be considered as the anterior end of the visceral sensory and visceral motor columns. It acts as the central regulator of the internal operations of the body and co-ordinates them with external events (Haymaker, Anderson, and Nauta 1969). It receives information from many other parts of the brain and its cells are also sensitive to changes in the composition of the blood. They act partly by producing the releasing factors that regulate the activities of the pituitary, which are carried in the hypophysial portal system down the stalk of the infundibulum, attached to the base of the hypothalamus (p. 442). The cells of the hypothalamus also evoke complicated patterns of behaviour, involving both visceral and somatic components, for example, the actions of feeding, sex, aggression, or going to sleep. Besides such actions

produced by mainly descending pathways, the hypothalamus also sends impulses upwards through the thalamus to the cortex. It is involved in some of the most complex cerebral activities, and indeed there is evidence that parts of it are essential for memorizing (p. 341). It provides the link by which the needs of the body ensure that what is recorded in the memory is appropriate and likely to improve chances of survival (see Adey and Tokizane 1967).

It is not too much to say that the operations of the hypothalamus lie close to the centre of the entity that we recognize as the 'personality' or 'character' of a man or animal. Some of the most significant and recently evolved features of man probably depend upon its special properties, for instance, our very long childhood, readiness to co-operate, and non-cyclic sexual impulses. A particular practical importance is the fact that anti-ovulatory contraceptive pills act upon its neurons (p. 478).

Corresponding to its central functions the hypothalamus lies in between higher and lower centres and is reciprocally connected with both. It can be regarded as a forward continuation of the reticular formation of the bulbar and mesencephalic regions and in turn continues forward to the septal nuclei of the medial walls of the hemispheres. This whole long stretch can be regarded as a single entity called the reticular formation or alternatively the 'isodendritic core' of the brain-stem (p. 271).

Studies by stimulation and extirpation show a considerable localization of centres in the hypothalamus. For instance, there is one making for more eating and another for less. It is a puzzle that in many places these centres do not correspond to any well-marked anatomical units. It may be that the neurons with particular functions are often intermingled. Indeed this may be the very feature that ensures their interaction to produce effective action patterns. Nevertheless there are some well-defined nuclei (Fig. 32.1) and the following are the chief subdivisions.

Anterior hypothalamus
Anterior hypothalamic area
Preoptic nucleus
Suprachiasmatic nucleus
Supraoptic nucleus
Paraventricular nucleus

Middle hypothalamus
Dorso-medial nucleus
Ventro-medial nucleus
Arcuate nucleus

Posterior hypothalamus
Posterior hypothalamic area
Premammillary nucleus
Mammillary nucleus

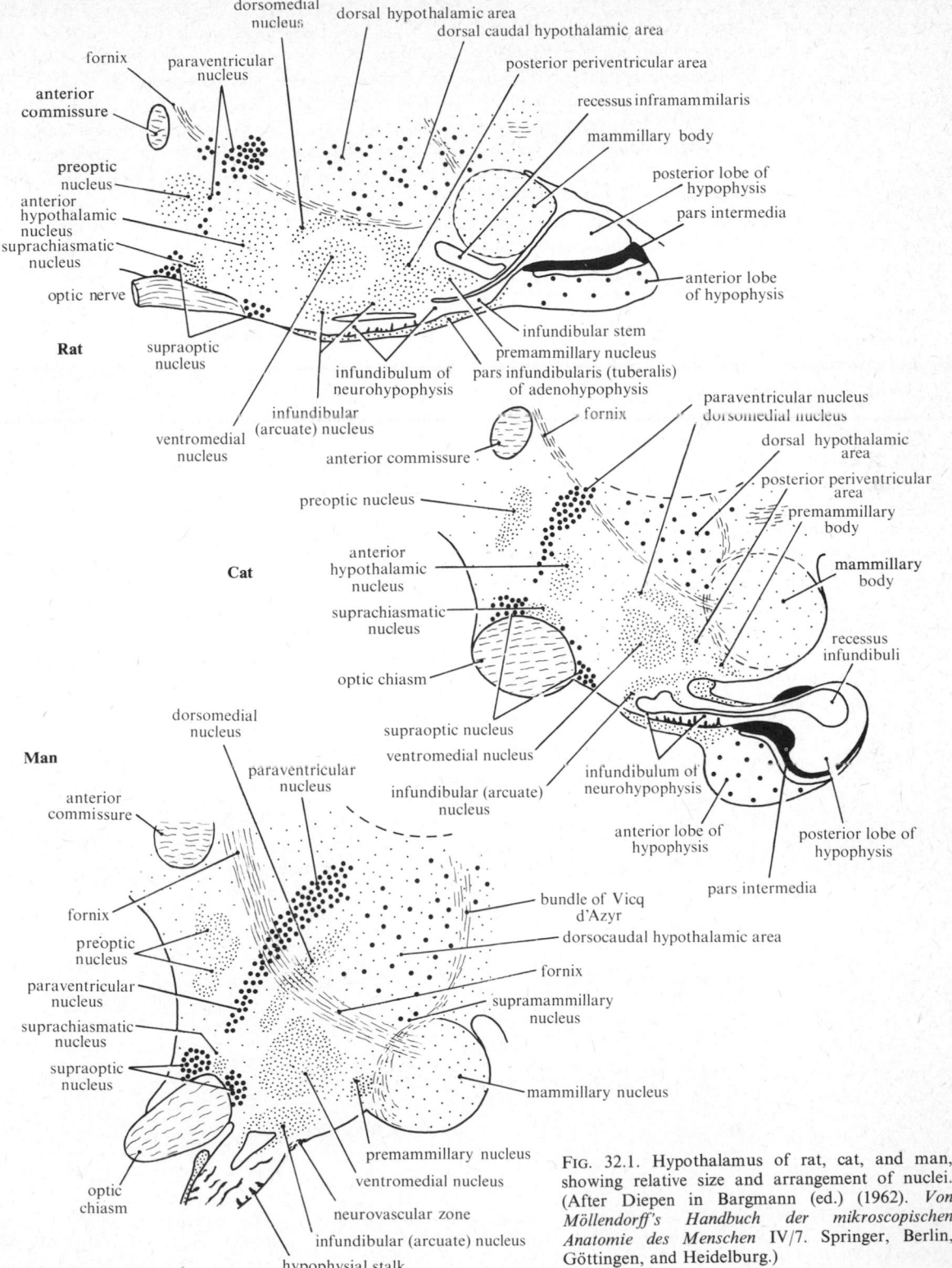

FIG. 32.1. Hypothalamus of rat, cat, and man, showing relative size and arrangement of nuclei. (After Diepen in Bargmann (ed.) (1962). *Von Möllendorff's Handbuch der mikroscopischen Anatomie des Menschen* IV/7. Springer, Berlin, Göttingen, and Heidelburg.)

4. Afferents and efferents of the hypothalamus

Direct fibre systems connect the hypothalamus with the midbrain tegmentum and central grey, the thalamus, the septum, and the hippocampus. There may be a direct connexion with the prefrontal cortex and there are certainly indirect ones through the thalamus. There is some evidence of a direct retino-hypothalamic pathway.

Many of these pathways conduct in both directions and the hypothalamus is both influenced by and influences many areas. The connexions can be summarized as follows:

(a) there are few, if any, direct connexions from any sensory pathways except the olfactory;

(b) the main connexions are reciprocal ones with the limbic forebrain (hippocampus and septum) and with the midbrain;

(c) there are efferent projections through the reticular formation to the visceral motor nuclei.

The functional state of the hypothalamus is thus continually influenced both from above and below, no doubt in highly specific and particular ways according to the situation. Its 'centres' are certainly affected by particular afferent signals, for instance, from the stomach in the regulation of hunger or the vagina in the control of ovulation (at least in some animals, p. 480). These signals arrive over complex polysynaptic pathways.

The influences upon the hypothalamus from above include the fibres from the hippocampus via the fornix (p. 318). These end especially in the mamillary body and from here the mammillo-thalamic tract leads to the anterior nucleus of the thalamus. This in turn is one of the many projections to the cingulate cortex. From there fibres reach through the entorhinal area to the hippocampus completing a circuit, known after its discoverer as the Papez circuit or the circuit of emotion. It is significant of the nature of hypothalamic function that injury to this circuit leads both to emotional disturbance and to failure to record in the memory (p. 343).

5. Regulation of endocrine activity

The hypothalamus plays a part in so many activities of the body that it is impossible to give a complete list. Moreover, it may enter into an activity either directly by its nervous output, indirectly through control of pituitary releasing factors or still more indirectly by the further effects of their hormones. Electrical stimulation of the hypothalamus produces many phenomena through the release of TSH, ACTH, or growth hormone (STH) (p. 439). Some of the most important actions of its cells are in the control of release of the hormones affecting reproduction, including lactation (p. 449). Through the posterior lobe of the pituitary it regulates the release of vasopressor and oxytocic hormones (p. 445).

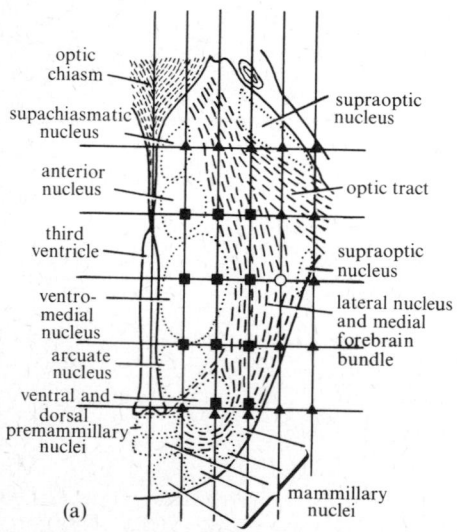

(a)

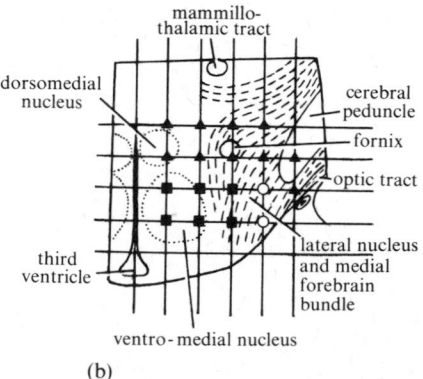

(b)

▲ Normal food intake
O No feeding behaviour
■ Hyperphagia

FIG. 32.2. Rat hypothalamus at the level of the ventromedial nucleus (a) in horizontal section, (b) in cross-section, showing feeding behaviour after small bilaterally symmetrical lesions in areas indicated. Horsley-Clarke co-ordinates superimposed at intervals of 0·5 mm. (After Krieg, after Anand and Brobeck (1951). *Yale J. Biol. Med.* **24.**)

6. Regulation of feeding and growth

The control of food and water intake depends upon the activation of specific neural mechanisms. Injury in the lateral parts of the hypothalamus leads to failure to eat or drink. Conversely, upon appropriate electrical, osmotic, or chemical stimulation in this region an animal will immediately begin to eat or drink. Cells in the ventro-medial hypothalamus have the opposite effect of inhibiting eating. After injury to them an animal will become very greedy and grow to an excessive size (Fig. 32.2). These centres probably regulate intake in normal life, basically by the genetically determined setting of their activity. This is then modulated by chemical signals, such as the level of glucose in the blood, and nervous signals such as are produced by the dryness of a thirsty mouth or contractions of an empty stomach. They are also, of course, open to the influence of external signals of smell and taste. All of these influences, and others, need to be considered in assessing the regulation of feeding behaviour.

7. Control of temperature

The hypothalamus is the central agent for the control of temperature (Fig. 32.3). Some of its cells have a genetically predetermined set-point at 37 °C (in man), and vary their output with changes in the temperature of the blood. After a lesion in the anterior hypothalamus an animal may lose the capacity to regulate its temperature downwards. Electrical stimulation here inhibits shivering, so that the temperature falls. Cooling of this region initiates the operation of heat productive actions and vice versa, and neurons have been found that change their firing rate with changes of 1 °C.

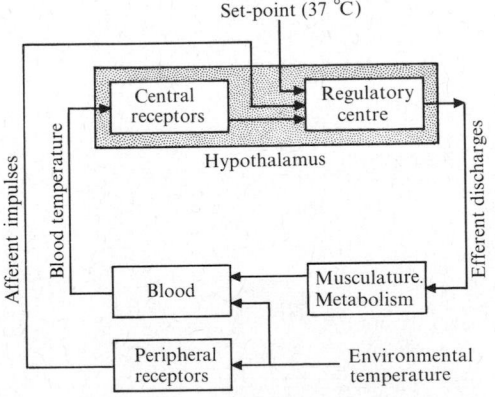

FIG. 32.3. The basic mechanism of thermoregulation. (Redrawn from Myers in Haymaker, Anderson and Nauta (ed.) (1969).) *The hypothalamus*. Courtesy of Charles C. Thomas, Publisher, Springfield, Illinois.

Stimulation in the posterior hypothalamus produces the opposite effects—that is, shivering is increased, but this area is not itself thermosensitive. If a warming device is inserted here the animal can be cooled to many degrees below normal. The receptors for heat and cold in the skin therefore contribute only in a minor way to this regulation, probably more by giving warnings that cause the animal so to behave as to avoid overloading the temperature regulating mechanism (p. 14).

8. Control of sleep

The biological significance of sleep remains uncertain, but it is a state into which nearly all birds and mammals periodically pass. Its regulation involves the hypothalamus and after lesions here, monkeys, cats, or rats will pass into a somnolent condition lasting for many days. After other lesions in rats there was complete loss of sleep periods, combined with excessive responsiveness to stimuli. These phenomena are probably due to interruption of a connexion between the lower levels of the reticular system and the hypothalamus (see Akert, Bally, and Scháde 1965; Oswald 1962).

By electrical stimulation in the hypothalamus a cat can be made to curl up and go to sleep within a few seconds. On the other hand, stimulation of the reticular activating system in the bulbar region produces arousal.

Sleep is accompanied by characteristic changes in the electroencephalogram (EEG). The alpha rhythm disappears and is replaced by characteristic slow waves. This is the condition of 'orthodox' sleep, which may be of various 'depths'. But there is another type called by various names such as fast, paradoxical, or rapid eye movement (REM) sleep (or dream sleep, because it is the phase when we dream before waking). It is not known how these two types are regulated, but in man periods of the two alternate through the night. If the dream periods are prevented the sleep is said to be unrestful.

9. Electrical stimulation of the hypothalamus

The representations of motor activities within the hypothalamus have been investigated by Hess (1949, 1957) by the method of implanting a small pair of electrodes in the head of a cat and activating these by remote control through an induction coil. This was one of the first investigations in which complicated behaviour patterns were produced in the unanaesthetized animal. The effects varied according to the position of the electrodes (Fig. 32.5). When they were in the periventricular grey matter of the caudal or dorsal parts of the hypothalamus (or the front part of the midbrain) visceral changes concerned with control of energy expenditure, such as rise or fall of blood pressure, alteration of heartbeat, or respiratory rate were seen. These changes are mediated through the sympathetic division of the autonomic nervous system, which prepares the body for attack or defence (p. 284). The hypothalamus may be regarded as the head centre controlling all the activities of that system (see Euler 1954). More elaborate preparatory actions were also seen. Following some stimulations the cat would snarl and its fur bristle and pupils dilate (Fig. 32.4). If approached when in this state it would readily scratch or bite. Many of the phenomena of this defence reaction are also controlled through the sympathetic system.

The hypothalamus also contains the head centre for the other division of the autonomic nervous system,

the parasympathetic, concerned with setting into action processes that conserve the body and build up its internal resources (p. 285). Electrodes placed more cranially, in the preoptic and supraoptic regions of the hypothalamus, produced movements involving groups of muscles concerned with the conservative functions mediated by the parasympathetic system. These include a whole range of activities of the gut, and others concerned with the finding and intake of food. Stimulation of different points produced defaecation, peristalsis, or vomiting.

From the most cranial parts of this region elaborate behaviour patterns were elicited, including sniffing, licking, and chewing, any of which might be seen after stimulation of points quite close together. They are similar to actions seen after stimulation of the hippo-

Fig. 32.4. Actions induced in cats by electrical stimulation in the hypothalamus. (a) Lapping; (b) licking; (c) and (d) sniffing; (e) dilation of pupils; (f) sleep; (g) snarling; (h) persistent gnawing. (After Hess 1957.)

campus (p. 319), which sends fibres through the fornix to the hypothalamus.

Evidently this whole region is concerned with the integration of activities for detection, intake, and digestion of food and with control of the amount taken. The olfactory system, one of the chief distance receptors for the finding of food, lies cranial to this region and is no doubt often responsible for activating first the sniffing actions and then those of eating. The gustatory receptors next assist with control of salivation (also elicited from stimulation of the hypothalamus) and at each of the later stages of digestion appropriate receptors come into action, for instance by providing information of the state of distension of the stomach or bowel.

Electrical stimulation shows that many other co-ordinated actions can be produced by the cells of the hypothalamus, for example, coughing, sneezing, and micturition. As we have seen stimulation at some sites in the hypothalamus has the effect of causing the animal to close its eyes and lie down to sleep, which may be considered as an extreme form of the parasympathetic conservative behaviour (Fig. 32.4).

These elaborate behaviour patterns evidently involve actions of somatic motor as well as visceral systems, and the fact that such complex movements can be elicited from activation of localized regions warns us that only an artificial separation can be made between internal and external actions.

Some of the actions of the hypothalamus are purely somatic, for example, Hess showed that following stimulation in the region shown in Fig. 32.5(c) the animal began to walk up and down and to make vigorous efforts to escape. These are somatic activities at a still higher level than those of the midbrain base (p. 300) and in life they are presumably under the direct control of the cerebral cortex.

The output of the cortical computing system no doubt produces its effects largely by stimulating appropriate sets of cells in the hypothalamus and other

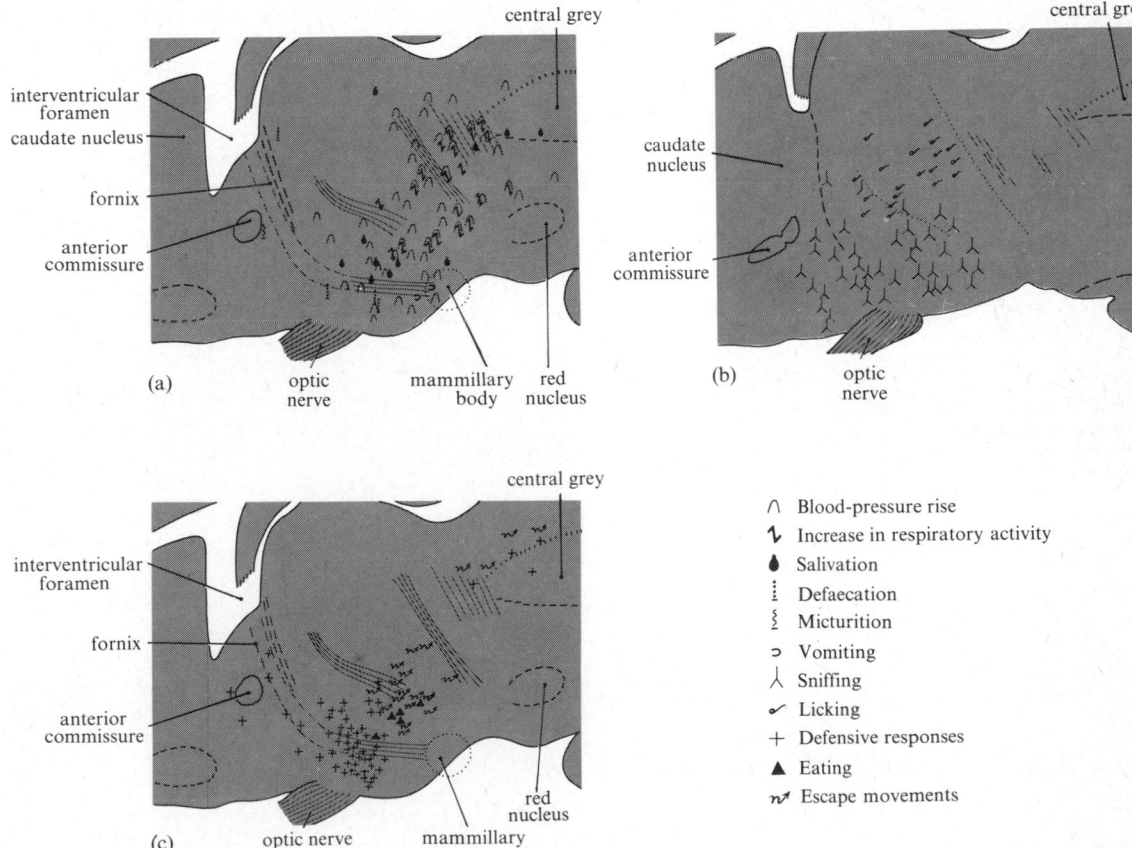

∩   Blood-pressure rise
ʅ   Increase in respiratory activity
🌢   Salivation
⋮   Defaecation
ʓ   Micturition
ͻ   Vomiting
ʌ   Sniffing
✓   Licking
+   Defensive responses
▲   Eating
𝑛ↄ   Escape movements

FIG. 32.5. Sites in the hypothalamus from which various responses were obtained on electrical stimulation. (After Hess 1957.)

higher motor centres. More complicated processes are also involved, for the hypothalamus sends fibres forwards to the thalamus and cortex.

Fig. 32.5 shows that many of the responses described can be elicited from stimulation of a wide area of the hypothalamus. This may be because they follow excitation of afferent or efferent pathways as well as of groups of cells. Yet the occurrence of such varied actions following stimulation of points so near together emphasizes the coupling action of these networks, by which the operations of so many motor systems are combined in diverse patterns.

10. The hypothalamus and emotion

The central nature of its activities gives the hypothalamus a special place in the generation of emotion and affect. Considered objectively it is the locus from which come the 'drives' that ensure the continuance of life whether by feeding, defence, or reproduction. Considered subjectively it is especially involved in the affective feelings of pleasure and pain, including hunger and its satisfaction, anger, and love and sexual feelings. Knowledge about these centres comes both from observation of human injuries and from animal experiment. People with lesions in the Papez circuit (p. 319) may show various conditions either of excessive emotion or of apathy and disinterest in life. All the

manifestations of anger can be provoked in a cat in which all the forebrain in front of the posterior part of the hypothalamus has been removed, but not if the cut is further back. The cat will hiss and growl, bare its teeth, arch its back, lash its tail, and put out its claws to scratch. These phenomena of 'sham rage' show that a whole pattern of action can be elicited from the hypothalamus alone. Of course the fact that rage can be elicited here does not mean that it is not normally influenced from centres further forward (p. 328). Nor does it tell us whether in any sense this can be considered as the locus of the subjective feeling of anger.

These questions are made even more difficult by the discovery of the possibility of electrical self-stimulation of the brain. Electrodes implanted, say, in the septal region or lateral hypothalamus of a rat can be made to deliver stimuli to the brain when the animal presses a lever. The animal will discover that pressing this lever is equivalent to a positive or pleasurable reward and will continue to press the lever until it is exhausted and falls asleep. There is evidence from stimulation of the brain in man under local anaesthesia that stimulation of the septal region produces pleasurable sensations ('I have a glowing feeling; I feel good'). These observations further emphasize the central position of the hypothalamus in the continuation of life (see Olds 1962; Trowill, Panksepp, and Gandelman 1969).

33 Basal ganglia and thalamus

1. The corpus striatum

THE lower parts of the cerebral hemispheres contain the regions known as the *basal ganglia*. They are concerned with the organization of the output of instructions from the cerebral cortex for the control of actions. There is a series of loops between the thalamus, cortex, and basal ganglia. These exert control partly by their influence through the motor cortex and the pyramidal tract, partly directly by descending pathways through the midbrain tegmentum (see Webster 1975).

The *corpus striatum* is the lateral part of the floor of each hemisphere (Fig. 37.15, p. 355). The evaginated wall of the hemisphere is here united with the side of the diencephalon. The striatum thus comes to lie immediately lateral to the thalamus (Fig. 33.2). The fibres of the internal capsule, proceeding to and from the cerebral cortex, run through the region and divide it

into several parts. The *caudate nucleus* and *putamen* together constitute the corpus striatum proper, which receives the input fibres and sends its output to the *globus pallidus,* from which in turn fibres pass to other parts.

The input to the striatum in mammals comes from three sources. (Kemp and Powell 1971) (Fig. 33.1).

(a) There is a topographically regular projection from all parts of the cerebral cortex, mostly of the same side, but partly from the other. The heaviest projection is from somato-sensory and motor cortices in agreement with the fact that the basal ganglia are associated with motor functions. There is little projection from primary sensory areas but there are also projections from association areas such as those in the depths of the superior temporal sulcus, which receive influences from the three major

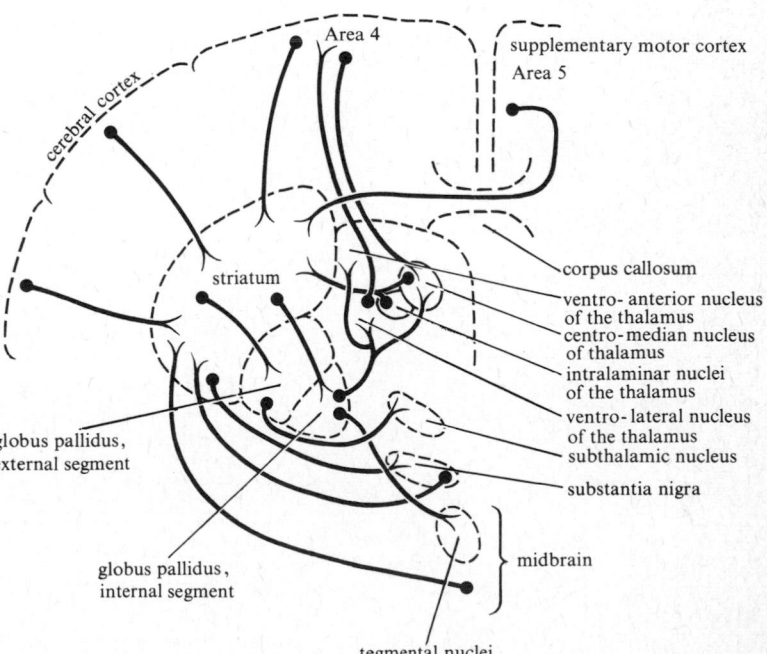

FIG. 33.1. Diagram of the main afferent and efferent channels of the corpus striatum and globus pallidus. Of the intralaminar nuclei of the thalamus only the centromedian component receives fibres from the globus pallidus but all these nuclei project to the striatum. (From Kemp and Powell 1971.)

sensory areas. The striatum thus allows convergence of influences from all parts of the cortex and in turn, as we shall see, influences the motor cortex. Projections from the various cortical areas overlap, so that no part of the striatum is influenced from one area only. Some of the efferent corticostriatal fibres are collaterals of fibres of the internal capsule proceeding to the pyramidal tract. It is not known whether this is a significant functional feature.

(b) There is a regular projection from the intralaminar nuclei of the thalamus to all parts of the caudate nucleus and putamen. These nuclei receive both ascending afferent and cortical projections (p. 314). Their overlapping projections to the striatum thus allow for further convergence of many influences.

(c) Numerous fibres project from the midbrain to the striatum, coming from the substantia nigra and the midbrain tegmentum near the red nucleus. These fibres are 'aminergic' (p. 289) and the striatum contains considerable quantities of dopamine.

The cells of the corpus striatum are stellate with numerous spreading dendrites, carrying spines. Endings from all three types of input fibre converge upon the same cell. Many of the axons of smaller striatal cells end within this centre itself. The efferent axons from the larger cells proceed to two destinations, the sub-

stantia nigra and the globus pallidus. From the former there are return connexions to the thalamus and to the striatum. The globus pallidus is the main output pathway. Its axons proceed to three destinations, the subthalamic nucleus, the intrinsic nuclei of the thalamus, and the midbrain tegmentum. The thalamic nuclei in question are all closely related to the motor cortex (for example, the ventral nuclei) and also receive fibres from the cerebellum. The strio-pallidum can thus sample the activity of the entire cortex, integrate this with the activity of intralaminar thalamic nuclei and midbrain tegmentum and, as a result, then send back influences to the motor cortex.

The efferent pathway from the pallidum to the midbrain may provide a more direct motor pathway to the spinal cord. Electrical stimulation in the region of the tegmentum where these fibres end gives facilitation of the gamma motor neurons, which influence movements by innervating the muscle-spindles (p. 266).

2. Thalamus and cortex

The cerebral hemispheres contain an elaborate set of centres providing the system by which much of behaviour is regulated. Various parts are usually recognized in them for reasons of convenience of description. In life they all co-operate, but there are distinct regions with different structures, and connexions which must

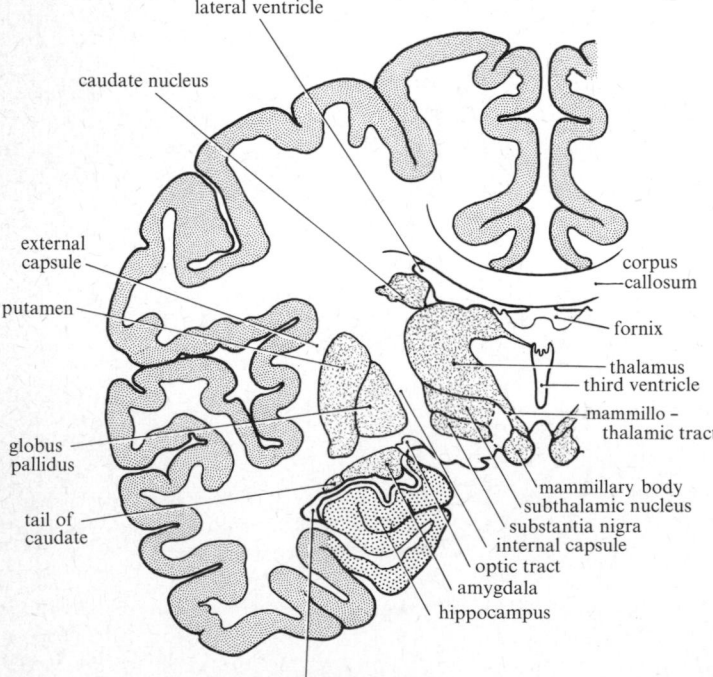

FIG. 33.2. Transverse section of human brain through the thalamus and corpus striatum.

contribute differently to the result. In a very general way we can say that the more dorsal parts of the hemisphere receive information from the periphery and compute the appropriate commands, which are then executed by the more ventral regions. But these are not precise statements and influences are continually passing both ways between sensory and motor to predict effective actions (see Kornhuber 1974).

The afferent pathways from all the receptors except smell make synapse in the thalamus, which is the thickened wall of the diencephalon (Fig. 33.2). From here fibres pass to the cerebral cortex in an orderly manner which is basically the same in all mammals (Fig. 33.3). Moreover, projections between thalamus and cortex proceed in both directions and the whole is one vast interacting reciprocating mechanism. It is by no means clear how it operates, but the thalamus must not be thought of as a mere relay station on the way to cortex. It is an integral part of the system for computing appropriate responses to the afferent input. The cortex (= rind, also known as pallium (= mantle)) became large when numerous afferent systems came to discharge into the hemisphere. Originally the cerebral hemispheres were purely olfactory centres (see Young 1962, p. 168).

In all mammals we can still recognize the original smell-brain (*rhinencephalon*), but it becomes progressively overshadowed in higher mammals by the non-olfactory part, the *neopallium*. Already in amphibians there is a lateral *pyriform cortex* and medial *hippo-campal cortex*, both receiving only olfactory fibres, and a dorsal general *somatic cortex*, which in addition receives from the thalamus projections concerned with impulses from other senses (Diamond 1967) (Fig. 33.4).

In mammals these three sorts of cortical tissue can still be recognized. The lower lateral portion, receiving fibres from the olfactory bulbs is the *pyriform cortex*, or paleocortex (p. 316). The lower medial part is the hippocampus or *archicortex*, no longer mainly olfactory (p. 317). These two regions have a relatively simple structure and are called *allocortex*, in contrast to the *neocortex*, the layered structure occupying the dorsal region or neopallium. The neopallium is subdivided in various ways by different authors (p. 320).

3. Thalamus

The thalamus is divided into numerous distinct areas or nuclei, each receiving input from a different source and connected by thalamic radiations with neocortex by the most direct route. These connexions are made in

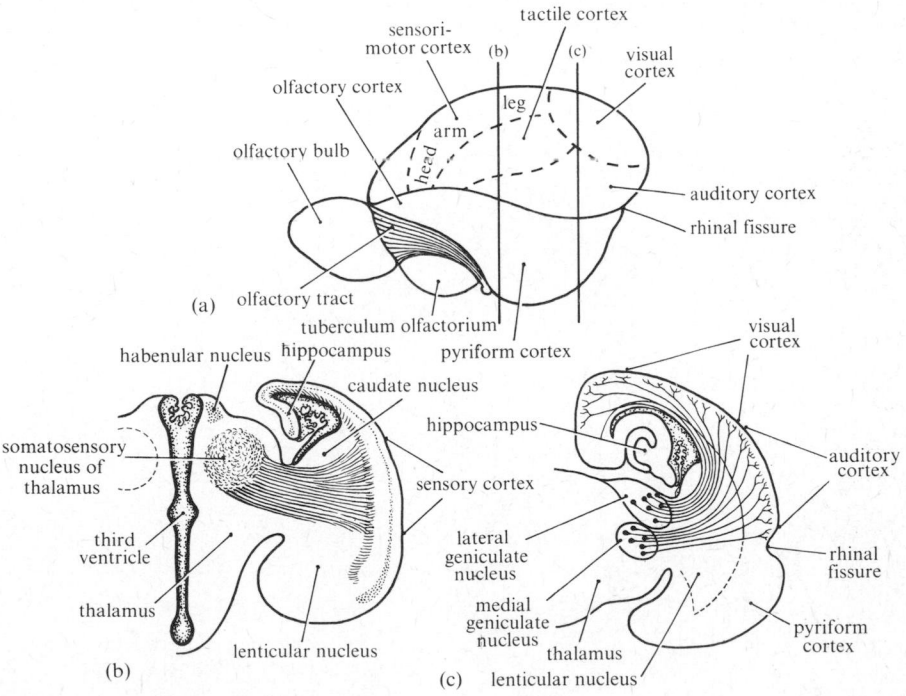

FIG. 33.3. (a) Cortical territories in the forebrain of a primitive mammal (platypus); (b) and (c) sections along lines indicated in (a). (After Elliot-Smith (1910). *Lancet* **88**.)

an orderly manner. That is, the receptor fields in the skin, eye, or ear are so projected that the topographical relations between parts of the sensory surfaces are preserved in the thalamus and reproduced on the cortex. However, the connexions are more general in lower mammals, such as the hedgehog, than in higher ones such as cat or monkey. In the former the projections overlap, so that impulses from each receptor reach to wide areas of the cortex. In the higher mammals specific areas of thalamus and cortex for each receptor can be recognized, though even here it is found that the cells can often be excited by sufficiently strong stimulation of receptors other than their specific ones. Thus cells of the visual cortex can be made to respond by loud noises. Furthermore, even in primates there are still large parts of the cortex whose cells receive a polysensory input, for example, those of the orbital gyrus (Fig. 29.2, p. 274; Brodman area No. 47).

Description of thalamo-cortical relations is obviously a very difficult matter, even if we restrict ourselves to the higher mammals, about which more is known. The thalamic nuclei can be classified in various ways, none wholly satisfactory. A major distinction is between *extrinsic nuclei*, which relay impulses from extra-thalamic sources to the cortex, and *intrinsic* nuclei, relaying impulses from other thalamic nuclei. The extrinsic nuclei may be divided into somatic or 'principal' nuclei, relaying from receptors of the muscles and joints, skin, eyes, and ears, and non-somatic, relaying

from other brain centres. In general the somatic nuclei are more posterior than the non-somatic (see Purpura and Yahr 1966).

4. Somatic relay nuclei

The somatosensory projection comes through the lemniscus bundles, ending in the posterior part of the more ventral thalamus (*posteroventral nucleus*) and projecting to the postcentral gyrus (Areas 1–3 of Brodman, see Fig. 29.2, p. 274). The *medial geniculate body* is a part of the back of the thalamus receiving impulses from the cochlear nuclei and inferior colliculus and projecting to the primary auditory cortical area in the superior temporal gyrus (Area 41). The *lateral geniculate body* lies lateral to it and receives fibres from the optic tract and projects to the primary visual cortex of the occipital lobe (Area 17).

5. Non-somatic relay nuclei

The *antero-lateral ventral nuclei* receive fibres from the cerebellum and the globus pallidus of the corpus striatum (p. 311). They are thus concerned with motor functions and project to the motor cortex in the pre-central gyrus (Area 4). The *anterior nucleus* of the thalamus receives the mammillothalamic tract and projects to the cingulate gyrus of the medial surface of the hemisphere (Areas 23–4). The medial or *dorso-medial nucleus* receives fibres from the hypothalamus and also

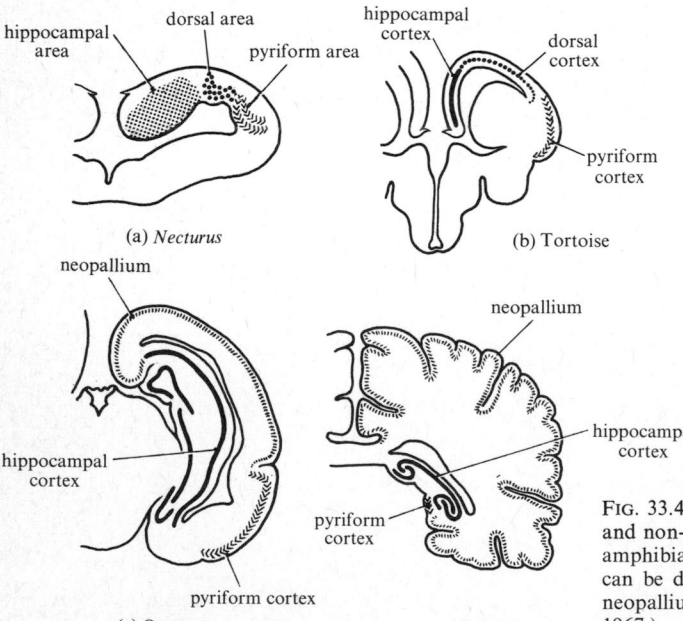

FIG. 33.4. A diagram of the relative extents of the olfactory and non-olfactory pallial fields in various vertebrates. In the amphibian and reptile it is all olfactory, but three divisions can be discerned. In the mammals there is a non-olfactory neopallium, enormously increased in man. (After Diamond 1967.)

from the spinothalamic tracts and projects to the frontal cortex (Areas 9 12) (p. 326).

6. Intrinsic nuclei

These lie laterally and include the *pulvinar* of primates, which receives from the midbrain tectum and probably from the lateral geniculate and projects to the visual areas 18 and 19, beyond the primary visual cortex. This pulvinar is a special development characteristic of the visual system of primates. Much more primitive sets of cells with intrinsic connexions are the *nuclei of the midline* of the thalamus, which project to the hypothalamus and corpus striatum, as well as to the medial surface of the cortex (cingulate gyrus, p. 343). This region is large in lower mammals and may be distinguished as paleothalamus from the neothalamic centres, which project only to cortex.

The axons of thalamic neurons thus mostly proceed laterally to the cortex, through a number of bundles, the *internal capsule*. But some of them pass forward to the striatum, others down to the hypothalamus or back to the tegmentum of the midbrain.

34   The cerebral cortex

1. General plan of the cortex

IN mammals centralization of the control of action has proceeded so far that almost every single act is influenced by the cerebral cortex. This region receives information directly or indirectly from nearly all the receptors in the body and every muscle and gland cell is influenced directly or indirectly by the impulses that originate in it. It may be considered as containing the most complete representation of the environmental conditions that is available to the animal. This representation is used in the operations of a vast computing system, which devises at each moment the behaviour that is most likely to provide for survival in view of the present situation and the experiences of the past, as recorded by heredity and individual memory (see Burns 1958; Sholl 1956).

The cerebral hemispheres, within which this computing system has developed, serve in lower vertebrates as purely olfactory centres (see Young 1962, p. 346). Sherrington suggested that their development into discriminating centres came about because the nose is the distance receptor that initiates the actions of hunting and finding a mate. If smell thus provides the 'drive' for behaviour it is obviously likely to be efficient for the 'smell-brain' to receive impulses from the optic, tactile, and other receptors that are furnishing the detailed information with which the chase is pursued. Moreover, chemical changes in the environment will often give valuable clues from past experience as to the presence of food, mates, and other conditions.

2. The smell-brain and amygdala

The olfactory nerve fibres (p. 410) end in the olfactory bulbs in tangled bundles, the *glomeruli*. As many as 26 000 olfactory axons end in each glomerulus and here they become associated with the dendrites of the *mitral cells* and *tufted cells* (Figs 34.1 and 44.6, p. 413). Each mitral cell makes connexion with a single glomerulus, but each glomerulus receives dendrites from 24 mitral cells and 68 tufted cells (in the rabbit). These mitral cells and tufted cells have intertwining collateral dendrites,

which may also be influenced by recurrent axonal collaterals. Moreover, there are periglomerular and granule cells that further complicate the neuronal chains (p. 414). The setting up of impulses in the axons of a given mitral or tufted cell presumably depends upon the stimulation of a particular set of olfactory receptors.

The mitral-cell axons proceed in the olfactory tract to the front part of the *pyriform lobe* (Fig. 34.1), occupying the lower lateral portion of the hemisphere. This palaeocortex (palaeopallium) has a structure simpler than that of the main cortex. It probably sends its output to higher cortical centres and is concerned in the more elaborate learned olfactory reactions. After ablation of this area olfactory conditioned reflexes that had been previously learned are lost.

The axons of the tufted cells pass to nuclei lying in the floor of the forebrain (amygdala), in the region of the anterior commissure. Experiments suggest that these fibres and centres mediate the simpler reflex responses to olfactory stimuli, probably by activating the hypothalamic centres that lie caudal to them. Electrical stimulation of the prepyriform cortex or the amygdala in animals gives rise to actions related to feeding, such as retraction of the lips, sniffing, licking, chewing, and salivation, which are also seen after stimulation of the hypothalamus (p. 309). However, the amygdala is much more than a simple olfactory centre. It receives impulses by various pathways from the various special senses and from many other parts of the brain. Its functions are not only the simple reflexes described above but also include influences upon the central motivations and drives of the individual. It is especially concerned with avoidance and defensive reactions, and electrical stimulation here in man produces a feeling of intense fear. After removal of the amygdala on both sides monkeys no longer show fear of snakes or humans and will pick up and taste every object presented. They may also show excessive movement and sexual activity. One way to consider this centre is as the most anterior part of the reticular

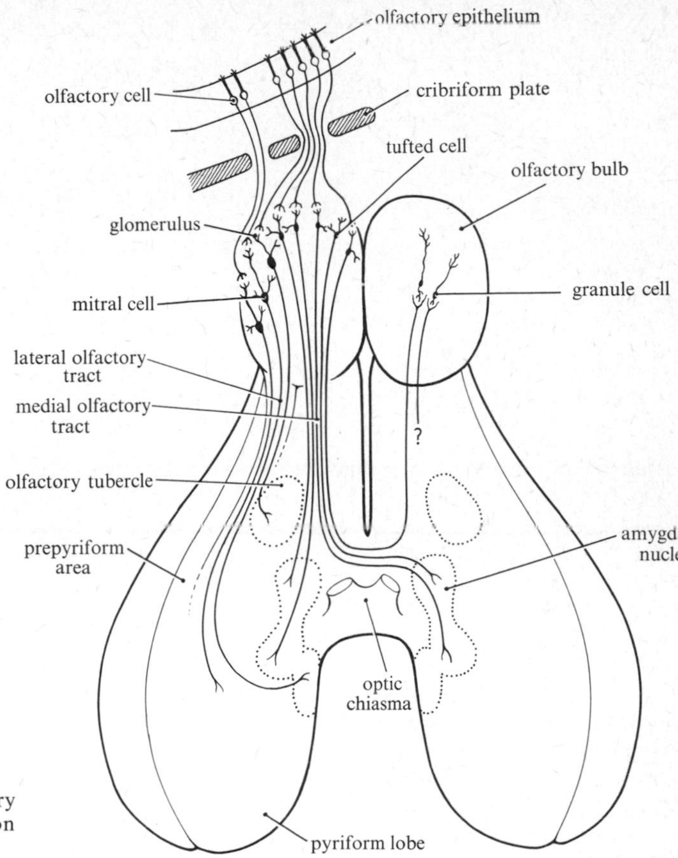

olfactory epithelium

olfactory cell

cribriform plate

tufted cell

olfactory bulb

glomerulus

granule cell

mitral cell

lateral olfactory tract

medial olfactory tract

?

olfactory tubercle

amygdaloid nuclei

prepyriform area

optic chiasma

pyriform lobe

Fig. 34.1 Diagram of the secondary olfactory pathways in a mammalian brain. (After Allison (1953). *Biol. Rev.* **28**.)

activating system (p. 271). It produces its effects by impulses passing to the several subcortical parts of the forebrain and also downwards in the medial forebrain bundle (see MacLeod 1971).

Fibre pass from the prepyriform region to other parts of the cortex, and it is probable that the function of this part of the olfactory system is more complicated than the simple initiation of feeding actions. The association of particular smells with situations useful or dangerous for the animal is one of the most important determinants of behaviour, especially in mammals, which were originally nocturnal animals. The apparatus for forming association between olfactory and other afferent impulses is therefore one of the most important regions of the brain. Even in man, where visual clues are more important than olfactory ones, we still find that particular smells may suddenly arouse complicated memories of past visual and other situations with which they have been associated, particularly if these situations were connected with the satisfaction of a strong biological need, such as food or sex. We express this in another way when we say that the olfactory sense

has strong emotional value. It is not unlikely that the action of these parts of the rhinencephalon has an important influence on the rest of the cerebral cortex, which has become developed, it must be remembered, in an area that was originally wholly devoted to olfactory functions.

3. The hippocampus

In this lower medial portion of the wall of the hemisphere, the cells have migrated away from the ventricle, but do not form complex layers and hence are called archicortex. This part of the cortex receives impulses from many other areas and sends signals to the hypothalamus. It may originally have been concerned with behaviour under olfactory influences but it persists and is indeed enlarged in microsmatic creatures such as whales and primates, where it is concerned, among other things, with memory. The continuity of the hippocampus with the neocortex can be seen in Fig. 34.2. The large cells form a layer of pyramids with apical dendrites proceeding towards the pial surface and basal dendrites and axons towards the ventricle

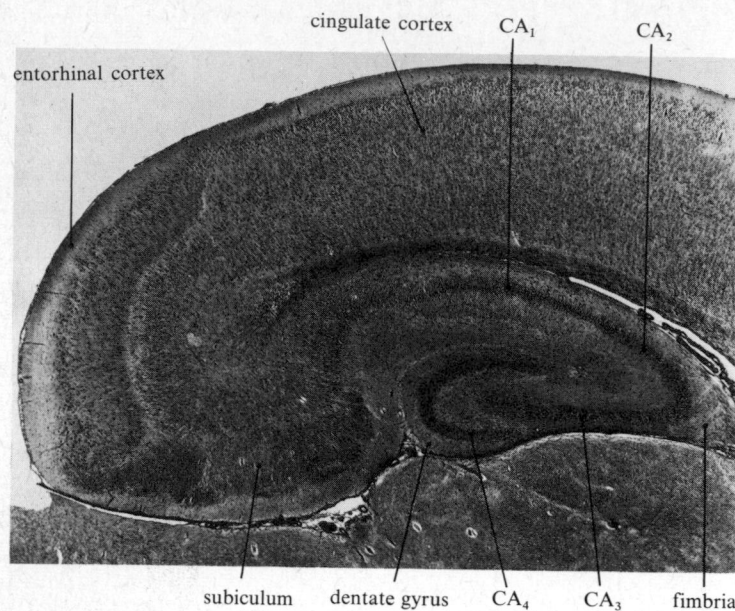

entorhinal cortex

cingulate cortex   CA₁   CA₂

subiculum   dentate gyrus   CA₄   CA₃   fimbria

FIG. 34.2. Cross-section of hind region of cerebral cortex of a rat.

(Fig. 34.3). The axons form bundles over the ventricular surface, the *fimbria*, and these join together to form a thick bundle, the *fornix*, running first backwards, then forwards and down in the lamina terminalis to the hypothalamus. Some of the fibres cross as the hippocampal commissure. Mixed with these pyramidal cells are smaller basket cells, with short axons, making synaptic connexion with the pyramids.

The hippocampus is turned upon itself and ends medially in a ridge characteristically indented, the dentate gyrus (Fig. 34.3). Input to the hippocampus is partly by fibres running in the fornix/fimbria, probably from the septum. But there is a large input all along its length from the medial and lower wall of the neocortex. This is known as the *entorhinal cortex* and receives olfactory fibres from the pyriform cortex (p. 316) and also impulses from a very wide cortical area through the cingulate cortex and cingulum (Fig. 34.3).

Axons of the cells of the entorhinal cortex reach to the *fascia dentata* by the *perforant pathway*, which crosses the original region of separation between the inturned cortical surfaces (Fig. 34.3). These fibres are excitatory to the cells of the fascia dentata, whose axons in turn enter the hippocampus proper as rows of *mossy fibres*, which make excitatory synapses with the apical dendrites of the hippocampal pyramids. These pyramids are divided into four distinct sets numbered CA₁₋₄ (Fig. 34.3). We can consider CA₃ and CA₄ together and they receive the mossy-fibre input. Their axons leave in the fimbria but also send branches, known

as the Schaffer collaterals, to excite the apical dendrites of the cells of areas CA₁ and CA₂. These collaterals also excite the basket cells, which in turn have an inhibitory effect on the pyramids of CA₃. The cells of CA₁ and CA₂ also receive a direct input from the entorhinal cortex. Their axons may run partly in the fimbria but some also run back to the subiculum.

We have here a system repeated many times over along the length of the hippocampus. Electrical stimulation of any region of the dentate gyrus activates cells in a corresponding slice of the hippocampus. There is also a longitudinal pathway, which must serve to integrate the actions of the various strips along the length.

This set of cells has various influences upon behaviour, which have been studied by the effects of ablation, electrical stimulation, and electrical recording. Rats deprived of both hippocampi show aberrations of search behaviour. In a maze they repeat the same actions over and over, as if lacking the normal tendency to show new behaviour to solve problems. They show a lack of power to inhibit a pre-potent response.

Humans lacking both hippocampi show defects of memory. Old memories are well preserved but there is great difficulty in setting up new ones, even for events occurring a few minutes previously. There seems to be especially a deficit in recognition of the situation, though with the aid of hints it may be recalled.

These data from animals and man suggest that the hippocampus is somehow concerned with searching

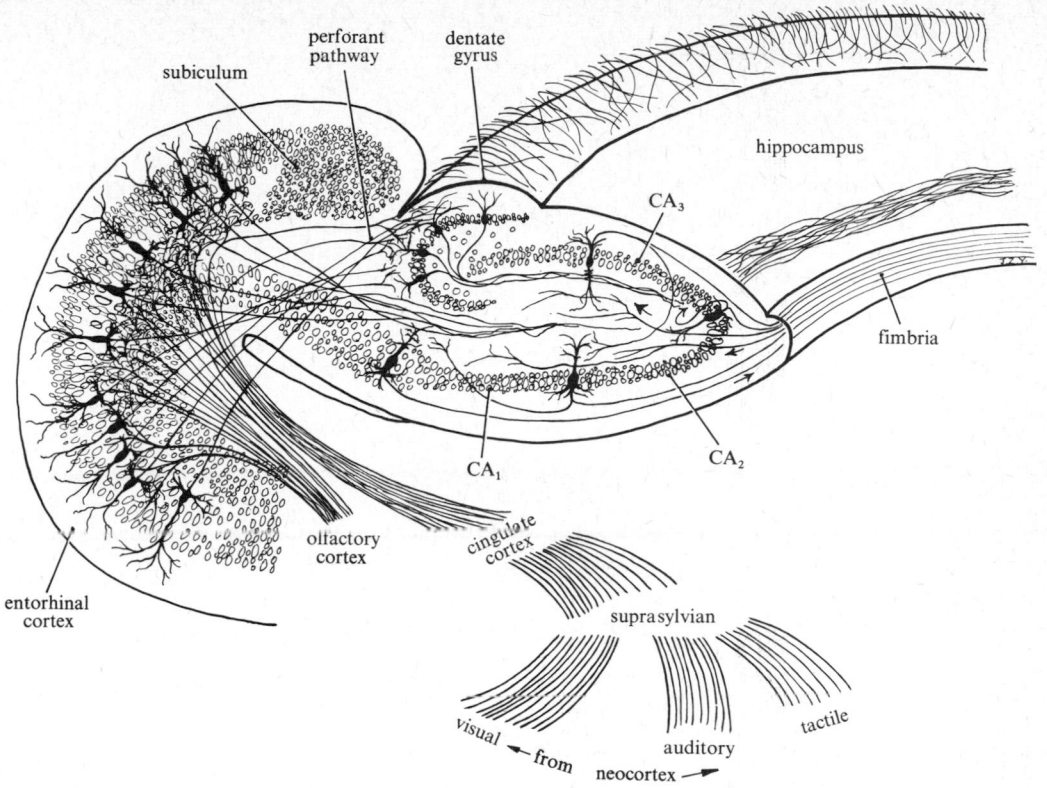

Fig. 34.3 Diagram of some of the connexions of the hippocampus. Based upon the rat, but with details from various mammals.

and the recognition of familiarity. It may be that it is involved in bringing together signals from the various receptor areas with those indicating reinforcement or rewards (Fig. 34.4, see p. 320). This would be an essential process in the formation of what we shall call a model in the brain by which memory records are stored (p. 341). Electrodes implanted in the hippocampus show special EEG potential changes (theta waves) during certain movements and perhaps during attention or arousal (see O'Keefe and Nadel, in preparation).

The efferent fibres of the hippocampus pass through the fimbria to the septum and hypothalamus, especially to the mammillary bodies. Damage to the Papez circuit (p. 306), for instance by the effect of alcoholism, produces not only emotional disturbances but also impairment of power to record in the memory. Clearly this part of the hemisphere though in a sense 'lower' is closely concerned with some of the most important activities of man and other mammals.

4. The neopallium

The special characteristic of the mammalian brain is the expansion of the mid-dorsal portion of the roof of the cerebral hemispheres, the region that receives ascending projection fibres from all parts of the body and is known as the general somatic cortex, *neopallium*, neocortex, or isocortex. The function of this tissue is hard to specify, but certainly includes formation of associations between patterns of signals coming from the various external receptors and those within the body that record the state of the various organs and hence the needs of the organism. It thus carries a store of information about the relevance of various patterns of events that have occurred. Subsequently, as new combinations of inputs arrive, it computes appropriate actions to suit each occurrence. It assembles and reacts to information from a variety of sources and provides a representation of the events that have occurred in the environment.

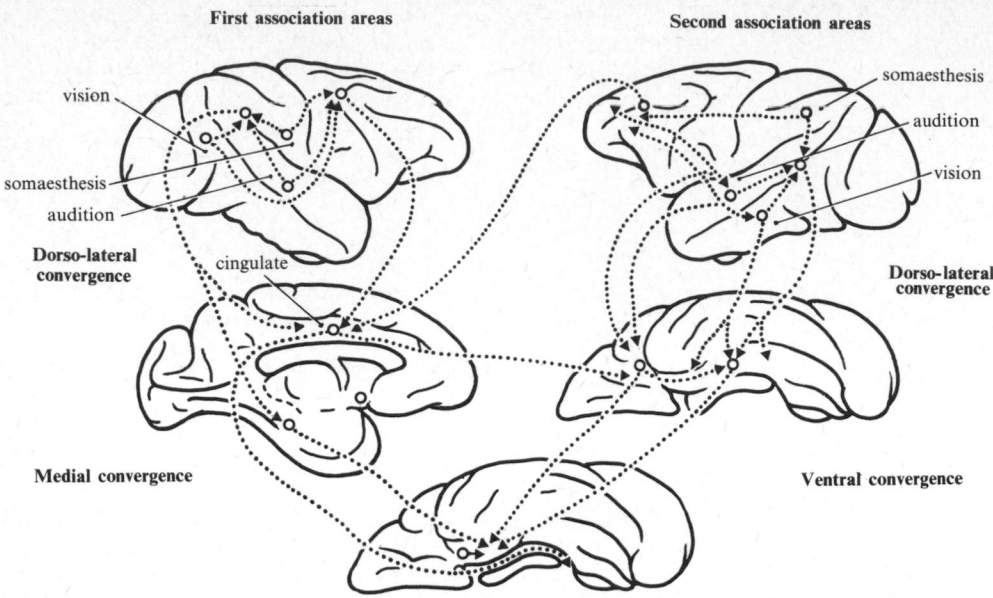

FIG. 34.4. The probable multisynaptic pathways by which the hippocampus has access to sensory information in the monkey, showing primary and secondary association areas for vision, somaesthesis, and audition. (After Van Hoesen, Pandya, and Butters (1972). *Science.* **175,** 1471–3. Copyright 1972 by the American Association for the Advancement of Science.)

The essential feature of the cortex is that a great number of pathways converge into it and the probability of a given response to input along any particular channel is initially low. The information of heredity provides in the cortex a system that is able to make predictive actions only after it has been trained. It produces few complete 'hereditary' responses to the environment, but has great capacities for building up representations as a result of the information transmitted by the receptors. Such information is of course a more 'up-to-date' and complete record of the environment than can be provided by heredity alone. It is the capacity to form this information store that gives to mammals their special ability to survive even in 'improbable' situations, that is to say those that are apparently unpropitious for life. Presumably in each animal species there are particular inherited capacities to acquire the sort of information that is likely to be useful.

5. The cerebral sulci and areas

The reaction to patterns of inputs, which is the characteristic feature of the activities of the cerebral cortex, seems to depend on the presence of a relatively thin sheet of tissue, within which the cells are arranged in columns and layers. The development of this extensive sheet leads in all larger mammals to folding of the surface of the cerebral hemisphere. The area available increases of course only with the square of the linear dimensions, whereas the volume of underlying white matter increases as the cube. The proportions are kept constant by folding of the surface to give the *sulci* and *gyri*; these are found in all large mammals. irrespective of affinities, whereas the hemispheres of their smaller relatives are smooth.

The pattern of the sulci depends on the way the stresses fall during the expansion of the brain roof. The basal ganglia form a solid mass along the floor of the hemisphere and expansion of the thinner pallium produces the *lateral (Sylvian) fissure*, the most constant of all the sulci (Fig. 34.5). Similarly the *rhinal fissure* laterally and *hippocampal fissure* medially separate the pyriform lobe and hippocampus from the neocortex above them.

The expanding sheet is attached to fixed points laterally and medially (the basal ganglia and corpus callosum) and as it enlarges it falls into the longitudinal folds that are commonly found in lower mammalian brains (Fig. 34.7). Lesser furrows form when the brain becomes still larger, and often these take triradiate patterns (Fig. 34.6), such as can be imitated by allowing contraction of the surface of a sphere, say, by coating

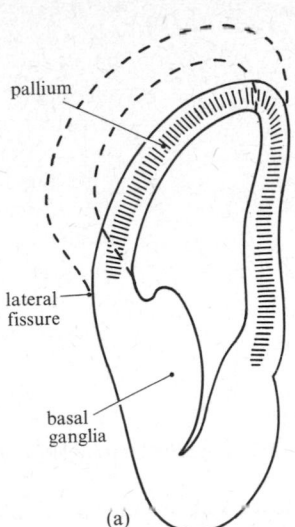
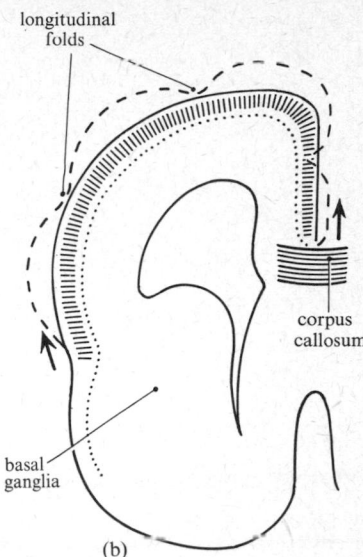

FIG. 34.5. (a) Diagram of the developing mammalian cerebral vesicle in transverse section showing how the lateral fissure is formed by expansion of the thin-walled pallium on the thicker basal ganglia. (b) The tendency to form longitudinal folds is produced by expansion of the pallium against the resistance of the corpus callosum and basal ganglia. (From Clark and Medawar (ed.) (1945) *Essays on growth and form.* Oxford.)

a balloon with gelatin or collodion and then allowing it to collapse. In primate brains the longitudinal folds are found only among the earlier types (lemurs), in the later forms the sulci are mainly transverse (Fig. 34.8). This is probably a result of the great expansion of the posterior (visual) and anterior (frontal) parts of these brains, which would have the effect of causing the crumpling to occur in the transverse rather than the longitudinal direction.

The longitudinal furrows of the earlier brains bear little relation to functional areas, but in primates some of the major sulci occur along lines that demarcate functionally distinct regions. Thus the *central sulcus* is a rather constant groove that divides the motor cortex in front from the sensory areas behind. In man there is a reduction in thickness of the cortex from 3·5 mm to 1·5 mm between these areas and this makes a line of weakness at which folding occurs.

The visual cortex surrounds the *calcarine fissure*, and the frontal gyri are separated by fissures that are

probably produced by similar factors. For purposes of description the surface of the hemisphere is divided into four main lobes, frontal, parietal, occipital, and temporal, and a finer division is made into numbered areas (Fig. 29.2, p. 274). Each of these areas was originally distinguished by Brodmann on a basis of supposed differences in neuronal structure. The numbers have been retained for the areas, although in some cases no exact criteria by which the neuronal structure of the areas can be recognized have been agreed (p. 322).

6. Localization within the cortex

The cortex is an organ for allowing association between signals arriving from many sources. Obviously in such a system there is in a sense no localization; it is the very fact of bringing together and allowing interaction that is important (Sholl 1956). Yet the cortex is far from being a uniform field in which all parts receive and give out similar influences. The input from each of

FIG. 34.6. (a) Patterns formed on the surface of a contracting balloon. (After Bull (1932). *Geol. Mag.* **69.**) (b) Sulcal patterns from the parietal lobe of a human brain, showing tri-radiate patterns, some of them linked, as in (a). (From Clark and Medawar *loc. cit.*)

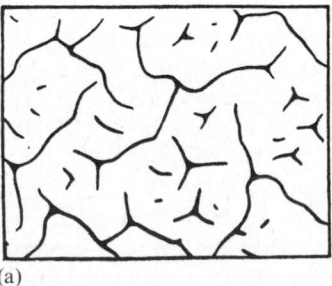
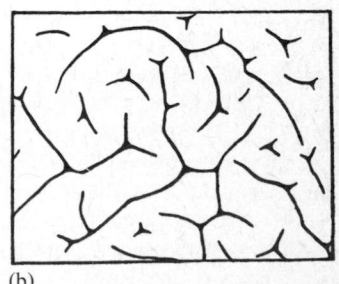

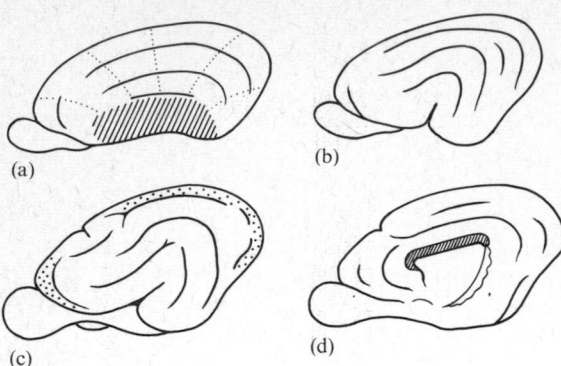

FIG. 34.7. (a) Diagram showing the tendency for longitudinal folds to develop in the cerebral cortex at right angles to the lines of stress (shown by dotted lines) that are set up by expansion of the thinner pallium on the thicker basal ganglia. (b) Effect of ventral flexure of the hemisphere in producing arcuate sulci. (c) and (d) Lateral and medial view of cat's cerebral hemisphere, showing arcuate sulci. The frontal and visual areas are stippled. (From Clark and Medawar *loc. cit.*)

the thalamic centres proceeds mainly to one particular part of the cortex. We can therefore recognize a *sensory cortex*, consisting of separate receptor areas where the impulses from the visual, auditory, tactile, and some other systems arrive. Efferent fibres proceed from many, perhaps all, parts of the cortex to lower centres, but the main pathway for direct control of the activities of the spinal cord is from the *motor cortex*. The parts of the cortex not directly identifiable as sensory or motor areas are sometimes called *association areas*: they receive fibres from the receptor areas and send them to the motor areas.

The areas of the primate brain were numbered by Brodmann about the central sulcus as a reference point (Fig. 29.2, p. 274). The system is arbitrary and the distinctions between the areas are not easy to define accurately by histological or other methods. Nevertheless, the divisions are convenient and the system of numbers is still widely used. In mammals other than primates there is no central sulcus and the terminology is not applicable, but the general distribution of functions is the same in all mammalian brains.

As we learn more about the cortex it becomes increasingly difficult to classify its parts in any simple way. The primary sensory areas are clear enough (Fig. 35.2, p. 331). The *somato-sensory cortex* concerned with the skin and muscles occupies the post-central gyrus of the parietal lobe, the *primary visual area* is the striate cortex of the occipital lobe (area 17, Fig. 29.2), and the auditory cortex occupies area 22. However, each sensory field is represented not once but two or more times,

with 'belt areas' around each primary sensory field. In a very general way we can say that the cells of the primary sensory areas respond to simple stimuli. Proceeding through the belt areas the information is serially processed so that the cells are activated only by particular patterns, perhaps temporal as well as spatial. The 'assocation areas' are therefore now seen to be largely concerned with greater elaboration of the information of the 'sensory areas'. At the highest levels, of course, the information from different senses interacts. Indeed in the region between parietal, occipital, and temporal regions Penfield identifies in man the '*association area of association areas*', where all interact.

Similarly on the output side there is in front of the *motor cortex* (area 4), which is concerned with the action of individual muscles, a second *pre-motor area* (6), which operates the action of whole limb segments. Further forward still is the *pre-frontal cortex*, which has subtle effects in regulating the behaviour of the whole individual (p. 324). The medial surface of the hemisphere (*cingulate cortex*) receives input from many other parts and is especially related with the hippocampus and perhaps concerned *inter alia* with the process of memory recording (p. 343).

As information about cortical structure accumulates it appears that all parts are based upon a common plan (p. 320). The differences in structure in various areas have been the subject of much study, the 'architectonics' being minutely described. For instance, the occipital cortex, which receives the visual projection, is characterized by a special band or stripe of medullated fibres and is hence known as the *striate area*; the motor cortex contains numerous large cells, and so on. Such broad differences undoubtedly exist, but there has been

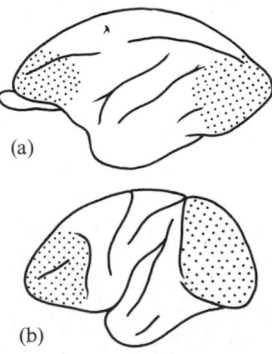

FIG. 34.8. Cortex of (a) *Lemur* and (b) the monkey *Cercopithecus* to show transition from pattern of longitudinal to transverse sulci as the frontal and visual areas became developed. (From Clark and Medawar *loc. cit.*)

much controversy about the division of the cortex into smaller areas. Lack of quantitative methods of analysis has made decisions about differences highly arbitrary.

7. The motor cortex

In attempting to understand cortical function it may be convenient to begin by study of the motor areas by which the cortex controls the rest of the central nervous system. Then we can proceed to investigation of the sensory areas by which the input of information is analysed. It is still not yet possible to understand how the whole system operates to produce appropriate patterns of action or how these are modified by experience. It is fortunate that we can to some extent recognize 'areas' with distinct 'functions', but it is important also to realize that in life most of them continually operate together (see Phillips 1973; Yahr and Purpura 1967).

The gyrus in front of the central sulcus contains the motor cortex (area 4) and that behind it the somatosensory cortex (areas 1–3). The motor and sensory areas are sharply distinguished by a change of thickness, indeed it is probably this change that determines the infolding of the cortex to form the central sulcus along the line between them during growth (p. 321). Nevertheless, the sensori-motor functions are not sharply distinct and there is continual interplay between the two areas. The post-central gyrus receives sensory inputs from muscle, joints, and skin and these provide some of the information directing the production of movements. However, this separation of 'sensory' and 'motor' is highly artificial. Both pre- and post-central areas send fibres to the spinal cord. Those from the pre-central area run to the motor neurons of the ventral part of the grey matter (p. 262). Those from the post-central cortex end in the dorsal columns and so influence the signals that are sent to the brain from the cord. The organism cannot therefore be regarded as a kind of puppet with a 'motor' system activated by its 'sense organs'. It is more a selector, choosing continually to 'attend' in one way or another to the changes going on around it. Nevertheless, different parts of the cortex act differently in this process of selection.

The motor cortex has long been recognized by the fact that stimulation of individual points within it produces contraction of isolated groups of muscles, at low thresholds. This is due to activation of the characteristic very large *Betz cells* of the area, whose large and fast-conducting axons enter the *pyramidal tract* (hence 'area giganto-pyramidalis'). There are some 3×10^4 of these cells in man, but the pyramidal tract contains over 10^6 fibres, many arising post-centrally, and many

of the smaller, slower ones running not to motor regions but to sensory ones, such as the trigeminal and dorsal column nuclei. Through these fibres the motor cortex and pyramidal tract regulate the input to the brain as well as the output from it. Fibres may pass down in pathways other than the pyramidal tract, known as the extrapyramidal system.

The pyramidal (corticospinal) tract is developed variously in mammals, determined both by their phylogenetic level and specialized activities. In all mammals it sends fibres either directly or as collaterals to the basal ganglia of the opposite side, including the red nucleus, susbstantia nigra, pons, and bulbar reticular nuclei. These are the connexions by which movement is regulated even in those mammals in which the pyramidal tract reaches only a short way down the spinal cord. Similarly, it always sends fibres to the trigeminal nuclei and those of the dorsal columns, and by these pathways regulates the afferent input from the face, trunk, and limbs. The tract reaches back to the spinal cord to various extents—hardly at all in hedgehogs, mainly to the ventral funiculi in other insectivores and monotremes. In ungulates it ends within the cervical cord. In many mammals it runs in the dorsal funiculi, reaching the cervical level in the phalanger, rabbit, and tree shrew, but to the lumbar region in the rat and in carnivores, though most of the fibres end further forward. The tract is much better developed in primates than in any other mammals.

In the cat and monkey the axons enter the base of the dorsal horn. By making lesions in the cortex and looking at the degenerating terminals it is found that the fibres from the more posterior somato-sensory areas go to the more dorsal parts of the horn (layers IV–VI of Rexed; Fig. 27.6, p. 262), where they can control the incoming sensory signals and their transmission to the brain. The pyramidal tract axons from the more anterior (motor) areas go to the more ventral layers of the cord, V–VII, where they influence cells controlling the ventral horn. In monkeys, apes, and man the pyramidal tract exercises a direct control over movements by terminations upon the motor neurons themselves, as well as in other parts of the cord. These cortico-motor-neuronal endings are most abundant in relation to the cells that innervate the distal muscles of the hand and foot. Interestingly enough, they are also found on the motor neurons that innervate the tips of the prehensile tails of spider monkeys. Evidently this direct control regulates fine movements. The more indirect pathways probably control the postural supporting muscles of the more proximal segments of the limbs, which are needed as bases for the fine manipulatory movements of the mouth, hands, feet, or tail.

8. Intrinsic cerebral networks of the pyramidal tract

The links of the cortex with other parts of the motor system are very complicated (Fig. 34.9). For example, collaterals of pyramidal tract fibres to the red nucleus may excite or inhibit cells there that conduct in the rubrospinal tract to the same motor neurons as are influenced directly from the motor cortex. Collaterals to the pons and inferior olive influence the cerebellum, which in turn influences the red nucleus and the motor cortex (through the ventro-lateral nucleus of the thalamus). Pre-central cells also send axons to the bulbar reticular formation, which by the ventral reticulospinal tracts controls the neurons of the motor horn that innervate the muscles of the trunk and proximal segments of the limbs.

It is only partly clear how all these centres interact to produce the patterns of movement. The effect of single cortical neurons on their neighbours can be studied by activating them antidromically by stimulation from below. It is found that the intra-cortical collaterals of the large fast fibres powerfully inhibit neighbouring Betz cells that have been penetrated with an electrode. Such collateral inhibition would have the effect of sharpening the focus of action. Other (slower) pyramidal tract fibres may facilitate their neighbours. Recurrent collaterals also inhibit cells of the cortico-rubral tract (Fig. 34.8).

It is still not known how the pyramidal tract (PT) cells are themselves controlled. They can be made to discharge impulses by a wide variety of peripheral stimuli, from light, sound, or touch or from the opposite cortex. Some respond only to stimulation of a local area of skin, or movement of a limb, or traction upon a tendon. In normal life individual PT cells may be activated separately. Recording from the pre-central gyrus of people under local anaesthesia, Penfield and Jasper showed disappearance of the EEG rhythm when the patient gripped an object or even when he was told to prepare to do so. The effect was not always restricted to the hand area. Recording from the median nerve it has been feasible to show the activation of a single motor neuron when a person makes a delicate movement of a finger. By recording from single PT cells of trained monkeys with a chronically implanted electrode the response can be followed when the animal presses a lever. The frequency of discharge increases when the movement is resisted (Fig. 34.10). The signal controlling the force of the movement is thus coded by the frequency of the impulses in particular fibres. Some cells were tonically active and their discharge was modulated during some aspects of the movements. Other cells were silent except in some phase of the action.

It is probable that much control of movement patterns is by discharge of impulses to the gamma motor neurons and hence alterations of the bias on the muscle-spindles (p. 265). Corticospinal fibres certainly influence these neurons either directly or through a polysynaptic pathway. The effect of cortical stimulation with repetitive electrical currents at say 40 Hz is to produce contraction of a particular group of muscles, say extensors of the elbow of a monkey, and relaxation of its antagonist. As Sherrington put it, the contracted biceps 'become suddenly soft, melting under the observer's touch'. Of course, in life, varying sequences of activations must be involved, for example, there are about 30 muscles in a hand and 'these are represented in the nervous centres in thousands of different combinations . . . just as many chords and tunes can be made out of a few notes' (see Phillips 1966).

The extent of cortex devoted to each part of the muscular system varies with the degree of precision of the movements involved rather than the bulk or power of the muscles. All parts of the body are represented in the cortex, with the lower parts nearer to the dorsal midline (Fig. 34.11). The hand and face have an especially large representation in man. Above the main motor area is a second one, extending over to the medial surface.

The degree of importance of the motor cortex for behaviour varies with the species. After cutting the pyramidal tracts in a cat or dog, locomotion is impaired only when the animal performs careful movements. In a monkey, independent movement of the fingers becomes impossible. Baby monkeys operated on at birth never learned separate use of the index finger and precision grip was permanently abolished. Monkeys trained to open problem boxes can still do so after removal of the motor cortex, but the actions are clumsy. This area is, as it were, the executor of fine movements, commands for which are developed elsewhere in the brain. In this connexion it is interesting that when Penfield elicited movements by stimulation of the motor cortex the patient could not prevent them and reported that he had not willed them to take place.

9. The pre-frontal cortex

The region in front of the pre-motor area (area 6, p. 274) can be regarded as the highest part of the brain, in the sense that it exchanges fibres not only with other parts of the cortex but also with the parts of the brain concerned with the internal state of the body. It is relatively larger in man than in any other animal and is involved in some of his most characteristic capacities, including social behaviour and voluntary actions. As

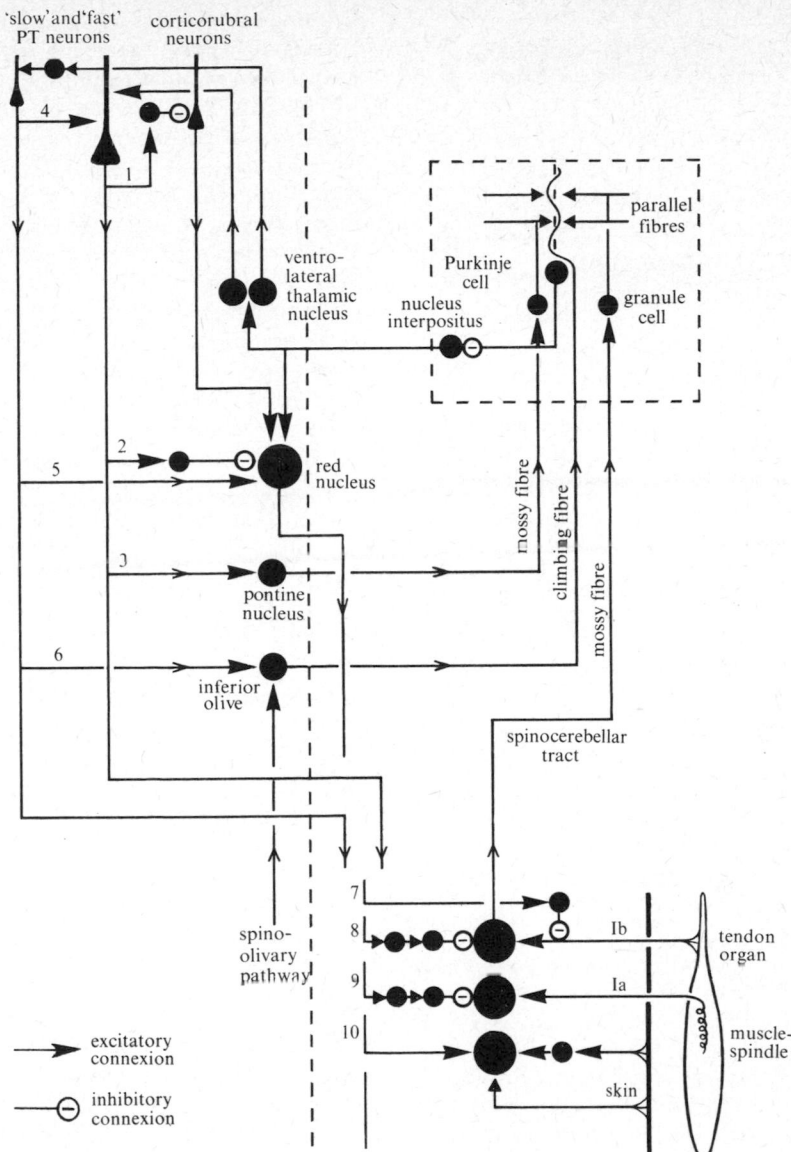

FIG. 34.9. Some of the connexions between the cat's pyramidal tract (PT) and cerebellum, greatly simplified. Midline indicated by interrupted vertical line.

'Fast' PT neurons: 1. Recurrent collaterals disynaptically inhibit corticorubral neurons. 2. Collaterals to red nucleus disynaptically inhibit cells of origin of rubrospinal tract. 3. Collaterals to pontine nuclei excite granule cells of cerebellar cortex of opposite side *via* mossy fibres.
'Slow' PT neurons: 4. Recurrent collaterals monosynaptically excite fast PT neurons. 5. Collaterals to red nucleus monosynaptically excite cells of origin of rubrospinal tract. 6. Collaterals to inferior olive which sends climbing fibres to opposite cerebellar cortex.

Spinocerebellar tract: Cells of origin are excited by inputs from skin, muscle-spindles (Ia) and tendon organs (Ib); and send mossy fibres into cerebellar cortex.
Actions of pyramidal tract on spinocerebellar inputs: 7. Presynaptic inhibition. 8. Disynaptic (postsynaptic) inhibition of Ib input. 9. Disynaptic inhibition of Ia input. 10. Monosynaptic facilitation of cutaneous input.
Purkinje cells of cerebellar cortex, excited by granule cells *via* parallel fibres and climbing fibres. Purkinje inhibit cells of cerebellar nuclei (nucleus interpositus).
Nucleus ventro-lateralis of thalamus excited by nucleus interpositus excites fast PT neutrons monosynaptically and slow PT neurons disynaptically.
(Figure kindly supplied by Dr. C. G. Phillips.)

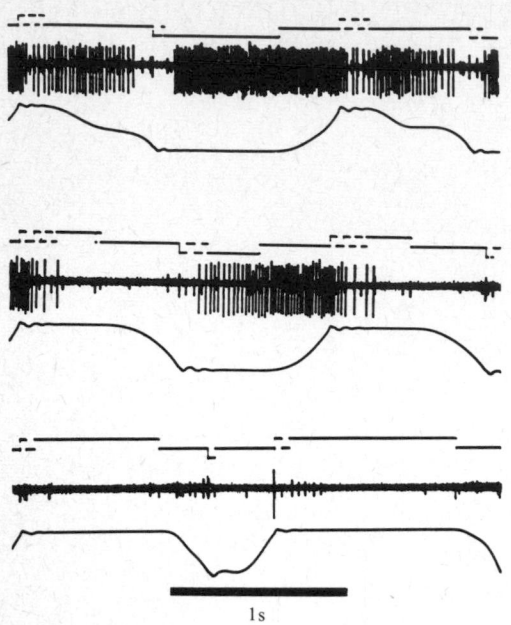

FIG. 34.10. Discharges of a monkey's pyramidal tract neuron recorded by an indwelling electrode during conditioned flexion of the wrist.
Wrist position is indicated by lowest trace in each set of three: upwards displacement represents flexion, downwards displacement extension. Middle trace is electromyograph. Top trace has three levels: upper for maximal flexion, lower for maximal extension, middle for all intermediate positions.
With unresisted movement (*centre*) the neuron discharged during flexion. When flexion was resisted by a weight (*top*) the neuron discharged more strongly during flexion and continued to discharge during extension, when paying out of weight was controlled by a diminishing flexor force. When flexion was aided by a weight (*bottom*) and controlled by an extensor force the neuron discharged one impulse only. (From Evarts (1968). *J. Neurophysiol.* **31**.)

might be expected such functions are hard to specify and define (Teuber 1972; Warren and Akert 1964).

The pre-frontal cortex of primates and carnivores covers the whole front end of the brain, and hence is often called the *frontal lobe*. In the rat it is divided into medial and lateral parts and does not reach the frontal pole. In primates this is a 'granular' cortex, containing many small cells, but in other mammals it is 'agranular' and indeed shows little division into layers. The thalamic region sending and receiving fibres from the frontal cortex is the medio-dorsal nucleus. This has lateral and medial parts corresponding to two cortical divisions which we can recognize: the dorso-lateral and orbital frontal cortices. The larger lateral medio-dorsal thalamus is a little-known region receiving a bundle of fibres arising in the medulla and perhaps conveying information about internal states (p. 314). The medial medio-dorsal

thalamus receives fibres from the olfactory (prepyriform) cortex and hypothalamus, also from the septum and interpeduncular region (Fig. 34.12) (Nauta 1972).

The frontal lobe also receives information from all other parts of the cortex through long, direct cortico-cortical pathways (Fig. 34.13). These paths come either directly from the 'belt zones' (never from the primary sensory areas) or indirectly via the lower temporal regions, which themselves receive from many sensory areas.

The efferent pathways of the frontal lobe are largely to the regions that provide the input (Fig. 34.14). The cortico-cortical projections come from the lateral surface. There is a large projection from the dorsal surface to the cingulate cortex, continuing round the back of the corpus callosum (the splenium) to the subiculum

FIG. 34.11. Motor areas of the cerebral cortex of the monkey. The 'simiusculi' show the areas from which movements of the parts concerned can be elicited by threshold stimulation with a 60 Hz a.c. stimulator and unipolar electrode 0·5 mm in diameter applied to the surface for 2 s. The main motor area lies in the precentral gyrus laterally and there is a secondary area on the medial surface of the cortex. The diagram is inadequate in so far as it does not indicate that areas activating neighbouring muscles overlap. (After Woolsey.)

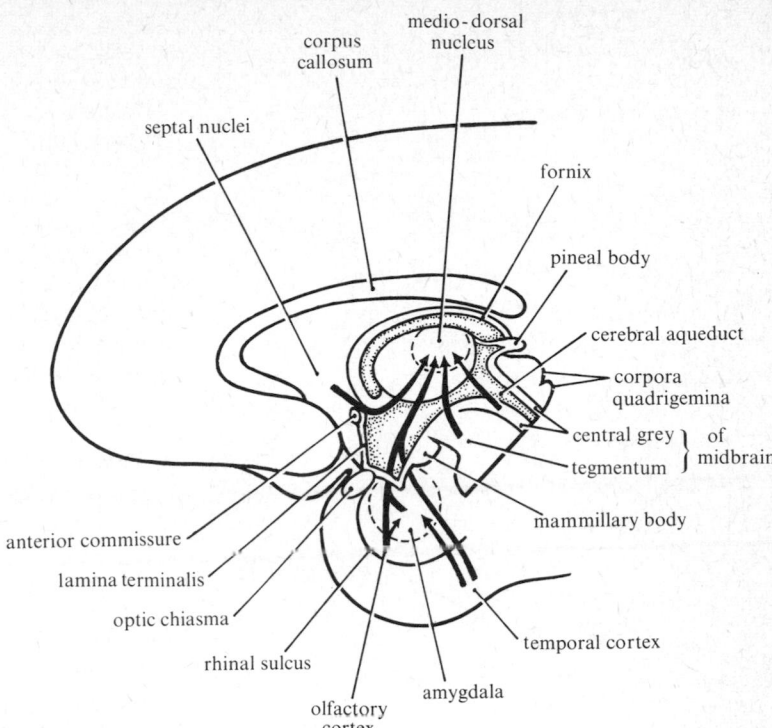

FIG. 34.12. Schematic diagram of medial transection showing some of the subcortical afferents to the thalamic mediodorsal nucleus. (After Nauta 1972.)

and so to the hippocampus. The frontal lobe also has extensive projections beyond the cortex, to the septum, striatum, hypothalamus, and midbrain base (tegmentum). These latter are regions that also receive fibres from the limbic system.

These connexions of the frontal lobe give some clues to its function. It combines afferent impulses from the cortical regions processing sensory information with those from the enteroceptive or viscero-endocrine regions of the brain. Moreover, it sends signals back to these same regions. Such '*efference copies*' may be used in various ways, including alteration of the input itself by selecting those features that are relevant to the internal state of the animal.

Experimental and medical evidence shows that the frontal lobe is not a single system, and we may distinguish three parts.

(a) *The frontal eye-fields*

Electrical stimulation in the region in front of the arcuate sulcus of primates produces conjugate movements of the eyes to the opposite side. After removal of this tissue a monkey at first fails to pay attention to events in the opposite visual field. But this is not simply a motor centre for eye movement. Recording from single neurons here shows that they are active

not before but *during* such movements. This behaviour may provide a clue to much frontal lobe function. Such signals may provide a 'corollary discharge' ((Sperry) see Teuber 1972), additional to the impulses passing down to motor centres. Reaching back to the more posterior regions of the cortex this 'efference copy' may re-set them to allow for the next steps consequential to the motor act. Such sequences are essential for what we call voluntary movements. After damage in this region the behaviour of a man or monkey may become compulsive or stimulus-bound. He cannot release objects grasped or may follow moving objects with hand or eye like a robot 'as if drawn by a magnet'.

Each neuron of these fields discharges only during one particular type of eye movement—say 'rapidly to the left and up 15°' or 'slow pursuit horizontally to the right'. They do not fire when movements are passively produced by electrical stimulation of motor nuclei. Units have even been found that fire only when the monkey is both looking at an object and reaching for it.

Further evidence that this region is concerned in 'voluntary' action comes from experiments in which monkeys are tested with spectacles that distort the image. A man or monkey soon adapts to this if he moves freely, but not if transported passively in a wheeled chair. After bilateral pre-frontal lesions a monkey is not able to adapt at all.

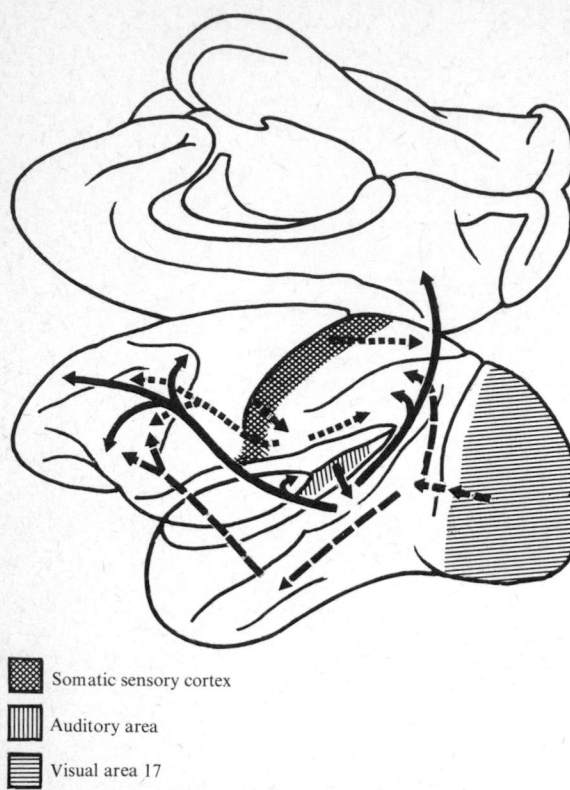

▨ Somatic sensory cortex

▥ Auditory area

▤ Visual area 17

FIG. 34.13. Schematic diagram showing some of the major association pathways from the primary sensory fields of the cortex. Solid arrows indicate afferents to frontal lobe carrying auditory information; broken arrows, ones carrying visual information; dotted arrows indicate somato-sensory afferents Pathways of all these sorts also converge on the inferior parietal lobule. (From Nauta, *loc. cit.*)

(b) *The dorso-lateral frontal cortex*

This large area is certainly not homogeneous, and its functions are complex. Lesions along the sulcus principalis (Fig. 34.14) produce defects in the monkey in tasks that involve a delayed response. The more posterior part is especially concerned with spatial problems, especially such as involve the position of the animal's own body. In man, lesions here also produce difficulties with such tasks, for instance, tracking of visual targets when the head and body are tilted. The hypothesis of corollary discharge might cover such defects and perhaps also explain more subtle changes, such as difficulty in making plans or in beginning or ending actions. In man, removals on the left side affect verbal fluency whereas those on the right influence visual memory, especially when it is a question of ordering of recency or categorizing cards by colour, form, or

number. The patient may say he knows what to do (for example, use colour as a clue to sorting cards) but cannot do it. These may be said to be defects of error utilization and could possibly also be produced by defective corollary feed-back.

A further difficulty is that results of removals vary with age. Removal of dorso-lateral frontal cortex from a young monkey, before the age at which it can do delayed response tasks, is not followed by a defect later. Moreover, there is no defect even in an adult following removal by stages. Evidently this cortex matures late and adjustments for lack of it can be made in other regions, and this is known to be true of other parts of the cortex too.

(c) *Orbital cortex*

Lesions to this cortex produce changes in emotional reactions and especially in social behaviour. For this reason surgical separation of it from the rest of the brain is sometimes used as a cure for persistent depression in man. The changes may be mild elevation (or sometimes depression) of mood or a state of euphoria and ignoring of social conventions. This is sometimes referred to as a failure of inhibition and the person may indeed be 'uninhibited', for example, about stealing food, sexual behaviour in public, or making personal comments that everyone thinks but doesn't say!

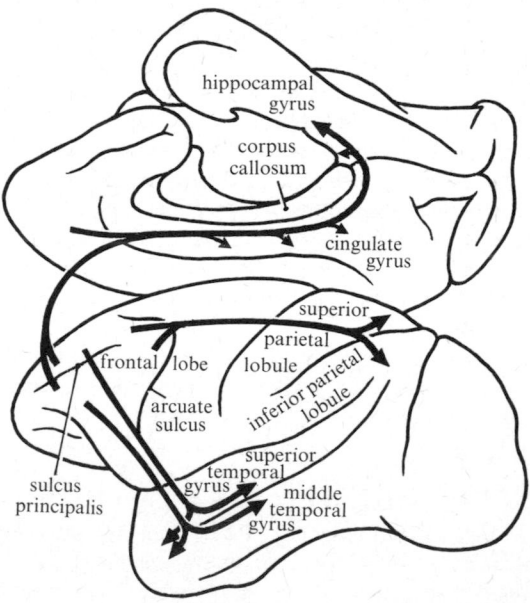

FIG. 34.14. Schematic diagram of the efferent connexions of the frontal lobe. The upper part of the figure is the medial aspect seen as if hinged and turned upwards, and therefore inverted. (From Nauta, *loc. cit.*)

The precise nature of the functions of this region remain obscure but it is reasonable to regard it as a higher inhibitory drive system, adjusting the organism to environmental factors and checking the overaction of internal drives. This would agree with its special role in man as a controller of social behaviour, which necessarily involves restraint.

The fact that the frontal lobes are so well developed in higher primates is of special interest. Higher primates, like other animals, are active and investigative animals, but they pursue their inquiries in a restrained and orderly way. One can often notice a marked difference between the ceaseless, if somewhat aimless, activity of the 'monkey house' at a zoo and the more dignified quiet of the corner occupied by chimpanzees, gorillas, or orang-outangs. Again, listening with several thousand other human beings to a speech or concert one cannot but be impressed at the astonishing restraint that characterizes human behaviour, which may be a result of the great development of the pre-frontal lobes.

35 Afferent projection areas of the cortex

1. Somato-sensory cortex

IMMEDIATELY behind the central sulcus is an area in the parietal lobe that receives projections concerned with the general somatic afferent systems for touch and pressure and probably also for heat and cold, taste, and proprioceptive senses. There is a detailed projection forward by fibres arising in the ventral nucleus of the thalamus and each connected with a particular area of the body surface. The arrangement resembles that of the motor area, the lower part of the body being represented on the upper portion of the hemisphere (Darian-Smith, Isbister, Mok, and Yokota 1966) (Fig. 35.1).

When a touch stimulus is given to any portion of the body surface electrical activity can be detected by electrodes placed in the appropriate part of these areas at the front end of the parietal lobe, which are numbered 1, 2, and 3 and on architectonic maps. This region thus constitutes the cortical receptor organ for touch, heat and cold, taste, and proprioception (Fig. 35.2).

An electrode passed through the primary somato-sensory cortex (SI) in a direction normal to the surface encounters a succession of neurons all activated by the same stimulus at one part of the body surface. In rats and mice the neurons of the face region are arranged in columns ('barrels') each forming the receptive field of one whisker. Caudal to the main somatic sensory area lies a second area (SII) in which electrical activity is also recorded after suitable peripheral stimulation. Each of the primary receptor areas of the cortex is accompanied by such a secondary area or 'belt area' (and a third and further sets may also be present). The neurons in the belt areas probably detect more complex features than those of the primary areas (p. 333).

Electrical stimulation of the primary somato-sensory cortex in a man under local anaesthesia produces reports of feelings (usually of touch), referred to the appropriate part. It is interesting that such procedure is not said to be painful, nor does it elicit referred pain (Penfield and Rasmussen 1950). The thalamocortical projection

system is concerned with the formation of associations between complex patterns of stimulation and with the computing of suitable responses. The impulses of the pain system operate primarily at a lower level, producing responses that are accurately predictable, and by a mechanism that is simpler than that of the cortex (p. 260). Yet the operations of the cortex certainly often

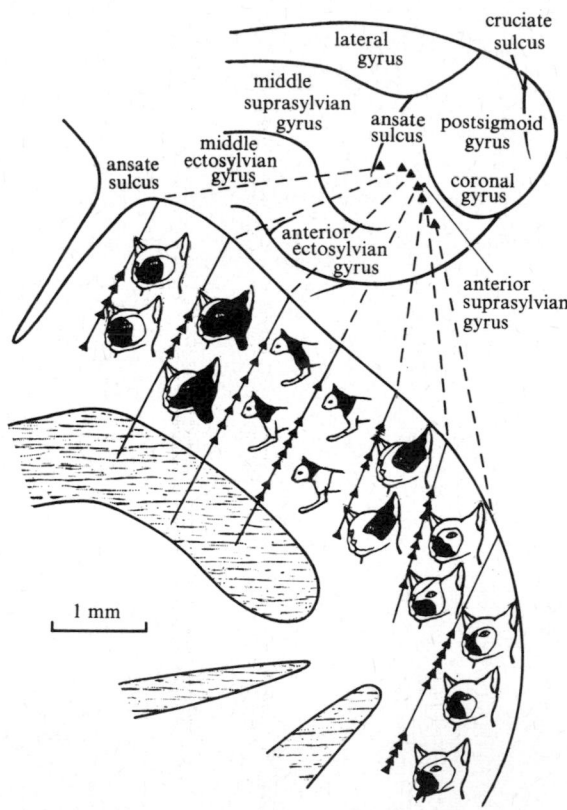

FIG. 35.1. Results of a series of electrode penetrations in the somato-sensory area of the cat. The neurons located are shown by triangles, with their associated receptive fields. These remain similar throughout the columns (except for the most ventral). (From Darian-Smith *et al.* 1966.)

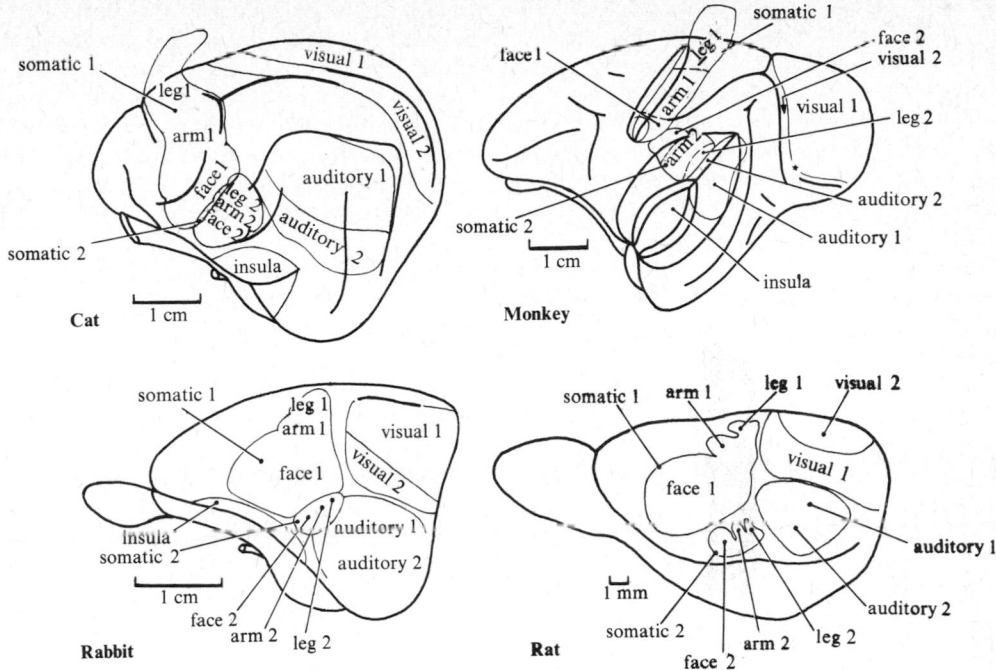

FIG. 35.2. Primary and secondary receptor areas of the cortex. It is likely that the areas giving auditory and somatic responses overlap in all cases. (After Woolsey (1952). In *The biology of mental health and disease*. Cassell, London.)

are influenced by pain impulses, acting as indicators of the undesirable nature of particular situations.

The area of the brain that is devoted to each part of the body varies with the importance of that part in the life of the animal. Thus in cats the limbs are used in varied ways and they have quite a large cortical representation, but in ungulates the limb area is small, perhaps because of the limited sensitivity of the hooves. In both horses and pigs the area concerned with the snout and lips is much larger than that of all the rest of the body (Fig. 35.3).

Behind S1, the equivalent of areas 1, 2, and 3 of Brodman, is a large part of the parietal lobe, known as areas 5 and 7, which is concerned in a less direct way with general somatic sensation. These areas receive projections from the association nuclei of the dorsolateral region of the thalamus, which are only indirectly connected with the ascending sensory system (p. 314). Injuries in areas 5 and 7 in apes produce temporary impairment of powers of localization of touch or discrimination of weights, and similar symptoms have been recorded in man.

There are two cortical areas for taste at the base of the parietal lobe in the region of the somato-sensory area for the mouth (Burton and Benjamin 1971).

2. Visual cortex

The fibres of the optic nerve in mammals end partly in the lateral geniculate body, partly in the superior colliculus of the midbrain (p. 314). The thalamus and

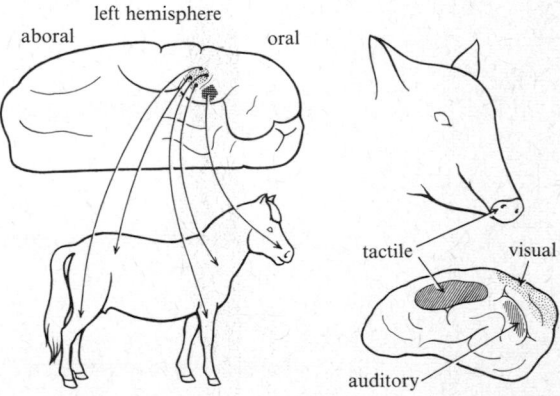

FIG. 35.3. Diagrams of the receiving areas in the brains of a horse and a pig. The anterior part of the receiving area in the horse brain is concerned entirely with the detailed representation of the nostril. In the pig the tactile area is connected with the opposite half of the snout. (After Adrian (1943). *Lancet* **121**; Adrian (1946). *Brain* **69**.)

colliculus both have two-way connexions with the cortex. A possible means of expressing the functions of the two sets of centres is that the colliculus is concerned with visual orientation and the thalamus and cortex with visual discrimination. But these are very broad statements, and the functions vary greatly in different mammals (see Jung 1973; Zeki 1974).

The occipital lobe of the cortex receives the visual radiation from the lateral geniculate body. In primates this is known as the striate cortex because of the presence of certain tangential bands of fibres as seen in section. This area (17 of Brodmann) is known physiologically as the primary visual cortex or visual area 1, VI. In front of it are other areas devoted to vision. These are known broadly as 'prestriate' areas (18 and 19) and can be subdivided by the properties of their neurons into second, third, and fourth visual areas. Still further areas concerned with vision are in the suprasylvian gyrus (called 'Clare–Bishop area' after its discoverers) and in the inferior temporal gyrus. This series of centres shows the great extent to which visual information can usefully be analysed. All these areas receive fibres from the striate cortex, directly or indirectly. The projection to visual area 1, VI, preserves the arrangement of the geniculate, and each part of the cortex thus corresponds to a particular part of the visual field. In visual areas 2 and 3, VII and VIII, there is also a topographical representation, but the arrangement is reversed spatially at each transition. The significance of this reversal is not known. Area 18 receives fibres from the pulvinar, a nucleus of the thalamus, which probably receives its input from the superior colliculus.

The relative amount of neocortex devoted to vision reaches its highest development in the primates. In the rhesus monkey it has been estimated from behavioural and physiological studies that something like one-third of the entire neocortex is concerned with vision. A very detailed analysis of the visual field is performed by these areas and different anatomical parts of the visual cortex specialize in analysing different aspects of the

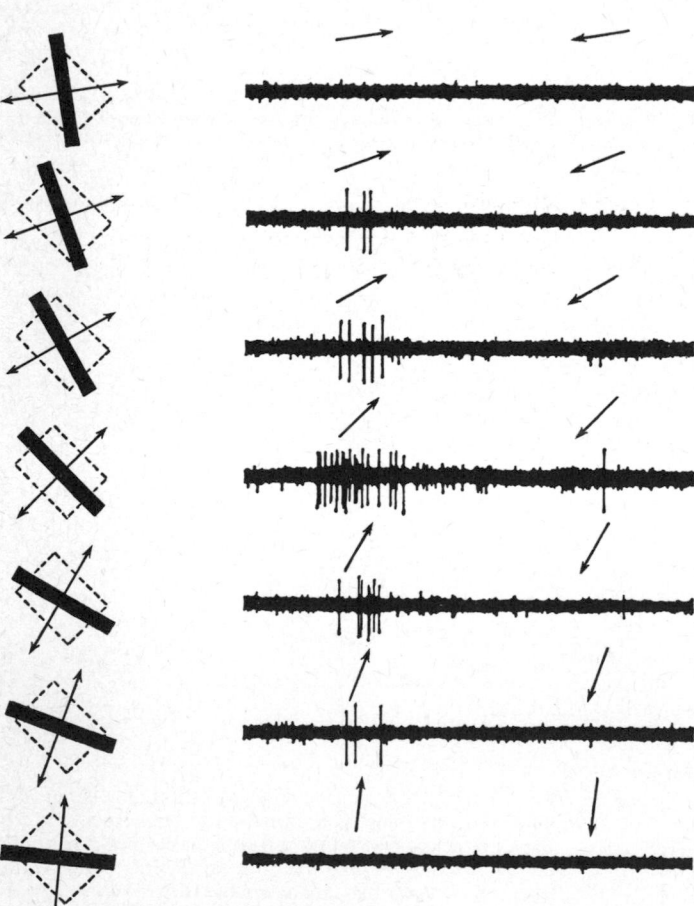

FIG. 35.4. Extracellular recordings from a microelectrode in single cells of a monkey's striate cortex in response to a black bar moving in the opposite visual field. Responses of a 'complex' cell in the right striate cortex. The receptive field in the left eye is indicated by the dashed rectangles. Whether or not the cell fires depends on the direction of movement and the angle of the bar. (From Hubel and Wiesel 1968.)

visual world. The cells of the lateral geniculate nucleus which project to the cortex have circular receptive fields divided into concentric excitatory and inhibitory regions. Even simple cells of the striate cortex have more complicated fields and respond best to edges of the appropriate length (Fig. 35.4). Thus, whereas the lateral geniculate nucleus registers difference in illumination, the striate cortex analyses the visual fields for different contours. These receptive field properties are presumably generated by convergent inputs to the cortical cells from the cells of the lateral geniculate (Hubel and Wiesel 1968).

The striate cortex deals with but one of the visual variables, namely a detailed analysis of contour. Other aspects of the visual world are analysed in detail in the other visual cortical areas. Visual areas 2 and 3 contain cells defined by Hubel and Wiesel as complex or hypercomplex, according to the variations of the contours to which they react. Other cells respond only to objects at a particular distance (Figs 35.4 and 35.5). Visual area 4, an area in front of these, contains cells that respond to particular coloured figures irrespective of their shape or direction of movement. The inferior temporal region carries the serial processing a stage further, receiving its input from the belt areas. Its cells respond to visual stimuli whose particular features are various and complicated. There are said to be cells in a monkey that respond to the outline of a monkey's hand, but not to a human one (see Blakemore 1974; Campbell 1974).

After removal of this inferior temporal area monkeys showed 'an insatiable visual curiosity', picking up objects they would ordinarily avoid, even snakes. They were very poor at visual learning tasks and in fact could be said to show 'psychic blindness'. Similarly patients with lesions to the temporal lobe on the right side were unable to remember faces or pictures, but had less trouble with words or numbers which are affected by lesions on the left (p. 336).

Evidently this region of the brain is concerned with visual learning, and it is interesting that it sends fibres to the entorhinal cortex and amygdala, which are concerned with learning (p. 343), although it also has many other outputs. There is thus a high degree of differentiation in the visual cortex in primates, with different areas specialized to analyse different features of the visual environment. It still remains a mystery how the information so analysed is subsequently brought together to give a unitary experience of vision. Removal of part or all of area 17 produces various degrees of blindness in different animals. In man complete removal of the area leads to complete blindness but in lower primates and other mammals form-discrimination alone is

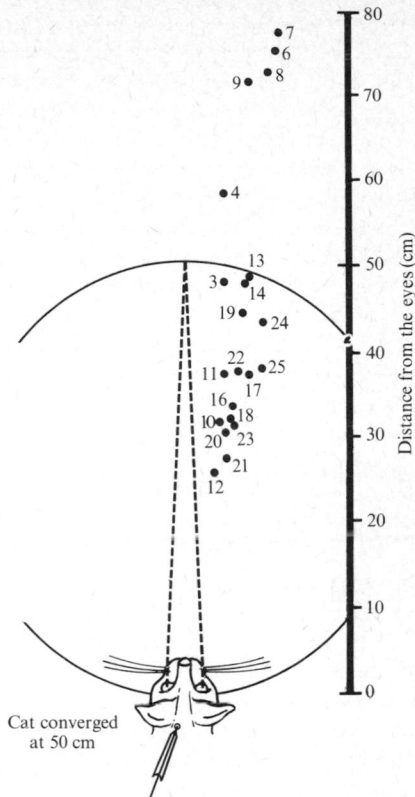

FIG. 35.5. Responses of single cells in cat primary visual cortex. Each numbered dot shows the horizontal position and depth in space which an object would have to occupy in order for its retinal images to fall on the binocular centres in the eyes for that unit. (From Barlow, Blakemore, and Pettigrew (1967) *J. Physiol.* **193**.)

affected and animals completely deprived of the occipital cortex can still react to changes in intensity of illumination.

In man a greater area of the brain is devoted to sight than to any other sense. The primary cortical (striate) area is much greater than that of the peripheral receptor itself. Thus, whereas the area of the pig's brain devoted to the snout is only a tenth of the receptor surface, the cortical receiving area for the fovea of the human eye is 10 000 times greater than the fovea itself.

3. The auditory pathway

The 24×10^3 fibres of each auditory nerve (in man) end in the cochlear nucleus. This contains about 10^5 neurons, but many are intrinsic to the nucleus and the number of axons leaving is about the same as entering. These fibres pass to three main destinations: the cerebellum; the superior olive, which is a nucleus in the medulla, and the nucleus of the lateral lemniscus, which can be considered its forward extension (Fig. 35.6). Some of these pathways run to the same, others to the opposite side. The superior olive sends its output largely to the nucleus of the lateral lemniscus, which thus receives input at either one or two neuron removes from the cochlea. Its output goes to the inferior colliculus, of the same or the opposite side, and this in turn sends fibres to end in the medial geniculate body of the thalamus, from which further fibres run to the auditory cortex.

There is a very large centrifugal pathway running in the reverse direction between these various centres and ending peripherally in the olivocochlear bundle of nerve fibres (Fig. 35.6) (Whitfield 1967).

It is not possible to give a fully rational account of the significance of this complicated set of centres along the pathways for hearing. It has been investigated by studying the effects on behaviour of interruption at various levels, or by electrical recording of their activity. As with other receptor systems it has been a problem to choose what auditory signals to use as tests. Investigators who have followed the model of physics and used pure tones were sometimes surprised to find that few units respond to these. But they might have thought that detection of pure tones is of little use to an animal.

More interesting results have been found when the sounds produced by other individuals of the species were used. The difficulty is that we understand how to analyse sounds in terms of frequency or pitch, because we have for so long related them to musical instruments. We have to learn to understand the 'musical' value of the innumerable sounds in nature, which are much more complex.

4. The analysis of cochlear signals

In the presence of a pure tone large numbers of auditory nerve fibres are activated (p. 406). What changes are accomplished along the pathway? At once we meet the difficulty that the changes at each level vary with the circumstances. They depend upon the depth of anaesthesia, or if there is none upon the descending influences, varying with other stimuli impinging on the animal. The difficulties are so great as to suggest to some people that analysis may be self-defeating. At the very first level, the cochlear nucleus, the effects are already varied and complex. Each neuron of the cochlear nerve may be distributed to a wide region of the basilar membrane (p. 405), and its central end may supply hundreds of cells of the cochlear nucleus. No simple statements can be made about the characteristic responses of each cochlear nerve fibre. Many of

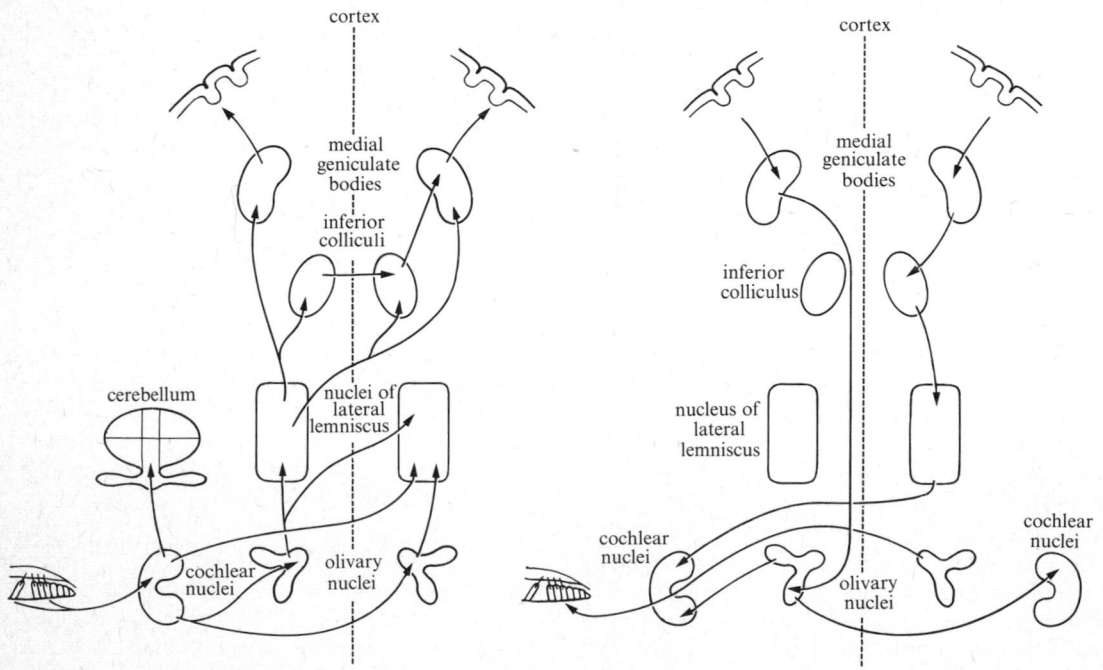

FIG. 35.6. The auditory pathway. Connexions to the cortex are shown in the left and those returning from it on the right. (After Whitfield 1967.)

them are spontaneously active at 10–200 Hz. Auditory stimulation usually increases the rate of discharge and the discharge may follow the stimulus, or bear an integral relation to it, especially at low frequencies. But pitch discrimination in mammals is not a period-time code. It involves the interaction of many fibres. The fibres that innervate many outer hair cells perhaps provide a low threshold system, while those supplying the inner ones give more discriminative responses. But owing to the spread of activity along the basilar membrane, numerous fibres are always activated (Fig. 35.7). Between just discriminable stimuli 99 per cent of the active fibres will be the same. With changing frequency of one semitone, perhaps 400 fibres are lost from one end of an array and added to the other. Changes with intensity may be about 200 fibres per decibel.

The fibres of the auditory nerve are distributed in a regular manner within the cochlear nucleus (*tonotopic distribution*). The frequency of greatest sensitivity of the neighbouring individual neurons shows a corresponding steady change. Spontaneous activity decreases along the auditory pathway. It is less in the cochlear nuclei than in the cochlear nerve fibres, and so on up to the auditory cortex. The frequency of responses also decreases along the pathway from some 250 Hz in the auditory nerve to 5–10 Hz at the cortex.

The transformation effected at the cochlear nucleus seems to be a sharpening by lateral inhibition. The fibres near the edge of the array that is excited by a tone are inhibited by a neighbouring one (Fig. 35.7). Stimulus frequency is thus signalled by the position of boundaries between active and inactive groups of units, and intensity is indicated by the number of units involved. Such effects are even more marked proceeding to higher levels (Whitfield 1967).

The superior olive and lateral lemniscus include the first point at which there is interaction between fibres from the two ears and it is probably concerned with location of sounds. Its cells are sensitive to time differences of as little as 10–20 μs, whereas arrival times at the ears of a cat differ by 250 μs. Some of the cells send axons to motor nuclei of the head and neck. This is also the lowest source of descending fibres to the olivo-cochlear bundle and it sends many fibres to the cochlear nucleus. However, the major portion of the cells of these nuclei project upwards to the inferior colliculus.

The cells of the colliculus show a very clear tonotopic localization. Some units are tuned to a narrow bandwidth, others more broadly. Many show binaural interaction and may be involved in sound localizations, but the idea that the colliculus is primarily concerned with auditory reflexes is now discredited. The main destination of its axons is to the medial geniculate of the thalamus and few fibres reach that nucleus without synapse in the tectum. The inferior colliculus is involved in simple tone discrimination probably by means of a distinct set of units, mainly monoaural. Frequency discrimination is still possible in cats after transection of the brain in front of the colliculus.

5. The auditory cortex

In the medial geniculate body the topology of the fibres is not understood. No clear linear tonotopic distribution has been revealed and the nature of the transformation effected to the auditory signals is not known. The fibres from this nucleus pass to the auditory cortex (area 22). This cortex lies in the temporal region and consists of a primary sensory cortex (AI in the cat), removal of which is followed by retrograde degeneration of the cells of the medial geniculate (p. 314). AII

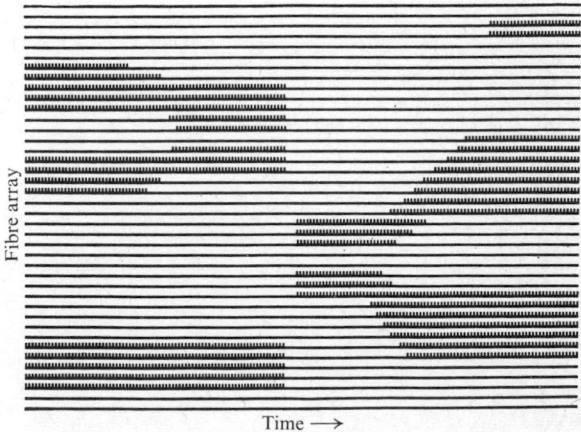

FIG. 35.8. Diagram illustrating distribution of groups of active fibres in total array. As the frequency structure of the signal changes with time so do the positions of the groups. (From Whitfield 1967.)

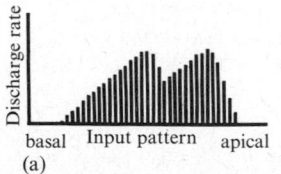 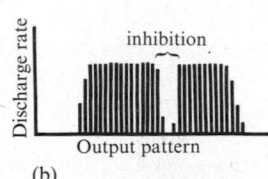

FIG. 35.7. (a) Discharge pattern in the auditory nerve on stimulation by adjacent tones. Each vertical line represents the mean discharge rate of 50–100 fibres. (b) Corresponding discharge pattern beyond cochlear nucleus. Mutual inhibition, by fibres usually stimulated by either tone, prevents fusion of blocks of discharges resulting from separate stimuli. (From Whitfield 1967.)

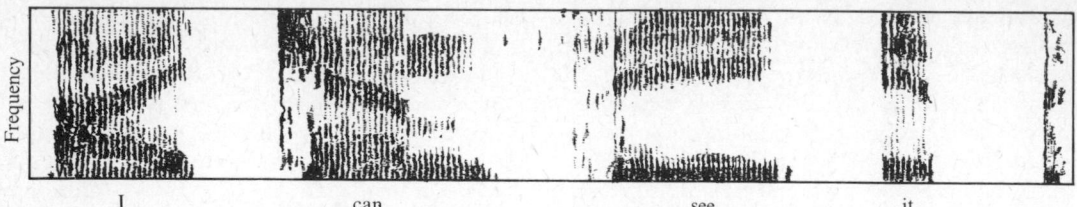

I	can	see	it

FIG. 35.9. Spectrogram showing frequency changes during speech. Different frequencies are represented by vertical levels with the highest at the top of the trace. The density of the trace records the intensity of a particular component. (After Potter, Kopp, and Green (1947). *Visible speech.* Van Nostrand, New York. Reprinted (1966), Dover, New York.)

and other areas near by are secondary ('belt') centres concerned with hearing. There are reciprocal connexions between the primary and secondary auditory areas, as in the visual and somato-sensory cortices. It is still uncertain how the projection to the auditory cortex is arranged. There is some evidence for a tonotopic distribution in the cat, with fibres for high-frequency tones in front. But the distribution of tone-specific fibres is by no means regular; indeed only 60 per cent of cells in the unanaesthetized cat respond to tones at all and many of these with a wide frequency range. About 10 per cent of cells respond only to frequency-modulated tones sometimes only to modulation in one direction. Other cells respond optimally to clicks and other complex noises e.g. the call sounds of other animals. The cells with a single type of response characteristic are probably arranged in columns (as in the visual cortex) but the distribution of the various types has not been mapped.

The auditory cortex is probably organized to detect temporal patterns of distribution of complex frequencies and intensities, rather than simple tones. The processes involved in the cortex are presumably the decoding of the significance of changing arrays of activity as in Fig. 35.8. It is interesting that the spectrogram of speech sounds shows rather simple arrays of this sort (Fig. 35.9).

The system would not be able to recognize all possible pattern sounds but only those of the high transition probabilities that have been learned by experience. The in-flow of information is increasingly restricted at higher levels. This may indeed be the function of the centrifugal pathway. Stimulation of the olivo-cochlear bundle reduces the response of auditory nerve fibres and similar effects may occur along the whole pathway. A mechanism for selective elimination of signals that are unrecognized and therefore unwanted would be a very efficient way of economizing channel capacity, making it relevant to the needs of the organism.

Electrical stimulation of area 22 in patients under local anaesthesia produces reports of roaring, and after its removal there is a reduction in auditory acuity (but not complete deafness); sometimes there are also defects of recognition of words and of speaking.

6. Speech centres

The analysis of speech sounds and the production of speech are unique among cortical functions in being limited to one side of the brain only, usually the left. If the left side is damaged in an infant, however, speech powers can develop on the right side. The analysis of speech sounds is no doubt very complex but an area specially involved lies behind the Sylvian (lateral) sulcus (Fig. 35.10). This is of course in between all the sensory areas and hence is an 'association area of association areas'. It is known as Wernicke's area, and after injury to it in man the patient is unable to understand speech or to produce intelligent speech, the condition of 'semantic aphasia' (see Liberman 1974; Penfield and Roberts 1959).

The actual production of speech sounds depends a region in front of the area of motor cortex devoted to the face and tongue (Fig. 35.10). After lesions to this Broca's area the patient can no longer speak properly, but can understand speech. This is a very crude division of the aphasias, which are clinically much more complicated, and indeed this highest of human functions is probably influenced by very many parts of the brain (see Lenneberg 1967).

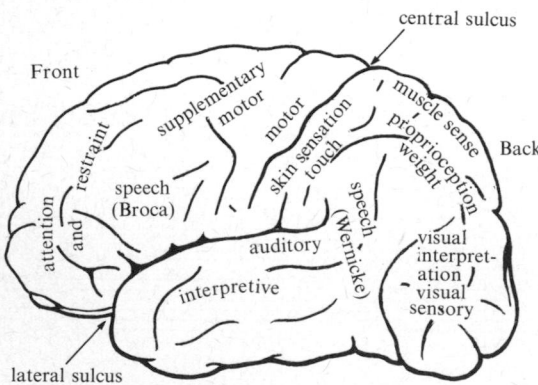

FIG. 35.10. The left (dominant) hemisphere of man showing various areas. (After Penfield and Roberts (1959). *Speech and brain-mechanisms.* Princeton University Press.)

36 Structure and functioning of the cerebral cortex

1. Structure of the cerebral cortex

IN the visual, somato-sensory, and auditory cortices, cells detecting each particular feature are in a column. Histological study shows this alignment of the cells (Fig. 36.1). Groups of cells and axons may lie together in columns, surrounded by cell-free areas of neuropil (Fig. 36.2). The barrel-like columns responsible for the information from the whiskers of rodents are especially conspicuous in this respect (p. 330) (Peters and Walsh 1972).

A second, equally striking feature of cortical organization is a tangential layering, shown especially by the fibres but also by the differing sizes of the cells in different layers. The pattern is conventionally described as consisting of six layers (Fig. 36.1). The cortical cells vary greatly in size and length of dendrites. The fact that cells of different sizes lie close together must mean that the size and length of branches of each cell have been determined by the specific activities in which it has been engaged rather than by some over-all graded control system.

Two main types are recognized: pyramidal cells and stellate cells (Fig. 36.3). The pyramidal cells have a large *apical dendrite* often reaching to the cortical surface and covered with fine dendritic spines. From its triangular cell body *basal dendrites* proceed in all directions. The axon passes to the white matter, giving collaterals upwards towards the surface, which may influence the cell's own dendrites or those of other cells within the same column. The ultimate destination of the axon may be elsewhere in the cortex or some subcortical station.

The stellate cells are often (but not always) small. They have many dendrites passing in all directions and not carrying spines. Some of them have axons that break up at no great distance from the cell body, while others show no obvious axon and possibly may be amacrine cells (microneurons) which function without propagating an action potential (p. 246). It may be that one function of the stellate cells is to sharpen contours of activity by inhibition. It is significant that these cells have a high concentration of rough endoplasmic

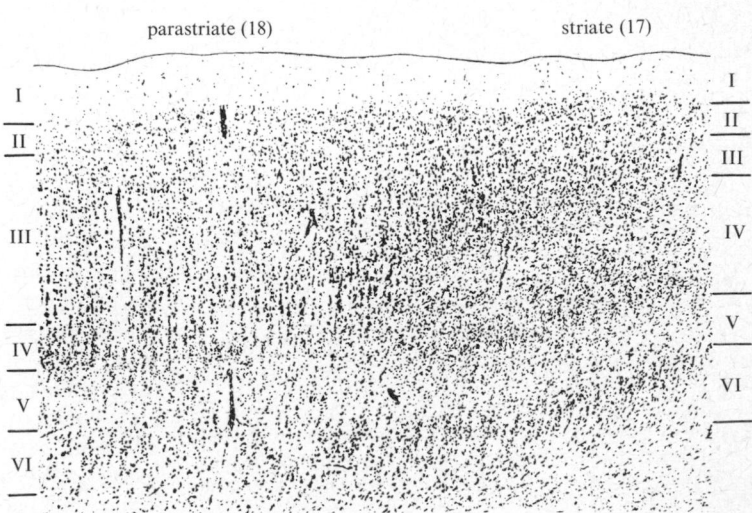

parastriate (18)   striate (17)

FIG. 36.1. The cells of the striate (visual) cortex (17) to right and parastriate (visuopsychic area (18)) to the left as seen after staining of the Nissl substance. (The layers are not all equally distinguishable.)

reticulum (Nissl substance). They are therefore active and synthetic cells, even though their axons are short or absent.

Electron microscopy shows two sorts of synapses in the cortex (Gray 1971). Type 1, the asymmetrical synapses, show a greater thickening of the membrane on the postsynaptic surface, whereas in type 2 (symmetrical) they are equal (Fig. 26.8, p. 252). The synaptic vesicles of the type 1 synapses are spherical whereas in type 2 they are oval ('flattened') and may be inhibitory. Nearly all the synapses on the pyramidal apical dendrites are made with the spines; they are type 1 and probably excitatory (Fig. 36.4). Each spine also shows

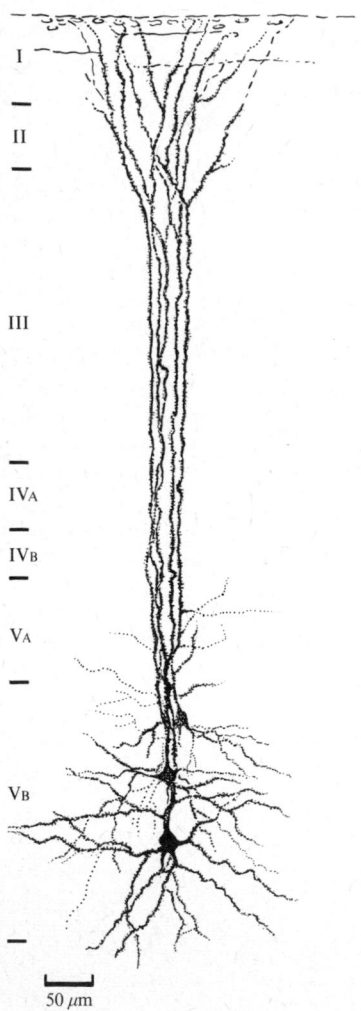

FIG. 36.2. Drawing of four layer V pyramidal neurons with apical dendrites ascending the cortex as a group, from a Golgi preparation of rat somato-sensory cortex. (After Peters and Walsh (1972). *J. Comp. Neurol.* **144.**)

clear channels (the 'spine apparatus') whose significance is not known. The synapses on the basal dendrites, on the other hand, are all of type 2 and may be the inhibitory endings coming from stellate cells. The synapses on the dendrites of the stellate cells are also largely of type 2, but a few are asymmetric and contain round vesicles. There are type 2 synapses on the axon hillocks and initial segments of both types of neuron (p. 252) (Jones and Powell 1970).

There are some tens of thousands of synaptic contacts on each cortical cell, perhaps as many as 6×10^4 (Cragg 1967). There are about 5×10^9 neurons in the cortex, so the total number of contacts is huge (10^{13}–10^{14}). It must be remembered however that each single knob is not necessarily the only connexion between two neurons. Indeed probably it never is. Unfortunately we do not know how many boutons there are on the end of any incoming axon. There are some indications that incoming fibres may climb alongside the apical dendrites each making synapse with many spines (Fig. 36.5). Whatever the method of branching and contact may be, in order to obtain the number of influences impinging on each cortical cell it would be necessary to divide the number of boutons per cell by an unknown factor, which may be 10^3 or even more.

The above interpretation of the synapses of pyramids and stellates is based largely on the fact that following interruption of the pathways leading to a piece of cortex there is degeneration of some endings on the apical dendrites (but not all). Conversely, degeneration of the synapses on basal dendrites is rarely seen, suggesting that they come from stellate and other cells nearby. The degeneration produced by lesions of commissural (callosal) fibres is concentrated in layers I, III, and IV (Fig. 36.3). Fibres passing between the somato-sensory cortical areas of the same side end mainly in layers III, IV, and V, but some run tangentially, close to the pial surface. The thalamocortical afferents end mainly in layer IV, overlapping into III and V. Thus all three sorts of afferents end in layer IV, and this has long been recognized as a receiving station. In the primary sensory and other receptive areas this layer contains many small neurons (hence 'granular cortex'). Layer V, on the other hand, contains very large cells, projecting to long distances, especially in the motor cortex where they are known as *Betz cells*.

Different parts of the cortex show great differences in structure. The archicortex and palaeocortex lack the granular and supragranular layers, but these are well developed in those primary sensory areas of the occipital, parietal, and temporal lobes that receive specific afferents. Conversely, the pre-central (motor) cortex and parts of the frontal cortex have fewer of

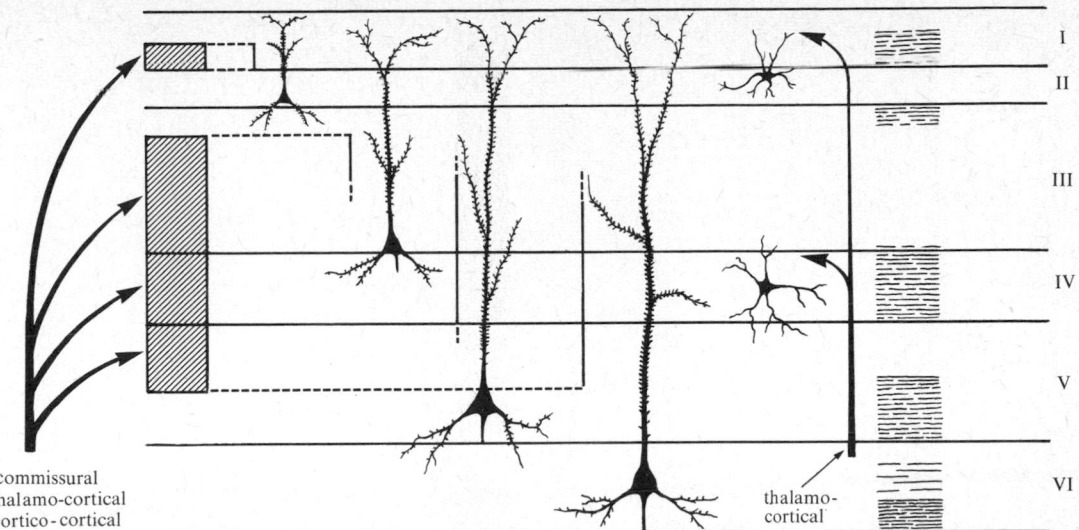

I
II
III
IV
V
VI

commissural
thalamo-cortical
cortico-cortical

thalamo-
cortical

FIG. 36.3. Diagram of possible main sites of termination of the main types of afferent to a region of the cerebral cortex. Stellate cells are shown on the right and may receive some

thalamocortical afferents. The position of the main horizontally running bands of intracortical association areas are also shown on the right. (After Jones and Powell 1970.)

the smaller cells and are called agranular. The division of the cortex into areas by Brodmann and others was made on a basis of differences in cells and layering, but the significance of these differences is still not understood.

2. Connectivity in the cortex

Each afferent fibre that arrives from the thalamus branches several times and the limits of its final branches lie as much as 650 μm apart in the visual area of the cat (Sholl 1956). Each single afferent fibre therefore

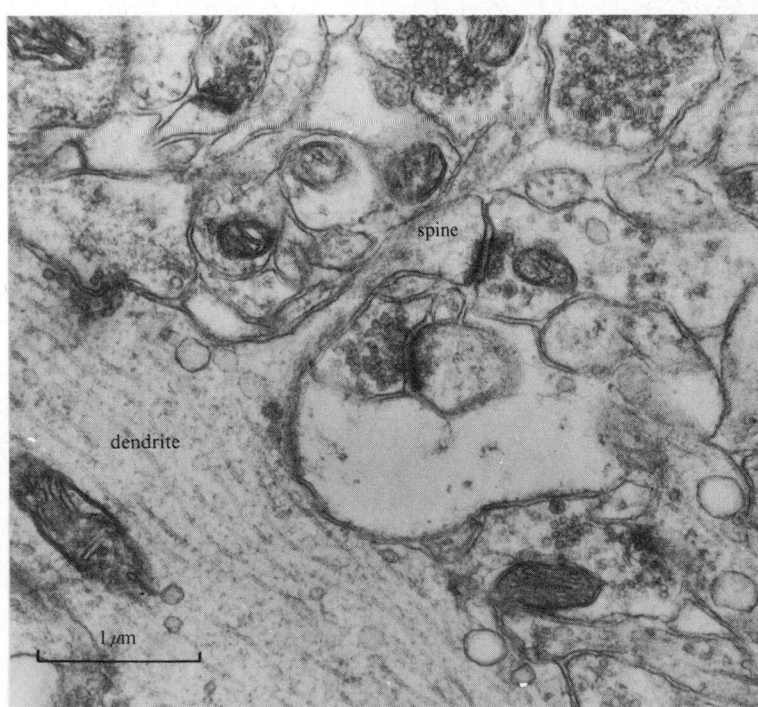

spine

dendrite

1 μm

FIG. 36.4. Electron micrograph of type 1 synapse on dendritic spine in rat visual cortex. (Figure kindly supplied by Prof. E. G. Gray.)

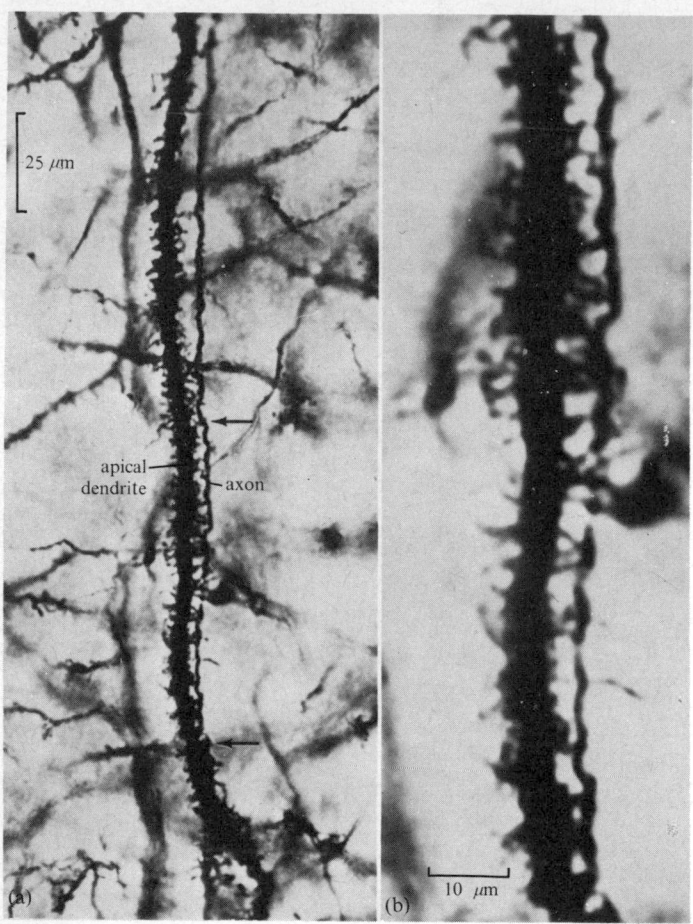

25 μm

apical—
dendrite   —axon

(a)   (b)   10 μm

FIG. 36.5. (a) Vertical axo-spinodendritic synapse from human neocortex between spines of the apical dendrite of a layer V pyramidal cell of the motor cortex and an axon of cortical layer III. (b) Detail of segment marked by arrows in (a). (Figure kindly supplied by Dr. M. Marin-Padilla.)

excites neurons lying within a considerable volume of cortex (about 0·1 mm³). Such a volume contains about 5000 neurons, most of them stellate cells (Fig. 36.6). There is therefore very great overlap between the volumes that are innervated by neighbouring afferent fibres. At present it is not possible to visualize the patterns in which the cortical neurons will be set into action by these overlapping inputs. The dendrites of each cortical cell extend over 250 μm and they share a volume of tissue with the dendrites of over 4000 other cells.

It is even harder to say how excitation will proceed away from the primary foci of excitation. The number of possible pathways provided by the axons of the stellate cells and the recurrent axons and collaterals of the pyramids is very great. Moreover, influences are continually reaching each section of the cortex from other cortical areas near and far, as well as from the thalamus. Many intra-cortical afferents (Figs 36.3 and 36.6) may reach to the surface layers and each divides

into branches extending over at least 150 μm. Each therefore may influence many cells, because the majority of the pyramidal cells send apical dendrites to the surface layers.

The total number of afferent fibres reaching any region of the cat's visual cortex is 25×10^3 mm², whereas there are 75×10^3 efferent fibres leaving it to proceed to other parts of the cortex or to noncortical regions (Fig. 36.6). At present we can form only a vague idea of how the output of impulses along these fibres is related to the input. In order to obtain a clearer picture we need more knowledge of the synaptic arrangements by which the incoming fibres influence the cortical cells. If we knew the means by which effective contacts are made in the intricate neuropil that occupies so much of the space within the cortex we might be able to understand how particular input patterns interact with the activities already going on and with whatever constitutes the cortical information store.

3. Feature-detecting neurons and responses to complex stimuli

The cortical end-stations for touch, hearing, and vision have all been shown to contain neurons each responding optimally only to certain limited specific environmental stimuli (p. 330). Some respond to simple features, for example a particular contour, others only to quite complex configurations, for example, to the outline of a whole monkey's hand (p. 333). There has been much debate as to whether such feature-detecting, 'gnostic' or 'pontifical' neurons constitute as it were a final stage of analysis whose firing patterns decide the behavioural output. Of course we can imagine that many neurons of each type are involved, but even with redundancy at this level some workers feel that the brain does not operate with such definite localized functional specificity. Instead it is suggested by some people that in life the discrimination of patterns is accomplished by the action of an ensemble of neurons, not activated on convergent or hierarchical principles. The question thus is whether the stimulus-specific template is provided by the properties of individual cells or as a statistical property of ensembles of them, the responses of individual cells perhaps changing with the circumstances. Indeed there has been some evidence that the evoked potential to a conditional stimulus is consistent although the firing of single units varies (John and Morgades 1969). One form of the 'ensemble' hypothesis is that the brain generates an internal activity that attempts to match or counterbalance what is coming in. This is known as analysis by synthesis and is a strategy adopted in some systems for pattern-recognition by computer (see MacKay and Gardiner 1972). Such a scheme may well operate, but still requires some form of template to make the initial hypothesis.

The 'gnostic neurons' certainly exist and they must surely be an important part of the mechanisms for form recognition by the brain. There is evidence that the generation of hypotheses is also necessary for complex pattern-recognition. Attempts at doing this by computer have shown that in even a simple object, say a cube, the actual edges and surfaces detected by a photocell will vary so much with the illumination that recognition is impossible. Some information must be provided about the geometrical and optical properties of the objects concerned. There is some evidence that persons born blind who have later been given sight are unable to distinguish shapes, at least until they have learned the 'rules for seeing'. Similar principles apply in the recognition of the complex sounds that constitute speech. Certain distributions of energy among frequencies must be recognized (formants) and at a higher level certain combinations constituting phonemes, syllables, words, and sentences. At each level expectations are generated and tested against the subsequent input before decisions are made as to the correct response to the sounds (Sutherland 1974).

4. Memory and the cortex

It is often considered that the cerebral cortex is the main seat of memory records in a mammal. This is true only if we interpret 'cortex' to include the basal parts such as the hippocampus and indeed the thalamus. Moreover, we know so little that it is not really clear in what sense memory records have a locus at all. Injury to particular parts causes difficulty with certain types of memorizing. For instance, memories of visual events are disturbed by lesions of the inferior temporal region in man or auditory ones by superior temporal lesions. Data from lesions, however, often show that a memory capacity is impaired but not totally lost. Items previously learned may be forgotten, but can be reacquired. It has seemed therefore to some authors that any given record is dispersed, perhaps quite widely

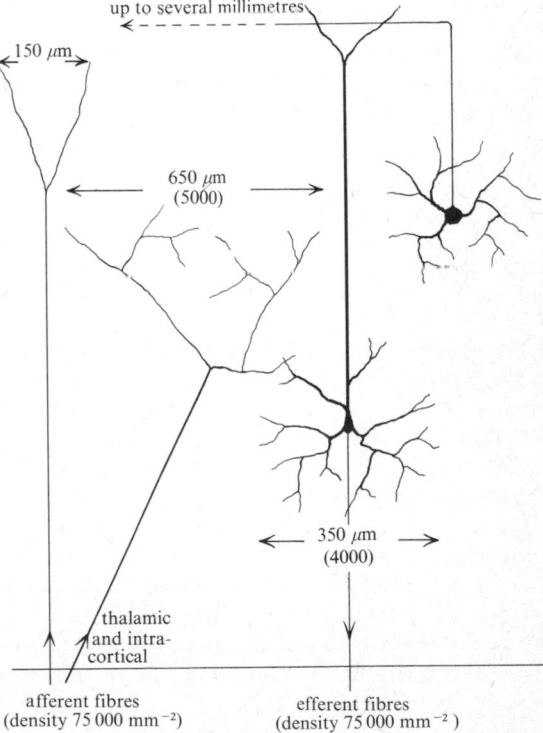

FIG. 36.6. Diagram to show the extent of the afferent axons and dendritic fields in the primary visual area of the cortex of the cat. (Figure kindly supplied by the late Mr. D. A. Sholl.)

through the cortical tissue (Lashley 1950). It has been suggested that the information may be dispersed as in a hologram. Yet careful tests with human beings after small head injuries have nearly always revealed some small defect in a particular capacity, which might be so unimportant that the person had not noticed it (Pribram 1971). Until we know more about the mechanisms of memory it is perhaps safest to assume that the record is localized in some parts of the brain at least for the individual senses, as it is known to be in an octopus (Young 1964).

It may be that we can distinguish between the memory record itself and the systems involved in 'reading in' to it and 'reading out' from it, as for a computer. On the other hand, we have continually to remember that information is carried in the brain by putting distinct items into separate channels, and so information is also presumably stored in the same way, by the properties of specific channels. How then do we find items in the memory? It is not by scanning some index to find the page, column, or row where the item is stored. Rather than such 'list addressing' there is probably a 'content addressing'. Many items are connected in the store by their contents so that when some fragment of information is presented a large set of items is recalled. This sort of view has been expressed by saying that memory storage consists in forming a model in the brain (Young 1964, 1966). It is significant that in people with hippocampal damage memories can be recalled if hints of their nature are given (Warrington 1971).

With such preliminary thoughts about the nature of memory we may consider a sample of the many experiments that have shown alterations in the activity of the brain as a result of learning. Changes of a memory type can be induced in a small volume of cortical tissue. Electrical stimulation by chronically implanted electrodes at say 6 Hz in a cat produces activity at that frequency in the mirror point on the other side. At first this ceases on both sides if the stimulus is turned off, but after some weeks activity at 6 Hz continues, even on the mirror side. (Morrel 1961; see John 1967). In a cat trained to press a bar to avoid a shock when a 4 Hz flicker appears, activity at this frequency appears in the visual cortex. If now a 10 Hz flicker is shown the cortex produces what seems to be a mixture of 4 Hz and 10 Hz. It may be that it first generates a play-back of the expected response. If 10 Hz is continued this frequency replaces the other (John 1967).

Whatever the memory change may be it must serve to alter the probability of producing one out of the several responses that are available to an animal that can learn. A simple plan of how this may be done is obtained by considering two possible alternative actions

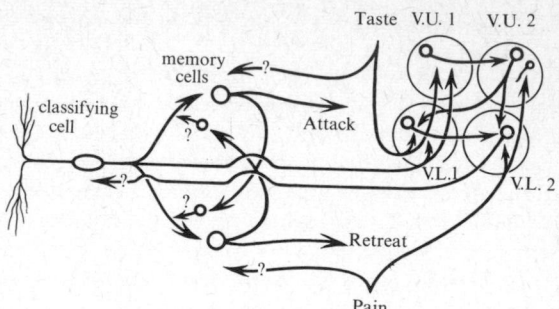

FIG. 36.7. The mnemon hypothesis. Diagram of parts of the visual memory system of an octopus.
The feature-detecting cells lie in the optic lobes and correspond to those of a mammalian brain. The one shown responds, say, to vertical contours. It is found experimentally that the animal can learn either to attack vertical rectangles or to retreat from them. If it receives, for example, a shock for an attack the 'pain' fibres will cause it to retreat, and it is suggested that collaterals then switch on the action of the small cells to block the unwanted channel. The centres to the right form a circuit, visual lower 1 and 2 (V.L. 1 and 2), with an upper circuit above (V.U. 1 and 2). Surgical interruption of these circuits impairs memory formation (see text).

consequent upon stimulation of a feature-detecting cell, which has been called the mnemon hypothesis (Fig. 36.7). The suggestion is that under appropriate conditions small cells begin the synthesis of an inhibitory transmitter, which switches off the connexions in an unwanted pathway. Memory would thus consist in learning what *not* to do, that is to say, the elimination of an initial redundancy. The memory may thus be said to be 'selective' rather than 'instructive', as is also the case of the system for producing antibodies (p. 191). Information storage consists essentially in selection among a set of possibilities. Heredity provides the sets of types of nerve cell as it does of types of lymphocyte. Experience then selects those that are appropriate to the particular individual.

There have been attempts to show that the memory record in the brain consists of synthesis of certain molecules—particularly of RNA. As the nature of the record is unknown it is impossible to deny this, but there is no hard evidence for such storage and it is most improbable given that the whole nature of information transmission in the nervous system depends upon putting each item into a distinct channel. Storage, therefore, is likely to consist of selecting particular channels. Of course the mechanism of selection is likely to involve synthesis of protein to act as a switch. And inhibition of protein synthesis does indeed interfere with memory. According to the mnemon hypothesis this is because it prevents the initiation of the production of an inhibitory transmitter. But various other forms of switch are feasible. The point is that the specificity of the memory item resides in the channel, not in the production of a particular molecule.

There is evidence that a mechanism for elimination of redundancy is at work in early learning by the mammalian brain. Kittens are born with cells each able to detect a particular contour (Hubel and Wiesel; see Gaze 1970). But if later the kitten sees only, say, horizontal lines it will be unable to respond to vertical ones and vice versa (Blakemore and Cooper 1970). The situation is essentially analogous to that suggested by the mnemon hypothesis, but the mechanism by which the unused cells are 'switched off' is not known. They may simply atrophy for lack of stimulation.

Of course in the cortex no simple scheme is operating, but at least we may be encouraged to seek for units of information storage, each conserving one bit (or mnemon), namely, the decision that a particular cell shall act in one of its possible modes or indeed simply whether is shall act or not. We already know that there are detectors responding to quite elaborate features (p. 333), and, by definition, in any brain that can learn, they must have more than one possible output. Moreover, the presence of certain small cells has been shown to be necessary for learning in the tactile memory of *Octopus* (Wells and Young 1965).

Whatever physical change is involved in memory it is certain that the process is influenced by the effects that the animal or man experiences as a result of its previous actions. Of course, in higher animals learning does not necessarily require a 'reward' such as food on every single occasion. Nevertheless the essence of a memory system is that it increases the probability that future actions will tend to ensure survival. This implies the operation at some stage of an internal monitor to indicate what should or should not be done. To give an analogy, when a man goes to higher altitude there is a receptor system indicating that more haemoglobin is required (p. 175). Memory in the nervous system is a special form of such adaptive responses and they must be basically monitored by internal receptors to ensure homeostasis. There must therefore be pathways by which signals of the rewards or results of action (reinforcement signals) are combined with signals from the receptor areas. The memory system of the octopus provides us with clues as to what to look for (Young 1965). In that animal there are lobes of the brain in which fibres bringing visual or tactile information are interwoven with others bringing signals of taste from the lips and probably pain from all over the surface of the body. These lobes not only allow the reward signals to meet those from the external world but in addition also spread the latter by branching. This allows for generalization, so that what is learned in one part of the receptive field is 'known' by it all.

There are similar 'mixing' lobes in the tactile memory system of the octopus and here we can prove their function experimentally. The octopus does not have to learn tactile tasks with each arm separately, eight times over: the information is exchanged between the arms by this system of branching and mixing. After removal of this 'mixer' lobe an octopus taught on the arms of one side cannot perform correctly on the other. Indeed after bisection of this part of the brain it can be taught in opposite directions on the two sides.

Of course, the memory system is much more complicated than this description indicates. We should not expect an instrument producing biologically 'correct' responses to a variety of situations to be simple even in an octopus. One complication is that a system is needed to maintain the information that a particular event has occurred until the action initiated by it terminates, say, in food or pain. In the octopus the circuit systems associated with the memory probably have this function. In fact there are double circuits, one above the other (Fig. 36.7). Interruption of the upper circuit, say by removing visual upper 2 (called the vertical lobe) makes it very difficult for the octopus to learn, especially to learn *not* to attack an object seen, or to draw in one that has been touched. It seems that these circuits provide the 'short-term memory' that keeps a record of what has happened and then combines it with the internal signals of the results of action. Perhaps maintaining the circulation, even for a time of the order of seconds, is necessary for the formation of a more long-lasting record.

The two memory systems in the octopus brain are certainly organized in approximately this way. The mnemon hypothesis may be wrong in some or many ways—for instance, in postulating an inhibitory transmitter—but it provides a logical schema of many of the features that must be present in memory systems, using the known anatomical features as a basis.

The experiments with *Octopus* thus show us some features that we should look for in any memory system and it may be that in mammals we find some of them in the cingulate cortex and hippocampus (p. 317). There is a set of aminergic fibres (p. 344), proceeding from the locus coeruleus of the medulla up through the midbrain and hypothalamus to the striatum and septum and so ultimately to the cingulate cortex (Fig. 36.8). The nucleus coeruleus lies at the head end of the set of nuclei in the medulla that receives taste signals by the nerves from the tongue and throat (VII and IX, p. 278). After removal of the nucleus coeruleus rats were found to be unable to learn to run a maze for a food reward (Crow 1973). With electrodes implanted in the nucleus a rat will press a lever repeatedly for reward.

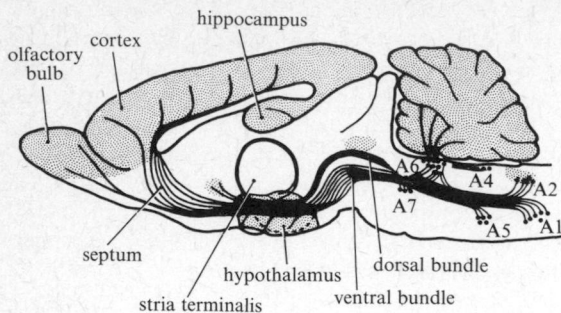

FIG. 36.8. Diagram of ascending aminergic pathways in rat brain, projected on to the midline. They run from various nuclei in the medulla labelled A1, A2, A4–A7. A6 is the locus coeruleus. It also sends fibres to the cerebellum. A3 is small and not shown. (After Ungerstedt 1971.)

The cingulate cortex that receives these aminergic fibres also receives fibres from all the main sensory areas (Fig. 36.8). It may be that part of the function of the cingulate–entorhinal–hippocampal system is to act as a 'mixer', combining signals from the exteroceptors with signals such as those of taste. A main output from the cingulate proceeds through the entorhinal cortex to the hippocampus, which has various functions (p. 317). It is connected in a circuit through the hypothalamus and anterior thalamus back to the cingulate cortex. Diseases interrupting this circuit lead to emotional disturbances and also to failures to form new memories, especially in man (the condition known as Korskakoff's syndrome). This connexion of memory formation with a circuit that is also involved in the internal states of the individual can hardly be accidental. It is possible that activity set up in this circuit serves to maintain a 'short memory' by re-excitation until the more permanent memory is 'printed' if a 'reward' arrives. But in the mammalian brain the situation is vastly more complicated than in the octopus and we can only make very tentative hypotheses as to its operations.

Electrodes indwelling in the hippocampus have detected cells that respond when a rat is in a particular part of a maze. It is suggested that they may provide the animal with a 'cognitive map' of the environment, telling it what is appropriate to do next when it finds itself in any given place (O'Keefe and Nadel, in preparation). This obviously could be achieved if the connexions of the cells produced suitable subsequent sequences of firing, but this is speculative. Besides the cells that respond to 'place' there are others in the hippocampus that respond only when particular objects are present. Still others respond to changes, that is to say, when old objects are absent or new ones present. It is also known that short bursts of electrical stimulation of the fibres of the penetrant pathway entering the hippocampus can produce changes that last at least for several days, showing that some enduring changes are possible there (Bliss and Gardner-Medwin 1973). When all this information about the hippocampus is brought together it should considerably advance our understanding of memory, but it must be admitted that at present we really know very little about the subject.

The capacity to form a memory record presumably varies with the way of life of the species. This may in time provide us with clues about the mechanisms involved. Each brain must have the power to store representations of those features of the world that are relevant for its possessor. This may be called building a model in the brain, which ensures appropriate action in the circumstances that the animal is likely to meet if future situations are similar to those previously encountered in the history of that species and of that individual.

37 Development of the nervous system

1. The origin of the nervous system

THE specialization of cells to respond to external changes first occurred on the outside of the body, and in mammals the entire nervous system still develops from ectodermal tissues. The receptive and conducting elements appeared many hudreds of millions of years ago by exaggeration in certain cells of the powers of responsiveness to environmental change. The 'response' that gives neurons their usefulness is not merely the setting up of a propagated action potential but also the power of outgrowth of axonal and dendritic processes. The nervous tissues today still require the stimulus of 'function' to achieve their full differentiation (p. 343). There has been a gradual elaboration of the response system during evolution so that the nervous tissues now become differentiated during development largely by response to stimuli (evocators) *within* the organism. The hereditary processes thus ensure that a nervous system is prepared during development and is ready to function at the appropriate time.

The evolution and development of the nervous system thus shows how heredity provides a recorded memory of past responses; the genes and developmental processes have been so selected as to provide a forecasting system that prepares the organism for the conditions it will meet. The activities of the nuclear materials and the differentiations that they allow after suitable evocation represent in a coded version the information received by the race during its past history.

2. The neural plate and neural tube

In a vertebrate the first response leading to the differentiation of nervous tissue is thus not to an external but to an internal stimulus. The chordamesodermal tissue that has passed in over the lips of the blastopore of an amphibian evokes changes in the overlying ectoderm, converting it into the *neural plate* (p. 486). There is probably a similar evocation in mammals, and in the chick the chordamesoderm invaginated through the primitive streak has a similar inducing action (Bellairs 1974).

The neural plate consists of cells taller than those of the remaining ectoderm and the plate proceeds to fold to form first a groove and then a complete tube (Fig. 37.1). This folding is an active process on the part of the cells of the neural tube and the neighbouring ectoderm, and is part of the movement of convergence by which the tissues of the embryo move dorsally. The folding may be a result of the action of microfilaments that run transversely across outer parts of the cells, pulling the apices together like purse-strings. At the same time the cells elongate, perhaps by action of the microtubules they contain. Neurulation can be prevented by vinblastin, a drug that disrupts filaments and tubules.

The margins of the neural groove meet first in the mid-region of the body and the folds then proceed to fuse in directions forwards and backwards from this (Fig. 53.6, p. 485). The ends of the tube remain open to the amniotic cavity for a while by the *anterior* and *posterior neuropores*. The posterior neuropore closes in such a way that the neural tube opens to the hind gut by a *neurenteric canal*.

All the nervous tissue throughout the body develops from cells within the neural tube and the neighbouring tissue, the neural crest (Jacobson 1970). The great majority of the neurons remain within the main axis of the brain and spinal cord, but the neuroblasts that will give rise to the neurons of the autonomic nervous system migrate out along certain of the nerves before they complete their differentiation. There have been reports of the formation of neurons in the wall of the gut from endoderm, but this has never been proved. The only other possible contributions to the nervous system come from the *placodes*, a series of patches of specialized ectoderm, which make contact with the cranial ganglia and may contribute cells to them (see Yntema 1943). The capacity to differentiate into nerve cells is thus almost wholly restricted to the region of the ectoderm that overlies the invaginated chordamesoderm.

3. The neural crest

The ectoderm along the edges of the neural plate forms a set of cells, the neural crest, along the

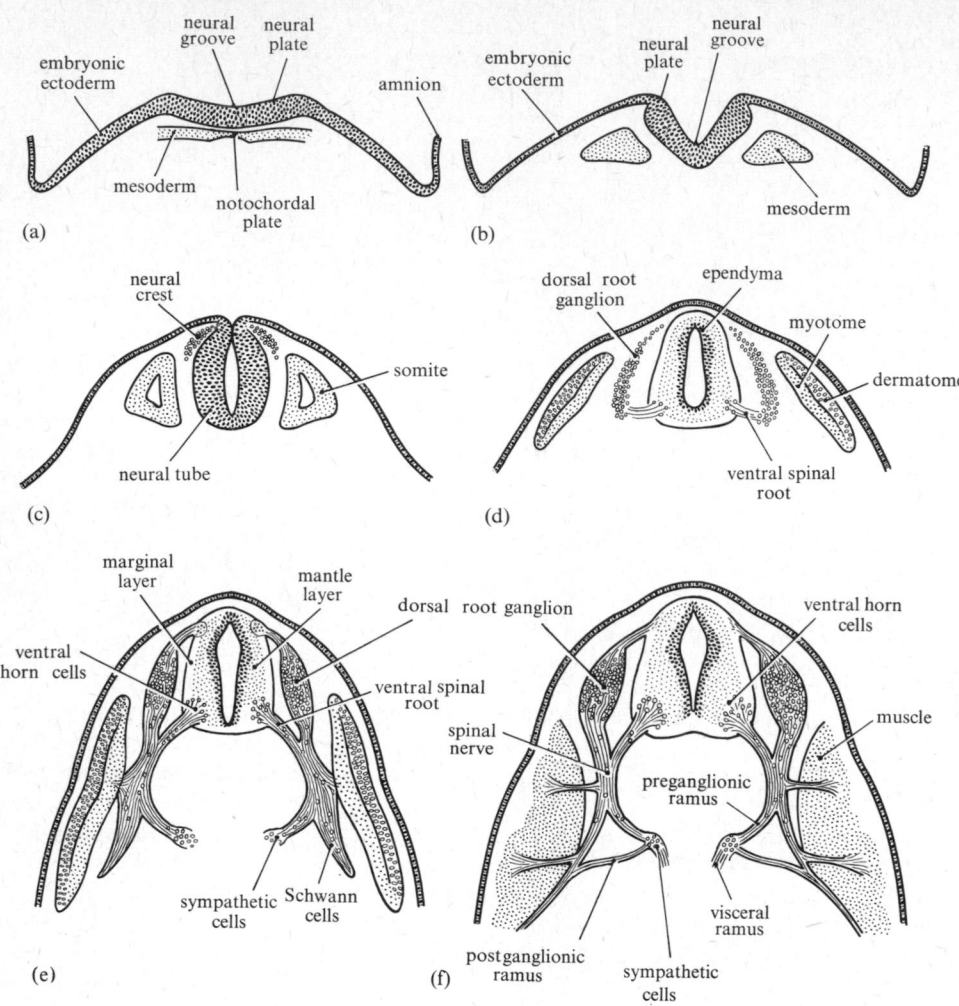

Fig. 37.1. Diagrams of transverse sections showing the development of the neural tube and the segmental nerves in a human embryo. (After Hamilton and Mossman 1972.)

whole length of the axis from the level of the mesencephalon backwards. These cells migrate laterally and produce the following tissues (Fig. 37.2): (a) the cells of the dorsal root ganglia, which form a segmental series, including the cranial ganglia V, VII, IX, and X, and all the spinal ganglia (the cells of the VIIIth nerve also belong to this series); (b) cells of the autonomic system, which migrate to their positions in the ganglia or viscera (p. 279); (c) cells of the adrenal medulla and other chromaffin tissue (p. 423); (d) the Schwann cells (neurilemmal cells) of all the peripheral nerves; (e) the cartilages of the branchial arches (p. 162); (f) the odontoblasts; (g) the melanoblasts of the skin; (h) the cells of the leptomeninges (pia and arachnoid).

The neural-crest cells evidently have special powers of movement and differentiation, which are presumably the result of their position at the margin of the region of influence of the chordamesoderm upon the ectoderm (see Horstadius 1950).

4. Development within the nervous system

The elaborate organization of the nervous tissue is arrived at by a series of processes: (a) proliferation; (b) cell death; (c) migration; (d) differentiation; (e) outgrowth; (f) functional differentiation; (g) learning.

The earlier of these processes take place mainly under the influence of heredity, the later require also the proper environmental stimuli. Together they lead to the pro-

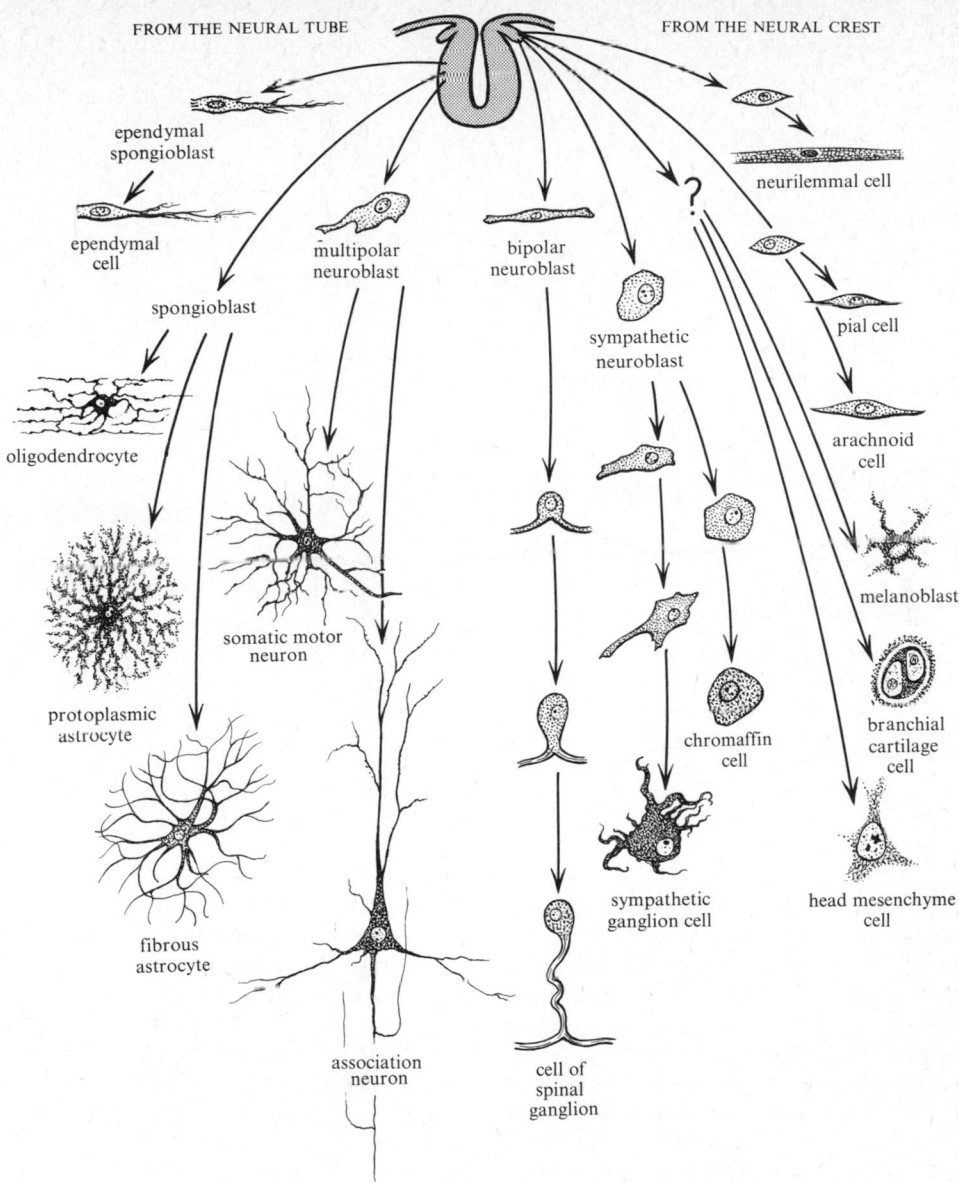

FROM THE NEURAL TUBE   FROM THE NEURAL CREST

ependymal
spongioblast

ependymal
cell

spongioblast

multipolar
neuroblast

bipolar
neuroblast

?

neurilemmal cell

sympathetic
neuroblast

pial cell

oligodendrocyte

arachnoid
cell

protoplasmic
astrocyte

somatic motor
neuron

melanoblast

fibrous
astrocyte

chromaffin
cell

branchial
cartilage
cell

sympathetic
ganglion cell

head mesenchyme
cell

association
neuron

cell of
spinal
ganglion

FIG. 37.2. Cell types derived from the neural tube and neural crest.

duction of a nervous system that is able to provide actions that are accurate forecasts because they are based upon the memories provided both by the hereditary system and by the experience of the individual.

Proliferation occurs by mitosis in the region of the neural tube near to the central canal and the resultant cells migrate outwards towards the surface, so that layers of cells are formed. In the wall of the tube it is possible to recognize three layers: an inner *ependymal layer*, middle *mantle layer*, and outer *marginal layer*.

The ependyma becomes the lining of the ventricles. The mantle layer contains the cells that become neurons, often called neuroblasts, though they cannot divide, and the glioblasts, which can divide further and become glia cells (Fig. 37.2). The marginal layer develops into the white matter.

The neural tube remains a single layer for some time, each cell being attached firmly at its ependymal end and more loosely at the surface. As each cell enters mitosis it loses the outer attachment and the nucleus

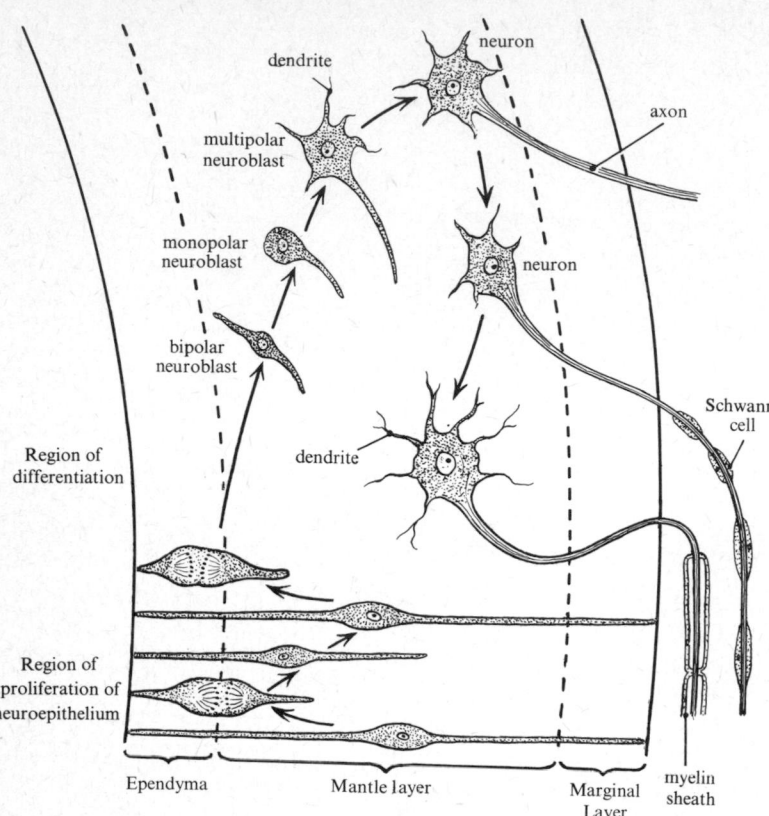

FIG. 37.3. Diagram showing stages in development of a motor neuron.

After a period of proliferation by mitotic division, some cells of the neuroepithelium lose their attachments to the other cells of the inner surface of the neural tube and migrate outward. The migration is accompanied by differentiation into a bipolar neuroblast by the growth of two processes, one of which is then absorbed while the other becomes the axon making the cell a monopolar neuroblast. Dendrites grow out, forming a multipolar neuroblast. The axon grows into the marginal layer and beyond, establishing the cell as a neuron. Schwann cells become attached to the axon and a myelin sheath is secreted.

moves close to the lumen (Fig. 37.3). After division the daughter nuclei move back. For this reason it was long thought that there is an inner layer of stem cells. After this early phase of multiplication most of the cells detach from the lumen. Some become neurons (p. 352); others become the glioblasts which may differentiate into astrocytes or oligocytes, but some of them retain the power of mitosis into adult life. Some capacity for proliferation remains in a subependymal zone, which may produce neurons even after birth and glia much later. The microglia are mesodermal phagocytes and originate outside the nervous system (p. 277).

The shape of the developing nervous system is determined by the fact that proliferation and migration do not occur uniformly throughout. Moreover, some of the early 'neuroblasts. fail to complete their differentiation, they undergo cytolysis and disappear. Local aggregations of cells are therefore not necessarily the result of local proliferation; they may be formed by migration, or by differential cytolysis and cell death.

The thickness of the wall thus comes to differ from place to place and bulges and foldings follow. Unfortunately we do not know in detail what determines the pattern of migration, proliferation, and cell death.

At an early stage in its differentiation the neural tube shows some sign of segmental structure. Three successive sets of these *neuromeres* appear, but their significance is not known. They do not correspond to the segments of mesoderm or nerve roots (Källén 1965).

5. Development of the nerve cells

The characteristic feature of the neurons is the presence of long conducting processes. The factors that control the growth and branching of the axoplasm to form the axons and dendrites are responsible for the whole organization of the nervous system. New nervous material is mostly synthesized within the nerve-cell body and the large amount of ribose nucleotides (Nissl substance) in the cytoplasm is connected with this activity. Cinematograph films of fibres growing in tissue culture show continual changes of shape. As the axons elongate they cross one another making a plexus pattern, which continually changes as the axons pull upon each other. No doubt similar changes take place during normal development within the body. Even in the adult we should not assume that the finer nerve plexuses are anatomically or functionally fixed. After injury to some of the fibres rapid re-adjustments occur in the remainder and it may be that there are changes with function, especially in the parts that are capable of learning.

The first sign of differentiation to form a neuron is that the cell detaches from the lumen and becomes bipolar (Fig. 37.3). It is then no longer capable of mitosis and is now '*determined*', that is, will continue to develop into a neuron even if moved to a foreign site. If a piece of the front of the neural plate is rotated through 180° its cranial cells develop into forebrain and grow out in the direction they would have proceeded normally. In the late neural plate neurons the genes for new neural development have been suppressed and their nuclei will no longer support normal development if placed in enucleated eggs, as those of late blastulae will do. From now on the genes for neural development express themselves. After a multipolar phase one process becomes the axon and others the dendrites.

The growing tip of the axon often carries a terminal club. This is usually called a 'growth cone', but there is evidence that only the pioneering fibres carry it and that it serves to find and recognize the appropriate end-place. Other fibres, without cones, then follow along the course of the pioneer (Lopresti, Macagno, and Levinthal 1973). In both cases the advance is by either an undulating membrane or by fine filopodia. These may extend and retract rapidly and are in constant motion. The rate of advance varies, with a maximum of 200 μm per hour in mammals. In each developing region the axons grow out in the direction of their final destination, but it is not known how this is determined. The formation of a long thin fibre must depend upon the physical properties of the contents, including the neurofilaments and neurotubules (Gilbert 1974). But there is probably also a turgor pressure, and if the axon is severed the part remote from the cell body breaks up into droplets (p. 239). Some axons contain RNA and are able to incorporate [³H]-leucine into protein, but this may be only in the mitochondria, which also contain DNA. Much new material is probably transported down from the cell. Some may also be taken up by pinocytosis at the periphery.

The factors controlling the direction and course of outgrowth of nerve fibres have been much debated. Three possible guiding influences have been considered (see Jacobson 1970):

(a) orientation of the fibres by electrical potential fields (Ingvar);

(b) chemical attraction of the fibres (*chemotaxis*— Cajal);

(c) orientation by contact with structures already formed or by 'ultrastructural' molecular characteristics of the medium (*contact guidance*—Weiss).

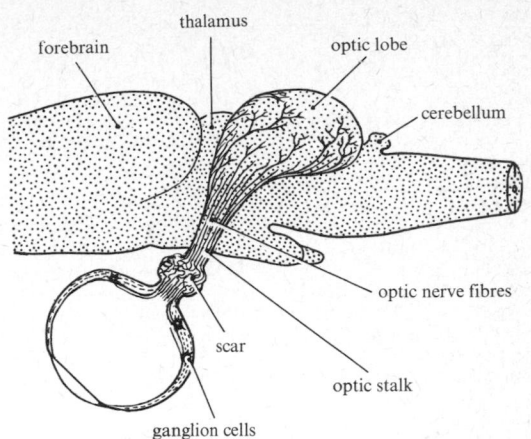

FIG. 37.4. Diagram showing the course of regenerating optic nerve fibres after section of the optic nerve of an amphibian. The optic nerve fibres grow out from ganglion cells in the ganglion-cell layer of the retina. They grow towards the brain in the optic stalk, becoming tangled at the region of section (the scar) and finally cross completely in the chiasma to the optic lobe of the opposite side. (After Sperry (1951). *Growth* **15**, *Suppl:* Symposium 10.)

Electrical potential fields can influence direction of nerve-fibre growth in tissue culture (Marsh and Beams 1946), but it is not certain that they do so *in vivo*. No definite proof of chemical attraction or repulsion of nerve fibres has ever been given, but it cannot be excluded. Fibres grow towards centres of active proliferation, but this is not necessarily evidence of a chemical influence. If a muscle is partially denervated the remaining fibres sprout and within a few days send branches apparently *direct* to the denervated end-plates. Nerve fibres that have been deflected from their proper course may nevertheless regenerate back to the right place. Thus regenerating optic nerve fibres of fishes or amphibians may find their way to the correct part of the midbrain tectum (Fig. 37.4), and the animals recover their normal power of snapping in the direction of an object moving in the visual field. During regeneration nerve fibres become much intermingled at the point of union of the two cut stumps (p. 240), and it is remarkable that sufficiently 'correct' connexions can be reformed to allow of such precise snapping movements (see Gaze 1970).

If the eyes are rotated through 180° an amphibian will make reversed or 'incorrect' snapping movements (Fig. 37.5). If after such rotation the optic nerves are severed and regeneration is allowed, the movements still take place in the reversed direction and continue to do so even after months of 'practice'. Several examples of such phenomena are known but may be due not to

(a)   (b)   (c)

FIG. 37.5. Diagrams showing the effect on visuomotor responses of rotation of the eyes of a toad through 180° and severance of the optic nerve with subsequent regeneration. (a) and (b) show the capture of a fly moving from left to right in front of a toad with normal vision. (c) shows the toad, whose eyes have been rotated 180°, failing to capture the fly because the movement made is appropriate to the retinal image of the fly prior to rotation of the eyes. (Data from Sperry, *loc. cit.*)

chemotropism but to the fact that nerve fibres branch out in many directions. Any that reach the 'correct' endings are 'recognized' and continue to develop; the others atrophy. It may be that there are specific chemical markers that allow fibres to recognize the correct pathway.

There are abundant signs that growing nerve fibres align themselves along nerve fibres already formed and along other structures. Contact guidance is therefore undoubtedly one of the main means by which the arrangement of fibres within the nervous system is produced. In a developing embryo the stresses set up by unequal rates of growth produce many surface boundaries within the nervous system and among the tissues. In addition, the large molecules of the materials in the intercellular spaces may become aligned by the forces to which they are subjected. Such ultrastructural organization would provide an orientating factor for the outgrowing nerve fibre.

The development of dendrites takes place later than that of axons and is perhaps dependent upon the latter. Dendrites grow by means of multiple filopodia and probably only those that receive correct axonal contacts persist. The form of the dendritic tree determines the function of the cell, but little is yet known about how this is achieved.

The organization of the various tissues of the embryo that is provided by the hereditary system thus determines to a considerable extent the pathways along which the nerve fibres grow, both in the central and peripheral nervous systems. Pioneering fibres grow, perhaps by contact guidance, along surfaces within the tissues. Later fibres follow these pioneers and innervate the developing tissues. Completion of the differentiation only proceeds, however, after effective connexions are formed (p. 242). There is some evidence that when a nerve fibre has made connexion with a muscle it becomes specifically changed in a fashion characteristic only for

that muscle. If a limb muscle of an amphibian larva is transplanted close to a normal limb, so that it receives nerves from the limb nerves, it is found always to contract at the same time as the muscle of the same name in the normal limb. If a whole limb is transplanted its actions will exactly copy those of the normal limb. This must mean that when a nerve fibre enters a given muscle its ganglion cell becomes specified by that muscle and functionally linked with other neurons supplying similar muscles, an effect known as *modulation* (Willier, Weiss, and Hamburger 1955).

The patterns in which the nerve cells are made active, however, are developed mainly under hereditary influences. A piece of spinal cord from the brachial region if isolated with a limb will produce co-ordinated movements, whereas with a piece of trunk cord the limb will make only irregular movements. The factors controlling the development of such *central action systems* appropriate to each region are unknown. They must be especially important in the brain.

6. Control of differentiation in the nervous system

Much of the differentiation during early stages depends upon a pattern of regional differences imposed upon the nervous system during the first stages of its evocation. Thus a portion of the spinal cord of a chick that would normally develop into the brachial region retains the characteristics of that part of the cord even if it is transplanted to other regions. In amphibians similar experiments show, however, that considerable regulation may occur.

During the later stages of development the mutual influences of neurons upon each other and of other tissues upon them play an increasingly important part (see Jacobson 1970). As in the case of so many other tissues the early development takes place under the control of hereditary instructions, but the later stages of differentiation of the details of structure depend upon interaction of the cells with each other and with the environment. Thus after removal of the developing limb-bud from an embryo of chick or frog the somatic motor cells of the ventral horn of the spinal cord are greatly reduced in size and number. Evidently when the outgrowing nerve fibres make contact with the muscles, they receive some 'stimulus' causing them to differentiate. Indeed, this contact with the periphery is necessary for the maintenance of the structure of the nerve fibre throughout life (p. 242). Conversely, overloading of a portion of the nervous system, for example, by implantation of extra limbs in a developing chick, may lead to the presence of increased numbers of neurons in the spinal ganglia. An interesting example of such a response is seen during regeneration of the tail of a lizard that has been thrown off (*autotomy*). The new tail is in-nervated entirely from the remaining proximal portion of the spinal cord and spinal ganglia, whose cells then grow to an exceptionally large size (Terni 1920).

Early development can proceed correctly in the absence of function. Tadpoles reared under anaesthetic develop well and move normally when placed in clean water. However, at later stages function is certainly necessary. Kittens at birth have cerebral cortical cells already organized to detect particular contours (p. 343). However, if an eye is covered for even a few days during a critical period between the fourth and seventh week the cells do not develop normally. It would be valuable to know whether there are such critical periods in man. Specific functional influences may be necessary. Thus kittens reared in an environment of horizontal lines are later unable to react to vertical ones or vice versa (Blakemore and Cooper 1970).

Thus the nervous system provides many examples of double-dependence, on factors from within and from the environment (p. 4). An elaborate system of in-fluences by chemical substances, by contact and by electrical factors may be at work. A substance known as nerve growth factor (NGF) that stimulates the growth of sympathetic nerve cells and fibres occurs (surpris-ingly) in the salivary glands of some animals. It is a protein of molecular weight about 40 000. Its relevance to the normal growth of nerves is not known (see Banks, Pearce, and Vernon, in preparation; Zaimis and Knight 1972).

7. Development of peripheral nerves

The nerve trunks thus develop as outgrowths of pro-cesses from the ventral horn cells, spinal root ganglia, and preganglionic autonomic fibres. The outgrowing fibres become aggregated into nerve trunks under hereditary influences. The pattern is essentially seg-mental, and even in a mammal it is rather regular along the whole length of the body (Fig. 37.6). In each segment there can be recognized a dorsal and a ventral root (Figs 37.7 and 37.8). The former contains afferent fibres, processes of the cells of the dorsal root ganglion, and, in the head region, also some special visceral motor and autonomic preganglionic fibres. The ventral roots contain somatic motor fibres and in the thoracolumbar region preganglionic sympathetic fibres (p. 281). Afferent fibres do not usually pass in ventral roots, except in those of the eye-muscle nerves and in the hypo-glossal nerve. The dorsal and ventral roots join in the spinal region, but not in the head.

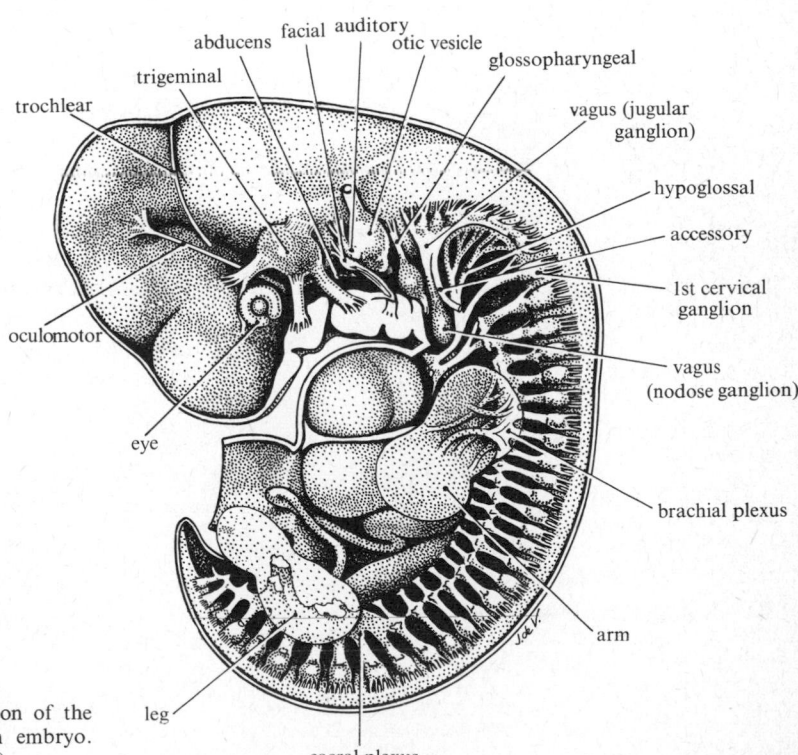

Fig. 37.6. Drawing of a reconstruction of the nervous system of a 10 mm human embryo. (After Streeter (1908). *Am. J. Anat.* **8.**)

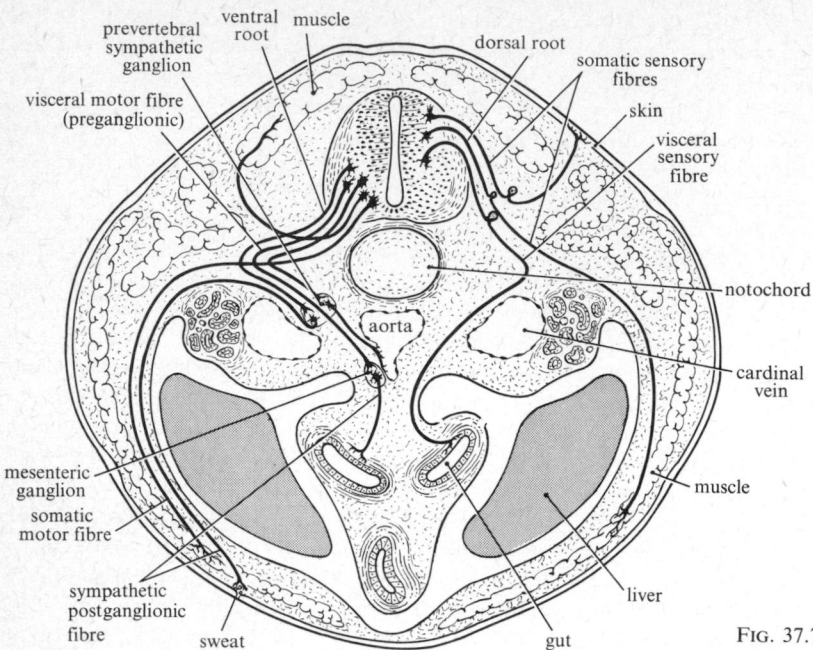

FIG. 37.7. Diagram showing the arrangement of the component fibres of spinal nerves.

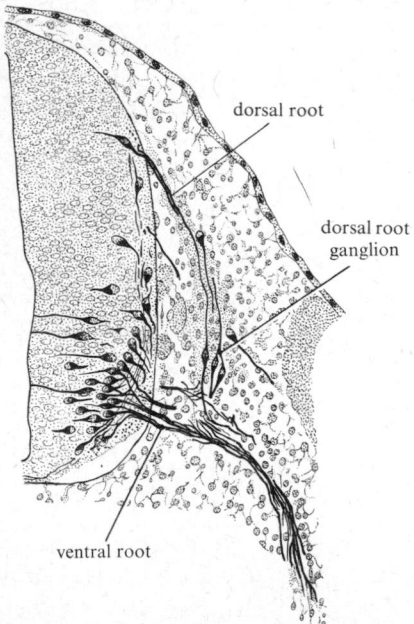

FIG. 37.8. Drawing of a transverse section of the neural tube of a 10 mm pig embryo (stained with silver to show axons). The dorsal root is formed from the processes of neuroblasts in the dorsal root ganglion whilst the ventral root is formed from the processes of neuroblasts in the cord. (After Held (1909). *Die Entwicklung des Nervengewebes bei den Wirbeltieren.* Barth, Leipzig.)

8. Development of the spinal cord

When the cells of the mantle layer differentiate the spinal cord comes to have thick walls and the narrow central canal is elongated vertically. Two distinct groups of cells can be recognized in the cord, the *alar plate* dorsally and the *basal plate* ventrally (Fig. 37.9). A groove, the *sulcus limitans*, marks the boundary between them. The dorsal and ventral midline tissues constitute the roof-plate and floor-plate of non-nervous tissue. The dorsal cells of the alar lamina receive the endings of the somatic afferent nerve fibres of the dorsal roots and thus constitute the *somatic sensory column* of the cord. Visceral afferent fibres end somewhat more ventrally, in the *visceral sensory column*. The cells of the basal plate constitute the motor columns, the *visceral motor*, lying more dorsally, and the *somatic motor* ventral to it.

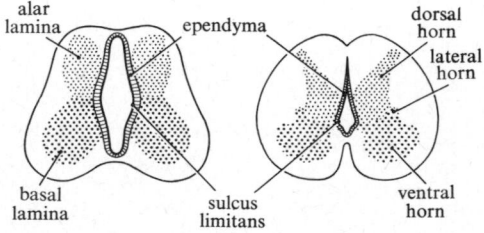

FIG. 37.9. Diagrammatic transverse sections of the spinal cord showing differentiation of the dorsal, ventral, and lateral horns.

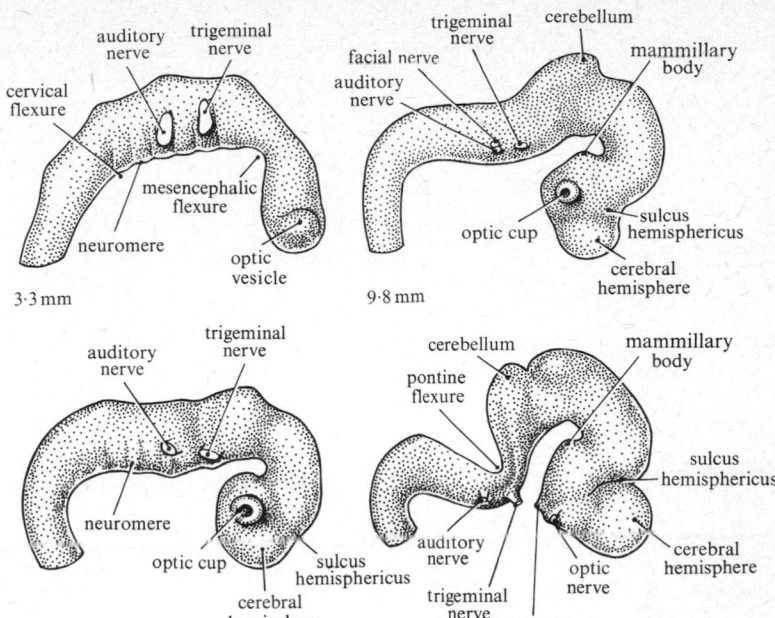

FIG. 37.10. Stages in the development of the brain of man. (Redrawn from figures of models reconstructed from sections by Hochstetter (1919). *Beiträge zur Entwicklungsgeschichte des menschlichen Gehirns.* Deuticke, Vienna and Leipzig.)

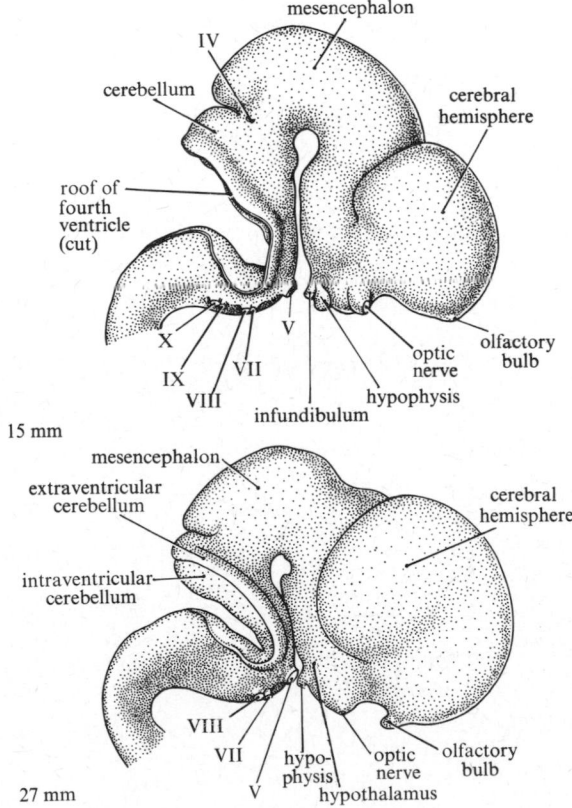

FIG. 37.11 Stages in the development of the brain of man (cont.).

These four functional columns can be traced throughout the length of the nervous system (p. 259), but it is not yet clear what developmental factors produce their segregation.

9. Development of the brain

The developing brain is at first a simple tube (Fig. 37.10) but the walls do not thicken to the same extent throughout, and thus the various flexures and outpushings are produced. At the front end of the spinal cord the central canal widens, the two sides of the mantle layer being, as it were, folded outwards, the roof-plate extending to form a thin non-nervous roof. This change of structure produces a sharp *cervical flexure* of the developing tube. A *mesencephalic flexure* in the same direction occurs in front of the future midbrain, at a point opposite to the front end of the notochord. This region corresponds to the front end of the series of segmental structures; it is the morphological front end of the body (p. 161). The first somite lies at about this level, and near here arise the trigeminal and oculomotor nerves, the dorsal and ventral roots of the first segment. The somatic motor column of the basal plate therefore ends near here and the whole of the forebrain can be regarded as an extension of the afferent centres of the alar plate. This excessive growth of the dorsal region of the tube produces the marked mesencephalic flexure.

As the cerebellum develops at the front end of the hind brain a further *pontine flexure* occurs in this region

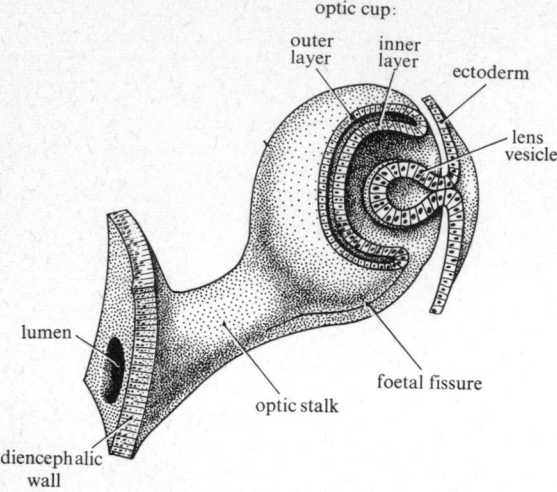

optic cup:
outer layer
inner layer
ectoderm
lens vesicle
lumen
foetal fissure
optic stalk
diencephalic wall

FIG. 37.12. Diagram of the optic cup and stalk of a 7·5 mm human embryo. (After a reconstruction by Mann (1928). *The development of the human eye.* Cambridge University Press.)

(Fig. 37.12). With the appearance of the cerebral hemispheres there is a marked *telencephalic flexure*.

The uneven thickenings of the walls produced by unequal cell proliferation and cell migration, and perhaps also by changes of cell shape, lead to development of the characteristic form of the brain and ventricles. The *optic vesicle* forms very early as a lateral pouch of what is at that time nearly the front end of the tube (Figs 37.11 and 37.12) but will become the diencephalon. The end of the optic vesicle rapidly becomes pushed in to make the optic cup and the lens differentiates from the overlying ectoderm (Fig. 37.13). The retina is thus morphologically a part of the lateral wall of the brain and the optic nerve represents a cerebral tract and not a true peripheral nerve.

The *cerebral* (*telencephalic*) *vesicles* appear later, in front of the optic cup. The hemispheres at first bulge forwards and remain open by a wide interventricular foramen to the unpaired between-brain (diencephalon) (Fig. 37.12). The lateral wall of the diencephalon becomes the thalamus, receiving the pathways from the

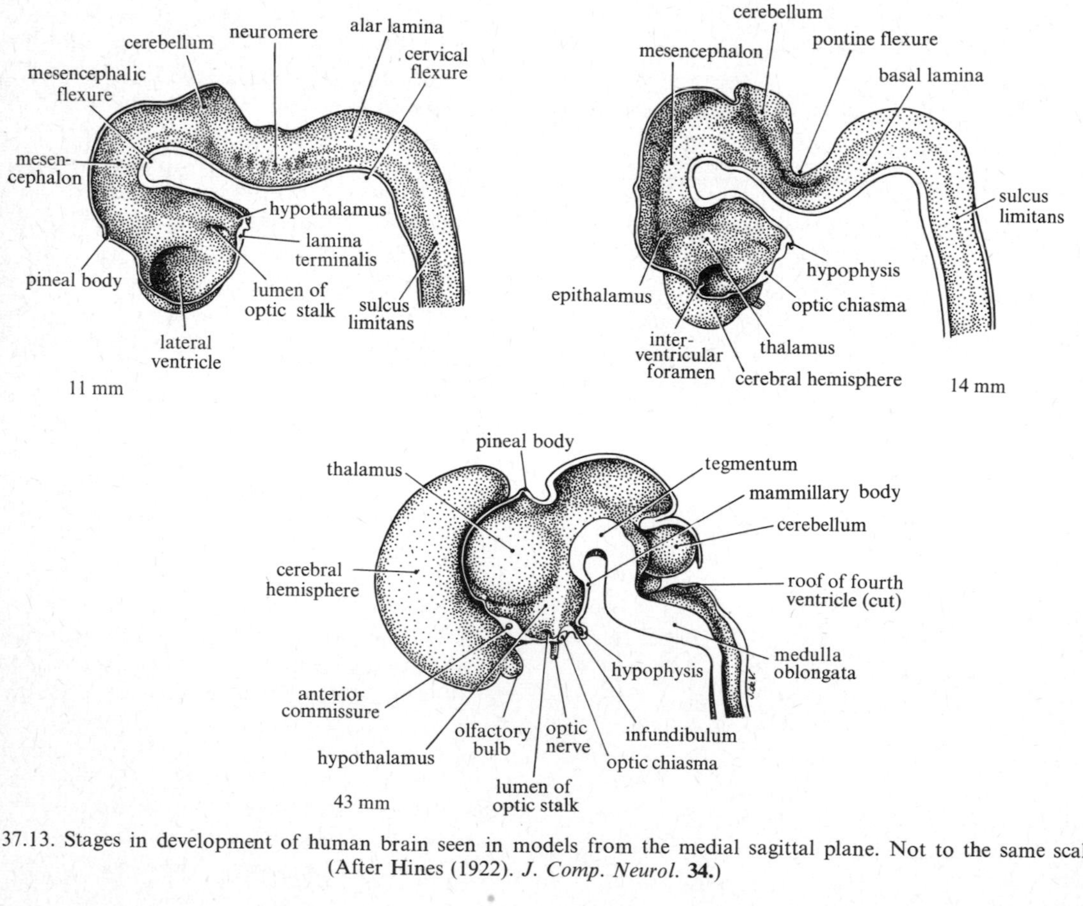

cerebellum
neuromere
alar lamina
cervical flexure
mesencephalic flexure
mesen-cephalon
hypothalamus
lamina terminalis
pineal body
lumen of optic stalk
sulcus limitans
lateral ventricle

11 mm

cerebellum
pontine flexure
mesencephalon
basal lamina
sulcus limitans
hypophysis
epithalamus
optic chiasma
inter-ventricular foramen
thalamus
cerebral hemisphere

14 mm

pineal body
thalamus
tegmentum
mammillary body
cerebellum
cerebral hemisphere
roof of fourth ventricle (cut)
medulla oblongata
hypophysis
anterior commissure
olfactory bulb
optic nerve
infundibulum
hypothalamus
optic chiasma
lumen of optic stalk

43 mm

FIG. 37.13. Stages in development of human brain seen in models from the medial sagittal plane. Not to the same scale. (After Hines (1922). *J. Comp. Neurol.* **34.**)

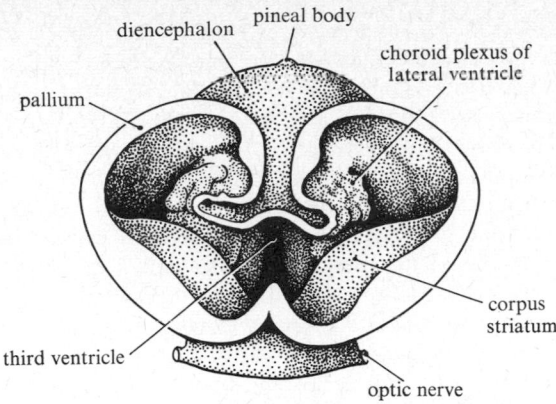

FIG. 37.14. Drawing of a model of the forebrain of a 17 mm human embryo. A cut has been made across the hemispheres and the view is from the front. (After Hochstetter, *loc. cit.*)

optic, auditory, and other afferent systems. This thalamic region is seen during development to be in direct continuity with the lateral wall of the cerebral hemisphere (Fig. 37.14), and the thalamic nuclei send fibres to the hemisphere and vice versa. The whole lateral wall of the forebrain, the original front end of the somatic sensory column, thus becomes developed into an elaborate memory system providing forecasts that control much of behaviour.

The ventral portion of the diencephalic region differentiates to form the hypothalamus (Fig. 37.15), perhaps representing the extreme anterior end of the visceral afferent and efferent columns (p. 259). The thalamus and hypothalamus are separated by a *hypothalamic sulcus* (Fig. 37.15), the forward extension of the sulcus limitans. The hypothalamus later acquires

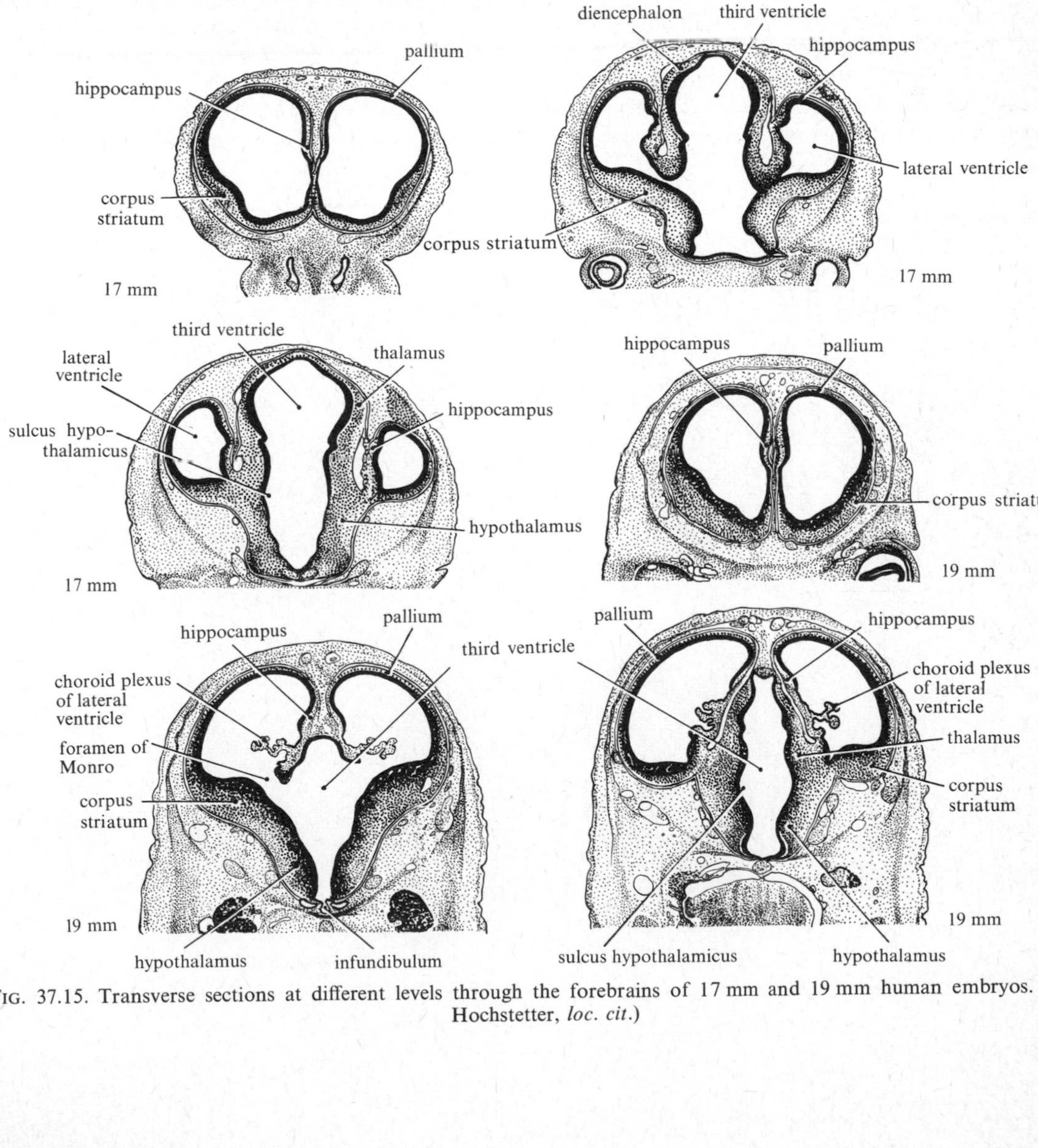

FIG. 37.15. Transverse sections at different levels through the forebrains of 17 mm and 19 mm human embryos. (After Hochstetter, *loc. cit.*)

connexions with the overlying thalamus and with the cerebral hemispheres (hippocampus) and plays an important part in determining the basic 'emotional' features of the action pattern of the forebrain.

The most ventral portion of the hypothalamus forms the infundibulum and this, with the stomodeal invagination makes the *pituitary body*. The fact that this gland develops at the point representing morphologically the extreme front end of the body is interesting in view of the dominant part it plays in the chemical signalling system that regulates development and adult life. The whole of the front region is evidently highly developed as a linked system of nervous and glandular tissue controlling the rest of the body.

10. Further development of the cerebral hemispheres

The evaginated cerebral vesicles extend both forwards and backwards and thus come to lie on either side of the unevaginated portion, the diencephalon. Where the medial wall of the hemisphere meets the lateral wall of the diencephalon the choroid plexus is formed, project-

ing into the lateral ventricle (Fig. 37.14). The interventricular foramina now become narrowed and the walls of the hemispheres proceed to differentiate. The more ventral portion becomes thickened early to form the *corpus striatum*. This is morphologically a direct forward continuation of the thalamus, but as the hemisphere grows backwards the corpus striatum comes to lie lateral to the thalamus (Fig. 37.14). It becomes continuous with the wall of the latter and fibres passing between the thalamus and the cortex grow through the corpus striatum as the *internal capsule*.

The dorsal portion of the wall of the hemisphere is the *pallium*, within which the cerebral cortex develops. The greater part of this becomes the *neocortex (neopallium)*, by migration of the cells away from the ventricle to give the characteristically layered structure. The differentiation first appears in the parietal region, concurrently with the arrival of the first thalamocortical fibres. The medio-dorsal wall of each hemisphere develops into a distinct type of cortex, the *hippocampus* (p. 317). The most lateral portion of the pallium also develops into the characteristic *pyriform cortex*, so called because this

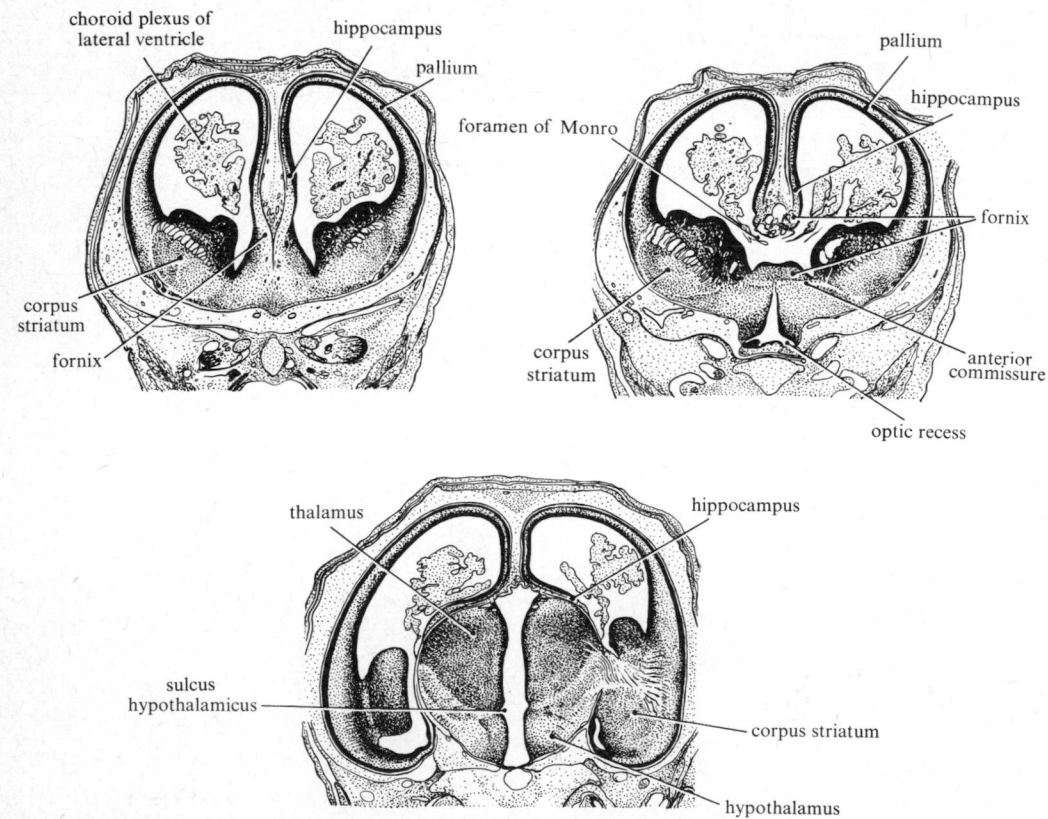

Fig. 37.16. Transverse sections at different levels through the forebrain of a 46 mm human embryo. (After Hochstetter, *loc. cit.*)

region is large and pear-shaped in mammals that have a well-developed olfactory system.

The *olfactory nerves* develop by outgrowth of processes of the cells of the olfactory epithelium of the nasal sac. These grow into the extreme front end of the developing cerebral hemispheres, where the *olfactory bulbs* then differentiate. The *olfactory tracts* join each bulb to the front part of the pyriform cortex, which is thus the part of the cortex most directly related to the olfactory function, which originally dominated all forebrain activities.

11. Cerebral commissures

The anterior end of the original neural tube (*lamina terminalis*) remains after the evagination of the cerebral hemispheres as the front wall of the diencephalon. In this lamina are formed the commissures between the two hemispheres. The *anterior commissure* is the more ventral and first to be formed (Fig. 37.16) and is followed by the *hippocampal commissure* above it. With the development of the neopallium there is a great further extension of commissural fibres, crossing in the *corpus callosum* and extending forwards and backwards far beyond the limits of the lamina terminalis. These crossing fibres may be said to unite the medial walls of the two hemispheres and they largely replace the hippocampus in the anterior region.

12. Later differentiation in the nervous system

The time at which the parts of the brain differentiate varies greatly; in general the control systems that are produced by heredity alone are laid down first. The parts responsible for acquiring memories during the lifetime of the individual differentiate later. At birth the mam-malian infant already possesses mechanisms for the regulation of digestion, respiration, and other functions. Even these basic regulation systems become markedly changed later. Temperature regulation, for instance, is imperfect in the new-born child (p. 17), and the operations of digestion later come under higher nervous control to a considerable extent.

Little is known about the factors that control the laying down of the detailed connexions within the nervous system. The fibres and cells certainly greatly influence each other and probably these influences depend upon the patterns in which the various cells are thrown into action by the stimuli they receive from the environment. In a chick after removal of the otocyst the cochlear centres differentiate normally up to 11 days, but the cells then undergo atrophy and mostly disappear. On the other hand, the vestibular nuclei differentiate almost normally (Hamburger 1952). The difference probably depends on the fact that the cochlear nuclei receive all their fibres from the ear, whereas the cells of the vestibular nucleus are influenced from many other sources.

There is further evidence from adults that cells whose input comes only from one source undergo atrophy if that source is removed. A most striking example of this *transneuronal degeneration* is the disappearance of the cells of the lateral geniculate body after removal of an eye or severance of the optic nerve.

It is probable, therefore, that the size reached by nerve cells and their processes depends upon the stimulation that they receive from the action of neighbouring nerve fibres. This stimulation may well be the factor that determines the details of axonal and dendritic branching and therefore of the synaptic relationships.

38   Receptor organs

1. Living measuring instruments

THE problem of finding terminology adequate to describe the actions of the sense organs is obviously especially difficult—for they themselves provide the information by which this and all other questions are decided. It has been usual to define the activities of these organs in terms of the sensory experiences to which they give rise. This is a form of the 'person language' for describing living activities (Young 1971), and it may be that for some purposes it is useful. However, following the principles laid down in Chapter 1, we shall try to describe the actions of the 'sense organs' in the language that is used for description of artefacts that have been devised to supplement them. The most obvious of these are *measuring instruments* (see Granit 1955). Every sense organ serves to measure some condition in the environment or to measure a change in condition. In everyday language we say that our sense organs 'tell us' (note the person language), say, about the intensity, of a light or that the intensity has changed, each constituting a 'stimulus'. Put otherwise we may say that one function of the receptors is to measure either the intensity or the extent of change of conditions. This may be a more useful way to start than by saying that 'sense organs respond to stimuli'. But the sense organs serve for more than simply measuring. The eye indeed measures intensity of light (brightness) and also its wavelength (colour), but in addition it records the pattern of distribution of light and so enables us to recognize a multitude of shapes. The skin measures mechanical movement and the ear loudness and, in a sense, frequency (pitch), but these senses too enable us to recognize complex shapes by touch or patterns of sound. We have not yet devised instruments that are able to recognize patterns as well as the body can, although simple pattern-recognition machines are made. It remains therefore difficult to describe these more complex activities of our receptors (see Adrian 1947).

However, the measuring instruments that man has invented provide us with powerful extensions of our own receptors and also with a new language for the description of living instruments. In the first place these artefacts give us a set of standards to which we can refer. We can speak much more precisely about weights, movements, or light changes if these are measured in c.g.s. or SI units by instruments than if they are assessed only by human receptors. Moreover, with very exact measuring instruments, such as electronic amplifiers, we can record the changes that go on within the body accompanying the operation of the receptors themselves. It is now known that the receptors all operate by sending to the brain distinct signals, the nerve impulses, in the afferent nerve fibres (p. 244). These nerve impulses are all alike in any one nerve fibre, but they differ in frequency according to the state of affairs at the receptor cell surface. We can therefore centre our study of all receptors around investigation of the conditions under which the receptor cells set up nerve impulses. Whether the animal or man actually responds by moving or speaking is, however, seldom determined by the actions of a single receptor fibre. The receptor surfaces are served by numerous nerve fibres, and the impulses in them interact at various levels as they pass through the nervous system.

We can therefore ask the following questions about any receptor system.

(a) Under what conditions does it set up nerve impulses in the nerve fibres leading away from it?

(b) What are the number and frequencies of the nerve impulses under various conditions?

(c) What are the spatial relations of the various receptors and their afferent nerve fibres, and what opportunities are provided for interaction of impulses within the nervous system?

(d) What features of the impulse patterns at different levels provide the relevant signals of information and how are they combined to ensure that choices are made between actions to ensure homeostasis?

As we become more adequately informed about these matters we begin to be able to give a satisfactory account even of the most complicated feats of measurement, such as the reaction to minute differences between sounds or shapes that we employ in hearing or reading.

2. Response to environmental change

Living organisms are composed of polyphasic physico-chemical systems whose steady states are responsive to changes in the external environment. Macromolecular colloidal solutions are easily changed by such factors as stirring, shaking, heating, illumination, or the actions of small amounts of other substances. Similarly even unicellular organisms are affected by changes in the mechanical, photic, or chemical conditions of their surroundings. The responses of organisms differ from those of non-living systems in that they are in the main such as will lead to a continuation of life; they are adaptive. During the course of evolution there has been produced from the responses of simple living systems an amazingly sensitive arrangement by which minute changes, occurring perhaps at a great distance from the surface of the animal, are measured and lead to readjustments that ensure its continuance. A hawk swooping to seize a lizard that has moved in the grass below shows one extreme development of the system by which the changes in the incident pattern of light have been linked up with the action system of the whole animal to produce forecasts that are effective in maintaining life.

An amoeba reacts to nearly all the types of change that affect a man. It is sensitive to heat, light, touch, vibration, and to chemical change. Yet light is a 'stimulus' for a man in a more extended sense than for an amoeba. The reaction of a protozoan when there is a change in intensity of some factor in its surroundings is usually limited to the setting in action of the motor system that operates to produce random withdrawal and advance in a new direction—avoiding-reactions which ultimately lead it to optimum conditions. The protozoan may be said to be a system that can receive only a minimal amount of information from its surroundings. If the only action that it can give is the avoiding-reaction then it is certain that every effective change in the surroundings will produce that response. The only alternative is whether or not the avoiding-reaction will be produced and only one unit or 'bit' of information passes through the system at a time.

Proceeding up the animal scale there is a great increase in the number of possible alternative actions and correspondingly an increase in the amount of information that can be transmitted. When a change occurs in the neighbourhood of a man he may move in various ways, either towards or away from the place where the change occurred. Instead of reacting only to a change in intensity he may respond in many different ways to differing patterns of intensities.

3. Communication theory

The problem of describing the action of the receptors in such complicated situations has been simplified by the techniques devised by mathematical engineers to help in the study of artificial aids to communication (Wiener 1948; Shannon and Weaver 1949; Cherry 1966). As has happened so often before the methods used to assist human functions have led to improved understanding of similar processes in the body. In order to improve communication by telephone, radio, and other means engineers need to measure the amount of information transmitted. This has led them to examine closely the concept of a communication channel, focusing on the fact that the function of transmission of information is to allow the sender to ensure that the receiver makes an appropriate choice between alternative possible actions. A first requirement for detailed specification of the system is therefore in theory to know what is the *set of actions* from which a given communication channel is to make selection (see Waterman 1968; Young 1971, p. 101).

In practice this requirement is seldom met either by the engineer or biologist. But in dealing with receptor organs it is important to find out at least something about the purpose for which their information is used. This often proves very revealing in showing up the limitations of our own narrow viewpoints. For example, the eyes or other photoreceptors of an animal have the function not only of choosing between such alternatives as whether an object is likely to be edible, dangerous or an obstruction, but also whether the days are getting longer and it will soon be time to breed. A mole's tiny eyes may not be much good for the first sort of choice but excellent for the second. Moreover they may serve him well simply by telling him to get *away* from light. There is a room for much more careful study of the alternatives between which choices are made by animals, and indeed by men.

More attention has been given to the *sets of signals* by which information is conveyed. Clearly these must be related to the sets of possible actions; there is no point in sending a lot more information than can be acted upon (or understood). However, this question obviously can be considered only when the possible actions are known, and in the meanwhile we mostly have to depend upon the analysis made by telephone (and other) engineers who are concerned with human speech, where the 'actions' of the receiver can be considered as recognition of the words of a language. They have emphasized that among the large set of words available to any person the words actually occur with very different frequencies. The amount of information conveyed by a word is inversely proportional to the frequency with which it is used.

In trying to apply such methods to the study of receptors we are again obstructed by the fact that we know neither the set of actions of the animal nor the set of code 'words', that are sent along the nerves from a receptor. Indeed neurologists are only now beginning to wrestle with the problem of identifying the significant signals sent by each type of receptor. It is known that the neural code consists of sequences of similar nerve impulses. We shall have to follow the evidence that has been discovered about the patterns of these impulses at various levels between input from the receptors and output of the muscles or glands, producing action. In few cases shall we be able to make the account even approximately complete.

We may, however, feel that our ideas are clarified by the attempt at such analysis. Instead of talking only about 'stimulus' and 'reaction' to it we can talk about *signals* and the *information* they provide, allowing *selection* between alternative actions. This will make us look for the patterns that constitute the signal among the *noise* that obscures it.

4. Receptors

Every species is provided with cells specialized in their sensitivity to the types of change that must be detected if the animal is to remain alive. The changes may be chemical, mechanical, photic, thermal, or electrical. Each receptor system contains a terminal apparatus sensitive to change of one of these types. Thus the receptors for touch, pressure, and sound all depend upon mechanical deformation. Chemical means of excitation are employed by the receptors for smell and and in certain internal organs sensitive to changes in the composition of the blood. Thermal energy changes stimulate the receptors that serve to measure temperature, and change of electromagnetic vibrations of a certain wavelength are the basis of vision, the light changes in that case serving to alter the condition of suitably photosensitive molecules of pigment in the rods or cones (see Carterette and Friedman 1974; Loewenstein 1971).

The next question is: how does the mechanism that provides the specific sensitivity of the end-organ to its adequate stimulus serve to set up nerve impulses? The sensory cells are known as *transducers*, they are activated by some energy change in the environment (say by light or sound waves) and this in turn activates nerve fibres, whose action potentials then serve to signal the occurrence of the change to the central nervous

system. The nerve impulses can thus be said to carry coded information about the change that has occurred (p. 247).

The act of transduction may occur either in the ends of the nerve fibres themselves or in specialized receptor cells, which have no axons but serve to activate nerve fibres in contact with them (Fig. 38.1). Examples of the first arrangement are the nerve terminals in the skin, which are the endings of fibres whose cell bodies are in the dorsal root ganglia (p. 257). The rods and cones of the eye, or hair cells of the ear are examples of the second arrangement. In both situations the effect of the external change is electrogenesis by the receptor membrane, which sets up a *generator potential*, and this then serves to initiate action potentials. The generator potential is graded as a linear function of the external change and persists at a new level. It is often a depolarization (but not always—in the retina it is a hyperpolarization). Its function is to act on an electrically excitable membrane to produce action potentials whose number and frequency vary with its strength and duration. In some fishes there are receptors directly responsive to electrical change, and of course nerve fibres can be electrically stimulated. The transducer membranes of other sense organs are, however, mostly not electrically excitable, at least not by small currents. They need some form of chemical amplification.

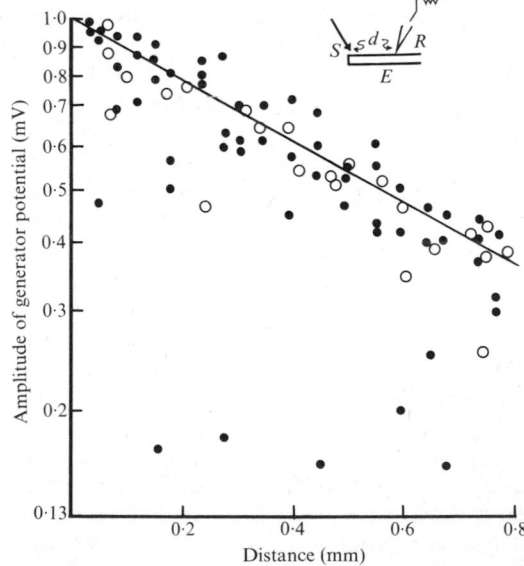

FIG. 38.1. Generator potentials of a Pacinian corpuscle.
A spot on the membrane is stimulated with mechanical pulses by a crystal-driven stylus. The potentials (solid circles) are recorded at various distances. The open circles are the passive (electrotonic) potentials recorded when the same spot is stimulated electrically by concentric microelectrodes. (From Lowenstein (1961). *Am. N.Y. Acad. Sci.* **94**.)

Pacinian corpuscles have been found to be a very suitable receptor for study of the generator potential. They are mechano-receptors detecting vibration and perhaps pressure. They occur in the human fingers but are more easily studied in the mesentery of a cat. Each corpuscle contains the ending of a dorsal root nerve fibre, wrapped in numerous sheets of tissue (Fig. 39.5, p. 369). If all these layers are removed the response of the surface of the nerve fibre can be studied by applying mechanical pulses by a fine stylus driven by an oscillating crystal (Fig. 38.1). The generator potentials produced can be led off the membrane and are found to decrease with distance from the spot stimulated. A current applied to the surface at the same spot is found to decrement in the same way. Such currents carried in tissues are said to be 'electrotonic', they are conducted passively without being boosted by any regenerative process such as is involved in an action potential (p. 246). The generator potential produced by various parts of the surface will, of course, summate, and if it reaches a suitable level may generate an action potential at an electrically excitable place on the attached nerve fibre. In the Pacinian corpuscle this is the constriction (node) before the fibre acquires its sheath of myelin (Fig. 39.4, p. 369). In an intact corpuscle the lamellae serve to cut out low mechanical frequencies and also to store energy during compression and so allow for an off response at decompression.

These studies of the generator potential were made in corpuscles that had been treated with the drugs procaine or tetrodotoxin, which prevent the actual initiation of an action potential. Without such treatment only a very small generator potential is detected before the action potential arises. Moreover, it is not certain that current must be conducted to a specific point such as the first node of Ranvier before action potentials can arise. It has been possible to record action potentials conducted along the terminal portion of the fibre.

However, there is little doubt that the first step in mechanoreceptor function is deformation of the membrane allowing entry of sodium ions, perhaps by creating pores. There is good evidence for some slowly adapting receptors that the deformation–receptor potential is linear for a sustained response, but for phasic (velocity dependent) displacements the mechanical conditions vary very much. The impulse frequency induced is a linear function of receptor potential over a wide range. The actual number and frequency of impulses generated varies widely for different receptors because of the mechanical conditions. Indeed it is absurd to expect any general relation since the varied signals provide part of the code by which information is sent to the brain.

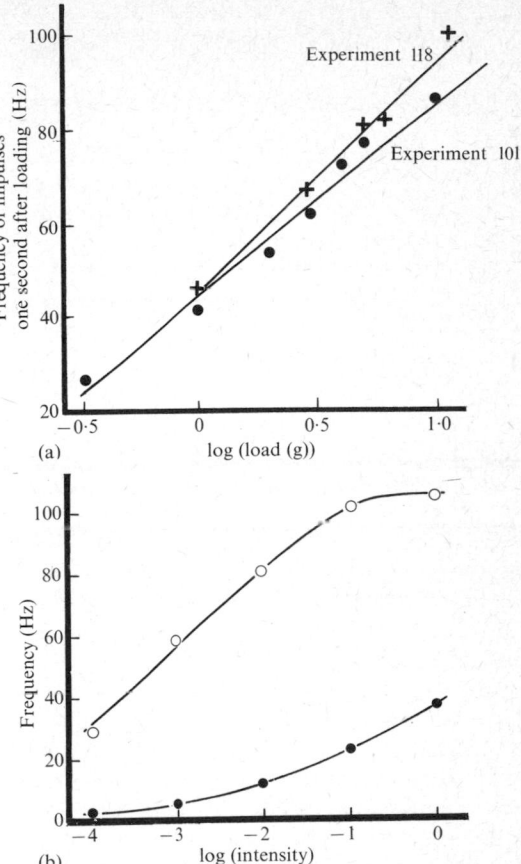

FIG. 38.2. (a) Discharge of frog muscle-spindle under different loads. (Matthews (1931). *J. Physiol.* **71.**) (b) Relationship between frequency of impulses and intensity of stimulating light in the frog. Upper curve, initial maximal discharge; lower curve, 3·5 s after illumination. The intensity is in arbitrary units. (Hartline and Graham (1932). *J. cell. comp. Physiol.* **1.**)

In receptors where the sensory cell is not directly continuous with the nerve fibre the connexion between the generator potential and the genesis of nerve impulses is more complicated (and not yet fully understood see p. 250). The function of the generator potential is to carry some signal for a short distance (the length of a retinal rod or hair cell) and then to initiate nerve impulses in the nerve fibre with which it is in contact but not continuity. This probably involves a chemical transmitter, although this has not yet been identified.

The pattern of impulses produced by each type of receptor organ is part of the code, being, of course, 'recognized' by the cells in the part of the brain concerned, which respond appropriately. Thus the muscle-spindle whose response is shown in Fig. 38.2 sets up a train of impulses whose frequency provides a measure of

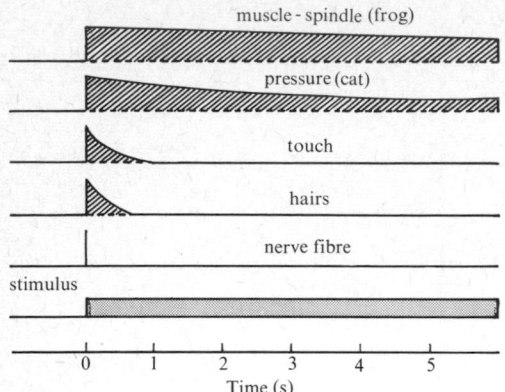

FIG. 38.3. Duration of firing of impulses by various types of end-organ, and a single nerve fibre, to a continuous stimulus. (Adrian (1928). *Basis of sensation*. Christophers, London.)

the load (or length) of the spindle. The frequency is proportional to the decadic logarithm of the load (Fig. 38.2). Many receptors discharge continuously in this way, the frequency serving to report the level or intensity of some external condition. In other receptors the discharge increases when the intensity at the receptor surface increases but the new high rate of discharge is not maintained. This decline in frequency is known as adaptation and is prominent, of course, in those receptors that signal a change of condition ('stimulus'). Following bending of a single hair the subject reports the sensation of touch only for a short period: the receptor system adapts rapidly. Conversely, the muscle-spindles discussed above adapt very slowly (Fig. 38.3).

It is not yet certain whether the phenomenon of adaptation indicates a change in the mechanism responsible for the specific sensitivity of the receptor (for example, the photochemical substance in the retina) or in the generator potential or other mechanism by which this specific change is transformed into a form of energy capable of setting up impulses in the nerve fibres. It may be that both are involved. Nerve fibres themselves show a process of accommodation, so that the impulse frequency declines under constant conditions. Each impulse is followed by a process that counteracts the setting up of a fresh impulse (p. 246). Yet a single stretch receptor may discharge at 300 Hz for many seconds. Clearly there is some mechanism in receptors that maintains firing in the nerve fibres in spite of accommodation. This is indeed necessary if the receptor is to provide a measure of a given level of intensity. One suggestion is that the numerous terminal branches that are often found at the ends of the afferent nerve fibres provide the mechanism. In a muscle-spindle each afferent terminal twig generates small non-propagating local potentials,

which summate at the points where they join (see Katz 1966). Even if there is accommodation of some of the nerve terminals under the generator potential, there are enough of them to allow repeated excitation at the junction point, on a statistical basis.

5. Stimulus and threshold

Receptor organs are thus able to signal a report of the condition in their neighbourhood or a change in that condition. To use the usual phrase, they show responses to stimuli. A stimulus may thus be defined as a change that is of sufficient magnitude to produce variation in the frequency of impulses in the nerve fibres connected with a receptor.

The lowest intensity at which a receptor will change its discharge is called the *threshold*, but the rate at which the intensity change takes place affects this level. Generally speaking the slower the change the less the stimulating effect. When operating under suitable conditions the sensitivity of receptors may be very great. The human eye can respond to a change of incident light from zero to only a few photons per second per square millimetre. One rod can probably be activated by a single photon and is thus at the maximum possible sensitivity for any light detector (p. 389). The ear is also able to respond to exceedingly minute forces (p. 406).

6. The frequency code

Nerve impulses have now been recorded in single fibres proceeding from muscle-spindles, cochlea, retina, touch organs, pain organs, and many other receptors. It is safe to say that receptors function by reporting extent of change in terms of a frequency code of nerve impulses. Changes in incident intensity are thus not recorded by the body as changes in the amplitude or duration of the signals set up by the receptor. The nerve fibre connected with a given receptor cell carries impulses that are essentially all alike and vary only in frequency. The maximum frequency reached may be 1000 Hz for some receptors recording vibrations, but is usually considerably less.

Each nerve fibre can carry only one sort of signal and therefore it is clear that response to a wide variety of types of change is achieved only by having many different types of receptor, each 'tuned', as it were, to respond to a particular change and sending signals in its own nerve fibre. This is the 'doctrine of specific nervous energies'; which, in simplest form, states that any given nerve fibre, however excited, gives rise to only its own characteristic type of activity in the central nervous system. The activation of each major receptor is indeed reported by man as yielding a characteristic sensation. When the elements of the retina are stimulated by a blow we record a sensation of light, not of pressure. It is gradually becoming apparent, however, that the particular response produced by the brain following arrival of impulses from any given receptor cell varies according to the activity in neighbouring cells, and is greatly influenced by the memory system that the organism has acquired (p. 341). The doctrine of specific nerve energies is true in so far as each single channel from the receptor cells is only normally activated by one form of external energy. But the multiplicity of channels and of their modes of combination is so great that it is not easy to forecast the response to activity of any one channel. Here, as elsewhere in biology, we have to find ways not only of acquiring knowledge about the activity of single units but also of forecasting the actions of many of them in combination.

Responsiveness to the details of the changes in the world around is ensured by the presence of several quantitatively different types of receptor within each of the main modalities. Thus the eye includes the rods, sensitive to low intensities of light, also cones of several types, with maximum sensitivity at various wavelengths, which enable us to distinguish colours (p. 391).

The specific effects of the discharge of the nerve fibre connected to a given receptor surface therefore depend on the central connexions rather than on any particular quality of its impulses. Yet there are characteristic differences between the impulses set up in different afferent nerve fibres. The peripheral nerve trunks contain fibres of a wide variety of sizes (p. 239) and the impulses of each sensory (and motor) system are conducted in fibres of a particular size and therefore travel at a characteristic velocity. The afferent impulses differ in other respects besides velocity, for instance, in duration. The characteristic patterns of response are produced by variation in the combinations that result when the impulses reach central neurons. The significant features of the impulse discharges from receptor surfaces are mainly the frequency and velocity and these, together with the spatial distribution of their connexions peripherally and centrally, determine the result of bombardment of the nerve centres following any particular peripheral change.

The properties of this code are still much debated. The question of the relation between the 'strength' of sensations has long been a chief problem of psychophysics. Gradations in magnitude from weak to intense constitute much of the information of our senses. According to the classical investigations of Weber and Fechner in the last century, the magnitude of the difference of intensity necessary to allow the stimuli to be distinguished increased with the intensity. Assuming

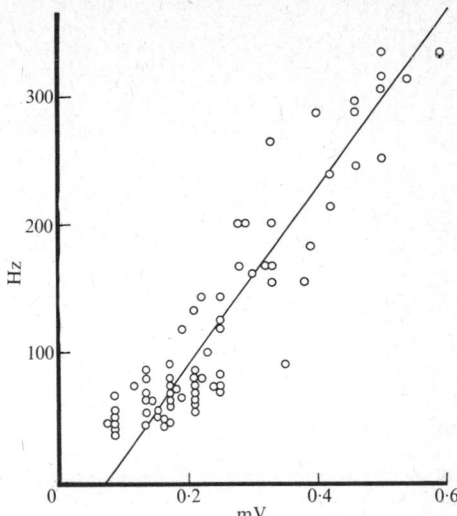

FIG. 38.4. Relation between frequency of impulses and local depolarization in frog muscle-spindle. (Katz (1950). *J. Physiol.* **111.**)

that each such just noticeable difference (j.n.d.) corresponds to an equal increment in sensation led to the conclusion that equal stimulus ratios produce equal stimulus differences, that is, the psychophysical law takes a logarithmic function.

Much evidence has been produced that seems to support this theory. Thus in a muscle-spindle Katz has recorded a depolarization within the end organ itself, which increases with the load. The number of impulses per second in the associated nerve fibre is proportional to this depolarization (Fig. 38.4). As we have seen the frequency of impulses is proportional to the logarithm of the load on the spindle. Presumably, therefore, the generator potential is also proportional to the logarithm of the load, though this has not been demonstrated experimentally.

However, in recent years the logarithmic law has been increasingly questioned. There is evidence that in many sensory systems equal stimulus ratios produce equal sensation ratios. In fact as a function of stimulus intensity s, the perceived intensity p grows as a power function

$$p = k(s-s_0)^b,$$

where s_0 is the threshold and k depends upon units. The exponent b varies with the modality and also with such parameters as the state of adaptation. s_0 is often so small as to be negligible, but for some senses, such as temperature, this must be taken as a neutral point (Stevens 1966).

The question can be tested in many ways. Human observers may be asked to assign numbers to a series of

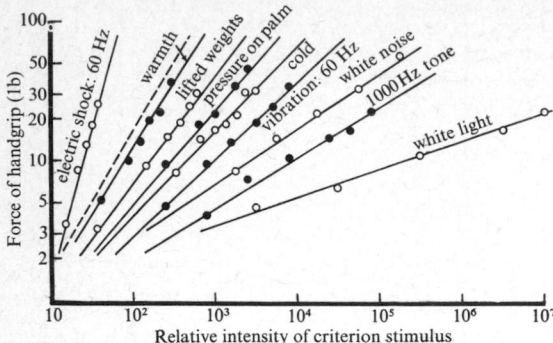

FIG. 38.5. Equal-sensation functions obtained by matching force of handgrip to various criterion stimuli. Each point stands for the median force exerted by ten or more observers to match the apparent intensity of a criterion stimulus. The relative position of a function along the abscissa is arbitrary. The dashed line shows a slope of 1·0 in these coordinates. (1 lb = 0·454 kg.) (From Stevens (1966). In *Touch, heat and pain* (ed. de Reuck and Knight). Ciba Foundation Symposium. J. & A. Churchill, London.)

stimuli of different intensity. A more objective way of doing this is to ask them to squeeze a handgrip with a force that matches the intensity of the stimulus (Fig. 38.5). The results that are given are surprisingly constant between observers and enable one to determine exponents for different modalities, ranging from 0·33 for visual brightness to 3·5 for electric current through the fingers. Even more conveniently the exponent found for two such 'matching functions' can be used to predict a third. Thus loudness has been matched to ten other continua.

Following these psychophysical investigations physiologists have been able to find situations in which aspects of the neural response follow the same law (for instance, in intensity of taste or loudness of sound). It may be that the transducer process itself is the critical factor that determines a linear relation between events in the outside world and the whole chain of signals through the nervous system.

7. Overlap and the information content of the channels

A characteristic feature of the organization of the nervous system is the overlap between conducting pathways, both peripherally and centrally. Thus many retinal elements are connected with a single optic nerve fibre. In man there are about 6×10^6 cones and 125×10^6 rods, but only 10^6 optic nerve fibres. At the other end of the system each nerve fibre arriving in the visual cortex may stimulate any among some 5000 neurons in whose dendritic fields its terminations lie (p. 341), Clearly the operation of the nervous system does not depend upon a technique of reproduction similar to that of photography or television, where the image is trans-

mitted piecemeal. The overlapping would be highly deleterious to such a system.

In the simpler parts of the nervous system a motor response may be produced when impulses are set up in relatively few afferent fibres. Moreover, the magnitude of the response may be determined by the frequency of the impulses. Thus the amount of contraction produced in the flexor reflex (p. 256) depends on the numbers of motor neurons activated and the frequency of their firing, which in turn depend upon the number of afferents excited and *their* frequencies. In this case, therefore, the organism may be said to measure the stimulus intensity by interpreting the frequency code in the afferent fibres.

This occurs in the parts of the nervous system in which the afferent fibres are connected with relatively few efferent channels. Impulses in the afferent channels thus decide only the extent to which a given reflex mechanism with which they are connected is operated. In other words the channels carry relatively little information, that is they control decision between few alternatives. The more elaborate receptor systems, such as the eye, may influence many effector systems. Any one of many responses may be made following input to the C.N.S. from the retina. We can therefore say that the optic nerve fibres carry a great deal of information, because they participate in the making of decisions between many possible alternatives.

This greater complexity of response to environment is made possible by the specialization of certain parts of the body of the metazoan for response to particular stimuli and the linking of the changes produced in these receptor organs with an elaborate pattern of activity already going on in the brain of the animal. Thus the light from a diamond interacts with the optical and nervous system of a man, causing him to react in devious and subtle ways such as shall lead him to acquire the jewel.

8. The probability of a response and the rhythm of life

Such specialization of response has been achieved by the setting aside of areas of tissue that have high sensitivity to certain of the changes around the organism. At the same time there has been a development of suitable central mechanisms allowing response to elaborate patterns of change. This appears clearly in the case of the set of signals that accompany pain. The hereditary system provides receptors of rather simple structure, which discharge when conditions occur that grossly damage the tissues (p. 367). If these conditions continue, the reaction of the body is such as to produce, if possible, a cessation of the trauma. These reactions often consist in a series of sudden jerks and writhings in various directions, often at random, comparable with the avoiding reaction of protozoans. When impulses signalling the occurrence of trauma arrive at the central nervous system the result is that the part or limb concerned is drawn away from the source of trauma. The probability that this avoiding response will be given is so high that we can forecast accurately that it will be shown by any intact individual of the species. The amount of information conveyed by the nerve fibres under these conditions is therefore low. Yet in higher animals the response is not completely invariant and may depend upon the circumstances, especially in man. In social animals sounds may be produced when a pain stimulus is applied and the sounds may serve to warn other individuals and in the case of man to bring others to the rescue. It is interesting that trauma is not accompanied by sound production in species that lack the appropriate social organization, for example, grasshoppers or frogs. In races of men with well-developed communication systems the 'response' to pain even includes discussions and the writing of books on the problem of pain, thus helping in the preparation against future damaging agents.

In higher animals life consists of a continual series of readjustments, many of them rhythmic. Study of these sequences of action is the essential requisite for investigation of the action of the receptors and measurement of the amount of information that they provide. Unfortunately we have at present few investigations of this sort and little knowledge of the rhythms of animal lives or of the way in which differences between them may be classified. Receptor action heralds each new phase, for example, nerve impulses are set up in the walls of an empty stomach. These impulses serve to initiate the search for food, but since this is a regular occurrence we may say that these impulses, like those of pain, provide only a small amount of information. The way in which food is sought may also follow a regular pattern, but there is probably more variability here than in the recurrence of food-seeking. The great amounts of information that ensure stability in spite of this variation come from the receptors that signal the presence of objects likely to provide food even when they are at a distance from the body. As we study species with more and more complex ways of life we find greater and greater variation in the way the rhythm of life is controlled. More and more information is used to maintain stability by minor or major changes in routine. It becomes more and more difficult to forecast the course of behaviour of any given individual.

As each reaction runs its course, for example, as the animal seeks its food and eats it, the initial discharges die down and are replaced either by inactivity or perhaps by discharges that lead the organism to remain in the condition it has reached. Such impulses could be called those of 'pleasure', but to remain in a favourable state is in general to be inactive and there is a connexion between pain and change, the stimulants of activity. There is a contrast between these and pleasure and repose, the negative state of inactivity, in which the organism is satisfied. Life tends to those conditions in which its continuance is most easily and completely ensured and the degree of activity is adjusted to this end; the organism reacts to threats to its continuance, produced either from within or from without.

It must be remembered that the receptor systems themselves are not passive agents that are set into action only in response to external change. Many of them discharge trains of impulses continually into the nervous system, for example, receptors both in the eye and the ear are continually active in this way and the effect of a 'stimulus' is to change their activity rather than to initiate it. The recognition of this 'spontaneous' activity of the receptors considerably alters one's view of the nature of the whole system. Together with the 'spontaneous' activity within the brain (p. 291) this may be the basis of much of the 'drive', which, as psychologists have long recognized, comes from within the organism. Indeed, there is now direct physiological evidence that the discharges of the receptors operate along specific pathways in the reticular system of the brain (p. 271) to 'arouse' the cortex. Cases have been reported of people whose afferent input became limited to very few sources (say one eye and one ear). When these channels were not stimulated the person fell asleep.

A still further complication is that there are *efferent* fibres running from the central nervous system to the receptors. A well-known example is the muscle-spindles, whose sensitivity is varied by discharges reaching them along efferent nerve fibres (p. 265). But there are efferent fibres also in the optic nerve and they have been shown to alter the sensitivity of the retina in some animals, as do fibres in the auditory nerve to the cochlea (p. 335).

If the receptors activate the C.N.S. and the latter sensitizes the receptors we obviously have a very complicated set of loops to consider. The effect of any pattern of external change will vary greatly according to the internal condition of the system; a fact with which we are all familiar. It is indeed hard to forecast the behaviour of living systems, but we can now begin to see the principles upon which they operate.

39 Receptors in the skin and viscera

1. Information from the body surface

IT is obviously important for the body to provide appropriate reactions to events occurring close to the surface. Some of these are likely to be traumatic and a system of 'pain' nerve fibres is present, whose central connexions, probably mainly laid down by heredity, determine that when such events occur there is withdrawal of the part or avoidance behaviour by the whole animal. On the other hand, some types of light touch produce movements that maintain the contact, for example, in animals that follow solid contacts in the dark, or in some forms of social and sexual behaviour. Moderate warmth may also elicit movement towards the source. The portions of the limbs that are in contact with the ground or other means of support provide information that is important for the maintenance of posture. Responses to the detailed pattern of contact are made possible by the special sensitive areas on the limbs (and sometimes tail) of primates and perhaps by other areas, such as the lips. Such reactions to the 'shape' of the object depend upon a more elaborate central organization than is involved in the responses to stimulation of pain receptors (see Iggo 1973).

2. Receptors in the skin

The classification of the skin sensations and identification of the receptors involved is unfortunately still in an unsatisfactory state (Sinclair 1967; de Reuck and Knight 1966; Catton 1970). In man it is usual to recognize a number of different conscious 'sensations' such as touch (light pressure), deep pressure, tickle, vibration, warmth, cold, quick pain, and slow pain. Histologists find a variety of types of nerve ending in the skin, but it is still uncertain whether each serves to signal the occurrence of one specific type of physical change. It is more likely that under many conditions several types of ending send signals and that the nervous system interprets each different pattern of signals as indicating a particular type of event and thus the need for appropriate action.

The best statement of the anatomical facts is that there is a spectrum of nerve endings in the skin extending from very simple free nerve endings to highly organized encapsulated endings in which a complicated structure surrounds the nerve terminal and determines the type of physical change to which it reacts (Fig. 2.1, p. 11). These encapsulated end-organs are particularly evident in those parts of an animal that perform specialized tactile functions, for instance the whiskers (vibrissae) of a cat or the array that makes up the Eimer's organ in the snout of a mole (Fig. 39.1). In man the Meissner's and Pacinian corpuscles of the fingers provide a comparable array, though their actions are not well understood.

Nerve fibres reach the skin through elaborate plexuses in the dermis, ensuring that each part of the skin is served by the overlapping areas of branching of many individual fibres. The afferent fibres vary in diameter from non-medullated C fibres through a range of diameters of small medullated B fibres to the largest A fibres (p. 267). Probably each nerve fibre supplies only one type of ending, and this alone suggests that there must be some degree of specificity in the signals that it carries. There is some evidence that particular regions of the skin constitute spots sensitive to warmth or cold, and certainly individual hairs signal touch. In this sense there are specific localized receptors for different sensations. Nevertheless the overlapping multiple innervation of many areas suggests that the varieties of stimulus change are transduced into patterns of impulses that are not all strictly specific (though some may be). Aspects of these patterns are then detected by the properties of the central cells such as their thresholds and capacity for spatial and temporal summation and adaptation (Melzack and Wall 1962).

Many of the nerves of the skin are in a dynamic state in the sense that they continually degenerate and reform. This is certainly true of those in the epidermis. Moreover if a piece of epidermis is stripped off the human skin with adhesive tape the diameter and position of all the nerve fibres in the dermis changes within 24 hours. Yet the sensory capacity does not change.

So the nervous system must recognize patterns of impulses set up by rather unspecific transducers.

3. Free nerve endings

To bring some order into the subject we may consider the known types of end-organ and the evidence as to their properties. The so-called free nerve endings occur throughout the dermis and epidermis of all mammals. They make fine branched trees of beaded fibres sometimes ending in slight swellings with accumulations of mitochondria. They are not really 'free' because they are included within Schwann cells while in the dermis and in the epidermis are wrapped in the cells (at least in the cornea in man). They are thus never freely exposed to the intercellular fluids, but it is not clear whether the wrappings have any specific effects upon their function. They are often the terminals of non-medullated nerve fibres, but may spring from small medullated fibres.

The function classically assigned to these endings by Frey was the signalling of pain and they certainly have some connexions with this sense. During regeneration after injury to a nerve the first detectable sensation returning to a denervated area is pain, and the fibres that can be found there are then all free nerve endings. Other sensations become possible as medullated fibres mature. This led the neurologist Head to his very influential theory that two distinct modes of sensation can be distinguished, a vaguer more diffuse and painful *protopathic sense* served by non-medullated fibres and a more discriminating *epicritic sense* served by larger fibres. Unfortunately the facts of regeneration are far more complicated and the theory cannot stand in its simple form. Nevertheless there is some truth in it, in that small fibres are often associated with pain and larger with fine discriminations, but in neither one is the association exclusive.

There is abundant evidence that free nerve endings can mediate sensations other than pain. Thus in the cornea they are the only type of nerve ending present. The first response to contact with the cornea is pain, but with care it is possible to recognize touch. Again the hairy skin contains no encapsulated end-organs (though the hairs are innervated) but it readily recognizes all qualities of sensation.

4. Pain

The capacity to detect damage is the most widespread of all sensory powers and is found in nearly all parts of the body. In man we regard it essentially as 'subjective' yet it is a sense with very clear functional correlates. When the networks of free nerve endings in a rabbit's ear are stimulated with a needle the animal moves its head in avoidance and in such situations we can speak of the *nociceptive response*. The free nerve endings are acting as '*nociceptors*'. The capacity to detect damage or trauma is obviously a fundamental need for an

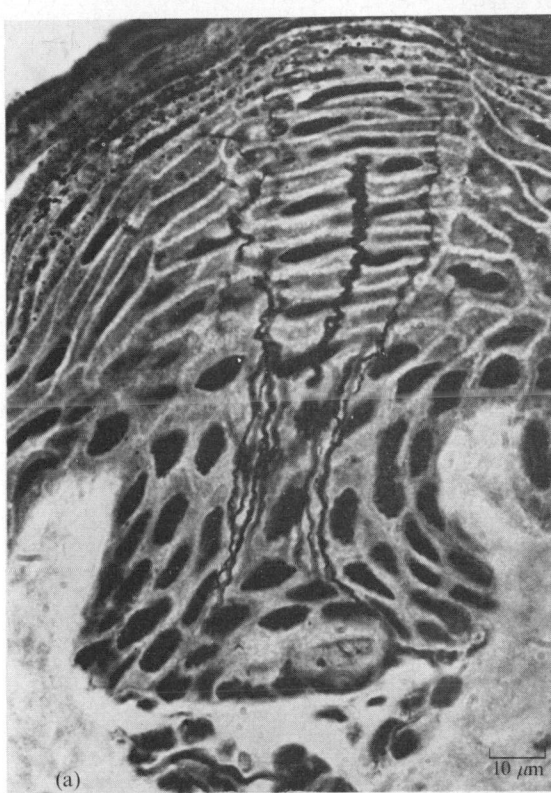

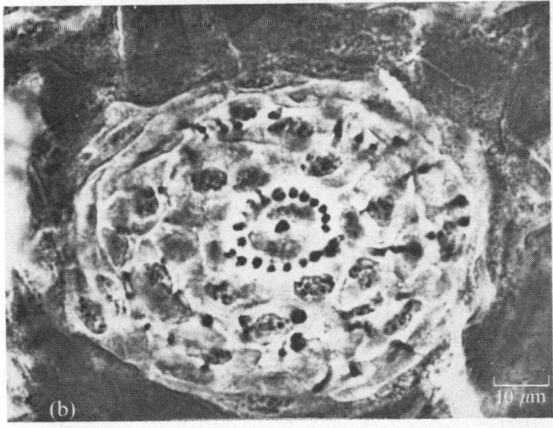

FIG. 39.1. Light-micrographs of Eimer's organ from snout of mole. (a) Vertical section. Axial and peripheral nerve terminals ascend from the dermo-epidermal junction to the stratum spinosum. (b) Transverse section. The single bare axial terminal is contained in a slender column of epidermal cells which itself is surrounded by a ring of peripheral terminals. (Figure supplied by Dr. T. A. Quilliam.)

organism and the nociceptors are actually distributed in places where following injury the signals produced can provide actions that may prevent further damage. Thus the cornea is exquisitely sensitive to 'pain'. All parts of the body surface are 'painful' when cut, burned, or otherwise traumatized. Visceral organs are painful when pulled or distended but not when cut. Muscles are painful when they lack blood (cramp). Damage to the surface of the brain itself does not produce pain, but some forms of 'headache' are probably caused by spasm in the cerebral blood-vessels.

Evidently nerve endings in these various regions transduce different traumatic events into appropriate signals. There are several theories as to the nature of painful stimuli and a complication is that excessive stimulation of almost any receptor gives rise to pain. It may be that pain results from some chemical intermediary, produced by degradation of nerve endings. Radiant heat produces lasting pain at about the level at which tissue damage begins (45 °C). However, under other conditions, much lower temperatures (37 °C) produce a temporary pain after which adaptation abolishes it. Again, water at 0 °C is both damaging and painful, but as vasodilation occurs the pain lessens. Controlled cooling can produce an anaesthetic effect.

Many chemicals cause pain when placed on a raw skin surface, but the most interesting are substances known to be liberated by tissue damage, such as histamine, serotonin, and hydrogen ions or potassium ions. Pain-producing substances have been isolated from the fluids of blisters.

The sensation of pain in man has been the subject of a great variety of observation and speculation. It is widely believed that two sorts of pain can be distinguished, a quick, prick-pain, and a slower, longer-lasting pain. They are probably related to two sorts of 'pain fibres', medullated and non-medullated.

Obviously in man the signals set up by traumatic events reach to higher cerebral levels and become associated with the visual, auditory, and other attributes of the situation so that they may be avoided in the future. The word 'painful' may be used to describe such situations. There is no specific 'centre' for pain in the cortex, but obviously the signals of nociceptors must reach to it, for we readily learn which auditory or visual sensations promise pain. There is a curious rare condition in which a person is totally insensitive to pain from any part of the body, with no other obvious disturbance of function (see Sinclair 1967).

5. Hairs and encapsulated end-organs

All the endings in the skin other than free nerve endings are associated with some form of non-nervous appara-

tus, which presumably plays a part in controlling the type of stimulus to which the ending responds. These organs vary from the tiny *Merkel's discs* composed of one or a few cells to the elaborate *Pacinian corpuscles* up to 1 mm in length. The functions of most of them are still poorly understood.

The *hair follicles* are perhaps the most universal. Each hair follicle is supplied by several large nerve fibres forming sheaths running around or along them (Fig. 2.1, p. 11). Each dorsal root fibre sends branches to 100 or more hairs and each hair receives innervation from at least four root fibres. There is no doubt that movement of hairs can give rise to a sensation of touch. The point of transduction seems to be where the hair leaves the skin. If this region is immobilized with collodion, moving of the hair base gives no sensation. On the other hand an artificial hair of wire cemented to the skin gives a sensation exactly like a real touch. But it is easy to show that movement of some hairs arouses no sensation at all. Conversely the hairless skin of the fingers feels the sensation of touch and so even does the cornea.

Following movement of a hair the nerve fibres connected with it produce a discharge that stops within a few milliseconds if the hair is held fixed in a new position. A small displacement may produce only one

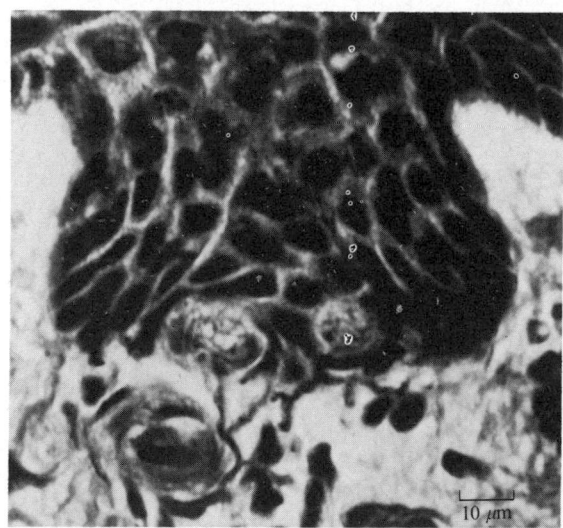

Fig. 39.2. Light-micrograph showing two Merkel's discs at the base of an Eimer's organ. The innervation of the left-hand disc is by an unmyelinated collateral arising from a myelinated stem fibre in the dermis. The surface of the discs is covered by an anastomosing system of fine nerves. A lamellated receptor lies in the dermis. (Figure supplied by Dr. T. A. Quilliam.)

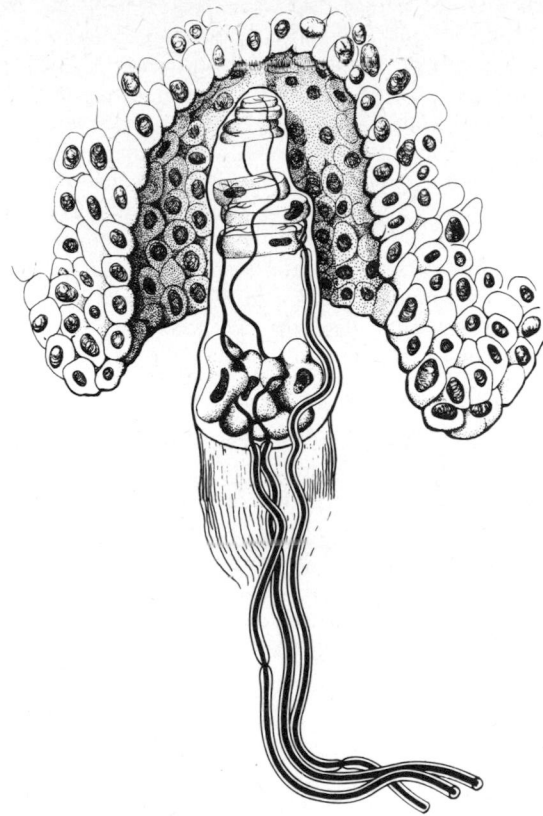

FIG. 39.3. Schematic drawing of a Meissner corpuscle from the human finger tip. (Figure supplied by Dr. T. A. Quilliam.)

Evidently deformation of the surface is the relevant change for the sensation of touch and this can be detected by other nerve endings with different structural characteristics. The *corpuscles of Meissner* in the finger are the most obvious of these (Fig. 39.3). The stratum corneum of the skin is here folded into ridges and is soft over the crests but harder in the valleys. From each crest a core of keratin extends down into the dermis. The corpuscles lie in these ridges and each is a column of flattened cells interleaved by nerve fibres (Fig. 39.3). Each corpuscle shares its nerve supply with many others, exactly as for the hair follicles.

Presumably these arrangements of the finger provide for the special sensitivity of the fingertip. If the soft skin of the ridge is readily deformed it may act as a magnifying lever. The Merkel's discs may provide for high sensitivity, the more protected Meissner's corpuscles for localization. The power of 'two-point' discrimination is so great that contacts only a few millimetres apart can be distinguished by the fingertips, whereas the minimum distance increases to a centimetre or more on the hairy skin.

impulse and continued displacement at a constant rate may produce a regular discharge at constant frequency, stopping when movement ceases.

In contrast to these rapidly adapting afferent units the skin contains two main types of slowly adapting cutaneous afferents (Iggo 1966). Those known as Type I show a discharge of very high frequency (>1000 Hz) and are associated with small dome-shaped structures in the skin. Small deformations of the dome produce a persistent discharge related both to the rate of change and amplitude of displacement, so that the frequency is maintained at a new level. Whereas the dome is very sensitive, pressure on the skin nearby has no effect. These properties are probably due to the relative stiffness of the elements in the organ. There is a dermal plug of collagen at its centre which insulates the receptors from displacements at the side. The receptors themselves are probably of the type known as Merkel's discs. Each of these is a single flattened cell wrapped around the fine termination of a myelinated nerve fibre (Fig. 39.2).

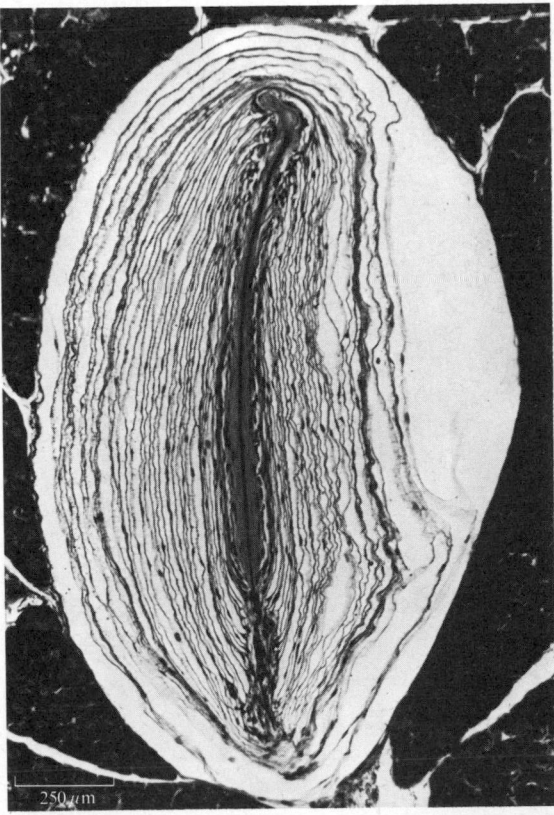

250 μm

FIG. 39.4. Light-micrograph of a Pacinian corpuscle from cat pancreas. (Figure supplied by Dr. T. A. Quilliam.)

The nature of the discharge from Meissner's corpuscles is not known but there is another set (Type II) of slowly adapting mechano-receptors. These organs have a regular resting discharge whose frequency increases smoothly with mechanical deformation and is subsequently maintained at a new level. These units differ from Type I in that they are very sensitive to lateral stretching of the skin. The histological appearance of these receptors is not known.

6. Pacinian and other lamellated corpuscles

A large class of mechano-receptors consist of a terminal nerve fibre around which are wrapped a large series of lamellae (Fig. 39.4). The Pacinian corpuscle is the best known of these. They occur in situations as varied as the mesentery of the cat and the human fingers. Somewhat similar structures are the corpuscles of Herbst and Grandry in the bill of birds and perhaps the genital corpuscles in the penis and clitoris of many mammals. All of them are mechano-receptors, probably sensitive to vibration.

Each Pacinian corpuscle is 1 mm or more long. The outer lamellae are fibroblastic but the inner consist of special cells, perhaps Schwann cells, which may be attached to the nerve-fibre surface by desmosomes. Each corpuscle is innervated by a single myelinated nerve fibre, usually a large rapidly conducting A fibre. They are readily stimulated by low-frequency vibration applied over quite a wide area of the outer surface of the ending and they adapt rapidly. They may thus be able to follow vibration to quite high frequencies (1000 Hz). Their functional effects are uncertain. Probably in the limb they serve to detect external vibration. They are often found in relation to blood vessels and may serve to monitor the flow, for example at arterio-venous anastomoses.

7. Temperature receptors

The immediate stimulus for thermal sensations is probably simply the temperature of the end-organs concerned. Thermosensitive fibres of the cats tongue were found to discharge for long periods at constant frequency even at a constant temperature. There seem to be distinct end-organs detecting 'cold' and 'warm', but in spite of much controversy the histological appearance of these has not yet been identified.

1. Sensitivity to radiant energy

THE eye of man is specialized to provide the body with information about changes in the flux of radiant energy between wavelengths of about 380 nm and 760 nm. It is interesting to speculate why organs for dealing with this particular range should have become well developed. Radiant energy of greater and lesser wavelengths is abundant around land animals, yet changes outside the visible range are either undetected by the human body or else produce only poorly localized responses (as to heat) or slow alterations such as the pigmentation of the skin by ultraviolet light.

Organs for detecting radiation in the visible range have probably become developed because there is here a larger variety of easily collected relevant information than would be available elsewhere in the spectrum. Radiation of longer or shorter wavelength is also significant for some animals. Organs for accurate detection of the direction of heat radiation are found; for example the special receptors on the head of pit vipers are used to detect the warm-blooded prey. Man-made direction-finding devices that employ a reflected beam of waves of 1–10 cm length show that this radiation could be 'useful' if the body had means for producing and detecting it; but these wavelengths are not continuously present in the sun's radiation.

At the other end of the spectrum ultraviolet radiation of wavelengths less than 300 nm is abundant in nature and is used for direction-finding by bees, fishes, and other animals, but so far as is known not by any mammal. Whether a particular type of change provides 'information' for an animal type depends upon whether such change provides a basis for selection among the various actions that the animal can make.

Changes that occur within the visible range have the advantage that they act as signals for many events that are important for the survival of an animal. Detection of the mere difference between light and darkness or of the intensity of illumination is itself of value, providing, for example, signs that can be used for the selection of appropriate habitats or niches out of the many that

are available. It is obviously important for many animals to change their behaviour in phase with the rhythms of day and night and of the seasons. In some animals there are photoreceptors specialized for detecting such changes with longer period (p. 452) but in mammals changes at different rates are all detected by the eyes. Clearly the eyes and brain are organized to allow more information than the mere intensity of illumination to be extracted. Often they enable the animals to react specifically according to the position of a change in the visual field, for example, to snap at the movement of some lighter or darker spot. The periphery of the human retina is used in this way when we see dimly 'in the dark' (p. 388). In some animals different reactions can be performed according to the wavelength of the light, in other words colour can be discriminated (p. 391). In a few species the eye and the associated parts of the brain allow 'form vision', the performance of distinct reactions chosen in relation to a combination of many features of the incident flux of radiation. This is especially true of man, who is able to respond differently to minute differences between objects having light-reflecting surfaces, that is to say recognizing their texture and shape. We probably extract more information with our eyes than do any other animals and it is largely on a basis of this visual information that the great variety of human actions is based. The eye is the chief means by which man makes in the brain a 'representation' of the occurrences around him that is useful for determining appropriate reactions to situations as they occur. This faculty more than any other provides us with the 'picture' of a world filled with objects, and this is the basis of much of our communication system and of our technology and way of life (Duke-Elder 1958).

Eyes usually operate by translating a change in the incident radiation into nerve impulses in the fibres of the optic nerve (Granit 1947, 1955). These are then 'analysed' by the brain to produce appropriate reactions. In studying the eye, therefore, we have to inquire how the light-sensitive surface is arranged and how the changes in it come to set up nerve impulses. As Rushton (1972) puts it, 'the impalpable optical image on the retina is turned into a more substantial picture of chemical change. The next stage is the way that one or two molecules of chemical change can cause nerve excitation. Following this is endless processing of nerve signals at various levels of the visual system'.

So in studying the eye we can investigate first the methods by which the energy is collected and focused by a lens system, then how the whole eye is moved to enable the individual to scan over the world around and report upon the changes going on there. Next we must investigate the photochemical changes in the sensitive elements and how these set up signals in nerve cells. Finally we should study how these signals interact to allow appropriate distinct reactions to different visual situations.

There are marked differences between the eyes of different animals according to the type of information required; that is to say the type of decision that the animal can make between different actions. Nocturnal animals require a high sensitivity, which would be a disadvantage in very strong light. Hawks require to localize minute movements and men can act differently according to the shape of small objects such as printed letters. The hereditary mechanism of each species provides the means of collecting the appropriate

information. By studying the differences we learn to recognize the significant features of the visual process (Walls 1942).

2. Development and structure of the eye

Embryologically we may distinguish three main parts in the eye, (a) the optic vesicle, which grows out from the brain; (b) the mesodermal tissues that later surround the outgrowth; and (c) the lens, formed from the overlying ectoderm (Fig. 37.12, p. 354). The wall of the diencephalon becomes evaginated and then makes a cup; the inner wall of the cup differentiates into the retina. The lens separates two chambers, the outer filled with a liquid *aqueous humour*, the inner with a more viscous substance, the *vitreous humour* (Fig. 40.1). The outer wall of the cup becomes the pigment layer covering the retina, and where the two layers are continuous at the margin of the cup there differentiate the pigmented inner side of the iris and the iris muscles (Fig. 40.3) (Duke-Elder and Wybar 1961).

The mesoderm around the retina makes a vascular layer, the choroid, corresponding to the pia arachnoid. The choroid layer expands at the front of the eyeball to form the ciliary body, containing the ciliary muscles, which operate the mechanism of accommodation (p. 374). The mesodermal layer is then continued to make the outer layer of the iris, whose inner lining is the

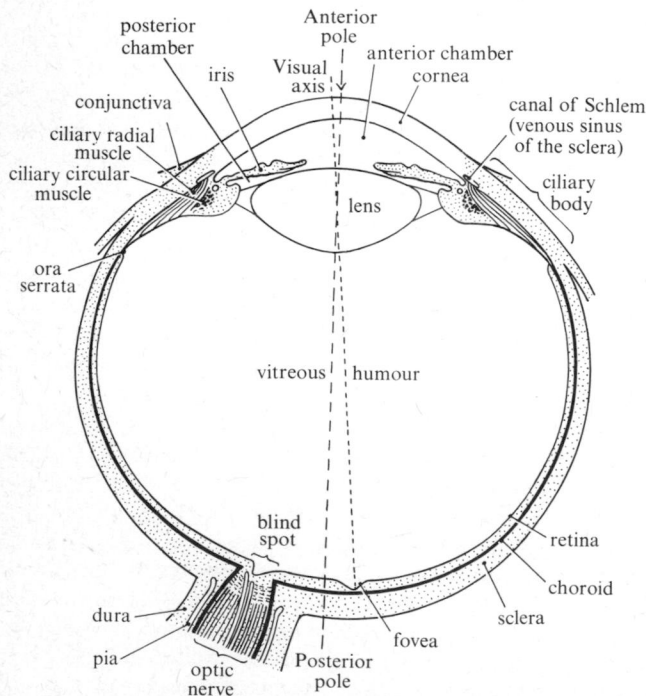

FIG. 40.1. Diagram of a horizontal section of the human eye.

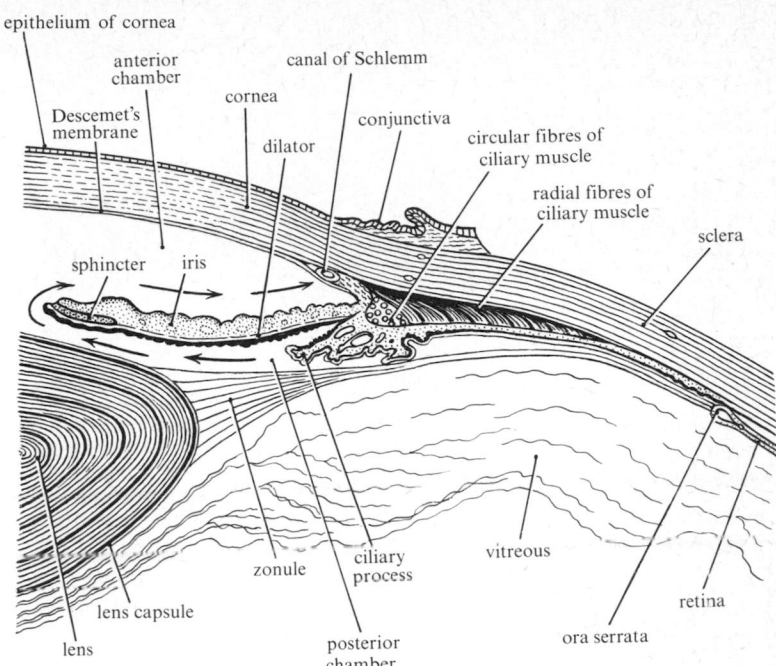

epithelium of cornea
anterior chamber
canal of Schlemm
cornea
Descemet's membrane
conjunctiva
dilator
circular fibres of ciliary muscle
radial fibres of ciliary muscle
sclera
sphincter   iris
zonule
ciliary process
vitreous
lens capsule
lens
posterior chamber
ora serrata
retina

Fig. 40.2. Diagrammatic section through meridian of eye of man.

ectodermal tissue of the optic vesicle. Outside the choroid is a tough coat of modified connective tissue, the *sclera*, showing in front as the white of the eye. The part of the sclera in front of the lens forms the outwardly convex, transparent *cornea*. The cornea is covered on its outer surface by a multilayered epithelium continuous with the *conjunctiva*, the lining of the lids. Over the surface of the cornea, the skin is very sensitive to pain, stimuli here serving to elicit lid reflexes and a flow of tears, by which foreign bodies are removed. The tears contain some organic matter and probably assist in the nourishment of the cornea, which contains no blood-vessels of its own.

Behind the cornea there is a split in the mesodermal tissues, making a space, the *anterior chamber of the eye*, lined in front by a single-celled mesothelial layer (sometimes called the 'endothelium') whose outer basement layer forms the distinct Descemet's membrane, covering the back of the cornea (Fig. 40.2). Internally the mesodermal tissue unites with the margin of the optic cup to form the *iris*. The iris is therefore a fold of tissue closing the front of the optic cup except for a circular aperture, the pupil, whose diameter is adjusted by circular (sphincter) and radial (dilator) muscles. These are formed by the development of muscular processes at the outer side of the epithelial cells of the outer ectodermal layer of the iris (Fig. 40.3). The *sphincter iridis* muscle is a circular ring of smooth

muscle fibres, which can be thrown into contraction by the action of postganglionic nerve fibres of the short ciliary nerves. These arise from the cells of the ciliary ganglion activated by the preganglionic fibres leaving the brain in the oculomotor nerve (p. 277). The *dilator muscles of the iris* consist of numerous radial bundles receiving their motor innervation from the sympathetic system (p. 281). Preganglionic fibres leave in the ventral roots of the upper thoracic region and run cranially to make synapse in the superior cervical ganglion with cells whose processes pass with the internal carotid artery to join the fifth cranial nerve, whence they pass to the eyeball in the long ciliary nerves. There is some evidence of reciprocal inhibitory innervation of these two muscles.

Behind the iris lies the *posterior chamber of the eye*, communicating with the anterior chamber around the margin of the pupil. The ectodermal layers at the periphery of the iris are thrown into a series of ridges, the *ciliary processes* (Fig. 40.2), projecting into the posterior chamber. Both chambers are filled with the *aqueous humour*, a solution resembling cerebrospinal fluid and containing 98 per cent water, some sodium chloride, and traces of protein. It is either secreted or filtered from the surfaces of the ciliary processes and is removed into a series of spaces leading to the circular *canal of Schlemm* (venous sinus of the sclera), which runs round parallel to the outer edge of the cornea. The

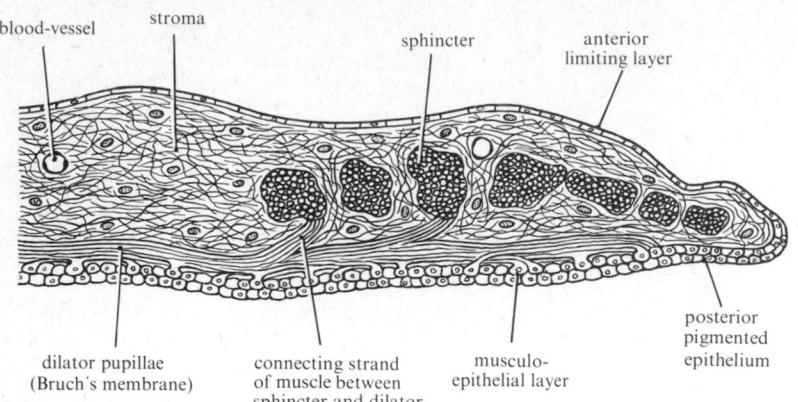

blood-vessel   stroma   sphincter   anterior limiting layer

dilator pupillae (Bruch's membrane)   connecting strand of muscle between sphincter and dilator   musculo-epithelial layer   posterior pigmented epithelium

FIG. 40.3. Semi-diagrammatic radial section through pupillary part of mammalian iris.

cells lining this canal are penetrated by channels leading to a system of trabeculae and veins beneath. These channels control the absorption of aqueous humour and hence the intra-ocular pressure, which determines the shape of the eyeball. Any obstruction of this circulation leads to excessive intra-ocular pressure, the condition of *glaucoma*.

3. The lens and accommodation

The lens of the eye is a biconvex body produced by the metamorphosis of the cells of the original ectodermal lens vesicle into lens fibres within an enclosing lens capsule. The anterior surface of the lens is less curved than the posterior. The lens is attached at the margin to the edges of the optic cup, along a line known as the *ciliary ring*, by a suspensory ligament, the zonule of Zinn, composed of radially arranged fibres. The elasticity of the sclera causes this ligament to pull upon the lens all round and hence to flatten it. Its curvature, and thus the focus of the whole lens system, can be altered by the contraction of the ciliary muscles resisting the elasticity of the sclera. It was shown by Thomas Young in 1801 that accommodation does not involve changes in the length of the eyeball. He fixed the anterior and posterior poles of his own eye with rings clamped together. He then observed no change during accommodation in the phosphene (false image) produced by the pressure. He also showed that there was no change in the curvature of the cornea, and that accommodation is not possible in an eye whose lens has been removed because of disease. The lens must therefore change its shape, which depends partly on the arrangement of the lens fibres and partly on the action of the elastic lens capsule, which is thicker in front than behind.

The *ciliary muscle* consists of unstriped fibres running in various directions, usually said to consist of two

parts, one with fibres directed radially (Brücke's muscle) and the other circularly (Müller's muscle). Contraction of either muscle relaxes the pull on the suspensory ligament producing focus on nearer objects. The radial fibres run from the ciliary ring of the lens to the scleral spur, a thickening close to the union of sclera and cornea. Their action is thus to draw the lens outward, relax the tension on the suspensory ligament, and hence allow the capsule of the lens, which is as elastic as rubber, to mould the contained plastic material to a more rounded form, giving the lens a shorter focal length. The circular fibres of Müller's muscle exert a similar effect by their sphincter action. The ciliary muscle is caused to contract by parasympathetic fibres coming from the ciliary ganglion. There is probably no sympathetic innervation (except of the blood vessels).

The refraction by which an image is thrown on the back of the retina takes place at the various surfaces across which the light passes. The anterior surface of the cornea, in contact with the air, provides a large part of this refraction. The cornea and aqueous humour have similar refractive indexes but the lens differs in refractive index from both humours.

The measure of the power of a lens is the *diopter*, which is the reciprocal of the focal length; a lens with focal length 1 m has, therefore, a refracting power of 1 diopter. The total refraction of the human eye is about 58 diopters.

When the eye is accommodated for distant vision the ciliary muscle is relaxed and the suspensory ligament pulls on the lens, reducing its curvature. The act of accommodation for near vision consists in contraction of the ciliary muscle, taking the tension of the ligament off the lens and allowing the latter to become more curved. The central part of the lens becomes more curved than the peripheral and the pupil also narrows

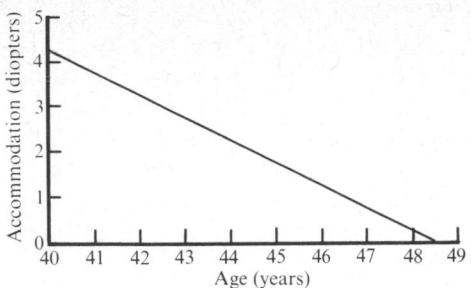

FIG. 40.4. Graph showing the decrease of human accommodation with age.

thus masking off the flatter outer portion of the lens, avoiding distortion of the image due to differing curvatures (spherical aberration). The range through which accommodation is possible thus depends on the power of the muscles and the elasticity of the lens. These factors become reduced with age, especially the second. The lens capsule loses its elasticity, and the substance of the lens its plasticity, due to an increase of insoluble proteins. The whole therefore fails to round up when the tension upon it is slackened, and the subject finds himself unable to focus upon near objects. This change takes place in nearly all people with advancing age. The full range of accommodation is about 16 diopters at puberty but falls to about 5 diopters between 30 and 40 and is as low as 1 diopter in most persons 60 years of age (Fig. 40.4) (Hofstetter 1965).

The mammalian method of accommodation has presumably been derived from some arrangement such as that found in amphibia, where the lens does not change its shape but is moved forward by protractor lentis muscles attached to the periphery of the zonule fibres (Young 1962, p. 351). With the development of a soft lens and an elastic capsule, the muscle would, with modifications, be able to take on the mammalian condition. Unfortunately we obtain little evidence about the evolution of this mechanism from the living reptiles. Like the birds they have striped ciliary muscles, acting by changing the shape of the soft lens by squeezing it.

The degree of accommodation varies greatly in different mammals and is high only in some ungulates, most carnivores, and in primates. Many rodents (mice) and ungulates such as the pig, sheep, and horse have little or no accommodation: the eye is permanently long-sighted. Special arrangements are found in amphibious and aquatic mammals to increase the refracting power of the lens in order to focus on the retina under water. In otters there is a very strong ciliary muscle and a powerful sphincter iridis which squeezes the lens. In seals and sea-lions there is a large round lens and powerful iris sphincter, and the pupil is reduced to a slit when the eye sees in the air. This condition, accompanied by a marked astigmatism of the cornea, enables the eye to focus on the retina both in air and in water without undue change in the shape of the lens; the performances of sea-lions show that their vision in air is quite efficient. The circular fibres of the ciliary muscle (Müller's muscle) are well developed in seals, whales, some ungulates, and the higher primates, especially man, all being animals that require extensive accommodation, though for different reasons.

4. Eye movements

The part of the retina most used in daylight occupies only a very small area, the fovea (p. 390). The eyes are continually being moved by the six extra-ocular muscles in such a way as to allow examination of objects by the fovea. We may recognize five types of movement in man (Alpern 1969).

(a) *Movements during fixation*
Even when the eyes are steadily fixating an object ahead they make repeated involuntary movements. These include a high frequency tremor ($\sim$50 Hz) covering less than 1′ of arc, as well as slow drifts up to 5′ and rapid flicks or saccades (see later, p. 376) of about 10′, made about once a second. These movements together make the image of a point object wander around a region 100 μm in diameter. The saccadic movements probably bring the image repeatedly back to one point. An image is found to fade away if it is 'stabilized' by a feed-back from a mirror on the cornea so that it falls continually on the same area. The fine movements prevent this adaptation.

(b) *Reflex eye movements*
(i) *Static reflexes*. For each position of the head the eyes move to return towards a 'normal' forward-directed position.

(ii) *Stato-kinetic reflexes*. These result from the influences of angular acceleration on the semicircular canals (p. 401). To illustrate them fix the eyes on a stationary target (say a pencil) and shake the head at about 4 times per second. The object remains visible. But if now the head is held still and the object moved at the same rate as before it will be blurred. In the first case the rapid responses of the semicircular canals provide the information for compensatory movements, whereas the visual tracking responses from the eyes are too slow to do so.

(iii) *Opto-kinetic nystagmus*. If the gaze is fixed on a slowly moving object (as when viewing an object in the middle-distance from a train) there is a slow deviation of the eyes as far as they can go and then a quick flick back in the opposite direction (nystagmus).

(c) *Saccadic eye movements*

These involve the sudden change in fixation from one object to another, or a return when the eye has 'wandered' (p. 375). A new object can be examined after an interval of about 200 ms. The movement is ballistic, that is, once started it cannot be stopped. In reading we make a series of saccades with fixation pauses of 200–300 ms.

(d) *Pursuit eye movements*

These produce continuous following of a slowly moving target.

(e) *Vergence movements*

These produce change in the angle that the lines of sight of the eyes make with each other to allow binocular fixation at different distances. Errors of these movements produce strabismus (squint).

41 The retina

1. The rods and cones

THE inner layer of the optic cup is composed of the light-sensitive cells themselves, together with several layers of nerve cells (Fig. 41.1). The retina is not strictly a peripheral receptor but an evaginated portion of the brain. The photochemical changes go on in the rods and cones, which are the outermost cells of the retinal layer and therefore lie with their outer (sensitive) ends in contact with the pigment-cell layer derived from the outer wall of the original optic vesicle, the space between the two layers being obliterated (Fig. 37.12, p. 354). The black pigment of the outer layer actually surrounds the outer portion of the rods and cones, separating the individual cells from each other. If the two layers become separated the retina is said to be 'detached' and does not function properly (p. 379).

The distinction between rods and cones is in the photosensitive pigments they contain and their function, as well as their shape. The rods contain rhodopsin (p. 378) and are long, thin, and sensitive to changes of light at low intensity. The cones contain various photolabile pigments by which colours are discriminated (p. 391). They are mostly short, and sensitive only at relatively high light intensity.

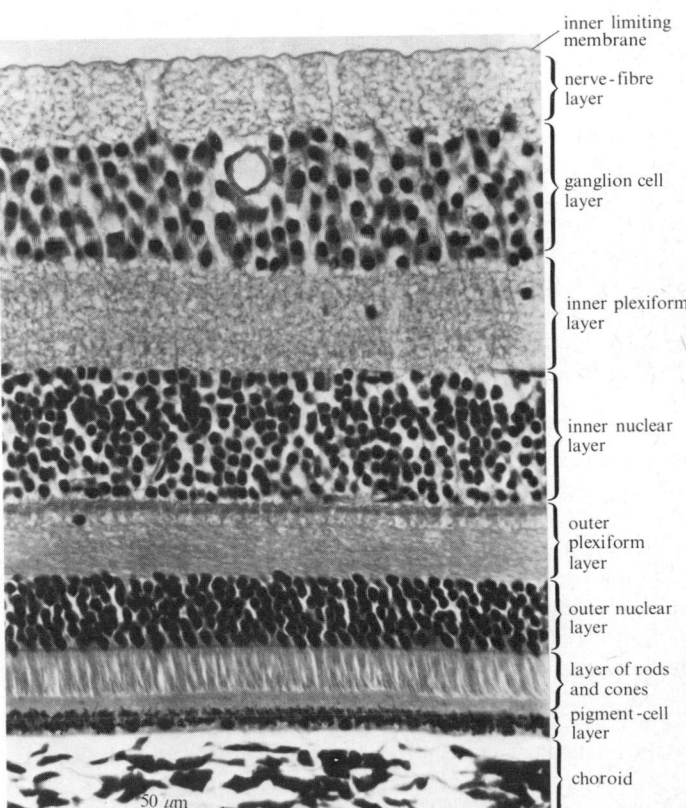

inner limiting membrane

nerve-fibre layer

ganglion cell layer

inner plexiform layer

inner nuclear layer

outer plexiform layer

outer nuclear layer

layer of rods and cones

pigment-cell layer

choroid

50 μm

FIG. 41.1 Light-micrograph from monkey eye showing the retina and part of the choroid. (From a preparation made by Mr. C. E. Bond.)

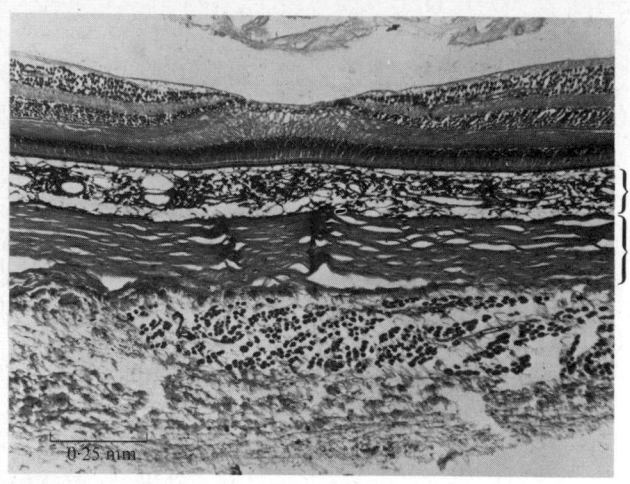

choroid

sclera

FIG. 41.2. Light-micrograph of the fovea from a monkey retina. Note reduction in thickness at this region of maximum visual acuity (see Fig. 42.2, p. 390). (From a preparation made by Mr. C. E. Bond.)

Near the centre of the retina is a circular area, the *macula*, with a pit, the *fovea centralis*, at its centre (Fig. 41.2). This pit is occupied only by cones and is responsible for fine discrimination. The cones here are larger than elsewhere. Rods become more numerous passing away from the fovea, but there are some cones throughout the retina (of primates). Rods and cones may be considered to have four parts (Fig. 41.3): (a) a long outer segment, which is the receptive portion; connected by a narrow neck, the cilium, with (b) an inner segment; which is in turn joined (in the case of the rod via a narrow 'rod fibre') to (c) a rounded cell body containing the nucleus; from which arises (d) a cone pedicle or inner rod fibre with terminal rod spherule, pointing towards the centre of the eye and making synaptic contact with the nerve cells of the retina (Fig. 41.6).

The outer segment has parallel sides in the rods but is usually expanded at its base in the cones, giving them their characteristic shape, but many of the cones of the fovea are elongated and shaped like rods. The outer segments of rods and cones have numerous layers on which are displayed the photolabile pigments (p. 389). Electron microscopy shows that the structure of a rod consists of a pile of discs, each with a double membrane at its upper and lower surfaces (Fig. 41.4). These membranous discs are derived in development from the cell surface. They have the typical three-layered structure of unit membranes. In the adult rods the membranes are no longer joined to the surface, so that the sacks are closed. In cones the continuity may be maintained (Fig. 41.4).

The visual pigment of the rods, rhodopsin, makes up a great part of their membranes. Rods are continually being formed at the base, and injected labelled amino

acids appear first there and move distally and disappear during the next few weeks. The distal ends of the rods are actually phagocytosed by lysosome bodies of the pigment epithelium. Moreover there is continual interchange between rods and epithelium. The rhodopsin is

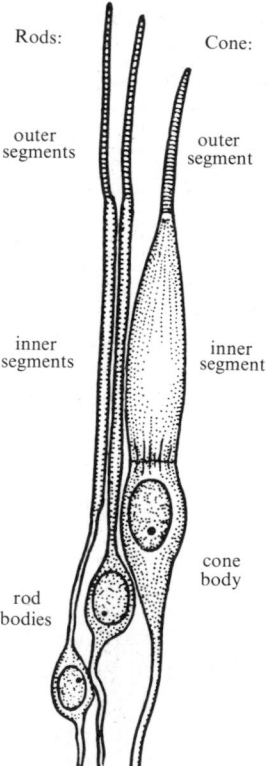

Rods: Cone:

outer   outer
segments   segment

inner   inner
segments   segment

                    cone
                    body
rod
bodies

FIG. 41.3. Diagrammatic views of two rods and a cone from the eye of rhesus monkey (*Macaca*).

largely broken down by light (p. 389) and after illumination for an hour 80 per cent of its vitamin A is found in the pigment cells. During dark adaptation it returns to the rods.

The outer segment of each receptor is joined to the inner by a narrow stalk having the structure of a cilium.

It contains nine fibrils each terminating in a basal body (centriole). The two filaments at the centre of the ring of a motile cilium are missing, but there is no doubt that the rods and cones are formed by modifications of cilia originally lining the central canal.

The inner segment is packed with very elongated

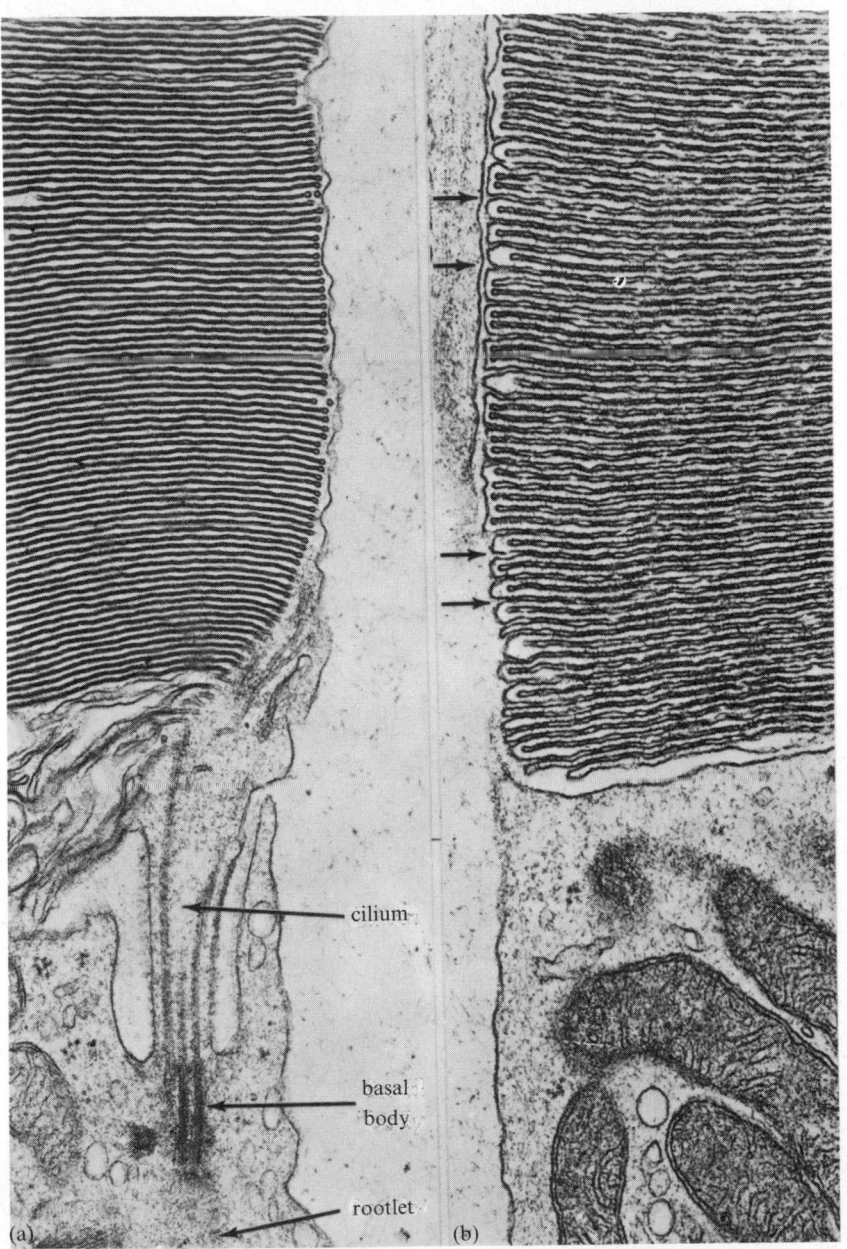

FIG. 41.4. Electron micrographs of visual cells from rat retina. (a) Part of the outer and inner segment of a rod showing the connexion in the form of a modified cilium. (b) The corresponding region of a cone. Arrows indicate lamellae open to the extracellular space. (Figure by courtesy of Dr. T. Kuwabara. In Bloom and Fawcett (1968). *A textbook of histology*. W. B. Saunders, Philadelphia.)

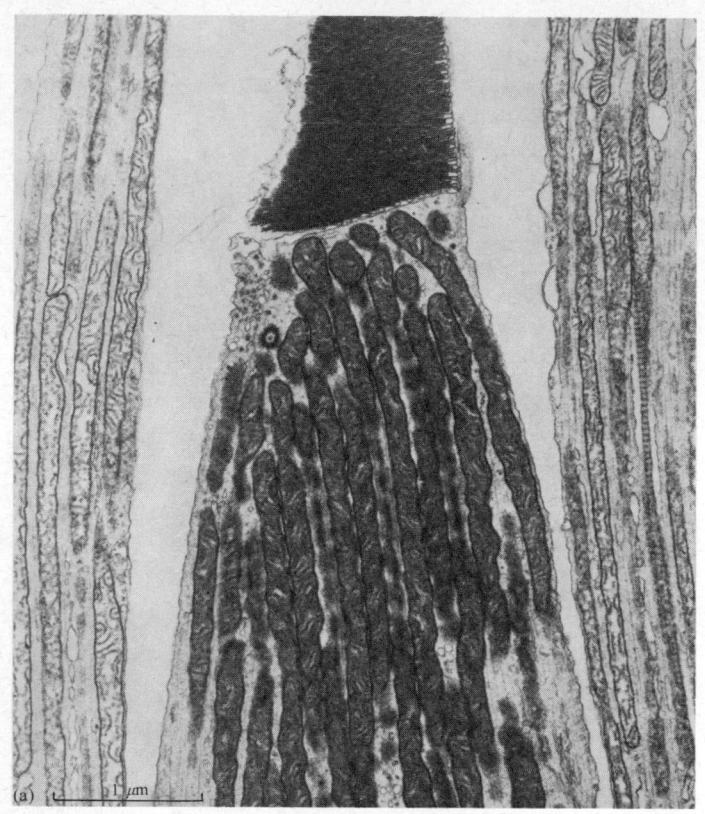

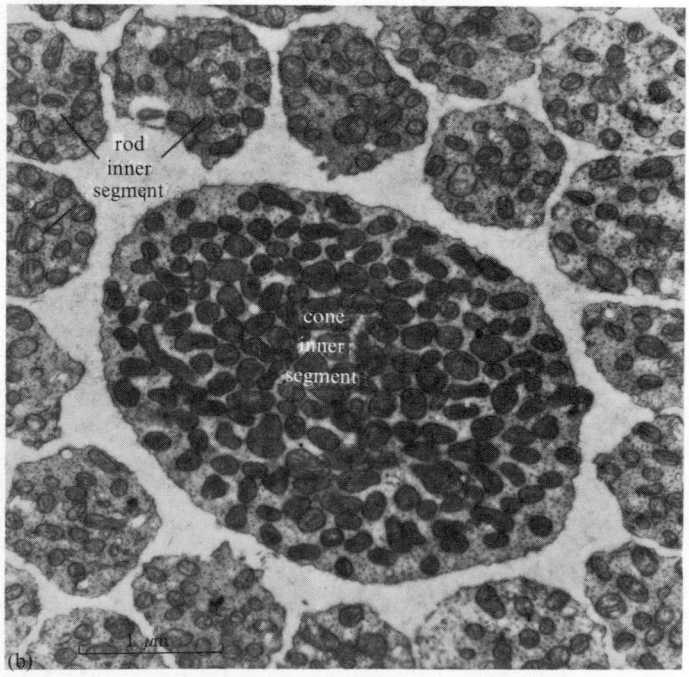

FIG. 41.5. Electron micrographs of visual cells. (a) Vertical section from human retina showing dense array of mitochondria in inner segment. (Figure by courtesy of Dr. T. Kuwabara.) (b) Transverse section through inner segments of one cone and several rods from rat retina. (Figure by courtesy of Dr. V. Marchesi. In Bloom and Fawcett (1968). *A textbook of histology*. W. B. Saunders, Philadelphia.)

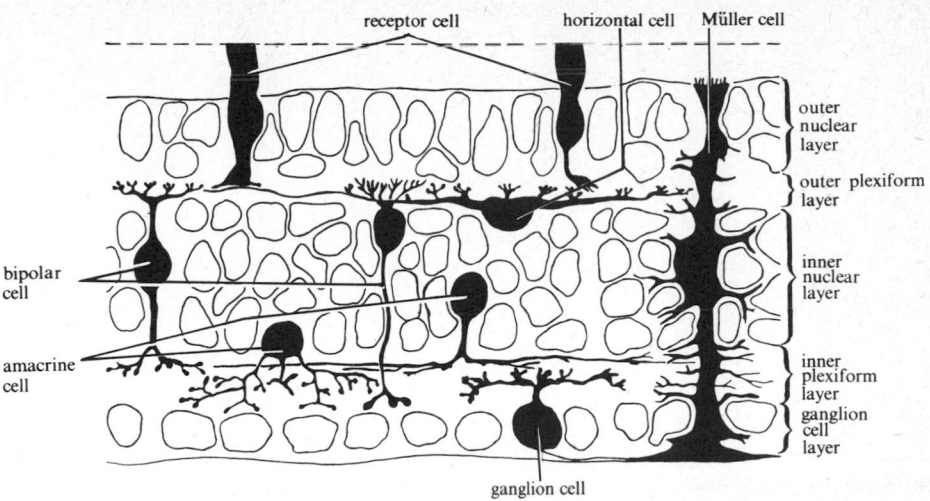

FIG. 41.6. Principal cell types of a vertebrate retina based on Golgi preparations of mud-puppy (*Necturus*) retina. The perikarya are mostly in defined layers as indicated. The Müller cells extend vertically through the thickness of the retina and their nuclei are usually found in the middle of the inner nuclear layer. (From Dowling 1970.)

mitochondria, presumably concerned with the amplification process by which a small change in the photopigments produces a signal (Fig. 41.5, also p. 382). Cross-striated rootlets extend from the centrioles among the mitochondria. At the base of the inner segment the rods (and cones) seem to be joined by an 'outer limiting membrane'. This is in fact a row of junctional complexes between the photoreceptors and the glia cells. The glia cells are here known as *Müller cells* and they run radially through the retina and surround all the other cell types Fig. 41.6.

2. Neurons of the retina

The retina contains four main types of neurons, with cell bodies arranged as inner and outer nuclear layers with inner and outer plexiform layers between, containing the synapses (Dowling and Boycott 1966; Boycott and Dowling 1969) (Fig. 41.7). The arrangement of the neurons allows the first stages of processing of the visual information. The number of channels is thereby reduced from the 120 million rods and 5 million cones to 1 million fibres in each optic nerve. It is obviously efficient for this economy to be made as near to the periphery as possible rather than carrying the full information over long optic nerves.

The synapses of the receptor cells contain an electron-dense ribbon surrounded by a regular array of synaptic vesicles (Fig. 41.8). These ribbon synapses make contacts each with three profiles, two of horizontal cell dendrites and one of a bipolar cell. This arrangement is known as a triad. There are two sorts of horizontal cells (in the monkey or human retina): one makes dendritic contact with numerous rods, the other with numerous cones. The axon proceeds tangentially for a long distance and makes contact with the bases of numerous

other rods or cones (Fig. 41.9). The cones make contact either singly with midget bipolars or several of them with a flat bipolar (Fig. 41.9). Numerous rods make contact with a single rod-bipolar (Boycott 1974).

The inner ends of the bipolar cells terminate in the inner plexiform layer, making synapses with the dendrites of amacrine cells and ganglion cells. The amacrines have branches wholly restricted to this layer; they therefore have no axon and their method of transmission of information is uncertain (see p. 385). The unistratified amacrines have three main branches extending for as much as a half a millimetre in the outer part of the layer (Fig. 41.9). The diffuse amacrines have numerous branches, either at one level or restricted mainly to the upper, middle, or lower third of the layer.

The ganglion cells are of two main types, monosynaptic ones, each synapsing with a single midget bipolar, and polysynaptic ones of various forms (Fig. 41.9). The midget bipolars and midget ganglion cells thus make relatively simple pathways from single cones. The other bipolars are under the influence of many rods or cones and the ganglion cells may be under the influence of both sorts of bipolar. The horizontal and amacrine cells ensure further interaction.

Each synapse of the inner ends of the bipolar cells contains ribbons and each makes contact both with an amacrine cell process and a ganglion cell dendrite (Fig. 41.9). This arrangement is known as a dyad. The amacrine cell process may make a reciprocal synapse with the bipolar terminal.

In the fovea the number of cones is exactly equal to that of bipolars and ganglion cells. However, even here there is much lateral interaction. In the peripheral retina the rods are more than a hundred times more numerous than the ganglion cells.

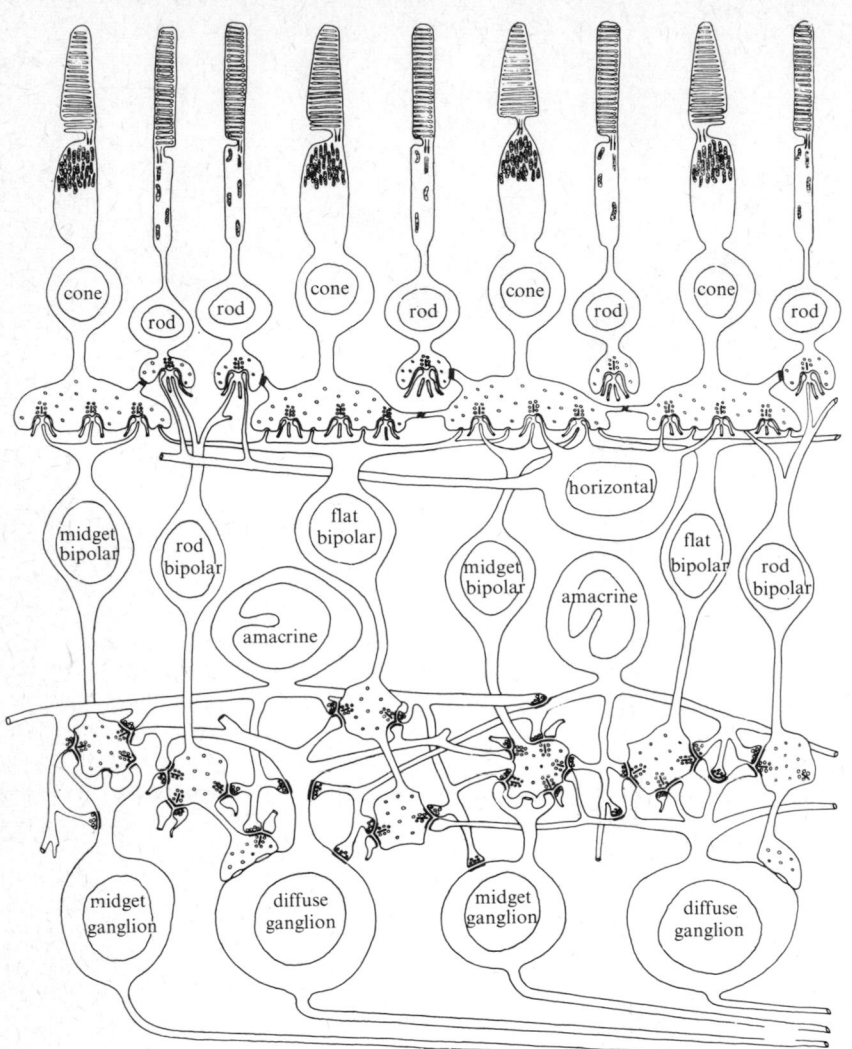

FIG. 41.7. Diagrammatic summary of contacts between principal cell types in the human retina. (From Boycott 1974.)

3. Signalling by the retina

The chemical changes produced by breakdown of the photolabile pigments are converted to electrical signals for transmission onwards. The sequence of electrical changes can be followed in the large cells of the urodele *Necturus*, the mud-puppy (Dowling 1970) (Fig. 41.10). An electrode within a cone records a resting potential of $-30\ \mu V$ to $-40\ \mu V$. Following illumination the cone hyperpolarizes and its membrane resistance rises. This signal is probably transmitted passively to the cone pedicle without the assistance of any regenerative process (i.e. there is no action potential). Indeed it is continuously graded according to the strength of the stimulus.

The operations within the retina seem to be conducted mainly by such *amplitude variant signals*, and these are satisfactory for accomplishing the appropriate interactions between the nearby receptor cells, horizontal cells, bipolar cells, and probably also amacrines. All of these respond to changes of illumination by amplitude variations (Fig. 41.10). Only when the signals reach the ganglion cells are they coded into *frequency variant nerve impulse signals*, which, being regenerative, provide for safe transmission over the optic nerve to the brain (see Rushton 1972).

Rods and cones all hyperpolarize upon illumination but cones have differing wavelength sensitivities (Fig. 41.11). The bipolar cells are arranged in pairs, giving a push–pull system. The members of a pair respond in opposite directions, one hyperpolarizing, the other depolarizing. These are responses to opposite visual events, one when the light goes on, the other at off;

one to illumination of the centre of the field, the other to its periphery; one to red another to green and so on. These opposite effects are combined various ways.

The horizontal cells may serve to operate a gain control, like that of an automatic camera, ensuring that the mean intensity is in the middle of the sensitivity range. Each of these cells receives signals from a large area of retinal surface and scales down the signals in the bipolars to the mean level. Thus, if all the red-cone signals of an area are attenuated in proportion to their output the contrast between the more active ones and their neighbours is increased. But it is not yet known how the activities of the various receptors are combined. There are probably many different pairs, balancing light against dark, red against green, and so on. The cells in the lateral geniculate nucleus, where the axons of the optic nerve end have many centre–surround organizations; red versus green, on versus off, and so on for all sorts of combinations.

In such ways the chemical changes in the rods and cones are converted into a code consisting of frequencies of impulses in specific neural pathways. Such a system is very different from the methods of coding and transmission used in our artefacts such as television, in which a great deal of information is carried in each channel. In the nervous system each channel carries only one sort of information, for example that there has been a particular degree of contrast between yellow and blue in a particular part of the visual field. We have as yet little means of knowing how the information in the many thousands of such channels is combined to allow recognition that, say, this is a cat and that a canary.

4. Electrical activity of the retina

The *electroretinogram* (ERG) is a series of potential changes recorded between an electrode on the cornea and an indifferent one elsewhere, say in the mouth. It usually consists of four phases (Fig. 41.12); an initial rapid corneal negativity, the *a* wave, followed by a large positive *b* wave, passing into a slowly rising *c* wave; when the light is switched off there is a further positive deflection, the *d* wave.

With a very bright flash the *a* wave is preceded by an

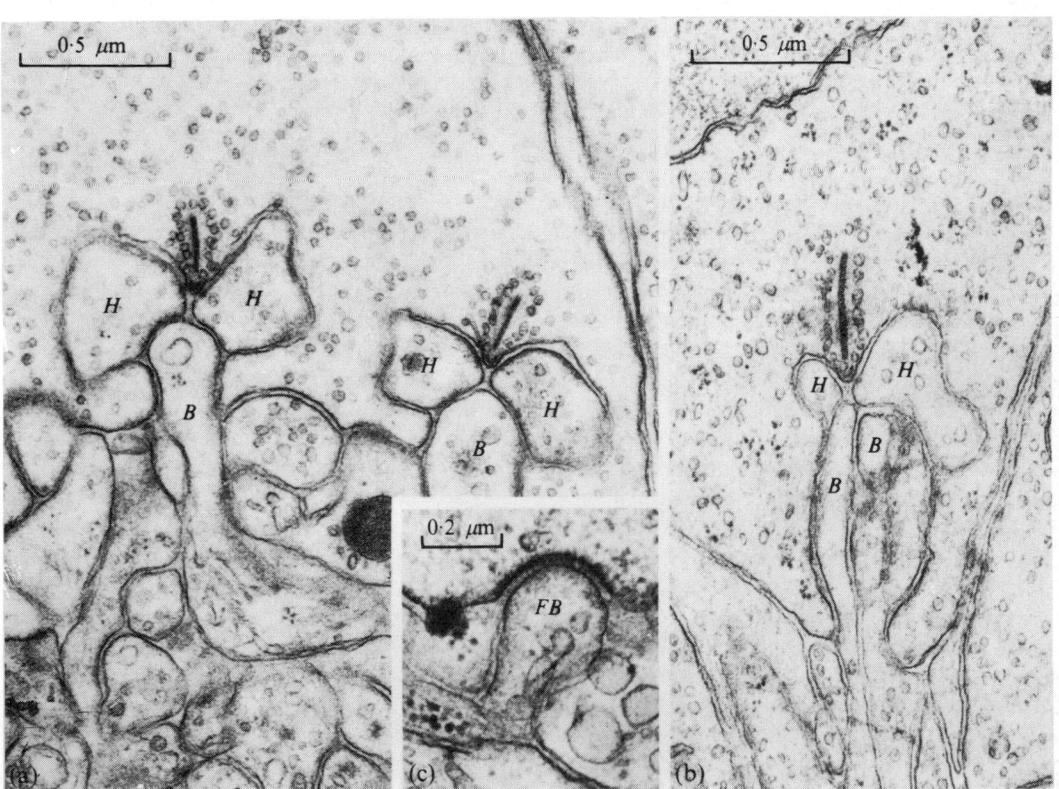

FIG. 41.8. Electron micrographs of synapses of receptor terminals. Typical ribbon synapses of (a) cone and (b) rod terminals with three and four processes entering each invagination respectively. *H*, horizontal cell process; *B*, bipolar cell dendrite. (c) Superficial contact of flat bipolar (*FB*) on receptor terminal. (After Dowling 1970.)

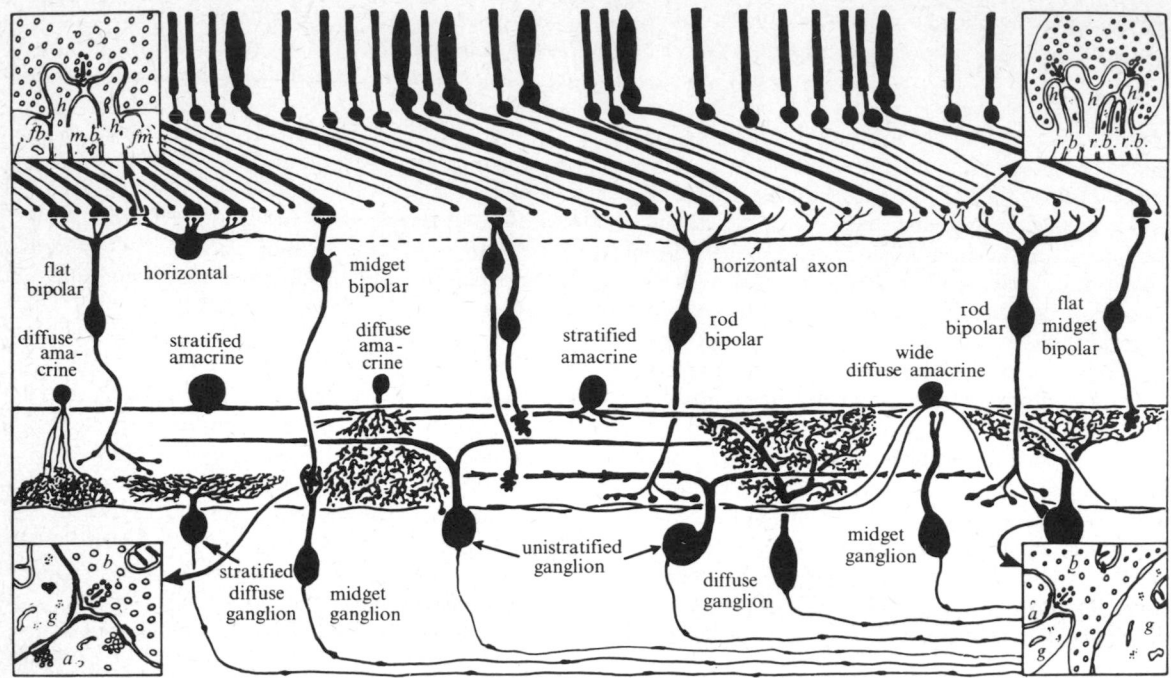

FIG. 41.9. Diagram of the types of nerve cell in the fovea of the monkey and man. Inset (*top left*), the synaptic relations at the cone pedicle, with *h*, terminals of horizontal cell; *f.b.* flat bipolar; *m.b.,* midget bipolar; and *f.m.,* flat midget cell dendrites; (*top right*), at the rod spherule the horizontal cell processes lie deeper and laterally relative to the bipolar cell dendrites, *r.b.*; (*bottom left and right*), various synapses between *b.,* bipolar; *g.,* ganglion and *a,* amacrine cells. (From Boycott 1974.)

early receptor potential occuring with no detectable latency. This is probably due to movements of charge in the molecules of rhodopsin during the photochemical reaction. Its amplitude is proportional to the amount of visual pigment bleached. On heating it disappears at the temperature at which the orientation of the rhodopsin is lost. This early potential is biphasic but it is not clear where is the asymmetry that causes current to flow. It may be due to the fact that some but not all of the sacks are open to the extracellular space (p. 378) and this would agree with the evidence that this potential depends mainly on cones.

It remains uncertain whether this early receptor potential is related to the signalling power of the rods and cones. The time-course of the hyperpolarization in a cone is exactly similar to the first wave (*a*) of the ERG. The other electrical changes indicated by the ERG (Fig. 41.12) are connected with activities within the retina. The second, *b* wave, matches responses seen in the glia cells of the mud-puppy. These, known as Müller cells, give slow sustained depolarization, lasting throughout illumination. Since these cells receive no synapses it may be that this response is produced by increased intracellular K^+, and perhaps spreads by the junctional complexes that they make (Fig. 41.13). The *c* wave of the ERG is associated with the metabolic regenerative processes produced by the pigmented epithelium; it is well-marked in frogs but less so in mammalian eyes. The *d* wave at 'off' is probably due to the swing of the negative *a* wave back to the base line, superimposed on the *c* wave (Dowling 1970).

Fig. 41.10 gives some idea of how these responses are related when a flash of light illuminates, on the left. The receptors operate independently and there is no feedback to them from horizontal cells. The small response of the cell on the right is due to stray light. Horizontal cells summate inputs from a wide area. The central response of bipolar cells is probably mediated by the direct contacts from receptors, the surround area corresponds to the spread of the horizontal cells. The transient response of the amacrines may be produced by their reciprocal synapses turning off the excitation from the bipolar cells. There is evidence that the amacrine cell interactions may determine the fact that some ganglion cells respond to movement of light only in a certain direction. There are also efferent fibres in the

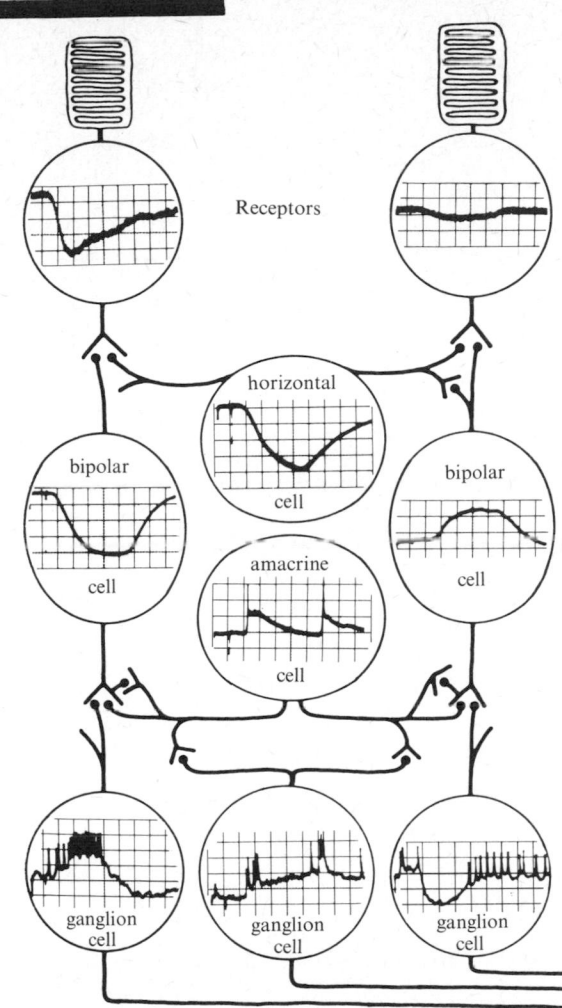

FIG. 41.10 Summary diagram relating synaptic organization of retina with intracellularly recorded responses in the mud-puppy. The experimental conditions are such that the right-hand receptor has continuous dim illumination whilst that on the left receives a bright flash. In general distal neurons respond with slow graded hyperpolarizing potentials whilst the proximal ones respond with depolarizing, mostly transient potentials. (From Dowling 1970.)

optic tract, at least in some mammals, ending within the retina. Their function is not fully known but they can alter the responses following illumination and thus provide a feed-back or reciprocating system (see Granit 1955).

5. Impulses in optic nerve fibres

The ganglion cells of the retina of a decerebrate cat in the dark generate frequent, irregularly spaced impulses.

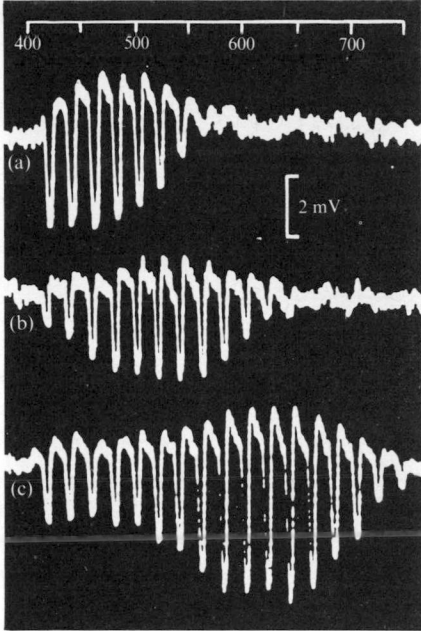

FIG. 41.11. Intracellular records from single cones of a carp. Each was stimulated in succession through a rotating wheel by lights of differing wave length but same light energy. Cone (a) responded mainly to blue, (b) to green, and (c) to red. (From Tomita *et al.* (1967). *Vision Res.* **7.** Reprinted with permission.)

These are probably due to thermal decomposition of retinal pigment, producing noise in the rods and cones. In cold-blooded animals the optic nerve fibres are generally silent in the dark. Individual fibres of the frog's optic nerve are of three kinds (Fig. 41.14). About 20 per cent show a burst of impulses when light is switched on and a sustained discharge while illumination is continued. About half of all the fibres give bursts of impulses at 'on' and 'off', but none in between (on–off fibres) and the remainder respond only when the light is turned off.

The receptive field of a single frog fibre is a circle of 0·5–1 mm diameter, most sensitive in the centre. The ganglion cells of the frog show further variety in that some respond only to special features such as linear stimuli (illuminated slits) with particular orientation and direction of movement.

In the cat the fields of the cells mostly have central and peripheral regions with opposite properties. Thus the centre of the field responds either to 'on' or to 'off', and cells with on centres have off surrounds and vice versa. Few or no cells of the cat or monkey have directional or orientational sensitivity such as those of the frog, but there are some in the rabbit.

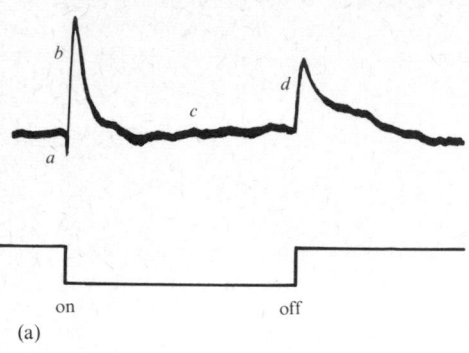

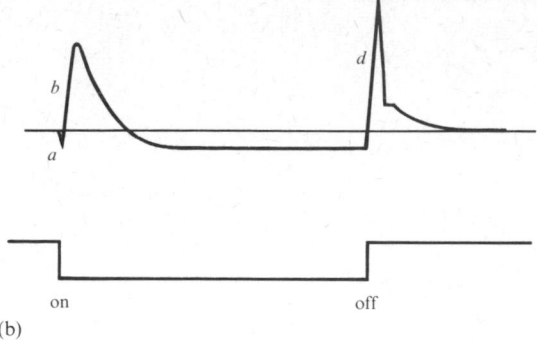

(a)   (b)

FIG. 41.12. (a) A typical electroretinogram showing main components: *a* wave (a brief corneal negativity); *b* wave, a large positive wave passing into the slowly rising *c* wave; and *d* wave, an on–off effect on removal of stimulus. (b) Electroretinogram of horned toad lacking *c* wave.

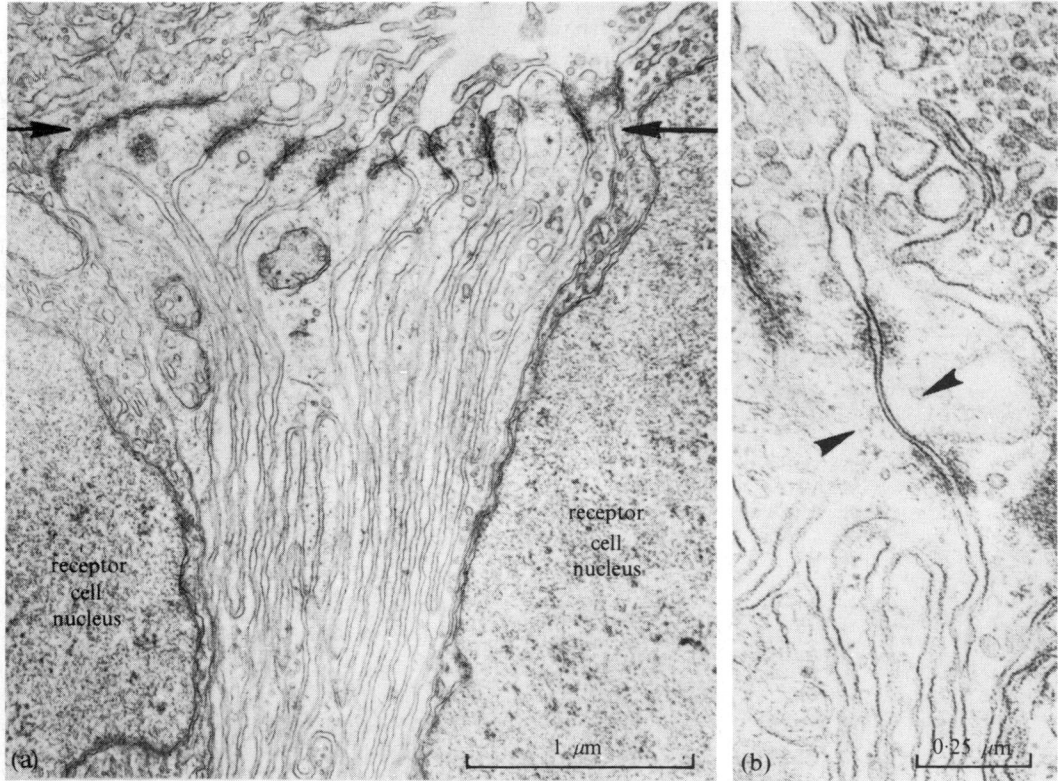

FIG. 41.13. (a) Electron micrograph of junctional complexes (*arrows*), mainly between Müller cell processes, at the level of the external limiting membrane. (b) Detail of complex showing gap junction between desmosomes. (From Dowling 1970.)

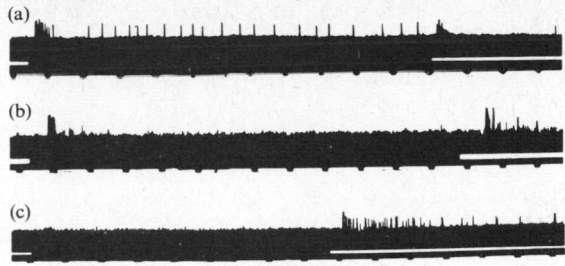

(a)

(b)

(c)

FIG. 41.14. Responses of the three types of intraocular nerve fibre of the frog. (a) On-fibre, responding with an initial burst of impulses followed by a maintained discharge during illumination with no response at 'off' (the apparent off-response in the record is partly due to the retinal potential and partly to the reaction of the fibre). (b) On–off-fibre, responding only to onset and cessation of illumination. (c) Off-fibre, responding only to cessation of illumination. Time marking: $\frac{1}{5}$ s. Illumination during interruption of white line at bottom of each trace. (From Hartline (1938). *Am. J. Physiol* **121**.)

Evidently the retina is truly part of the brain in that its cells compute responses to complex patterns of input from the receptors, varying according to the type of information needed by the habits of the animal. Further complications arise when we consider the effects of altering the intensity or colour of illumination. Single ganglion cells rarely increase their frequency of discharge steadily with increased intensity of stimulation. Some actually decrease and others change behaviour in complex manners. In dark-adapted eyes the spectral sensitivity curve of all animals with numerous rods agrees with the absorbtion spectrum of rhodopsin (or porphyropsin, see p. 389). When the eyes are light-adapted the sensitivity of each ganglion cell changes in ways that show that it is now connected to other receptors in addition to, or instead of, rods. In the diurnal ground squirrel there are clear-cut classes of ganglion cells corresponding to the three classes of cones (p. 391).

42 Vision by day and night

1. The duplicity theory of vision

ALTHOUGH the rods and cones are fundamentally alike, their particular characteristics enable man to use the eye in two ways so different that they almost be said to provide two distinct receptor organs. It was first noticed by Schultze in 1866 that the eyes of nocturnal animals contain mainly or only rods, whereas in related diurnal forms there are either both rods and cones or cones alone. According to the *duplicity theory*, which he put forward, the rod retina is responsible for *scotopic* or dim-light vision, having a high sensitivity but poor powers of discrimination, whereas the cones are responsible for *photopic* or bright-light vision, involving fine discrimination, including that of colour and shape.

2. Scotopic vision and dark adaptation

It is easy to find evidence of this double function in our own eyes. Everyone is familiar with the fact that the eyes gradually become 'accustomed to the darkness'. This process of dark adaptation can be investigated by first making the eyes fully 'light-adapted' by looking at a bright light and then testing the sensitivity of the retina by finding the weakest light that can be seen after various subsequent times in the dark (Fig. 42.1). By this method it is found that the threshold of the eye falls rapidly; after half an hour in the dark the eye responds to light 10^4 times less intense than at first; it has become 'dark-adapted'. The curve of dark adaptation shows a sharp kink and the usual interpretation of this is that during the first six minutes the cones are rapidly increasing in sensitivity; the threshold of the rods falling only later but continuing for a much longer period. Investigation of many visual processes shows kinks of this sort, providing evidence in favour of the duplicity theory.

Anyone can confirm that the greater sensitivity of the retina is in its peripheral part; when fully dark-adapted and using low intensities of illumination we can only vaguely discern the presence of objects, but we see them better if we do not look directly at them;

that is to say by using the periphery of the retina rather than the fovea.

As the visual threshold drops during the dark-adaptation the visual acuity falls: it is a commonplace that when things are seen in a dim light they are indistinct. Moreover, they then have no colour; only very bright moonlight allows the cones to function, hence in dim light we see things in shades of grey or silver. Red light has no effect on the rods, a fact that allows us to use the two parts of the visual system separately when we want both to make fine discriminations (say for reading) and yet remain able to use the eye at low threshold. Instruments on the dashboard of a motor car or an aeroplane, if illuminated with red light, can be read easily at night-time by use of the cones, allowing the rods (unaffected by red light) to develop their maximum sensitivity for peering out of the vehicle into the darkness.

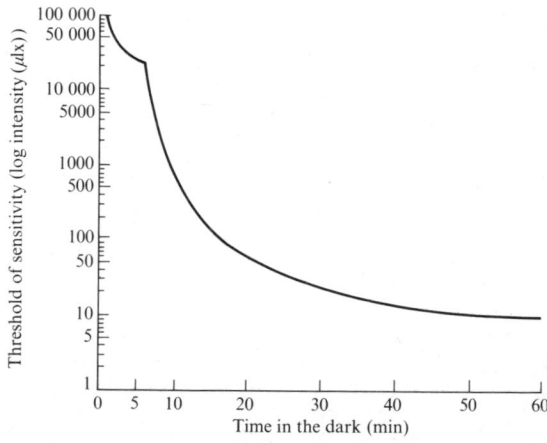

FIG. 42.1. Course of dark-adaptation in the human eye. The cones are completely adapted after about six minutes. The rods then begin to be adapted. The change over from cone adaptation to rod adaptation is marked by the sharp change in direction of the curve. (After Wright and Granit (1938). *Br. J. Ophthal.,* Monogr. Suppl. No. 9.)

3. Visual purple (rhodopsin)

It is not certain whether the rod or cone type of vision appeared first in the vertebrate series, but probably it was the scotopic (rod) type; this is somewhat the simpler and may be described first. The incident light sets up signals in the rods by a process involving change of the substance *visual purple* (rhodopsin). This substance is found in the retina of man and most animals except those that have only cones. Some freshwater fishes have a related pigment, porphyropsin. A fully light-adapted retina is colourless but retinae removed as dark-adaptation proceeds show a purple tint, which disappears quickly on exposure to light. This can be readily verified with frogs' eyes. Rhodopsin constitutes as much as 80 per cent of the protein of the membranes of the outer rod segment. The molecules are not packed randomly. In plane-polarized light, rods are dichroic, absorbing the light more strongly if its electric vector is perpendicular to their long axes than if it is parallel to them. This means that the chromophore groups of the rhodopsin are orientated in a plane parallel to the long axes of the rods.

Rhodopsin has a molecular weight of about 40 000 and is composed of a protein, opsin, associated with the carotenoid substance 11-*cis*-retinaldehyde, or retinene, derived from vitamin A (retinal). Deficiency of this vitamin therefore produces night-blindness. The maximum sensitivity of rhodopsin is to blue–green light (500 nm) and this is approximately (but not exactly) the region of maximum sensitivity of the dark-adapted eye (510 nm). The pigment is not decolorized by red light (see Wald 1951, 1959).

The simplest hypothesis is that in scotopic vision the light changes rhodopsin to an unstable substance that breaks down into opsin and retinene, the process of breakdown stimulating the rod to the activity, the provision of a generator potential, by which the next link is activated. Dark adaptation consists in the enzymic resynthesis of rhodopsin. But the increase in sensitivity proceeds much faster than the resynthesis. To explain this it has been suggested that the presence of bleached rhodopsin either lowers the sensitivity of the rods or causes them to give signals even in the dark, so that the effect of illumination is lessened (Barlow 1964).

4. The absolute threshold of vision

The remarkable nature of the photochemical process is demonstrated when we consider the extraordinarily high sensitivity of the retina. Much energy is absorbed by the refracting media as light passes through the eye. Therefore in order to determine the minimum light energy needed to initiate the visual process it is necessary to find what is the smallest amount falling upon a given corneal area that will stimulate. After calculating how much of this light is absorbed by the lens and so on, we can find how much actually reaches the retina. Hecht and his colleagues (1942) have made careful determinations with minute spots of light illuminating an area containing 500 rods. They found that a very small light source is reported as seen at 60 per cent of presentations if it contains between 54 and 148 quanta. Estimates of the amount of light lost by reflection and absorbtion before it reached the retina lead to the conclusion that for the brightest of these flashes not more than 14 quanta would be absorbed by rhodopsin. From this it is obvious that the probability that one of the 500 rods receives more than one quantum is only 0.178, which is much less than 0.6, the probability of seeing the flash. This result therefore means that every quantum received by a rod causes it to transmit a signal that can be used in deciding whether a flash has occurred. Since the results show that about 5 quanta must be absorbed evidently about 5 rods must be activated.

If this is so it must follow that a very faint light, near the threshold, is sometimes seen and sometimes not. This phenomenon can in fact easily be observed by asking subjects to say whether or not they see a faint flash. As the light intensity is increased with an optical wedge the frequency with which the flash is reported increases in exactly the manner that would be expected as the chances rise that enough rods will be struck by single quanta. No setting of the wedge can be found such that at lower intensities the light is *never* reported while at higher ones it is *always* reported. These fluctuations in the apparent threshold had previously been supposed to be due to changes in the eye, or in the observer. Hecht's work shows that in reporting that a very dim flash is only sometimes visible an observer is directly recording the quantal nature of light. A retinal rod reaches the absolute limit of sensitivity set by the quantum and molecular theories. Observers of the visual threshold had therefore noticed that light is a quantal or discontinuous phenomenon long before that fact was indirectly established by physical analysis. That the facts had not forced themselves on the attention of physicists is an interesting comment on modern man's distrust of his own recording apparatus. The importance of the clues provided by faint light sources in deciding human actions agrees with the discovery that we are provided with a system capable of collecting information from the smallest possible change in the emission of radiant energy at these wavelengths (see Pirenne and Denton 1952).

One quantum of light energy can transform only a single molecule of visual purple or other receptor substance. There must therefore be some mechanism by which the products of this molecule can affect others and produce a sufficient change to fire off activity in the rod. The only suggestion at present about how this occurs is that it is related to the elaborate internal organization of the rods, as shown by their cross-striation and birefringence. Perhaps a change in any part of this system leads to extensive reorientations throughout. The great length of the rods and large number of their plates, especially in nocturnal animals, presumably increases the chance that any photon will be effective in disturbing the state of the whole.

5. Photopic vision

The cones of the eye provide a mechanism of low sensitivity but high acuity, suitable for obtaining information in the bright light of day-time. Cones are found in the retinae of nearly all diurnal animals, either mixed with rods or in a special central fovea. In some purely diurnal creatures (squirrels, many birds, many lizards, snakes, and turtles), they are the only visual cells in the eye. The differences between the photochemical processes in the two types of visual cell are not well understood. Cones contain special pigments responsible for their photosensitivity (p. 391). The shortness of the outer segment of typical cones is presumably connected with their low sensitivity. In some animals they perform photomechanical movements opposite to those of rods, becoming elongated in brighter, contracted in duller illumination.

Vision at high luminosities permits collection of much information and development of elaborate behaviour patterns. First, there is the possibility of using great resolving power to discriminate fine detail and small movements, thus of taking correspondingly detailed action. Secondly, the eye and brain together can recognize and react to a great variety of significant patterns and shapes from the varied changes of light intensity around. Thirdly, photopic vision is often associated with a high degree of wavelength discrimination, producing the phenomenon we know subjectively as colour vision.

6. Resolving power

A greater degree of resolution is possible at high than at low luminous intensities because there is so much light that the summation mechanism present in the rod system can be dispensed with. Fine movements can then be discriminated and remarkable feats of resolution are performed by animals such as birds and primates when they are picking out small seeds or insects, or detecting movements at great distances (Pirenne 1967). These achievements depend on the presence of a high density of cones. The central fovea of man is a circle about 500 μm across, containing 30 000 narrow cones. The cells are so densely packed that not even capillaries intervene. Outside this area is an outer fovea, where there are some rods and capillaries. The cones of the central fovea are elongated and thus resemble rods anatomically, though not in sensitivity. They are so arranged that the light reaches them directly from the vitreous, their connexions with bipolar cells being made round the margin of the fovea. The retina of the central fovea is therefore slightly thinner and that around it thicker than the retina elsewhere. The amacrine layer associated with the foveal region is especially thick.

The maximum acuity of human vision is found only in a very minute region, less than 35′ of arc, within the rod-free fovea (Fig. 42.2). All our critical seeing is done with this tiny area, the eye being of course continually moved to scan a much larger field (p. 375).

The resolving power of the eye is defined as the least angle at which two contours can just be discriminated, and may be as little as 10″ of arc subtended at the eye. The acuity varies with the intensity of illumination. In the fovea the resolving power is near that set by the mosaic of cones and the imperfections of the optical image due to spherical and chromatic aberration. It is interesting that the eye is so economically designed that all these factors are equally limiting.

7. Form vision

The power of form recognition is one of the most developed human attributes but its basis is little under-

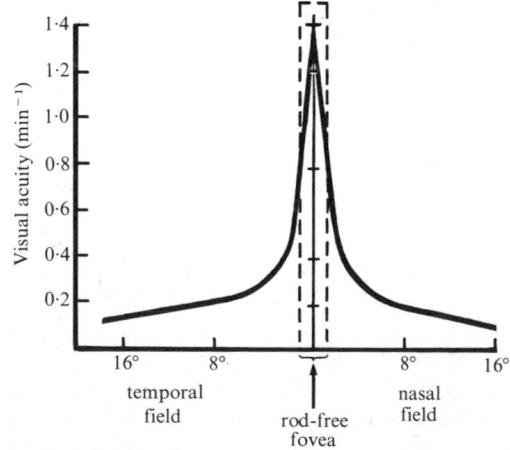

FIG. 42.2. Graph showing the visual acuity in the horizontal meridian. Data from two subjects with normal eyes. (After Alpern (1969). In *The eye* (ed. Davson), Vol. 3. Academic Press, New York.)

stood. Human actions are controlled by information about the great variety of objects provided by the natural environment and by our artefacts. This information allows a much more detailed set of choices than does the information about flashes, patches of light and shade, or hints of movement that are available to other animals.

The recognition of shape depends on the picking out of certain significant outlines from the many that are presented in the visual field. Some shapes seem to force themselves upon us, for instance, the startle pattern of round eye-spots shown by many animals: yet to a large extent we have to learn to recognize shapes (p. 333). Probably the process goes on in the cerebral cortex and the necessary requirements as far as the eye are concerned are that there should be a system able to register a pattern of light and shade and to project it to the cortex, where its features can be matched against those stored in the memory system as a result of previous experience (p. 341). The significant processes presumably occur at the boundaries between areas that are illuminated to different extents. It is possible that events in the retina itself play a part in detecting these boundaries, perhaps by sharpening them or increasing their contrasts. Retinae rich in cones (and the central areas of mixed retinae) are associated with very thick inner retinal layers and especially with numerous amacrine cells. It is significant that these pathways that carry large amounts of information also allow much overlap (see Granit 1955).

8. Colour vision

Any animal whose photosensitive cells have a range of maximum sensitivities could respond differently to illumination by light of different wavelengths. In practice only eyes working at high luminous intensities allow colour vision and only a limited number of animals make use of this information. Accurate investigation of colour vision is difficult because it must take account of the possibility that the differential sensitivities of the animal enable it to distinguish wavelengths as if they were of different intensities. Many animals, for instance dogs, show themselves able to discriminate one colour from another but will yet confuse each colour with some tint of grey. True colour vision, the discrimination of wavelength independently of brightness, is known with certainty to occur in some teleostean fishes, in turtles, lizards, birds, and primates, all being diurnal animals with high visual acuity. It may be present in some other groups but is probably absent from elasmobranchs and amphibians, and from most mammals.

Human reaction to colour consists in giving a series of names to light between wavelength about 760 nm (red) and 380 nm (violet). All the visible wavelengths presented together, or certain combinations of them, we call colourless, white, or grey; removal of certain wavebands leaves the remainder coloured. Thus coloured paper reflects only some wavelengths and coloured liquid or glass transmit only some. Absorption of all wavelengths produces black. It is found that white, grey, and all the colours visible in the spectrum can be obtained by mixing three 'primary' colours (trichromacy). Within limits any three colours can be chosen, provided they are from widely separated parts of the spectrum. For mixing lights it is usual to take blue, green, and red as 'primaries', but with pigments, blue, yellow, and red are chosen.

The phenomena of trichromacy would be explained if the path of information from the retina is somewhere restricted to three colour channels, each expressible as a continuous variable. This was the trichromatic theory first clearly expressed by Thomas Young in 1802. We have already seen that the rod-pigment rhodopsin has a wide frequency spectrum with maximum sensitivity at 502 nm. It has recently been shown that there are in addition three further photolabile pigments, probably located in the cones. By experiments that measure the amount of light reflected from the human fovea, Rushton (1965, 1972) found the red and green pigments and called them erythrolabe and chlorolabe. Observations of these pigments in single cones have since been made by Marks, Dobelle, and MacNichol (1964) and Brown and Wald (1964) (Fig. 42.3). They have also found some cones that seem to contain the blue pigment (cyanolabe), which for technical reasons cannot be shown by Rushton's technique of ophthalmoscopic densitometry.

The three-colour hypothesis is therefore in the main confirmed. In the human fovea there are at least three photolabile pigments having the expected spectra and located mainly in three separate sets of cones (Fig. 42.4). It cannot be rigorously excluded, however, that there are no further complications, for instance other pigments, perhaps only slightly photolabile. Nor is it certain that the pigments are each confined to one type of cone. Although the evidence for the pigments is quite good they cannot be extracted from mammalian retinae by any such technique as is used for rhodopsin. The eyes of diurnal birds such as the domestic fowl or pigeon do, however, yield, in addition to rhodopsin, smaller quantities of a second pigment, iodopsin, absorbing maximally at longer wavelengths. This pigment can also be synthesized from 11-*cis*-retinaldehyde. The mammalian cone pigments are probably not very dissimilar as is suggested by the fact that in the para-

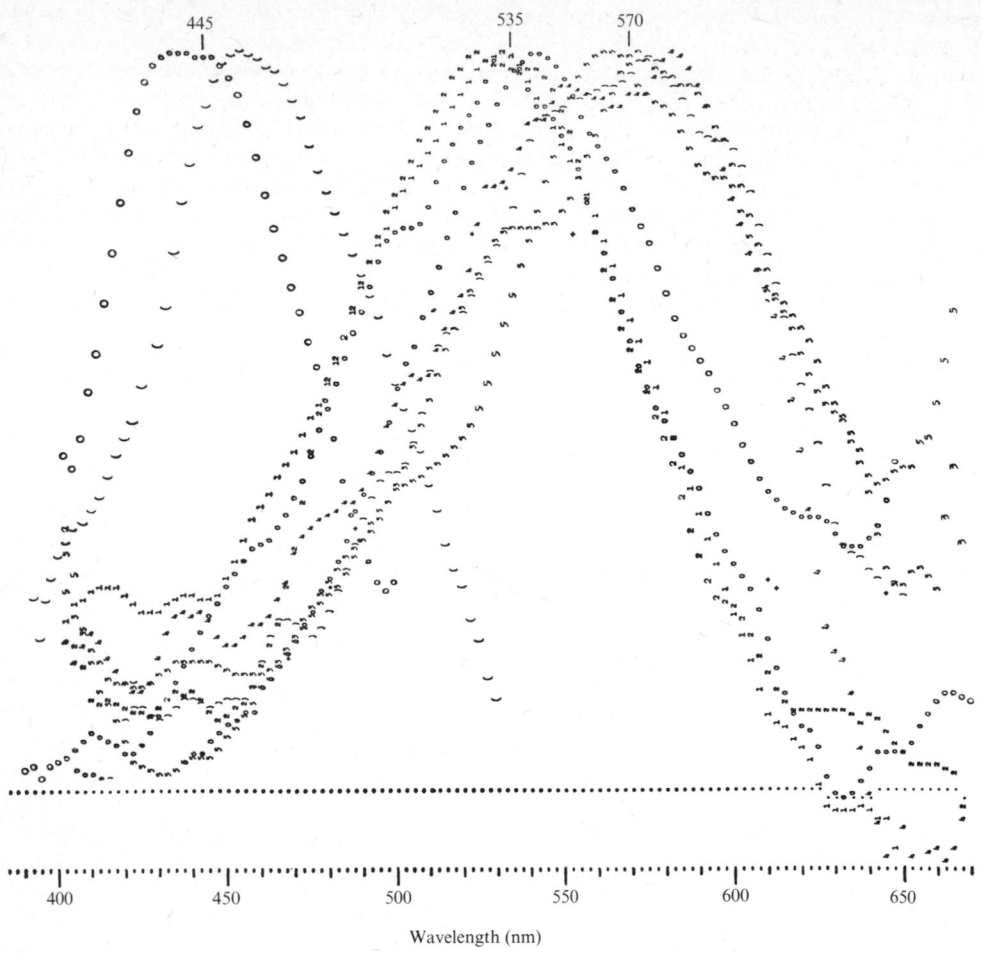

445   535   570

400   450   500   550   600   650

Wavelength (nm)

FIG. 42.3. Difference spectra of human (parentheses) and monkey (*Macaca*) cones (numerals). (From Marks, Dobelle, and MacNichol (1964). *Science, N.Y.* **143**, 1181–3. Copyright 1964 by the American Association for the Advancement of Science.)

foveal region, where rods and cones are equal in abundance, the dark adaptations of rods is slowed if the cone pigments are bleached, so that presumably they compete for the retinal.

Although it is reasonably certain that colour vision is a property of three types of cone it is only partly understood how the action of these is combined either within the retina or beyond (see p. 387). In the ground squirrel (*Citellus*), whose eyes contain only cones, about one-fourth of the fibres of the optic nerve respond characteristically to different wavelengths (Michael 1968). Some are excited by green and inhibited by blue and others the opposite. Both sorts are little excited by white, because of the antagonism. Presumably these colour cells are connected to two groups of bipolar cells one receiving synapses from green-sensitive and the other group from blue-sensitive cones.

9. Colour blindness

Abnormalities of colour vision are inherited as sex-linked recessives and are much more common in men (8 per cent) than women (0·4 per cent). Monochromats, that is people unable to distinguish any colours, are very rare. Dichromats, lacking one of the three pigments, are of three sorts. Protanopes and deuteranopes both have poor powers of discrimination at the red end of the spectrum but differ in other respects. They are each about 1 per cent of the male population. Rushton showed that protanopes lack erythrolabe and deuteranopes lack chlorolabe. Tritanopes are people who have little ability to discriminate blue from green and they probably lack cyanolabe, though there is little evidence of this. The condition is not more frequent in men, but is rare (< 1 in 10^4).

The great majority of colour-blind men are called

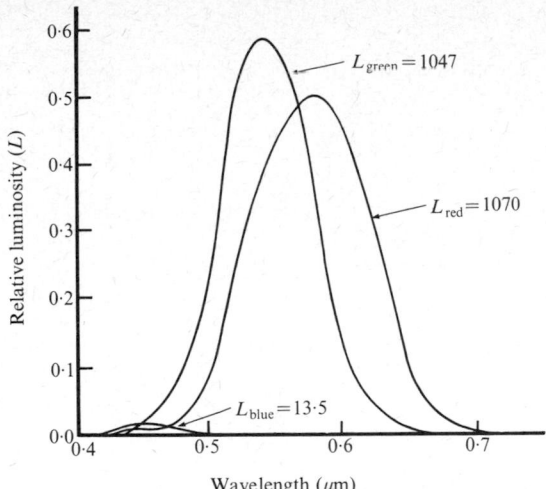

FIG. 42.4. Fundamental response curves for an equal energy spectrum. (After Pitt (1944). *Proc. R. Soc. B.* **132**.)

anomolous trichromats (6 per cent). To match all colours with the three primaries they need to see them in unusual proportions. Of the various different forms the commonest is red–green colour blindness in which the capacity to discriminate is reduced in the red–green–yellow region.

The existence of such common hereditary abnormalities raises the question of the usefulness of the power of colour discrimination for men now and in the past. It is obviously useful to be able to distinguish red (blood), green (plants), and blue (sea and sky), though it is perhaps hazardous to suggest how these particular colours were selected during evolution until we know more of the power of discrimination of other primates and other animals. Each type of animal may have photolabile pigments suited to its need, if it is chemically practicable to provide these. Thus deep-sea fishes have pigments with sensitivity maxima between 475 nm and 490 nm, near to the wavelength of the light reaching them. Humans, of course, can with advantage distinguish objects with a wide variety of colours beyond the simple red, green, and blue. It has been suggested that the genes for male colour blindness were selected because they enabled their possessors to defeat camouflage devices by prey animals. But populations of modern hunter-gatherers do not include these genes in large numbers (see Young 1971, p. 553).

10. Special adaptations for nocturnal vision

The relative degree of emphasis on photopic and scotopic vision varies greatly in different species. Many animals that are mainly nocturnal are also able to achieve quite good visual discrimination in the daytime, for instance the cat. These animals have cones as well as rods in the retina. Some completely nocturnal creatures have abandoned cone vision altogether and developed rod retinae, with a very high sensitivity to light. The retina is completely free of cones in bats, armadilloes, and probably in hedgehogs, lemurs, guinea-pigs, whales, and seals; rats and other nocturnal rodents have only very few of them. As many as 1000 rods connect with each bipolar cell in such rod retinae. The other cells of the retina are reduced and the whole is therefore thinner than in diurnal animals, and the optic tract is slender. Many other special features can be recognized in these night eyes. They are usually very big, allowing for a wide pupil and necessitating a large anterior chamber and large round lens. This may give the eye a tubular form, which may restrict its movement within the orbit to such an extent that the animal can only change its line of sight by means of a mobile neck. It is no accident that owls as well as tarsiers can turn round and look backwards (see Duke-Elder 1958).

Protection of the very sensitive retina from bright light is afforded in nocturnal animals by the presence of pupils capable of wide excursions. Animals that are mainly nocturnal but also come out in the day often have a pupil that closes to a narrow slit and the edges of the iris may even come together to leave one, two, or in the case of the gecko, several tiny apertures. Such pupils may be horizontal as in the dogfish or vertical as in cats. Presumably the narrow cracks allow the formation of pin-hole ('stenopaic') images independently of the use of the lens. In lower vertebrates the rods may also be protected by appropriate migration of pigment around them and by photomechanical movements of the 'myoid' segments of the rods themselves, pushing the sensitive outer segment into the pigment layer. It is not certain whether such movements also occur in mammals.

The sensitivity of the eye at low intensities is increased in many animals by a reflecting or diffusing arrangement, the *tapetum lucidum*. This usually lies in the choroid but may be formed in various ways, for instance, from a layer of connective tissue shining as does the tissue of a tendon (ungulates), or from a special layer of cells filled with granules of guanin, as in carnivores, prosimians, and especially in elasmobranch fishes. In the semi-diurnal *Mustelus* and other sharks the guanin can be covered or uncovered by movement of pigment cells of the choroid. This device may be used by animals living in dimly lit situations, as well as by those that come out at night. Thus there are tapeta in the eyes of seals and of whales and they have been developed inde-

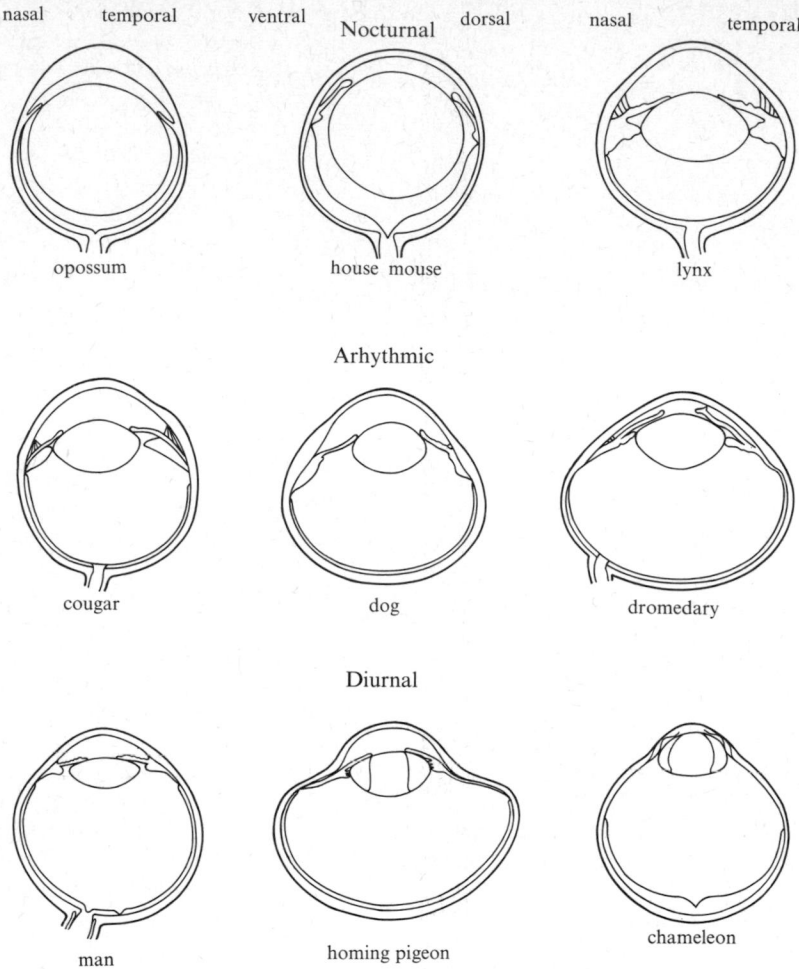

nasal temporal   ventral   Nocturnal   dorsal   nasal   temporal

opossum   house mouse   lynx

Arhythmic

cougar   dog   dromedary

Diurnal

man   homing pigeon   chameleon

FIG. 42.5. Diagrams showing the relations between intra-ocular proportions and environmental light intensity. (After Walls 1942.)

pendently in various species of fish living in the turbid shallow waters of lakes.

11. Special adaptations for diurnal vision

The eyes of diurnal animals are mostly arranged to allow fine discrimination at the expense of a high level of sensitivity; they are cone rather than rod eyes. The high acuity is differently used according to whether the animal is insectivorous (swallows), or feeds on seeds (many birds), or is a predator, such as the hawks, or a shy herbivore, like the ungulates, which use their large eyes to keep a look-out for enemies. Animals that move swiftly also have large eyes, allowing high resolution for the avoidance of obstacles.

Eyes used during the day-time show various characteristics that appear to be associated with their use for fine discrimination. They are often large, allowing a large retinal image. In small diurnal animals this may make the eye relatively large (Fig. 42.5) but it is the absolute size that is important and large animals may have relatively small eyes ('Haller's Law'), which are yet very effective, as in the horse. The cornea–lens system usually lies farther from the retina in diurnal than in nocturnal eyes, allowing for a larger image (Fig. 42.5). Correlated with this, the cornea and lens are less sharply curved in diurnal animals, giving the necessary longer focal length. The pupil is often small in the eyes of purely diurnal animals.

In the retina it is found that the proportion of cones to rods is relatively higher the more completely the animals keep to day vision. Completely rod-free eyes are found in lizards and snakes, many birds, and squirrels. The presence of rods would, of course, be a disadvantage if it interferes with the photopic system by interposing blind spots between the cones. Human sensations in very bright light show the disadvantage of possessing a high-sensitivity system in day-time and it is not surprising that sun-loving reptiles, such as lizards and snakes, possess only the less sensitive cones. Presumably it is partly for this reason that where, as in man, the eye is used both by day and night, the rods occur at the periphery and there is a central area of cones.

The effect of the blind spot might well be serious in the absence of binocular vision, which ensures in man that no part of the visual field falls on both blind spots. In squirrels and some fishes the optic tract leaves the eyeball not at one point but as an elongated line, so that the blind spot is a thin horizontal stripe along the upper border of the eye and does not interfere with vertical discrimination of the approach of enemies from above (Fig. 42.6). Nocturnal squirrels have the more usual round blind spot. In birds the blind spot is associated with a vascular membrane, the pecten, but the method of working of the arrangement is uncertain (see Young, 1962, p. 487).

The central area is developed to varying degrees. The depression (fovea) is not always present and it has been suggested that since the vitreous humour and the retina differ in refractive index the steep curvature of the walls of the foveal pit may produce a magnification of the image in some reptiles and birds, such as the swallows, and some primates (marmosets). More probably, as Pumphrey suggests, the aberration produced by this curvature gives changes in the apparent size or other characteristic of an image as it crosses the fovea, thus allowing fixation and assisting in providing the high visual acuity found in these animals (Young, 1962, p. 485). The presence of a central area of special visual cells does not necessarily mean that the animal is diurnal. In crocodiles there is a patch of cells probably of high sensitivity rather than high acuity.

The yellow pigmentation that gives the human central area its name of *macula lutea* is probably a colour filter, providing an increase of visual acuity by eliminating chromatic aberration. The images of objects reflecting the various parts of the spectrum occur at different levels and although the great length of the central cones reduces the disadvantages of this effect, a yellow filter, cutting out especially the violet end of the spectrum, should act to reduce the blurring. Many birds have yellow or orange oil droplets in the cones, and the cornea and lens are distinctly yellow in some diurnal animals (especially in squirrels and to some degree in man).

Colour filters in the eye may also serve to increase contrast in important regions of the spectrum, for instance, yellow filters by eliminating the blue element would assist in distinguishing between various greens. In pigeons the lower part of the retina has yellow droplets, increasing the contrast of objects seen against the sky, whereas the upper part, with mostly red droplets, is suited to see objects against the green of the world beneath the bird. The red droplets in the cones of turtles may serve to reduce glare and it is perhaps significant that they are also abundant in the retina of kingfishers.

It is difficult to be sure of the exact significance of these as of many other details of the optic system, but it can hardly be an accident that the lenses of squirrels and some primates are yellow, that pigment granules are found in the sharp-sighted birds, and that colour

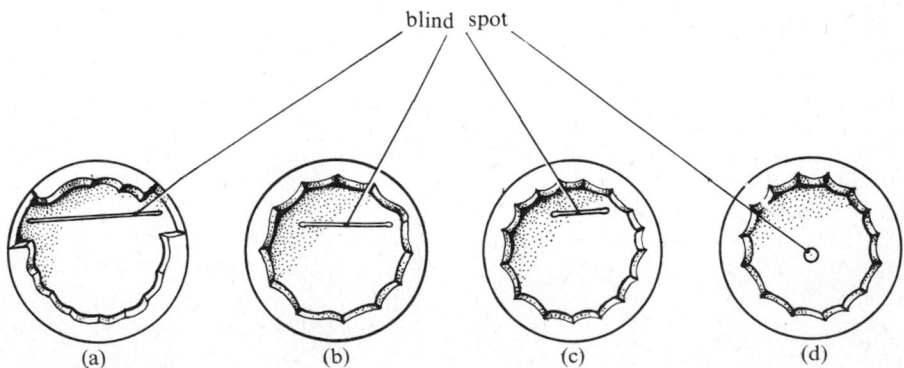

FIG. 42.6. Diagrams showing the blind spot in various members of the squirrel family. (a) Prairie-dog, *Cynomys* (inhabits open spaces, very bright light); (b) wood-chuck, *Marmota* (inhabits less bright spaces); (c) grey squirrel, *Sciurus* (inhabits dense woods); (d) flying squirrel, *Glaucomys* (nocturnal, with a nearly pure-rod retina). The drawings are not to the same scale and the anterior segments have been cut away. (After Walls 1942.)

filters are completely absent from the mainly nocturnal or crepuscular ungulates and carnivores. The yellow pigment of the human macula has a significant effect in absorbing blue light (when it is especially dense it may make its possessor blue blind). Riflemen and others have long known that the use of extra-ocular yellow filters produces a significant increase in visual acuity.

Thus the eye shows a great range of variation in different animals and many of the variants are correlated with the habits. Even after discounting all the dangers of speculation on such themes it can hardly be denied that the structure of the eyes is appropriate to the way in which they are used, by whatever process this result may have been achieved.

43 Hearing and receptors for orientation in space

1. Hearing and orientation in space

THROUGHOUT the vertebrates the apparatus of the inner ear shows a remarkably constant organization, providing receptors that respond to slight mechanical deformation and give information (a) about the position of the head in relation to gravity; (b) on the angular acceleration of movement of the head in various directions; and (c) on the direction and frequency of incident sound-waves (de Reuck and Knight 1967; Eldridge and Millar 1971).

The first two functions, mediated by the utricle and saccule and the semicircular canals, are universal in vertebrates; the third function, hearing, occurs from fishes to mammals but is absent in some animals, for instance tortoises. The connexion of gravity receptors with those for hearing depends on the fact that both rely on detection of mechanical deformation by small forces through their effects on the movement of fine hairs. Yet the information is used in very different ways. The impulses originating in the gravity receptors serve mainly for reflex control of posture. Changes in the pattern of incident sound-waves provide signals for the occurrence of a wide variety of events associated with the animal's food, enemies, and mates. Full use of this information involves its assessment by the elaborate computing system of the higher nerve centres, therefore responses to sounds vary with past experience and are not always forecastable for all individuals of a species.

2. Parts of the ear

The mammalian inner ear consists of three distinct receptor systems (Fig. 43.1): (a) the semicircular canals and their cristae, giving response to rotational movements of the head in space; (b) the *maculae* of the utricle and saccule, providing tonic gravitational receptors; and (c) the *organ of Corti* for measuring direction,

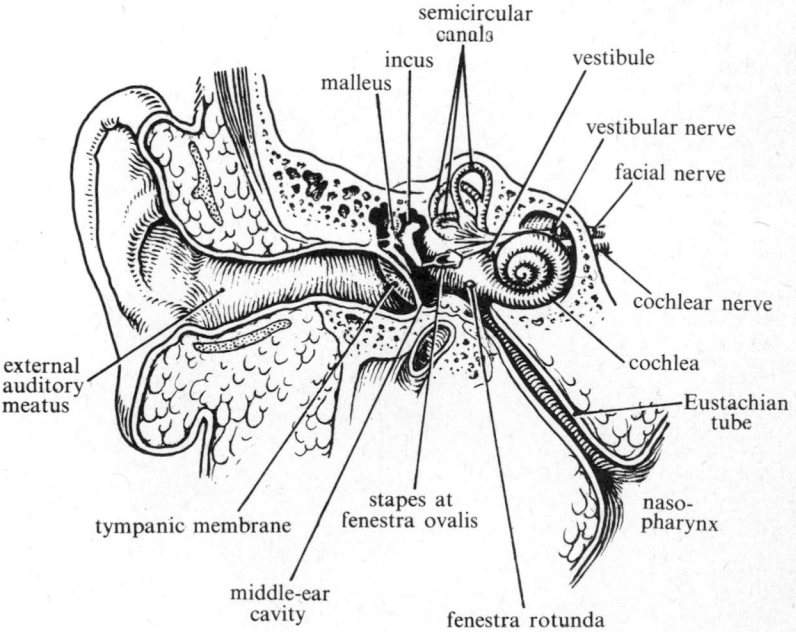

FIG. 43.1. Composite drawing of the ear of man.

intensity, and frequency of incident sound-waves between limits, in man, of about 20 Hz and 15 000 Hz. All of these receptors are formed by division of an original ectodermal sac, the otic vesicle. This becomes embedded in the otic bone (petrous portion of the temporal bone as it is known in man). The reception of sound-waves is mediated by the outer and middle ears and the tympanum, parts not directly connected with the otic vesicle but developed from the first branchial (hyoidean) gill slit (p. 161). The tympanic membrane stretched across the canal marks the outer end of the middle ear. The region external to the tympanum is the outer ear, terminating in the pinna, which is variously adapted to reflect sound-waves and to assist in the localization of sound.

3. The middle ear

The middle ear communicates with the pharynx by the *Eustachian tube* (pharyngo-tympanic tube). This is normally closed but is opened during swallowing by the tensor palati muscle allowing equalization of pressure on the two sides of the ear drum.

The *tympanic membrane* consists of a thin core of connective tissue, covered externally by skin of the outer ear, internally by squamous epithelium of the middle ear. It is not round and its shape is such as to make it nearly aperiodic, that is to say, to reduce its tendency to vibrate especially readily at certain frequencies. Vibrations are transmitted from the large ear drum (85 mm^2) to the inner ear by the three *auditory ossicles*, *malleus* resting on the drum, *incus* in the middle, and *stapes* touching a membrane covering a small hole, the *fenestra ovalis* (oval window) (3·2 mm^2)

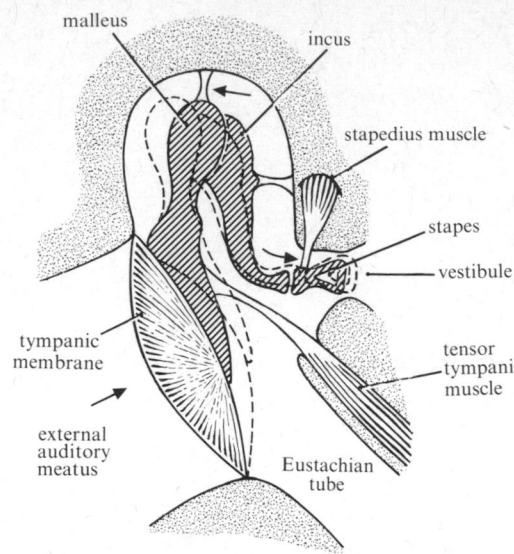

Fig. 43.2. Diagram of middle ear showing the ossicles at rest and after inward displacement of the tympanic membrane. The superior of the three malleus ligaments and the ligament of the incus are also shown.

in the petrous temporal bone. The ossicles, which are supported by tiny suspensory ligaments, are connected by diarthrodial joints, and the whole system ensures that every inward movement of the centre of the tympanic membrane produces a movement of about one-third the extent in the foot plate of the stapes (Fig. 43.2). The pressure at the stapes is twenty-two times larger than at the drum. A similar reduction apparatus is used in man-made recording devices

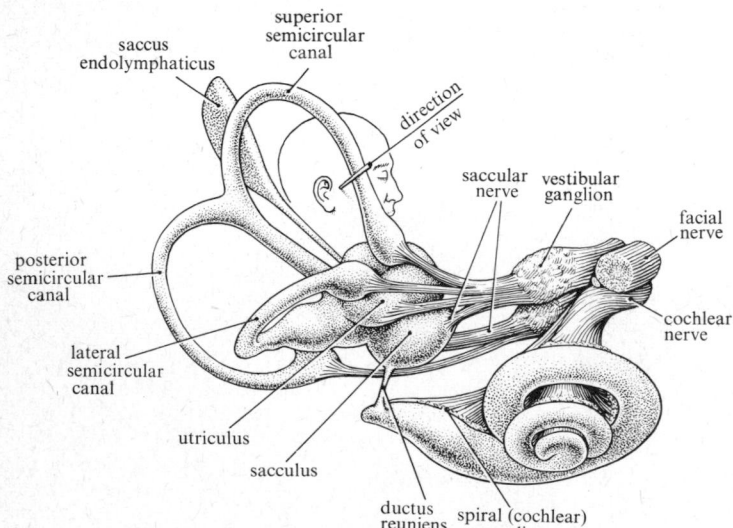

Fig. 43.3. Distribution of the auditory nerve to the membranous labyrinth. (After Hardy (1934). *Anat. Rec.* **59.**)

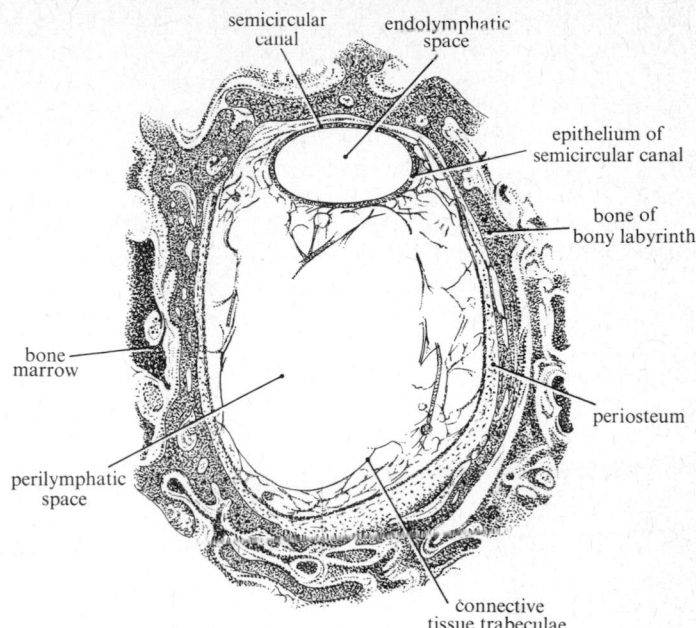

semicircular canal

endolymphatic space

epithelium of semicircular canal

bone of bony labyrinth

periosteum

connective tissue trabeculae

bone marrow

perilymphatic space

Fig. 43.4. Transverse section of adult human lateral semicircular canal.

whenever a lightly damped membrane activated by small forces and large displacements of the air has to produce larger forces and small displacements in a more heavily damped liquid system (impedance matching) (see Naftalin 1965).

There are two muscles in the middle ear, the *tensor tympani*, which pulls the drum inwards, and the *stapedius*, which pulls the stapes away from the oval window. These muscles show a reflex contraction when sound falls on the ear and the increased tautness they

produce provides protection against loud sounds by decreasing the amplitude of the vibrating systems. Their action is thus analogous to that of the iris of the eye.

4. The inner ear

The otic sac forms a series of cavities, the membranous labyrinth, containing receptors for the separate functions performed (Fig. 43.3). There are two sacs, the *utricle* and *saccule*; three *semicircular canals*, set in

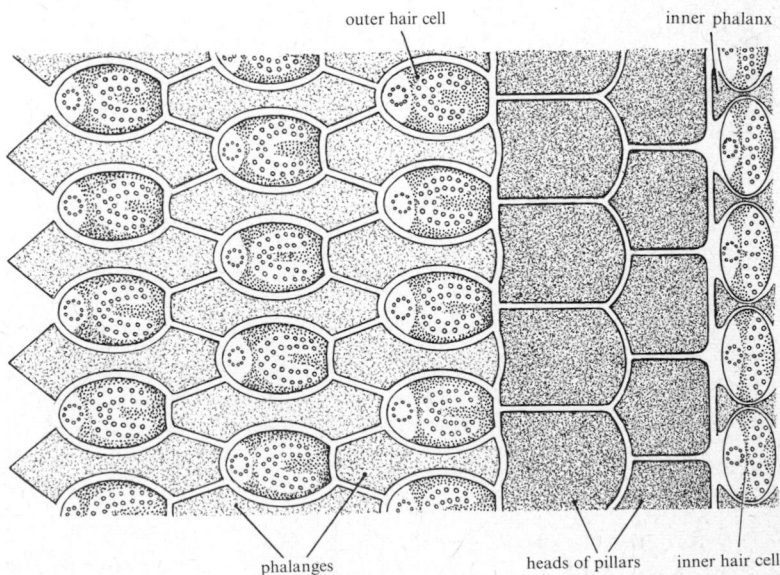

outer hair cell

inner phalanx

Fig. 43.5. Diagrammatic surface view of the organ of Corti. The centrioles of the hair cells face the stria vascularis. (After Wersäll, Flock, and Lundquist (1965). *Cold Spring Harbor Symp. quant. Biol.* **30.**)

phalanges

heads of pillars

inner hair cell

different planes; and a coiled tube, the cochlea. The sensory passages derived from the otic vesicle contain a fluid, the *endolymph*. They are surrounded by a series of spaces in the mesodermal tissues, which separate the labyrinth from the bone. The membranous labyrinth is therefore in contact with the periosteum of the bone only at a few places (Fig. 43.4). For the most part it is surrounded by quite large irregular cavities communicating with the subarachnoid spaces by a perilymph duct. These cavities are crossed by trabeculae connected with the periosteum and filled with a fluid, the *perilymph*.

Altogether there are six sensory areas in the labyrinth. Three of these, the *cristae ampullares*, occupy swellings one at one end of each of the three semicircular canals. The organs of static or tonic balance are the two *maculae*, one each in the utricle and the saccule. Finally, the auditory organ is a long ridge running up one side of a coiled tube, the *cochlea*.

5. The hair cells and their nerves

The various organs of the inner ear are innervated by the eighth nerve (auditory or vestibulo-cochlear). This is essentially a dorsal root, part of the facial nerve, modified to innervate the invaginated portion of the skin that makes the otic vesicle. As in other dorsal roots the cell bodies lie outside the central nervous system, forming the vestibular and cochlear ganglia. The peripheral branches of these nerve cells run to sensitive patches in the inner ear, where they end in contact with specially modified *hair cells* (Fig. 43.5). These are ectodermal cells whose modified cilia are the mechano-receptors that follow the motions of the fluid endolymph.

The finest afferent branches of the eighth nerve fibres end as swellings pressed against the bases of the hair cells. These endings contain few or no vesicles. There are often synaptic vesicles within the hair cells and they are arranged around a characteristic flat ribbon (Fig.

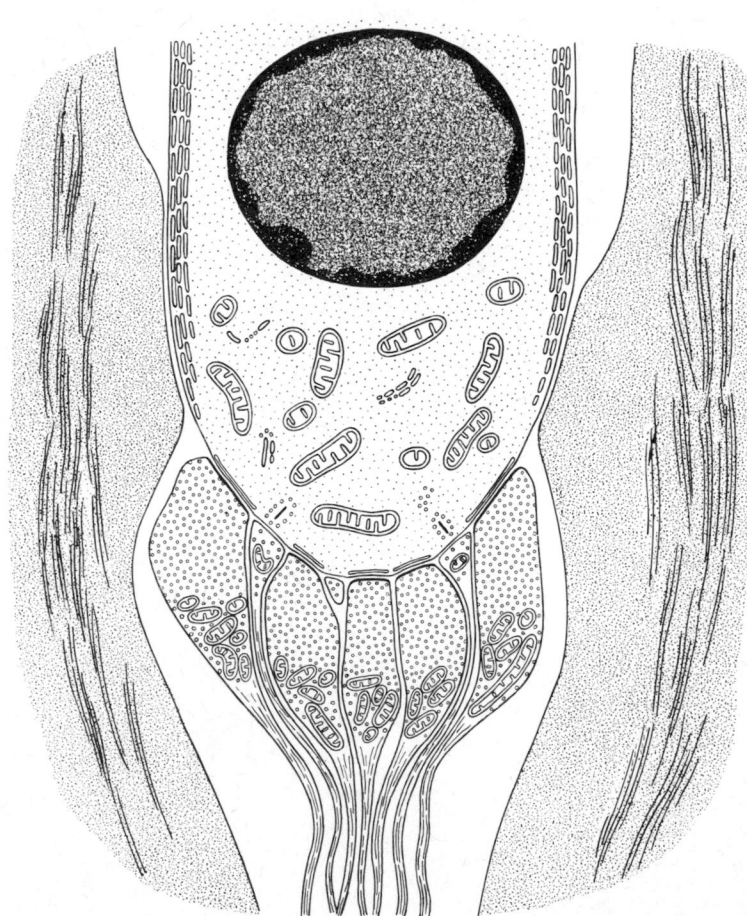

FIG. 43.6. Diagrammatic view of basal part of an outer hair cell with supporting Deiters' cells seen by electron microscopy. Two types of synapses are present. Cochlear nerve endings have associated synaptic ribbons, and the efferent endings contain many synaptic vesicles. (After Smith and Sjöstrand (1961). *J. Ultrastruct. Res.* **5**.)

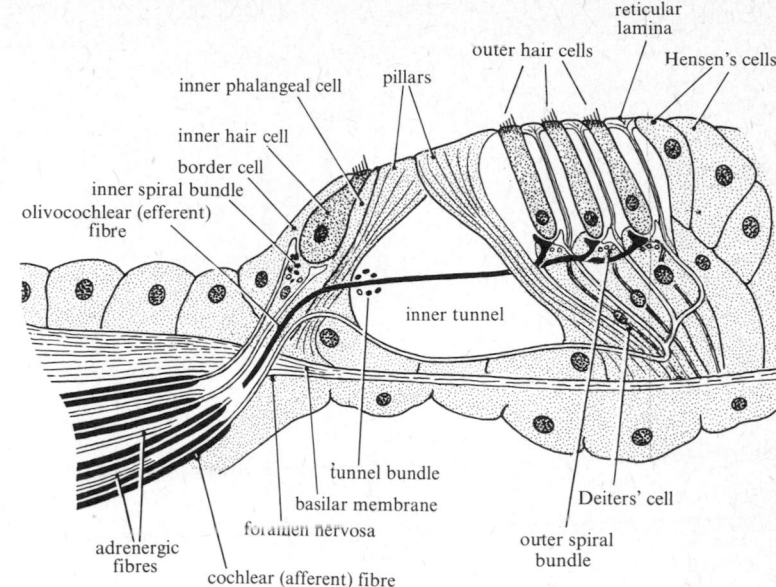

FIG. 43.7. Diagrammatic transverse section of organ of Corti. The cochlear nerve provides the afferent innervation (*white*). Myelination ends at the basilar membrane. Some fibres run straight to the inner hair cells, others take spiral courses in the inner spiral bundle before termination and the remainder terminate ultimately on the outer hair cells. The efferent innervation (*black*) derives from the olivo-cochlear nerve. No definite terminations on the inner hair cells have been found; most fibres join the inner spiral and tunnel bundles and cross the tunnel to terminate on the outer hair cells. (After Spoendlin 1966.)

43.6) similar to that in the bases of the retinal rods and cones (p. 381). The hair cell probably excites the nerve ending by chemical means, but the transmitter is not known for certain. There is a concentration of acetylcholinesterase near some of these endings. In addition to these afferent synapses many of the hair cells also bear the endings of efferent nerve fibres (Fig. 43.7). At these endings the efferent nerve fibres are presynaptic to the hair cells and sometimes also to the afferent nerve fibres. The terminals of the efferents contain synaptic vesicles.

6. The vestibular apparatus

This provides an important part of the information by which the balance and posture of the body is maintained (p. 301). About 19 000 nerve fibres supply its receptor organs in man and carry impulses to the vestibular nuclei in the medulla (Fig. 31.4, p. 293). From here impulses pass to spinal motor and oculomotor centres, to the cerebellum, and probably also to autonomic centres and to the cerebral cortex too. The influence of the vestibular receptors is thus very widespread but probably operates mostly along high-probability channels, fixed by heredity. The relatively small number of vestibular nerve fibres allows the transmission only of small amounts of information, producing similar effects in all members of the species (see Wersäll and Flock 1965).

7. The semicircular canals

The three semicircular canals are tubes forming almost complete circles and set approximately at right angles

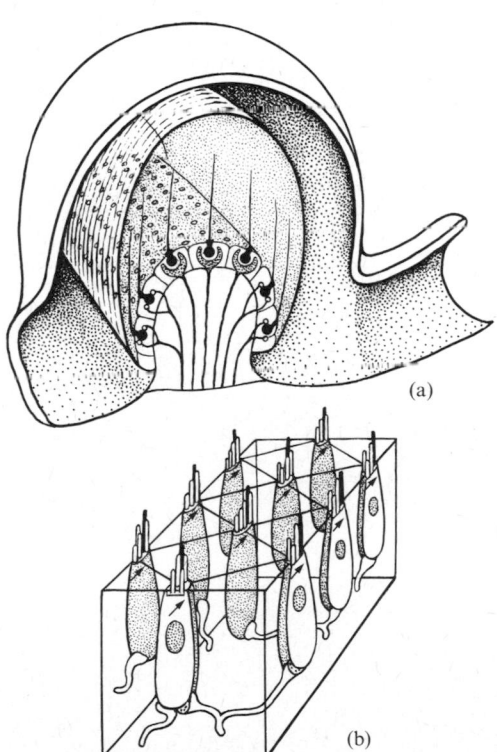

FIG. 43.8. (a) Diagram of crista ampullaris of the vertical canal of *Raja clavata* with (b) schematic enlargement of receptor cells in sensory epithelium showing orientation of sensory hair bundles. The kinocilium (*black*) always occurs on the same side of the bundle and is orientated away from the utriculus in the vertical canals but towards it in the horizontal canal. Cupula omitted from (b). (After Flock and Wersäll (1962). *J. Cell Biol.* **15**.)

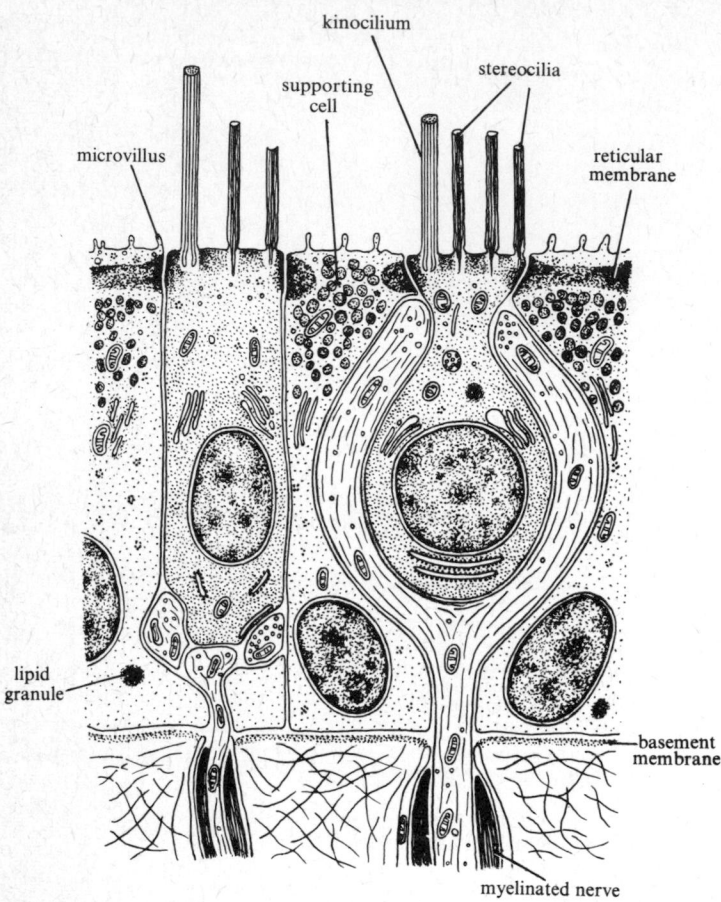

FIG. 43.9. Section through the sensory epithelium of the crista ampullaris showing Type I (*left*) and Type II (*right*) hair cells. (After Wersäll (1965). *Acta Oto-lar. Suppl.* **126.**)

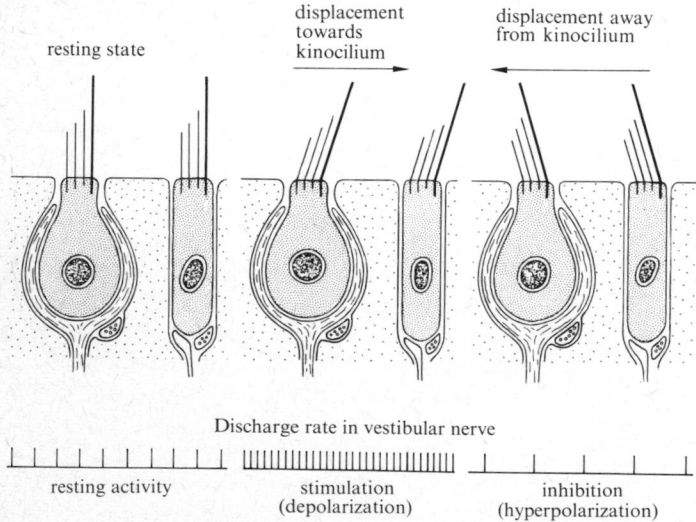

FIG. 43.10. Schematic illustration of the relation between hair-cell orientation and the discharge rate on displacement of the sensory hair. (After Wersäll, Gleisner, and Lundquist (1967). In *Myotatic, kinesthetic and vestibular mechanisms* (ed. de Reuck and Knight). Ciba Foundation Symposium. J. & A. Churchill, London.)

to each other (Fig. 43.3). They spring from a larger central sac, part of whose wall is evaginated to form the utricle. Each tube carries an ampulla at one end, and in this lies the sensory crista (ridge) (Fig. 43.8). This carries a number of sensory hair cells and supporting cells (Fig. 43.9). The hair cells are embedded in a gelatinous material, the *cupula*. When fluid moves in the canal it moves the cupula and this activates the hair cells and alters the discharge of impulses by the vestibular nerve fibres that end at their bases. The hairs

on each cell are of two sorts, several non-motile stereocilia and a single motile kinocilium (Fig. 43.8). The groups of hairs are similarly orientated on all the cells and the response of the hair is determined by the direction from which the stimulus approaches the bundle of hairs (Fig. 43.10). Displacement of the cupula in the direction from the stereocilia towards the kinocilium leads to a drop of d.c. potential across the hair cell (p. 407) and an increased discharge of the afferent nerve fibre and vice versa for displacement in

FIG. 43.11. Differentiation and polarization in the macula sacculi. (a) The macula is shown in its natural vertical orientation. The largest crystals of the statoconial membrane are in the antero-inferior part and the smallest at the striola. The underlying sensory cells are polarized (*arrows*) away from the line dividing the pars interna from the pars externa. (b) Transverse section showing variation in crystal layer thickness and crystal size, distribution of type I and type II cells, and polarization of sensory cells in middle of striola. The arrangement ensures a constant flow of impulses from the cells of lower part, which are polarized to respond to movement downwards, and are responsible for postural reflexes. (After Lindeman (1969). *Ergeb. Anat. EntwGesch.* **42.**)

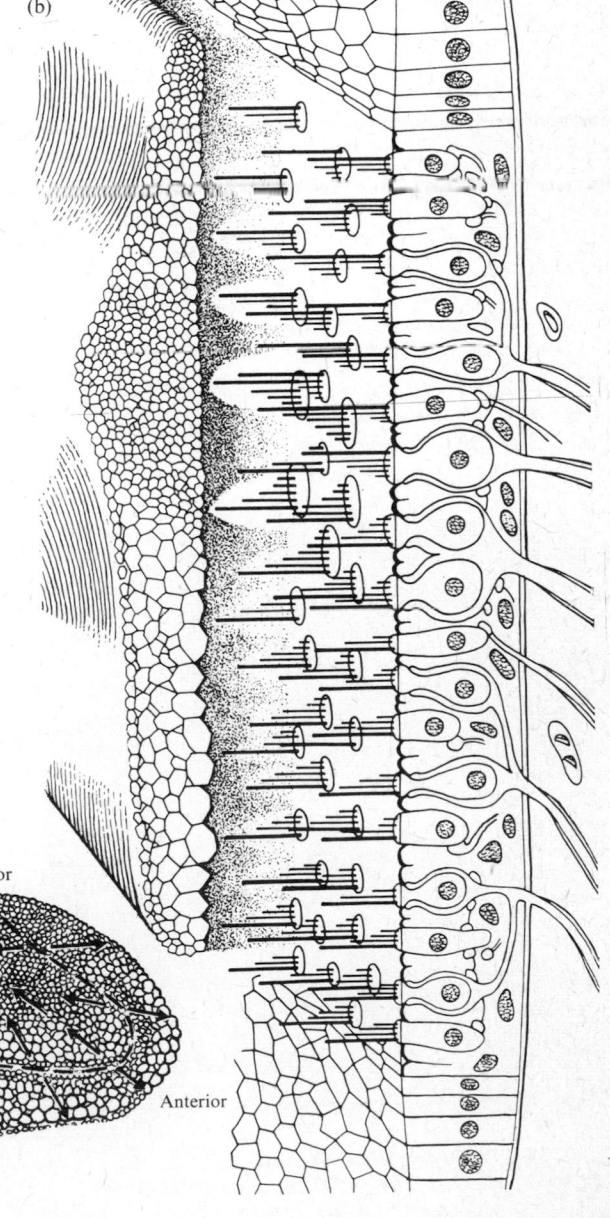

(b)

(a)

Superior

Posterior

Anterior

Inferior

striola

the opposite direction. The nerve fibres probably send impulses continually to the vestibular nuclei of the brain (Fig. 31.4, p. 293), and the information provided by the changes in rates of discharge of the fibres from the ampullae of the three canals produces movements of the neck and back, limbs, and eye muscles to compensate for angular rotations. The system is very sensitive; it is calculated that in fish (where the movements of the cupula can be watched directly) a change of pressure of 0·05 mm water is sufficient to produce movement. Another way of testing the sensitivity is to record the minimum head movement that gives rise to compensating movements of the eyes. In rabbits movement has been found to follow angular accelerations of as little as 0·09 degrees per second per second.

8. Utricle and saccule

The utricle communicates with the saccule by a narrow duct, from which springs a long tube, the *ductus endolymphaticus,* representing the original canal connecting the otic vesicle with the surface during development. The maculae of the utricle and saccule consist of patches of sensory epithelium, containing, like the ampullary cristae, supporting cells and hair cells. The hairs are not actively motile but embedded in a gelatinous membrane bearing many tiny calcium carbonate

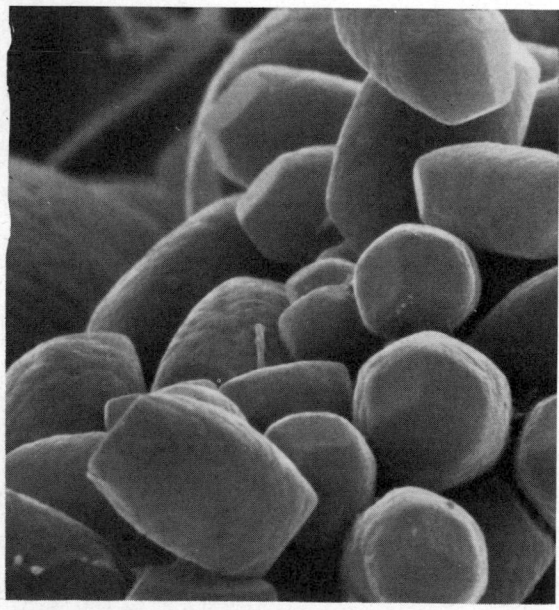

FIG. 43.12. Scanning electron micrograph of human otoconia. Field width 12·5 μm. (Figure kindly supplied by Dr. D. J. Lim.)

crystals, the *otoliths* or ear dust (Figs 43.11 and 43.12) (Lim 1969; Lindeman 1969). The membranes are thus weighted and since the two maculae are placed vertically and at right-angles to each other the hair cells are pulled upon in a characteristic manner for each position of the head. The receptor organs of the maculae therefore provide steady streams of impulses, varied in pattern for each position and providing the information by which the motor nerve centres maintain the muscles in an appropriate state. The maculae are thus organs for tonic, the cristae for phasic, maintenance of posture. There is also some evidence that the saccule is an auditory receptor for vibrations of low frequency.

9. The cochlea

The auditory receptor of man lies in a tube leading from the saccule and is coiled in a spiral with $2\frac{1}{4}$ turns, the *ductus cochlearis* or scala media (Fig. 43.13). This narrow tube, containing a viscous fluid, the *endolymph,* is enclosed in a much wider tube of *perilymph,* the whole making the coils of the cochlea. The cochlear duct does not lie at the centre of the perilymph space but is attached to one side and rests on a partition, the spiral lamina, which divides the whole perilymphatic space into two, a lower scala tympani and upper scala vestibuli.

The vibrations of the stapes, the innermost of the three ossicles, are not transmitted direct to the cochlear duct but to the perilymphatic fluid at the base of the scala vestibuli. Here the outer wall of the labyrinth comes into contact with a hole in the bony wall, the *fenestra ovalis,* into which the end of the stapes fits. The fluid in the vestibular scala is thus set into vibration. The whole cochlea is enclosed in bone, except at the lower end of the scala tympani, where there is a second hole in the petrous temporal bone, the *fenestra rotunda.* Every inward movement of the membrane covering the oval window must therefore be followed by an outward movement of the round one and the fluid system between these two is so arranged that its vibrations set in motion the *spiral lamina,* the membrane on which the sensory hair cells of the cochlea are placed. The scala vestibuli communicates at the tip of the coil with the scala tympani by a minute pore, the helicotrema, but this is too small to allow movements of the frequency of auditory vibrations to be transmitted through the fluid. Accordingly every time the pressure rises in the scala vestibuli the membranes separating this cavity from the scala tympani are moved, including the spiral lamina, which forms the base of the cochlear duct.

The lamina is of complicated structure and includes the *basilar membrane,* a strip of tissue about 3 cm long, containing a number of *auditory strings* or basilar

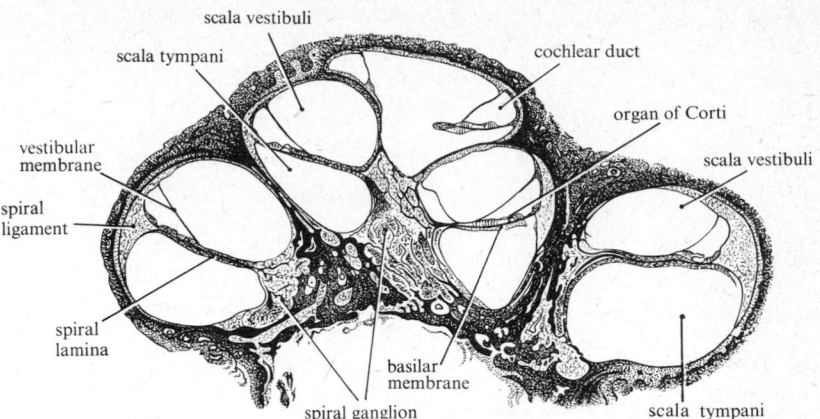

scala vestibuli

scala tympani

cochlear duct

organ of Corti

scala vestibuli

vestibular
membrane

spiral
ligament

spiral
lamina

basilar
membrane

spiral ganglion

scala tympani

FIG. 43.13. Section of human
cochlea.

fibres, placed transversely to the axis of the cochlea. On these rests the sensory ridge, forming part of the lining of the inner cochlear duct itself. This sensory ridge, the *organ of Corti*, consists of supporting cells and sensory hair cells (Fig. 43.14). The supporting cells are of several sorts, including *inner* and *outer pillars* and *Deiters' cells*; the two former are filled with filaments (not collagenous) and they enclose a space, the tunnel of Corti. The sensory hair cells are correspondingly divided into a row of inner hair cells and three to five rows of outer hair cells. The Deiters' cells lie between the hair cells and also contain filaments. There are about 25 000 hair cells in each cochlea in man (Engstrom, Ades, and Anderson 1966; Iurato 1967; Spoendlin 1966).

The *sensory hair cells* all have the same basic organization. Each has an apical zone carrying up to 150 non-motile stereocilia, arranged in a regular pattern. There is no kinocilium but a basal body is present. Below this is an intermediate zone containing much endoplasmic reticulum and mitochondria, and a nucleus. Finally the basal zone makes contact with afferent nerve fibres and here there are synaptic vesicles. There are 30 000 afferent nerve fibres in the cochlear nerve (cat). Only 5000 of these go to the outer hair cells, each sending branches to many receptors. The remaining 25 000 innervate the inner hair cells, each going to two or three hair cells. All nerve fibres are unmyelinated within the organ of Corti, but acquire sheaths outside it.

There are also some 500 efferent fibres in the nerve, the olivo-cochlear bundle of Rasmussen (p. 336). These end in synapses making contact with the bases of the outer hair cells themselves but only with the terminal nerve fibres of the inner hair cells. These efferent fibres contain abundant cholinesterase.

In addition, it has been shown by fluorescence microscopy that there are adrenergic nerve fibres to the cochlea. These do not extend beyond the habenula perforata, which is the level at which the afferent fibres acquire myelin sheaths. Some of these adrenergic fibres are vascular efferents and degenerate after removal of the superior cervical ganglion. However, others degenerate only after section of the cochlear nerve and therefore presumably originate in the brain. They may influence the setting up of nerve impulses in the afferent fibres at the point where the latter acquire their sheaths (Spoendlin 1966).

The hairs are embedded in a gelatinous *tectorial membrane* and the sensitivity to sound depends upon the shearing movements between the two. The hairs are moved by the basilar membrane, which, according to the resonance theory of Helmholtz (1863) consists of a series of strings arranged like piano wires. If they resonate according to the incident sound wave they would excite the hair cells by pressing them against the tectorial membrane or pulling them away from it. The basilar membrane has indeed some 24 000 transverse fibres, increasing from 60 μm in length near the base to 500 μm in length at the tip of the spiral. Injuries at the base impair hearing of high notes and vice versa. At the apical end there are about 470 strings per octave and the accuracy of discrimination suggests about $2\frac{1}{2}$ strings for each interval. However the 'strings' are not under tension and it is a variation in stiffness along the membrane that allows differentiation of tones.

The actual movements in the cochlea have been observed by Békésy by sprinkling fine particles on the membrane and observing stroboscopically with a microscope. This only partially confirms the resonance theory. Because of the variation in stiffness, any mechanical disturbance sets up a travelling wave moving at decreasing velocity from base to apex. Low frequency disturbances travel further than high and the maximum amplitude shifts towards the apex as frequency de-

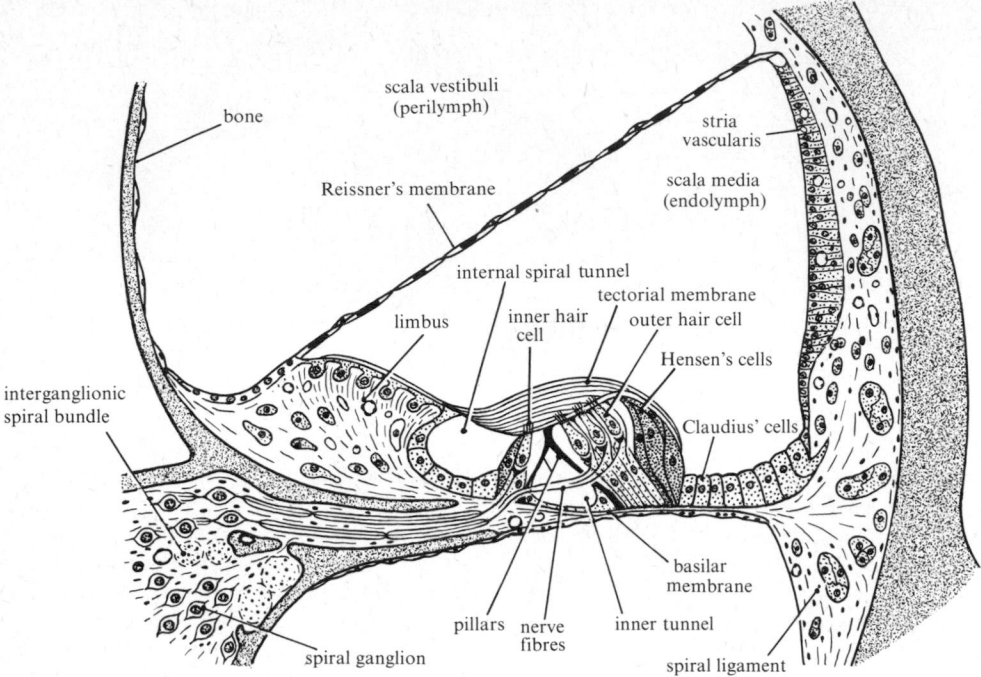

FIG. 43.14. Diagrammatic section of one turn of mammalian cochlea. (After Davis 1965.)

creases. Even at high frequences a large number of hair cells are involved, not a few specific ones as in a simple resonance theory. By studying the capacity of cats to distinguish tones after cochlear lesions it has been estimated that about 3000 hair cells are activated by a tone of 1000 Hz at 40 dB and 2940 of these are also activated by the adjacent distinguishable tone (see p. 334).

It is still uncertain precisely how the stimulation of the hair cells occurs. Transducers can be either passive if the impinging energy is simply transferred into, say, electrical energy, or active if the transducer receives its power from some other source so that the output energy is controlled by the entering energy but does not depend upon it. Greater sensitivity can be achieved in this way and there is reason to think that this is how the ear operates. A possible source of energy is a remarkable positive potential of 80 mV between the cochlear duct and the perilymph. This potential is maintained by a specialized tissue, the *stria vascularis*, running along the cochlea duct (Fig. 43.14). This secretes the endolymph, which is unique among extracellular fluids in containing a very high concentration of potassium and a low concentration of sodium. This is presumably connected with the positive potential (Fig. 43.15) (Davis 1970).

The input energy to the ear at the threshold of hearing is exceedingly small. At 1000 Hz it is calculated to be the equivalent of a pressure of 2×10^{-9} N cm^{-2}. This is far too small to move even a part of the basilar membrane through any distance greater than atomic dimensions, moreover it is less than the energy of Brownian movement, which would drown any signals in noise. The solution of this problem is not known but may lie in the relation of the hair cells to the tectorial membrane. The latter is a very watery substance (90 per cent) containing protein and calcium and other ions. It is suggested that as a result of bending of the hairs the conductivity of the hair-cell membrane may be altered and thus the current flow powered by the endolymphatic potential of $+80$ mV is modulated. However, this is not generally agreed and another suggestion is that the tectorial membrane contains metallo-protein or mucopolysaccharide compounds so organized as to produce a piezo-electric effect. By whatever means, it seems clear that at threshold amplitudes of movement of atomic dimensions or less cause a change in ohmic resistance proportional to amplitude (or pressure). This causes a change in leakage current so that the energy of the signal is transduced and amplified.

During auditory stimulation a series of a.c. potential

changes can be recorded, the cochlear microphonics. There is good evidence that these potentials depend upon the hair cells (they disappear if the latter are destroyed) and they may well be produced by current flow following changes in membrane resistance produced at the contact between the tectorial membrane and the stereocilia driven by the sum of the endolymphatic potential and the intracellular potential of the hair cell ($+60$ mV). Such currents may serve to excite the nerve endings by release of chemical transmitter. Generator potentials of up to 20 mV have indeed been found by intracellular recordings. These potentials would summate to set up action potentials in the auditory nerve fibres probably at the place where they become myelinated. The 5000 fibres, each innervating many outer hair cells, have a lower threshold than the 25 000 each innervating one or a few inner hair cells and allowing for finer discrimination.

There is extensive feed-back throughout. Stimulation of the olivo-cochlear fibres raises the auditory threshold potential and attenuates the cochlear microphonics. These fibres fire continuously and must play a large part in regulating audition. There are extensive feed-backs between the cerebral auditory centres.

Small differences in sound patterns are faithfully transmitted over considerable distances and are therefore used by many animal species as the clues for social and sexual behaviour. Fishes and birds use the auditory system in this way and the recognition of speech patterns provides the main basis for the detailed human communication system. Little is known of the means by which the fine differences are discriminated either by the peripheral or central apparatus (p. 335). The number of nerve fibres in the cochlear nerve is not high, considering the wide range of distinct actions they allow.

10. Echolocation

Echolocation has been shown to occur in bats, toothed whales, shrews, and some birds. The emitted sounds are produced by the larynx in bats, probably clicks amplified by the air sacks of the head ('melon') in whales, and as clicks of the tongue in shrews and birds.

For echolocation, high frequencies and wide band width give good resolution of direction and range. In some species of bats the larynx produces short (1–2 ms) frequency-modulated pulses of pure tones at up to 100 kHz and 100–200 pulses per second. Others use pulses of 30–100 ms of constant wavelength and several harmonics. Other mammals may use high-frequency sounds, for instance young mice communicate with their mother by squeaks at up to 90 kHz.

In bats echolocation is the main source of information about the environment. It was shown by Spallanzani in 1793 that blinded bats can avoid fine wires, but not if their ears have been blocked. They can also detect the direction and distance of insects while in flight and can be trained to recognize an insect from objects of other shapes thrown into the air. To do this they must detect a faint echo, recognize it, determine its distance and direction in both vertical and horizontal axes, and also whether it is moving and in what direction: all this in spite of noise.

The external ears are often huge and complicated with a large tragus, and may be moved in synchrony with the pulses (in the horseshoe bats, Rhinolophidae). The tympanic membrane is small and the ossicles light

FIG. 43.15. Modified 'resistance microphone' theory of cochlear excitation with the primary 'battery' in the hair cell and an accessory battery in series in the stria vascularis. The stria vascularis has been positively identified as the generator of the positive potential of the scala media. The leakage current through the hair cells is assumed to be modulated by an increase and decrease of ohmic resistance due to bending of the cilia or to shearing forces applied to them. A change in leakage current is assumed to cause liberation of chemical mediator which excites the nerve ending (synapse). The electrical pathways through Hensen's cells and basilar membrane, and Reissner's membrane and limbus represent all the shunting pathways from scala media to scala tympani. (Figure kindly supplied by Dr. H. Davis.)

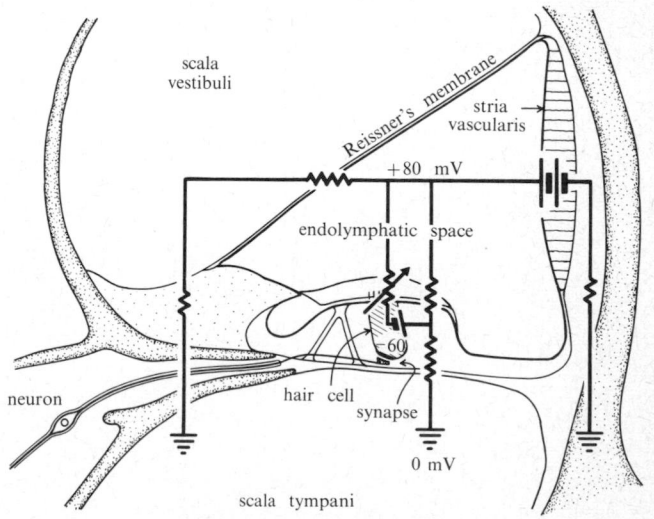

and tightly coupled. The middle-ear muscles are very large and contract about 5 ms before the emission of each pulse. This alters the cochlear microphonics, which attenuate as the sound is emitted and then recover over the next 10 ms. The bat is thus protected against the outgoing sound but ready to receive the echo (see Sales and Pye 1974).

The cochlea shows thickening of the basilar membrane in certain regions, mainly in the basal turn, but nearer the apex in species that use lower frequencies. These thickenings may therefore be responsible for narrowly tuned responses.

Analysis of the echoes seems to be mainly by the lower auditory centres of the brain. The cochlear nucleus, olivary complex, and inferior colliculus are large, but the cortical auditory areas not specially developed. The responses of cells of the inferior colliculus are narrowly tuned; each has a best frequency and best intensity. Bats make use of inhibitory interactions to restrict the response areas of individual units. There may be a change of 20 dB for a 1 per cent change in frequency and even at best frequency there is an optimum intensity, often high, with sharp gating. There may be changes in response with less than 0·5 dB change of intensity. There are units that respond only to constant wavelength pulses and not to frequency modulation or vice versa. Others respond in ways that depend on the direction, rate, and intensity of the frequency-swept pulse. Units may show a marked change with the direction of the sound, the threshold varying 9 dB per degree.

The units show remarkable temporal precision. Frequency can be discriminated on only 2–3 cycles. Recovery is complete in 2 ms. These capacities appear better in the collicular than the cochlear nucleus or auditory nerve. There seems to be a capacity to emphasize information about second or subsequent signals arriving at the third-order level.

Distance is presumably estimated by measuring the intervals of firing of neurons responding to both the outgoing and return sounds. Direction can be determined from only one or two echoes, probably by comparison of binaural intensities at several frequencies. By comparing those at three frequencies, preferably widely separated, the bat could theoretically determine the one point in space from which they would all have the observed binaural differences.

The special properties of these neurons provide an excellent example of how in each species the brain is adapted to receive and analyse the information that is relevant to its way of life. The capacities of whales for echolocation have been less fully investigated. Dolphins can discriminate in this way between spheres differing by only 10 per cent in diameter. They can avoid metal wires as little as 0·2 mm in diameter. The sounds reach the ear through the lower jaw and the air sacks in the 'melon' of the head.

44 Chemo-receptors

1. Chemical changes near and far

IT is of great importance for organisms to be able to identify the chemical nature of substances that they meet around them, distinguishing the valuable from the dangerous. Chemical cues are particularly useful for higher animals that have elaborate systems for computing associations and for making predictions; food, enemies, companions, and many other factors are associated with characteristic chemical conditions. It is significant that the cerebral cortex, the most elaborate and successful of all computers and predictors, arose from the part of the brain concerned with the discrimination of smells. Chemo-receptors are found in many animals and it is interesting that the substances to which they are sensitive are usually organic compounds such as are likely to be produced by other organisms. Elements and inorganic compounds are less often identified (see Ohloff and Thomas 1971; Pfaffmann 1969; Wolstenholme and Knight 1970).

The chemo-receptors of mammals are the organs of taste and smell, between which there is a sharp distinction anatomically and functionally. The sense of smell is in the main a distance chemo-receptor, whereas that of taste only comes into operation when substances are in close contact with the body, often only after they are in the mouth. Corresponding to this difference in distribution is a wide difference in biological functions, the sense of smell serves to detect clues that guide the animal to its food or mate. The sense of taste enables it to discriminate only between 'desirable' and 'undesirable' food. This explains why the smell system is connected to the elaborate forebrain computor and produces responses based upon memory, whereas the afferent impulses in taste fibres enter the medulla oblongata and are mainly connected directly with relatively simple motor systems.

Correspondingly there is also a great difference in thresholds. Quinine can be tasted at a molar concentration of 4×10^{-7}, sugar only at 2×10^{-2} M, but organic sulphur compounds, such as amyl thioether, can be smelled in solutions as dilute as $5 \cdot 8 \times 10^{-9}$ M. It is said that musk can be smelled when there is as little as 4×10^{-5} mg l^{-1} of air (Pfaffmann 1951).

The taste receptors allow for the collection of only a small amount of information, enabling us to distinguish the four qualities sweet, salt, acid, and bitter; whereas by smell an animal or man can recognize a great many substances, especially organic compounds. The 'tastes' of food are their smells as produced by the effect of small quantities reaching the olfactory epithelium from the mouth. Responses to the true taste receptors are produced with a high degree of probability over relatively simple pathways containing few fibres. The cortical centres for taste, if present at all, are small. Responses to smell, on the other hand, involving discrimination of many distinct situations, are largely learned. The number of olfactory receptors is very high and the central pathways are complicated.

2. Taste

Throughout the vertebrate series the capacities for smell and taste are always localized in separate regions. In man the receptors for taste are the papillae of the tongue, at the centre of each of which is a taste bud containing some 50 cells (Fig. 44.1). These are elongated epithelial cells which are continually renewed from basal cells; thus all stages of the one type may be seen. Labelling with tritiated thymidine, in a rat, shows that the cells live for about 10 days (see Beidler 1971b).

Each papilla is innervated by fibres from two sources; those of the lingual nerve mediate the sense of touch in the outer parts of the bud. Only some 50 unmyelinated fibres enter the central taste bud. These come through the chorda tympani branch of the facial nerve to the front and from the glosso-pharyngeal for the back of the tongue. Each nerve fibre innervates several taste buds, and makes contact with many cells within each bud. In spite of these connexions careful study shows that in man a single papilla suitably stimulated responds only to one of the four tastes. There are differences in distribution, the tip of the tongue being usually said to be more sensitive to sweet, the sides to salt and sour,

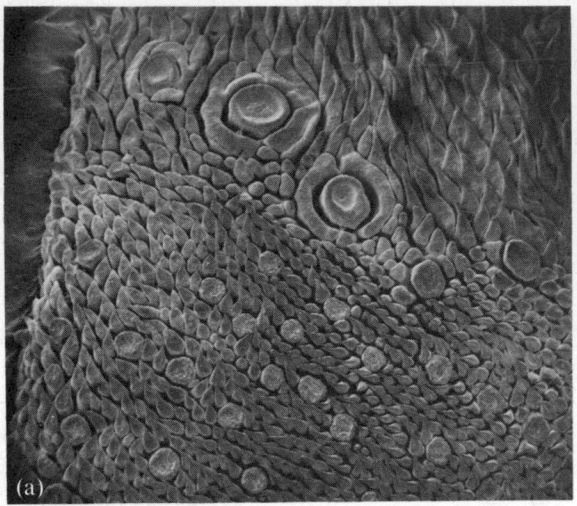

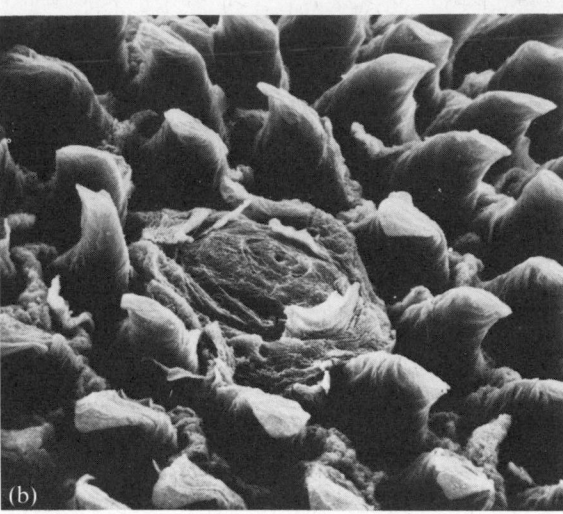

FIG. 44.1. Scanning electron micrographs of tongues. (a) Puppy. In addition to the tapering filiform papillae and fungiform papillae, large circumvallate papillae are present. Field width 4 mm. (b) Rat. Many filiform papillae surround a fungiform papilla at the centre of which is a pore indicating the presence of a taste bud. Field width 0·2 mm. (Figures kindly supplied by Dr. P. P. C. Graziadei.)

and the back to bitter. But all parts are sensitive to all four qualities.

Each taste cell carries numerous microvilli projecting into a pit filled with amorphous material at the centre of the bud (Fig. 44.2). This apical pole also contains numerous large secretory droplets. The base of the cell contains synaptic vesicles and is enclosed by branches of the nerve fibres.

Little is known about the transduction and coding processes in taste buds. There are presumably specific receptor substances and indeed a 'sweet' protein which complexes with sugar and saccharin has been isolated from ox tongues (see Beidler 1970). A 'bitter' protein also has been found. The active principle of 'miracle fruit' of West Africa makes sour substances taste sweet and is a protein that probably binds near the sweet site and stimulates it by a conformational change in the presence of acid.

It has been suggested that the sweetness of a molecule depends upon an AH—B grouping where AH is a proton donor and B an acceptor. AH and B groups may find complementary sites on the chemoreceptor surfaces whose properties allow hydrogen bonding (Shallenberger 1971). When the basic taste stimuli, sodium chloride, quinine, hydrochloric acid, or sucrose, are placed successively on the tongue most of the nerve fibres were found to respond to two or more of them (Fig. 44.3). Fibres responding to two of them were the most common, then those responding to one, next those responding to three, and only a few to all four.

It is not known whether this is due to multiple sensitivity of the individual transducers or to the branching of the nerve fibres innervating taste cells with different specific sensitivities (Pfaffman 1970). There is known to be multiple innervation and there is also evidence that each gustatory cell is part of the receptive field of more than one afferent fibre.

These overlaps are probably an essential feature of the mechanism for discrimination between different substances. Although only four taste qualities are recognized there are great differences in the sweetness of different sugars, and so on. Indeed there is really a continuous gustatory spectrum in some respects. There is certainly a close relation between the tastes of salt and sweet and many salts taste sweet at threshold concentrations. How the pattern of peripheral innervation and central action produces such discrimination is not known. It may perhaps show features similar to the discrimination of touch (p. 366).

3. Smell

The receptive surface for distance chemo-reception (smell) is the olfactory epithelium of the nose. The receptors are primary sensory cells, that is to say cells lying in the epithelium, whose axonic processes pass as the olfactory nerve fibres through the cribriform plate of the ethmoid bone to the olfactory bulb (Fig. 34.1, p. 317). Neurosensory cells of this type (neuroepithelial cells) are not found in any other part of the body of any chordate above *Amphioxus*. The receptor cells have a spindle-shaped body whose cytoplasm contains much granular endoplasmic reticulum (Nissl bodies) (Fig. 44.4). The peripheral end or 'dendrite' carries a terminal knob with 12 or more olfactory cilia, up to 200 μm long. Each has the usual $9+2$ microtubules, except that a few micrometres from the base the cilium becomes very narrow, with the number of microtubules decreasing. These long, fine ends of the cilia lie parallel to the surface, embedded in mucus (Fig. 44.5). They are certainly motile in some animals. Between the receptor cells are supporting cells bearing microvilli and joined to each other and to the receptor cells by a web of desmosomes. They also contain a yellow-brown pigment, which gives a characteristic colour to the olfactory epithelium (see Beidler 1971a).

The olfactory glands (of Bowman) are simple tubes secreting the mucus. In addition to the unmyelinated olfactory nerve fibres, myelinated fibres of the trigeminal nerve reach the epithelium and form fine terminal fibres.

It is often supposed that these provide for a sense of touch, but it has been shown that they also discharge impulses in the presence of odorous substances, usually (but not always) with a higher threshold than the true olfactory receptors. They probably serve to operate 'olfactory reflexes', one of which restricts the nasal passage by sympathetic action on the blood-vessels. Strong smells produce constriction of the vessels and may arrest the breathing. These reflexes are thus similar to those of the iris and middle-ear muscles. The axons of the olfactory cells are exceedingly fine, some of them as small as 0·1 μm in diameter. There is a total of about 100 million of them in the rabbit.

The olfactory cells are all histologically identical; yet animals, and men, are capable of distinguishing by smell between a wide range of substances. There is no complete theory that will account for this capacity. There cannot be as many types of peripheral receptor as there are qualities of smell and it is difficult to see

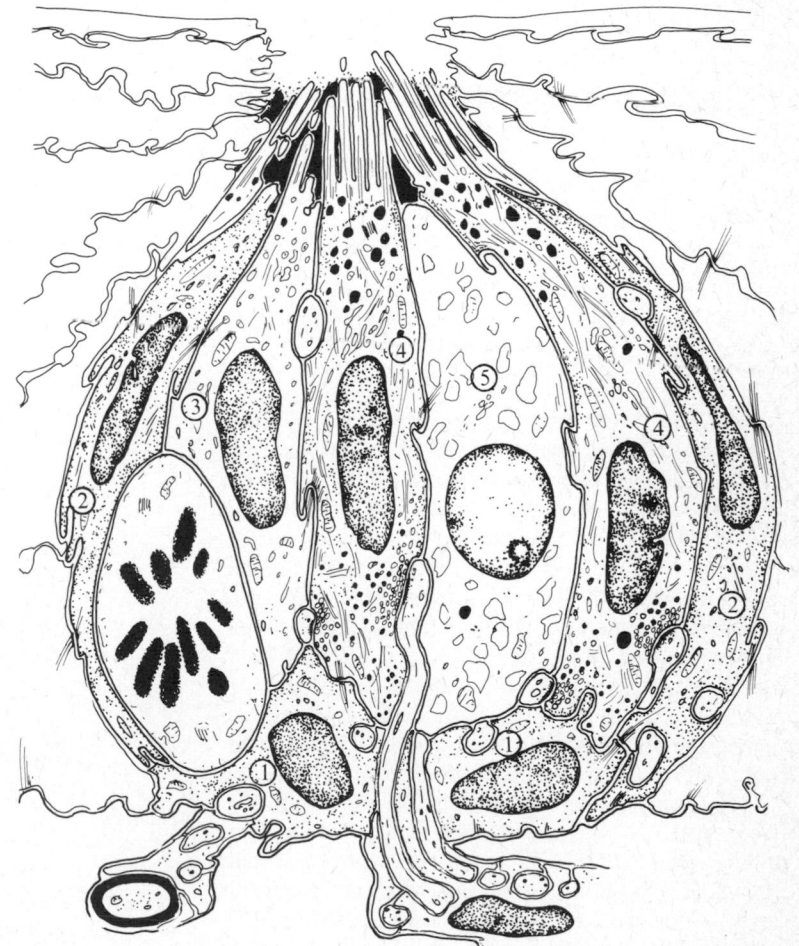

FIG. 44.2 Drawing of vertical section through taste bud from rabbit foliate papilla. There are five recognizable stages: Basal cells proliferate (1) and elongate (2) until the pore is reached and form a few taste microvilli (3). At maturity many microvilli and secretory drops are present at the apical pole (4). Finally dying cells autolyse (5) and are shed. (From Andres (1970). In discussion, in *Taste and smell in vertebrates* (ed. Wolstenholme and Knight). Ciba Foundation Symposium. J. & A. Churchill, London.)

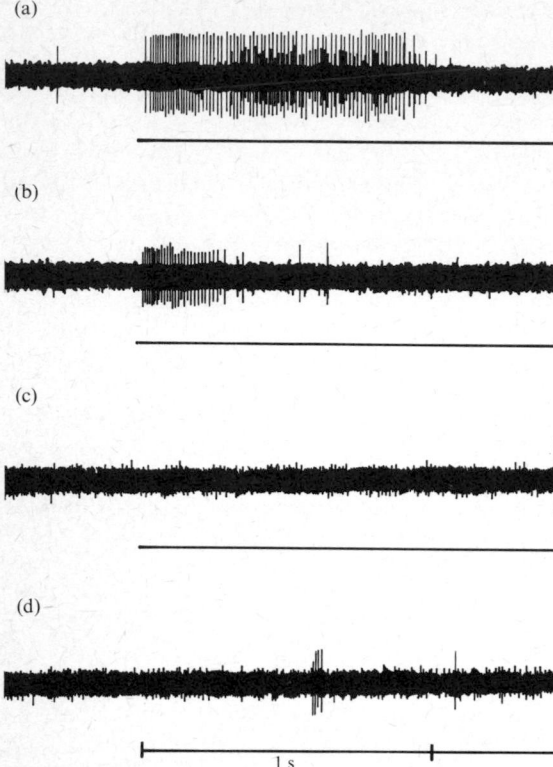

FIG. 44.3. Electrophysiological responses of rat glosso-pharyngeal fibres to basic taste stimuli. Recordings from two fibres are represented by spikes of different height. (a) 0·3M sucrose (b) 0·01N HCl (c) 0·3M NaCl (d) 0·001M quinine hydrochloride. (Redrawn from Frank and Pfaffman (1969). *Science* **164**, 1183–5. Copyright 1969 by the American Association for the Advancement of Science.)

how a single set of receptors could be variously stimulated by different substances so as to produce the different impulse frequencies necessary to allow the discrimination. The difference from taste clearly lies in the multiplicity of qualities that can be recognized. Presumably the basis of this capacity is the fact that the olfactory receptor cells and their nerve fibres are exceedingly numerous. They run as bundles of minute non-medullated fibres to end in the olfactory bulb. Before entering the glomeruli of the latter the fibres interweave in a complicated manner so that each glomerulus receives fibres from different areas, though not from very distant parts of the receptor surface. The olfactory fibres make synapse in the glomeruli (Fig. 44.6), each of which is connected with the dendrites of many mitral cells or tufted cells (p. 316). These cells have accessory dendrites that interweave in a plexiform layer and may make synapse with each other (Fig. 44.7). There are also periglomerular granule cells with short

axons among the glomeruli. These granule cells are apparently activated synaptically by the dendrites of the mitral cells and then they in turn inhibit large fields of mitral cells (Rall, Shepherd, Reese, and Brightman 1966). There are thus indications in the olfactory system of the plexiform and overlapping arrangement that is found elsewhere in systems capable of distinguishing between many distinct combinations, the inhibition serving to sharpen distinctions between neighbours.

4. Interaction in the olfactory pathways

There are widespread possibilities for interaction between the influences coming from distinct olfactory cells (Allison 1953a, 1953b). In the rabbit each glomerulus receives impulses from no less than 26 000 olfactory receptors and sends them to 24 mitral cells and 68 tufted cells. The former then send axons to the olfactory cortical system, the prepyriform cortex (p. 317). The tufted cells send impulses to centres in the amygdala and related regions at the base of the forebrain, serving probably for rather simple reactions. Thus 1000 first-order axons enter the olfactory bulb for each second order axon of the lateral olfactory tract.

It is not at all clear how olfactory discrimination is achieved. Electrical recording shows that each single olfactory nerve fibre is not odour-specific. That is, it does not discharge only when a particular chemical and its related compounds are wafted into the nose. Indeed almost every odour seems to effect every receptor one

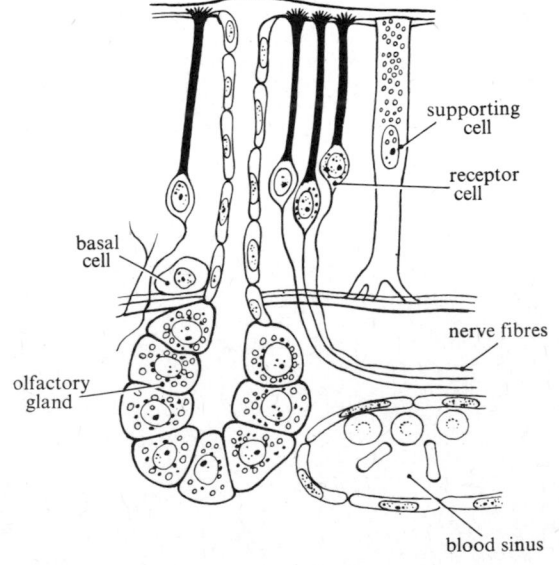

FIG. 44.4. Section of olfactory epithelium from the rabbit showing an olfactory gland. Receptor cells send their axons into the olfactory nerve in a bundle. (From Allison 1952b.)

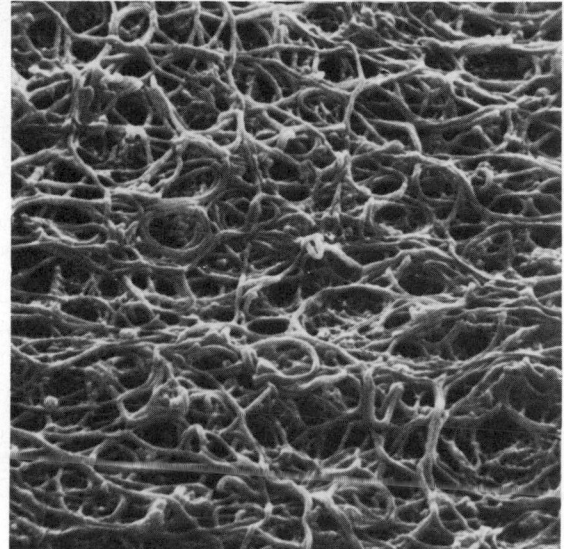

FIG. 44.5. Scanning electron micrograph of frog olfactory mucosa. The pattern of cilia is similar in most vertebrates. Field width 15 μm. (Figure kindly supplied by Dr. P. P. C. Grazeidei.)

way or another (Lettvin and Gesteland 1965). Most axons fire even when a stream of pure moist air is present and the addition of a given chemical may increase or decrease the firing rate (Fig. 44.8). Any set of compounds will affect different cells to different extents. Any two substances that smell alike may stimulate one cell similarly and another in opposite directions (Shibuya 1969).

Since there is no synapse between the receptors and their axons this bewildering set of responses must be due to varied combinational effects of the odorous molecules on the receptor traps. It is as if every axon expresses a point of view with respect to all compounds—and each axon has a separate point of view. In fact so far it has not been possible to understand the principles by which information about chemical composition is coded in the olfactory nerve and decoded by the brain. Presumably there are specific receptor molecules somewhere in the surface of the olfactory cells, and these react with particular chemical groupings to cause the cell to produce first a generator potential and then nerve impulses. It is probable that the olfactory cilia carry at least many of the receptor molecules (Ottoson 1970), though some olfactory cells lack cilia (in the vomero-nasal organ of Jacobson).

The amount of substance necessary may be very

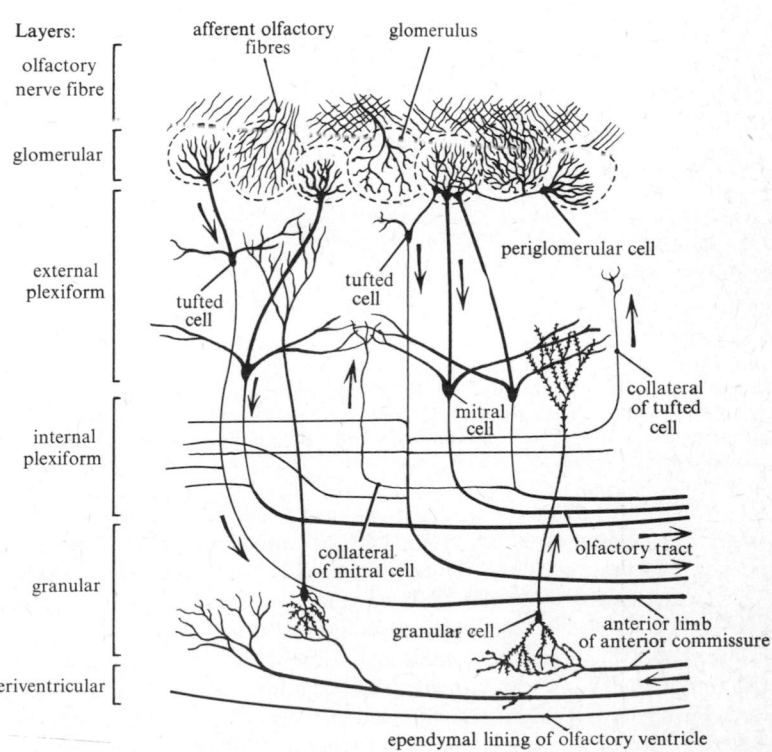

FIG. 44.6. Diagram showing the main pathways for impulses within the olfactory bulb in rabbit. (After Allison 1953a.)

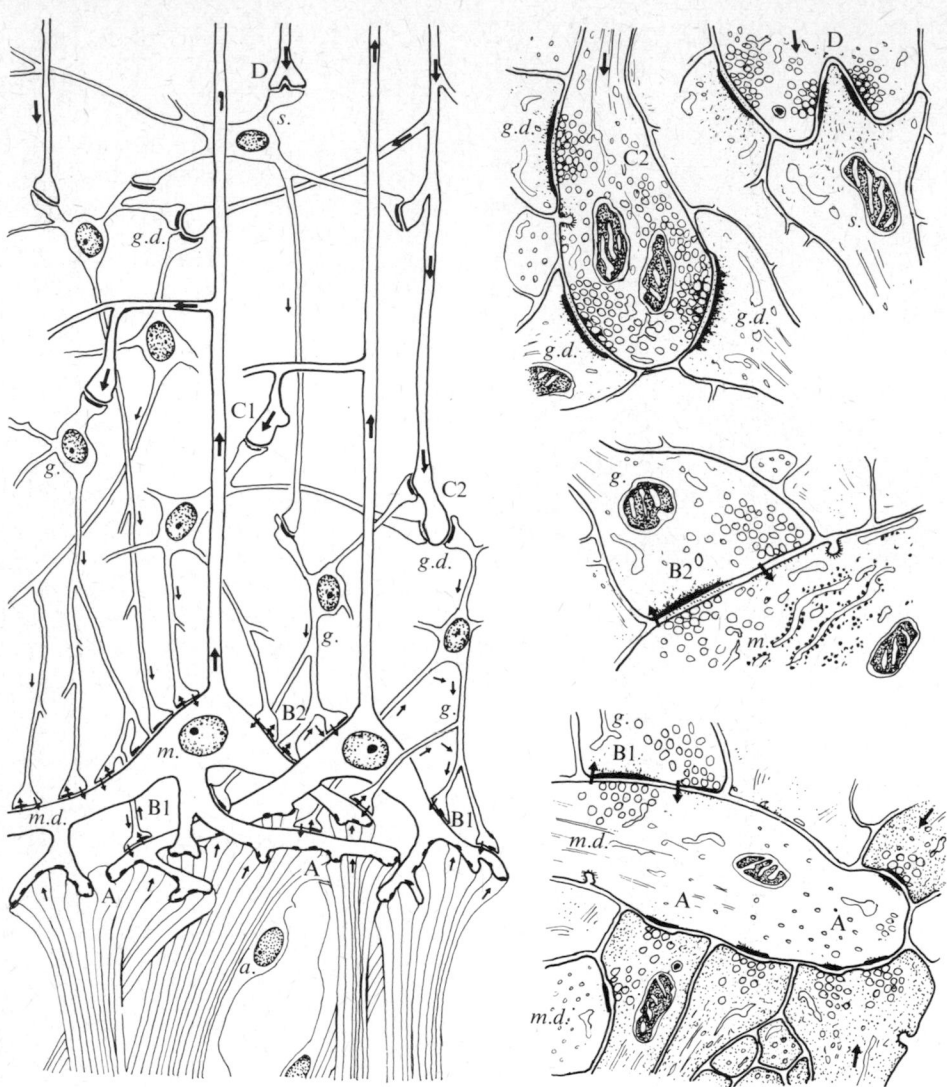

FIG. 44.7. Diagram of synaptic connexions in the fish olfactory bulb. Hypothetical presentation of neuronal connexions (*left*). Detailed drawings of types of synapse (*right*). Synapses between fila olfactoria and mitral cell dendrites (A), reciprocal granule cell ending on mitral cell dendrite (Bl), on the soma of a mitral cell (B2), mitral cell collateral endings (C1) and central bulbopetal fibre endings (C2) on granule cell dendrites; synapses with interdigitated membrane complex in the periventricular zone, probably on stellate cells (D). C1 and C2 form morphologically similar synapses. *g.*, granule cell; *m.*, mitral cell; *s.*, stellate cell; *a.*, astrocyte; *d.*, dendrite; *g.d.*, ganglion dendrite; *m.d.*, mitral cell dendrite. The arrows indicate the direction of transmission of the stimulus. (From Andres *loc.cit.*)

small. The human nose can respond to quantities of odorous substance such as β-ionone as small as 10^{-14} g. This includes many molecules (108) but is far less than can be weighed in the best analytical balance. It is not known how the combination serves to activate the cell membrane and allow the setting up of a current by ionic leakage, but it is likely that the action potential is generated where the cell narrows to form the nerve fibre. The electro-olfactogram (EOG) is a record of potential change from an electrode on the olfactory epithelium when an odorous substance is applied there (Fig. 44.8), and probably it is at least partly a generator potential. It shows certain regularities distinctive of particular substances. But no such regularities have been found at the level of single fibres. Again, recording from different parts of the olfactory bulb of the rabbit

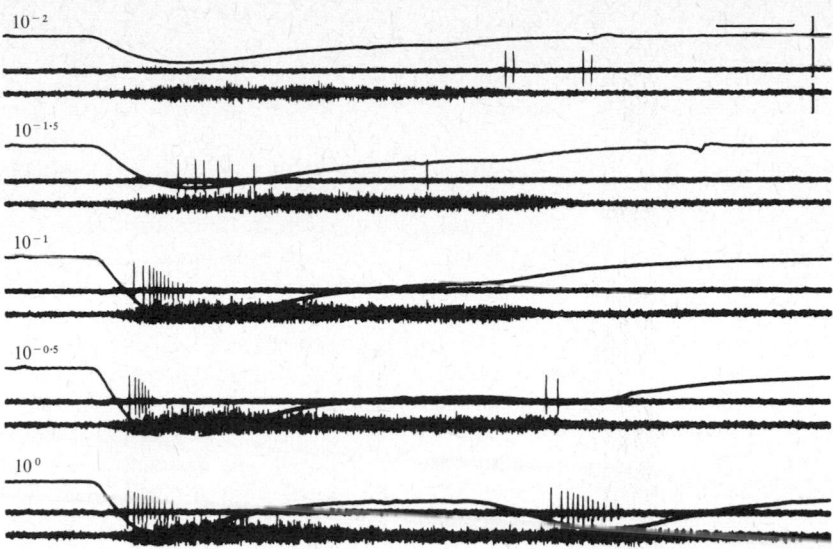

Fig. 44.8. Simultaneous recordings of electro-olfactogram spike discharges, and olfactory nerve twig discharges in response to various concentrations per saturation of amyl acetate. Calibration 2 mV, 1 mV, 100 μV from above. (From Shibuya. In Pfaffman 1969.)

shows that some compounds are more effective in stimulating particular regions, e.g. heptane stimulates the hinder end, and amylacetate the front (Adrian 1951). It may be that there are differentially responsive patches in the epithelium or different classes of receptor each connected to certain part or parts of the bulb. Presumably every animal has receptors or combinations of them for substances important for it and indeed Adrian found a region in the cat's olfactory bulb that was very sensitive to trimethylamine, a substance abundant in fish.

The means by which the olfactory centres perform the 'analysis' of smell remains uncertain. The cells of the olfactory bulb give rhythmic discharges, which can be recorded in the absence of olfactory stimulation. Impulses arriving in the olfactory nerve fibres interact with these rhythmic discharges, both having access to the mitral cell pathways (p. 413). When a smell is first applied the olfactory signals reduce the rhythmic discharges that are transmitted through to the olfactory tract but 'before long the continuous discharge builds up again and swamps the olfactory effect' (Adrian 1951).

The olfactory tract ends partly in the pyriform lobe of the cortex and studies in the hedgehog (where this lobe is large and readily accessible) show waves of electrical activity at each inspiration. This is probably the result of the effect on the olfactory system of the substances inevitably carried in a stream even of 'pure' air. When smelly substances are added the regular cortical waves no longer appear and their place is taken by small irregular waves. The presence of these small waves suggests that a particular pattern of activity is being set up and is the basis of the discrimination (Adrian 1942).

5. The classification of smells

Special difficulties in the physiological analysis of smell arise from the way in which our language handles the problem (Lettvin and Gesteland 1965). Everyone recognizes a great variety of smells, but we can only describe them by simile. There are no obvious and accepted units of smell. Even experts such as wine-tasters or perfumers use metaphors, such as 'floral top-notes' or 'spicy overtones', using the analogy of sounds. Others use suggested similarities between smells and colours. But in the end each substance simply has its own smell. Moreover in general smells do not blend as colours do, we certainly can often recognize the components of a mixture. Yet small constituents may give mixtures special new properties, as musk does to perfumes or glutamate to soups. However there must be units of some sort as is shown by people with partial anosmia; some are unable to smell cyanides, others butyric acid, and still others cannot detect the smell of asparagus in the urine, but otherwise their sense of smell is normal. It can hardly be that we have receptors adapted only to molecules that have been met in evolutionary history, because newly synthesized molecules have characteristic smells.

There have been many attempts to discover some limited number of primary odours and to relate these to chemical structure. Henning (1926) recognized six primary types, flowery, fruity, foul, spicy, burnt, and resinous. He considered all smells to be combinations of these, which he represented as lying at the corners of a prism. Zwaardemaker tried to identify organic radicles that are 'osmophores', thus nitriles give a pleasant smell (euosmophores) mercaptans an unpleasant one (kakosmophores). There may be curious similarities, as in the substances employed to make an old car smell like new, or bread as if freshly baked, but no one knows why they have this effect. An interesting fact is that many smelly substances absorb infrared radiation and produce Raman shift, that is, altering the frequency of light reflected from them. These properties depend upon oscillations within the molecules at high frequencies (10^{13}–10^{14} Hz) and are so specific that they are used by chemists to identify molecules. It may be that such oscillations are somehow connected with activation of the olfactory receptors. It is claimed that there are correlations between particular odours and specific combinations of frequency of molecular vibration (Wright and Burgess 1970).

A further suggestion for the classification of smells is that there are certain groupings associated with distinctive common molecular dimensions or reactivities (Amoore 1965). Various lines of evidence have suggested that there are seven basic classes of odour, ethereal, camphoraceous, musky, floral, minty, pungent, and putrid. These are the names most commonly found in the descriptions of chemical compounds, others such as lemon or garlic are rarer. Amoore found that for the first five of these the shapes and sizes of the molecules having the smell fall within restricted limits. The other two smells are characteristic of molecules with a particular electronic character. However, the molecules associated with particular smells do not always have characteristics that are similar within reasonable limits. We are really still in ignorance of the events by which chemical substances specifically activate these highly sensitive cells of the nose (see Amoore 1971; Moulton 1971).

45 Endocrine organs. The thyroid and parathyroid

1. Methods for the study of hormones

ONE of the great advances made in biology during the twentieth century has been in knowledge of the regulation of the life of the body by substances produced in endocrine glands and released into the blood. The relation of this system of chemical signalling to other means of maintaining the integrity of the body will be considered after the various endocrine organs and their effects have been described. Then we shall be in a position to consider the class of *hormones*, specific chemical signalling substances, produced in one region and acting elsewhere (p. 454).

It is necessary first, however, to say something about the methods by which the ductless glands have been studied. Recognition that a particular organ has an endocrine nature has come in various ways, sometimes from clinical observation of the results of abnormality, often from experiments with animals. Full demonstration of the existence of a gland of internal secretion can only follow a detailed experimental study. If it is alleged that an organ produces a specific substance acting at a distance it must be shown first that removal of the organ is followed by absence of the distant effect. Secondly, extracts of the organ introduced into the blood-stream must be shown to be able to produce these effects and to compensate for the removal. Thirdly, it should be shown that the active substance is produced in effective concentrations during normal life, and the control of its release described. Fourthly, if possible, the substance should be isolated in the pure state and chemically identified. Finally, it is very convincing if excessive or reduced secretion, produced either by disease or by stimulation or suppression of the action of the gland, can be shown to have the expected effects.

In practice achievement of a full understanding of the action of the endocrine glands has been a gradual process, depending largely on careful study of the methods of preparation of extracts. Extracts from many tissues contain substances that have powerful actions upon other tissues and it cannot be alleged that an organ functions as an endocrine gland, that is to say

produces chemical signals, simply because extracts prepared from it have certain effects. For example, histamine, which is found in aqueous extracts of most tissues, has a powerful vasodilatator action. A fall in blood-pressure following injection of a tissue extract is therefore likely to be due to the presence of histamine and not to a principle specific to that tissue. A rise in blood-pressure is more significant. Any results produced by extracts have to be considered in relation to the methods used in preparing them. The fact that a particular substance can be extracted from a gland does not prove that this material is normally released by that gland into the blood-stream.

There has been great progress in the isolation and recognition of specific hormones and many of them, but not all, have been chemically identified. They belong to many different classes of substance. The hormones of the pituitary gland are probably proteins, those of the gonads and adrenal cortex are steroids (p. 426), while the active principles of the thyroid and adrenal medulla are smaller organic molecules. This diversity warns us that the hormones cannot be recognized as a class having any common chemical characteristic. If the word hormone is to have a clear meaning it must be shown that the substances included under it, though chemically unlike, act as signals in the body, by controlled release and transport to a distance.

Much confusion has arisen from giving fancy names to substances alleged to be contained in tissue extracts. All names added to scientific nomenclature should be attached to clearly defined and recognizable entities. The only completely exact procedure is to obtain the chemically pure substance before naming it. Only then can the substance strictly be said to be born as an individual scientific entity. If an earlier christening is insisted upon the name should be applied to an extract prepared in a way that is carefully described. Unfortunately endocrinologists have been no less anxious than other systematists to achieve immortality by additions to the nomenclature, and the variety of names used for extracts of the glands has produced much confusion. It is not very instructive to learn that the adrenal cortex contains 'cortin', especially when it is found that more than twenty chemically distinct steroid substances are included in the extract. Yet it may sometimes be helpful to adopt the convention that such a name shall be applied to extract prepared in a particular way from certain parts of a gland. The standardization of nomenclature and the adoption of agreed methods of assay has occupied a large part of the time of endocrinologists.

2. The thyroid gland

The thyroid, like so many other endocrine organs, is a region where certain metabolic processes that go on throughout the body take place with particular intensity. In this case the process is the uptake of iodine by the cells and its combination to make substances that stimulate respiration. Possibly this uptake originally took place directly from the exterior. The thyroid is derived from the endostylar feeding groove of the pharynx of the earliest chordates, an organ by which mucus is produced for the entanglement of minute organisms. In the lampreys (cyclostomes) the change in function takes place during ontogeny. The larval lamprey lives in the mud and has an open endostylar sac, which secretes mucus; at metamorphosis this sac turns into a series of typical thyroid vesicles (Young 1962, p. 117).

In higher vertebrates the thyroid gland separates from the floor of the pharynx during development; it takes its raw material from the blood and gives its product back to the blood again. This transformation of a region performing a general function in relation with the outside world into a special endocrine organ may first have been dictated by a change of habits. The result has been to allow detailed control of a particular aspect of life by centering that type of activity in one organ. The earliest craniates, feeding on micro-organisms, possessed a large pharynx and perhaps collected their iodine there. When they came to take in large pieces of food the iodine had to enter the blood stream lower down the alimentary canal, and perhaps it was present only in small amounts. The deficit may actually have played a part in the stimulation of the cells of the pharynx that were previously concerned with its uptake to develop into an active thyroid. The endostyle is a mucus-secreting gland but it is concerned with iodine metabolism, for *Amphioxus* or larval lampreys placed in water containing radioactive iodine show a concentration of this substance in the endostyle.

Whatever may have been the origin of the thyroid the secretion of the gland has become of such import-ance that in mammals its presence in too great or too small amounts influences the functioning of every part of the body. The cells of the mammalian thyroid are arranged in characteristic vesicles. They give the appearance of secretory alveoli but are actually closed sacs (Fig. 45.1). Each vesicle is lined by cuboidal epithelium, whose cells are the active secreting agents of the gland. The vesicles contain a material known to histologists as the 'colloid'. This substance consists of the effective product of the gland and is a reservoir from which the material can be passed back through the cells into the blood-stream. Between the vesicles lie blood and lymph capillaries and nerves, the latter controlling the blood-flow but not actually providing secreto-motor fibres to the cells. The thyroid has an exceptionally rich blood-supply. Besides the vesicles the gland contains groups of *parafollicular* cells, with clear cytoplasm. These are now known to produce the hormone *calcitonin*, which lowers the amount of calcium in the blood and is also formed in the para-thyroids (p. 422). These cells are formed from the ultimobranchial body, derived from the fifth branchial arch (p. 123).

The evidence that the thyroid functions as a gland of internal secretion is complete along all the five classic lines: (a) symptoms follow its removal; (b) the symptoms can be prevented by treatment with extracts; (c) it has been proved that the gland emits an active substance into the blood-stream; (d) the active principle has been isolated, identified, and synthesized; (e) diseased conditions involving both reduced and excessive secretion are known.

Historically our knowledge of the gland has grown gradually and recognition of the more superficial symptoms of thyroid dysfunction preceded understanding of its general role in the control of metabolism. There lingers still a tendency to describe its activities in terms of certain more obvious symptoms of its removal, for instance, retardation of growth. It is now possible, however, to specify thyroid activity more exactly. The gland produces an iodine-containing substance that acts throughout the body by promoting the activity of the oxidizing systems, and also of protein synthesis (see Pitt-Rivers and Trotter 1964; Wollman 1969).

The iodine is stored in the gland in combination with a glycoprotein (thyroglobulin). The hormone that is released into the blood is a mixture of the tyrosine derivative triiodothyronine:

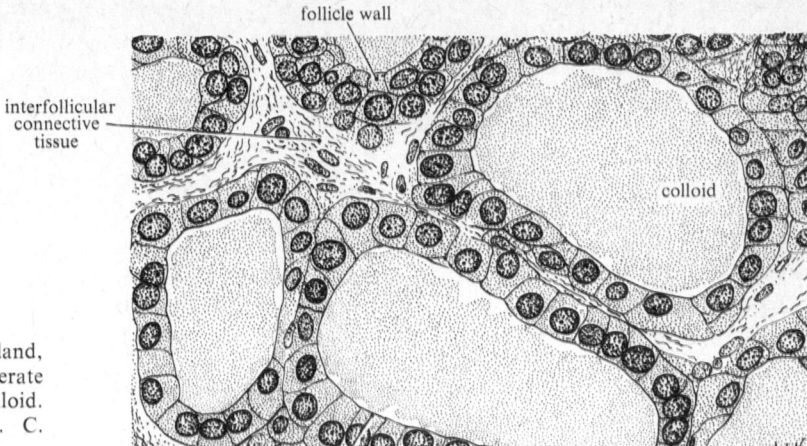

follicle wall

interfollicular
connective
tissue

colloid

FIG. 45.1. Adult human thyroid gland, showing cubical epithelium and moderate distension of the follicles with colloid. (From a photograph by Mr. K. C. Richardson.)

apical cytoplasm of cubical epithelial cell

and the similar substance with four iodine atoms (thyroxine). These are powerful stimulants of the action of respiratory enzymes. Injected into the body they cause a marked rise in the basal metabolic rate. There can be no doubt that this is the chief part that the thyroid plays in life; presumably by virtue of some special property of the iodine atom.

The thyroid tissue has thus become a specialized region for producing metabolic stimulants. The gland contains 20 per cent of all the iodine in the body and attracts this element when it enters the blood. The protein part of the thyroglobulin is produced in rough endoplasmic cisternae towards the outer parts of the follicle cells. It is transferred to the Golgi apparatus lying more centrally and probably receives its carbohydrate there and is then discharged into the lumen. Iodine taken into the cell is rapidly transferred to the inner margin of the vesicle and there added to the thyroglobulin. Release of hormone is probably produced by the breakdown of the colloid by the microvilli lining the follicle. These become longer when the gland is stimulated by the thyrotropic hormone of the pituitary (TSH), which probably acts by stimulating the enzyme adenyl cyclase as a second messenger. This is the means by which several hormones activate their target organs. The TSH binds specific receptors in the membrane of the thyroid follicular cell, activating adenyl cyclase. This then catalyses the conversion of ATP to cyclic $3',5'$-adenosine monophosphate (cAMP) which in turn activates the various specific enzyme systems of the cell. By this mechanism the signal for activity is transmitted inwards from the surface. Changes in the ultrastructure of the follicle cells can be seen within a few minutes after injection of TSH and continue for days (Dumont 1971).

3. Control of thyroid secretion

There is a close relationship between the demands of the tissues and the activity of the thyroid itself. The effect of partial removal or underactivity of the gland may be to stimulate growth and activity of the remaining thyroid tissue. It is therefore difficult to say whether an enlarged condition of the gland (*goitre*) is an index of reduced or excessive activity. The sulphur-containing substances thiourea and thiouracil have the effect of preventing synthesis of the hormone. Following their administration the gland becomes enlarged, although at the same time symptoms of hypothyroidism develop. Goitres can also be produced by treatment with cyanides and other substances that depress the respiratory systems of the cells, the thyroid responding by increased output.

The amount of hormone secreted by the gland is therefore related in some way to the 'demand' for it by the tissues and we get a glimpse of the complicated sets of relationships by which the activity of the body promotes the production of substances that themselves make further activity possible. Thyroid secretion is also controlled by more specific factors; the nervous system may exercise a regulatory influence through control of the blood-supply, and the TSH stimulates discharge of the colloid vesicles and produces symptoms of increased thyroid functioning. Conversely if the level of thyroid hormones in the blood falls, the pituitary is stimulated to produce increased TSH, which falls off again when the thyroid hormones increase. Being influenced by so many factors the gland fluctuates considerably in size and such changes can often be observed in man, for instance, during puberty, menstruation, or pregnancy (see Brown–Grant 1966; Dobyns 1964).

4. Hypothyroidism

The thyroid hormone is necessary for the normal metabolism of most if not all cells of the body and it is not surprising that any defect in the gland produces profound and widespread effects. Complete removal is not fatal, but produces retarded growth in young animals and in the adult a lowered metabolic rate, apathy, thickening and other changes in the skin, and many other effects on the nervous system, gonads, pituitary, adrenals, and other organs. Administration of thyroid extract or of thyroxine can compensate for these defects. The basal metabolic rate of an individual (BMR) is determined by taking a form of breathing test while at rest (see MacGregor 1964).

Symptoms indicating serious reduction in thyroid activity in man are known as *cretinism* when they occur during development and as *myxoedema* in the adult. A cretin shows slow growth, with many abnormalities, for instance pot-belly, gross intellectual deficiency, and retarded sexual development. The influence of the thyroid in controlling growth and differentiation is seen in the fact that it is responsible for producing metamorphosis in amphibian tadpoles, acting by promoting the growth of some cells, the atrophy of others. The gland is also necessary for the periodic moulting of reptiles, birds, and some mammals; probably it influences all processes of continual replacement of cells, such as proceed in the skin, the intestine, and the blood vascular system.

The name myxoedema refers to a characteristic thickening of the subcutaneous tissues of the skin, but this is not a true fluid oedema. In any case the name is inadequate to describe the complete changes in appearance and personality that are produced by thyroid deficiency. The characteristic feature of the condition is the reduction of the basal metabolic rate, which may fall to 60 per cent of the normal, associated with lowered cell respiration throughout the body and serious impairment of many activities. Treatment with extracts of the gland, or with thyroxine, produces marked improvement, first an increase of basal metabolic rate and then a reversal of the changes in the nervous system, skin, and elsewhere throughout the body.

5. Goitre

It would be expected, perhaps, that the symptoms of myxoedema should be associated with shrinkage or small size of the thyroid, actually the reverse may be the case; the gland is often enlarged. The increased TSH production causes hypertrophy of the thyroid but its cells fail to produce their hormone and the pituitary continues its high level of production. A further complication is that the symptoms of thyroid deficiency and enlargement of the gland may follow from a lack of iodine in the diet. It has long been known that goitre and cretinism are common in continental and mountainous districts remote from the sea. Such 'endemic' goitre occurs in Switzerland, in Derbyshire, and in the middle-west of America. Iodine is abundant in sea-water and the presumption is that in these continental districts men and animals fail to get the small amounts that become available closer to the sea. The use of dried seaweed and sponges was a Hippocratic remedy for goitre. Man requires 0·1 mg of iodine a day, which would be obtained from two servings of fish a week. The other main sources are milk and eggs, but these contain little when produced in inland areas.

It is not entirely clear how the circulation of iodine proceeds in nature and there may well be complicating factors that make the iodine unavailable, for instance, the presence of fluorine or other ions. Moreover, there are undoubtedly hereditary forms of hypothyroidism and the dietary factor cannot too easily be assumed to be wholly responsible for the large number of goitres in any area. The two factors may, of course, go together, a low level of inherited activity becoming apparent only when associated with low iodine intake. Treatment with thyroid preparations may be equally helpful in either case; it has been shown by large-scale experiments in the United States and Switzerland that addition of iodine to drinking-water, or provision of iodized chocolate or salt, can reduce endemic cretinism and goitre.

6. Hyperthyroidism

A true condition of excessive action of the thyroid occurs with the symptoms that would be expected, high metabolic rate, fast pulse, warm skin, raised nervous excitability, and various disorders of metabolism. Such symptoms may appear either because of hereditary enlargement of the gland or from the development of a tumour of its cells (thyroid adenoma). These conditions of hyperthyroidism are not always easy to distinguish from their opposites. In the forms known as exophthalmic goitre the eyes protrude, probably because of the deposition of fat in the orbits. A further paradox of thyroid behaviour is that administration of iodine may temporarily relieve the symptoms of hyperthyroidism! Understanding of the iodine balance of the body is indeed still far from complete (see Morgans 1964).

Study of the thyroid thus shows clearly how difficult it is to describe the factors that control the activity of any part of a living organism. Conventional biological analysis attempts to recognize a 'body' provided by 'heredity' and acted upon by its 'environment'. If this scheme were adequate we could forecast the result of alteration in any part of the system; hereditary over-development of the thyroid would produce one result, deficiency of iodine (change in the 'environment') another, and removal of part of the gland a third. Such predictions are falsified by the interactions of the parts of the body, and the tendency to self-regulation of the whole. Deficiency of iodine may be followed by increase in the size of the gland and a goitre may be a sign either of excessive or of reduced 'hereditary' activity. We need a new method of thought, which instead of throwing emphasis on the fixed material body shall describe the activities of each life and the factors that control the steady states that they maintain.

7. The parathyroids

The pharyngeal wall, besides providing the thyroid cells specialized to regulate the metabolic rate, provides cells that perform the important function of regulating the calcium exchanges of the body. The parathyroid glands (see Arnaud, Tenenhouse, and Rasmussen 1967) are four minute masses of tissue arising from the endoderm of the third and fourth branchial pouches, close to the thymus (p. 186). They are usually partly embedded in the thyroid gland and are liable to be removed with it. They are necessary for the maintenance

of life and it is therefore easy to see how early experimenters were led to suppose that thyroid removal is fatal in some animals. The glands weigh altogether only about 0·1–0·2 g in man. They consist of closely packed cells, only exceptionally arranged as vesicles. Numerous vascular sinusoids lie between the cells.

After removal of the parathyroids the animal or man within a few hours develops abnormal muscular contractions and convulsions, leading to death by spasm of the laryngeal and respiratory muscles. This tetanus results from a changed excitability in the muscle fibres and nerve centres and is now known to be only one sign of the fall of calcium that is the result of lack of the parathyroid secretion. Normally in man there is 10–12 mg of calcium and 4–5 mg of phosphorus per 100 ml blood. Following parathyroidectomy the calcium may fall to 6 mg or lower and the inorganic phosphorus rise to 9 mg. Relief of the symptoms can be obtained by injection of calcium salts or of an acid extract of the gland, which is known as parathormone or, from its discoverer, Collip's hormone. The hormone is a polypeptide of molecular weight $\sim 10\,000$, with 84 amino-acid residues (Fig. 45.2). Automated procedures for degradation of proteins and identification of the fragments have allowed the full sequence of residues to be determined for parathyroid hormone and calcitonin (Fig. 45.3). The hormone has a direct effect upon bone.

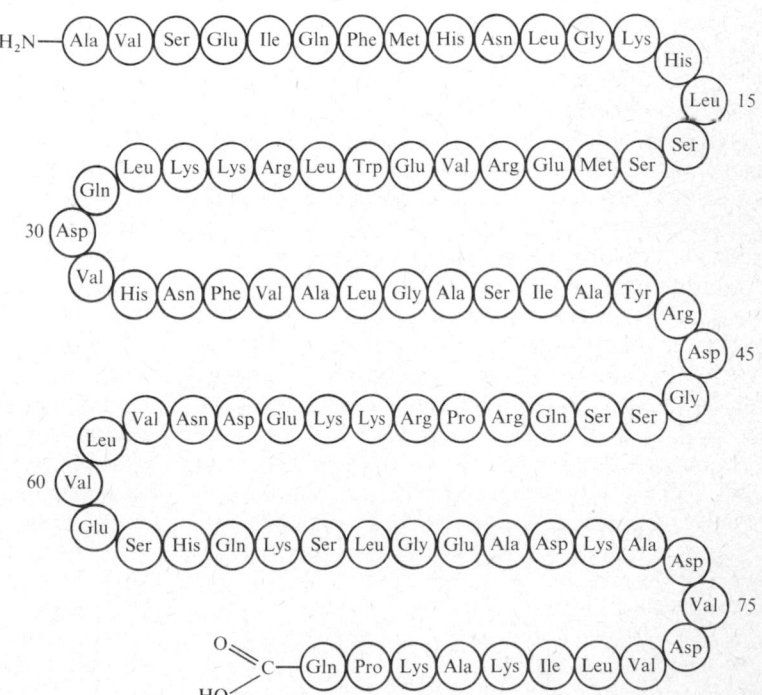

FIG. 45.2. Amino-acid sequence of bovine parathyroid hormone. (After Potts *et al.* 1971.)

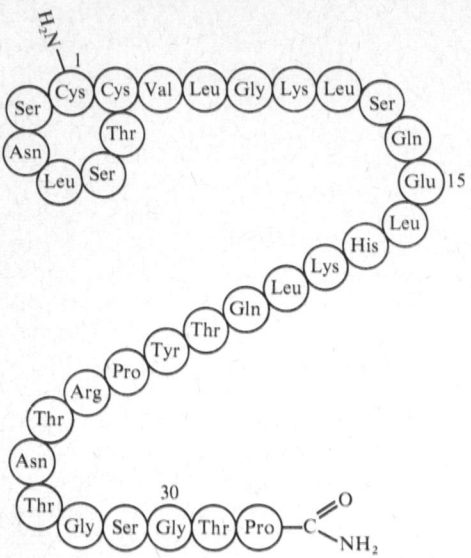

FIG. 45.3. Amino-acid sequence of salmon calcitonin I. (After Potts *et al.* 1971.)

If a piece of the gland is grafted onto the brain the bone nearby is absorbed. The hormone also causes the kidney to excrete more phosphate, leading to a removal of calcium phosphate from bone. The hormone has been synthesized and the NH_2-terminal sequences 1–34 and 1–29 shown to act directly upon the adenyl cyclase activity of kidney and bone cells respectively. The sequence 1–20 alone is inactive.

The function of the secretion of the parathyroids is thus to control the amount of calcium circulating in the blood. In performing this function the glands co-operate with the vitamin D introduced with the food or synthesized by the action of sunlight in the skin, this vitamin being necessary for the deposition of calcium in the bones. Injection of large amounts of the hormone produces solution of calcium from the bones and a high calcium level in the blood (as much as 20 mg per 100 ml), which may be followed by calcification of soft parts, for instance, of the walls of the arteries.

The secretion of the parathyroids is regulated by the calcium level in the blood, not directly by any other hormone. Perfusion of the glands of a goat with blood containing 14 mg per 100 ml of calcium led to an 80 per cent decrease in hormone within 30 minutes. Later perfusion with low calcium led to a sevenfold increase. If the diet is deficient in calcium the glands become enlarged and they also increase during lactation, when much calcium is being mobilized for transfer to the milk.

The regulation of blood calcium also involves calcitonin, a hormone antagonistic to parathormone, produced by the parafollicular cells of the thyroid (p. 418) (MacIntyre 1968; Potts, Kentmann, Niall, and Tregear 1971). If the amount of calcium in the fluid perfusing the thyroparathyroids of a dog is raised there is a fall within a few minutes in the systemic calcium. This fall is said not to occur when the parathyroid only is perfused (this can be done in the goat). Calcitonin is a protein with 32 amino acids and minimum molecular weight 3600 (Fig. 45.3). It acts by inhibiting bone resorption. Calcium is constantly leaving the plasma to be deposited in bone or excreted by the kidneys or gut. Conversely, calcium constantly enters the plasma by resorption of bone and intestinal absorption. The level is maintained by alteration in the levels of parathyroid hormone and calcitonin, whose release from the glands varies with the blood calcium level. It is not clear whether the calcitonin is produced by both thyroid and parathyroids or only by the former.

Calcitonin has been synthesized and study of fragments has shown that the whole sequence must be present. Fragments or analogues shorter by even one residue are totally inactive. The amount of calcium affects the membrane properties of every cell, and especially of the mitochondria. It is obvious that the stabilization of this ion is of the first importance for a balanced life. In fish-like animals this is probably achieved at the same time as regulation of the total salt content of the blood, by the special chloride-secreting cells in the gills, which are able to pass salt inwards or outwards as required by the state of the surrounding water (Young 1962, p. 203). Both elasmobranchs and teleost fishes have large ultimobranchial bodies and indeed salmon calcitonin has an especially high potency. Presumably the parathyroids of land vertebrates have developed from these cells; they continue to perform a function of the gills that was still needed after aquatic respiration ceased. As the parafollicular cells arise from the 4th branchial arch and ultimobranchial body, the calcitonin must have a similar origin.

46   The adrenal glands

1. Chemical signals ensuring response to stress

THE adrenal bodies of mammals (often called supra-renals) are a pair of organs of orange–yellow colour, lying near the kidneys. They contain two distinct types of cell, the cortical, arising from the coelomic epithe-lium, and the medullary, which develops with the sympathetic ganglion cells from the neural crest. The significance of the association of the two types of tissue is still not clear.

The two tissues both serve to send signals that elicit, in various parts of the body, reactions that are appro-priate to conditions of stress. Adrenaline and nor-adrenaline, secreted by the medulla, produce immediate preparations for attack or defence, by changes in the distribution of the blood, in the level of blood sugar, and in other ways (p. 426). The secretions of the adrenal cortex are concerned in many ways with long-term preparations to resist adverse conditions and infection, making it possible for the various organs to undertake the slower regulatory adjustments by which they assist in maintaining homeostasis. An animal without its adrenal cortex is therefore apt to fail when any of its tissues are called upon to adapt themselves to changed conditions. This organ is therefore of central importance in the processes that ensure that the various tissues remain adequate to meet the conditions of the environ-ment.

2. Evolution of the adrenal glands

As with several other endocrine organs we can trace the phylogenetic history from the condition in which the adrenals consisted of cells scattered widely through the body. Even in mammals there remains a consider-able amount of adrenal tissue that is not incorporated within the gland. In *Amphioxus* no tissue of definitely adrenal nature has been identified and in lampreys and hag-fishes (cyclostomes) there are no compact adrenal bodies but some cells found scattered along the cardinal veins have been claimed as cortical and medullary tissue. It is probable that in these early types of verte-brate the individual tissues throughout the body make

their own responses to conditions of stress, without the assistance of secretions by a centralized adrenal system.

In elasmobranch fishes there are two separate sets of tissue, the *interrenal glands*, representing the adrenal cortex, and the *suprarenals* the medulla. Both tissues are found spread over many body segments, the supra-renals being a long series of pairs, extending from the level of the heart to the hind end of the coelom (Young 1962, p. 165). In bony fishes the two parts are also separated, but in all tetrapods they are combined to form a single organ, the interrenal tissue (cortex) more or less enclosing the suprarenal tissue (medulla). In urodeles the organ is spread out over many segments, but in tetrapods it develops mainly in a few segments, although small groups of cells of both sorts occur elsewhere. In mammals the sympathetic ganglia con-tain medullary tissue and there are masses of cortical cells in various parts of the abdomen. Centralization of the adrenal cells into one organ has therefore proceeded gradually throughout vertebrate history and is not com-plete even in mammals. The hormones evidently operate locally, as well as for the body as a whole under central control.

3. The adrenal medulla

The adrenal medulla occupies the centre of the gland, forming an irregular parenchyma of cells secreting substances that assist the sympathetic system in making the body ready for attack or defence. The cells contain powerful reducing substances and show chemical reactions that are rare elsewhere in the body, from which they get the name of *chromophil cells*. The tissue has a characteristic open structure on account of the large vascular sinusoids between the cells (Fig. 46.1).

The adrenal medulla is richly innervated and the nerve fibres make contact with the gland cells them-selves and not merely with the blood-vessels (Fig. 46.2). The nerve fibres reach the glands from the splanchnic nerves (p. 281) and since they carry a myelin sheath it was suggested by Elliott that they are preganglionic fibres (p. 279). This has been proved to be correct by

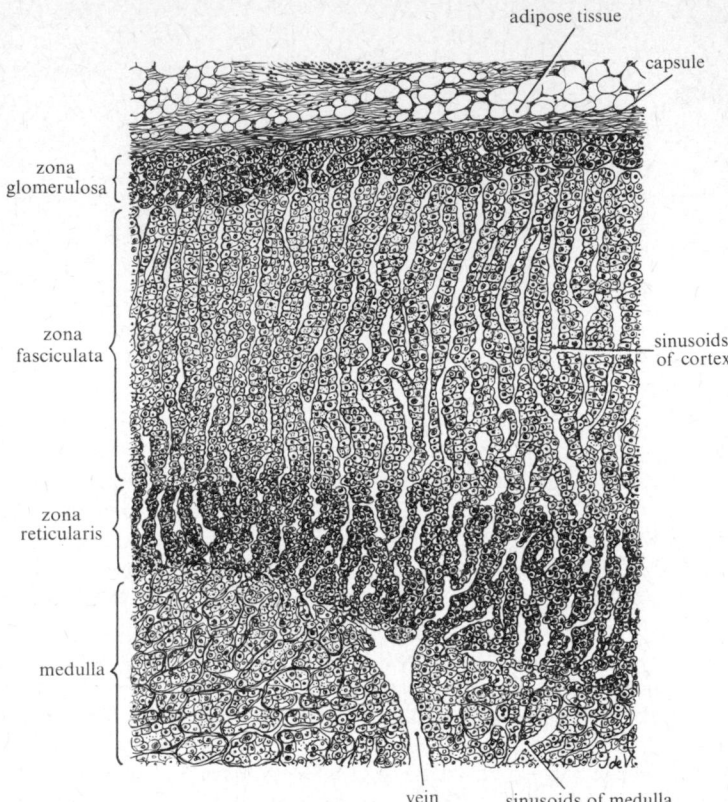

adipose tissue

capsule

zona
glomerulosa

zona
fasciculata

sinusoids
of cortex

zona
reticularis

medulla

vein   sinusoids of medulla

FIG. 46.1. Adrenal gland of monkey
(*Macaca*) showing zones of the cortex.
(Drawn from a photograph supplied by
Mr K. C. Richardson.)

the demonstration that the fibres in the gland degenerate
after the ventral roots of the hinder thoracic and upper
lumbar region have been severed (Fig. 46.3).

Like other preganglionic nerve fibres those of the
adrenal produce their effect by the liberation of acetylcholine close to the surfaces of the adrenal medullary
cells (p. 287). No other visceral tissue is innervated in
this way, without the interposition of a postganglionic
autonomic nerve cell, and it is clear that the chromophil cells of the adrenal themselves represent the postganglionic neurons. The interest of this relation is
greatly increased by the discovery that the sympathetic
nerve fibres as well as the adrenal cells act by the
production of adrenaline, or a similar substance (p. 289).
The hormone manufactured in the gland is poured into
the blood-stream and acts at a distance, that produced
at the ends of sympathetic nerve fibres is liberated
where the nerve fibres run close to the smooth muscle
cells or gland cells that they co trol (Levine 1969).

It was shown by Oliver and Schäfer in 1894 that the
adrenal (like the pituitary) contains a substance that
on injection produces rise of blood-pressure. The
adrenal substance was identified in 1901 and synthesized soon after. Like thyroxine, *adrenaline* (= epine-

phrine) is a derivative of tyrosine, but in this case a
relatively simple one, its composition being *l*-1-[3,4-
dihydroxyphenyl]-2-methylaminoethanol:

$$\text{HO}-\text{C}_6\text{H}_3(\text{OH})-\text{CH(OH)CH}_2.\text{NH(CH}_3).$$

Salts of adrenaline are readily soluble in water but are
quickly oxidized in alkaline solution to a pink substance. Large amounts of a reducing substance, probably ascorbic acid (vitamin C), are present in the gland.

Besides adrenaline the medulla also contains the
homologous non-methylated aminoethanol base, *noradrenaline*:

$$\text{OH}-\text{C}_6\text{H}_3(\text{OH})-\text{CH(OH).CH}_2.\text{NH}_2.$$

The actions of this substance are similar to those of
adrenaline, though there are quantitative differences.

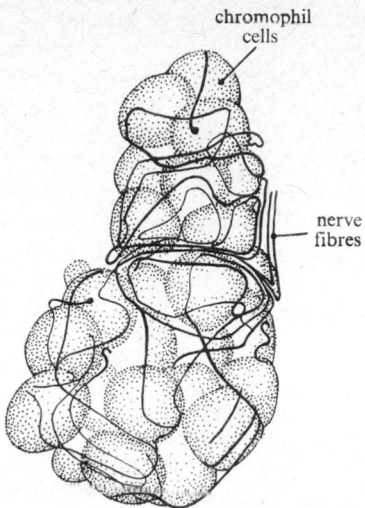

FIG. 46.2. Diagram of the innervation of the adrenal medulla of a guinea-pig showing the nerve fibres ramifying around the chromophil cells. (After Willard (1936). *Q. Jl. micr. Sci.* **78.**)

The adrenaline-like substance produced at sympathetic nerve endings throughout the body (p. 285) is mainly noradrenaline (see Euler 1951).

4. Actions of adrenaline

Injection of adrenaline produces effects throughout the body that mimic those of stimulation of the sympathetic nerves (p. 284); bringing into play the mechanisms that make the animal ready to fight or run. There is constriction of the arterioles of the skin and of most of the visceral organs, but dilation of those of the muscles and especially of the heart, producing a diversion of blood into these organs. Adrenaline has an accelerating action on the beat of the isolated heart if the vagi have been cut; when they are intact the rise of blood-pressure that follows the adrenaline injection may act upon the receptors of the carotid sinus to produce a reflex fall of heart-rate. An adrenaline concentration of 2×10^{-3} mg per kg body weight is sufficient to raise the blood-pressure when injected into a decerebrate cat and this probably represents about the amount released per minute by the animal under experimental conditions (see below). Adrenaline inhibits the contraction of the smooth musculature of most of the viscera, including that of the stomach and urinary bladder, but it excites contraction of the gall-bladder, ureter, and sphincters of the gut. Its action is thus in general to stop visceral activities, although it causes contraction of the uterine muscle, especially during pregnancy.

Several of its actions promote the efficiency of oxygen transport and of the supply of fuel to muscle. It produces relaxation of the muscles of the bronchioles of the lungs (advantage is taken of this in the use of adrenaline for the relief of asthma) and contraction of the spleen, causing an increase in the number of circulating red corpuscles. Acting upon the liver, adrenaline causes mobilization of sugar and thus has effects

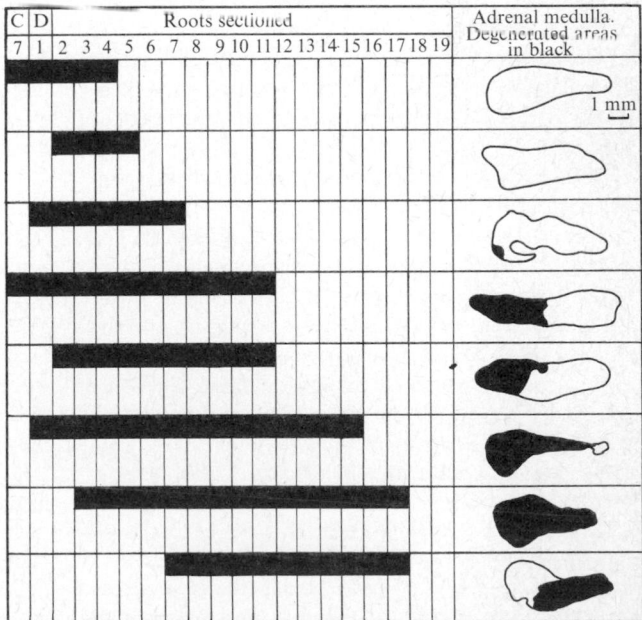

FIG. 46.3. Diagram showing degeneration in the adrenal medulla following section of spinal roots in the cat. The black rectangles show the nerves cut at operation. The figures to the right show the area of degeneration that results (Young (1939). *J. Anat.* **73.**)

opposite to those of insulin (p. 433), producing hyper-glycaemia and even glycosuria. The clotting time of the blood is decreased. Other particular actions of adrena-line produce further assistance in defence, varying with the animal species. Thus by causing contraction of the dilator muscle of the iris the pupil is opened wide. Action on the muscles of the hairs produces 'bristling' of the fur in cats, salivation and lachrimation occur in some species, and sweating in many ungulates, though not in man.

This is an imposing list of actions; adrenaline, a substance produced normally in the body, is evidently a very potent drug. Related substances such as ephe-drine, tyramine, and benzedrine are mostly somewhat less powerful in action than adrenaline itself but are used instead of it as coagulants, for the relief of asthma, to constrict inflamed vascular membranes in the nose, or for their general 'tonic' and stimulating effects. Being more stable than adrenaline, they can be taken by the mouth.

5. Amount of adrenaline secreted

It would be of great interest to know the conditions and rate of secretion of adrenaline in the intact body and various techniques have been devised for estimating this. Cannon and his colleagues used cats whose hearts had been denervated by severing the vagi and sym-pathetic nerves. They found that the heart accelerated when such animals were excited or hurt. For instance, when a dog was placed in front of a cat the increase reached as much as 40 beats a minute and occurred with-in 10 seconds. If the adrenal glands were then removed these increases were no longer seen and this provides a convincing demonstration of the 'emergency' function of the gland. Electrical stimulation of the great splanch-nic nerves produced acceleration of the heart-rate of intact cats.

Direct assay of the amount of adrenaline produced has been performed by using the vena cava, into which the adrenal veins discharge, as a collecting pocket. The amounts recorded by this technique are small when the animal is at rest (about 2×10^{-3} mg kg^{-1} min^{-1}) but they increase nearly ten times during splanchnic stimulation. It has been suggested that the adrenaline has a tonic rather than an emergency action and some workers have even denied that it plays any part at all in the normal life of the body. This would be strange indeed for such an active substance. There is little doubt that adrenaline is secreted to provide a specific set of signals that rapidly reach all parts of the body when an animal or man is frightened or angry and also during violent exercise, pain, and perhaps in other conditions such as cold, anaesthesia, or asphyxia. We must, however, avoid such oversimplifications as the statement that 'the emotion of fear is due to the secretion of adrenaline'; our new knowledge of these matters is interesting, but not yet sufficient to encompass a complex human emotion. The controversy emphasizes the importance of detailed quantitative knowledge about endocrine organs. After the general outlines of the function of an organ have been discovered by experi-ment much quantitative investigation is required before we are able to acquire satisfactory understanding and control of its actions during normal life (see Malmejac 1964).

6. Structure of the adrenal cortex

The adrenal cortex is described as containing three parts (Fig. 46.1). Immediately beneath the capsule is a *zona glomerulosa*. Within this in many animals the cells are arranged in columns, the *zona fasciculata*, con-stituting the major part of the cortex. Within it lie further irregular cords of cells the *zona reticularis*. The gland is richly supplied with blood and the capillaries form sinusoids. The only nerve fibres in the gland run to the blood-vessels. The secretion is therefore controlled by hormones, in particular, the adrenocorticotropic hormone (ACTH) of the anterior pituitary (p. 440).

The adrenal cortex is specially concerned with the metabolism and production of steroid hormones. These act upon cells throughout the body. In particular, they influence two types of activity, the regulation of water and electrolyte balance, and the metabolism of carbo-hydrates, including the mucopolysaccharides of the interstitial spaces. These two actions are mediated by distinct types of hormone, the mineralocorticoids (the most potent of which is aldosterone) and the glucocorticoids (cortisol and corticosterone) (see Bransome 1968; Mulrow 1972).

There has been much discussion about the relation of the various hormones to different parts of the cortex. One theory is that cortical cells are continually being formed beneath the capsule in the zona glomerulosa and pass inwards through the fasciculata to be dis-charged in the reticularis. This view is supported by the fact that regeneration can occur from a small outer piece of tissue, but not from inner ones. However, it is also generally held that the different zones produce different hormones—mineralocorticoids in the glomeru-losa, glucocorticoids in the fasciculata, and androgens in the reticularis.

The adrenal cortex is very large in embryonic mammals and reduces rapidly after birth. The part concerned lies at the inner side of the cortex and is known as the X zone. Its significance is not understood.

7. Cytology of the adrenal cortex

The biosynthesis of steroids is accomplished by smooth endoplasmic reticulum and Golgi membranes. There is very little rough reticulum in the cells of the adult cortex. There are many 'cytoplasmic vacuoles' of 30–600 nm diameter. These are formed either by pinocytosis at the cell surface or by 'dictyogenesis', i.e. formation from the dictyosomes of the Golgi apparatus. There are also very many 'liposomes', structures containing one or more types of lipid.

The steroid hormones are synthesized from cholesterol, which can itself be synthesized from acetate. The enzymes concerned are partly localized in the mitochondria, which in addition to the usual respiratory system contain a second one on the inner membranes, which is concerned in hydroxylating the steroids (11β-hydroxylase) (Mulrow 1972).

The synthesis is largely dependent on the presence of ACTH and is almost absent after hypophysectomy. The ACTH binds to some receptor in the surface of adrenal cortical cells and within 10 seconds of adding it to a preparation of cortical cells they begin to produce adenyl cyclase. This in turn produces cyclic AMP which acts as the intracellular messenger, setting off the steroid synthesis.

8. Functions of the adrenal cortex

The cortex makes up the major part of the adrenal: its secretions are essential for life. If both adrenals are removed the animal dies within a week or two (unless accessory adrenal tissue is present). The symptoms tell us much about cortical function. They include fall in blood sugar, loss of blood volume, and excess concentration of salts and other substances in the blood. These changes are followed by fall of blood-pressure, renal failure, and other disturbances. Moreover, animals without adrenals are much more sensitive than normal to the stress of unfavourable circumstances such as large temperature changes, wounding, or starvation (see Eisenstein 1967).

The adrenal cortex is mainly concerned with the production of steroid hormones. It is obvious from the above list of symptoms that these substances affect the life of a great proportion of the cells of the body. No doubt they enter into cellular metabolism in many different ways. The symptoms produced by errors in each type of cell depend upon its normal function, which is disturbed by the absence of the steroids.

Some of the effects of adrenalectomy can be corrected by compensating for the functional deficiency. Life can be prolonged after the removal by allowing rats to drink 3% NaCl, which they will take in great amounts. This treatment does not help with other symptoms, such as the fall in blood sugar and low resistance. These can be corrected, however, by administration of cortical steroids, after which glucose is again formed from protein and resistance to stress is increased.

9. Cortical steroids

The demonstration of the endocrine function of the adrenal cortex began with the discovery that extracts ('cortin') are able to prevent death following adrenal removal. The substances involved are steroids, so called because of their derivation from waxy materials such as cholesterol (Greek *stereos* = solid). These are based on the four-ring structure as shown:

Steroid molecule skeleton   Cortisol

Corticosterone   Aldosterone

The three chief steroids produced by the cortex are known as corticoids (or corticosteroids). In addition, the gland produces small amounts of the sex hormones, which are also steroids.

The corticoids are of two sorts. The glucocorticoids, mainly cortisol (= hydroxycorticosterone) and corticosterone, are concerned with metabolism, especially of carbohydrates. The mineralocorticoids, mainly aldosterone, influence the metabolism of sodium and potassium. Other steroids found in the gland such as cortisone are probably intermediate stages of synthesis.

10. Glucocorticoids

Corticosterone and cortisol have a large part of their action on the liver, where they promote the production of glucose (from protein and other sources (gluconeogenesis). They also inhibit the effect of insulin in allowing the tissues to use glucose (p. 433). The effects of injection of these steroids are felt in many parts of the body. For instance they may promote breakdown of tissue protein in lymphoid tissue. This may have the effect of reducing the inflammation produced by bacterial or other damage and injections of corticoids may be used clinically for this purpose. These are very powerful substances, however, and their use may produce many side-effects.

Glucocorticoids are not stored in large amounts in the gland. Their production is controlled by the ACTH and the amount of them that is produced increases within minutes of injection of ACTH. This is also the mechanism by which they are increased following stress. The half-life of ACTH is brief (a few minutes) and the output of the corticoids is completely dependent on this. The hypothalamus is sensitive to the amounts of circulating glucocorticoids and this in turn controls the output of ACTH releasing factor.

The glucocorticoids circulating in the blood are mainly reversibly bound to an α-globulin and are therefore not able to diffuse freely into the cells. This bound supply of hormone constitutes a circulating store providing a supply of these hormones to some extent independent of rapid change in production, but the details of the control system are obscure.

11. Mineralocorticoids

Aldosterone is controlled very differently from the other corticoids. It is not carried bound in the blood, nor is it regulated by ACTH, although this has some effect in increasing it. There is no negative feed-back effect of the steroid in preventing its own production by the gland. Probably the cells of the adrenal cortex are directly sensitive to the levels of sodium and potassium in the blood. They may also be influenced by the hormone angiotensin liberated by the kidney (p. 222) (see Edelman and Fimognari 1968).

12. Effects of corticoids in the body

We are far from a complete understanding of the significance of the steroids for the metabolism of cells in general but it may be that the activities of the adrenal cortex represent a specialization of steroid exchanges that occur in all cells throughout the body. The cortical hormones influence many tissues and it is uncertain what common factor, if any, is involved. The effect of these substances is to assist the mechanisms by which each type of cell mobilizes energy for the performance of its characteristic type of work. There are indications that the adrenal steroids influence the permeability of the surfaces of cells and perhaps also of mitochondria and of other surfaces within cells at which enzymic actions proceed (p. 456). Among other effects the cortical steroids influence carbohydrate metabolism and water and salt exchanges, which are among the most fundamental of all cell activities. Adrenal de-

ficiency leads to change in concentration of carbohydrates and salts in the blood, and especially to a fall in blood sodium. The intracellular potassium is raised as the sodium of the intercellular fluids falls. Further effects follow these changes, especially loss of water (see Feigelson and Feigelson 1964; Renold and Ashmore 1960).

It may be that the fundamental effect of the corticosteroids is on the energy-producing systems of the cells. Control of these would, of course, influence many other metabolic activities, including protein metabolism, fat production, and salt exchanges. There are indications that all the adrenal and gonadal steroids act in this way on the fundamental metabolic activities of the cells. The actions of the different steroids differ in the details of their effects and are thus able to serve as signals that produce specific influences on the tissues.

In man there is a daily rhythmic variation in the level of cortisol in the blood. The maximum is at 6 a.m. and there is a decline throughout the waking hours. The ACTH varies correspondingly being almost absent from the blood at 6 p.m. This circadian rhythm seems to ensure appropriate levels for an active day. If the individual changes his day to two 8-hour waking periods separated by sleep then after a few days two pulses of plasma cortisol appear. The rhythm probably arises in the hypothalamo-pituitary system.

After removal of both adrenals there is loss of appetite, vomiting, muscular weakness, lowering of blood-pressure and of temperature. Lesions appear in the linings of the gut and the actions of the renal tubules are affected. Death follows within 2 weeks after removal of the adrenals in most animals, but survival is prolonged during pregnancy, perhaps on account of the presence of progesterone from the corpus luteum. Injection of progesterone will prolong life after adrenalectomy, but complete protection is obtained by injection of cortical extract or cortisone, especially if accompanied by a diet low in potassium but rich in sodium. Doses of large quantities of sodium ions are of considerable benefit in maintaining life after adrenalectomy.

13. Reaction of the adrenal to stress

The term 'stress' is used for any condition that tends to upset the normal homeostatic equilibrium. Since such upsets are inevitable for every creature the body is provided by heredity with specific mechanisms for bringing it back to its steady state. After an animal or man has undergone stress, say by trauma, there are changes in glucose tolerance and nitrogen excretion and other signs of alteration of the fundamental processes of energy interchange and protein metabolism by the cells. The adrenal cortical secretions play a part in these adjustments. For example, adaptation of the blood sugar and liver glycogen to low atmospheric pressure does not occur after adrenalectomy. The adrenal cortex is very sensitive to the demand for its secretions by the tissues. If one or more of these target organs uses increased amounts there follows a hypertrophy of the adrenal cortex. Response by this gland therefore plays a large part in the processes by which the organism adapts itself to particular conditions.

Both parts of the adrenal are thus involved in the actions in which the body reacts to situations that are liable to produce serious disturbance of its activities. Selye and others have recently emphasized the wide variety of changes that are involved in this *general adaptation syndrome* in response to stress. Stress may be set up in various ways, for example, by cold, fatigue, infections, or toxic agents. In response to any of these the body shows a common syndrome, which includes discharge of adrenal hormones, hypertrophy of the adrenal cortex, involution of the thymus and lymphoid apparatus (p. 188), and disturbances of the gastrointestinal tract, kidneys, reproductive organs, and others. This syndrome includes both immediate defence reactions that are produced mainly through the sympathetic adrenal medullary system and slower adaptive reactions, which, as Selye puts it, 'comprise the "*learning*" of defence against future exposure to stress' (Selye 1956).

The control of the adaptation syndrome is complicated. The hypothalamus plays a central part. Its neurons are activated by disturbances, whether these are nervous or metabolic. The hypothalamic neurons liberate the releasing factor that leads to increased production of ACTH (p. 440). The adrenal cortex responds to this substance during the immediate, or *alarm reaction*, stage by an output of steroid hormones and other substances. This is seen histologically within a few minutes of stressing (say, by exercise or haemorrhage) as a loss of lipid granules and cholesterol from the cells and a reduction of the normal high ascorbic acid content of the gland. There follows a stage in which the body shows increased powers of resistance to stress and during this period the adrenal cortex increases in width and there is an increase in the lipid granules, and in cholesterol and ascorbic acid. At the same time there is an increase in the number of mitotic figures in the cortex, especially in the outer layers (zona glomerulosa). Stress of a continuous type, such as is produced by fasting, low temperature, or pregnancy, produces a gradual increase in the amount of ACTH secreted and in the size of the adrenal gland. Following removal of the adrenals animals show a greatly reduced

capacity to resist the damaging effects of various environmental conditions, such as cold, burning, anoxia, or infection.

The cortical steroids probably exert their protective effects in different ways under various conditions. It has been suggested that the gland serves to neutralize toxic products of shock, such as histamine, but that has never been proved. Injections of ACTH or corticoids lead to stimulation of the activity of the lymphoid tissue and to the breaking up of great numbers of lymphocytes. If these cells constitute a reserve of the serum globulins responsible for antibody production (p. 191) we should see here the explanation of the activity of the gland in resisting infection.

Injections of ACTH and cortisone have a marked influence upon a series of conditions such as rheumatoid arthritis, which may collectively be called *collagen diseases*. It is not clear how these responses are related to other aspects of the adaptation syndrome, or indeed whether there is any such connexion. The influence may be directly upon the fibroblasts and other mesenchymal tissues or upon some aspect of the antigen–antibody relationship.

An interesting example of the effect of different environmental conditions is seen in the adrenals of rats of wild and laboratory stocks described by Richter (Fig. 46.4). The male rats whose adrenals are shown both weighed 390 g, but the adrenal weighed 93 mg in one and only 18 mg in the other. Correspondingly it was found that the domesticated rats showed a much lower resistance to various poisons than did the wild

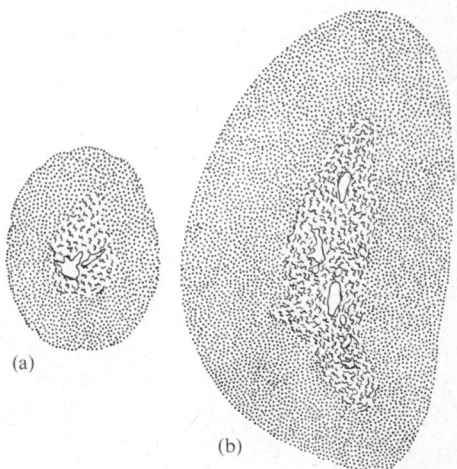

(a)

(b)

FIG. 46.4. Diagrams of sections of the adrenal glands of (a) a domesticated rat and (b) a wild rat, showing the effects of environmental conditions on the size of the adrenal glands. (After Richter (1952). *Ciba Fn. Colloq. Endocr.* **3.**)

strain. On a deficient diet the laboratory animals had fits and died within 5–15 days, whereas the wild ones showed few fits and never succumbed. The behaviour of the two strains is also markedly different, the wild animals being fierce and aggressive. In captivity they are always ready to escape and they feed and breed poorly; animals of the laboratory stock rarely try to escape and they breed well. After adrenalectomy the laboratory rats were readily kept alive by suitable provision of salt solutions, whereas the wild ones always died under this therapy and often did not survive even with large doses of deoxycorticosterone.

Thus the condition of the adrenal system and indeed of the whole defence mechanism corresponds to the environmental conditions under which the animals live. It is not certain how much of the difference is due to hereditary influences and how much to memories established during the lifetime of the individual. Norway rats have become domesticated for laboratory use only during the last 100 years at most. Yet it may be that selection of genes has produced large changes even during this relatively short time. Richter observes that selection would operate especially severely at times of mating and lactation. Wild rats are so fierce and 'suspicious' that only the tamest of them will mate in captivity. Similarly the mothers are so 'apprehensive' that at the least disturbance they kill the entire litter. Laboratory conditions could rapidly provide selection for the tamer types. Richter reports that attempts to tame young wild rats are only partly successful. They remain nervous and ready to bite or escape and their adrenals are large. Yet, it may be that a large part of the difference depends upon the amount of stressing received during the lifetime and hence on the different conditions of the adrenals and other organs.

14. The adrenal glands and sexual development

The adrenal cortex sends signals that assist in the control of sexual differentiation and the maintenance of the oestrous cycle in the female. After removal of the ovaries cyclical changes may continue in the uterus and vagina, but they cease if the adrenals are removed and the animal is maintained on a uniform dose of cortical extract. This suggests that rhythmic changes in the adrenal may be an essential part of the basic timing mechanism of the oestrous cycle.

Abnormal activity of the adrenal cortex is associated with abnormalities of sexual development and functioning. These are usually in the direction of masculinity, producing precocious development in boys and masculinization of girls. The overactivity, which may be due to a tumour, leads to production of unusual amounts of androgenic steroids. The fact that such changes can so greatly alter the time of onset of sexual maturity makes plausible the suggestion that mutation and selection may have operated through the adrenal to produce new races. If, as has been suggested, man has developed by the neoteny of a race of apes, the necessary genetic changes may have affected the functioning of the adrenal cortical tissue, perhaps related to a change in the action of the pituitary gland.

47 The pancreatic islets and the control of carbohydrate metabolism

1. Diabetes

IT was discovered at the end of the nineteenth century that removal of the pancreas from a dog or cat produces a characteristic set of symptoms involving disturbance of the carbohydrate metabolism. The amount of sugar (glucose) present in the blood, normally about 0·12 per cent, rises after the operation to 0·2 per cent and even 0·4 per cent (hyperglycaemia), and sugar appears in the urine. Meanwhile sugar is produced in great amounts, mainly from protein, and there are also characteristic disturbances of fat metabolism. The animal shows muscular weakness, wasting, and other abnormalities and usually dies within three weeks. This set of symptoms resembles that seen in the human condition known as *diabetes mellitus*, in which sugar is found in the urine. The condition diabetes (Greek = a siphon, from *diabainein*, to pass through) was so named by Areteus relative to the symptom of excessive urination—'the epithet diabetes has been assigned to [this] disorder being something like the passing of water by a siphon'. Willis who in 1670 discovered the sweet taste of urine in one form of the disease designated it diabetes mellitus (literally, honey-sweet), to distinguish it from diabetes insipidus, in which excessive flow of urine is now known to be due to disturbance of the antidiuretic hormone of the posterior pituitary (p. 445).

2. The pancreatic islets

The pancreas, besides producing digestive enzymes, in its secretory acini, also contains groups of cells that have a rich blood-supply but no secretory ducts to the exterior. The veins drain into the portal vein, allowing the hormones to act immediately on the liver. As early as 1909 it was suggested that these cells produce a substance, *insulin*, that is responsible for preventing the conditions seen after removal of the pancreas. In 1922 Banting and Best provided satisfactory evidence that extracts of the pancreas are able to alleviate the symptoms of diabetes. The final proof that the islet cells produce the hormone has been complicated by the fact that they are mixed with the enzyme-secreting acinar tissue, and by the presence of two main types of cell within the islets. The β-cells are the most numerous and contain granules soluble in acid alcohol whereas those of the α-cells are insoluble in it (see Krahl 1974). It is now considered that the β-cells are the source of the hormone insulin, for the following reasons.

(a) In bony fishes the islets and enzyme-producing tissues can be separated and the former are found to contain insulin but not the latter. (b) Ligation of the pancreatic duct in a mammal is soon followed by degeneration of the acinar tissue but the islets remain for a while and during this period the insulin content of the pancreas remains high. (c) If part of the pancreas is removed the β-cells of the remainder undergo changes that can be interpreted as hypertrophy, especially if the animal is kept on a high carbohydrate diet. These changes do not, however, occur if injections of insulin are given. (d) When there is a tumour that increases the number of the β-cells, abnormally large amounts of insulin appear, moreover, growths of the islet tissue in abnormal parts of the body by metastasis of the tumour produce insulin. (e) Conversely if the β-cells (but not the α-cells) are destroyed by injection of certain anterior pituitary extracts or of the substance *alloxan* then the insulin content of the gland is found to fall proportionately.

The α-cells produce *glucagon*. This is a single-chain polypeptide hormone which increases breakdown of glycogen and formation of glucose in the liver and hence increases the level of glucose in the blood, an effect opposite to insulin. Glucagon also has a potent effect on the β-cells, increasing the production of insulin.

3. Structure of insulin

Insulin has now been isolated and crystallized in the form of combinations that it makes with zinc or other metals. It is a protein, of minimum molecular weight about 6000 but normally polymerized. All insulins consist of an A, or glycine, and a B, or phenylalanine chain, of 21 and 30 amino acids, respectively, inter-

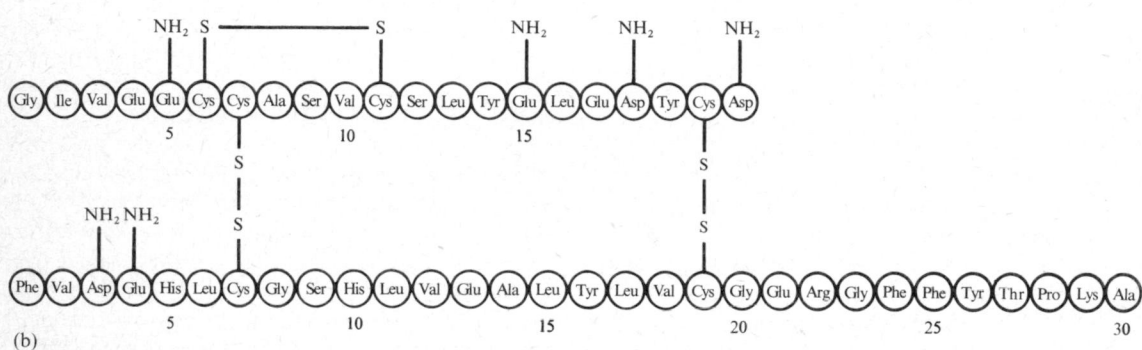

FIG. 47.1. Amino-acid sequence of insulin. (a) Invariant residues of 20 animal insulins; the residues left blank differ in insulins from different animals. (From Grant and Coombs (1970). *Essays Biochem.* **6.**) (b) Beef insulin (From Sanger, Thompson, and Kitai (1955). *Biochem. J.* **59.**)

connected by two disulphide bridges (Fig. 47.1). The sequences may vary considerably in different species, mainly at amino acids $A_{8\ 10}$ under the disulphide bridges, without impairing activity. Insulin forms crystals in the presence of zinc, in which three insulin dimers make a spheroid around two zinc ions. The A_{19} tyrosine and the asparagine at the COOH-terminal of the A chain are essential for the biological activities, and disruption of the disulphide bonds removes all activity. Insulin was the first protein to be synthesized, independently by Katsoyannis and Meienhofer in 1963.

4. Secretion of insulin

Insulin is produced in the β-cells of the islets by ribosomes attached to the rough endoplasmic reticulum. It is transferred via the Golgi bodies to membrane-bound vesicles in which it is stored as granules which are electron-dense, by virtue of their zinc (Grodsky 1970).

Release of the insulin is by movement of the granules to the surface and fusion of the vesicle membrane with the plasma membrane (Fig. 47.2). The movement may

be produced by actomyosin-like filaments that are present, and is inhibited by colchicine, which destroys microtubules.

Insulin is liberated when the cells are stimulated by glucose. The first effect is a rapid release within one minute from a small labile 'compartment', probably the vesicles. The entering glucose is rapidly phosphory-lated to glucose 6-phosphate and via the cyclic-AMP mechanism may perhaps activate the filaments. This phase is not affected by inhibition of protein synthesis with puromycin. Only up to 3 per cent of the pancreatic content is released in this way. After a few minutes of continuous stimulation the rate of secretion falls, due to feed-back inhibition. After several further minutes of glucose stimulation there is then a secondary rise in insulin output. This is puromycin-sensitive and there-fore involves synthesis. Secretion stops at once if the stimulus is removed, but there may be excess storage if stimulation has been prolonged, leading to excessive release in later renewed stimulation. Insulin injected into the blood is rapidly removed; 90 per cent has disappeared after 20 minutes (in man).

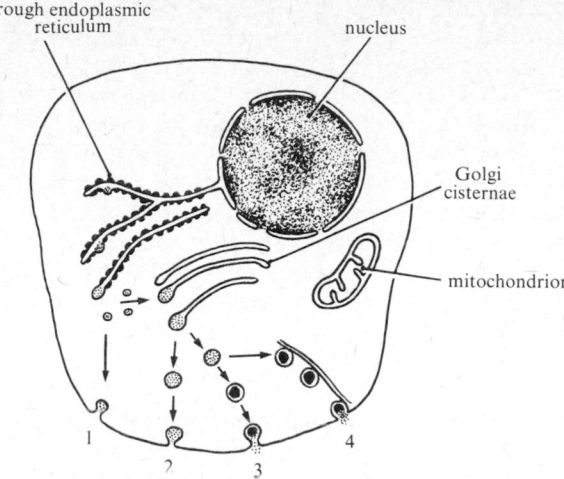

rough endoplasmic
reticulum

nucleus

Golgi
cisternae

mitochondrion

1   2   3   4

FIG. 47.2. Possible methods of insulin release.
The rough endoplasmic reticulum of β cells contains a pale
grey amorphous substance presumably containing the proinsulin polypeptide chains formed by the attached ribosomes.
Direct release may occur (1) but most proinsulin is transferred
to small Golgi cisternae and packaged in membrane bound
granules in which transformation to insulin probably occurs,
either in the Golgi or just after vesicle formation. Release may
also take place at this stage (2), however, the primary storage
site of finished insulin is the β granule and these move to the
surface and fuse with the plasma membrane, rupture and
liberate their contents (3). It has been suggested that microtubules play a role in the movement to the surface (4).

Other factors that stimulate the secretion of insulin
are pituitary growth hormone, adrenal and gonadal
steroids, thyroid hormone, and amino acids, fats, and
other metabolites. The action of most of these on the
pancreas is indirect however, the main direct influence
being the action of glucose which enters the cells
directly.

5. Actions of insulin

The effect of injection of insulin is to produce rapid
and complete reversal of the hyperglycaemia and other
symptoms of pancreatic removal: the blood-sugar level
falls and glucose disappears from the urine, protein
breakdown is reduced, abnormalities of fat metabolism
are corrected, and glycogen begins to be laid down in
the liver. These results suggest that insulin has an effect
on the oxidation of sugar by the tissues and on the
utilization of glucose in the liver for conversion to
glycogen. Like so many other hormones, therefore,
insulin affects one of the fundamental metabolic processes that occur in every cell throughout the body;
in this case the oxidation of carbohydrates by which
energy is provided for doing work (see Levine and
Goldstein 1955; Wool, Stirewalt, Kurihara, Low,
Bailey, and Oyer 1968).

Insulin is usually described as an anabolic hormone.
It increases the uptake and oxidization of glucose by
muscle and in fact increases the ability of tissues to use
sugar and fats. It also increases the incorporation of
amino acids into protein and this is probably a direct
effect, independent of glucose. The regulation of the
normal growth of many tissues is the result of the co-
ordinated action of several hormones, especially insulin
and somatotropin (p. 439).

6. Sugar metabolism

Various sugars can be utilized by the body but before
being oxidized they are converted to glucose and we
may therefore consider only the metabolism of this
substance. Glucose is a mixture of two isomers in
which the α form predominates. The amount of sugar
in the blood of man is kept at between 0·08 per cent and
0·16 per cent. This sugar serves to provide fuel for
combustion by the muscles and other tissues to provide
energy. The breakdown of the glucose, like many other
of the metabolic operations of carbohydrates, involves
phosphorylation. The glucose is converted by a series
of enzymes, first to hexose phosphate, and then to
triose phosphate, which in turn is converted to pyruvic
acid, and this, if there is abundant oxygen available
through the cytochrome and other oxidase systems,
breaks down to carbon dioxide and water (the Krebs
cycle).

Alternatively, under anaerobic conditions, the
pyruvic acid forms lactic acid; in either case the effect
is that when glucose is present in large amounts the
series of reversible enzyme reactions proceeds in the
direction of glucose, with liberation of energy. The
whole process is accelerated by insulin and takes place
less readily in its absence, but the exact points of the
cycle at which the insulin acts have not been determined.

The carbohydrate of the blood is kept constant by
ingestion from the alimentary canal and by formation
of glucose in the liver, either from the carbohydrate
reserve (glycogen) or if necessary by production of
glucose from other sources, such as protein or fat
(gluconeogenesis). Following ingestion of carbohydrate
the blood glucose may rise above 0·2 per cent but the
excess is rapidly removed from the circulation by conversion to glycogen in the liver and muscles. Glycogen,
like starch, is a polysaccharide, formed from a number
of glucose units $(C_6H_{10}O_5)_n$. It is formed from glucose
after phosphorylation to give a hexose phosphate, from
which by further conversions the polymer is produced.
The reaction is reversible and the influence of insulin is
probably to push it in the direction of increased
glycogen production. Adrenaline has the reverse effect
and its liberation during muscular activity therefore

assists in maintaining stable conditions by ensuring mobilization of glucose from the liver. An example of this action is the apocryphal case of the diabetic who, having taken an overdose of insulin, suffered from hypoglycaemia and was arrested as a drunk. His anger at the injustice of the treatment given him by the police was accompanied by adrenaline production sufficient to mobilize sugar, which effected a cure that astonished the police surgeon.

The human liver contains about 100–200 g of glycogen, which would provide for less than a day's supply if the body were left without food. Provision of glucose to the tissues is essential for the maintenance of life; the nerve cells in particular are soon damaged in its absence. The results of hypoglycaemia (reduced blood sugar) are first of all a hyperexcitability and desire for food and then in extreme stages convulsions and coma, due to impairment of the actions of the nerve cells.

7. Control of blood sugar

Claude Bernard long ago showed that damage to certain parts of the medulla oblongata, by the insertion of a needle, produces irregularities of carbohydrate metabolism, his so-called '*diabetic puncture*'. We have little further information about this 'centre' or its method of action, presumably it is through the vagus. The hypothalamus probably acts as a still higher centre influencing carbohydrate metabolism.

The rate of glucose metabolism, like so many other activities of the body, is thus subject to the influence of many factors and it is difficult to say with certainty that it is 'controlled' by any single organ. The amount of sugar ingested and the demands made for fuel by the activities of the body directly influence the blood-sugar level. If these factors operated alone there would be fluctuations in the amount of glucose available for the tissues, whereas a constant supply is needed for their proper working, especially for the brain and the heart. Two means are available to ensure this (a) a store of carbohydrate in the liver, in the form of glycogen, which can be readily converted into glucose; and (b) an enzyme system that is able to produce sugar from protein sources if needed. To ensure that these systems are called into action at the right times and to the right extent there is an elaborate hormonal and nervous control (see Cahill, Ashmore, Renold, and Hastings 1959).

Rapid mobilization of reserves is largely produced by secretion of adrenaline. Increased utilization of excess carbohydrate depends on increased secretion of insulin. Adrenal cortical and perhaps thyroid hormones stimulate the process of gluconeogenesis, by which the extra sugar is formed when necessary, and these glands operate under control of the anterior pituitary, which also has direct effects by producing a substance antagonistic to insulin. Regulation of the pituitary itself is a complicated affair, probably dependent partly on actions of the central nervous system and partly on the reciprocally stimulating effects of the pituitary on the other endocrines and the influence of the level of blood sugar and other metabolites on these organs. A characteristic feature throughout this, as so many other systems in the body, is the tendency of the action of a set of cells to be increased when the demand for its product rises.

8. Evolution of blood-sugar control

This elaborate balance of activities regulating the blood-sugar level is probably a recent mammalian acquisition, but we have too little information about the conditions in lower vertebrates to be able to trace its history in detail. As in other endocrine systems we find a key position occupied by a substance that is produced in a gland of internal secretion, the pancreas, which, within the chordate series, has become differentiated from a tissue that was earlier concerned with ingestion from the outside world. It is interesting that in lampreys the islet cells still form part of the intestinal wall. The cells lining the intestine have thus become regulators of carbohydrate metabolism by their actions at the source. The balance of sugar between these cells and the blood probably originally depended on the demands made by other tissues. A fall in blood sugar would stimulate within the intestinal cells the processes of sugar absorption and production, which are presumably common to all tissues.

In this way we can imagine that the process of producing the substance that is able to stimulate carbohydrate metabolism became especially active in these intestinal cells. Those animals with sets of genes ensuring that this activity was especially pronounced would be particularly successful and the intestinal cells would come to have their effect not only by actively producing sugar for the blood but by liberating insulin into the blood to stimulate appropriate processes elsewhere in the body. By some such method a substance related to a particular metabolic process came to be produced at a single central site, the pancreas, by a specialization of activities that are common to tissues throughout the body.

The adrenal cortex plays an equally prominent part in the control of carbohydrate metabolism, acting upon the intracellular metabolic activities throughout the body. Thus many influences affect the blood-sugar level, some mainly at the points of entry and production of carbohydrates, others controlling their use. The interaction of this elaborate set of factors to control the blood-sugar level in a mammal presumably serves to maintain a constancy that would not be possible with a simpler system. The demand for sugar by the tissues sets in action a series of processes that not only release carbohydrate from reserves but also stimulate its production and probably activate the animal to search for fresh supplies. No detailed analysis of this elaborate homeostatic system is yet available but it serves as an excellent example of the way in which the living organization maintains its constancy.

48 The pituitary gland

1. The master gland

THE pituitary body (hypophysis cerebri), like the other ductless glands, is derived from tissues that previously performed other functions. Its activities are a special development of a fundamental part of the control system of all cells, namely, that which regulates metabolism and growth. The effects of the pituitary hormones influence the processes of synthesis of the very substance of the cells, including the metabolism of proteins, carbohydrates, and fats. The substances produced by the gland are themselves proteins and some of them act upon the enzyme systems of cells throughout the body that affect synthesis and growth. Many other activities are also influenced by the pituitary and thus proper co-ordination of these with the metabolic and growth processes is ensured (see Harris and Donovan 1966; Haymaker, Anderson, and Nauta 1969).

On account of the multiplicity of its effects the pituitary is therefore accurately described as a master gland; it is the central computor of the chemical co-ordinating system, acting upon the other endocrines and in turn acted upon by them so that it provides a focal point at which much of the bodily activity is adjusted (see Ganong 1974; Ganong and Martini 1969).

The gland weighs only about 0·5 g in man and lies in the sella turcica of the sphenoid bone, close to the optic chiasma and attached by the infundibular stalk to the floor of the diencephalon. In the regulation of many processes it works with the overlying nervous centres of the hypothalamus, which constitute the central neural mechanism for the control of the internal activities of the body.

2. Phylogenetic history of the pituitary

The gland develops from two distinct rudiments, an inpushing of the buccal ectoderm, known as *Rathke's pouch* (hypophysis in the narrow sense), and a downgrowth from the brain, the *infundibulum*. The buccal portion perhaps corresponds to Müller's organ in *Amphioxus*, which is a patch of thickened ciliated and glandular ectoderm, probably concerned with feeding. In the sea squirts (Tunicates) there is a ciliated pouch, the *subneural gland*, lying in a similar position. Carlisle has shown that this pouch is a receptor organ, serving to control feeding. It also detects the presence of eggs or sperms in the water current and causes liberation of sperms, or ovulation, by production of a chemical substance that stimulates the nearby ganglion. The substance may be related to the gonadotropin of higher chordates and injection of mammalian chorionic gonodotropin into a sea squirt is followed by liberation of gametes.

The two parts of the pituitary are found constantly throughout the Craniata but with some differences of position. The cavity of Rathke's pouch remains large and open to the exterior in lampreys and in the fish *Polypterus*, but becomes much reduced in many vertebrates and in adult man consists only of a few separated spaces. The cavity of the infundibular stalk likewise varies from a large space in cyclostomes to none at all in man.

3. Parts of the pituitary

The major parts of the adult pituitary (Fig. 48.1) are the *pars anterior* (= *distalis*), formed by folding of the epithelium of the front face of Rathke's pouch, the *pars intermedia*, developed from its hinder face, *pars tuberalis* from its upper portion, and *pars nervosa* from the infundibulum. From the position of the rudiments it follows that, in mammals, the pars intermedia lies against the pars nervosa and the pars tuberalis surrounds the infundibular stalk and thus lies close to the tuber cinereum of the brain. It has been common in the past to distinguish between anterior and posterior lobes of the pituitary, with a dividing line at the pituitary cleft (remains of the cavity of Rathke's pouch), but it is not certain that the parts of the posterior section act as one unit (see Harris and Donovan 1966).

These four parts can be recognized in nearly all vertebrates (Green 1951), but their relative positions in

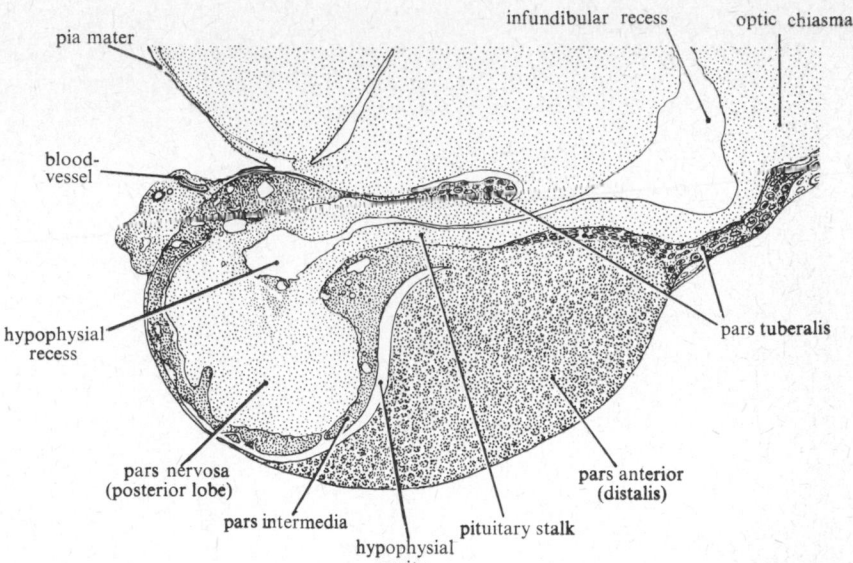

Fig. 48.1. Longitudinal section of the hypophysis of a cat. (From a photograph by Mr. K. C. Richardson.)

the adult may vary, for instance, in elasmobranchs the pars 'anterior' is behind the others and in anurans the tuberalis forms a separate region. Experimenters have taken advantage of such special conditions for the independent removal of the parts. The parts are not developed to the same extent in all vertebrates, for instance, there is said to be no intermedia in birds, whales, sea cows, or armadilloes. Again, the pars nervosa is a single-layered epithelium in lampreys but a relatively massive solid organ in man. Much may be learned from such differences; it is to be expected that an organ so closely bound up with the growth and life of the animals would change as the life changes.

4. The hypophysial portal system

The blood-supply of the pituitary is peculiar in that it comes from the hypothalamus through a portal system. Capillaries within the tissue of the infundibulum unite to form vessels that pass down the pituitary stalk and break up again to form sinusoids within the adenohypophysis itself. From here blood is then collected into hypophysial veins, discharging into the jugular system. This portal circulation is the means by which the pituitary secretions are controlled. There are no secretomotor fibres to the cells of the pars distalis. Instead they are regulated by the action of a set of *releasing factors*, substances secreted by the nerve cells of the hypothalamus and discharged in the infundibulum and carried in the portal circulation to activate (or inhibit) the cells of the pituitary (p. 442).

The interrelations of the various parts of the hypothalamus, infundibulum, and adenohypophysis are so close that difficulties arise in naming them. The hypothalamus proper may be held not to include the stalk of the infundibulum, which has a distinct structure and is best regarded as part of the neurohypophysis. In particular, its upper part, the *median eminence* of the tuber cinereum is highly vascular and is the place where capillary loops of the hypophysial portal blood-vessels take up the releasing factors for carriage to the adenohypophysis (Figs 48.2 and 48.3). This is done by the blood-vessels along the stalk or stem of the infundibulum, which lead to the adenohypophysis. The neural lobe (pars nervosa) itself is separately supplied by blood from the inferior hypophysial artery. It receives the nerve fibres of the hypophysial tract, which regulate the secretions of its hormones (p. 445) (Xeureb, Pritchard, and Daniel 1954*a*, 1954*b*).

5. Pars anterior or pars distalis

The secretions of the pars anterior probably affect every cell in the body. They are beginning to be quite well known chemically and six distinct hormones can be recognized.

> Somatotropin = growth hormone, GH.
> Prolactin = lactotropic hormone—mammotropin, MH.
> Thyrotropic hormone, TSH.
> Follicle stimulating hormone, FSH.
> Luteinizing hormone, LH.
> Adrenocorticotropic hormone, ACTH.

Simple methods of staining distinguish three types of cell in the anterior pituitary, with acidophil granules, basophil granules, and non-stainable granules (chromophobic). The identification of the cells responsible for

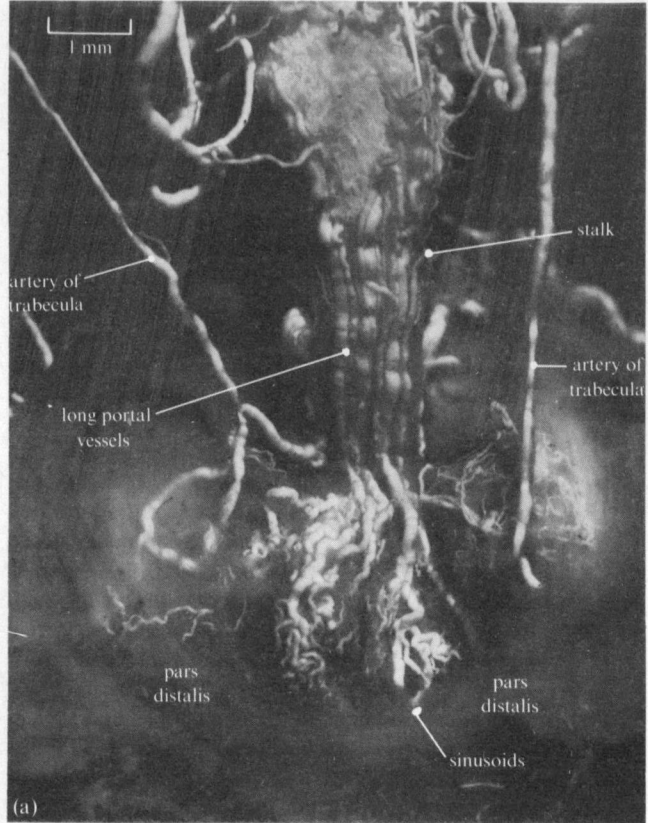

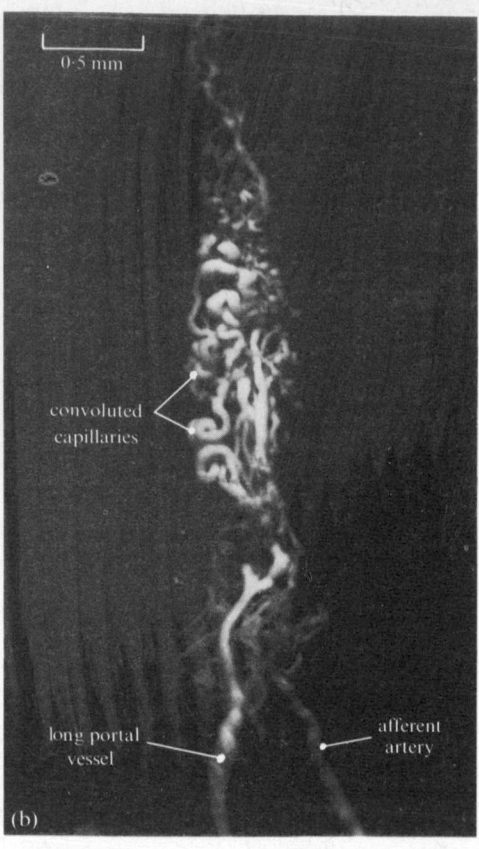

FIG. 48.2. (a) Human pituitary gland seen from in front and above. The blood-vessels have been injected with neoprene latex and the tissues partially macerated, showing the long portal vessels running down the stalk and breaking up into sinusoids of pars distalis, the anterior lobe. (b) Neoprene cast of a long spike of convoluted capillaries taken from an injected human pituitary stalk (all tissue has been macerated), showing the afferent artery to this capillary complex, and the long portal vessel into which it drains. (Figures by courtesy of Professor Peter Daniel.)

the six hormone types found in the pars anterior has been complicated by the variety of staining methods used and the differences between species. A simple general statement is that there are three main cell types, acidophils, basophils, and chromophobes. In some species the basophils occupy a separate zone and it was early found that cells from this zone of the ox stimulate metamorphosis of tadpoles and hence presumably include the thyrotropes. Cells of the acidic zone caused growth but no metamorphosis and hence include the somatotropes.

After much further study it has been shown, using Herlant's stain, that there are two sorts of acidophil, somatotropes (aurantiphils), staining orange, and mammotropes or lactotropes (erythrophils), staining red with carmine, which produce prolactin.

The basophils have caused even greater difficulty but are now shown to include three sorts. The thyrotropes stain blue with aldehyde-fuchsin, hence cyanophils. The gonadotropes stain red but are said by some to be of two sorts, those with masses of flocculent granules secrete FSH, whereas the LH cells have dispersed granules.

Finally, the corticotropes are the chromophobes, without stainable granules.

Acidophils	Somatotropes,	aurantiphils
	Lactotropes,	erythrophils
Basophils	Thyrotropes,	cyanophils
	Gonadotropes,	amphophils
	Folliculotropes,	flocculent granules
	Interstitiotropes,	dispersed granules
Neutrophils	Corticotropes	

The cells can also be distinguished with the electron microscope, mainly by the sizes of their granules. Thus mammotropes have very large irregular granules (6–900

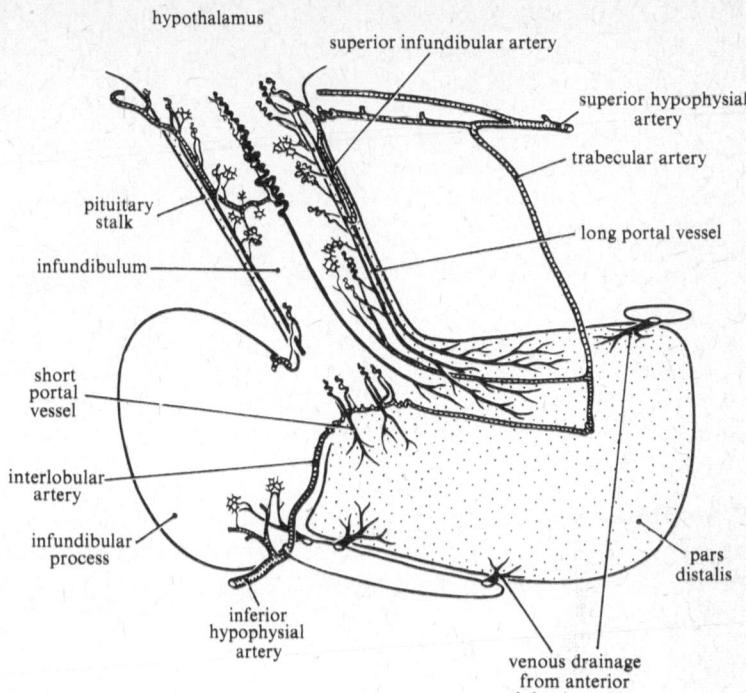

FIG. 48.3. Diagram of the portal blood-supply of the human pituitary. Both the superior and inferior hypophysial arteries break up into capillaries in the infundibulum. From these capillary beds the portal vessels conduct to the sinusoids of the pars distalis. These drain into veins in the capsule. (After Xeureb *et al.* 1954.)

μm), somatotropes smaller ones (350 μm), gonadotropes smaller still (200 μm), and thyrotropes only 150 μm. These granules can be collected in separate pellets by ultracentrifugation and shown to contain the appropriate hormones.

6. Growth hormone

The connexion between the pituitary gland and growth was recognized by clinicians and anatomists long before the concept of hormones was established. In the condition of gigantism or *acromegaly* the bones are very long and the pituitary fossa in the skull is found to be enlarged. The proof of the existence of a growth-promoting factor was given by Evans and Long in 1921. They made intraperitoneal injections of saline pituitary extracts into rats, which proceeded to grow to twice the size of controls. The substance responsible has been prepared in highly purified form from at least six species, including man. There are differences between animals but the hormone is a protein with molecular weight ranging from 21 500 to 48 000. The amino-acid composition has been determined but not the sequence.

Growth hormone has various metabolic actions. It probably stimulates growth by increasing protein synthesis. Injections of it cause a fall in the level of amino nitrogen in the blood. Dwarf races of mice lack acidophil cells in the pituitary and conversely they are numerous in giants. Human growth hormone can be used to increase the growth of dwarfs.

Somatotropin has numerous other metabolic effects. It may cause the deposition of fat in the liver. It mobilizes sugar from the liver, acting in the opposite sense to insulin in producing hyperglycaemia (p. 433). It may have effects on the islet tissue of the pancreas. It influences electrolyte balance, causing retention of sodium, potassium, and phosphorus and disturbance of the calcium/magnesium balance. It is not known what general actions lie behind these effects but clearly somatotropin is one of the most important controlling agents of metabolism.

7. Lactogenic hormone, prolactin or mammotropin

The development of the breasts and production of milk are complex processes affected by sex hormones as well as by both anterior and posterior lobes of the pituitary. The growth of the gland and its active production of milk are distinct though related processes. The anterior pituitary hormone, prolactin, or mammotropin, is a protein produced by a special type of acidophil cell, the mammotrope. Injection of this substance will produce milk secretion even in a rabbit oophorectomized when adult, though it will not cause growth of the mammae of an immature animal. The substance also has the effect of causing secretion of 'milk' by the crop of a pigeon. Perhaps the clue to this curious connexion is the similarity of the metabolic processes involved. The hormone is a protein with molecular weight about 23 000. The amino-acid composition is known but not the sequence (see Wolstenholme and Knight 1972).

The actual release of milk under the tactile stimulus of suckling is produced by a hormone of the posterior pituitary, oxytocin (p. 445).

8. Thyrotropic hormone, TSH

This is produced by specific basophilic cells of the pars anterior. It causes both increased production of thyroid hormones and their release. Ten minutes after a single intraperitoneal injection of TSH into a rat, droplets of colloid are seen being taken up by pinocytosis from the lumen of the thyroid follicles into the cells (p. 419). The activity of TSH is assayed by the response of the thyroid, or by its effect on the thyroid of hypophysec-tomized tadpoles. The only other action of TSH is to produce release of free fatty acids from adipose tissue.

The hormone has proved difficult to purify. It is a protein with molecular weight between 28 000 and 40 000. There are considerable differences in the amino-acid composition in different vertebrates and it is pre-sumably for this reason that eel TSH, for example, has little effect in rats. The level of TSH has been studied in human plasma in various conditions. It is increased in people with untreated thyroid deficiencies (cretinism or myxoedema) and lowered in hyperthyroidism.

9. The gonadotropic hormones

The effects of the pituitary upon the gonads were dis-covered by P. E. Smith in 1926. Already by 1931 two substances were separated, one causing follicular growth in the ovary and the other rupture of the follicles and discharge of the ovum and formation of a corpus luteum. These two, follicle stimulating hormone (FSH) and luteinizing hormone (LH) are produced by basophil cells and following castration these cells increase in the pituitary. Although it is claimed that two distinct sorts of basophils are involved there is still some uncertainty about the distinctness of the two hormones and their effect. Neither has been prepared in a form that is pure and without the effects of the other. The situation is complicated by the fact that they work together ('synergistically'). Thus LH can cause discharge only of follicles that have been ripened under the influence of FSH.

Nevertheless there is no doubt that two at least partly distinct hormones are involved. FSH can be assayed by its effect in causing growth of the ovarian follicles of hypophysectomized rats. It also causes increase of weight of the testes of similarly treated

males, but LH has the same effect. FSH also causes discharge of spermatozoa by *Xenopus* and this effect is specific, but not a very sensitive test. Preparations of partly purified FSH have a molecular weight of about 30 000. They contain carbohydrate as well as amino acids. FSH can be assayed in human plasma and shown to increase at menarche and especially after the meno-pause. It can be separated from human urine and, here again, increases in older women.

Luteinizing hormone, besides its action on the follicles, causes differentiation of the interstitial cells of the testis, and the discharge of androgen. The hormone is protein-like with a molecular weight between 40 000 and 100 000. LH is also produced by the placenta (p. 492).

The amounts of FSH and LH in the blood are con-trolled by distinct releasing factors, FSHRF and LHRF (p. 444). The regulation is complicated because there are feed-back effects from both pituitary and gonadal hormones. Thus FSH and LH inhibit the production of the releasing factors in the median eminence. In addition there are inhibitory effects of oestrogens and progesterones, probably also acting on the hypothalamus. These effects are important in the regulation of the menstrual cycle and the action of contraceptive steroids (p. 478) (Schally and Kastin 1970).

There are also the reverse effects of the absence of steroid hormones. Moreover, there are important positive feed-back effects. Thus the secretion of oestro-gen stimulates the release of LH, probably by an action on the median eminence.

10. Adrenocorticotropin, ACTH

More is known about ACTH than any other pituitary hormone; its sequence of amino acids has been analysed and it has been synthesized. It is now known that it is produced by the chromophobe cells (p. 438). Its effect is to produce growth of the adrenal cortex, especially proliferation of the cells of the zonae fasciculata and reticularis. It also increases release of the hormones by several times, especially the glucocorticoids (cortico-sterone in the rat or cat, cortisol in dog or man, p. 427). As a result the ascorbic acid and cholesterol content of the adrenal is reduced after injection. ACTH also has several extra-adrenal effects, especially expansion of melanocytes (pigment cells) of amphibia and dissolution of fat (lipolysis). The former effect is presumably due to the presence of the same sequence of seven amino acids in ACTH and MSH (melanocyte stimulating hormone) (p. 444).

The structure of ACTH has been determined by the method of breaking it up into fragments, first applied to insulin by Sanger. Table 48.1 (pp. 442–3) shows the essential steps in the process of identifying the positions of the 39 component amino acids. The structure has been determined for extracts of the pituaries of human, cow, sheep, and pig. The only difference found was of one amino acid in the pig. The entire sequence of 39 amino acids has been synthesized by Schwyzer and Sieber (1963) in Switzerland. The complete sequence is not necessary for activity, which seems to depend upon the first 24 residues from the NH_2-terminal. Residues 4–10 constitute the heptapeptide in common with MSH.

The amount of ACTH in the blood is low but can be assayed. It shows a 24-hour rhythm being greatest at about 6.00 a.m., when the secretion of adrenal steroids is at a maximum. In nocturnal animals the maximum is in the afternoon. This suggests a preparation for the daily activities and it is indeed the function of the adrenal to provide this. In the presence of any especially stressful event the ACTH level increases to provide for the emergency. Conversely ACTH is quickly removed from the blood; its half-life is only 5 minutes.

As with other hormones the amount of ACTH in the blood depends partly on the demand from the target organ. It is markedly raised by adrenal insufficiency, for instance in Addison's disease.

11. The pars intermedia

This is formed from the epithelium of the hinder face of Rathke's pouch and is therefore closely applied to the pars nervosa. It has the single function of producing the melanocyte stimulating hormone (MSH) (also called intermedin) and is reduced in man, where this hormone is probably little used (Fig. 48.4) (see Harris and Donovan 1966, Vol. 3; Wingstrand 1966).

The cells of the pars intermedia are basophil and contain glycoprotein granules; they are supplied by nerves from the pars nervosa. The amount of MSH released can be varied rapidly in fishes, amphibians, and reptiles which show marked colour changes for purposes of concealment or temperature regulation. The changes are produced by expansion and contraction of the pigment granules in the melanocytes. It is not certain whether this is produced by migration of the pigment within the cells or by amoeboid movement. In some fishes the melanocytes are controlled by sympathetic nerve fibres, but in most animals the regulation is by variations in the amount of MSH. The function of MSH in mammals (and birds) is uncertain. The intermediate lobe is quite large in some ungulates (e.g. giraffes). The skin of mammals of course contains melanocytes, though these do not usually vary their state of expansion (p. 23). However, injection of small amounts of MSH (8 mg per day) produces marked darkening of the skin of man. It is possible that MSH has other actions that are more important.

MSH has been obtained in the pure state and analysed and synthesized. It includes two slightly different polypeptides. αMSH contains 13 amino acids

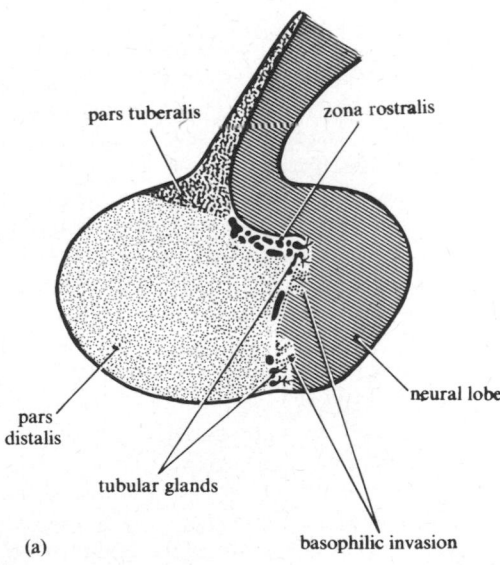

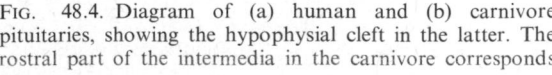

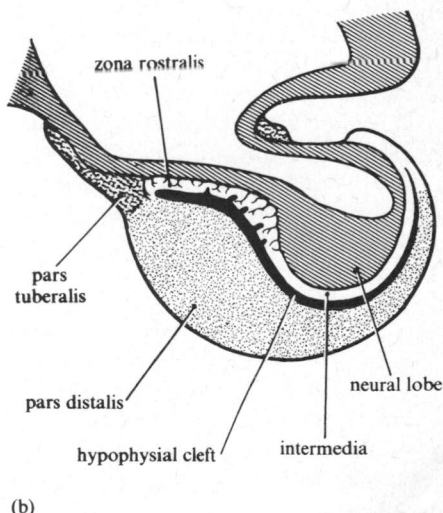

FIG. 48.4. Diagram of (a) human and (b) carnivore pituitaries, showing the hypophysial cleft in the latter. The rostral part of the intermedia in the carnivore corresponds to a zone of cysts in the human. Here there are also tubular glands and areas of basophil cells. (After Wingstrand 1966.)

T.

The sequence of amino
(from Li, D

Agents employed†	Peptide fragments‡		Amino
FDNB	H	Ser	
Ct	C5c	Ser-Tyr	
PITC	H	Ser-Tyr-Ser-Met-Glu	
Ct	C5b	Ser-Met-Glu-His-Phe	
Ct	C4c	Arg-Try	
T	T5	Ser(Tyr, Ser, Met, Glu, His, Phe)Arg	
T	T8	Try-Gly-Lys-Pro-Val-Gly	
Ct	C2a	Gly-Lys-Pro-Val-Gly	
T	T11b		
Ct	C2b		
T	T9a		
Ct	C7a		
Ct	C6a		
T	T2		
Acid-T2	PAH3		
Acid-T2	PAH1		
Acid-T2	PAH2		
Ct	C5e		
Cp	H		
Ct	C7d		

Complete sequence: Ser-Tyr-Ser-Met-Glu-His-Phe-Arg-Try-Gly-Lys-Pro-Val-Gly
                      1   2   3   4   5   6   7   8   9   10   11 12 13 14

† FDNB, fluorodinitrobenzene; Ct, chymotrypsin; PITC, phenylisothiocyanate; T, trypsin; Cp, carboxy-pe
‡ H, hormone; other peptide fragments, derived from H by digestion techniques indicated.

and βMSH 18. The sequence of αMSH is identical with that of the NH$_2$-terminal tridecapeptide of ACTH (p. 441). All three hormones contain the sequence Met-Glu-His-Phe-Arg-Try-Gly. αMSH is more active than βMSH and both are many times more active than ACTH. The pentapeptide His-Phe-Arg-Try-Gly has an effect on melanophores, but is one million times less active than αMSH.

12. Releasing factors

The main agents controlling the production and release of the hormones of the adenohypophysis are the releasing factors. These are peptide hormones produced by the cells of the hypothalamus and carried to the median eminence, where they are released into the portal vessels and passed to the adenohypophysis. It must, however, be remembered that there is a direct effect of circulating hormones on the pituitary, for example, adrenal corticoids suppress production of ACTH (see Geschwind 1969).

The search for the relasing factors was stimulated by the failure to find nerve fibres innervating the cells of the pars distalis. The crucial evidence for the existence of the factors was that if the pituitary was removed from young rats and transplanted back near the median eminence the animals reproduced. If it was placed under the temporal lobe of the brain they did not.

Much further evidence has come from other experiments. After section of the pituitary stalk most anterior pituitary functions are reduced, but may recover if the blood-vessels are allowed to regenerate. The proof that it is the blood-vessels and not the nerves which are involved is that electrical stimulation of the hypothalamus or median eminence will evoke discharge of pituitary hormones but stimulation applied directly to the anterior lobe is ineffective.

The precise sites of origin and means of transport of the releasing factors are not yet known. Almost certainly the mechanism is one of transport of neurosecretory granules, similar to that which is best known in the fibres controlling the pars nervosa (p. 447). Nerve fibres from various parts of the hypothalamus make a

ine adrenocorticotropin
Chung 1961)

nce

g
g-Arg-Pro-Val   NH$_2$
g-Arg-Pro-Val-Lys |
      Lys(Val,Tyr,Pro, Asp,Gly,Glu,Ala, Glu,Asp)Ser-Ala-Glu
      Lys(Val,Tyr,Pro, Asp,Gly,Glu,Ala, Glu,Asp,Ser,Ala,Glu,Ala, Phe)
         Val-Tyr-Pro -Asp(Gly,Glu,Ala, Glu,Asp,Ser,Ala,Glu,Ala, Phe, Pro,Leu, Glu, Phe)
         (Val,Tyr,Pro, Asp,Gly,Glu)Ala
             Gly-Glu(Ala, Glu,Asp,Ser,Ala)
              Ala-Glu-Asp
                              Ala-Phe-Pro-Leu
                                   Leu-Glu-Phe
                                       Glu-Phe

                           NH$_2$
                           |
g-Arg-Pro-Val-Lys-Val-Tyr-Pro-Asp- Gly-Glu-Ala-Glu-Asp-Ser-Ala-Glu-Ala-Phe-Pro-Leu-Glu-Phe
 18   19   20   21   22   23   24   25   26   27   28   29   30   31   32   33   34   35   36   37   38   39

cid-T2, partial acid hydrolysis of peptide T2 obtained from tryptic digestion.

plexus among the neuroglial cells of the median eminence. Presumably the releasing factors are discharged when the neurons giving rise to the fibres of this plexus are activated by appropriate external or internal stimuli.

Releasing factors for all the six hormones have now been identified and many of them purified by chromatography so that they act when injected in nanogram doses. None has yet been prepared pure or synthesized (McCann and Porter (1969)—but see Yates, Russell, and Maran (1971)). All are polypeptides inactivated by proteolytic enzymes, and of low molecular weight (1000–(?)2500). They are heat-stable at 100°C for 10 minutes.

The growth hormone releasing factor (GRF or GHRF or SRF, somatotropin releasing factor) has been isolated from the hypothalamus, and lesions of the hypothalamus produce dwarfs. Injection of the extract causes a visible increase within five minutes in granules in the somatotropes of the pituitary. Later there is release of granules at the membrane and increase of rough endoplasmic reticulum and Golgi membranes (Fig. 48.5). The factor is a polypeptide with a molecular weight of about 2000.

The other releasing factors act similarly. The corticotropin releasing factor (CRF) causes increased ACTH secretion 1–2 minutes after injection, this is also seen at about the same time after stressing (p. 429). Two slightly different peptides have been found to have CRF activity. Their amino-acid composition resembles that of melanophore stimulating hormone (MSH, p. 441) and there seems to be some relationship of the pars nervosa to control of adrenal cortical function. The negative feed-back action of cortisol on ACTH production may be mediated by suppression of CRF production, perhaps by uptake of the steroid from the cerebrospinal fluid in the ependyma of the third ventricle.

Thyrotropic releasing factor (TRF) has a similarly prompt action and has been isolated from the hypothalamus. TSH increases in the blood 3 minutes after injection of TRF. Electrical stimulation in the anterior

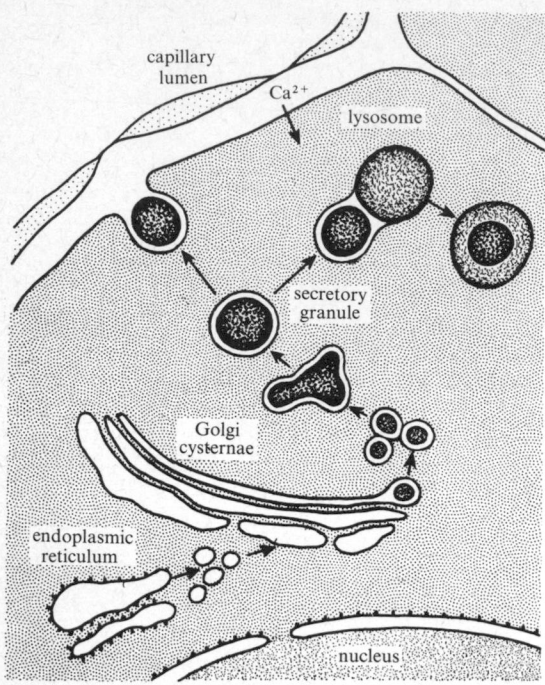

FIG. 48.5. Scheme showing mechanism of a releasing factor. Specific depolarization of the pituitary cell membrane due to the primary action of a releasing factor is followed by uptake of Ca^{2+} and the consequent release of storage granules; further granules move to the surface and discharge causing the synthetic machinery to start. (After McCann and Porter 1969.)

menstrual cycle is regulated by a complex feed-back system (p. 478).

MSH is controlled by both releasing and inhibiting factors (MRF and MIF) in the animals that change their colours.

The sites of production of these releasing factors are known only approximately. Presumably they are formed in nerve cells and there is evidence that those concerned with the gonadotropic hormones are in the tuber cinereum, while those for MRF are in the paraventricular nuclei. The nerve cells are in turn presumably controlled by both nervous and hormonal influences. For example, there is evidence that cortisol is taken up by the cells of the hippocampus, which may be the sensitive area for the inhibition of secretion of CRF and hence ACTH following response to stress. There are presumably very important and complex interactions of somatic response with nervous and 'psychological' factors.

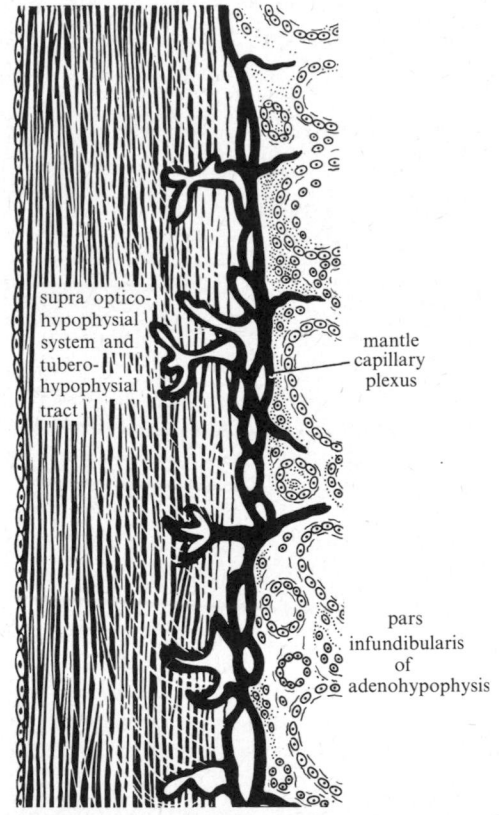

FIG. 48.6. Suggested course and termination of fibres of tuberohypophysial tract on the capillary loops and at the neuro-adenohypophysial contact zone along the edge of the infundibulum. (Redrawn from Haymaker (1969). In *The hypothalamus* (ed. Haymaker, Anderson, and Nauta). Courtesy of Charles C. Thomas, Publisher, Springfield, Illinois.)

hypothalamus or the median eminence of rats caused an increase of three to four times in TSH. Injection of thyroxine into the 'thyrotropic area' of the hypothalamus inhibits TSH release, even in minute doses. By making the thyroxine radioactive it can be shown that it does not reach the pituitary directly. Injections of pituitary extract produced an even more rapid inhibition, so there is a double negative feed-back. TRF is a peptide, probably with 12–15 residues. The releasing factors also increase synthesis of the appropriate hormones in the pituitary. Increase of circulating thyroid hormones inhibits production of TRF; production increases after thyroidectomy.

Production of prolactin is controlled by an *inhibiting factor* (PIF). When a pituitary gland is grafted to the median eminence secretion of prolactin is *less* than when it is grafted elsewhere.

Regulation of the gonadotropic hormones is by separate releasing factors (LHRF and FSHRF). The surge of luteinizing hormone preceding ovulation is probably produced by an increase of LHRF. The

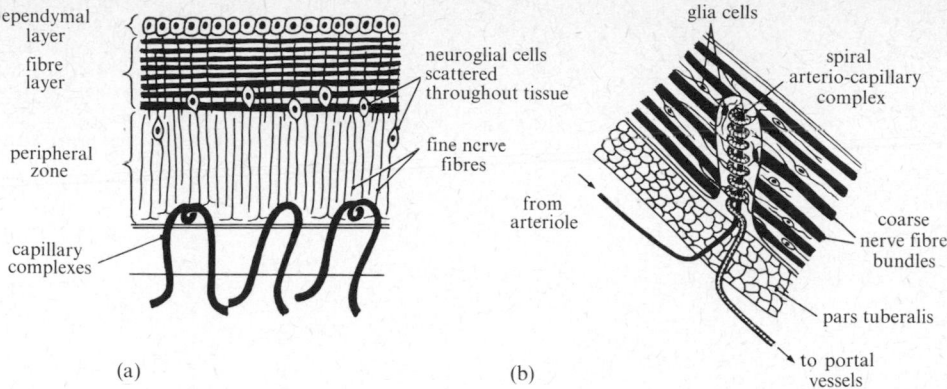

ependymal layer
fibre layer
peripheral zone
capillary complexes

neuroglial cells scattered throughout tissue
fine nerve fibres

(a)

glia cells
spiral arterio-capillary complex
from arteriole
coarse nerve fibre bundles
pars tuberalis
to portal vessels

(b)

FIG. 48.7. (a) Sagittal section of median eminence as it is in most mammals showing how the fibres of the hypophysial tract are related to a peripheral layer of neuroglia and nerve fibres and this in turn to the portal capillaries. (b) Details of the relationship of the portal capillaries to fibres of the hypophyseal in man. (From Sloper 1966.)

The releasing hormones are probably produced within the hypothalamic neurons and carried down the nerve fibres to be released in the median eminence. The pituitary stalk contains several layers (Fig. 48.6). It also contains numerous glial cells of special structure and loops and blood-vessels of the hypothalamo-hypophysial portal system. In many animals these vessels lie in a peripheral layer of the stalk, but in man they are included in special inpushings into the tissue of the median eminence (Fig. 48.7). The significance of this arrangement is not known but it may be that special arrangements for the controls of pituitary secretions are responsible for some of the particular features of human life (Haymaker 1969; Sloper 1966).

There has been much discussion of how releasing factors affect the pituitary cells. There is some evidence that they produce a depolarization. This may have an effect on Ca^{2+} uptake, producing the discharge by a mechanism similar to that of muscular contraction. Of course the factors also influence longer-term synthesis of the hormones.

13. Tanycytes

Cells of the ependyma of the hypothalamus and median eminence also play a part in the control of the pituitary secretion. They are known as tanycytes and are elongated cells with one pole facing the ventricle and the other attached to the wall of a capillary of the portal system. The inner end of the cell carries microvilli and there is some evidence that these take up material from the cerebrospinal fluid and carry it to the portal vessels. The cells form many complex loops not only around vessels but also around neurons of the median eminence (Fig. 48.8). The evidence suggests that this pathway may be especially concerned in the secretion of luteinizing hormone and in particular with the sudden release of this that precedes ovulation. The same pathway has also been suggested for the passage of TRF. It is still not clear how these substances reach the cerebrospinal fluid. It may be that they are secreted by distant regions of the brain, perhaps the temporal lobes (see Bleier 1972).

14. Hormones of the neural lobe

It was first discovered by Oliver and Schäfer in 1895 that extracts of the posterior lobe of the pituitary increase the blood-pressure. Subsequently it was found that they also have a special effect on the blood-vessels of the loop of Henle in the kidney, preventing the secretion of urine. A third action of such extracts is to cause contraction of the pregnant uterus at parturition and later the ejection of milk from the mammae—actions that are called 'oxytocic' because they are achieved by very small doses (*oxy* = sharp). There has been a long debate as to whether there are three separate hormones for these vasopressor, antidiuretic, and oxytocic actions. It is certainly possible to obtain a protein material with a molecular weight of about 30 000 that has all the properties. Nevertheless, it is also possible to split this into two fragments, one with oxytocic and the other vasopressor–antidiuretic activity. It is still not clear whether this is an artificial separation into 'hormone-fragments'. The two 'pure' hormones are much smaller molecules and both have molecular weights of about 1000. They can be separated by electrodialysis from the inactive protein component, which is called *neurophysin*. Both have been synthesized.

Oxytocin is a peptide with 9 amino-acid residues,

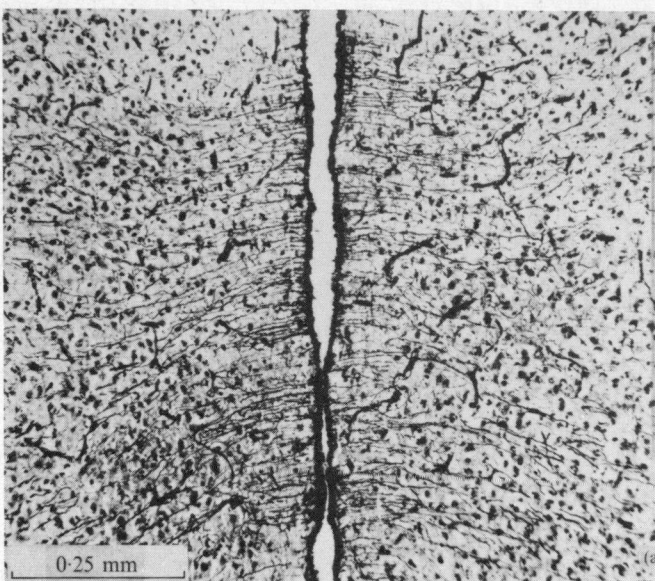

Fig. 48.8. Ependymal processes in the hypothalamus of the rabbit. (a) Tuber cinereum in horizontal section at level of ventro-medial nucleus. Many basal processes extend peripherally into the medial hypothalamus from ependymal cells lining the third ventricle. (b) Branched process from the ependymal layer (*lower left*) connecting with neurons of the dorso-medial nucleus. (Figures kindly supplied by Dr. Ruth Bleier.)

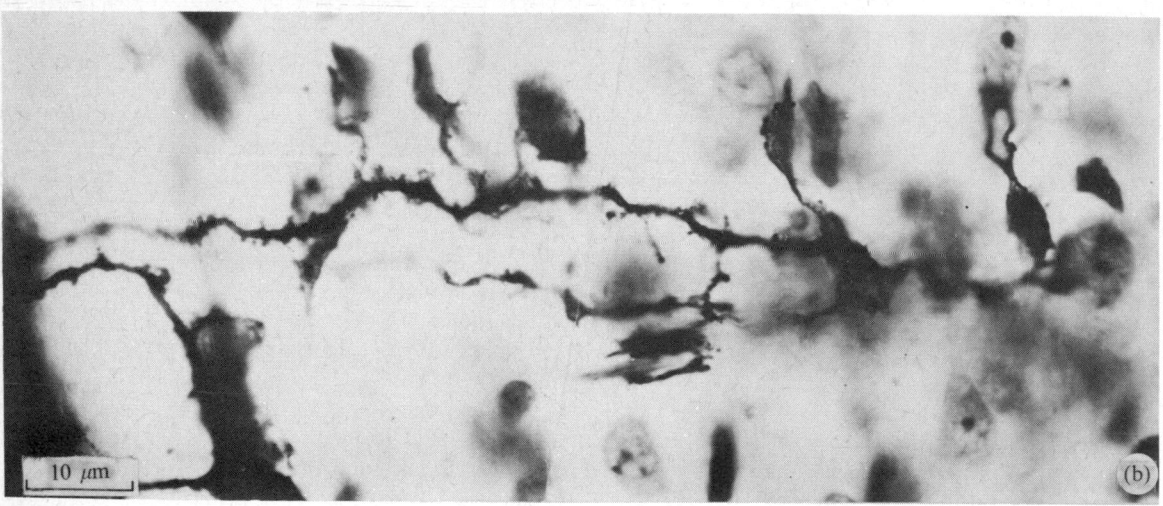

10 μm

including two cystines. Vasopressin or antidiuretic hormone (ADH) also has eight amino acids, six being the same as those of oxytocin.

Oxytocin

S————————S
| |
Cys-Tyr-Ile-Gln-Asn-Cys-Pro-Leu-Gly-NH$_2$

Arginine vasopressin (human)

S————————S
| |
Cys-Tyr-Phe-Gln-Asn-Cys-Pro-Arg-Gly-NH$_2$

Arginine vasotocin

S————————S
| |
Cys-Tyr-Ile-Gln-Asn-Cys-Pro-Arg-Gly-NH$_2$

The vasopressin of the pig and some other animals has lysine instead of arginine. In lower vertebrates the corresponding substance is arginine vasotocin, showing other small differences. Study of the distribution of these active principles suggests that they have all been evolved from the arginine vasotocin of fishes by a series of steps involving change of one amino acid at a time. The adaptive advantages of the changes are not yet clear, and indeed many animals have active prin-

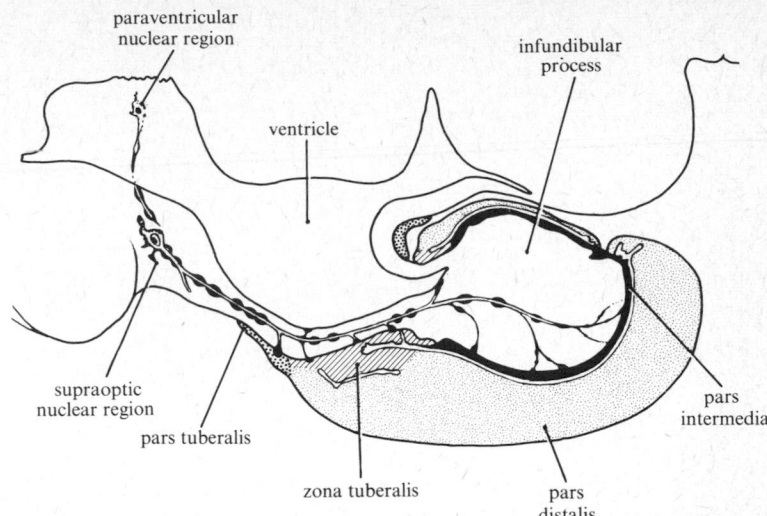

Fɪɢ. 48.9. Sagittal section of hypothalamus and pituitary of cat. (From Sloper 1966.)

ciples for which they seem to have no use (Sawyer 1966). For example there is as much oxytocin in the male as in the female pituitary of mammals. The posterior pituitary functions of fishes are presumably concerned with regulation of their ionic internal environment. When the amphibia came on land it was obviously advantageous to develop an antidiuretic mechanism. The development of oxytocin involved a further modification of this method of control and its application to

produce uterine contraction at parturition and then secretion of milk (see Follett and Heller 1964*a*, 1964*b*).

15. Neurosecretion

Although vasopressin and oxytocin can be extracted from the posterior lobe of the pituitary it is probable that they are not produced there but in the nerve cells and fibres of the hypophysial tract. This is a set of nerve fibres arising from the cells of the supraoptic and

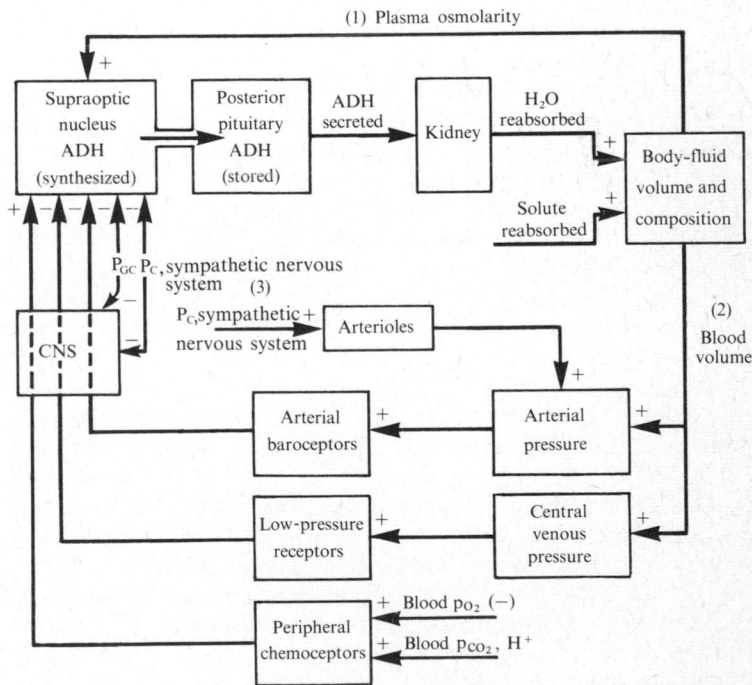

Fɪɢ. 48.10. Control of vasopressin secretion and the regulation of blood volume. P_{GC}, plasma glucocorticoid concentration; P_C, plasma catecholamine concentration. Increase or stimulation ($+$), decrease or inhibition ($-$). (After Share 1969.)

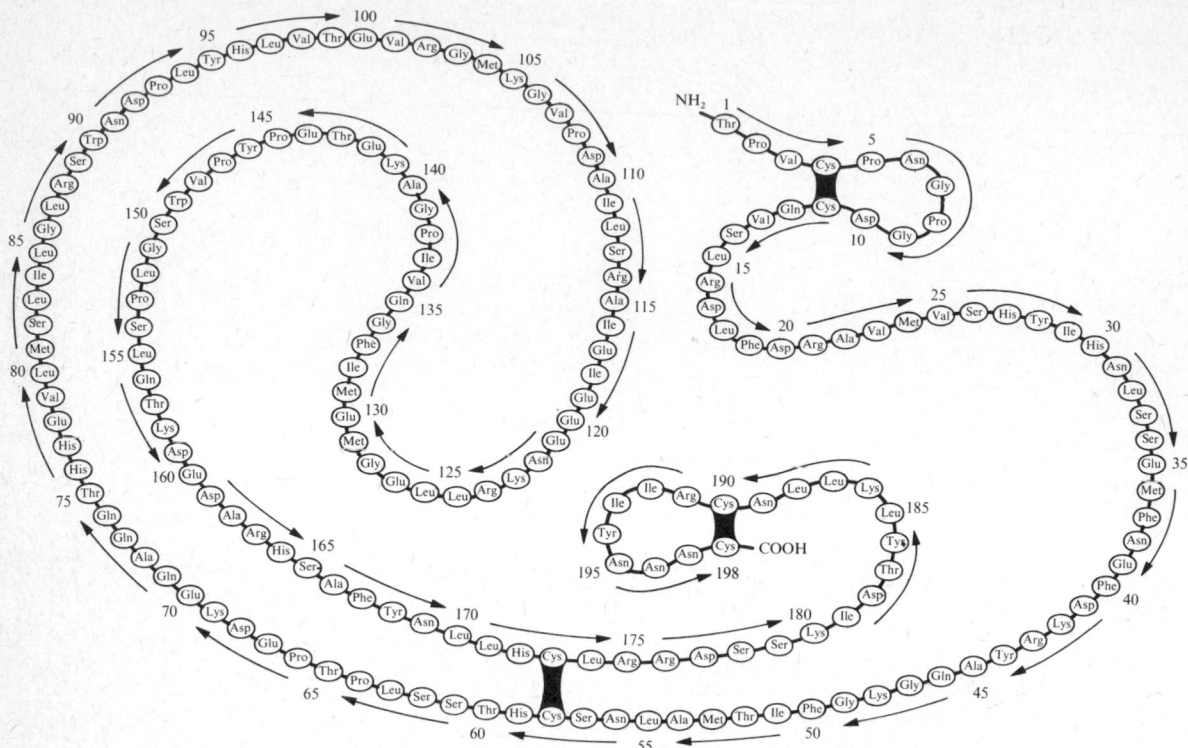

FIG. 48.11. Amino-acid sequence of ovine prolactin. (From Li (1972). In *Lactogenic hormones* (ed. Wolstenholme and Knight). Ciba Foundation Symposium. Churchill Livingstone, Edinburgh and London.)

paraventricular nuclei (Fig. 48.9). Their processes pass down the infundibular stalk and end partly at least within the neural lobe. These nerve fibres are peculiar in containing rather large membrane-bound particles about 100–300 μm in diameter, the *neurosecretory granules*. They have characteristic staining reactions, in particular they stain with chrome–alum haematoxylin (Gomori stain) and with stains for disulphide groups. By section of the stalk it can be shown that they are formed in the nerve cells and carried downwards in the axons to the pituitary. If a ligature is tied around the stalk the granules become dammed up above it. The granules can be isolated by centrifugation of sucrose homogenates of the neurohypophysis, and this fraction contains the two hormones, associated with protein as *neurophysin*.

The hypothalamic nuclei contain ADH, but it is not certain whether all the hormone is manufactured there and carried downwards, or whether some is made in the nerve terminals or in the cells of the neural lobe. These cells are called pituicytes and may be a modified form of neuroglia. Their precise relation to the nerve endings remains to be determined. It may be that the

hormones can be stored in the neural lobe and not always released directly into the blood.

The cells of the supraoptic and paraventricular nuclei are certainly sensitive to the changes in the composition of the blood. A micropipette can be inserted into the nucleus and fluids of greater or lesser osmotic concentration introduced. With a strong solution there is immediate increase in ADH and reduction of the urine. The nerve cells thus function as osmometers and they in fact contain rather large vacuoles, presumably connected with this function.

Dehydration, produced by withdrawal of fluid or injection of hypertonic solution produces a decrease in the neurosecretory granules and in the hormone content of the gland. There is therefore little doubt that the granules contain hormone, though it is uncertain how it is released. The nerve fibres of the hypophysial tract can also carry nerve impulses and stimulation of them electrically or by acetylcholine causes release of hormone.

The regulation of blood volume by ADH is a good example of a homeostatic control mechanism (Fig. 48.10). There are three feed-back loops. The first is the

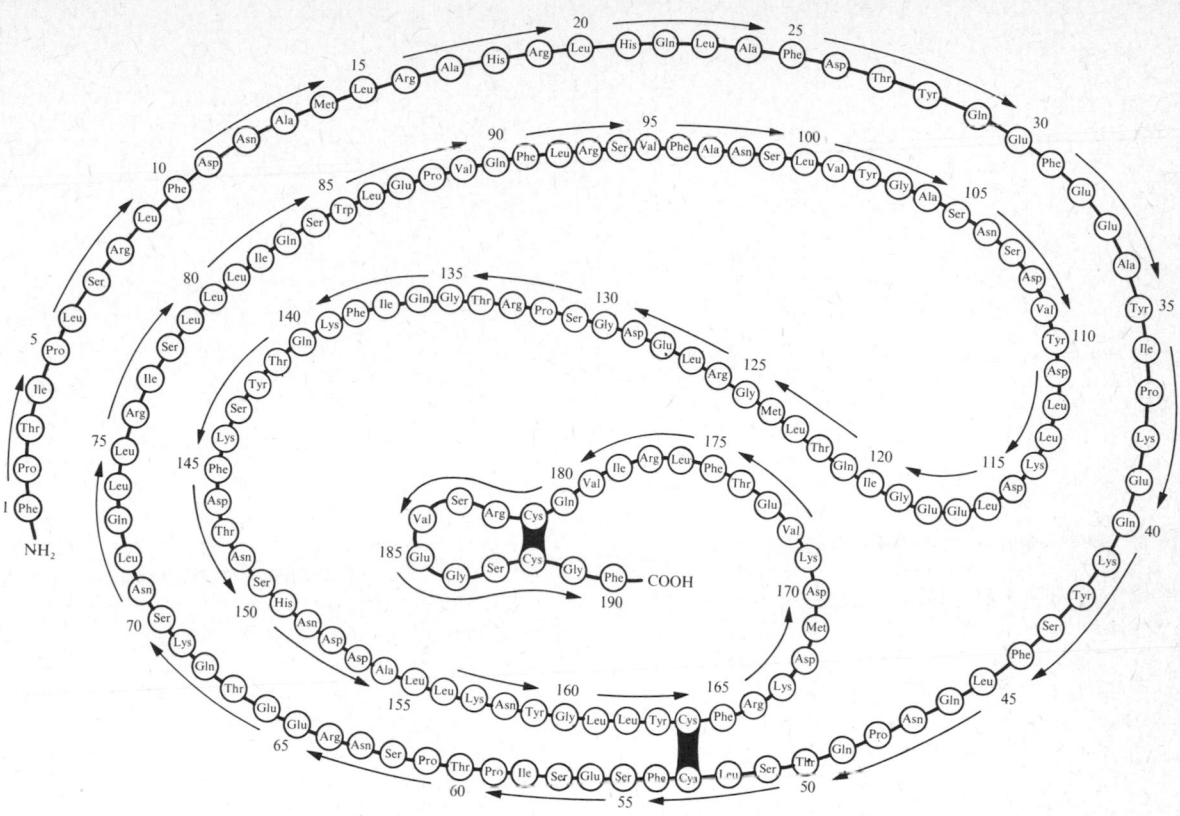

FIG. 48.12. Amino-acid sequence of human growth hormone. (From Li *loc. cit.*)

direct measurement of osmolarity by the supraoptic nucleus; the second is a measure of changes in blood volume assessed by comparison of signals from arteries and veins; and the third loop acts by changes in arterial pressure produced by changes in catecholamine levels. The diagram also indicates that the ADH secretion is affected by other factors such as blood oxygen and carbon dioxide and the levels of gluco-corticoids and catecholamines. We can thus obtain some impression of the complexity of the regulatory mechanism, but at present there is complete lack of quantitation of the stimuli and responses. There is not even a precise method for measuring the concentration of ADH in the blood (see Share 1969).

There is the further problem of separate regulation of production of ADH and oxytocin. Suckling or milk-ing produces 100 times more of the latter than the former. It is possible that there are two sorts of cells in the nuclei, stimulated by different environmental or internal situations. The hormones rapidly disappear from the blood, with a half-life of only a few minutes.

The secretion of milk thus has a double pituitary control. Prolactin from the anterior lobe promotes its production. Then tactile stimulation of the teats elicits secretion of oxytocin, which causes contraction of the myoepithelial cells of the mammae and ejection of the milk (p. 20). This process is known as 'let-down'.

It is probable that there is also a secretion of oxytocin as a result of stimulation of the vagina during copula-tion. This may serve to produce contraction of the uterus and Fallopian tubes and thus transport of spermatozoa.

16. Prolactin

Prolactin is a hormone of the anterior pituitary found in all vertebrates and with diverse actions. It induces development of the mammary gland and secretion of milk in mammals and is necessary for proper gonadal function in some rodents. It influences maternal behaviour in mammals and egg incubation in birds also secretion of 'milk' from the crop of pigeons. In many vertebrates it acts as a growth stimulant and in fishes

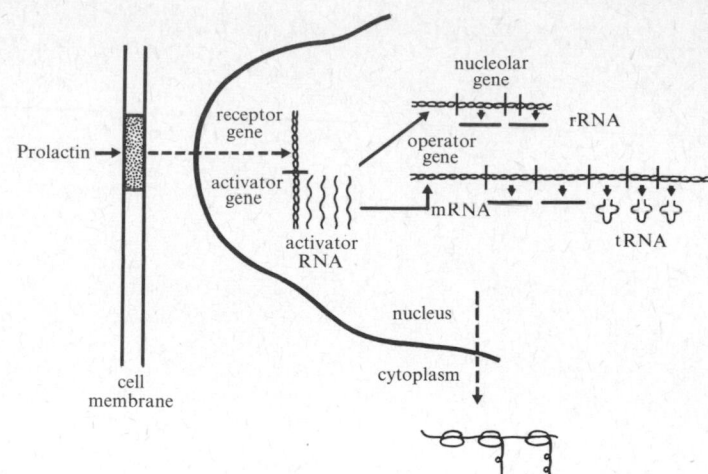

FIG. 48.13. Scheme of the sequence of hormone actions and cellular transitions during mammary epithelial cell differentiation *in vitro*. Chorionic somatomammotropin ≡ placental lactogen. (From Turkington (1972). In *Lactogenic hormones* (ed. Wolstenholme and Knight). Ciba Foundation Symposium. Churchill Livingstone, Edinburgh and London.)

and some aquatic amphibians it influences the control of electrolytes in the blood. It is therefore a uniquely versatile hormone, with a broad spectrum of actions. A list of no less than 82 different actions has been made for it (Nicoll and Bern 1972). Many other actions are related to growth, metabolism, and protein synthesis. It is not surprising therefore to find that prolactin is similar in structure to growth hormone (Figs 48.11 and 48.12) (Li 1972). A similar hormone is produced by the placenta ('chorionic somatomammotropin' or 'placental lactogen').

Prolactin is produced by specific cells of the anterior pituitary which are stained bright red by erythrosin in Herlant's differential stain (p. 438). The granules in the cells become numerous in the pituitary of a lactating rat separated for a few hours from the young, but become depleted in a female allowed to nurse them for

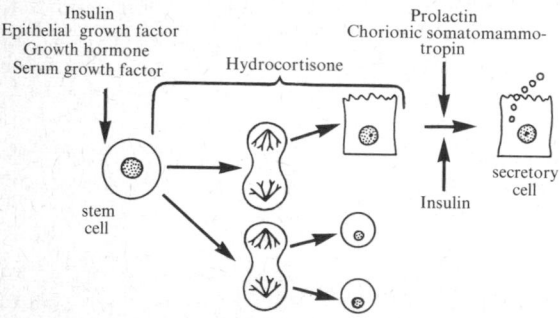

FIG. 48.14. Proposed scheme of the response of the mammary alveolar cell to prolactin showing: hormone sensing; processing of hormonal 'information' in the cytoplasm and nucleus; transcription of multiple classes of RNA for synthesis of specific milk proteins, and other induced proteins in the cytoplasm. (From Turkington *loc. cit.*)

30 minutes before killing. The prolactin is released in the form of granules. Secretion is regulated by a releasing factor secreted by the median eminence but this is inhibitory (prolactin inhibiting factor PIF). Bilateral lesions in the median eminence produce a rise of 10 times in serum prolactin within 30 minutes and the high level is maintained indefinitely. The control of PIF secretion is apparently from sites at the thalamo-hypothalamic border in the medial hypothalamus, stimulation of which elicits prolactin secretion (Fig. 52.7, p. 479). The exteroceptive stimuli of suckling have a direct effect in inducing secretion, as also does coitus. The level of prolactin in the blood has a direct feed-back effect in depressing prolactin secretion.

The mode of action of the hormone has been followed in detail by study of its effects on organ cultures of mouse mammary glands (Turkington 1972) (Figs 48.13 and 48.14). The multiplication of cells of the mammary gland is induced by a complex of synergistic hormonal actions by insulin, growth hormone, and an epithelial growth factor (in rodents). Before prolactin can produce its full effects there must be a pre-treatment with hydrocortisone. The stimulus to produce milk depends upon the interaction of the prolactin with the mammary cell membranes, isolated preparations of which specifically bind prolactin and not other pituitary glycopeptides. The effect is to stimulate production of 45S and 32S pre-ribosomal RNA by the nucleus. The method by which the membrane change is translated into a signal to the nucleus is not known but cyclic AMP-binding protein and protein kinases are involved. The altered transcription is then followed by production of multiple classes of RNA required for synthesis of milk proteins, and by secretion of milk products, which also require prolactin for completion as well as other factors.

1. Photoreceptor or gland?

THE pineal or epiphysis of man is a body about 5 mm in diameter attached by a stalk to the non-nervous choroidal roof of the diencephalon. Descartes assigned to it the supreme function of seat of the soul, supposing that it acted as a valve to regulate the flow of animal spirits through the ventricles. There was no serious advance in understanding of it until 1957 when Lerner and his co-workers at Yale set out to isolate the active principle *melatonin*, for which they used 250 000 bovine pineal glands (Lerner *et al.* 1958).

The pineal develops as a sac-like outgrowth of the roof of the diencephalon, at first often paired. In many lower vertebrates the outgrowth then forms an eye, with well-formed photoreceptors in the wall of a sac. In mammals and birds the walls soon become greatly folded to form follicles and then a nearly solid parenchyma. The function of this pineal 'gland' is now beginning to be clear. Although no longer a photoreceptor it is part of the organism's mechanism for producing adjustment of reproductive and other functions to long-term changes in illumination, such as those of the seasons (Kelly 1962; Stebbins and Eakin 1958; Wurtman, Axelrod, and Kelly 1968).

In the adult mammalian pineal the follicles of cells contain only traces of lumen and are mingled with a stroma of mesenchyme, with fibroblasts and blood-vessels. The characteristic cell is the pinealocyte, with long, branching processes. These contain the usual organelles and some vesicles, but show few special signs of secretory activity. There are also much-branched astrocytes. Sometimes small lumina lined with ependyma remain at the centre of follicles (see Wolstenholme and Knight 1971).

2. Innervation of the pineal

The pineal of mammals has no nervous connexion with the brain, but receives numerous postganglionic nerve fibres from the superior cervical ganglion by way of the nervi conarii (Fig. 49.1). These certainly approach very

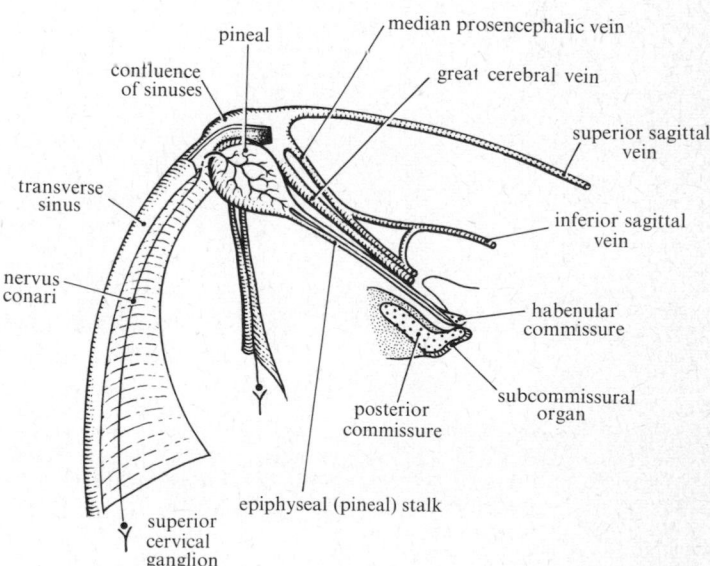

FIG. 49.1. Schematic drawing showing the autonomic innervation of rat pineal gland. The nervi conarii run along the tentorium cerebelli to the superior part of the pineal. (After Kappers 1965.) *Progress in Brain Research*, **10**.)

photoreceptor cell

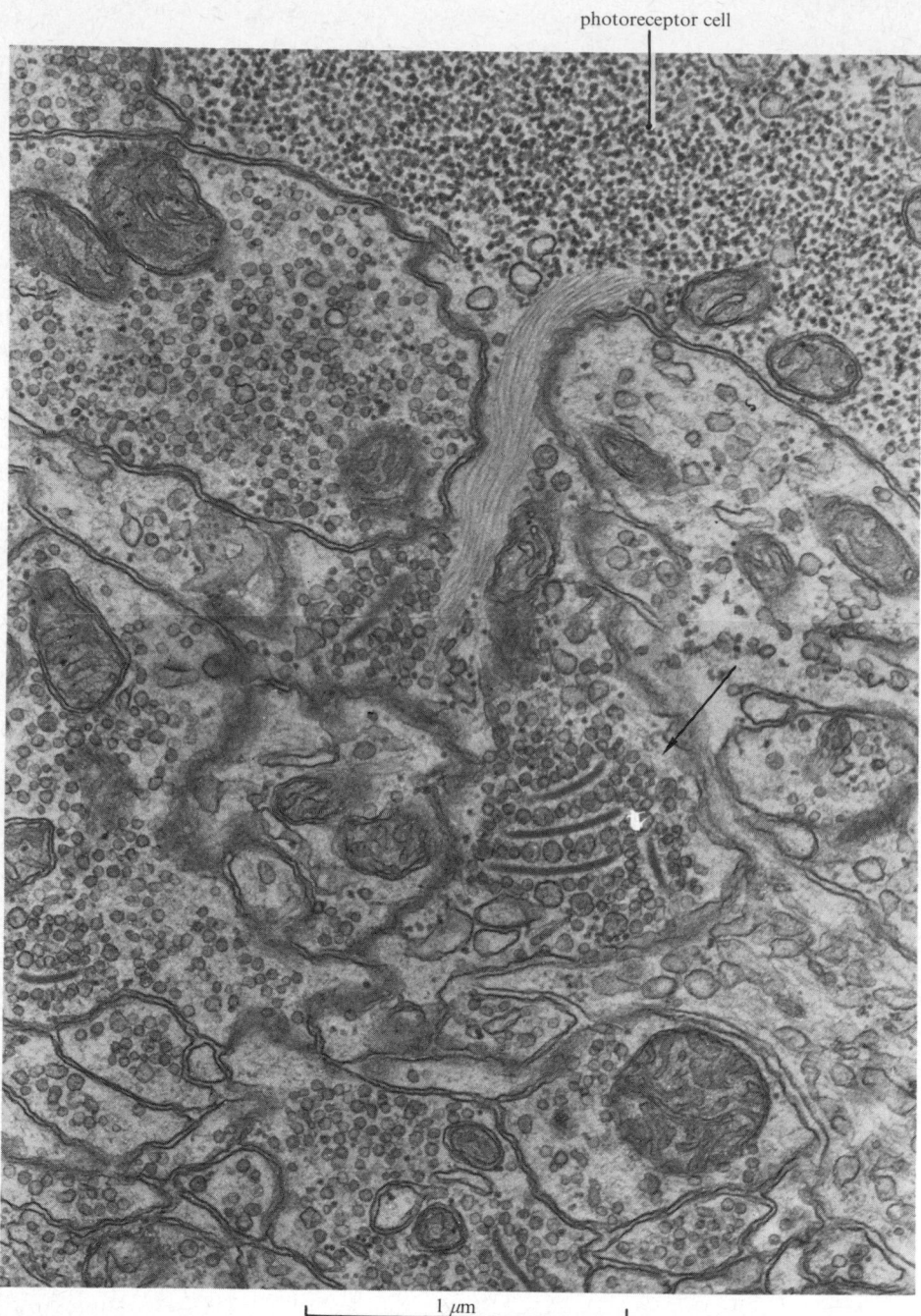

1 μm

FIG. 49.2. Electron micrograph showing part of the synaptic region in the wall of the pineal of a larval newt. The basal process from a pineal photoreceptor cell terminates in enlargements containing synaptic ribbons and presynaptic vesicles (*arrow*). These terminations presumably synapse with dendrites of neurons forming the pineal tracts and nerves. (Figure kindly supplied by Prof. D. E. Kelly.)

close to the pinealocytes and probably make synaptic contacts with them (Fig. 49.2). The sympathetic endings contain the dense-cored vesicles characteristic of catecholaminergic synapses. Such vesicles have also been reported in the pinealocytes. Within the pineal cells there are numerous synaptic ribbons. These are organs otherwise found in the presynaptic cytoplasm of the endings of receptors of the eye and ear (p. 400) or in the photoreceptive pineal of lower vertebrates. The finding of them here is not only a fascinating sign of the ancestry of the pineal but may help to illuminate the function of the ribbons.

3. Metabolism of the pineal. Melatonin

The metabolism of the pineal shows several curious features, not all of which are understood. It has a turnover of phosphorus ten times that of any other part of the brain. It becomes heavily calcified at puberty in man by granules of calcium phosphates and carbonates (brain sand), a fact of great value in the interpretation of X-ray pictures, but not yet explained.

It was discovered long ago that frogs fed on extracts of bovine pineals became light in colour. The significance of the response is not clear, but it provides a means of assay. The characteristic product of the pineal is the indole compound melatonin (*N*-acetyl-5-methoxytryptamine). The enzyme responsible for its production (hydroxyindole-*O*-methyl transferase, HIOMT) is found only in the pineal. Other substances present in considerable amounts in the pineal are serotonin (5-hydroxytryptamine), noradrenaline, and histamine. This is the only part of the body whose sympathetic fibres contain serotonin.

4. The pineal and reproduction

Although the precise means of physiological action of these substances is not known they are certainly connected with the function of the pineal in relation to light. There are diurnal fluctuations in all of these amines as there is in the weight of the pineal itself. Moreover rats kept in the light for some weeks show a reduction in weight of the gland of up to 25 per cent compared to others kept in the dark. This response of the pineal to light is undoubtedly connected with the control of the maturation of the gonads. Administration of pineal extracts decreases the weight of the gonads. It is suggested therefore that in the dark the pineal synthesizes a product that depresses the gonad. Indeed its cells show much greater basophilia in animals kept in the dark than in the light. This inhibition probably operates at various time-scales. In particular it may provide the mechanism by which many birds and mammals come into reproductive activity with increasing length of day.

The control of the pineal activity is almost certainly through the superior cervical ganglion. After removal of the ganglia in a ferret, light no longer stimulates maturation of the gonads. This is not a result, as was first thought, of the closure of the eyelids (ptosis), which is a consequence of the operation because it denervates the muscles of the lid, since it occurs if the lids are removed, whereas maturation still occurs if they are sutured.

Removal of the pineal also prevents the effect of light on the maturation of the ferret. A connexion of the gland with reproduction has been suspected for many years. Tumours of the pineal have been found to be associated with precocious maturity in man, again suggesting an inhibitory influence. Possibly the pineal plays a part in the inhibition of the maturation of the gonads during the long childhood of man (see Young 1973). This and the absence of cyles of sexuality would explain its atrophy in the adult. In hamsters kept in the dark or blinded for 30 days the weight of the testes fell from 2·2 g to 0·8 g. No fall occurred if either the pineal or the superior cervical ganglia were removed. The pathway from the eyes to the superior cervical ganglion passes through the hypothalamus and can be interrupted by lesions that damaged both median forebrain bundles (in the rat).

These effects are almost certainly produced by melatonin. The enzyme HIOMT that produces it is found in increased amounts in animals kept in the dark. The increase does not occur in animals treated with puromycin or actinomycin D to block DNA-directed protein synthesis.

It is not known how melatonin produces its effects. Administration of synthetic melatonin will produce a fall in weight of the gonads. It is taken up both by the gonad and in the brain and pituitary and its effect may therefore be either central or peripheral or both.

5. The subcommissural organ

The subcommissural organ is an enigmatic tissue lying in the floor of the brain below the pineal system. In all vertebrates except birds and mammals its cells produce a mucoid material that pours into the ventricle and is condensed to form a long flowing strand, Reissners fibre, which travels the whole length of the spinal cord. Its function is quite unknown. In mammals the subcommissural organ forms follicles near the pineal organ, which perhaps release a secretion into the blood.

50 Internal secretions and homeostasis

1. History of the hormone concept

HAVING surveyed the various endocrine organs we may now proceed to say something more about the 'hormone concept'. The idea that action is produced at a distance by the transport of chemical substances came rather late to biology, perhaps because such chemical signalling is remote from the classical scheme of causation that was applied to living things by the seventeenth-century physicists. Incidentally, the development of other chemical ideas has suffered in the same way; it is a curiosity of the growth of science that descriptions of the motions of the planets became quite precise at a time when ideas about the composition and combining powers of terrestrial matter remained very imperfect. Similarly, physics was applied to biology long before chemistry was and the integration of the activities of the various parts of the body was ascribed to the working of the nerves, by the conduction along them of the vital spirits.

Yet the specific effects that certain organs exercise upon distant parts of the body has been known since antiquity; for instance, that castration of men produces eunuchs, with changes in the body-form, hair, and reproductive organs. In 1849 Berthold showed by the transplantation of the testes of cockerels to various parts of the body that they are glands, producing effects by pouring substances into the blood-stream.

Knowledge of chemical correlation developed little further until the end of the nineteenth century, in spite of some attention to the subject by Claude Bernard, in connexion with his concept of the maintenance of the stability of the *milieu intérieure*. Clinical observers proceeded in advance of physiologists, correlating symptoms seen in patients with abnormalities of the thyroid, adrenal, and pituitary glands. Then Oliver and Schäfer showed in 1895 that the adrenal and pituitary bodies contain substances that raise the blood-pressure, and adrenaline was isolated and chemically identified in 1901. Even after this advance progress was still slow; the greater part of our knowledge about hormones has been acquired since 1920. Before that date little was known of the functions of the thyroid and sex glands, less of the pituitary and pancreatic islets, and nothing at all of the parathyroids or adrenal cortex.

The name hormone was first used by Starling in 1905 to describe the substance secretin (p. 146) that he and Bayliss had discovered in 1902. In 1914 Starling defined a hormone as 'any substance normally produced in the cells of some part of the body and carried by the blood-stream to distant parts which it affects for the good of the body as a whole'. This definition includes not only specific excitants, such as adrenaline or secretin, but also substances such as carbonic acid, which is produced by the tissues during their respiration and acts upon the respiratory centre of the medulla (p. 156). Indeed, with this definition it is difficult to exclude many other substances, for example, the amino acids that are poured into the blood-stream by the cells of the intestine and produce definite effects in the liver.

The study of hormones is clearly closely bound up with investigation of how the cells of the body of a metazoan animal influence each other for their common good. The specific hormones, produced by special cells, are only performing in an exaggerated manner actions that are common to all cells. The movement of molecules and ions is the basis of all integrated activity in the body. In the mainly aqueous medium they move rather quickly, by diffusion or, for charged particles, along electrical gradients. Cell to cell chemical communication may be common (Loewenstein and Kanno 1967; Loewenstein and Penn 1967). Moreover there are specialized long-distance chemical transport systems in the blood and lymph streams and along axons (p. 232). The classical hormones are the most sophisticated members of this chemical communication system. They are produced in very small organs, highly specialized for this function. Their output is multi-controlled, but with a rather simple feed-back mechanism. Incidentally they are fairly similar in all vertebrates, both in structure and effects. It is not convenient to use the term hormone so widely as to include all stimulants. The word is now usually reserved for the specific excitants, the term *autocoid* suggested by Schäfer for this purpose not having gained general acceptance. The word *parahormone* is available for the non-specific excitants such as carbonic acid, but it is not in common use.

The essence of a hormone, in the exact sense, is that it constitutes a specific chemical signal. We have seen that the concept of signalling implies a system with a set of instructions to act in a certain manner (p. 247). The signals ensure correct action because they are sent out in a controlled pattern or code, transmitted to a distance, and 'decoded' by receivers by the selection of some of the possible actions of the latter. The glands of internal secretion are able to act in this way because the hereditary instructions of the genes ensure that release of their products is so controlled that it occurs under appropriate conditions. The signals are decoded by certain tissues that are sensitive to them and are hence known as *target organs*.

Chemical signalling has certain clear advantages over signals carried along the specific channels of the nerve fibres. It provides a means of sending messages to all parts of the body (like the smells used as alarm signals in a coal-mine). Widespread effects are particularly important for the regulation of metabolism. The presence of different reactions to a single hormone allows the integrated operation of several processes (a sort of pleiotropic effect). Various ways of ensuring the possibility of negative feed-back by which material produced under the influence of a hormone circulates to reduce the activity of the initiator (for example, the pituitary and sex hormones). Endocrine activity is often initiated by the nervous system but the chemical signals provide a means of producing an effect sustained over some long time. In fact we can recognize various time-scales of control from the very rapid actions of large medullated nerve fibres and motor end-plates, through the slower conduction of small non-medullated post-ganglionic fibres, secreting transmitters by diffuse endings in smooth muscle, say of the gut, through to hormones circulating in the blood to produce still longer lasting actions.

For full understanding of the working of such a system we need quantitative information about the controlled release of the hormone, about the amounts of it that circulate in the blood, and about the effects that it produces in various tissues.

2. Vascular hormones and tissue hormones

Even with this definition the distinction between hormones and other stimulants remains difficult, especially since in recent years it has become increasingly clear that much signalling in the body is achieved by the transport of chemical substances for short as well as for long distances, and not necessarily through the blood-stream. The classic experiments of Loewi (1921) showed that nerves exert their effects upon muscles by the liberation of specific excitatory substances. The hearts of two frogs were perfused in such a way that the Ringer's fluid passing in and out of them became mixed. Stimulation of the vagus nerve to one of the hearts then produced slowing of both. It is now certain that such specific *chemical transmitters* ('neurohumors', see Parker (1932)) are released not only by the nerves to smooth muscle but also by those of skeletal muscle and at interneuronal junctions throughout the nervous system (p. 251). One of the substances involved is the ester *acetylcholine*. This is produced by the 'cholinergic' nerve fibres, which include preganglionic autonomic fibres, postganglionic parasympathetic and some sympathetic fibres, and the nerve endings in striated muscle. Are we to call such substances that are liberated in the tissues 'hormones'? Adrenaline provides an especially interesting case since it is produced by the cells of the adrenal medulla and a similar substance ('sympathin', noradrenaline) is liberated at nerve endings of sympathetic postganglionic nerve fibres (p. 285).

Many other instances of the transport of stimulating substances for long or short distances through the tissues are known. The evocating action of invaginated chordamesoderm on the overlying ectoderm might be called hormonal, but there is not complete evidence that it depends upon the transport of a substance. Similarly, in the development of the gonad the medulla and cortex mutually inhibit each other (p. 226). When two newt tadpoles of unlike sex are grafted together (parabiosis) the 'medullarin' produced by the gonad of the male diffuses into the twin and suppresses the proper development of the *nearer but not the further ovary* (see Witschi 1939). This case therefore closely resembles that of the bovine free-martin (p. 226), except that in the tadpoles the hormones diffuse through the tissues and not in the blood-stream. In other species of amphibia the effect is blood-borne, so that in parabiosis *both* ovaries are influenced (see Burns 1961).

These facts make a strong case for recognizing, besides the classical *vascular hormones*, also a class of *tissue hormones*, and we can even speak of *intracellular hormones* including the 'second messengers' discussed below and specific substances produced in the nucleus and carried to other parts of the cell to control differentiation and daily metabolic activities. This was suggested long ago on general grounds (Young 1934), and is of course the essence of the method of transmission of the genetic code. The value of such discussion of terms is that it focuses attention on the way the homeo-stasis of the body is ensured, the products of cells or parts of cells being carried to a distance and serving as specific signals that control metabolism elsewhere.

3. Target cells, receptors, and second messengers

The specificity of the responses to chemical signals is ensured by the presence of specific 'receptors' or 'receptor sites' on the target cells. The receptors bind the hormone with a specificity dependant on comple-mentarity of molecular conformation (stereospecific affinity, see Finar (1968)). The interaction induces a conformational change in the receptor (allosteric reaction) leading to the appearance of enzymic activity either in the receptor itself or in cellular proteins interacting with it.

For many hormones there is evidence that the inter-action of hormone with receptor stimulates or inhibits the enzyme adenyl cyclase which is present in the cell membrane and may be part of the receptor molecule. This catalyses the conversion of ATP to adenosine $3',5'$-cyclic monophosphate (cAMP), and stimulates the cell's particular synthetic activity. The cyclic AMP thus functions as a 'second messenger' to the hormone (the 'first messenger') (Rasmussen, Goodman, and Tenen-house 1972) (Fig. 50.1).

Such a mechanism may be at work not only in endocrines but also in control of exocrine secretion, synaptic transmission, and other cellular activities. There is evidence that cAMP applied outside the cell will mimic the effect of the hormone (e.g. of adrenaline on liver glycogenolysis, p. 425) and that it is synthesized when the hormone is applied to the tissue. However the situation is not simple and the 'second messenger' hypothesis is not fully accepted. Cyclic AMP is probably not the sole messenger, and change in the cellular distribution of calcium in particular is often, perhaps always, also necessary. The effect of the first messenger may be to allow an influx of calcium ions. The cAMP may sometimes act by regulating the calcium flux across mitochondrial membranes, allowing increased $[Ca^{2+}]$ in the cytosol, rather as occurs in the activation of muscular contraction (p. 61). It may be that both act as messengers and the actions of each is to reduce that of the other, which would constitute the means by which the signal is quenched, obviously a very important cybernetic consideration.

The multiple effects of cAMP are probably the

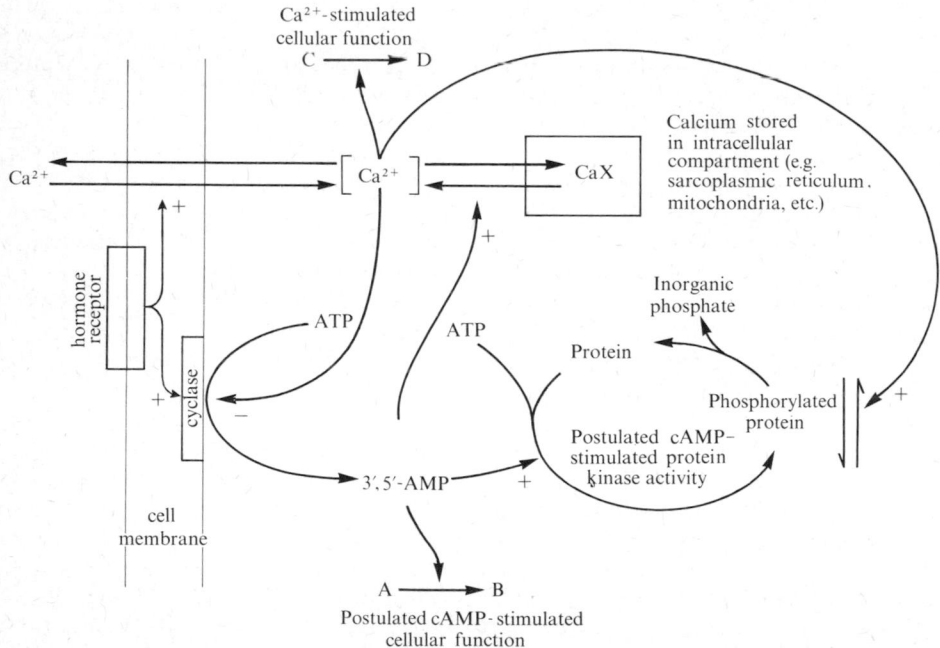

FIG. 50.1. A general model of cell activation with cyclic $3',5'$-AMP and Ca^{2+} acting as second (intracellular) messengers, and each regulating the concentration of the other in the cytosol. cAMP also controls the protein kinase group of enzymes and the phosphoprotein products some of which are regulated by Ca^{2+}. Each intracellular messenger may also have independent effects on cell function ($A \rightarrow B$; $C \rightarrow D$). (After Rasmussen *et al.* 1972.)

result of its capacity to activate or inactivate enzymes that catalyse key reactions in metabolism. There is some evidence that it regulates transcription at least in bacteria and perhaps also translation. Many of its effects are based upon the regulation of activity of enzymes, by action upon other enzymes, known as protein kinases, which convert the substrate protein by phosphorylation. The means by which it does this are not fully understood. One hypothesis is that it affects the interaction of the sub-units of proteins (Jost and Rickenberg 1971). Its multiple effects would thus depend upon the functions of the particular proteins present in the cell. There are probably numerous suitable proteins in a cell, with varying affinities for the nucleotide, which would produce a subtle regulatory system.

In any case there is evidence that hormones such as catecholamines or prostaglandins produce an actual physical change in mammalian red cells—a decrease in deformability. This is accompanied by an increase of cAMP in the cells (of rats, but not of humans). We can be fairly sure that first messengers act by producing membrane changes that release the specific activity characteristic of the cell by the intermediacy of cAMP and calcium ions.

4. Relation of endocrines to cell metabolism

Even if we consider only the more familiar blood-borne hormones we are recalled at many points to their relation to the ordinary metabolism of the tissues. The earlier investigations of each hormone often seemed to indicate that it exerts only one specific action upon some other part or tissue of the body, for instance, adrenaline on blood-pressure, 'pituitrin' on the uterus, or sex hormones on the secondary sex characters. Further study has shown that in nearly every case the substance secreted influences not only certain distant cells but some special aspect of the functioning of many cells throughout the body. Thus the growth-promoting hormone of the anterior pituitary not only 'stimulates growth of the long bones', as earlier accounts suggested, but modifies protein synthesis throughout the body (p. 439).

Similarly, the thyroid secretion stimulates the respiration of many tissues of mammals, though not of fishes, where its action may be primarily on osmoregulation and thus secondarily on metabolism. The parathyroid hormone controls the exchanges of calcium and phosphorus. The adrenal cortical steroids and those of the gonads affect cell metabolism throughout the body, while insulin, adrenaline, and the pituitary substances have varied actions on carbohydrate metabolism. The difference between these effects and those produced by

the non-specific 'parahormones', such as amino acids secreted by the intestinal cells, is that the amount of 'true' hormone secreted is regulated by nervous and other means. The hormone thus serves as a specific signal, part of the internal control system. The parahormone fluctuates under external influence. The essence of a 'signal' is that it is part of a 'code' for transmitting information within a given system and in this it differs from an outside 'influence' subject to random fluctuations (p. 459).

There is evidence from the phylogenetic history of the endocrine organs that in the earlier stages of vertebrate evolution there were no definite endocrine glands or systems of specific chemical signals (Barrington 1963, 1964). No ductless glands have been recognized in *Amphioxus*. No aggregations of cells comparable to the adrenal cortex or medulla have been found in lampreys (but cortisol and corticosterone occur in hag fishes, e.g. *Myxine*). It may be that all the specific endocrine organs have arisen by control in certain cells of a type of metabolism that was at first more general throughout the tissues. Thus the appearance of iodothyronines in the endostyle is a special case of the widespread occurrence of the binding of iodine by tyrosine. It is not clear whether iodinated proteins played a special part in cell physiology before they became specialized as hormones.

In the case of the steroid hormones, however, we are dealing with substances whose metabolism is fundamentally important, since they are components of the membranes around all cells and their organelles. It may well be that by-products of steroid metabolism served as messengers between neighbouring cells before they became blood-borne signals. It is interesting that regulation of water and salt balance and other metabolic activities is the basic function of one of the steroid hormones (aldosterone), while most of the others are concerned with the fundamental function of reproduction, but of course at a higher level than that of cells alone.

The catecholamine hormones are another set of tyrosine derivatives, substances found very widely in living organisms. We do not know when they first began to serve as messengers but they do so already as synaptic transmitters in many invertebrates as well as in the vertebrate autonomic and central nervous systems (pp. 287, 251). Their use as blood-borne hormones was certainly an early special development of vertebrates.

When we come to consider the polypeptide hormones of the hypothalamus and pituitary we are dealing with substances of the type that must have been responsible for the control of growth and metabolism from the earliest stages of evolution. Again it is interesting that still today regulation of growth and sexual development is by such substances. But of course regulation is no longer by their crude action as metabolites and they have become differentiated during evolution into an extensive set of specific excitants for diverse targets. The cells of these organs have thus become endocrine glands by control of the release into the blood-stream of substances that act all over the body or on some special organ. In this way some feature of cell metabolism that was at first at the mercy of fluctuations in the environment, or of the influences of all the cells upon each other, came to be more exactly regulated.

The origin of the endocrine pathways for control of digestion provides another example (p. 146). In lampreys the exocrine pancreas is represented by secretory cells at the anterior end of the intestine and the endocrine part by groups of cells, the follicles of Langerhans around the junction of the foregut and intestine, but not communicating with its lumen. It is reasonable to suppose that as with the digestive epithelium, concerned with the transport of future metabolites from the gut, some of its cells come to vary their activity with the level of metabolites in the blood. This could have led to changes in the discharge of their own products and their evolution into hormones (Barrington 1964).

5. Evolution of the endocrine system

It is a striking fact that the ductless glands have nearly all been produced by a metamorphosis of function during evolution, so that a set of cells that was under the direct influence of the outside world became cut off from it. In several instances an organ dealing with some raw material has become separated from its external source of supply and then converted into an endocrine gland. Thus the thyroid gland is derived from the endostyle, which at an earlier stage took up iodine directly from food or water and passed it to the tissues (p. 418). The parathyroids are probably derived from the cells of the gills of fishes that are concerned with salt transfer (p. 422). The organs corresponding to the pituitary in earlier chordates are concerned with the uptake of food (Müller's organ of *Amphioxus*) or with receptor functions (subneural gland of tunicates, p. 436). The pancreatic islet tissue is in lampreys actually dispersed in the intestinal wall (p. 434).

We are fortunate in having, in the tunicates, *Amphioxus*, and the lampreys, survivors of the early stages of so many of these endocrine systems, and the picture that they show is too general to be accidental. It seems that in the later vertebrates there has been a change in the use of these tissues. For instance, the cells of the parathyroid region, instead of dealing locally with salts presented to them by the environment and by the blood, came to secrete regulating substances into the blood. Similarly for the thyroid, pancreas, and perhaps other glands. The transformation involves the conversion of an organ whose products are subject to the fluctuations of the environment into one that releases controlled amounts to stimulate cells elsewhere. We can speak of this as the evolution of a signalling system, indicating that the gland becomes specialized to emit hormones and certain tissues become specialized to react to them, so that information is transmitted in a code.

The picture of the actual transformation of cell functions remains, unhappily, more vague. We have, however, some further hint in the fact that the endocrine glands, like other cells, are responsive to the need for their products. Poisoning of the respiratory enzymes of the cells with cyanide leads to hypertrophy of the thyroid (p. 419); castration is followed by increase in the production of gonadotropic hormone by the pituitary (p. 440); increase of blood sugar leads to increased secretion of insulin (p. 434); and so on. Admittedly these reactions to demand are themselves little understood but they certainly occur and they suggest that, following a change in habits, the products that accumulate because of lack of some substance, say iodinated amino acids, may become the morphogenetic stimulus that produces a hypertrophy of the cells that previously provided the material. This may eventually lead to the production of a new endocrine organ. Evidently here, as elsewhere, we cannot separate studies of evolution from those of development, nor of development from physiology and biochemistry. Our task is to try to have in mind as much as possible of all the past web of interactions that has made existing vertebrates what they are.

6. Evolution of homeostatic control

In any organism an elaborate balance is continually struck between various metabolic activities and the environment, the end-result being that the action of the whole system maintains itself. Presumably, in the simplet metazoans, including the early chordates such as *Amphioxus*, every cell performs all the necessary functions for itself and co-ordination is at a low level. There is little provision for maintaining a steady level of concentration of essential substances in the body. Violent oscillations would be expected in such a system and as a corollary there is no capacity to maintain the delicate balance of activities that is necessary to make use of a complicated organ such as the cerebral cortex. In particular, storage of the effects of past experience is obviously likely to be poor in a system that is subject to violent fluctuations.

Stability in all living organizations is ensured by the double dependence of the tissues (p. 4). They react to the information received through their hereditary mechanism and to the information that is provided by the events occurring around them. In lower organisms this two-way communication goes on within each tissue, fluctuations are liable to result and no very elaborate ('improbable' or 'higher') organization can result. The organism remains relatively similar to its environment and fluctuates with it.

Evolutionary change has consisted largely in the development of specialized systems of control by a code of signals and with feed-back, allowing more steady adjustment of the various phases of life. In such higher animals each individual differs more widely from the environment than primitive organisms do, but the whole organization provides an elaborately coded representation of the changes that are likely to occur in the surroundings. By means of this representation anticipatory actions are taken that enable the animal to remain constant in spite of wide external fluctuations.

Evolution of the endocrine organs of vertebrates provides beautiful examples of this development. For instance, the increasingly elaborate organization of the organs that affect carbohydrate metabolism has produced a steadier blood-sugar level (p. 434). The details of the way in which the glands achieve such homeostasis in the body is a very attractive field for research, especially of a quantitative and mathematical nature. For instance, it seems likely that something could be done to express the mutual interactions of rate of ingestion of sugar, rate of insulin secretion, level of liver glycogen and blood glucose, rates of secretion of the anterior pituitary, adrenal cortex, and medulla, and thyroid glands, as well as the level of muscular activity. All these processes and levels influence each other and the blood sugar, and it is possible to imagine a set of equations to express these relations by considering the transmission of the information that determines the life of the animal. Unfortunately we still lack knowledge about the amounts of hormone in the blood and their control and rate of change, which is needed before any such expressions can be made exact.

The pituitary gland evidently provides a focus which, influencing the organs and influenced by them, allows a balance to be struck. It may well be that control or feed-back systems are involved here, as in the nervous system. Many endocrine organs become enlarged when 'demand' for their product is increased. There is also some evidence that excess of a hormone in the blood may stimulate the production of an antihormone. For instance, after repeated injection of thyrotropic pituitary extracts rats become refractory to them. Several of the endocrines are known to be involved in the elaboration of specific substances as a defence against infection and the connexions between the endocrine and lymphatic systems are seen in the similar effects on the thymus of adrenal cortical injections and infective fevers (p. 429).

In order to understand the metabolic life of the body we must not make sharp distinctions between different phases—anabolism catabolism, respiration, excretion, and so on. The workings of the various types of enzyme system remain closely linked, even when they proceed mainly in specialized organs. The maintenance of the life of the organism depends on the balancing of these actions against environmental changes. The endocrine glands, as regulators of the balance, perhaps more than other systems determine the course of the whole life. This is conspicuous in the effects, for instance, of thyroid withdrawal, which reduces the level of activity of cells throughout the body and transforms the 'character' of the individual. The existence of such effects has led to speculative attempts at classification of human and animal types, active persons as hyperthyroid, sluggish as hypothyroid, large and bony ones as acromegalic, and so on.

Following this reasoning change in the relative rates of development of the various endocrines has been held responsible for many evolutionary developments— for example, those involving neoteny, such as are expressed by saying that 'Man is a foetal ape.' There is not enough evidence at present about the actual amounts of hormones circulating in the blood to allow much detail in such analysis but it probably contains a germ of truth. The secretions of the glands are no doubt of outstanding importance in determining the characteristics of each type of life, as are also many other factors, for instance, the proteolytic enzyme make-up or the arrangement of the cerebral cortex. The differences between species of mammals or races of men as sets of polyphasic systems cannot yet be expressed by any simple set of factors, but it is a satisfactory sign of the progress of biology that the task can be envisaged and that attempts to deal with it have begun.

51 The reproductive tract of mammals

1. Reproduction and long-period homeostasis

THE reproductive system is the part of the body concerned with maintenance of the organization of the race rather than with that of any single creature. Yet it is a mistake to try to separate these two types of control too sharply. The organization that is preserved by homeostasis is not the property of any one 'individual', but of the whole race. The persistance of living things is due to the fact that they have control mechanisms operating with different time-scales. Ordinary physiological activities serve to make rapid adjustments, and reproductive activities allow for more gradual ones (p. 9).

Each separate creature plays its part in maintaining the race for a while, but much of its activity is directed to ensuring that others shall continue after its death. This is achieved by the production of numerous and varied offspring and the placing of these in situations where their equipment is likely to allow them to survive. The homeostatic functions of reproduction therefore depend upon efficient action (a) of the mechanisms of mating, ensuring variety among offspring, and (b) of the mechanisms for caring for the young while it develops and for enabling it to start its life under suitable conditions. The production of gametes is thus only a part of reproductive activity. Many of the actions of a mammal are directed towards increasing the chance that the young shall survive; social, family, and individual life are so closely related that they are hardly distinguishable.

The function of the reproductive system being to maintain the stability of the race over very long periods, no system of rapid feed-back of signals can ensure that its functions are efficiently performed. The control of the reproductive functions is mostly by the hereditary instructions—they are, as we say, innate or instinctive. To ensure that reproduction occurs at times of the year when the young are likely to survive, the slowly operating chemical signalling of the endocrine system is used (Chapter 52). Control through the nervous system is important, however, for the performance of reproductive actions in relation to other members of the species and in all mammals there are communication signs by which individuals recognize others of the same or opposite sex and elicit the appropriate mating reactions from them (Hafez 1971).

The reproductive system is therefore elaborately organized and includes not only the ovary and testis, which produce the gametes, but also the ducts by which these are carried to the exterior. Further, there are *secondary sexual characters* providing signals ensuring that the sexes meet and pair, and mammary glands and special behaviour mechanisms for the care of the young. The whole reproductive system is co-ordinated by an elaborate chemical communication system, centred upon the hypothalamus and the pituitary gland.

2. The reproductive tracts

The cells that will form the gonads arc set aside very early in development (p. 225). They come to lie in organs that develop in the mesoderm and in all vertebrates the gonads have a close relationship to the coelom. In lampreys both the eggs and the sperm are shed into the coelom and pass to the exterior through coelomic funnels (Young 1962, p. 93). The female reproductive system is still arranged essentially on this plan in mammals but in both sexes the system of ducts has become complicated to allow fertilization and development within the body.

The arrangements by which this is ensured are further developments of those that allow the meeting of eggs and sperm in fishes and other vertebrates whose eggs are laid in the water. The sperms must act in considerable concentration and they can swim only for a short time, therefore special nuptial ceremonies for bringing the sexes together are found in many fishes, and there may be elaborate copulatory organs and internal fertilization. Such mechanisms became further developed in the early terrestrial reptiles.

If the eggs are fertilized internally it is a short step to retain them within the mother, perhaps at first mainly as a means of protection against desiccation.

Yet the viviparous habit characteristic of mammals probably did not develop directly from the changes that finally emancipated the vertebrates from the water. There was an intermediate stage in which eggs with shells were laid and the monotreme mammals, platypus and echidna, still remain in this stage. In the marsupial and placental mammals, however, the whole system of genital ducts in both sexes is highly developed for

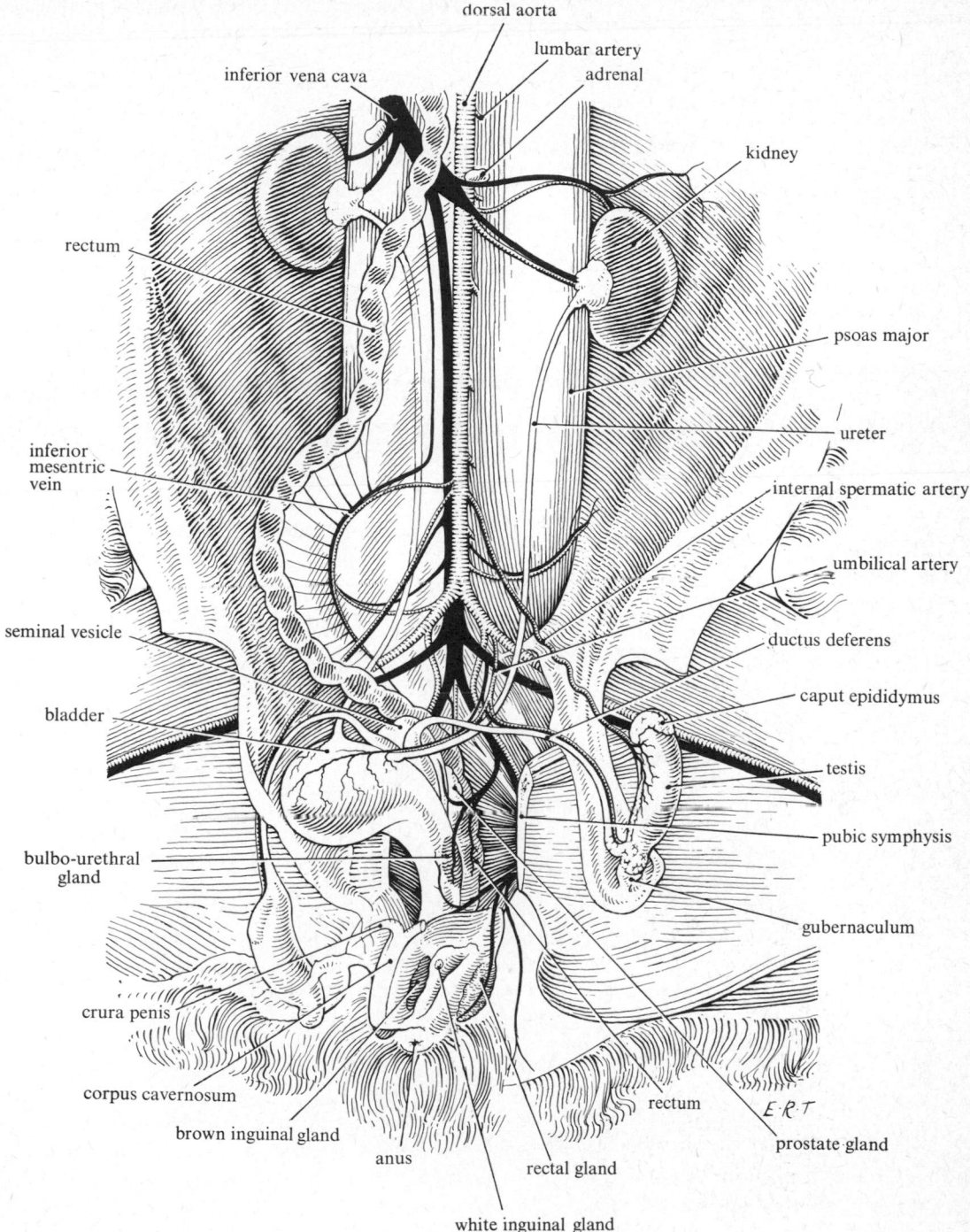

FIG. 51.1 The urogenital system of the male rabbit.

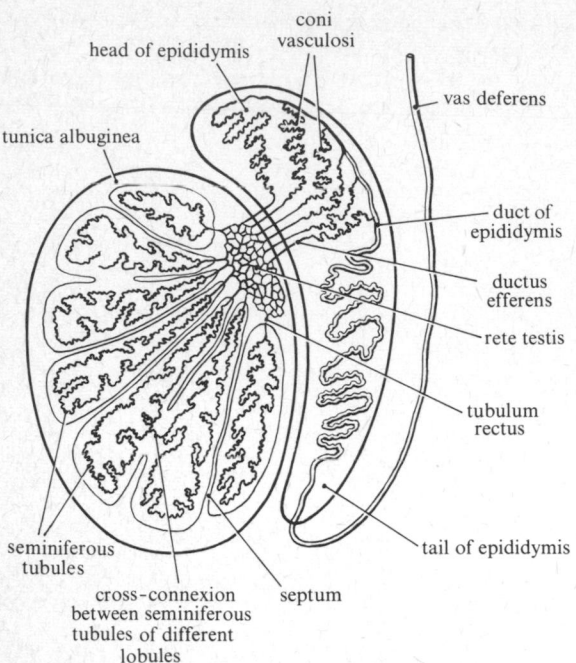

FIG. 51.2. Diagram of a testis and epididymis showing general arrangement of seminiferous tubules and excretory ducts.

internal fertilization and viviparous development. The ducts are maintained in their proper functional state by the action of sex hormones produced by the gonads under the control of the pituitary gland and the adrenal cortex (Chapter 52). These hormones also regulate the manifestation of the secondary sexual characters by which the sexes are attracted to each other and stimulated to the activities of mating.

3. The testis and scrotum

The *testis* of mammals (Fig. 51.1) is peculiar in that in nearly all species it descends from the body cavity into a special portion of the coelom, the *scrotal sac*, apparently because spermatogenesis is not able to proceed at the high temperature of the body cavity. It is not known why this should be necessary, since in birds and a few mammals (elephants) the testes are retained in the body cavity and therefore have a high temperature.

The scrotum is a pouch of the abdominal cavity and in the rabbit the communicating passage, the inguinal canal, is wide and the testis can readily be pushed up into the abdomen. In man the communication is by means of the narrow *spermatic cord*, containing the spermatic artery and vein and the vas deferens. The lining of the sac is continuous in development with that

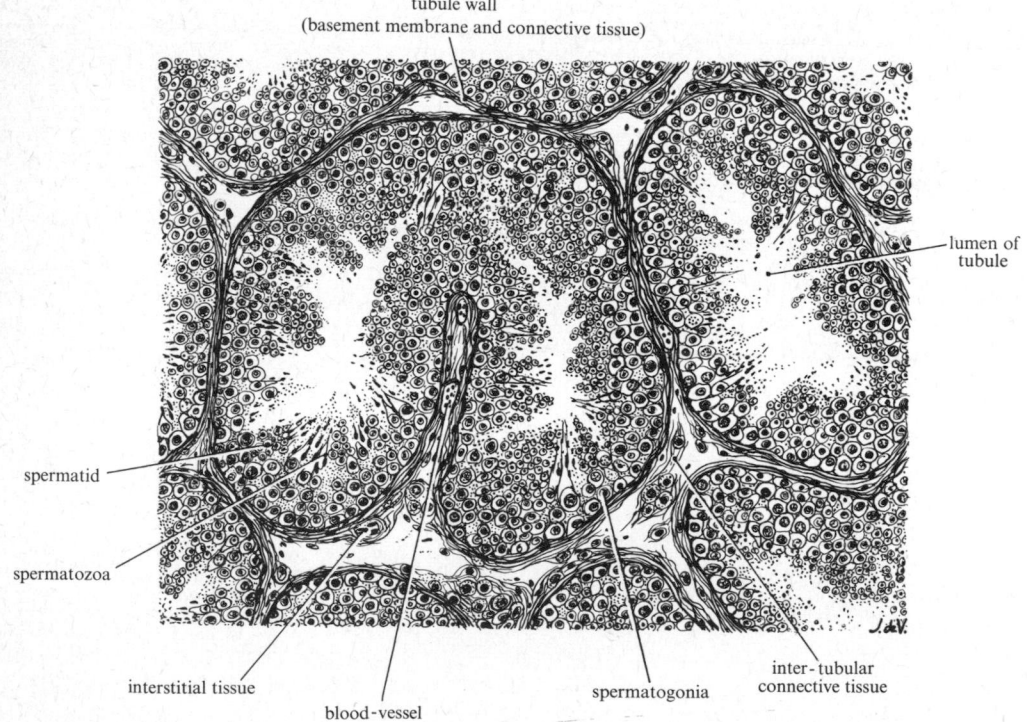

FIG. 51.3. Low power view of human testis. (From a photograph by Mr. K. C. Richardson.)

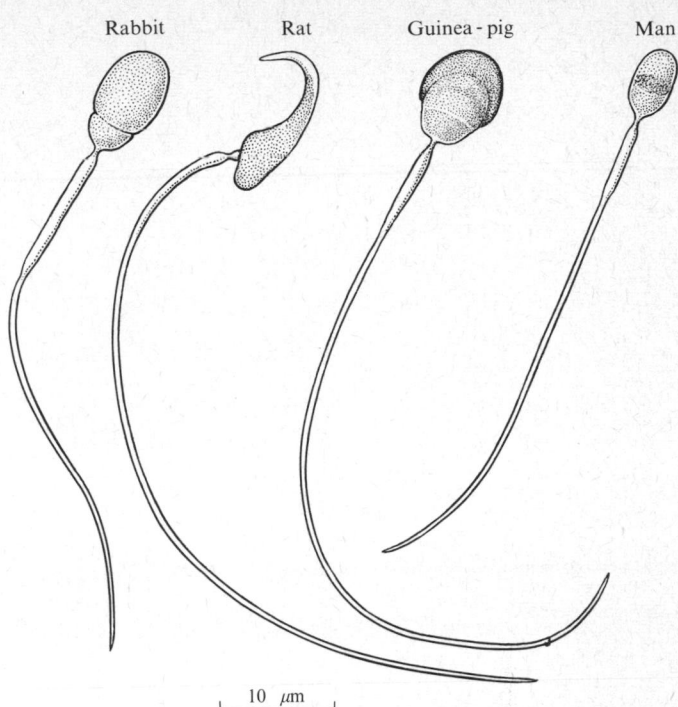

Rabbit   Rat   Guinea - pig   Man

FIG. 51.4. Drawings of mammalian sperm.   10 μm

of the abdominal coelom and its muscle is continuous with those of the abdominal wall. The testis is attached to the base of the scrotum by a ligament, the gubernaculum. Fixation by this cord is at least partly responsible for the descent of the testis during development (p. 228).

The testis itself (Fig. 51.2) consists of a number of exceedingly long, coiled seminiferous tubules, whose walls produce the spermatozoa (the process of spermatogenesis) (Fig. 51.3). The total length of the tubules in a man is said to be 250 m. In each seminiferout tubule there is an outer layer of spermatogonia, with nuclei of moderate size, which by dividing produce the primary spermatocytes and further spermatogonia. The spermatocytes have large nuclei and pass through the meiotic divisions, in which the chromosone number is halved, to form secondary spermatocytes with more compact nuclei. These then become the spermatids, which by the process of spermiogenesis differentiate into spermatozoa. Throughout spermatogenesis the spermatids are attached to nutritive Sertoli cells, whose nuclei have large nucleoli, suggesting that they conduct active protein synthesis and transfer the products to the developing spermatozoa. The process of spermatogenesis is under the control of the follicle stimulating hormone of the pituitary (FSH) and requires the presence of vitamin E.

4. Spermatozoa

These vary in shape in different mammals, but all have two main parts, a head and a tail (Fig. 51.4). The head is composed mainly of the nucleus, usually flattened. The DNA is densely packed and markedly birefringent; presumably the molecules are highly orientated. Spermatozoa have a short life (in the active state) and it may be that the DNA, conveniently packed in this way, does not influence the sperm metabolism. There is no RNA. Spermatozoa do not grow or divide and rapidly undergo 'senescence'.

The nucleus is covered by a cap, the acrosome, derived from the Golgi apparatus and containing enzymes (perhaps including hyaluronidase) that dissolve the cumulus cells and corona radiata around the ovum.

The tail contains four regions, neck, mid-piece, main-piece, and end-piece. Throughout the tail runs a core or axoneme containing characteristic microtubule doublets, arranged as a ring of nine around a pair of single microtubules (Fig. 51.5). These contain subfilaments and provide the contractile machinery. Around the axoneme there is a ring of nine coarser outer fibres. They are prominent in the mid-piece and main-piece in rodent sperms, but taper away in those of humans. It is still uncertain how this apparatus of the axoneme is related to movement. The tail has an axis running

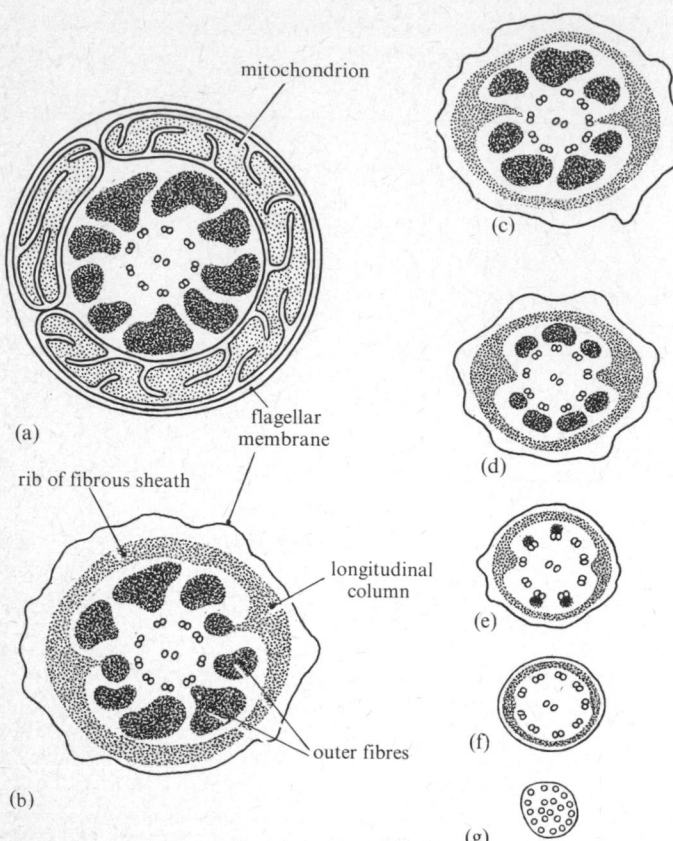

FIG. 51.5. Semi-diagrammatic series of transverse sections of mammalian sperm tail as seen by electronmicroscopy. (a) Mid-piece; (b–f) successively posterior sections through the main piece; (g) near tip of end piece.

through the plane of the two central microtubules, which corresponds to the plane of the thickenings of the sheath (see below). The neck contains a basal apparatus of three granules of centriolar origin similar to that of other cilia or flagella. The mid-piece is a thickened region containing a spiral of mitochondria. The main piece is the largest part of the tail and here the filaments are bounded by a protein sheath with helical strands. These are thickened on two sides to form ribs or fins (Fig. 51.5). In the end-piece the membrane is simpler. It is probable that the tail bends in a plane perpendicular to the fin (as do cilia), though others believe that the wave is helical. The beat normally passes downwards from the mid-piece and drives the sperm forwards in a fairly straight path, rotating along its longitudinal axis. The velocity, for bull sperm, is about 100 μm s^{-1} under certain conditions. The energy for the movement comes from oxidative phosphorylation in the mid-piece, with endogenous phospholipid as substrate. ATP is involved in the chemo-mechanical coupling (see Bishop 1962; Fawcett 1962).

Sperms are produced in vast numbers, perhaps 200 000 000 in the 2–3 ml of a normal human ejaculate, of which 10 per cent or more show various abnormalities. Only one fertilizes the egg, but a man with a concentration of less than 20 000 000 per ml is usually infertile.

Since the male is the heterogametic sex, the spermatozoa must be of two sorts—X- and Y-bearing. In spite of many efforts these have never been distinguished by their size or other features or separated by centifrigation or other methods. In many mammals far more males are conceived than females, so at some stage there must be differential behaviour. There is a popular belief that acidity assists female-producing spermatozoa and vice versa, but there is no convincing evidence of this.

5. Male sex hormones

The production of spermatoza is continuous in the testes of man after puberty but is affected by nutritional and other environmental conditions. In species with a restricted mating season the spermatogenesis occurs only at that time. The activity of the testis is controlled by secretion of the anterior lobe of the pituitary (p. 440). Besides producing spermatozoa the testis is also the seat of formation of the male sex hormones, *testosterone* and other similar androgenic steroids (see

Testosterone

Austin and Short 1972), whose effect is to produce and maintain the secondary sex characters, by which males are recognized, the female is stimulated to copulation, and the social and family organization is maintained (p. 446). These hormones are produced by the *interstitial cells*, groups of which lie between the tubules (Fig. 51.3). These contain masses of smooth endoplasmic reticulum presumably containing the enzymes for the synthesis of androgens.

The production of androgens is controlled by the luteinizing hormone of the pituitary (LH) which is also known as interstitial cell stimulating hormone, ICSH The cells of the hypothalamus are sensitive to the level of androgen and when it falls they emit the releasing factor LHRF and vice versa (p. 442). We thus have the situation that fertility (capacity to produce effective sperms) is controlled by FSH, but potency (secretion of accessory glands and capacity to copulate) by LH. Tying off the vas deferens (vasectomy) prevents the emission of sperm but the interstitial tissue remains and potency need not be affected.

6. Male genital ducts

The products of the testis are carried by a system of tubes derived from the mesonephric (Wolffian) ducts (p. 227). The seminiferous tubules open into a network of tubes, the *rete testis*, from which 15–20 *vasa efferentia* lead to the long, much coiled *epididymis*, which carries the sperms to the vas deferens (Figs 51.1 and 51.2). The cells lining the vasa efferentia are ciliated but those of the epididymis and the vas are columnar, with partial superposition of layers (hence, pseudo-stratified) and some of them are glandular. The total length of the epididymis reaches several metres in man and this long tube serves as a means of storing sperms. Injection of the radioactive form of a DNA precursor, [³H]histidine, into the testis shows that it takes 2 months or more from the beginning of spermatogenesis to the liberation of spermatozoa.

The vas deferens has a thick muscular wall, giving it a characteristic firm consistency. It enters the abdomen

from the scrotal sac through the inguinal canal and curves medially to open into the urethra. Before reaching this the vas passes through a system of glands and sacs whose secretions, together with the products of the testis and epididymis, make up the semen (Fig. 51.1). The *vesicula seminalis* is a long folded pouch opening at the expanded lower end (ampulla) of the vas. Its cavity is much divided and its wall is secretory. The *prostate* is a large mass of glandular tissue lying around the upper part of the urethra, into which is poured its secretion by numerous ducts. *Cowper's glands* (bulbourethral) are a pair opening into the urethra lower down. The secretions of the seminal vesicles and prostate make up the major part of the volume of the semen and the contraction of their walls assists in ejaculation. Receptors in their walls perhaps arouse the sex urge when they are full.

7. Spermatozoa within the male genital tract

After formation by the germinal epithelium spermatozoa are probably washed away down the tubules by the continuous flow of fluid from the testes. The passage through the epididymis is by muscular contractions and has been followed by injection of indian ink. It takes several days or even weeks and is faster if the animal is allowed to copulate. Spermatozoa are stored in the tail of the epididymis and vas deferens (not in the vesicula seminales) and are sufficient to allow repeated fertile ejaculations (36 in an hour from a bull). Spermatozoa are kept inactive within the male probably mainly by the unavailability of oxygen, perhaps by an inhibiting factor.

8. The penis

One of the biological problems of life on land is to ensure fertilization in the absence of an external watery medium in which sperm and eggs can be mixed. In early stages of tetrapod life this was brought about by the close apposition of the genital regions, which even in fishes are often brought together, and are swollen at the time of breeding by the dilation of their blood-vessels. In reptiles and mammals the method has been perfected by development of a tube, the *penis*, formed during evolution from the wall of the cloaca. The penis becomes rigid by filling of the blood-vessels of the erectile tissue, which has developed from the original vascular network.

The penis thus consists of a tube containing the terminal portion of the urethra and three vascular columns, a pair of *corpora cavernosa* and a median *corpus spongiosum* around the urethra. These are sacs divided into intercommunicating spaces. During erection they become filled with blood and so enlarged and stiffened. This is produced by relaxation of the walls of coiled helicine arteries, allowing increased inflow. This action is produced through the special parasympathetic fibres of the nervus erigens, which also supply the ischio- and bulbo-cavernosus muscles. The vas deferens, seminal vesicles, and prostate are supplied separately, by the hypogastric nerves from the sympathetic chain.

The reflexes of erection and copulation are elicited mainly by the tactile sensitivity of the penis, assisted by olfactory and other associated stimuli. The tip of the penis (glans) is covered by a smooth sensitive skin containing the genital corpuscles, modified organs of touch, and Pacinian corpuscles. Impulses from them through the spinal cord produce erection and then the reflexes of ejaculation. These involve contractions first of the musculature of the epididymis and vas deferens, then of the prostatic walls and urethral glands. The final discharge is by rhythmical contractions of the bulbo- and ischio-cavernosus muscles by which the base of the penis is attached to the pelvic girdle. The secretions of the prostate are probably emitted first, then the spermatozoa which are finally washed out by the secretions of the seminal vesicles. It is not certain whether these latter play any part in nourishing the sperm, but they contain much fructose. Spermatozoa swim by wave-like undulations of the tail rotating about the long axis. They progress at a rate of several millimetres per minute, but not necessarily in one direction (see p. 464) (see Metz and Monroy 1967; Monroy 1965).

9. Female reproductive tract

The mammalian female genital tract (Fig. 51.6) is built on the plan of other vertebrates, namely a pair of tubes (Müllerian ducts) leading from the coelom to the exterior (p. 228). The special mammalian features are the development (a) of part of the tubes as one or a pair of uteri, regions specialized for nourishing the young, and (b) of the terminal part as a vagina for copulation (see Lamming and Amoroso 1967).

The ovaries have a wide outer cortex and smaller central medulla (Fig. 52.2). The cortex is the main part and is occupied by numerous follicles containing ova. The outer surface is covered by a cuboidal epithelium known as the germinal epithelium though it is no longer thought to give rise to germ cells. The various stages of development of the follicles are described on p. 471. The cells of the follicle walls produce the female sex hormones, first oestrogens, while they are ripening, then progesterone after ovulation. The remaining tissue of the cortex includes a stroma of connective tissue fibres and stromal cells which differentiate into inter-

stitial cells that secrete steroids in some mammals, but not in the adult woman.

Close to the ovary is the fimbriated funnel, whose folded, ciliated lips receive the eggs when shed. The first part of the female tract is the Fallopian tube (uterine tube), of narrow diameter, whose walls are ciliated, glandular, and muscular. Fertilization occurs in the upper end of this tube (p. 472) and the ova are carried along it by ciliary and muscular action. In many mammals, for instance the rabbit, the tract remains paired and there are thus two uterine horns; in the human the lower parts are fused to make a single *uterus*. This is the part of the tract that can react to the presence of an embryo in such a way as to allow placentation to occur (p. 486). Its walls are glandular and muscular and its blood-vessels have special characteristics. These features enable the uterus to react to the presence of a foetus in forming the *placenta*, by which the foetus is nourished (p. 487). The whole uterus lies in a fold of peritoneum, the mesometrium or *broad ligament*.

The lower end of the single or paired uteri narrows to form a canal, the *cervix*, which communicates with the lowest portion of the Müllerian duct, the *vagina*. This is a tube lined by a variable, often squamous epithelium (p. 473), whose wall contains muscular and erectile tissue. Its exterior opening is protected by two pairs of folds of skin, the *labia majora* and *labia minora*, enclosing a vestibule into which open the *bulbo-urethral* or *Bartholin's glands*, comparable to the Cowper's glands of the male, serving to lubricate the opening to the vagina. At the dorsal side of the vestibule lies the *clitoris*, consisting, like the penis, of paired corpora cavernosa of erectile tissue, with a highly sensitive tip.

10. Secondary sex characters and mating

The coming together and mating of males and females is ensured by an elaborate system of signals, receptors, and effector mechanisms whose importance is shown by the large part they play in controlling the appearance and habits of animals and man. The initial bringing together of the sexes is largely a matter of olfactory signals in many mammals (see p. 20) but visual signals also play a large part, especially in primates, including man. The *secondary sexual characters*, differences between the two sexes in parts of the body other than those directly concerned with reproduction, serve mainly to provide signs for recognition and attraction.

The *sex drives* that impel mating behaviour depend on the internal condition of the individual, produced by action of the endocrine system (especially the hormones of the gonads) on the nervous system. In those mammals that have a breeding season the nervous system of both sexes becomes sensitized at certain times by the endocrine secretions, so that it responds to stimulation by the secondary sexual characters of the opposite sex.

In man the sex drives fluctuate less markedly with the seasons, but they become modified and lessened after removal of the influence of the gonads. We understand little of these interactions of glands and brain and in man the great elaboration of the nervous system makes sexual behaviour varied and difficult to forecast. Much of the work of psychologists has been directed to analysis of the various ways in which the

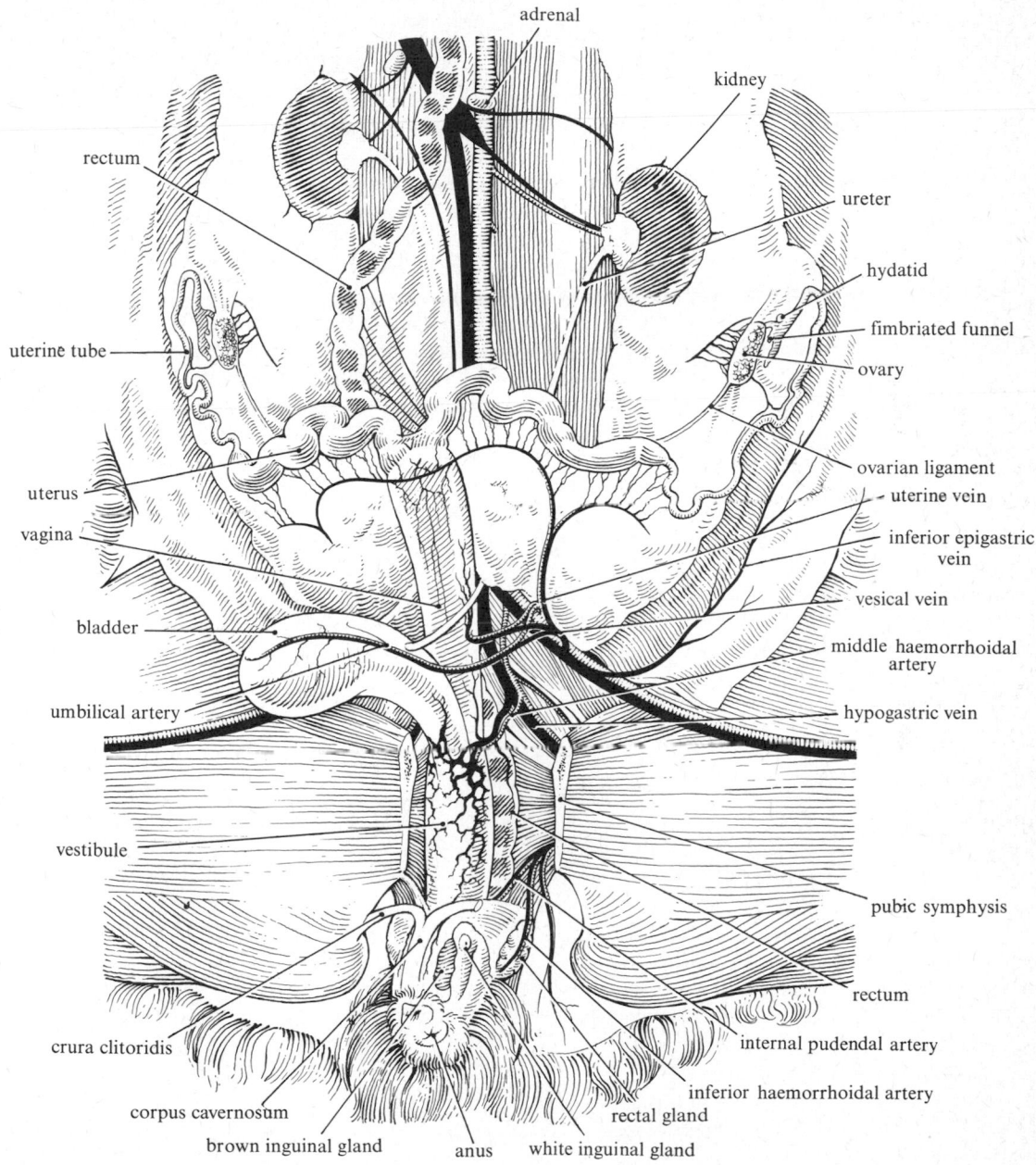

FIG. 51.6. The urogenital system of the female rabbit.

sex drives operate, especially when they are frustrated by the restrictions imposed by social organization.

The completion of copulation requires co-operation between the male and female, which is ensured by the responses to the sexual signals and the sensory equipment of the penis and vagina. Under proper reciprocal stimulation there is erection of the penis, lubrication of the vestibule and vagina by secretion of Bartholin's glands, and finally in the orgasm of both sexes the ejaculate of the male is received into the upper end of the vagina and passes through the cervix into the uterus and uterine tubes.

11. The control of reproduction

The effectiveness of reproduction is judged only by the persistence of the race and therefore no system of signals within the body can ensure that it is adequately controlled. The afferent and efferent impulses involved in orgasm serve to ensure that the eggs and sperm are brought together, but this is only the beginning of the process of reproduction and there can be no feed-back control within the body to guarantee its completion. The further nervous and endocrine systems involved in reproduction operate mainly under hereditary and not under cerebral cortical control. The reproductive tracts receive nerve fibres from the autonomic nervous system and these play a part in regulating the movements of the uterus, vas deferens, and other organs. The parts of reproductive behaviour that involve reaction by the whole individual are of course under the control of the brain, but responses are in the main to hereditarily determined olfactory, auditory, and visual signs. The elaborate computing system of the cortex is only indirectly involved, although in primates and especially in man it comes to play a large part in reproductive as in all other behaviour.

Memories acquired by the tissues during the lifetime of the individual clearly cannot provide useful forecasts about the effectiveness of the reproductive systems. We cannot say that the ovary, testis, or uterus carries memory stores in the sense that we can identify these in the femur or the muscles. The reproductive organs change with use, but the changes are not under the influence of any immediate selective factors that make them more effective as forecasters of efficient reproductive behaviour. The higher nervous centres enter into mating behaviour and there is evidence of changes with learning in the copulatory behaviour of carnivores and primates. This is even more conspicuous in the mechanism for the care of the young and particularly in man, where the information learnt by the individual as to how to teach subsequent generations is one of the most striking characteristics.

In spite of this control by acquired memories it remains true that the hereditary instructions play an outstanding part in the control of reproduction even in man. The stability of the race through thousands of years is ensured by the operations of the gonads and reproductive tracts under the influence of spinal reflex centres and endocrine mechanisms, which develop and function with relatively little intervention by the highest parts of the nervous system. Even the signals for sexual recognition and stimulation are probably largely inherited, though subject to modification by learning, especially in man.

52 The oestrous cycle of mammals

1. Preparation of the uterus

THE protection, nourishment, and care of the young within the uterus is ensured in placental mammals by the close integration of ovulation and fertilization with events that prepare the uterus to receive the embryo. The necessary changes in the female constitute the *oestrous cycle*, culminating in ovulation, which in many species is timed to occur close to a period at which the female will receive the male, thus ensuring fertilization. There is a surprising variety in the means by which this timing is achieved in various mammals, notably in the relationship between ovulation and copulation and hence in the time of survival of the sperms and probably also of the ova.

In the mouse, where we have full information on this point, ova remain fully 'active' for only about 6 hours after ovulation and sperms are not found motile in the oviduct later than 12 hours after copulation (p. 481). On the other hand in bats copulation occurs in the autumn, but fertilization not until the following spring. It is obvious that in this field conclusions must not lightly be carried over from one species to another; exact statements can be made only about phenomena known to take place by careful observation of each type of animal. Unfortunately, since mammalian development is internal, accurate data can be obtained only by the slaughter of a large number of animals, which cannot easily be undertaken in the large domestic mammals. In man the operations of obstetricians provide a partial substitute, but we are still without detailed information about some of the most important aspects of human reproduction, such as the length of life of the eggs and sperm.

Yet the present state of our knowledge of these matters contrasts favourably with the ignorance of a short time ago. Considering the vital importance of the subject, especially in man, it is astonishing that 30 years ago there was no reliable information about the time of ovulation of women, indeed the erroneous view that this usually occurs during or soon after the menstrual flow was very common. Ingenious and patient investigations have now given a satisfactory general idea of the nature of the female sexual cycle in mammals, and also much information about their hormonal control (see Parkes 1960; Short 1972). The effects of this information, together with the further knowledge of details that will no doubt accumulate to supplement it, will be of great importance in determining reproductive practices and moral codes in the future.

In spite of the variety in the sexual cycles of various female mammals we can recognize a pattern common to all. Several influences control the female sex cycle, one or another predominating in each species, but probably all act to some extent in every female mammal. Thus in the doe rabbit ovulation only occurs after copulation, whereas in women it usually occurs as a result of endocrine factors on or about the fourteenth day after the beginning of the previous menstrual flow, irrespective of mating. But we may not from these facts conclude that endocrine cycles have no influence on the ovary of the rabbit or that copulation is always without influence on that of a woman.

2. The oestrous cycle

In mammals, as in other vertebrates, the secondary sexual characters and condition of the genital ducts are regulated by secretion from the ovary and this is influenced by pituitary secretions, which are in turn themselves influenced by the ovarian hormones. The female reproductive system, therefore, does not remain at a steady level of activity but undergoes cyclical changes. The period of oestrus or heat is the time of sexual activity of the female, occurring either rhythmically throughout the year or at one or more seasons. At the time of oestrus the eggs are shed from the ovary and the females desire and will receive the males. The word oestrus is derived from the Greek name for the gadfly or warble fly, whose sting drives cattle crazy. Its use in the present sense was already suggested by ancient pastoral mythology. The goddess Hera, having discovered that her husband Zeus had lain with another goddess, Io, sent a gadfly to sting her rival, who was driven into many strange places by the torment of this 'oestrus' (see Asdell 1964; Bullough 1961; Perry 1971).

At the time of oestrus the desire and attractions of the females, reacting with those of the males, may produce very violent behaviour in individuals and in herds of mammals. Human social organization is possible because men and women instead of these sweeping changes of desire, experience attractions that are more persistent if somewhat less insistent. Man and old-world monkeys are peculiar in this respect; in most mammals mating only occurs at the times when oestrus overtakes the female. Thus the cat and bitch come on heat two or three times a year and these periods of oestrus are separated by long periods of *anoestrus*, in which the reproductive system is quiescent and the female will not mate. In other mammals the oestrous periods follow each other in regular cycles, either throughout the whole year (man) or during a certain breeding season (many rodents). Animals of the type with single periods of heat are said to be *monoestrous*, those with a succession of cycles are *polyoestrous*. The condition is not rigidly fixed for any one species and the surroundings and nutrition alter the sexual cycles

markedly. Conditions of domestication seem to increase the number of periods of oestrus, thus wild sheep and goats only have a single mating season, in the autumn, and but a single oestrous period in that season. Domesticated sheep and goats may show several oestrous periods in the autumn and winter, also an extra breeding season in the spring or even, in the case of Australian merino sheep, a regular series of cycles throughout the year.

In spite of these variations we can give the following general statement about the breeding seasons of the various mammalian orders. Marsupials breed once or twice a year and are monoestrous or polyoestrous. Insectivora, the most primitive placentals, are polyoestrous; thus in the shrew, cycles a few days in length succeed each other throughout a long breeding season, in the course of which several litters may be born. The same condition is found in rodents and primates, groups that remain close to the insectivoran stock. Thus the rat shows cycles 4–5 days long. Old-world monkeys, like man, have cycles of about 28 days, with a special condition of menstrual bleeding, discussed below. In some bats copulation occurs in the autumn

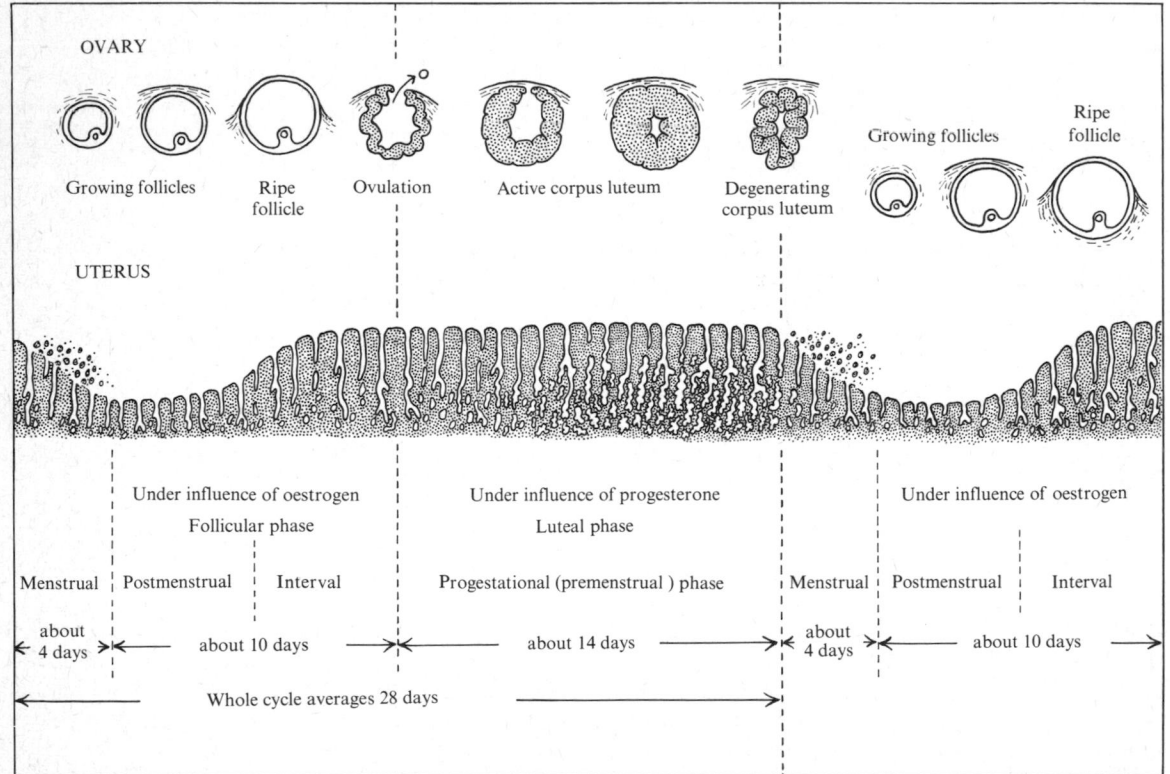

FIG. 52.1. Diagram showing the sequence of events in the menstrual cycle. (After Corner (1946). *The hormones in human reproduction*. Princeton University Press.)

but fertilization is delayed until the spring. In armadilloes, some carnivores, and roe deer there is delayed implantation, the blastocyst lying dormant in the uterus.

Perissodactyla are mainly polyoestrous. The mare in domestication experiences a series of cycles throughout the spring and early summer, each lasting 19–23 days. In the wild state the season is probably shorter but includes several cycles. Artiodactyla are frequently monoestrous in the wild state, and it is in these animals, especially those living in herds, that the 'rutting season' produces the greatest disturbance in the lives of both sexes. The males experience a season of rut, corresponding to the oestrus of the females, and animals such as the bison congregate into great herds of fighting and copulating individuals.

Carnivora are also monoestrous in the wild state, but the fact that they, like the artiodactyls, become poly-oestrous in captivity suggests that the primitive state for mammals is one in which at the appropriate time of year several successive periods of oestrus occur. We have therefore to look for the basic changes that produce this rhythmicity.

3. Proestrus—the follicular phase

The various types of cycle do not differ fundamentally and we can recognize a series of events common to all mammals (Fig. 52.1). In monoestrous species there is a period of *anoestrus* during which the female reproductive organs are quiescent. Then follows a gradual ripening of follicles in the ovary, quickening under the influence of external stimuli such as light, acting through the pineal gland (p. 453). Thus anoestrus passes into *proestrus*, in which the ova in the follicles of the ovary ripen rapidly. This is the period of 'coming into heat' which precedes the true oestrus.

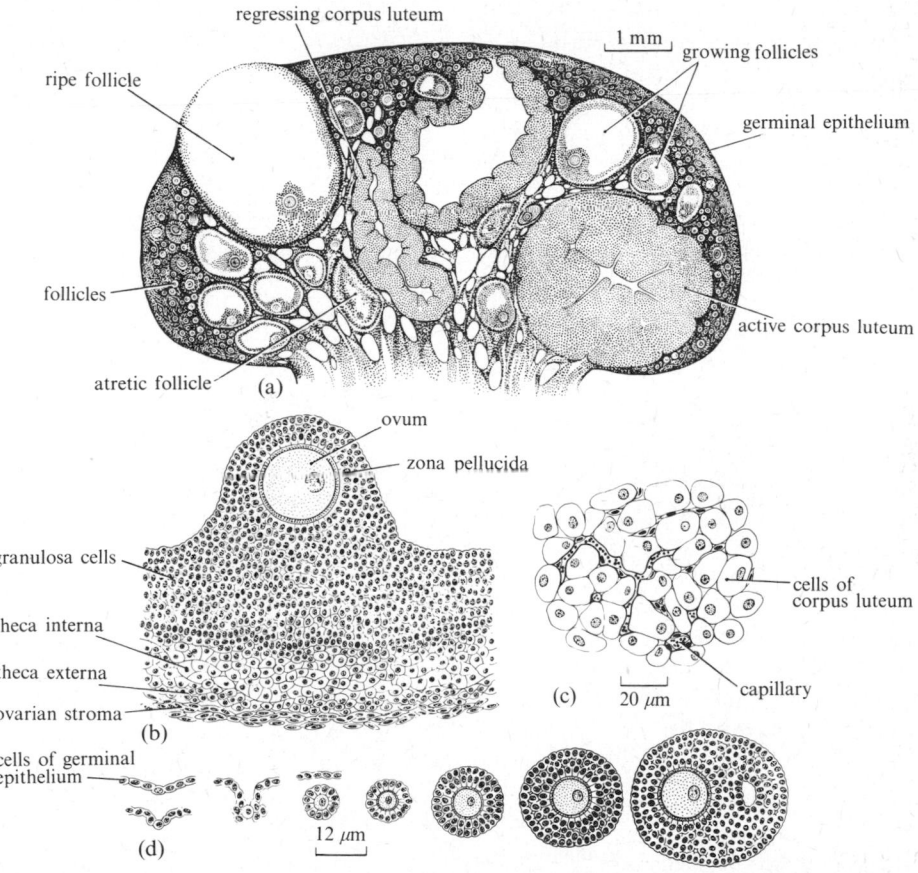

FIG. 52.2. (a) Schematic drawing of the mature ovary of a monkey (*Macaca*); (b) cells of the wall of a ripe follicle; (c) cells of the corpus luteum; (d) suggested stages in the development of a follicle from the germinal epithelium, but it is doubtful whether ova are formed by sinking in the manner shown. (Partly after Corner *loc. cit.*)

4. The ripening of ovarian follicles

Each ovary of a new-born child contains some 2 000 000 oocytes, which have migrated into it from their origin in the embryonic endoderm. Their number decreases throughout life. Thus, for example, by the age of seven years the number has decreased to 300 000. Later some mature and are shed, but far more undergo degeneration (*atresia*). Each oocyte is at first surrounded by a single layer of follicle cells (Fig. 52.2). The oocytes in these primary follicles by the time of birth have all entered the prophase of their first meiotic division and they only complete this at ovulation. Maturation consists in mitosis of the follicle cells to give several layers surrounding a fluid-filled cavity, forming a structure which is known as a Graafian (ovarian) follicle.

When about six layers are present the surrounding stromal cells become differentiated to form a case, the *theca*, around them. The inner layers are now known as the *membrana granulosa* and they show little further development during the first or follicular phase of development, lasting in a woman for about 11 days. During this time, however, the thecal cells undergo rapid development and produce the oestrogens characteristic of this phase of development. Their secretion thickens the wall of the vagina in many mammals, in preparation for copulation (Fig. 52.3). They also begin the preparation of the uterus to receive the embryo. The thecal part of the follicle is very vascular but the

vessels do not penetrate to the granulosa cells because of a thick *membrana propria*, separating the two regions. The granulosa cells probably produce progesterone, which is present only in small amounts during the follicular phase. After ovulation blood vessels penetrate the membrana propria and the granulosa cells form the corpus luteum (p. 474).

The oocyte grows meanwhile and acquires a non-cellular special membrane, the zona pellucida. As the follicle becomes bigger still spaces appear within it, filled with a liquor folliculi. In the ripe follicle the oocyte is suspended near the centre by a cord of cells, the cumulus. Usually only one follicle ripens each month in a woman, the process taking 10–14 days. The ripe follicle bulges at the surface, and finally bursts, discharging the oocyte into the coelomic cavity, where it is taken up by the fimbriated funnel. The remains of the cumulus cells form a corona radiata around the oocyte, which completes its maturation and becomes an ovum.

5. The oviduct

The oviduct is specialized to receive the ovum and allow its fertilization and transport to the uterus. The outer end of the oviduct, the ampulla, or fimbriated funnel, consists of elaborate folds, covered with cilia. The tube itself is also ciliated and in addition contains glands which in the rabbit provide the egg with an

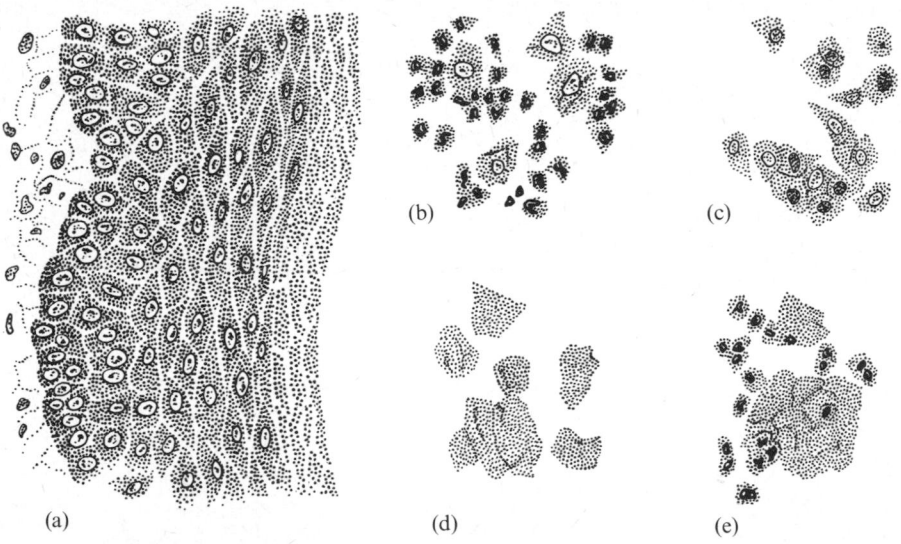

(a)   (b)   (c)   (d)   (e)

FIG. 52.3. The vaginal cycle in the white rat.
(a) Part of the vaginal wall at the time of ovulation (oestrus). The cells of the inner surface at the right, are cornified and are without nuclei. (b), (c), (d), and (e) show cells shed from the inner surface of the vaginal wall; (b) at dioestrus in the vaginal cavity there are epithelial cells (pale nuclei) and leucocytes (dark nuclei); (c) just before oestrus the epithelial cells swell; (d) cornified cells as in (a) after being shed; (e) after oestrus, leucocytes return and cornified cells disintegrate. (After Long and Evans (1923). *Mem. Univ. Calif.* **6**.)

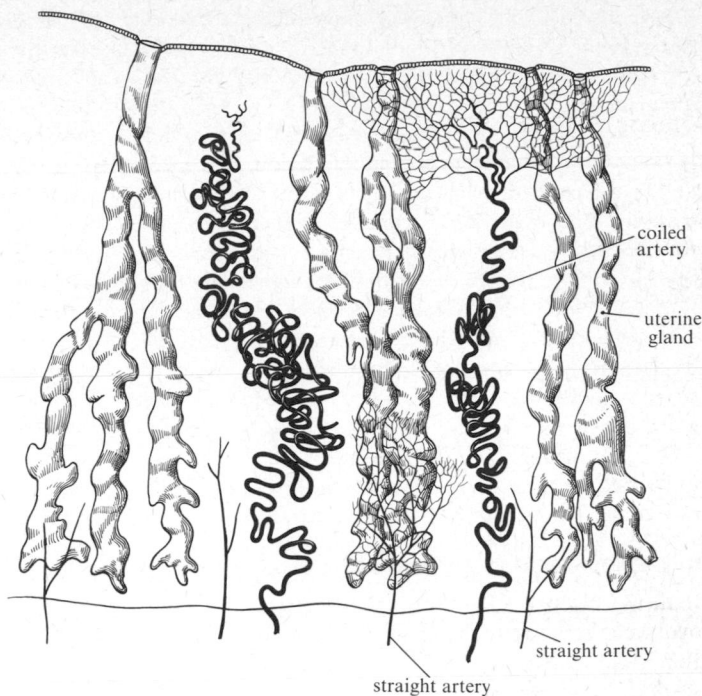

coiled
artery

uterine
gland

straight artery

straight artery

FIG. 52.4. Arteries of the endometrium of
the rhesus monkey (*Macaca*). (After
Daron (1936). *Am. J. Anat.* **58.**)

albuminous coat and, in monotremes and some marsupials, also a shell. There are coats of smooth muscle
and at the time of ovulation the oviduct shows active
movements. At this time the blood-vessels of the funnel
are engorged so that its fimbriae embrace the ovary. The
cilia beat towards the uterus and they and the movements transport the ovum. The tubes must carry the
sperms in the opposite direction but it is not known how
the reversal occurs. The tubes receive autonomic nerve
fibres.

6. The uterus

The wall of the uterus contains a thick muscle coat, the
myometrium, and a glandular mucous membrane, the
endometrium. The latter undergoes a cycle of change
throughout each monthly period in a woman. It consists of a surface columnar epithelium invaginated as a
series of simple glands (Fig. 52.4). During the follicular
phase of the cycle these glands proliferate by mitosis
and the endometrium increases 2–3 times in thickness,
and becomes more vascular.

7. Changes in the vagina

During the prooestrous (follicular) period there are
also changes in the vagina, especially marked in rodents.
In anoestrous or castrated animals the vaginal epithelium consists of only one or two layers; the only cells
to be found in the contents of the organ are leucocytes,

which can pass readily through the thin wall. Under
the influence of oestrogens, however, the vaginal
epithelium becomes much thickened and keratinized
and some of the superficial cells are shed into the
lumen and can be recognized if a 'smear' preparation
is made (Fig. 52.3). These changes provided the means
by which early investigators tested for the presence of
oestrogenic hormones. In women the vagina is lined by
a stratified epithelium. The cells do not usually kera-
tinize but become loaded with glycogen, especially at
the time of ovulation. This may provide nutrient for
sperm and allows fermentation to lactic acid and the
development of a characteristic bacterial flora. The
cells show a slight tendency to desquamate at the time
of ovulation.

8. Oestrus

Proestrus is succeeded by the climax of the cycle,
oestrus proper. During this time ovulation usually
occurs, either spontaneously or, in some species (rabbit,
cat, ferret), only if there is copulation. The distended
Graafian follicle bursts, washing out the enclosed ovum,
still surrounded by some of the cumulus cells. The ovum
is thus shed close to the opening of the oviduct, whose
fimbriated funnel is provided with cilia. The eggs pass
down the oviduct partly by ciliary, partly by peristaltic
action. It is during this period of oestrus that the
female of many species is receptive to mating. The

vaginal wall is now thick and often highly cornified and a smear preparation of the vaginal contents shows cornified epithelial cells that have been shed, but no leucocytes.

Marked changes of electrical potential, as recorded between the abdominal wall and the vagina, take place at the moment of ovulation. This sign has served to confirm that ovulation occurs in women near the fourteenth day after the beginning of the previous menstrual flow. At this time pains ('Mittelschmerz') are sometimes felt, some women experience characteristic sensations and claim that they know the time of ovulation; it is also sometimes alleged that desire is then increased. A change in body temperature also occurs at the middle of the human menstrual cycle. There is a drop on about the fourteenth day, followed by a rise, so that the temperature is generally higher during the second half of the cycle than the first.

There is considerable variation in these matters and clearly the climax of oestrus is less marked in women than in most mammals. In some monkeys the time of ovulation is indicated by sharp changes in the coloration of the face and buttocks and in other secondary sexual characters. In the baboon these changes were early used in demonstrating that ovulation occurs at the middle of the menstrual cycle (Zuckerman 1932). In other mammals various phenomena of receptivity are seen at the height of oestrus, the vaginal lips become tumid and special secretions are formed that attract the males, for instance to a bitch on heat. In some animals there is slight bleeding from the vagina either just before or at the height of oestrus. It was natural to compare this with menstrual bleeding, leading to the erroneous view that ovulation occurs soon after menstruation. Bleeding in the middle of the month sometimes occurs in women and would correspond to this oestrous bleeding of animals.

The facts thus suggest that in women the follicles begin to ripen after the menstrual flow. This period may therefore be characterized as the follicular phase of the cycle. The oestrogens produced by the developing follicle cause proliferation of the wall of the uterus and cornification of cells of the vagina. This part of the cycle culminates in ovulation, usually on about the fourteenth day after the beginning of the previous menstrual flow.

9. The corpus luteum. Pseudo-pregnancy

The events following oestrus vary considerably according to the type of mammal and as to whether ovulation, fertilization, and placentation have occurred. The ruptured ovarian follicle rapidly becomes filled with gland cells derived from the granulosa cells, forming the *corpus luteum* (Fig. 52.2) and producing the steroid hormone *progesterone*. The luteal cells are characteristic of steroid-secreting glands, pale-staining and with vacuoles in the cytoplasm after treatment with lipid solvents. They contain yellow pigment giving the gland its name. If there is fertilization and pregnancy the corpus luteum grows further, stimulated by the chorionic gonadotropin (p. 492). After either menstruation or parturition the gland regresses to leave a white scar, the *corpus albicans*.

In women the uterine lining continues to grow after ovulation, under the influence of the progestational hormones of the corpus luteum, even in the absence of fertilization (the progestational period or *dioestrus*). The changes in the uterine wall during this period consist of a great further development of the glands, which become coiled and begin to secrete. The arteries near to the surface of the endometrium acquire a pecular coiled structure (Fig. 52.4). Their function is to provide a rich blood-supply just under the endometrial surface at the time the developing ovum is ready to implant. At the height of the progestational period the endometrium acquires a marked sensitivity to mechanical stimuli, which induce in it changes similar to those of early placentation (p. 486) and known as a '*decidual reaction*'.

If fertilization has not occurred the progestational period lasts for about 14 days in women. At the end of this time the corpus luteum regresses and the thickened endometrium breaks down, producing the menstrual flow. The coiled arteries constrict, so that the endometrial surface becomes ischaemic; they then dilate individually for short periods, leading to the loss of blood. Some tissue also sloughs off and the endometrium becomes much thinner. The changes of menstruation are therefore due to contraction of the coiled arteries, but some slight menstrual flow occurs in New World monkeys, which have no coiled arteries. Evidently this type of blood-supply is a specialization superimposed on an older mechanism.

The factors that produce development of the corpus luteum in the absence of conception vary in different mammals. In some species (rabbit) sterile copulation may be followed by *pseudo-pregnancy*. If there is no mating, however, the corpus luteum exercises little or no influence and there is a gradual regression of the uterine thickening (*metoestrus*), without sloughing of tissue or bleeding. There is, however, some bleeding at this time in the bitch and cow. In all cases metoestrus leads to a return to the condition with fewer glands and smaller muscles and smaller blood-supply to the uterus. The vaginal wall at this time also shows degenerating epithelial cells and leucocytes. In animals with interrupted breeding periods the metoestrus is followed by anoestrus, but in those with a regular rhythm the uterus, after a longer or shorter period, again begins to develop. A new epithelium rapidly covers the bare uterine mucosa and mitotic divisions begin in the cells of the stroma and glands.

10. The menstrual cycle

The regular menstrual rhythm of a woman is therefore a series of oestrus periods (Fig. 52.1). Following the flow, proestrus, the follicular phase of the cycle, prepares the body for ovulation, which occurs about 14 days after the beginning of the previous flow. The remaining 14 days of the cycle are a luteal (progestational) phase in which the uterine growth continues, making preparation for the foetus and ending only if the latter fails to arrive (see Vande Wiele, Bogumil, Dyrenfurth, Ferin, Jewelewicz, Warren, Rizkallah, and Mikhail 1970).

This interpretation of the menstrual cycle as a combination of follicular and luteal phases has been challenged on the ground that in monkeys and perhaps also in the adolescent human, menstrual cycles without ovulation may occur. The fact that a cycle of uterine development and bleeding can be accomplished without formation of a corpus luteum emphasizes the similarity of the functions of the undischarged and the ruptured follicles, to which we shall refer later (p. 476).

11. Oestrogenic hormones of the ovary

The ovary itself is a major source of oestrogenic hormones (female sex hormones). The two chief active substances obtained from the ovary are *oestrone* and *oestradiol*. There are also many other naturally occurring and synthetic substances that have oestrogenic activity. *Stilboestrol* is one of these; it is not destroyed by digestion and therefore can be taken by mouth.

Oestrone

Oestradiol

Stilboestrol

Though the formula of stilboestrol as written has some resemblance to that of naturally occurring oestrogens the substance is not chemically very similar. Its action may be compared with that of a skeleton key and the similarity is certainly suggestive, but the mode of action of synthetic oestrogens remains uncertain.

In mammals the oestrogenic hormones play a major part in the control of the secondary sexual characters. Following removal of the ovaries there is a loss of the sexual instinct of the female, and indeed a loss of activity in general, as measured in rats in a revolving cage. The operation is also followed by reduction in size of the uterus and vagina, reduction of the glands of the uterus, reduction in the mammary glands, and changes in many of the soft parts and skeleton. Of special interest is the fact that removal of the ovary produces changes in the anterior lobe of the pituitary, which is one of the main influences controlling the ovary (see below). Injection of ovarian extracts or oestrogens is able to reverse all of these changes. The uterus becomes hyperaemic and distended with secretion, following multiplication of the gland-cells by mitosis, while the vaginal epithelium becomes thickened and cornified.

The changes characteristic of the proestrous phase of the oestrous cycle are produced by the oestrogens secreted by the ovary. The main source is the cells of the theca interna of the ripening follicle, large cells containing fatty substances and with a rich blood-supply. The oestrogenic substances are present in considerable amounts in the liquor folliculi. Small amounts of oestrogens are probably produced elsewhere in the ovary but the theca interna is the main source and the phase of the cycle in which it is active is therefore appropriately referred to as the follicular phase.

Oestrogens also have many other effects throughout the body, maintaining the secondary sex characters peculiar to the female. In some species certain characters become specially accentuated at the height of oestrus, serving as signals to attract the males. Thus the colouring of the buttocks, genitalia, and face of some female monkeys reaches its full development just before ovulation, of which indeed it provides an excellent sign (p. 474).

The secretion of oestrogens is also responsible for the later stages of development of the genital ducts before puberty. The vagina of the rat remains a solid cord of cells until maturity, which occurs at two months after birth. It can be made to open at one month or earlier by injection of oestrone. The constancy of this reaction throughout vertebrates is remarkable, oestrone also causes opening of the genital ducts of immature lampreys (Young 1962, p. 96).

12. Hormones of the corpus luteum

Following release of the egg from its follicle the cells of the granulosa develop into luteal cells and produce the related steroid *progesterone*, whose effects are partly like those of the oestrogens, partly distinct. If

Progesterone

fertilization has occurred and an embryo is present in the uterus the corpus luteum continues to develop and its secretions effect the changes in the female uterus and metabolism that are necessary for the support of the foetus. The change-over of function at ovulation is not sudden and indeed small amounts of progesterone are secreted in the later days of the follicular phase of the cycle. Oestrogens continue to be produced in the progestational phase, from persisting cells of the theca interna.

Progesterone has a powerful action upon the uterus, causing further growth of both endometrium and myometrium, especially if the organ has already been stimulated by oestrogens. There is a strong influence on the uterine glands, muscles, and blood-vessels. Other effects throughout the body prepare the female for pregnancy by various changes in metabolism.

The corpus luteum is therefore the 'pregnancy gland' and is essential for the earlier stages of pregnancy though it may be removed later in women (but not in rat or cow) without leading to abortion. The relaxation of the ligaments of the pubic symphysis before birth depends in some species at least (guinea-pig) on a substance produced by the ovary, but this 'relaxin' is probably not identical with progesterone.

13. Artificially induced oestrous cycles

The effect of the two hormones on the endometrium has been studied by the ingenious technique of planting small portions of the uterine epithelium into the anterior chamber of the eye in monkeys (Markee 1940). The piece becomes supplied with blood from the vessels of the iris and remains as well-differentiated uterine tissue for years. Its monthly fluctuations follow those of the uterus and can be studied by direct observation through the cornea. Using this technique as a test it has been found possible to produce by appropriate injection and withdrawal of hormones an almost complete replica of the menstrual cycle in animals from which the ovaries had been removed. When oestrogen only was given daily the fragment implanted in the eye became progressively larger and more vascular and coiled arteries could be seen within it, though they never reached close to the surface. If after 28 days the injections of oestrogen were stopped, then 2 or more days after the last injection bleeding of the fragment began. This shows that menstrual bleeding can occur following a change in the level of oestrogen only.

Changes equivalent to those of ovulatory menstrual cycles were obtained by injecting the castrated monkeys with oestrogen throughout the month, together with progesterone on days 15–28. The grafted pieces of endometrium now showed their full development. Coiled arteries grew close to the surface and formed characteristic terminal loops and curls. Both hormones were discontinued on the twenty-eighth day and 1–3 days later (more regularly than after oestrogen only) bleeding from the fragment (and from the vagina) began.

These experiments confirm all the other evidence in showing that the endometrial changes are produced by the effect on the uterus of the hormones secreted by the ripening follicles and the corpus luteum. The development of the endometrium in the later part of the cycle is due to the influence of progesterone added to that of oestrogens, and the final breakdown and flow of the blood follows the withdrawal of the hormones and consequent constriction of the coiled arteries, which depend on them. The ischaemic endometrial surface then partly breaks away and some blood escapes.

14. Changes in hormone levels during oestrous cycles

The control of reproduction of all vertebrates is operated by a complex system if interactions involving the hypothalamus, pituitary, and gonads (Chapter 47). Some hypothalamic lesions or pituitary removals prevent maturation of the gonads of young mammals and are followed by atrophy of those of an adult. Other lesions or implants of pituitary tissue cause

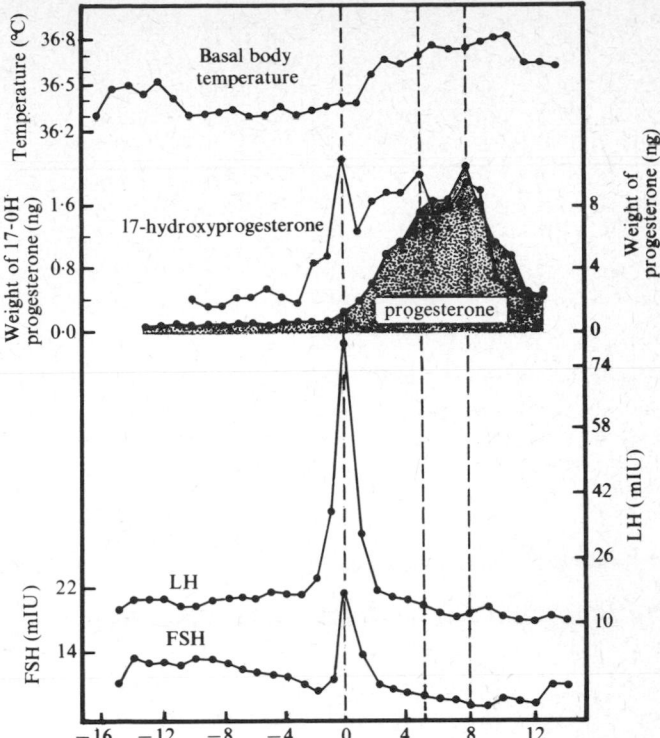

FIG. 52.5. Mean daily basal body temperature, daily plasma 17-hydroxyprogesterone and progesterone concentrations and mean daily plasma LH and FSH concentrations during 16 presumed ovulatory cycles synchronised around the day of the LH mid-cycle peak. (From Ross *et al* (1970). *Recent Prog. Horm. Res.* **26**.)

precocious development of the gonads. Conversely the action of oestrogens or androgens upon the hypothalamus are necessary for the proper development of adult sexual functioning.

Knowledge of the complicated interactions of these factors-in control of oestrous cycles has increased with the development of methods for assay of hormones in the blood (see Ross *et al.* 1970). These immune assays depend basically on using a pure gonadotropin as an antigen injected into a rabbit. Serum from this animal will then contain antibodies that will cause coagulation of latex particles which have been impregnated with the gonadotropin to be assayed. The changes have been followed in detail through the menstrual cycle of normal women. Plasma FSH and LH are measured by radio-immune assay and expressed by reference to a standard of International Units (I.U.). The steroids are measured by competitive protein-binding assays with an exactitude reaching to a coefficient of variation of 8 per cent at 30 ng ml^{-1} of plasma. Such exact measurements make it possible to construct computer models of the possible interacting effects of the hormones. Besides helping understanding of these processes this greatly assists in controlling them (Fig. 52.5).

The FSH increases in the first part of the follicular phase, but falls at the end of it. The secretion of

oestrogens increases throughout the follicular phase slowly at first, then rapidly reaching a peak on the day before the LH peak. This is paralleled by the rise in 17-hydroxyprogesterone, the steroid measured in the study shown in Fig. 52.5, whose function is not known. Meanwhile the LH remains constant and low. Then just before ovulation there is a rise in FSH and a very large one in LH. These rises are presumably a result of some action of the increasing oestrogenic steroids upon the hypothalamus, overcoming its inhibition of the production of releasing factors. The joint rise in LH and FSH is probably necessary to trigger ovulation, perhaps by causing secretion of fluid into the follicle. After these surges both hormones fall to low levels, the FSH in particular. The LH now causes development of the corpus luteum and secretion of progesterone. In the absence of pregnancy and hence stimulation by chorionic gonadotropin the corpus luteum begins to regress after about 8 days and as the progesterone falls the uterine mucosa breaks down and the cycle terminates in menstruation. The fall in progesterone seems to be the factor that triggers the next rises in FSH and LH, which begin before the onset of the menses, perhaps by removal of an inhibition of the production of FSHRF and LHRF.

As an example of attempts to model these pheno-

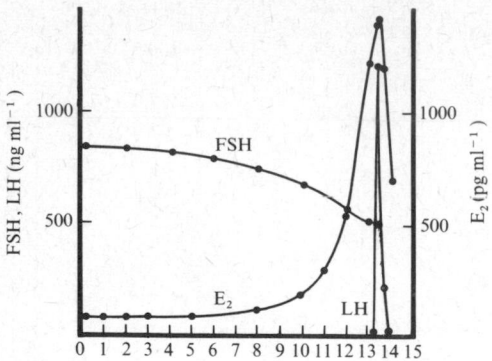

FIG. 52.6. Computer output for concentrations of FSH, oestradiol (E_2), and LH during preovulatory period. Threshold for release of pituitary LH was set at a plasma concentration of 500 pg oestradiol per millilitre. (From Ross *et al. loc. cit.*)

mena, the output of a computer programme based on another set of data is shown in Fig. 52.6.

15. The hypothalamus and regulation of oestrous cycles

The basic control of the oestrous cycle is operated from the hypothalamus, where there are sex centres responsible for regulation both of the release of gonadotropic hormones and for production of appropriate behaviour. The anterior hypothalamic region of the pre-optic and suprachiasmatic nuclei contains a clock mechanism which induces cyclic ovulation about midnight every fourth day in a rat (Fig. 52.7). This is done by some action of the hypothalamus 8 hours earlier. Injection of dibenamine on the previous day blocks the ovulation, but only if given between about 2.00 and 4.00 p.m. The clock operates to produce a surge of release of LH and the development of such a clock is a property of the genotype of both sexes, but is suppressed in the male by testosterone secreted just after birth. If male rats are castrated in the few hours after birth then when adult they produce cyclic surges of LH, which will induce ovulation in implanted ovaries (Arimura and Findlay 1971).

The area responsible can be activated by electrical stimulation in the rabbit lasting only 3 seconds. Presumably the natural stimulation of the vagina during copulation produces a nervous bombardment of this focus. The hypothalamic centre in turn bombards the median eminence and changes in the nerve terminals around the blood vessels have been seen there. The LH released produces ovulation 9 hours later (in the rabbit). The region in question probably also requires stimulation by oestrogen, which is found to accumulate there (after injection) with accompanying changes in electrical activity of its cells.

These nuclei in turn stimulate the hypophysiotropic nuclei of the ventro-medial hypothalamus. This area is responsible for maintenance of gonadotropin secretion, and can do this when isolated from the rest of the brain. The cyclical events and ovulation, however, depend upon stimulation from the anterior area. The hypophysiotropic nuclei produce their effects in turn by acting upon the centres in the median eminence in which releasing factors are stored.

The regulation of the cycle of course varies with the species but is probably essentially similar in all. As puberty develops the FSH begins to stimulate the ovary. Oestrogen probably provides a necessary ovulatory background to the anterior hypothalamus even in those species that ovulate spontaneously, the nervous reflex following copulation is needed in some species (rabbit and cat) and may play a part in rat and man (p. 480). Section of the pre-optic tuberal pathway (Fig. 52.7) abolishes ovulation.

The influence of progesterone during pregnancy checks further ovulation and this action is probably directly upon the median eminence. The evidence for this is that implants of progesterone prevent ovulation if they are made here, but not when made elsewhere. This is therefore the probable site of action of contraceptive pills containing substances that resemble progesterone in their action and which can be taken by mouth.

16. Factors that control times of reproduction

The organization of the ovarian and pituitary secretions of mammals therefore ensures a cyclical hypertrophy of various tissues of the uterus and vagina, so timed as to allow fertilization to take place when the uterus is ready to receive the embryo. These cycles may go on regularly and continually throughout the reproductive life of the individual, as in the case of women. There are, however, two influences that intervene most powerfully in this rhythm in most mammals, namely, the seasons and the males. In nearly all animals there are one or more breeding seasons, during which complete cycles are produced. For instance, in macaque monkeys menstrual cycles continue throughout the year, but for much of the time they occur without ovulation; only in a limited breeding season can conception occur. In many rodents oestrous cycles occur only at certain seasons and, as we have seen, in monoestrous species, such as the wild sheep, there may be only a single heat period in each year (Bullough 1961).

Various factors in the environment produce these seasonal changes. Nutrition certainly plays a part, both by the total quantity of food taken and the presence of essential vitamins. Inadequate nourishment readily

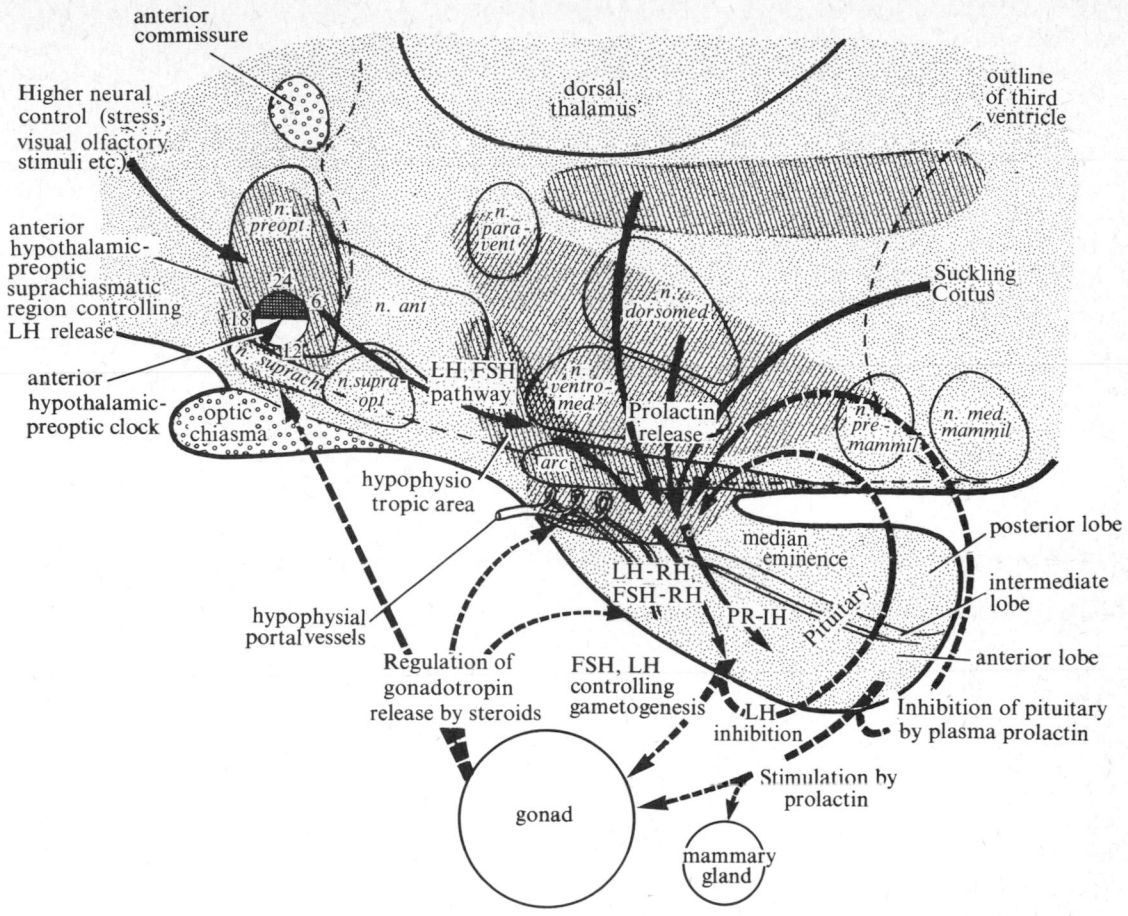

FIG. 52.7. Schematic diagram showing the regulation of gonadotropin release via the hypothalamus.
Hatched areas are those principally concerned in gonadotropin control. Solid arrows indicate neural communication; broken arrows indicate hormonal communication. Nuclei (*n.*) shown are anterior hypothalamic; arcuate, dorso-medial, medial mammillary; paraventricular, premammillary, preoptic, supra-chiasmatic, supraoptic, and ventro-medial. (After Arimura and Findlay 1971.)

disturbs the menstrual cycles of women, and species that are monoestrous in nature become polyoestrous with the better food supplies available in domestication.

Change in the length of day is a dominant factor responsible for producing the onset or regression of breeding. The light acts by altering the secretion of melatonin by the pineal (Chapter 49). However, even under conditions where change in length of day is slight, animals may show a sharply marked breeding season. For instance, bats in the New Hebrides all become pregnant once a year, though their island is only 16° from the equator and shows a nearly constant rainfall, while they themselves live all day in dark caves with almost constant temperature. Right on the equator, however, they breed all the year round, although groups within the colony show synchronized breeding, perhaps by social stimulation. There is still much to be learned about the control of breeding seasons.

17. Influence of males on the female cycle

The presence of males seems to initiate new oestrus cycles in rats. When they are introduced to a group of females previously kept alone, then mating begins on the third day. There is evidence that olfactory stimuli are involved. The odour of a strange male will prevent implantation of the eggs fertilized by another male four days before. Members of a group of female rats may tend to cycle synchronously and similar effects have been reported by groups of women. Substances producing such hormone-like effects outside the body are called *pheromones.*

'Psychic' factors of various sorts deeply affect the mating behaviour of both sexes. Courtship displays may play a part in the induction of ovulation in mammals as they do in birds. Isolated female pigeons will ovulate if a mirror is put in the cage.

The influence of copulation on the female cycle is marked in some species. In the cat and the ferret follicles ripen spontaneously but only rupture if copulation occurs during oestrus. In the rabbit the follicles do not fully mature without copulation, which is therefore followed by ovulation some 10 hours later. Here the details of the process have been explored and it has been shown that stimulation of the nerves of the female genitalia sends nerve impulses via the spinal cord to the brain, where they cause the hypothalamus to liberate FSHRF to activate the pituitary to produce follicle stimulating hormone. It is not clear why this mechanism should be present in only some species; it has been suspected that something similar may occur in other mammals and it is obviously important to know whether it is present in man. Unfortunately we have no clear evidence on the point. It is certain that ovulation usually occurs spontaneously at about the middle of the cycle in a woman, but this does not prove that copulation is without influence in controlling the movement of ovulation near mid-cycle, or even at other parts of the cycle. Evidence from the young human embryos that have been carefully investigated suggests that conception normally occurs near mid-cycle. Statistics of pregnancies following alleged isolated copulations show the middle of the cycle as the time of maximum fertility.

53 The development of mammals

1. Early development of the mouse

THERE are more data about the fertilization and early development of the mouse than are available for other mammals. The average length of the oestrous cycle is about 9 days, dioestrus lasting 1–11 days, proestrus $\frac{1}{2}$–$1\frac{1}{2}$ days, oestrus 1–3 days, and metoestrus 1–5 days. Search for ova in the tubes of animals killed at various periods showed that ovulation will occur spontaneously at some time during oestrus even in the absence of the male. There is, however, a connexion between ovulation and copulation because most females killed shortly after copulation at various parts of oestrus have (in the oviduct) freshly, ovulated eggs. It is still not certain whether this concurrence is due to the fact that the female will only accept the male near to the time of ovulation or to a stimulation of ovulation by copulation. The complexity of the stimuli necessary for completed mating is shown by the fact that some of the females would not mate throughout oestrus and yet many of these ovulated spontaneously.

The sperms pass rapidly up the oviduct and some have been found near its upper end 15 minutes after copulation. They probably remain mobile for less than 12 hours and the ova are only able to be fertilized within 6 hours after ovulation. The margin of time available for fertilization is therefore rather narrow, probably it usually occurs within 2 hours after copulation. The maturation divisions of the ovum also occur during the hours immediately after ovulation, the second division following fertilization.

The ovum itself at this stage is about 70 μm in diameter† and contains granules of various sorts, including yolk. There is, however, no recognizable animal pole; the nucleus lies near the centre of the ovum. The freshly ovulated ovum is surrounded by a number of follicle cells, *corona* or *cumulus cells*, which leave the ovary with it (Fig. 53.1) and serve to cement

---

† Other mammalian ova are larger, 125 μm in the rabbit, 140 μm in man. There is no simple relationship between the sizes of the ovum and of the adult.

together all the ova ovulated from one ovary. Within the corona cells lies a transparent membrane, the *zona pellucida* or 'oolemma' (Fig. 53.1), inside which the ovum undergoes its development. This is a layer of mucopolysaccharide about 3 μm thick. The sperms have to penetrate the corona cells and their power to do so probably depends on the presence of the enzyme hyaluronidase in the ejaculate. Although one sperm alone enters each egg a considerable amount of semen is needed to make fertilization possible; the probability of effective fertilization by any one sperm is small, and this explains the fact that large numbers of sperms are produced. Even if a sperm collides with an ovum it is not certain to achieve fertilization. Penetration depends on the lytic action of enzymes produced by the sperm upon the egg investments. After entry the sperm head rotates through 180° and the egg and sperm nuclei then fuse.

Cleavage takes place as the ova pass down the oviducts and by the time the embryo reaches the uterus, about 3 days after copulation, it has reached the early blastula stage and is composed of a hollow ball of cells, known as a *morula*. Cleavage is complete but not quite equal, yet even at the morula stage the cells differ so little in size that no clear distinction of animal and vegetative poles is possible (Lewis and Wright 1935).

The total volume of the embryo has so far remained about that of the ovum but during the fourth to sixth days it swells, apparently by the secretion inwards of fluid. At this stage it is known as a *blastocyst*, a hollow sphere whose wall, the *trophoblast*, is thin over most of the circumference but thickened at one point, the *inner cell mass* (Fig. 53.1). The trophoblast persists throughout gestation as the outer covering of the embryo, in contact with maternal tissues. The increase of diameter causes the embryo to press against the zona pellucida, which becomes thin and finally ruptures, liberating the embryo into the uterus.

Embryos can be collected from the oviducts by washing, and cultivated *in vitro* up to the blastocyst stage. They then can be re-implanted into the uterus of

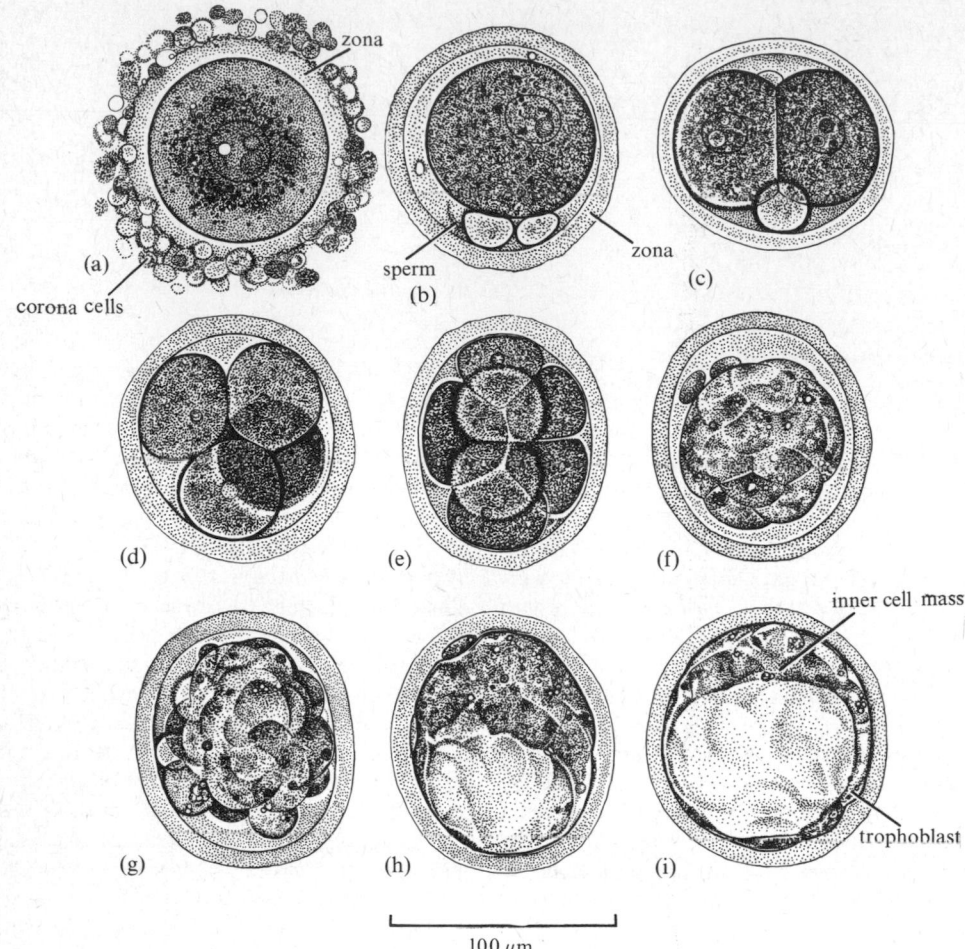

100 μm

FIG. 53.1. Stages in early development of the egg of the mouse, recovered from Fallopian tubes and uterus at known times after ovulation and copulation. (After Lewis and Wright 1935.)

(a) Ovarian egg, obtained 7–10 hours after copulation in late-oestrus, surrounded by corona cells. Note accumulation of dark yolk about nucleus, two nucleoli, clear peripheral zone and fuzzy surface of zona. (b) Normal egg, 20–20½ hours after copulation. Two polar bodies and sperm in perivitelline space. Zygote nucleus somewhat diagrammatic. (c) 2-cell stage 24 hours after copulation. (d) 4-cell stage 44½ hours after copulation. Nearly spherical cells each in contact with the other three by small contact areas. In this and subsequent figures only the nucleoli are seen. (e) 7 cells, 48 hours after copulation. (f) Morula from uterus, 72 hours after copulation. (g) Young blastocyst from uterus 82 hours after copulation, loosely arranged cells on surface. (h) Blastocyst formed from same specimen as (f), kept for 4 hours *in vitro*. (i) Blastocyst 82 hours after copulation. Vitellus completely fills zona. Note large amount of fluid.

other individuals and allowed to develop to birth. With such techniques many observations of early development have been made (Bellairs 1971). Mammalian embryos differ from those of lower vertebrates in that ribosomal RNA is synthesized during cleavage. This is perhaps related to the scarcity of yolk, but the cells of embryos of marsupials and the platypus contain nucleoli and presumably synthesize RNA.

The cells of the morula are all equipotential and can regulate. If one of the first two blastomeres is killed a whole animal develops and two eggs can be fused to make a single animal. Indeed as many as 16 whole morulae can be united. Such mosaics can be chimaeras of male and female or of different strains.

The mouse embryo reaches the blastocyst stage, ready to implant in the uterus, about 6 or 7 days after fertilization. This time-interval is similar in a wide variety of mammals, irrespective of the length of the uterine tubes. Implantation occurs after about a week in the rabbit and human, as well as in the mouse. In the macaque it occurs a little later (ninth day); in the pig, dog, and cat after about 2 weeks. In deer, however, the blastocyst lies unattached in the uterus for several weeks and in the badger for as long as eight months ('delayed implantation'). The embryos of rodents, primates, and carnivores remain spherical at this stage, but in the artiodactyl ruminants an extraordinarily rapid elongation occurs, the pig's blastocyst being nearly 30 cm long by the time of implantation (see Boyd and Hamilton 1952).

2. Gastrulation in mammals

Until the blastocyst stage the development of placental mammals consists of a cleavage not differing essentially from that of other chordates with little yolk, such as *Amphioxus*. The blastocyst stage itself and the subsequent stages of gastrulation are less easy to compare with those of lower vertebrates; moreoever, they vary widely in different mammals. The inner cell mass is the definitive embryonic region, the remainder of the blastocyst wall constitutes the trophoblast or extra-embryonic ectoderm, whose function is to establish intimate relationship with the tissues of the mother.

The inner cell mass gives rise very early to endoderm and often to some precocious mesoderm. The endoderm cells grow out in all directions from the edges of the embryonic disc formed by the inner cell mass and thus come to line the trophoblast, converting it into a *yolk sac*, though this contains no yolk. In some embryos, presumably of the more primitive mammals and including the insectivora and the rabbit, this endoderm may be in close contact with the trophoblast and the yolk sac is large (Fig. 53.2). In others, scattered 'mesenchymal' cells separate endoderm and trophoblast, so that the yolk sac forms a small separate inner vesicle. This primary mesoderm early arranges itself to form inner and outer mesodermal sheets, with an extra-embryonic coelom between them (Fig. 53.3).

3. Embryonic membranes

The *amnion* is formed very early, either by the production of folds covering the embryonic disc (rabbit, Fig. 53.2(a)), or as a hollowing out of the inner cell mass, without the formation of folds (man, Figs 53.2(b) and 53.3). The effect of the variations in early development is to produce embryos that appear at first sight to differ considerably. In those mammals that have departed less from the reptilian condition the appearance at this stage is very like that found in an animal with yolky eggs (see Amoroso 1952).

The embryo is covered by amniotic folds of ectoderm and somatic mesoderm and rests on a relatively large (though empty) yolk sac. In the more modified types, such as man, at the corresponding stage the embryo itself is less developed and lies as a germinal disc between two sacs, an upper amniotic and a lower yolk sac (Fig. 53.3). The outer trophoblast, now lined by mesoderm, is the *chorion*, which at first is of course avascular. Blood-vessels are provided to allow communication between the embryo and the trophoblast. In the primitive mammals this communication is provided (as in reptiles and birds) when the *allantois* grows out into the extra-embryonic coelom. In other mammals mesoderm formation anticipates this out-

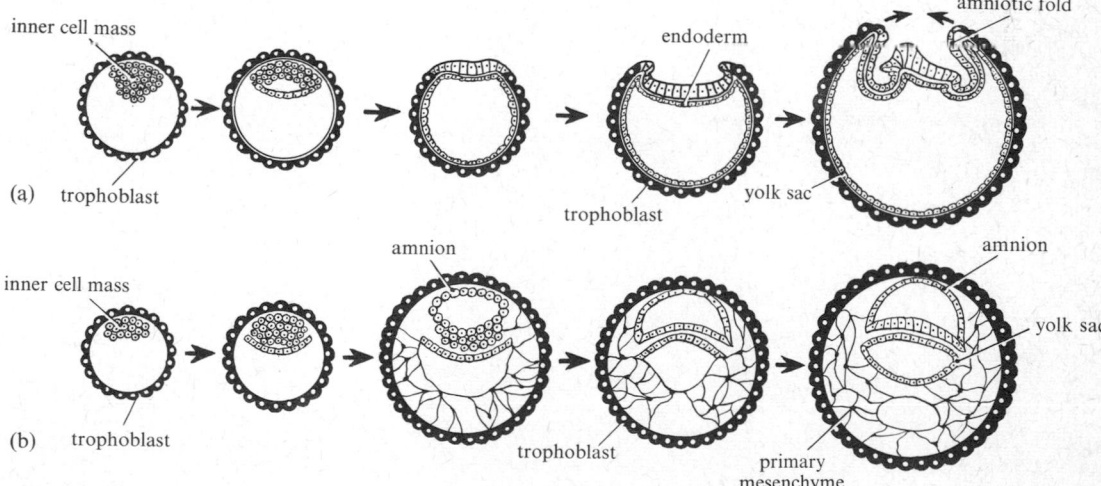

FIG. 53.2. Development of the amnion in mammals: (a) in the rabbit by formation of folds; (b) in man by hollowing out the inner cell mass. (After Boyd and Hamilton 1952.)

growth; the allantoic cavity itself then never becomes large and the embryo is connected with the trophoblast by a strand of mesoderm. This strand represents both the meeting-points of the chorion and amnion and the outgrowing allantois of earlier forms (Fig. 53.6) and it becomes the umbilical stalk. In both modes of development the amniotic cavity eventually becomes large and the fully formed embryo, attached by its umbilical stalk, floats in a fluid-filled sac.

The amniotic wall consists of two layers, as in the

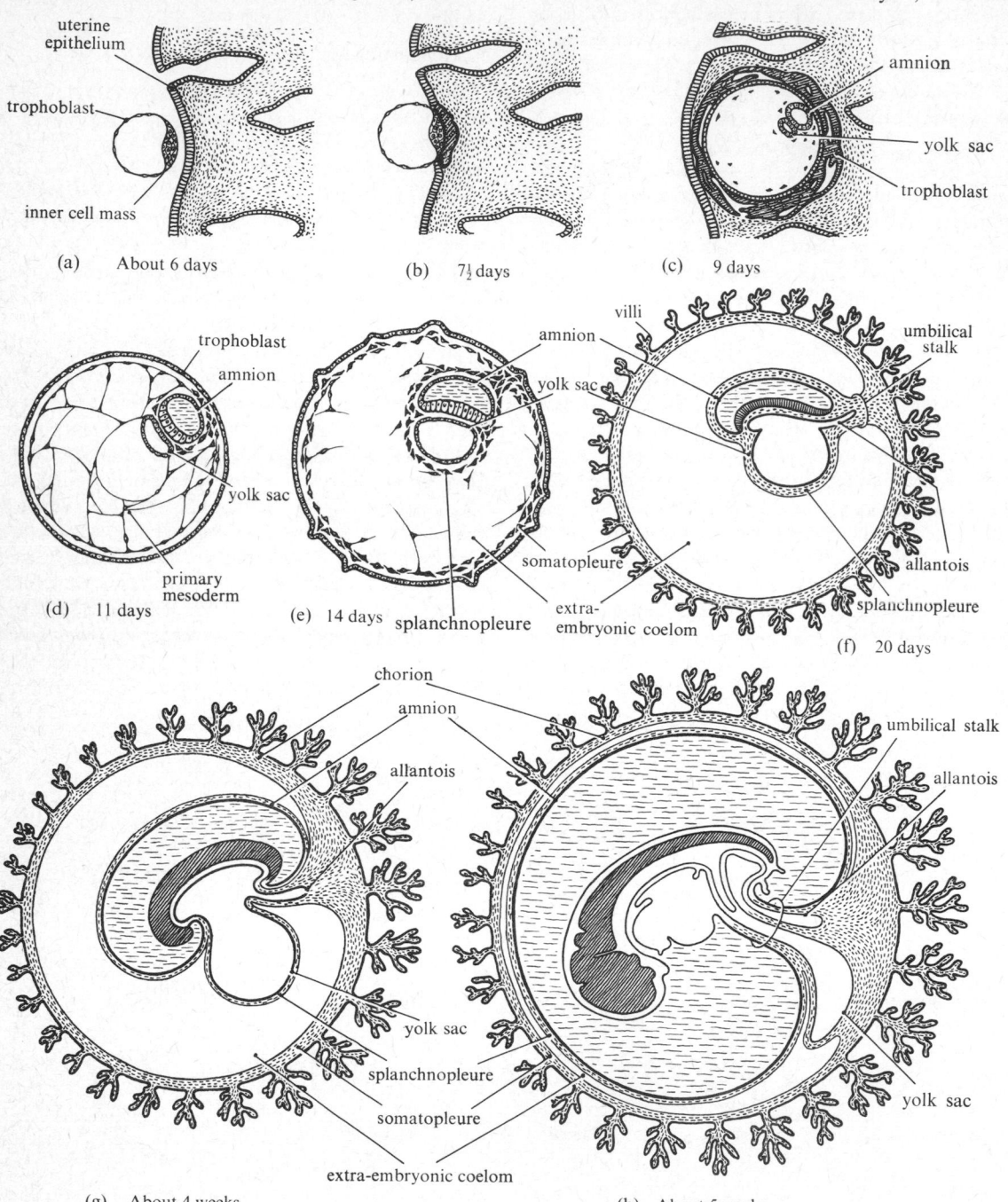

Fig. 53.3. Early stages of development in man.

chick, an inner true amnion and outer chorion (the original trophoblast) which, when vascularized, forms the most important link between embryo and mother. The amniotic fluid is not a stagnant pool but is very rapidly changed. In the human embryo at the end of gestation it has a half-life of only 95 minutes. This means that 3–4 litres of water are exchanged between mother and foetus every hour. Part of the interchange is due to swallowing of the fluid by the foetus, who drinks nearly 500 ml a day. The amniotic fluid contains various organic and inorganic constituents. The proportion of non-protein nitrogen increases during pregnancy, reflecting the secretion of urine by the foetal kidneys (Assali, Dilts, Plentl, Kirschbaum, and Gross 1968).

4. Formation of the embryo

The manner of development of the embryo is one of the most tantalizing parts of mammalian embryology (see Hertig and Rock 1941, 1945; Streeter 1942, 1945, 1948; Harrison 1961). We can discern enough to be sure that the processes are modified from those found in animals with yolky eggs, but few details are available of the actual morphogenetic movements in any mammal (Fig 53.4). The embryonic area ceases to be round and during the beginning of the third week of intra-uterine life in man (or second week in the rabbit) it can be seen at the bottom of the amniotic sac as an oval shield, consisting of ectoderm above and endoderm below. This shield can be compared with the blastoderm of a chick or reptile embryo and soon comes to show,

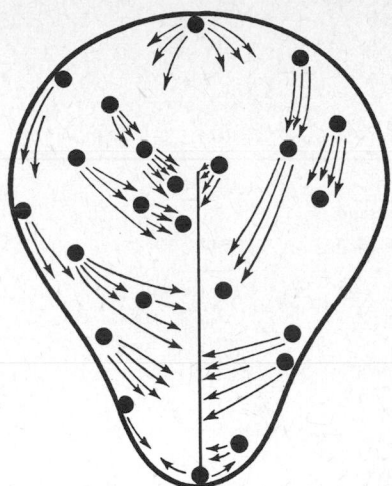

FIG. 53.4. Morphogenetic movements in the rabbit embryo during primitive streak formation. Black spots show regions marked with the neutral red; arrows show subsequent displacement of dye. (From Daniel and Olson (1966). *Anat. Rec.* **156.**)

towards its hind end, a primitive streak, provided with a primitive knot about the middle of the shield (Fig. 53.5). A head process appears as a thickening of the shield in front of the primitive knot, at about the sixteenth day in man.

These appearances suggest strongly that gastrulation and embryo formation proceed by morphogenetic movements resembling those of a reptilian ancestor.

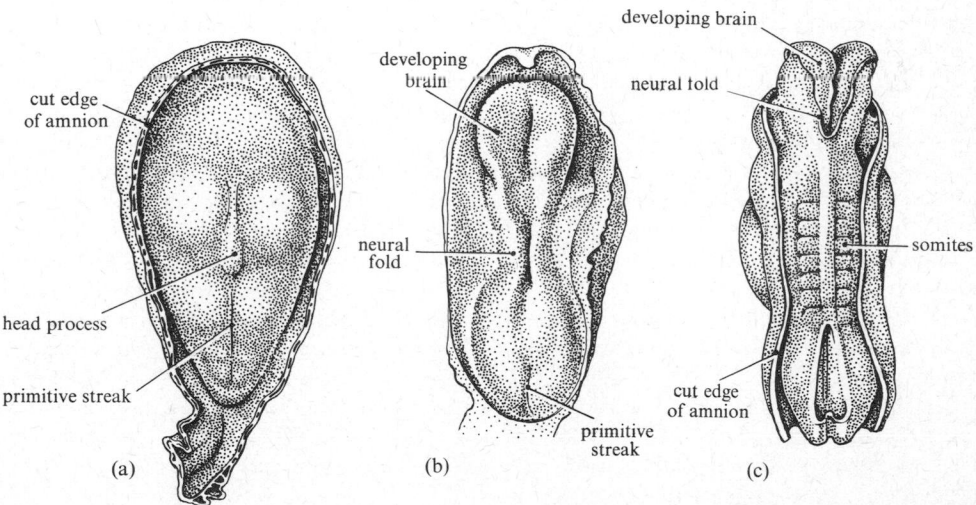

FIG. 53.5. Formation of the embryo in man. (a) 16 days, surface view of germ disc after removing the roof of the amniotic sac; (b) 18 days; (c) 20 days. (After Corner (1944). *Ourselves unborn.* Yale University Press, New Haven.)

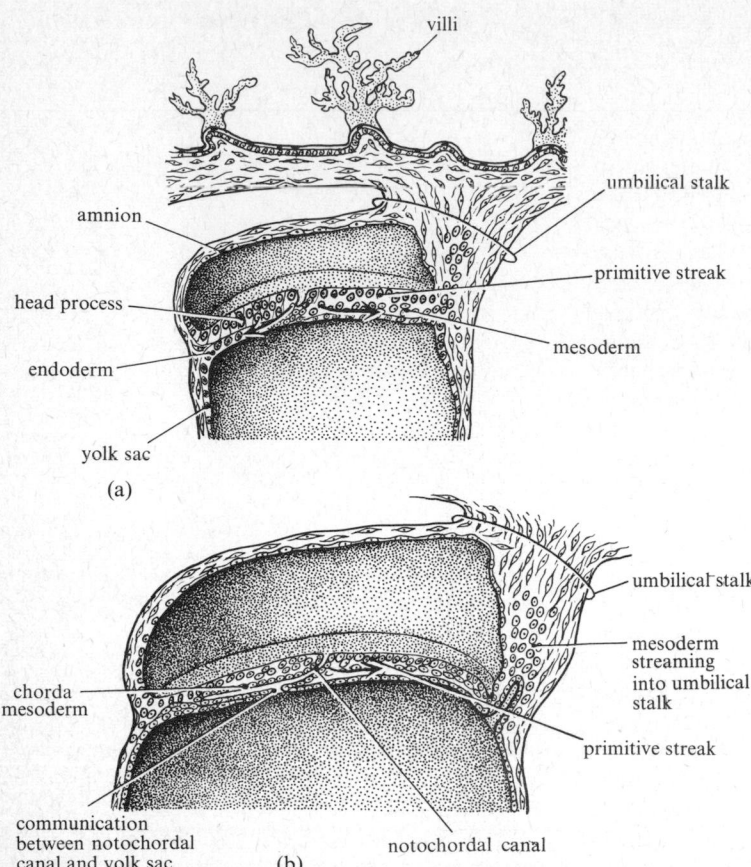

(a)

(b)

Fig. 53.6. Semi-diagrammatic views of human embryos cut in the sagittal plane. (a) 16 days and (b) a little later. (After Hamilton and Mossman 1972.)

It is probable that material streams towards the middle line and passes in at the primitive streak, then forwards and outwards to form the notochord and mesoderm. The primitive groove is continued forward as a 'notochord canal', representing the 'chordamesoblastic canal' of reptiles. This canal may actually open into the yolk sac (as in man, Fig. 53.6). Grafts of the embryonic axis of rabbits are able to induce formation of neural tissue from the ectoderm of a chick. We may therefore assume that in mammals there is the same sequence of induction by invaginated material as in other vertebrates.

Mesoderm (sometimes called secondary mesoderm to distinguish it from the earlier-formed 'mesenchyme') can be seen in connexion with the sides of the primitive streak, as would be expected if invagination is occurring. Everything suggests movements of this sort and the primitive streak moves backwards (extension) in man, just as in the chick, as the embryo forms in front of it. Unfortunately, owing to the conditions of intra-uterine development it has not yet been possible to observe any of these processes directly and we must for the present be content with the study of whole embryos and sections prepared at various stages.

By the eighteenth day in man a distinct embryo with neural folds and somites can be seen in front of the primitive streak. The neural folds begin to close a little later, first at the mid-dorsal level and then progressively forwards and backwards (Fig. 53.5). Meanwhile head and tail folds appear as in reptiles and birds, constricting off the embryo from the yolk sac by a narrow yolk stalk. The heart forms as in lower vertebrates and the blood-system early becomes connected with a system of blood islands that is established in the yolk sac. A pair of omphaloidean arteries formed in the yolk stalk joins the paired dorsal aortae, which arch round the pharynx as the first pair of branchial arches, joining the ventral aorta to the heart. This stage of a definitive embryo with beating heart is reached in the fourth week in man and after ten days in the rabbit.

5. Implantation of the embryo

We can now return to consider how the embryo establishes the connexions with the mother that are characteristic of mammalian development. Preparation to receive the embryo begins before ovulation with the effect of oestrogens on the uterus (Chapter 52). These cause increased development of the glands and muscles and an improved blood-supply. Preparation is continued in the progestational phase that follows, the glands becoming further developed and the arteries, in primates, approaching close to the surface (Fig. 52.4, p. 473).

As the blastocyst grows it often lies free in the cavity of the uterus, with the entire surface of the trophoblast in contact with the endometrium. In other species the blastocyst remains attached only at the site of first contact, leaving a free surface projecting into the uterine cavity. In yet other types, including man, the whole blastocyst sinks into the endometrium and becomes surrounded by maternal tissues, the *interstitial* form of placentation (see Amoroso 1952).

Whatever the means by which the first attachment is made, there soon develops an organ allowing for interchange between mother and foetus over part or all of the surface of the trophoblast. This organ, the *placenta*, consists of the intimately apposed or fused foetal and maternal tissues. On the foetal side the extra-embryonic ectoderm of the trophoblast (chorion) is connected with the blood-stream of the embryo either by contact with the wall of the yolk sac (*vitelline placenta*) or allantois (*allantoic placenta*). In the latter condition the outer covering is often known as the allanto-chorion and this name is used also when, as in man, the blood-vessels come from the precociously developed mesoderm of the allantois, although the cavity of the latter fails to develop. The events of early placentation in man have become more clear through the finding of embryos before and in process of embedding (see Harrison 1961). The trophoblast shown in Fig. 53.7 is greatly thickened on one side and its cells are so active in division that cell boundaries are not reformed and a *syncytiotrophoblast* results. This tissue, rapidly invading the uterine wall, has already destroyed the maternal epithelium and is proliferating in the endometrium beneath. The remainder of the surface of the blastocyst still projects into the uterine cavity and consists of a relatively thin cellular layer (*cytotrophoblast*).

Study of such a single stage does not, of course, show what has happened since the blastocyst entered the uterus, and we can only guess that the implantation has been produced by some action of the cells of the trophoblast on the uterine wall. It is not clear at what stage the mammalian embryo begins to draw nourishment from the mother, but it must be early, since the egg contains remarkably little yolk, less even than in other 'oligolecithal' eggs such as that of *Amphioxus*.

About four days pass before the egg arrives in the uterus. Being presumably an actively metabolizing tissue it must soon be in need of supplies. It is probable that these are provided by the mother even before the blastocyst implants (see p. 489).

In man the blastocyst rapidly sinks into the endometrium and by the eleventh day is surrounded on all sides. Although it is usual to ascribe this implantation to enzymatic invasion by the trophoblast it is possible that the uterine wall at this stage readily undergoes erosion on contact with any foreign object. In higher primates especially, the presence of the coiled arteries close to the endometrial surface gives the latter very peculiar properties; it may be regarded as an unstable system, ready to react. The '*decidual reaction*', a change of the endometrium produced by contact with an embryo, can be made to occur by mechanical stimula-

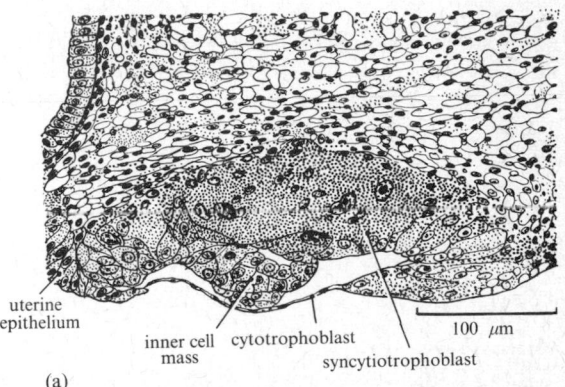

uterine epithelium

inner cell mass

cytotrophoblast

syncytiotrophoblast

100 µm

(a)

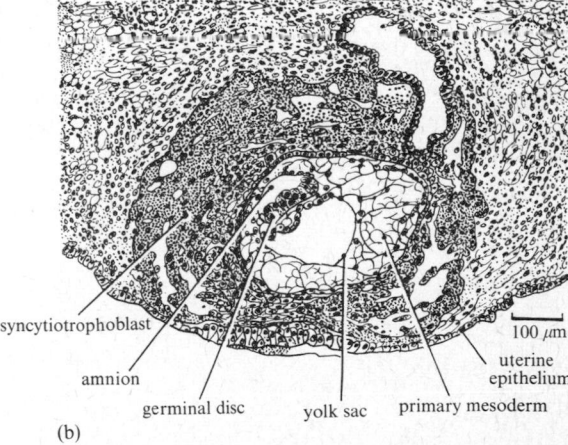

syncytiotrophoblast

amnion

germinal disc

yolk sac

primary mesoderm

uterine epithelium

100 µm

(b)

FIG. 53.7. Early stages of implantation in man. (a) $7\frac{1}{2}$ day-embryo. (After Hertig and Rock (1945). *Contr. Embryol.* **31**.) (b) $11\frac{1}{2}$ day-embryo. (After Hertig and Rock (1941). *Contr. Embryol.* **29**.)

tion of the progestational uterus by such objects as glass beads or suture threads. In other mammals the action of the trophoblast on the uterine tissues is less violent than in primates, but in all the blastocyst comes into contact with the uterine epithelium over part or the whole of its surface. The trophoblast might be expected to produce rejection by the uterus as a homograft. The mechanism for its protection is not known (see Kirby 1970).

The reaction of the uterus to the embryo is therefore an important part of the mechanism of placentation, but the trophoblast itself is very active. This has been shown by planting fertilized mouse ova into the anterior chamber of the eye and watching their development (Runner 1947). The trophoblast of this implant develops well for as much as three weeks, forming phagocytic trophoblastic giant cells, which penetrate the epithelia of the eye and produce extravasation of blood. In a developing embryo the trophoblast grows by division of the cells of the cytotrophoblast, which then fuse to make the syncytial masses. The cytotrophoblast covers the surface of the growing villi of the cotyledons for the first three months of pregnancy but then disappears, leaving each villus covered only by a thin syncytial layer. A thin connective tissue space and basement membrane is the only further separation from the foetal capillary endothelium. The surface of the syncytium is covered by numerous microvilli (Fig. 53.8).

6. The yolk sac

The next phase to be considered is the establishment of special means of connexion between the trophoblast as an organ of interchange with the mother and the developing embryo itself. This seems to have been effected in the earliest mammals by means of the yolk sac, which is still the only organ responsible in most marsupials, and plays a large part during the earlier stages of development in placentals. We have seen how the endoderm early spreads out over the inner side of the trophoblast to make the yolk sac. This double wall forms the non-vascular yolk-sac placenta that occurs in marsupials and in the early stages of many placentals. Later the 'primary mesoderm' extends between the trophoblast and the endoderm. In this mesoderm the first blood-vessels develop; becoming connected with the omphaloidean arteries and veins they are able to carry nourishment back to the embryo. Obviously in mammals it is the outer covering of the yolk sac that

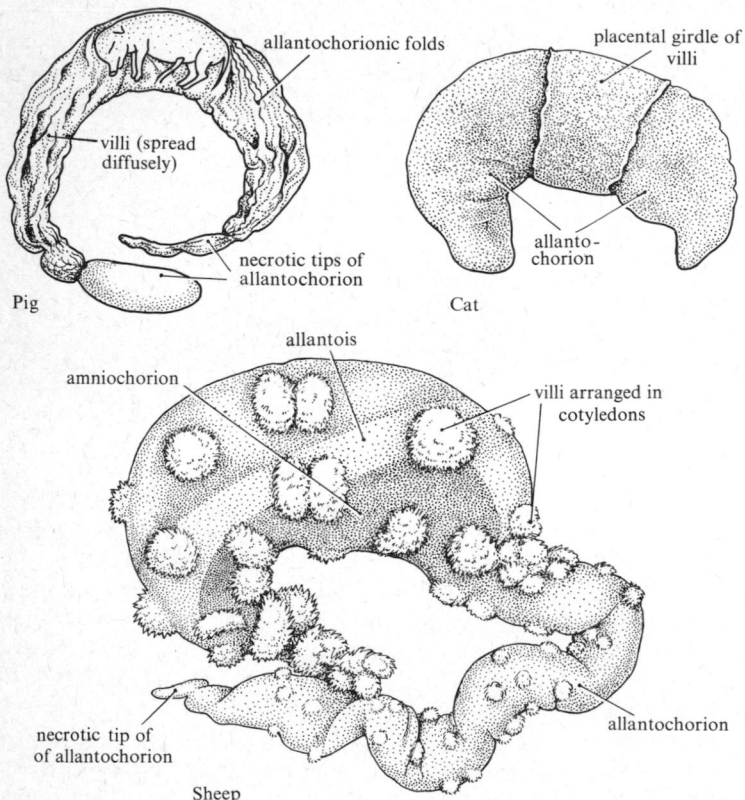

FIG. 53.8. Diagrams of some mammalian chorionic sacs showing differences in the distribution of villi. (After Amoroso 1952.)

is important rather than the inner endodermal layer that digests the yolk in a reptile or bird. Accordingly this outer trophoblastic layer becomes highly developed and folded into villi, while the endodermal sac becomes greatly reduced. In primates the yolk sac does not become applied to the outer wall of the trophoblast at all; when the yolk stalk is formed the sac remains only as a small dilatation at its distal end (Fig. 53.3). In the mammals that have a less-developed placenta the yolk sac is a more important organ. In marsupials it forms the organ of interchange throughout the whole of the rather short intra-uterine life. In rodents the vascularization of the yolk sac proceeds over only about half of its surface and the remaining portion may become eroded away, leaving the yolk-sac cavity open to that of the uterus. The *inner* wall of the yolk sac may then become applied to the maternal endometrium, a condition known as *inverted yolk-sac placentation*. Yolk sacs that have opened in this way may be used directly as digestive organs for the absorption of material from the uterus.

7. Histotrophe and haemotrophe

The material for nourishment of the embryo (*embryotrophe*) is derived from two sources. The *histotrophe* includes secretions of the uterine glands and breakdown products of the uterine mucosa and placenta, absorbed through the wall of the yolk sac. The *haemotrophe* is the nourishment supplied through the blood-stream. The secretion of the uterine glands is rich in fatty substances in many animals ('uterine milk'). This histotrophe probably provides an important source of nourishment in ruminants and other animals but is less important in man, where the association of foetal and maternal circulations soon becomes very close.

8. The allantois

The chief connexion between the absorbing surfaces and the embryo is provided not by the yolk sac but by the allantois. In reptiles and birds an outgrowth from the hind gut, covered by splanchnic mesoderm, provides a direct channel for the transport of oxygen to the foetus, besides fulfilling its primary function as a bladder for the storage of excretory products. In the early placental mammals, in addition to these two functions, it also serves to bring nourishment from the mother to the embryo. In most marsupials the allantois is a small sac growing out into the extra-embryonic coelom and forming no special contacts with the trophoblast. It presumably serves as a storage bladder. In the marsupial bandicoot (*Perameles*), however, it joins the chorion over a considerable area, villi being formed inter-digitating with those of the uterus.

This is essentially the arrangement in the placenta of the majority of eutherians. Insectivores, bats, artiodactyls, perissodactyls, and primitive primates, carnivores, cetaceans, and others all possess a medium or large allantoic bladder, filled with fluid, presumably excretory matter that cannot be returned to the mother across the placenta. In the horse and in artiodactyls crystals of calcium urate (*hippomanes*) occur in the fluid. These large allantoic sacs are found in the mammals in which the contact between embryonic trophoblast and maternal tissue is perhaps less close (p. 491). The allantois in these animals is thus in the condition found in reptiles, birds, and marsupials, a sac lying in the extra-embryonic coelom, and its splanchnopleuric mesoderm vascularizes the chorion. Its arteries, the umbilical arteries, become the main source of blood for the placenta and its veins (allantoic or umbilical veins) drain into the right atrium.

In primates (except lemurs) and rodents, which have a haemochorial placenta, conditions are much modified by the reduction of the cavity of the allantois. Thus in man the mesoderm corresponding to the splanchnopleure of the allantois forms very early (Figs 53.3 and 53.6). The allantoic cavity appears later as a small diverticulum extending from the hind gut into the strand of tissue that attaches the embryo to the trophoblast. It never becomes enlarged, presumably because the placenta serves from the earliest stages to carry away the excretory products, which it is well suited to do since few layers intervene between maternal and foetal circulations (p. 491). The portion of the allantois within the foetus becomes the bladder, and a cord, the *urachus*, connects this with the umbilicus and may contain a small allantoic cavity.

9. Villi and cotyledons of placenta

Correlated with these differences in the yolk sac and allantois are great differences in the mode of formation of villi by the trophoblast and the contact of these with the uterine wall. It is not clear what morphogenetic processes are responsible for villus formation. At its simplest it seems to be merely a wrinkling of the surface of the combined chorion and allantois. Thus in the pig the very long blastocyst contracts, forming a number of wrinkles over the surface (Fig. 53.9).

A similar *diffuse placenta* is found in the horse and other perissodactyls. In the carnivores a much more complicated arrangement of villi is formed and they are restricted to a band around the centre of the embryo, hence *zonary placenta*. There are found intermediate conditions between the diffuse and zonary arrangements, and also between the latter and the condition known as *cotyledonary*, in which only certain

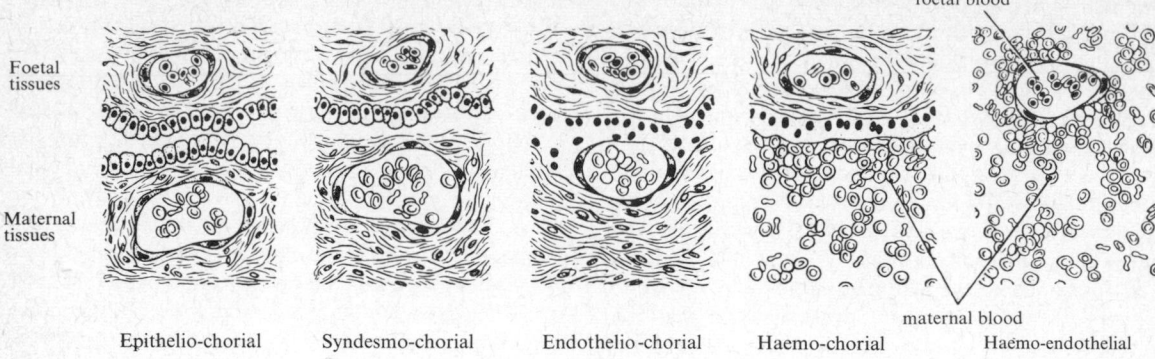

FIG. 53.9. Diagrams of the various types of placenta showing relations between foetal and maternal tissues. (After Amoroso 1952.)

parts of the placental zone of the allanto-chorion form villi. This is found in many ruminants (sheep and cow) though others may have a combined diffuse and cotyledonary placenta (Whitetail deer, *Odocoileus virginianus*) or even a purely diffuse one (*Moschus moschiferus*, the musk deer). The position of the cotyledons is apparently determined by the presence of specially differentiated areas in the wall of the uterus. In the insectivores, bats, rodents, and primates the villi are at first found over the whole trophoblast but sooner or later become restricted to a limited area, producing a *discoidal placenta*.

10. Types of placental union

More important than the distribution of the villi, which may be quite different in closely related forms, is their actual structure and the opportunities that they provide for interchange between mother and foetus. It has long been realized that the closeness of this connexion varies greatly in different mammalian orders. In some animals, such as the pig and horse, the chorionic villi are readily detached from the maternal tissues at parturition, so that no uterine material is lost. This is the *non-deciduate* condition, to be distinguished from the *deciduate* forms, in which some uterine tissues come away with the foetal membranes.

This is, however, a relatively gross way of classifying the conditions, which include various different arrangements of the foetal and maternal tissues. The villi in all cases are tubular folds of the foetal chorion, having a core of foetal mesoderm, richly supplied with blood-

TABLE 53.1
Placental types

Type of placenta	Maternal uterine mucous membrane				Foetal chorion				Examples
	Maternal blood	Maternal endothelium	Maternal mesodermal connective tissue	Maternal epithelium	Foetal epithelium	Foetal mesodermal connective tissue	Foetal endothelium	Foetal blood	
Epithelio-chorial	+	+	+	+	+	+	+	+	Horse, pig, donkey
Syndesmo-chorial	+	+	+	−	+	+	+	+	Cow, sheep, goat
Endothelio-chorial	+	+	−	−	+	+	+	+	Dog, cat
Haemo-chorial	+	−	−	−	+	+	+	+	Primates, some rodents
Haemo-endothelial	+	−	−	−	−	−	+	+	Some rodents and rabbits

vessels. These villi make branching structures of various forms, invading the maternal tissues in the area in which the placenta is developing. In the diffuse placenta of the sow and mare the epithelium of the villi is in contact with the maternal uterine epithelium, and interchange must take place between foetal and maternal blood across all the intervening tissues, that is to say foetal endothelium, connective tissue and epithelium, and maternal epithelium, connective tissue, and endothelium (Table 53.1 and Fig. 53.9). Such a type of placenta is known as *epithelio-chorial*. A similar condition is found in the lemurs. In many higher placentals, however the intervening barrier is reduced by the erosive action of the foetal trophoblast on the uterine tissues. In the ruminants, such as cow and sheep, the maternal epithelium is removed in this way, producing the type known as *syndesmo-chorial*. In carnivora the erosion goes farther, so that the villi are in contact with the maternal endothelium, hence, *endothelio-chorial*. In insectivores, rodents, bats, and higher primates the maternal endothelium is eroded so that the foetal villi are bathed in a lake of maternal blood, an arrangement known as *haemo-chorial*. In the rabbit and rat the chorionic epithelium itself disappears, producing a still closer relationship (*haemo-endothelial*). This classification was devised by Grosser and it still widely used, but presents many difficulties and must not be applied too rigidly. Placentae often show different structure as they develop and various conditions may be found at any one stage. Moreover the number of layers between mother and foetus is not the main important feature.

11. Efficiency of the placenta

It was assumed by Grosser that the fewer the layers separating the foetal and maternal blood the greater the efficiency of interchange. However placental transfer does not depend wholly or even mainly on diffusion but on the activity of the cells, which often contain very numerous mitochondria and much glycogen, making energy available for transport. Numerous microvilli are present at the absorbing surface.

Nevertheless the closeness of foetal and maternal circulations is found to be correlated approximately with: (a) the size of the molecules that can cross the barrier; (b) the rate of diffusion of ions; (c) the presence of a large allantois; (d) the amount of the histotrophe secreted by the uterine glands; and (e) the richness of the colostrum, the secretion of the mammary glands immediately after birth before the flow of milk begins (Huggett 1961; Assali *et al*. 1968). Yet it must be noted that the so-called 'less efficient' placentae may result in

the production of very well developed and active young, for example we may contrast the foal with the new-born human baby. Attempts to decide the 'efficiency' of placentae of various types leads to much confusion and uncertainty. Nearly all the mammals that produce large and well-developed young have placentae of the allegedly less efficient type, with a large allantois.

12. Evolution of the placenta

It is equally difficult to correlate placental structure with our knowledge of the affinities of the mammalian orders based on other evidence. Common sense would suggest that the diffuse and epithelio-chorial type of placentation is primitive; but it is found among the most specialized mammalian orders. The insectivores, shown both by their anatomy and geological history to be an ancient group, are said to have discoidal haemo-chorial placentation! If our ideas of affinity were based on placentation alone we should include lemurs with 'ungulates', because their placentae are epithelio-chorial, and the conies (*Hyracoidea*) with the insectivores, primates, and other haemo-chorial types. A classification based on placentation would put the sloths among man's nearest relatives. It may be that such absurdities are the result of imperfect study and classification of placentae, and it is difficult to believe that the haemo-chorial placental type is the primitive one. Probably the earliest eutherians possessed epithelio-chorial placentae.

13. Metabolism of the placenta

The cells of the placenta perform more varied functions than any others in the body. They act simultaneously as lungs, liver, intestine, and kidney and produce both steroidal and protein hormones. Electron microscopy shows that many types of cell are present to control these activities and the interchange between mother and foetus (see Wynn 1968).

In general the placenta allows the passage of substances of small molecular weight such as salts, sugars, urea, amino acids, simple fats, and some vitamins and hormones, but not the larger molecules of proteins, nor red blood corpuscles or other cells (Assali *et al*. 1968). Small molecules pass more readily than larger ones, and this process resembles ultrafiltration. Changes on one side of the barrier usually lead to corresponding changes on the other but there may be persistent differences on the two sides that are difficult to explain by diffusion theory. Thus in some epithelio-chorial placentae the blood sugar is higher on the foetal side. Discrete particles and non-pathogenic bacteria never pass the placental barrier.

The transfer of oxygen and carbon dioxide has been

studied in the sheep and is obviously fundamental. Transfer is mainly by diffusion but the actual P_{O_2} of the foetal umbilical vein is, however, much lower than the maternal arterial P_{O_2} and does not rise greatly even if the mother breathes oxygen. Placental membranes are freely permeable to water and electrolytes and the foetal and maternal bloods have identical osmolarity and concentrations of sodium, potassium, and chloride ions. Studies of the transfer of carbohydrates are complicated since these are utilized by the tissue of mother and foetus. Labelled glucose injected into the mother (rabbit) is first retained in the placenta and then released to the foetus. Quite large molecules such as ^{14}C-labelled cholesterol and steroid hormones can cross the placenta as can fat-soluble vitamins.

Amino acids can be transferred rapidly, as can small polypeptides. These transfers are not by simple diffusion but involve association with membranes and may require specific energy-dependent enzymatic carrier mechanisms. Still larger macromolecules may be transferred by pinocytosis.

One of the most important transfers is obviously of immunity and the mechanism for this varies with the number of layers separating maternal and foetal blood (Table 53.2). There is no passage of γ-globulins across the placenta in the pig (6 layers), ruminant (5 layers), or carnivore (3 layers). Instead another method is used; the new-born ruminant absorbs lactoglobulins present in the colostrum. In rodents (3 layers) there is some transfer by a route involving the amniotic fluid and the splanchnopleure of the yolk sac. In primates, with a separation by only the two layers of the haemochorial placenta, γ-globulins can cross the placenta and there is no intestinal absorption after birth. The human placenta is further exceptional in that it also allows the mutual exchange of leucocytes and blood platelets.

TABLE 53.2

Time of transmission of passive immunity in mammals

Species	Transmission of passive immunity	
	Pre-natal	Post-natal
Ox, goat, sheep	0	+ + + (36 hours)
Pig	0	+ + + (36 hours)
Horse	0	+ + + (36 hours)
Dog	+	+ + (10 days)
Mouse	+	+ + (16 days)
Rat	+	+ + (20 days)
Rabbit	+ + +	0
Guinea-pig	+ + +	0
Man	+ + +	0

In addition to acting as an organ of transfer the placenta has many other important biochemical properties, for instance it stores much glycogen. Various special types of cell develop in the placenta, presumably in connexion with these metabolic processes. The *decidual cells* are connective tissue cells of the endometrium that become loaded with glycogen or lipoids. They are usually prominent early in pregnancy but are lacking altogether in some placentae. The giant cells, which may be mono- or multinuclear, develop from the trophoblast; their function is unknown but they may be phagocytic.

The placenta is also the seat of production of a variety of hormones (Simmer 1968). The substances known as *chorionic gonadotropins*, which resemble pituitary hormones, are present in large quantities and are excreted in the urine during pregnancy, at least in some species. Their presence is the basis of a test for pregnancy in women. During the second month large amounts of a substance able to stimulate ovulation are excreted and one test consists of injecting urine samples into rabbits and later examining the ovaries. Another modification uses the ovulation of the aquatic frog *Xenopus*. The chorionic gonadotropin stimulates the corpus luteum of the ovary, at least during the early part of human pregnancy. However the removal of the ovaries even in the early months does not necessarily produce abortion and the function of the gonadotropin is therefore obscure. In some other species (rats) the ovaries are necessary throughout pregnancy. The placenta also produces a lactogenic hormone (see Wolstenholme and Knight 1972).

A variety of steroids can be isolated from the placenta and some of these are produced there. Progesterone is the most abundant and has various possible functions. It inhibits contraction of the uterine muscles and promotes the growth of the breasts. Large amounts of oestrogen are also produced and one of their effects is to stimulate growth of the myometrium. The smooth muscle cells increase in length more than 10 times during pregnancy and new cells are added by mitosis. Injection of labelled oestrogen shows that it is bound within a few minutes to the muscle cell nuclei, where it presumably sets in action the production of messenger and ribosomal RNA.

14. Parturition

Parturition at the end of gestation is evidently a more serious matter in those mammals in which there is a close interfusion of foetal and maternal tissue than in the non-deciduate types in which the allanto-chorion simply comes away from contact with the uterine surface. It is still not clear exactly what stimulus initiates parturition. The posterior pituitary produces a secretion (*oxytocin*) that has a powerful effect on the smooth muscle of the uterus, but during the earlier part of pregnancy some agent protects against its action. The effective stimulus to parturition is thus perhaps the withdrawal of this protection, allowing uterine contractions to begin. The progesterone secreted by the corpora lutea (and perhaps later by the placenta) may provide the inhibition. If a rabbit is made to ovulate by injection of pregnancy urine on the 25th day of pregnancy, birth does not occur normally on the 32nd day but is delayed until the 40th, when the induced corpora lutea degenerate. The villi of the human placenta are very rich in acetylcholine, which may play a part in stimulating the uterine musculature. Very complicated processes must be involved in the separation of foetal and maternal tissue, and in the healing of the latter, if the placenta is deciduate.

Bibliography

General works

BALINSKY, B. I. (1970). *An introduction to embryology* (3rd edn). W. B. Saunders, Philadelphia.

BARCROFT, J. (1934). *Features in the architecture of physiological function.* Cambridge University Press.

BARTLEY, W., BIRT, L. M., and BANKS, P. (1968). *The biochemistry of tissues.* John Wiley, London.

BAYLISS, L. E. (1959–60). *Principles of general physiology* (5th edn), Vols 1 and 2. Longmans, London.

BELL, G. H., DAVIDSON, J. N., and EMSLIE-SMITH, D. (1972). *Textbook of physiology and biochemistry* (8th edn). Churchill Livingstone, Edinburgh and London.

BELLAIRS, R. (1971). *Developmental processes in higher vertebrates.* Logos Press, London.

BLOOM, W. and FAWCETT, D. W. (1968). *A textbook of histology* (9th edn). W. B. Saunders, Philadelphia.

CAMERON, G. R. (1952). *Pathology of the cell.* Oliver and Boyd, Edinburgh.

CLARK, W. E. LE GROS, and MEDAWAR, P. B. (ed.) (1945). *Essays on growth and form; presented to D'Arcy Wentworth Thompson.* Clarendon Press, Oxford.

Cunningham's Textbook of anatomy (1972). (12th edn, ed. G. J. Romanes). Oxford University Press, London.

DAVSON, H. (1970). *A textbook of general physiology* (4th edn), Vols 1 and 2. J. & A. Churchill, London.

DAVSON, H. and EGGLETON, M. G. (ed.) (1968). *Starling and Lovatt Evans' Principles of human physiology* (14th edn). J. & A. Churchill, London.

EBERT, J. D. and SUSSEX, I. M. (1970). *Interacting systems in development* (2nd edn). Holt, Rinehart and Winston, New York.

FAWCETT, D. W. (1966). *The cell: its organelles and inclusions. An atlas of fine structure.* W. B. Saunders, Philadelphia.

GOODRICH, E. S. (1930). *Studies on the structure and development of vertebrates.* Macmillan, London. (Reprinted (1958). Dover, New York.)

GOSS, R. J. (1964). *Adaptive growth.* Logos Press, Academic Press, London.

GRASSÉ, P.-P. (ed.) (1966–). *Traité de Zoologie: anatomie, systématique, biologie,* Tome XVI, *Mammifères* (7 in 8 fascicules). Masson, Paris.

Gray's Anatomy (1973) (35th edn, ed. R. Warwick and P. L. Williams). Longman, London.

GRAY, J. (1968). *Animal locomotion.* Weidenfeld and Nicolson, London.

HAM, A. W. (1974). *Histology* (7th edn). J. B. Lippincott, Philadelphia and Toronto.

HAMILTON, W. J. and MOSSMAN, H. (1972). *Hamilton, Boyd and Mossman's Human embryology; prenatal development of form and function* (4th edn). W. Heffer, Cambridge.

Handbook of physiology: a critical, comprehensive presentation of physiological knowledge and concepts (1959–). American Physiological Society, Washington, D.C.

HOCHACHKA, P. W. and SOMERO, G. N. (1973). *Strategies of biochemical adaptation.* W. B. Saunders, Philadelphia.

KÜHN, A. (1971). *Lectures on developmental physiology* (2nd edn, transl. R. Milkman). Springer-Verlag, Berlin.

LANGMAN, J. (1969). *Medical embryology; human development—normal and abnormal* (2nd edn). Williams & Wilkins, Baltimore.

LEHNINGER, A. L. (1970). *Biochemistry: the molecular basis of cell structure and function.* Worth, New York.

LOEWY, A. G. and SIEKEVITZ, P. (1969). *Cell structure and function* (2nd edn). Holt, Rinehart and Winston, New York.

Marshall's Physiology of reproduction (1952–66), Vols 1–3. (3rd edn, ed. A. S. Parkes). Longmans, London.

MATTHEWS, L. HARRISON (1969–71). *The life of mammals,* Vols 1 and 2. Weidenfeld and Nicolson, London.

MOORE, K. L. (1973). *The developing human: clinically oriented embryology.* W. B. Saunders, Philadelphia.

NEEDHAM, J. (1931). *Chemical embryology,* Vols 1–3. Cambridge University Press.

—(1950). *Biochemistry and morphogenesis.* Cambridge University Press.

— (1959). *A history of embryology* (2nd edn). Cambridge University Press.

NELSEN, O. E. (1953). *Comparative embryology of the vertebrates.* Blakiston, New York.

OPPENHEIMER, J. M. (1967). *Essays in the history of embryology and biology.* M.I.T. Press, Cambridge, Massachusetts.

PARKER, T. J. and HASWELL, W. A. (1972). *Textbook of zoology* (7th edn, ed. A. J. Marshall and W. D. Williams), Vols 1 and 2. Macmillan, London.

PATTEN, B. M. (1968). *Human embryology* (3rd edn). McGraw-Hill, New York.

PORTER, K. R. and BONNEVILLE, M. A. (1968). *Fine structure of cells and tissues* (3rd edn). Lea & Febiger, Philadelphia.

RANSON, S. W. and CLARKE, S. L. (1959). *The anatomy of the nervous system: its development and function* (10th edn). W. B. Saunders, Philadelphia.

REITH, E. J. and ROSS, M. H. (1970). *Atlas of descriptive histology* (2nd edn). Harper & Row, New York.

RHODIN, J. A. G. (1974). *Textbook of histology: a text and atlas.* Oxford University Press, London.

ROMER, A. S. (1962). *The vertebrate body* (3rd edn). W. B. Saunders, Philadelphia.

SELKURT, E. E. (ed.) (1971). *Physiology* (3rd edn). Little, Brown, Boston.

SHERRINGTON, C. S. (1947). *The integrative action of the nervous system* (revised edn). Cambridge University Press.

SINGER, C. and UNDERWOOD, E. A. (1962). *A short history of medicine* (2nd edn). Clarendon Press, Oxford.

Starling and Lovatt Evans' Principles of human physiology (1968). (14th edn, ed. H. Davson and M. G. Eggleton). J. & A. Churchill, London.

STEVENS, S. S. (1951). *Handbook of experimental psychology*. Chapman & Hall, London.

TRINKAUS, J. P. (1969). *Cells into organs: the forces that shape the embryo*. Prentice-Hall, Englewood Cliffs, New Jersey.

WADDINGTON, C. H. (1956). *Principles of embryology*. Allen & Unwin, London.

WALKER, E. P. (1964). *Mammals of the world*, Vols 1–3. Johns Hopkins Press, Baltimore.

WEBER, R. (ed.) (1965–7). *The biochemistry of animal development*, Vols 1 and 2. Academic Press, New York and London.

WHITE, A., HANDLER, P., and SMITH, E. L. (1973). *Principles of biochemistry* (5th edn). McGraw-Hill, New York.

YOUNG, J. Z. (1951). *Doubt and certainty in science*. Clarendon Press, Oxford.

— (1962). *The life of vertebrates* (2nd edn). Clarendon Press, Oxford.

— (1971). *An introduction to the study of man*. Clarendon Press, Oxford.

Chapter 1. Communication, control, and the continuation of life

ADOLPH, E. F. (1968). *Origins of physiological regulations*. Academic Press, New York and London.

ASHBY, W. ROSS, (1960). *Design for a brain* (2nd edn). Chapman & Hall, London.

CANNON, W. B. (1932). *The wisdom of the body*. W. W. Norton, New York.

KALMUS, H. (ed.) (1966). *Regulation and control in living systems*. John Wiley, London.

LANGLEY, L. L. (ed.) (1973). *Homeostasis: origin of a concept*. Dowden, Hutchinson & Ross, Stroudsburg, Pennsylvania.

MACHIN, K. E. (1964). Feedback theory and its application to biological systems (with a commentary by B. R. Wilkins). *Symp. Soc. Exp. Biol.*. **18**, 421–46.

SOMMERHOFF, G. (1974). *Logic of the living brain*. John Wiley, London.

WATERMAN, T. H. (1968). Systems theory and biology—view of a biologist. In *Systems theory and biology* (ed. M. D. Mesarović), Proceedings of the Third Systems Symposium at Case Institute of Technology, pp. 1–37. Springer-Verlag, New York.

WILKINS, B. R. (1966). Regulation and control in engineering. In *Regulation and control in living systems* (ed. H. Kalmus), pp. 12–28. John Wiley, London.

YOUNG, J. Z. (1946). Effects of use and disuse on nerve and muscle. (Sydney Ringer Memorial Lecture, University College Hospital Medical School.) *Lancet* (2), 109.

— (1971). *An introduction to the study of man*. Clarendon Press, Oxford.

Chapter 2. The skin and control of temperature

BLIGH, J. (1973). *Temperature regulation in mammals and other vertebrates*. North-Holland, Amsterdam and London.

Handbook of Physiology (1964). Section 4: *Adaptation to the environment* (ed. D. B. Dill). American Physiological Society, Washington, D.C.

HARDY, J. D. (1961). Physiology of temperature regulation. *Physiol. Rev.* **41**, 521–606.

KELLY, D. E. (1966). Fine structure of desmosomes, hemidesmosomes, and an adepidermal globular layer in developing newt epidermis. *J. Cell. Biol.* **28**, 51–72.

MARTIN, C. J. (1902). Thermal adjustment and respiratory exchange in monotremes and marsupials; a study in the development of homoeothermism. *Phil. Trans. R. Soc. B* **195**, 1–37.

MONTAGNA, W. (1962). *The structure and function of skin* (2nd edn). Academic Press, New York.

MROSOVSKY, N. (1971). *Hibernation and the hypothalamus*. Appleton-Century-Crofts, New York.

SCHMIDT-NIELSEN, K. (1964). *Desert animals: physiological problems of heat and water*. Clarendon Press, Oxford.

SPEARMAN, R. I. C. (1966). The keratinization of epidermal scales, feathers and hairs. *Biol. Rev.* **41**, 59–96.

WHITTOW, G. C. (ed.) (1970–3). *Comparative physiology of thermoregulation*, Vols 1–3. Academic Press, New York and London.

Chapter 3. Epidermis and dermis. Appendages and appearances

BILLINGHAM, R. E. and MEDAWAR, P. B. (1948). Pigment spread and cell heredity in guinea-pigs' skin. *Heredity* **2**, 29–47.

COTT, H. B. (1940). *Adaptive coloration in animals*. Methuen, London.

FOX, H. MUNRO and VEVERS, G. (1960). *The nature of animal colours*. Sidgwick & Jackson, London.

FRASER, R. D. B., MACRAE, T. P., and ROGERS, G. E. (1972). *Keratins: their composition, structure and biosynthesis*. Thomas, Springfield, Illinois.

HOOF, J. A. R. A. M. VAN (1962). Facial displays in higher primates. *Symp. Zool. Soc. Lond.* **8**, 97–125.

MATTHEWS, L. HARRISON (1969–71). *The life of mammals*, Vols 1 and 2. Weidenfeld and Nicolson, London.

PORTMANN, A. (1952). *Animal forms and patterns: a study of the appearance of animals* (trans. H. Czech). Faber & Faber, London.

RUDALL, K. M. (1968). Intracellular fibrous proteins and the keratins. In *Comprehensive biochemistry* (ed. M. Florkin and E. H. Stotz), Vol. 26, part B: *Extracellular and supporting tissues* (continued), pp. 559–91. Elsevier, Amsterdam.

Chapter 4. The connective tissues

CARR, I. (1973). *The macrophage: a review of ultrastructure and function*. Academic Press, London.

FITTON JACKSON, S. (1964). Connective tissue cells. In *The cell: biochemistry, physiology, morphology* (ed. J. Brachet and A. E. Mirsky), Vol. 6, pp. 387–520. Academic Press, London.

— (1968). The morphogenesis of collagen. In *Treatise on collagen* (ed. B. S. Gould), Vol. 2, part B, pp. 1–66. Academic Press, London.

FRANZBLAU, C. (1971). Elastin. In *Comprehensive biochemistry* (ed. M. Florkin and E. H. Stotz), Vol. 26, part C: *Extracellular and supporting tissues* (continued), pp. 659–712. Elsevier, Amsterdam.

GRANT, M. E. and PROCKOP, D. J. (1972). The biosynthesis of collagen. *New Engl. J. Med.* **286,** 194–9.

GREENLEE, T. K., Jr, and ROSS, R. (1967). The development of the rat flexor digital tendon, a fine structure study. *J. Ultrastruc. Res.* **18,** 354–76.

HARKNESS, R. D. (1961). Biological functions of collagen. *Biol. Rev.* **36,** 399–463.

— (1964). The physiology of the connective tissues of the reproductive tract. *Int. Rev. Connective Tissue Res.* **2,** 155–211.

— (1966). The structure of collagen. In *Progress in the biological sciences in relation to dermatology,* 2 (ed. A. Rook and R. H. Champion), pp. 3–23. Cambridge University Press.

KULONEN, E. and PIKKARAINEN, J. (ed.) (1973). *Biology of fibroblast.* Academic Press, London.

LAURENT, T. C. (1972). The ultrastructure and physical-chemical properties of interstitial connective tissue. *Pflügers Arch. ges. Physiol.* **336** (Suppl.), 521–42.

MILLER, E. J. and MATUKAS, V. J. (1974). Biosynthesis of collagen: the biochemist's view. *Fedn. Proc. Fedn. Am. Socs exp. Biol.* **33,** 1197–204.

PEACOCK, E. E. and VAN WINKLE, W. (1970). *Surgery and biology of wound repair.* W. B. Saunders, Philadelphia.

RAMACHANDRAN, G. N. (ed.) (1968). *Biology of collagen,* Vol. 2, parts A and B. Academic Press, London.

RHODIN, J. A. G. (1967). Organization and ultrastructure of connective tissue. In *The connective tissue* (ed. B. M. Wagner and D. E. Smith), pp. 1–16. International Academy of Pathology Monograph. Williams & Williams, Baltimore.

ROSS, R. and BORNSTEIN, P. (1969). The elastic fiber. I. The separation and partial characterization of its macromolecular components. *J. Cell. Biol.* **40,** 366–81.

VERNON-ROBERTS, B. (1972). *The macrophage.* Cambridge University Press.

VIIDIK, A. (1973). Functional properties of collagenous tissues. *Int. Rev. Connective Tissue Res.* **6,** 127–215.

WEISS, P. (1961). The biological foundation of wound repair. *Harvey Lect.* **55,** 13–42.

Chapter 5. Skeletal tissue: cartilage and bone

BARNETT, C. H., DAVIES, D. V., and MACCONAILL, M. A. (1961). *Synovial joints: their structure and mechanics.* Longmans, London.

BARRETT, A. J. (1968). Cartilage. In *Comprehensive biochemistry* (ed. M. Florkin and E. H. Stotz), Vol. 26, part B: *Extracellular and supporting tissues* (continued), pp. 425–74. Elsevier, Amsterdam.

Ciba Foundation Symposium II (n.s.) (1973). *Hard tissue growth, repair, and remineralization.* Elsevier, North-Holland.

FITTON JACKSON, S. (1964). Connective tissue cells. In *The cell: biochemistry, physiology, morphology* (ed. J. Brachet and A. E. Mirsky), Vol. 6, pp. 387–520. Academic Press, London.

FREEMAN, M. A. R. (ed.) (1973). *Adult articular cartilage.* Pitman Medical, London.

GLUCKSMANN, A. (1942). The role of mechanical stresses in bone formation *in vitro. J. Anat.* **76,** 231–9.

HALL, B. K. (1970). Cellular differentiation in skeletal tissues. *Biol. Rev.* **45,** 455–84.

HANCOX, N. M. (1972). *Biology of bone.* Cambridge University Press.

MCLEAN, F. C. and URIST, M. R. (1968). *Bone: fundamentals of the physiology of skeletal tissue* (3rd edn). University of Chicago Press.

PRITCHARD, J. J. (1961). Ossification. *Scott. Med. J.* **6,** 177–87.

— (1963). Bone healing. *The scientific basis of medicine annual reviews* 1963, pp. 286–301.

VAUGHAN, J. M. (1970). *The physiology of bone.* Clarendon Press, Oxford.

ZIPKIN, I. (ed.) (1973). *Biological mineralization.* John Wiley, New York.

Chapter 6. The framework of the body

BASSETT, C. A. L., PAWLUK, R. J., and BECKER, R. O. (1964). Effect of electric currents on bone *in vivo. Nature* **204,** 652–4.

EVANS, F. GAYNOR (1973). *Mechanical properties of bone.* Thomas, Springfield, Illinois.

FROST, H. M. (1964). *The laws of bone structure.* Thomas, Springfield, Illinois.

HALL-CRAGGS, E. C. B. (1969). Influence of epiphyses on the regulation of bone growth. *Nature* **221,** 1245.

HOYTE, D. A. N. and ENLOW, D. H.

(1966). Wolff's law and the problem of muscle attachment on resorptive surfaces of bone. *Am. J. phys. Anthrop.* **24,** 205–14.

MURRAY, P. D. F. (1936). *Bones.* Cambridge University Press.

RYDELL, N. (1966). Intravital measurements of forces acting on the hipjoint. In *Studies on the anatomy and function of bone and joints* (ed. F. Gaynor Evans), pp. 52–68. Springer-Verlag, Berlin, Heidelberg and New York.

SIJBRANDIJ, S. (1963). Inhibition of tibial growth by means of compression of its proximal epiphysial disc in the rabbit. *Acta anat.* **55,** 278–85.

Chapter 7. Muscles

BOURNE, G. H. (ed.) (1972–3). *The structure and function of muscle* (2nd edn), Vols 1–3. Academic Press, London and New York.

ECCLES, J. C. and SHERRINGTON, C. S. (1930). Numbers and contraction-values of individual motor-units examined in some muscles of the limb. *Proc. R. Soc. B* **106,** 326–56.

ELFTMAN, H. (1941). The action of muscles in the body. *Biol. Symp.* **3,** 191–209.

FATT, P. and KATZ, B. (1953). Chemo-receptor activity at the motor end-plate. *Acta physiol. Scand.* **29,** 117–25.

FUCHS, F. (1974). Striated muscles. *A. Rev. Physiol.* **36,** 461–502.

HALL-CRAGGS, E. C. B. (1974). Mammalian skeletal muscle and its innervation. In *Essays on the nervous system: a Festschrift for Professor J. Z. Young* (ed. R. Bellairs and E. G. Gray), pp. 106–30. Clarendon Press, Oxford.

HILL, A. V. (1927). *Living machinery.* G. Bell, London.

— (1950). The dimensions of animals and their muscular dynamics. *Proc. R. Instn Gt Br.* **34,** 450–71.

HUNT, C. C. and KUFFLER, S. W. (1954). Motor innervation of skeletal muscle: multiple innervation of individual muscle fibres and motor unit function. *J. Physiol., Lond.* **126,** 293–303.

HUXLEY, H. E. (1957). The double array of filaments in cross-striated muscle. *J. biophys. biochem. Cytol.* **3,** 631–47.

— (1972). Molecular basis of contraction in cross-striated muscles. In *The structure and function of muscle* (2nd edn, ed. G. H. Bourne),

Vol. 1, Part 1: *Structure*, pp. 301–87. Academic Press, New York and London.

KATZ, B. (1966). *Nerve, muscle, and synapse*. McGraw-Hill, New York.

PROSSER, C. L. (1974). Smooth muscle. *A. Rev. Physiol.* **36**, 503–35.

Chapters 8–10. The vertebral column. The forelimb. The pelvic girdle and hind limb

BARNETT, C. H. and NAPIER, J. R. (1953). The form and mobility of the fibula in metatherian mammals. *J. Anat.* **87**, 207–13.

DARCUS, H. D. (1951). The maximum torques developed in pronation and supination of the right hand. *J. Anat.* **85**, 55–67.

GRAY, J. (1944). Studies on the mechanics of the tetrapod skeleton. *J. exp. Biol.* **20**, 88–116.

— (1968). *Animal locomotion.* Weidenfeld and Nicolson, London.

HAINES, R. W. (1953). The early development of the femoro-tibial and tibio-fibular joints. *J. Anat.* **87**, 192–206.

JOSEPH, J. (1960). *Man's posture: electromyographic studies*. Thomas, Springfield, Illinois.

THOMPSON, D'A. W. (1942). *On growth and form* (2nd edn). Cambridge University Press.

Chapter 11. The head of mammals

DE BEER, G. R. (1937). *The development of the vertebrate skull*. Clarendon Press, Oxford.

Chapter 12. Development of skeleton, muscles, and coelom

DE BEER, G. R. (1937). *The development of the vertebrate skull*. Clarendon Press, Oxford.

GILBERT, P. W. (1957). The origin and development of the human extrinsic ocular muscles. *Contr. Embryol.* **36**, 59–78.

GOODRICH, E. S. (1930. *Studies on the structure and development of vertebrates*. Macmillan, London.

Chapter 13. The intake of food: mouth and teeth

FRANK, R. M., SAUVAGE, C., and FRANK, P. (1972). Morphological basis of dental sensitivity. *Int. dent. J., Lond.* **22**, 1–19.

GREGORY, W. K. (1934). A half century of trituberculy, the Cope–Osborn theory of dental evolution. *Proc. Am. phil. Soc.* **73**, 169–317.

HERVEY, G. R. (1959). The effects of lesions in the hypothalamus in parabiotic rats. *J. Physiol., Lond.* **145**, 336–52.

KENNEDY, G. C. (1953). The role of depot fat in the hypothalamic control of food intake in the rat. *Proc. R. Soc.* **B140**, 578–92.

Orban's Oral histology and embryology (1972) (7th edn, ed. H. Sicher and S. N. Bhaskar). C. V. Mosby, St. Louis.

ROMER, A. S. (1966). *Vertebrate paleontology* (3rd edn). University of Chicago Press.

STEVENSON, J. A. F. (1969). Neural control of food and water intake. In *The hypothalamus* (ed. W. Haymaker, E. Anderson, and W. J. H. Nauta), pp. 524–621. Thomas, Springfield, Illinois.

Chapter 14. Digestion

GREGORY, R. A. (1965). Secretory mechanisms of the digestive tract. *A. Rev. Physiol.* **27**, 395–414.

ROUILLER, CH. (ed.) (1963–4). *The liver: morphology, biochemistry, physiology*, Vols 1 and 2. Academic Press, New York and London.

TAVILL, A. S. (1972). The synthesis and degradation of liver-produced proteins. *Gut* **13**, 225–41.

Chapter 15. Respiration

CORDA, M., EULER, C. VON, and LENNERSTRAND, G. (1965). Proprioceptive innervation of the diaphragm. *J. Physiol., Lond.* **178**, 161–77.

EULER, C. VON (1973). The role of proprioceptive afferents in the control of respiratory muscles. *Acta neurobiol. exp.* **33**, 329–41.

LLOYD, T. C. (1971). Respiratory gas exchange and transport. In *Physiology* (3rd edn, ed. E. E. Selkurt), pp. 451–70. Little, Brown, Boston.

MERRILL, E. G. (1974). Finding a respiratory function for the medullary respiratory neurons. In *Essays on the nervous system: a Festschrift for Professor J. Z. Young* (ed. R. Bellairs and E. G. Gray), pp. 451–86. Clarendon Press, Oxford.

SCHLAEFKE, M. E., SEE, W. R., and LOESCHCKE, H. H. (1970). Ventilatory response to alterations of H^+ ion concentration in small areas of the ventral medullary surface. *Resp. Physiol.* **10**, 198–212.

Chapter 16. Development of the gut and respiratory system

BOYD, J. D. (1950). Development of thyroid and parathyroid glands and the thymus. *Ann. R. Coll. Surg.* **7**, 455–71.

DU BOIS, A. M. (1963). The embryonic liver. In *The liver: morphology, biochemistry, physiology* (ed. Ch. Rouiller), Vol. 1, pp. 1–39. Academic Press, New York and London.

PATTLE, R. E. (1969). The development of the foetal lung. In *Foetal autonomy* (ed. G. E. W. Wolstenholme and M. O'Connor), pp. 132–42. J. and A. Churchill, London.

TOWERS, B. (1968). The fetal and neonatal lung. In *Biology of gestation* (ed. N. S. Assali), Vol. 2: *The fetus and neonate*, pp. 189–223. Academic Press, New York and London.

Chapters 17 and 18. The internal environment. Blood-vessels and capillary circulation. The cells of the blood

BECKER, A. J., MCCULLOCH, E. A., and TILL, J. E. (1963). Cytological demonstration of the clonal nature of spleen colonies derived from transplanted mouse marrow cells. *Nature* **197**, 452–4.

BOGGS, D. R. (1966). Homeostatic regulatory mechanisms of hematopoiesis. *A. Rev. Physiol.* **28**, 39–53.

CANNON, W. B. (1932). *The wisdom of the body*. Kegan Paul, London.

OWEN, J. A. (1967). Effects of injury on plasma protein. *Adv. clin. Chem.* **9**, 1 41.

PANTIN, C. F. A. (1964). Homeostasis and the environment. *Symp. Soc. exp. Biol.* **18**, 1–6.

PERUTZ, M. F. (1969). The haemoglobin molecule. *Proc. R. Soc.* **B173**, 113–40.

SEEGERS, W. H. (1969). Blood clotting mechanisms: three basic reactions. *A. Rev. Physiol.* **31**, 269–94.

WEYER, E. M. (ed.) (1968). Erythropoietin. *Ann. N.Y. Acad. Sci.* **149**, 1–583.

WILT, F. H. (1967). The control of embryonic haemoglobin synthesis. *Adv. Morphogen.* **6**, 89–125.

Chapter 19. Protection and defence of the body

COHN, Z. A. (1968). The structure and function of monocytes and macrophages. *Adv. Immunol.* **9**, 163–214.

HIRSCH, J. G. (1965). Phagocytosis. *A. Rev. Microbiol.* **19**, 339–50.

WARD, P. A. (1968). Chemotaxis of mononuclear cells. *J. exp. Med.* **128**, 1201–21.

Chapter 20. Antibodies

BILLINGHAM, R. E., BRENT, L., and MEDAWAR, P. B. (1956). Quantitative studies on tissue transplantation immunity. III. Actively acquired tolerance. *Phil. Trans. R. Soc.* **239**, 357–414.

CARR, I. (1970). The fine structure of the mammalian lymphoreticular system. *Int. Rev. Cytol.* **27**, 283–348.

CLARK, S. L. (1963). The thymus in mice of strain 129/J, studied with the electron microscope. *Am. J. Anat.* **112**, 1–33.

— (1964). The penetration of proteins and colloidal materials into the thymus from the blood stream. In *The thymus* (ed. V. Defendi and D. Metcalf). Wistar Institute Symposium Monograph No. 2, pp. 9–31. Wistar Institute, Philadelphia.

DAVIES, A. J. S. (1969). The thymus and the cellular basis of immunity. *Transplant. Rev.* **1**, 43–91.

EDELMAN, G. M. and MÖLLER, G. (1973). The immune response as a model for cellular maturation and differentiation. *Neurosci. Res. Progr. Bul.* **11**, 87–154.

EVERETT, N. B. and TYLER, R. W. (1967). Lymphopoiesis in the thymus and other tissues: functional implications. *Int. Rev. Cytol.* **22**, 205–37.

GREAVES, M. F., OWEN, J. J. T., and RAFF, M. C. (1973). *T and B lymphocytes: origins, properties and roles in immune responses.* Elsevier, Amsterdam.

HUMPHREY, J. H. and WHITE, R. G. (1970). *Immunology for students of medicine* (3rd edn). Blackwell, Oxford and Edinburgh.

KUECHLER, E. and RICH, A. (1969). Sequential synthesis of messenger RNA and antibodies in rabbit lymph nodes. *Nature* **222**, 544–7.

LENNOX, E. S. and COHN, M. (1967). Immunoglobulins. *A. Rev. Biochem.* **36**, 365–406.

LEVEY, R. H. and MEDAWAR, P. B. (1966). Nature and mode of action of antilymphocytic antiserum. *Proc. natn. Acad. Sci. U.S.A.* **56**, 1130–7.

NOSSAL, G. J. V. (1969). The cellular basis of immunity. *Harvey Lect.* **63**, 179–211.

ROITT, I. M. (1971). *Essential immunology.* Blackwell Scientific Publications, Oxford.

WARREN, K. H. (ed.) (1968). *Differentiation and immunology.* Academic Press, New York and London.

WOODRUFF, M. (1969). Lymphocytes and antilymphocytic serum. *Endeavour* **28**, 65–7.

Chapter 21. The heart and circulation

GREENFIELD, A. D. M. (1965). The peripheral circulation. *A. Rev. Physiol.* **27**, 323–50.

GUYTON, A. C., COLEMAN, T. G., and GRANGER, H. J. (1972). Circulation: overall regulation. *A. Rev. Physiol.* **34**, 13–46.

KORNER, P. I. (1971). Integrative neural cardiovascular control. *Physiol. Rev.* **51**, 312–67.

ROSS, G. (1971). The regional circulation. *A. Rev. Physiol.* **33**, 445–78.

Chapter 22. Development of the blood vascular system

ASSALI, N. S., BEKEY, G. A., and MORRISON, L. W. (1968). Fetal and neonatal circulation. In *Biology of gestation*, Vol. II: *The fetus and neonate* (ed. N. S. Assali), pp. 51–142. Academic Press, New York and London.

GOODRICH, E. S. (1930). *Studies on the structure and development of vertebrates.* Macmillan, London. (Reprinted (1958). Dover, New York.)

Chapter 23. Excretion and the control of water balance

BULGER, R. E. (1965). The shape of rat kidney tubular cells. *Am. J. Anat.* **116**, 237–55.

MOREL, F. and ROUFFIGNAC, C. DE (1973). Kidney. *A. Rev. Physiol.* **35**, 17–34.

ORLOFF, J. and BURG, M. (1971). Kidney. *A. Rev. Physiol.* **33**, 83–130.

SMITH, H. W. (1953). *From fish to philosopher.* Little, Brown, Boston.

WINDHAGER, E. E. (1969). Kidney, water, and electrolytes. *A. Rev. Physiol.* **31**, 117–72.

Chapter 24. Development of the urogenital system

BURNS, R. K. (1961). Role of hormones in the differentiation of sex. In *Sex and internal secretions*, (3rd edn, ed. W. C. Young), Vol. 1, pp. 76–158. Baillière, Tindall & Cox, London.

DU BOIS, A. M. (1969). The embryonic kidney. In *The kidney* (ed. Ch. Rouiller and A. F. Muller), Vol. 1, pp. 1–59. Academic Press, New York and London.

FRASER, E. A. (1950). The development of the vertebrate excretory system. *Biol. Rev.* **25**, 159–87.

GOODRICH, E. S. (1935). The study of nephridia and genital ducts since 1895. *Q. Jl. Microsc. Sci.* **86**, 113–392.

GRUNEWALD, P. (1952). Development of the excretory system. *Ann. N.Y. Acad. Sci.* **55**, 142–6.

MINTZ, B. (1960). Embryological phases of mammalian gametogenesis. *J. cell. comp. Physiol.* **56**, Suppl. 1, 31–47.

PHARRISS, B. B. and SHAW, J. E. (1974). Prostaglandins in reproduction. *A. Rev. Physiol.* **36**, 391–412.

VERNIER, R. L. and SMITH, F. G., Jr, (1968). Fetal and neonatal kidney. In *Biology of gestation* (ed. N. S. Assali), Vol. 2: *The fetus and neonate*, pp. 225–60. Academic Press, New York and London.

WADDINGTON, C. H. (1938). The morphogenetic function of a vestigial organ in the chick. *J. exp. Biol.* **15**, 371–6.

WILLIER, B. H. (1955). Ontogeny of endocrine correlation. In *Analysis of development* (ed. B. H. Willier, P. A. Weiss, and V. Hamburger), pp. 574–619. W. B. Saunders, Philadelphia.

— WEISS, P. A., and HAMBURGER, V. (1955). *Analysis of development.* W. B. Saunders, London.

Chapter 25. The nervous system

ARIËNS KAPPERS, C. U., HUBER, G. C., and CROSBY, E. C. (1936). *The comparative anatomy of the nervous system of vertebrates, including man.* Macmillan, London.

BARONDES, S. H. (ed.) (1967). Axoplasmic transport. *Neurosci. Res. Progr. Bull.* **5**, 307–419.

BOURNE, G. H. (1968–72). *The structure and function of nervous tissue.* Vols 1–6. Academic Press, New York and London.

CURTIS, B. A., JACOBSON, S., and MARCUS, E. M. (1972). *An introduction to the neurosciences.* W. B. Saunders, Philadelphia.

DAVISON, A. N. and PETERS, A. (1970). *Myelination.* Thomas, Springfield, Illinois.

GRAFSTEIN, B. (1967). Transport of

protein by goldfish optic nerve fibers. *Science* **157**, 196–8.

KASTEN, F. H. (1966). Charles Marc Pomerat: experimental biologist and humanist. *Med. biol. Illust.* **16**, 78–88.

OCHS, S. (1972). Rate of fast axoplasmic transport in mammalian nerve fibres. *J. Physiol., Lond.* **227**, 627–45.

RAMÓN Y CAJAL, S. (1909–11). *Histologie du système nerveux de l'homme et des vertébrés* (Trans. L. Azoulay), Vols 1 and 2. A. Maloine, Paris.

SCHMITT, F. O. (ed.) (1970). *The neurosciences: second study program*. Rockefeller University Press, New York.

— and WORDEN, F. G. (ed.) (1974). *The neurosciences: third study program*. M.I.T. Press, Cambridge, Massachusetts and London.

SINGER, M. and SCHADÉ, J. P. (eds) (1964). Mechanisms of neural regeneration. *Progress in Brain Research* **13**.

—and—(eds) (1965). Degeneration patterns in the nervous system. *Progress in Brain Research* **14**.

WAXMAN, S. G. (1972). Regional differentiation of the axon: a review with special reference to the concept of the multiplex neuron. *Brain Res.* **47**, 269–88.

YOUNG, J. Z. (1945). The history of the shape of a nerve fibre. In *Essays on growth and form presented to D'Arcy W. Thompson* (ed. W. E. Le Gros Clark and P. B. Medawar), pp. 41–94. Clarendon Press, Oxford.

Chapter 26. Signalling in the nervous system

CREED, R. S., DENNY-BROWN, D., ECCLES, J. C., LIDDELL, E. G. T., and SHERRINGTON, C. S. (1932). *Reflex activity of the spinal cord.* Clarendon Press, Oxford. (Reprinted (1972) with annotations by D. P. C. Lloyd.)

ECCLES, J. C. (1964). *The physiology of synapses*. Springer-Verlag, Berlin.

ERICKSON, R. P. (1974). Parallel "population" neural coding in feature extraction. In *The neurosciences: third study program* (ed. F. O. Schmitt and F. G. Worden), pp. 155–69. M.I.T. Press, Cambridge, Massachusetts and London.

FATT, P. and KATZ, B. (1952). Spontaneous subthreshold activity at motor nerve endings. *J. Physiol., Lond.* **117**, 109–28.

IVERSON, L. L. (1974). Biochemical aspects of synaptic modulation. In *The neurosciences: third study program* (ed. F. O. Schmitt and F. G. Worden), pp. 905–15. M.I.T. Press, Cambridge, Massachusetts and London.

LANGLEY, O. K. and LANDON, D. N. (1967). A light and electron histochemical approach to the node of Ranvier and myelin of peripheral nerve fibers. *J. Histochem. Cytochem.* **15**, 722–31.

KATZ, B. (1966). *Nerve, muscle, and synapse*. McGraw-Hill, New York.

ROBERTSON, J. D. (1964). Unit membranes: a review with recent new studies of experimental alterations and a new subunit structure in unit membranes. In *Cellular membranes in development* (ed. M. Locke), pp. 1–81. Society for the study of development and growth, 22nd symposium. Academic Press, New York.

WALL, P. D. (1965). Impulses originating in the region of dendrites. *J. Physiol., Lond.* **180**, 116–33.

WEIGHT, F. F. (1974). Physiological mechanisms of synaptic modulation. In *The neurosciences: third study program* (ed. F. O. Schmitt and F. G. Warden), pp. 929–41. M.I.T. Press, Cambridge, Massachusetts and London.

Chapter 27. The spinal cord

ECCLES, J. C. and SCHADÉ, J. P. (ed.) (1964a). Organisation of the spinal cord. *Progress in Brain Research* **11**.

— and — (ed.) (1964b). Physiology of spinal neurons. *Progress in Brain Research* **12**.

GELFAN, S. (1964). Neuronal interdependence. *Progress in Brain Research* **11**, 238–58.

REXED, B. (1964). Some aspects of the cytoarchitectonics and synaptology of the spinal cord. *Progress in Brain Research* **11**, 58–90.

ROMANES, G. J. (1964). The motor pools of the spinal cord. *Progress in Brain Research* **11**, 93–116.

WALL, P. D. (1967). The laminar organization of dorsal horn and effects of descending impulses. *J. Physiol., Lond.* **188**, 403–23.

Chapter 28. Servo-control of movement

BARKER, D. (1974a). The motor innervation of muscle spindles. In *Essays on the nervous system: a Festschrift for Professor J. Z. Young* (ed. R.

Bellairs and E. G. Gray), pp. 131–54. Clarendon Press, Oxford.

—(1974b). The morphology of muscle receptors. In *Handbook of sensory physiology*, Vol. 3, part 2, *Muscle receptors* (ed. C. C. Hunt), pp. 1–190. Springer-Verlag, Berlin.

— STACEY, M. J. and ADAL, M. N. (1970). Fusimotor innervation in the cat. *Phil. Trans. R. Soc. B***258**, 315–46.

EVARTS, E. V. (1967). Representation of movements and muscles by pyramidal tract neurons of the precentral motor cortex. In *Neurophysiological basis of normal and abnormal motor activities* (ed. M. D. Yahr and D. P. Purpura), pp. 215–51. Raven Press, Hewlett, New York.

MATTHEWS, P. B. C. (1972). *Mammalian muscle receptors and their central actions*. Arnold, London.

MERTON, P. A. (1951). The silent period in a muscle of the human hand. *J. Physiol., Lond.* **114**, 183–98.

PHILLIPS, C. G. (1970). An outline of recent work on the spinal cord of the cat. *Paraplegia* **8**, 86–100.

Chapter 29. The organization of the brain

GLEES, P. and MELLER, K. (1968). Morphology of neuroglia. In *The structure and function of nervous tissue* (ed. G. H. Bourne), Vol. 1: Structure 1, pp. 301–23. Academic Press, New York and London.

NAUTA, W. J. H. and KARTEN, H. J. (1970). A general profile of the vertebrate brain with sidelights on the ancestry of cerebral cortex. In *The neurosciences: second study program* (ed. F. O. Schmitt), pp. 7–26. Rockefeller University Press, New York.

POMPEIANO, O. (1973). Reticular formation. In *Handbook of sensory physiology*, Vol. 2, *Somatosensory system* (ed. A. Iggo), pp. 381–488. Springer-Verlag, Berlin.

RAMÓN-MOLINER, E. and NAUTA, W. J. H. (1966). The isodendritic core of the brain stem. *J. comp. Neurol.* **126**, 311–35.

Chapter 30. The autonomic nervous system

BURN, J. H. and RAND, M. J. (1965). Acetylcholine in adrenergic transmission. *A. Rev. Pharmac.* **5**, 163–82.

BURNSTOCK, G. (1972). Purinergic

nerves. *Pharmacol. Rev.* **24**, 509–81.

CHIEN, S. (1967). Role of the sympathetic nervous system in hemorrhage. *Physiol. Rev.* **47**, 214–88.

ERÄNKÖ, O. (1955). Histochemistry of noradrenaline in the adrenal medulla of rats and mice. *Endocrinology* **57**, 363–68.

FALCK, B. and OWMAN, Ch. (1965). A detailed methodological description of the fluorescence method for cellular localization of biogenic monoamines. *Acta. Univ. lund.*, sectio II, No. 7, 1–23.

KUNTZ, A. (1953). *The autonomic nervous system* (4th edn). Lea & Febiger, Philadelphia.

LANGLEY, J. N. (1921). *The autonomic nervous system*, Part 1. W. Heffer, Cambridge.

MATTHEWS, M. R. and RAISMAN, G. (1969). The ultrastructure and somatic efferent synapses of small granule-containing cells in the superior cervical ganglion. *J. Anat.* **105**, 255–82.

NEIL, E. (ed.) (1972). *Handbook of sensory physiology*, Vol. 3, part 1: *Enteroceptors.* Springer-Verlag, Berlin, Heidelberg and New York.

PATOŃ, W. D. and ZAR, M. A. (1968). The origin of acetylcholine released from guinea-pig intestine and longitudinal muscle strips. *J. Physiol., Lond.* **194**, 13–33.

Chapter 31. The control of posture and movement. The cerebellum and midbrain

BELL, C. C. and DOW, R. S. (1967). Cerebellar circuitry. *Neurosci. Res. Progr. Bull.* **5**, 121–222.

BRAITENBERG, V. (1967). Is the cerebellar cortex a biological clock in the millisecond range? *Progress in Brain Research* **25**, 334–46.

CREED, R. S., DENNY-BROWN, D., ECCLES, J. C., LIDDELL, E. G. T., and SHERRINGTON, C. S. (1932). *Reflex activity of spinal cord.* Clarendon Press, Oxford.

ECCLES, J. C., ITO, M., and SZENTÁGOTHAI, J. (1967). *The cerebellum as a neuronal machine.* Springer-Verlag, Berlin.

FOX, C. A. and SNIDER, R. S. (ed.) (1967). The cerebellum. *Progress in Brain Research* **25**.

ITO, M. (1974). The control mechanisms of cerebellar motor systems. In *The neurosciences: third study program* (ed. F. O. Schmitt and F. G. Worden), pp. 293–303. M.I.T. Press, Cambridge, Massachusetts and London.

JANSEN, J. and BRODAL, A. (1954). *Aspects of cerebellar anatomy.* J. G. Tanum, Oslo.

LLINÁS, R. (ed.) (1969). *Neurobiology of cerebellar evolution and development.* Proceedings of First International Symposium of the Institute for Biochemical Research. American Medical Association, Chicago.

MARR, D. (1969). A theory of cerebellar cortex. *J. Physiol., Lond.* **202**, 437–70.

NIEUWENHUYS, R. (1967). Comparative anatomy of the cerebellum. *Progress in Brain Research* **25**, 1–93.

PALAY, S. L. and CHAN-PALAY, V. (1973). *Cerebellar cortex: cytology and organization.* Springer-Verlag, Berlin.

SZENTÁGOTHAI, J. (1965). The use of degeneration methods in the investigation of short neuronal connexions. *Progress in Brain Research* **14**, 1–30.

Chapter 32. The hypothalamus

ADEY, W. R. and TOKIZANE, T. (ed.) (1967). Structure and function of the limbic system. *Progress in Brain Research* **27**.

AKERT, K., BALLY, C., and SCHADÉ, J. P. (ed.) (1965). Sleep mechanisms, *Progress in Brain Research* **18**.

EULER, U. S. VON (1954). Visceral functions of the nervous system. *A. Rev. Physiol.* **16**, 349–70.

HAYMAKER, W., ANDERSON, E., and NAUTA, W. J. H. (ed.) (1969). *The hypothalamus.* Thomas, Springfield, Illinois.

HESS, W. R. (1957). *The functional organization of the diencephalon.* (ed. J. R. Hughes). (Translation of *Die Zwischenhirn, Syndrome, Lokalisation, Funktion* (1949, 1954)). Grune & Stratton, New York and London.

OLDS, J. (1962). Hypothalamic substrates of reward. *Physiol. Rev.* **42**, 554–604.

OSWALD, I. (1962). *Sleeping and waking.* Elsevier, Amsterdam.

TROWILL, J. A., PANKSEPP, J., and GANDELMAN, R. (1969). An incentive model of rewarding brain stimulation. *Psychol. Rev.* **76**, 264–81.

Chapter 33. Basal ganglia and thalamus

DIAMOND, I. T. (1967). The sensory neocortex. *Contrib. sensory Physiol.* **2**, 51–100.

KEMP, J. M. and POWELL, T. P. S. (1971). The connexions of the striatum and globus pallidus: synthesis and speculation. *Phil. Trans. R. Soc.* **B262**, 441–57.

KORNHUBER, H. H. (1974) Cerebral cortex, cerebellum, and basal ganglia: an introduction to their motor functions. In *The neurosciences: third study program* (ed. F. O. Schmitt and F. G. Worden), pp. 267–80. M.I.T. Press, Cambridge, Massachusetts and London.

PURPURA, D. P. and YAHR, M. D. (1966). *The thalamus.* Columbia University Press, New York and London.

WEBSTER, K. E. (1975). Structure and function of the basal ganglia: a non-clinical view. *Proc. R. Soc. Med.* **68**, 203–10.

Chapter 34. The cerebral cortex

BURNS, B. D. (1958). *The mammalian cerebral cortex.* Arnold, London.

MACLEOD, P. (1971). Structure and function of higher olfactory centers. In *Handbook of sensory physiology,* Vol. 4, part 1: *Chemical senses: olfaction* (ed. L. M. Beidler), pp. 182–204. Springer-Verlag, Berlin.

MILNER, B. (1974). Hemispheric specialization: scope and limits. In *The neurosciences: third study program* (ed. F. O. Schmitt and F. G. Worden), pp. 75–89. M.I.T. Press, Cambridge, Massachusetts and London.

NAUTA, W. J. H. (1972). Neural associations of the frontal cortex. *Acta neurobiol. exp.* **32**, 125–40.

O'KEEFE, J. and NADEL, L. (In preparation). *The hippocampus as a cognitive map.*

PHILLIPS, C. G. (1966). Changing concepts of the precentral motor area. In *Brain and conscious experience* (ed. J. C. Eccles), pp. 389–421. Springer-Verlag, Berlin, Heidelberg and New York.

—(1973). Cortical localization and 'sensorimotor processes' at the 'middle level' in Primates. *Proc. R. Soc. Med.* **66**, 987–1002.

SHOLL, D. A. (1956). *The organization of the cerebral cortex.* Methuen, London.

TEUBER, H.-L. (1972). Unity and diversity of frontal lobe functions. *Acta neurobiol. exp.* **32**, 615–56.

WARREN, J. M. and AKERT, K. (ed.) (1964). *The frontal granular cortex and behavior, proceedings of a symposium*. McGraw-Hill, New York.

YAHR, M. D. and PURPURA, D. P. (ed.) (1967). *Neurophysiological basis of normal and abnormal motor activities*. Raven Press, Hewlett, New York.

Chapter 35. Afferent projection areas of the cortex

BLAKEMORE, C. (1974). Developmental factors in the formation of feature extracting neurons. In *The neurosciences: third study program* (ed. F. O. Schmitt and F. G. Worden), pp. 105–13. M.I.T. Press, Cambridge, Massachusetts and London.

BURTON, H. and BENJAMIN, R. M. (1971). Central projections of gustatory system. In *Handbook of sensory physiology*, Vol. 4, part 2: *Chemical senses: taste* (ed. L. M. Beidler), pp. 148–64. Springer-Verlag, Berlin.

CAMPBELL, F. W. (1974). The transmission of spatial information through the visual system. In *The neurosciences: third study program* (ed. F. O. Schmitt and F. G. Worden), pp. 95–103. M.I.T. Press, Cambridge, Massachusetts. and London.

DARIAN-SMITH, I., ISBISTER, J., MOK, H., and YOKOTA, T. (1966). Somatic sensory cortical projection areas excited by tactile stimulation of the cat: a triple representation. *J. Physiol., Lond.* **182**, 671–89.

EVANS, E. F. (1974). Neural processes for the detection of acoustic patterns and for sound localization. In *The neurosciences: third study program* (ed. F. O. Schmitt and F. G. Worden), pp. 131 45. M.I.T. Press, Cambridge, Massachusetts and London.

GROSS, C. G., BENDER, D. B., and ROCHA-MIRANDA, C. E. (1974). Inferotemporal cortex: a single-unit analysis. In *The neurosciences: third study program* (ed. F. O. Schmitt and F. G. Worden), pp. 229–38. M.I.T. Press, Cambridge, Massachusetts and London.

HUBEL, D. H. and WIESEL, T. N. (1968). Receptive fields and functional architecture of monkey striate cortex. *J. Physiol., Lond.* **195**, 215–43.

JUNG, R. (ed.) (1973a). *Handbook of sensory physiology*, Vol. 7: *Central processing of visual information*, Part 3A: *Integrative functions and comparative data*. Springer-Verlag, Berlin.

—(1973b). *Handbook of sensory physiology*, Vol. 7: *Central processing of visual information*, Part 3B: *Visual centers in the brain*. Springer-Verlag, Berlin.

LENNEBERG, E. H. (1967). *Biological foundations of language*. Wiley, New York.

LIBERMAN, A. M. (1974). The specialization of the language hemisphere. In *The neurosciences: third study program* (ed. F. O. Schmitt and F. G. Worden), pp. 43–56. M.I.T. Press, Cambridge, Massachusetts and London.

MONTERO, V. M., ROJAS, A., and TORREALBA, F. (1973). Retinotopic organization of striate and peristriate visual cortex in the albino rat. *Brain Res.* **53**, 197–201.

NADEL, L. and O'KEEFE, J. (1974). The hippocampus in pieces and patches: an essay on modes of explanation in physiological psychology. In *Essays on the nervous system: a Festschrift for Professor J. Z. Young* (ed. R. Bellairs and E. G. Gray), pp. 367–90. Clarendon Press, Oxford.

PENFIELD, W. and RASMUSSEN, T. (1950). *The cerebral cortex of man*. Macmillan, New York.

—and ROBERTS, L. (1959). *Speech and brain mechanisms*. Princeton University Press.

TEUBER, H.-L. (1972). Unity and diversity of frontal lobe functions. *Acta Neurobiol. exp.* **32**, 615–56.

WELKER, C. (1971). Microelectrode delineation of fine grain somatotopic organization of Sm I cerebral neocortex in albino rat. *Brain Res.* **26**, 259–75.

—and SINHA, M. M. (1972). Somatotopic organization of SmII cerebral neocortex in albino rat. *Brain res.* **37**, 132–6.

WHITFIELD, I. C. (1967). *The auditory pathway*. Arnold, London.

WOOLSEY, C. N. (1965). Organization of somatic sensory and motor areas of the cerebral cortex. In *Biological and biochemical bases of behaviour* (ed. H. F. Harlow and C. N. Woolsey), pp. 68–81. University of Winsconsin Press, Madison.

ZEKI, S. (1974). The mosaic organization of the visual cortex in the monkey. In *Essays on the nervous system: a Festschrift for Professor J. Z. Young* (ed. R. Bellairs and E. G. Gray), pp. 327–43. Clarendon Press, Oxford.

Chapter 36. Structure and functioning of the cerebral cortex

BLAKEMORE, C. and COOPER, G. F. (1970). Development of the brain depends on the visual environment. *Nature* **228**, 477–8.

BLISS, T. V. P. and GARDNER-MEDWIN, A. R. (1973). Long-lasting potentiation of synaptic transmission in the dentate area of the unanaesthetized rabbit following stimulation of the perforant path. *J. Physiol., Lond.* **232**, 357–74.

CRAGG, B. G. (1967). The density of synapses and neurones in the motor and visual areas of the cerebral cortex. *J. Anat.* **101**, 639–54.

CROW, T. J. (1973). Catecholamine-containing neurones and electrical self-stimulation: 2. A theoretical interpretation and some psychiatric implications. *Psychol. Med.* **3**, 66–73.

GAZE, R. M. (1970). *The formation of nerve connections*. Academic Press, London and New York.

GRAY, E. G. (1971). The fine structural characterization of different types of synapse. *Progress in Brain Research* **34**, 149–60.

JOHN, E. R. (1967). *Mechanisms of memory*. Academic Press, New York and London.

—and MORGADES, P. P. (1969). Neuronal correlates of conditioned responses studied with multiple chronically implanted moving electrodes. *Exp. Neurol.* **23**, 412–25.

JONES, E. G. and POWELL, T. P. S. (1970). An electron microscope study of the laminar pattern and mode of termination of afferent fibre pathways in the somatic sensory cortex of the cat. *Phil. Trans. R. Soc.* B**257**, 45–62.

LASHLEY, K. S. (1950). In search of the engram. *Symp. Soc. exp. Biol.* **4**, 454–82.

MACKAY, D. M. and GARDINER, M. F. (1972). Two strategies of information processing. See *Neurosci. Res. Prog. Bull.* **10**, 77–8.

MORRELL, F. (1961). Electrophysiological contributions to the neural basis of learning. *Physiol. Rev.* **41**, 443–94.

NADEL, L. and O'KEEFE, J. (1974). The hippocampus in pieces and patches: an essay on modes of explanation in physiological psychology. In *Essays on the nervous system: a Festschrift for Professor J. Z. Young* (ed. R. Bellairs and E. G.

Gray), pp. 367–90. Clarendon Press, Oxford.

O'KEEFE, J. and NADEL, L. (In preparation). *The hippocampus as a cognitive map.*

PETERS, A. and WALSH, T. M. (1972). A study of the organization of apical dendrites in the somatic sensory cortex of the cat. *J. comp. Neurol.* **144**, 253–68.

PRIBRAM, K. H. (1971). *Languages of the brain: experimental paradoxes and principles in neuropsychology.* Prentice-Hall, Englewood Cliffs, New Jersey.

SHOLL, D. A. (1956). *The organization of the cerebral cortex.* Methuen, London.

SUTHERLAND, N. S. (1974). Object recognition. In *Handbook of perception* (ed. E. C. Carterette and M. P. Friedman), Vol. 3: *Biology of perceptual systems*, pp. 157–85. Academic Press, New York and London.

UNGERSTEDT, U. (1971). Stereotaxic mapping of the monoamine pathways in the rat brain. *Acta physiol. scand.* **82**, Suppl. **367**, 1–48.

WARRINGTON, E. K. (1971). Neurological disorders of memory. *Br. med. Bull.* **27**, 243–7.

WELLS, M. J. and YOUNG, J. Z. (1965). Split-brain preparations and touch learning in the octopus. *J. exp. Biol.* **43**, 565–79.

YOUNG, J. Z. (1964). *A model of the brain.* Clarendon Press, Oxford.

— (1965). The organization of a memory system. *Proc. R. Soc.* B**163**, 285–320.

— (1966). *The memory system of the brain.* Oxford University Press, London.

Chapter 37. Development of the nervous system

BANKS, B. E. C., PEARCE, F. L., and VERNON, C. A. (In preparation). Nerve growth factors and their antisera. *Int. Rev. Biochem.*

BELLAIRS, R. (1974). Early differentiation of the nervous system. In *Essays on the nervous system: a Festschrift for Professor J. Z. Young* (ed. R. Bellairs and E. G. Gray), pp. 1–30. Clarendon Press, Oxford.

BLAKEMORE, C. and COOPER, G. F. (1970). Development of the brain depends on the visual environment. *Nature* **228**, 477–8.

GAZE, R. M. (1970). *The formation of nerve connections.* Academic Press. London and New York.

GILBERT, D. S. (1972). Helical structure of *Myxicola* axoplasm. *Nature New Biology* **237**, 195–8.

HAMBURGER, V. (1952). Development of the nervous system. *Ann. N.Y. Acad. Sci.* **55**, 117–32.

HÖRSTADIUS, S. (1950). *The neural crest.* Oxford University Press, London.

JACOBSON, M. (1970). *Developmental neurobiology.* Holt, Rinehart and Winston, New York.

KÄLLÉN, B. (1965). Early morphogenesis and pattern formation in the central nervous system. In *Organogenesis* (ed. R. L. DeHaan and H. Ursprung), pp. 107–28, Holt, Rinehart and Winston, New York.

LOPRESTI, V., MACAGNO, E. R., and LEVINTHAL, C. (1973). Structure and development of neuronal connections in isogenic organisms: cellular interactions in the development of the optic lamina of *Daphnia. Proc. natn. Acad. Sci.* **70**, 433–7.

MARSH, G. and BEAMS, H. W. (1946). *In vitro* control of growing chick nerve fibres by applied electric currents. *J. cell. comp. Physiol.* **27**, 139–57.

TERNI, T. (1920). Sulla correlazione fra ampiezza del territorio di innervazione e grandezza delle cellule gangliari. 2. Ricerche sui ganlï spinali che innervano la coda rigenerato, nei Sauri (*Gongylus ocellatus*). *Arch. ital. Anat.* **17**, 507–43.

WILLIER, B. H., WEISS, P. A., and HAMBURGER, V. (ed.) (1955). *Analysis of development.* W. B. Saunders, Philadelphia.

YNTEMA, C. L. (1943). An experimental study on the origin of the sensory neurones and sheath cells of the IXth and Xth cranial nerves in *Amblystoma punctatum. J. exp. Zool.* **92**, 93–119.

ZAIMIS, E. J. and KNIGHT, J. (ed.) (1972). *Nerve growth factor and its antiserums.* Athlone Press, London.

Chapter 38. Receptor organs

ADRIAN, E. D. (1947). *The physical background of perception.* Clarendon Press, Oxford.

CARTERETTE, E. C. and FRIEDMAN, M. P. (ed.) (1974). *Handbook of Perception*, Vol. 1: *Biology of perceptual systems.* Academic Press, New York and London.

CHERRY, C. (1966). *On human com-munication: a review, a survey, and a criticism* (2nd edn). M.I.T. Press, Cambridge, Massachusetts.

GRANIT, R. (1955). *Receptors and sensory perception.* Yale University Press, New Haven.

KATZ, B. (1966). *Nerve, muscle, and synapse.* McGraw-Hill, New York.

LOWENSTEIN, W. R. (ed.) (1971). *Handbook of sensory physiology*, Vol. 1: *Principles of receptor physiology.* Springer-Verlag, Berlin.

MOUNTCASTLE, V. B. (ed.) (1968). *Medical Physiology.* Vols 1 and 2. (12th edn), C. V. Mosby, St. Louis.

SHANNON, C. E. and WEAVER, W. (1949). *The mathematical theory of communication.* University of Illinois Press, Urbana.

STEVENS, S. S. (1966). Transfer functions of the skin and muscle senses. In *Touch, heat and pain* (ed. A. V. S. de Reuck and J. Knight), pp. 3–17. J. & A. Churchill, London.

WATERMAN, T. H. (1968). Systems theory and biology—view of a biologist. In *Systems theory and biology* (ed. M. D. Mesarović), pp. 1–37. Proceedings of the Third Systems Symposium at Case Institute of Technology. Springer-Verlag New York.

WIENER, N. (1948). *Cybernetics.* John Wiley, New York.

YOUNG, J. Z. (1971). *An introduction to the study of man.* Clarendon Press, Oxford.

Chapter 39. Receptors in the skin and viscera

CATTON, W. T. (1970). Mechanoreceptor function. *Physiol. Rev.* **50**, 297–318.

DE REUCK, A. V. S. and KNIGHT, J. (ed.) (1966). *Touch, heat and pain.* Ciba Foundation Symposium. J. & A. Churchill, London.

IGGO, A. (1966). Cutaneous receptors with a high sensitivity to mechanical displacement. In *Touch, Heat and Pain* (ed. A. V. S. de Reuck and J. Knight), pp. 237–56. J. & A. Churchill, London.

— (ed.) (1973). *Handbook of sensory physiology*, Vol. 2: *Somato-sensory system.* Springer-Verlag, Berlin.

MELZACK, R. and WALL, P. D. (1962). On the nature of cutaneous sensory mechanisms. *Brain*, **85**, 331–56.

SINCLAIR, D. (1967). *Cutaneous sensation.* Oxford University Press, London.

WALL, P. D. (1974). 'My foot hurts

me': an analysis of a sentence. In *Essays on the nervous system: a Festschrift for Professor J. Z. Young* (ed. R. Bellairs and E. G. Gray), pp. 391–406. Clarendon Press, Oxford.

Chapters 40–2. The eye and its movements. The retina. Vision by day and night

ALPERN, M. (1969). Types of eye movement. In *The eye*, Vol. 3: *Muscular mechanisms* (2nd edn), (ed. H. Davson), pp. 65–174. Academic Press, New York.

BARLOW, H. B. (1964). Dark adaptation: a new hypothesis. *Vision Res.* **4**, 47–57.

BOYCOTT, B. B. (1974). Aspects of the comparative anatomy and physiology of the vertebrate retina. In *Essays on the nervous system: a Festschrift for Professor J. Z. Young* (ed. R. Bellairs and E. G. Gray), pp. 223–57. Clarendon Press, Oxford.

— and DOWLING, J. E. (1969). Organization of the primate retina: light microscopy. *Phil. Trans. R. Soc.* **B255**, 109–76.

BROWN, P. K. and WALD, G. (1964). Visual pigments in single rods and cones of the human retina. *Science* **144**, 45–52.

DAVSON, H. (ed.) (1969). *The eye* (2nd edn), Vol. 1: *Vegetative physiology and biochemistry;* Vol. 3: *Muscular mechanisms.* Academic Press, New York and London.

— (1972). *The physiology of the eye* (3rd edn). Churchill Livingstone, Edinburgh and London.

DOWLING, J. E. (1970). Organization of vertebrate retinas. *Invest. Ophthalmol.* **9**, 655–80.

— and BOYCOTT, B. B. (1966). Organization of the primate retina: electron microscopy. *Proc. R. Soc.* **B166**, 80–111.

DUKE-ELDER, W. S. (1958). *System of ophthalmology*, Vol. 1: *The eye in evolution.* Henry Kimpton, London.

— and WYBAR, K. C. (1961). *System of ophthalmology*, Vol. 2: *The anatomy of the visual system* (ed. W. S. Duke-Elder). Henry Kimpton, London.

GRANIT, R. (1947). *Sensory mechanisms of the retina.* Oxford University Press, London. (Reprinted with corrections (1963). Hafner, New York and London.)

— (1955). *Receptors and sensory perception.* Yale University Press, New Haven.

HECHT, S., SHLAER, S., and PIRENNE, M. (1942). Energy, quanta and vision. *J. gen. Physiol.* **25**, 819–40.

HOFSTETTER, W. H. (1965). A longitudinal study of amplitude changes in presbyopia. *Am. J. Optom.* **42**, 3–8.

MARKS, W. B., DOBELLE, W. H., and MacNICHOL, E. F. (1964). Visual pigments of single primate cones. *Science* **143**, 1181–3.

MICHAEL, C. R. (1968). Receptive fields of single optic nerve fibers in a mammal with an all-cone retina. III. Opponent colour unit. *J. Neurophysiol.* **31**, 268–82.

PIRENNE, M. H. (1967). *Vision and the eye* (2nd edn). Science paperbacks, London.

— and DENTON, E. J. (1952). Accuracy and sensitivity of the human eye. *Nature* **170**, 1039–42.

RUSHTON, W. A. H. (1965). A foveal pigment in the deuteranope. *J. Physiol., Lond.* **176**, 24–37.

— (1972). Pigments and signals in colour vision. *J. Physiol., Lond.* **220**, 1–31P.

WALD, G. (1951). The chemistry of rod vision. *Science* **113**, 287–91.

— (1959). The photoreceptor process in vision. In *Handbook of physiology* Section 1. *Neurophysiology* (ed. H. W. Magoun), Vol. 1, pp. 671–92. American Physiological Society, Washington, D.C.

WALLS, G. L. (1942). *The vertebrate eye and its adaptive radiation.* Cranbrooke Institute of Science Bulletin 19.

YOUNG, J. Z. (1971). *An introduction to the study of man.* Clarendon Press, Oxford.

YOUNG, T. (1801). On the mechanism of the eye. *Phil. Trans. R. Soc.*, 23–88.

— (1802). On the theory of light and colours. *Phil. Trans. R. Soc.*, 12–48.

Chapter 43. Hearing and receptors for orientation in space

DAVIS, H. (1965). A model for transducer action in the cochlea. *Cold Spring Harb. Symp. quant. Biol.* **30**, 181–90.

— (1970). Anatomy and physiology of the auditory system. In *Hearing and deafness* (3rd edn, ed. H. Davis and S. R. Silverman), pp. 47–82. Holt, Rinehart and Winston, New York.

DE REUCK, A. V. S. and KNIGHT, J. (ed.) (1967). *Myotatic, kinesthetic and vestibular mechanisms.* Ciba Foundation Symposium. J. & A. Churchill, London.

ELDRIDGE, D. H. and MILLAR, J. D. (1971). Physiology of hearing. *A. Rev. Physiol.* **33**, 281–310.

ENGSTRÖM, H., ADES, H. W., and ANDERSSON, A. (1966). *Structural pattern of the organ of Corti. A systematic mapping of sensory cells and neural elements.* Almqvist & Wiksell, Stockholm.

IURATO, S. (1967). *Submicroscopic structure of the inner ear.* Pergamon, Oxford.

LIM, D. J. (1969). Three dimensional observation of the inner ear with the scanning electron microscope. *Acta Oto-lar. Suppl.* **255**, 1–38.

LINDEMAN, H. H. (1969). Studies on the morphology of the sensory regions of the vestibular apparatus. *Ergebn. Anat. EntwGesch.* **42**, Heft 1, 1–113.

NAFTALIN, L. (1965). Some new proposals regarding acoustic transmission and transduction. *Cold Spring Harb. Symp. quant. Biol.* **30**, 169–80.

SALES, G. D. and PYE, J. D. (1974). *Ultrasonic communication by animals.* Chapman & Hall, London.

SPOENDLIN, H. (1966). *The organization of the cochlear receptor.* Advances in oto-rhino-laryngology, Vol. 13. S. Karger, Basel.

WERSÄLL, J. and FLOCK, Å. (1965). Functional anatomy of the vestibular and lateral line organs. *Contrib. sensory Physiol.* **1**, 39–61.

WERSÄLL, J., GLEISNER, L., and LUNDQUIST, P.-G. (1967). Ultrastructure of the vestibular end organs. In *Myotatic, kinesthetic and vestibular mechanisms* (ed. A. V. S. de Reuck and J. Knight), pp. 105–16. J. & A. Churchill, London.

Chapter 44. Chemoreceptors

ADRIAN, E. D. (1942). Olfactory reactions in the brain of the hedgehog. *J. Physiol., Lond.* **100**, 459–73.

— (1951). Differential sensitivity of olfactory receptors. *J. Physiol., Lond.* **115**, 42P.

ALLISON, A. C. (1953a). The structure of the olfactory bulb and its relationship to the olfactory pathways in the rabbit and the rat. *J. comp. Neurol.* **98**, 309–53.

— (1953b). The morphology of the olfactory system in the vertebrates. *Biol. Rev.* **28**, 195–244.

AMOORE, J. E. (1965). Psychophysics

of odor. *Cold Spring Harb. Symp. quant. Biol.* **30**, 623–36.

— (1971). Progress towards some direct quantitative comparisons of the stereochemical and vibrational theories of odor. In *Gustation and olfaction* (ed. G. Ohloff and A. F. Thomas), pp. 147–60. Academic Press, New York and London.

BEIDLER, L. M. (1970). Physiological properties of mammalian taste receptors. In *Taste and smell in vertebrates* (ed. G. E. W. Wolstenholme and J. Knight), pp. 51–67. J. & A. Churchill, London.

—(ed.) (1971*a*). *Handbook of sensory physiology*, Vol. 4, part 1: *Chemical senses: Olfaction.* Springer-Verlag, Berlin.

—(1971*b*). *Handbook of sensory physiology,* Vol. 4, part 2: *Chemical senses: Taste.* Springer-Verlag, Berlin.

HENNING, H. (1926). Psychologie der chemischen Sinne. In *Handlbuch der normalen und pathologischen Physiologie* (ed. A. Bethe *et al.*) Vol. 11 *Receptionsorgane I*, pp. 393–405 Springer-Verlag, Berlin.

LETTVIN, J. Y. and GESTELAND, R. C. (1965). Speculations on smell. *Cold Spring Harb. Symp. quant. Biol.* **30**, 217–25.

MOULTON, D. G. (1971). Detection and recognition of odor molecules. In *Gustation and olfaction* (ed. G. Ohloff and A. F. Thomas), pp. 1–25. Academic Press, New York and London.

OHLOFF, G. and THOMAS, A. F. (ed.) (1971). *Gustation and olfaction.* Academic Press, New York and London.

OTTOSON, D. (1970). Electrical signs of olfactory transducer action. In *Taste and smell in vertebrates* (ed. G. E. W. Wolstenholme and J. Knight), pp. 343–54. J. & A. Churchill, London.

PFAFFMANN, C. (1951). Taste and smell. In *Handbook of experimental psychology* (ed. S. S. Stevens), pp. 1143–71. Chapman & Hall, London.

—(ed.) (1969). *Olfaction and taste III.* Rockefeller University Press, New York.

— (1970). Physiological and behavioural processes of the sense of taste. In *Taste and smell in vertebrates* (ed. G. E. W. Wolstenholme and J. Knight), pp. 31–45. J. & A. Churchill, London.

RALL, W., SHEPHERD, G. M., REESE, T. S., and BRIGHTMAN, M. W. (1966). Dendrodendritic synaptic pathway for inhibition in the olfactory bulb. *Expl. Neurol.* **14**, 44–56.

SHALLENBERGER, R. S. (1971). Molecular structure and taste. In *Gustation and olfaction* (ed. G. Ohloff and A. F. Thomas), pp. 126–32. Academic Press, London and New York.

SHIBUYA, T. (1969). Activities of single olfactory receptor cells. In *Olfaction and taste III* (ed. C. Pfaffmann), 109–16. Rockefeller University Press, New York.

WOLSTENHOLME, G. E. W. and KNIGHT, J. (ed.) (1970). *Taste and smell in vertebrates.* Ciba Foundation Symposium. J. & A. Churchill, London.

WRIGHT, R. H. and BURGESS, R. E. (1970). Specific physicochemical mechanisms of olfactory stimulation. In *Taste and smell in vertebrates* (ed. G. E. W. Wolstenholme and J. Knight), pp. 325–37. J. & A. Churchill, London.

— and — (1971). Molecular mechanisms of olfactory discrimination and sensitivity. In *Gustation and olfaction* (ed. G. Ohloff and A. F. Thomas), pp. 134–44. Academic Press, New York and London.

Chapter 45. Endocrine organs. The thyroid and parathyroid

ARNAUD, C. D., Jr, TENENHOUSE, A. M., and RASMUSSEN, H. (1967). Parathyroid hormone. *A. Rev. Physiol.* **29**, 349–72.

BROWN-GRANT, K. (1966). The control of TSH secretion. In *The pituitary gland* (ed. G. W. Harris and B. T. Donovan), Vol. 2, *Anterior pituitary,* pp. 235–69. Butterworth, London.

DOBYNS, B. M. (1964). The thyroid–pituitary relationships. In *The thyroid* (ed. J. B. Hazard and D. E. Smith), pp. 76–99. Williams & Wilkins, Baltimore.

DUMONT, J. E. (1971). The action of thyrotropin on thyroid metabolism. *Vitams Horm.* **29**, 287–412.

HAMBURGH, M. and BARRINGTON, E. J. W. (ed.) (1971). *Hormones in development.* Appleton-Century-Crofts, New York.

MACGREGOR, A. G. (1964). Hypothyroidism. In *The thyroid gland,* (ed. R. Pitt-Rivers and W. R. Trotter), Vol. 2, pp. 112–29. Butterworth, London.

MACINTYRE, I. (1968). Calcitonin: a review of its discovery and an account of purification and action. *Proc. R. Soc. B***170**, 49–60.

MORGANS, M. E. (1964). Hyperthyroidism. In *The thyroid gland,* (ed. R. Pitt-Rivers and W. R. Trotter), Vol. 2, pp. 151–70. Butterworth, London.

PITT-RIVERS, R. and TROTTER, W. R. (ed.) (1964). *The thyroid gland,* Vols 1 and 2. Butterworth, London.

POTTS, J. T., Jr, KEUTMANN, H. T., NIALL, H. D., and TREGEAR, G. W. (1971). The chemistry of parathyroid hormone and the calcitonins. *Vitams Horm.* **29**, 41–93.

WOLLMAN, S. H. (1969). Secretion of the thyroid hormones. In *Lysosomes in biology and pathology* (ed. J. T. Dingle and H. B. Fell), Vol. 2, pp. 483–512. Elsevier, New York.

Chapter 46. The adrenal glands

BRANSOME, E. D. (1968). Adrenal cortex. *A. Rev. Physiol.* **30**, 171–212.

COPE, C. L. (1972). *Adrenal steroids and disease* (2nd edn). Pitman-Medical, London.

EDELMAN, I. S., and FIMOGNARI, G. M. (1968). On the biochemical mechanism of action of aldosterone. *Recent Progr. Hormone Res.* **24**, 1–44.

EISENSTEIN, A. B. (ed.) (1967). *The adrenal cortex.* Little, Brown, Boston.

EULER, U. S. VON (1951). Hormones of the sympathetic nervous system and the adrenal medulla. *Br. med. J.* **1**, 105–8.

FEIGELSON, P. and FEIGELSON, M. (1964). Studies on the mechanism of cortisone action. In *Actions of hormones on molecular processes* (ed. G. Litwack and D. Kritchevsky), pp. 218–33. Wiley, New York.

IDELMAN, S. (1970). Ultrastructure of the mammalian adrenal cortex. *Int. Rev. Cytol.* **27**, 181–281.

LEVINE, R. J. (1969). The adrenal medulla and catecholamines. In *Duncan's Diseases of metabolism* (ed. P. K. Bondy), pp. 886–903. Saunders, Philadelphia.

MALMEJAC, J. (1964). Activity of adrenal medulla and its regulation. *Physiol. Rev.* **44**, 186–218.

MULROW, P. J. (1972). The adrenal cortex. *A. Rev. Physiol.* **34**, 409–24.

RENOLD, A. E. and ASHMORE, J. (1960). Metabolic effects of adrenal corticosteroids. In *Diabetes* (ed. R.

H. Williams), pp. 194–215. Hoeber, New York.

SELYE, H. (1956). *The stress of life.* McGraw-Hill, New York.

Chapter 47. The pancreatic islets and the control of carbohydrate metabolism

CAHILL, G. F., ASHMORE, J., RENOLD, A. F., and HASTINGS, A. B. (1959). Blood glucose and liver. *Am. J. Med.* **26**, 264–82.

GRODSKY, G. M. (1970). Insulin and the pancreas. *Vitams Horm.* **28**, 37–101.

KRAHL, M. E. (1974). Endocrine function of the pancreas. *A. Rev. Physiol.* **36**, 331–60.

LEVINE, R. and GOLDSTEIN, M. S. (1955). On the mechanisms of action of insulin. *Recent Progr. Hormone Res.* **11**, 343–80.

STEINER, D. F. (1966). Insulin and the regulation of hepatic biosynthetic activity. *Vitams Horm.* **24**, 1–61.

WOOL, I. G., STIREWALT, W. S., KURIHARA, K., LOW, R. B., BAILEY, P., and OYER, D. (1968). Mode of action of insulin in the regulation of protein synthesis in muscle. *Recent Progr. Hormone Res.* **24**, 139–213.

Chapter 48. The pituitary gland

BLEIER, R. (1972). Structural relationship of ependymal cells and their processes within the hypothalamus. In *Brain–endocrine interaction. Median eminence: structure and function* (ed. K. M. Knigge, D. E. Scott and A. Weindl), pp. 306–18. S. Karger, Basel.

EVANS, H. M. and LONG, J. A. (1921). The effect of the anterior lobe of the hypophysis administered intraperitoneally upon growth, and the maturity and oestrus cycles of the rat. *Anat. Rec.* **21**, 61–3.

FOLLET, B. K. and HELLER, H. (1964*a*). The neurohypophysial hormones of bony fishes. *J. Physiol., Lond.* **172**, 74–91.

—and—(1964*b*). The neurohypophysial hormones of lungfishes and amphibians. *J. Physiol., Lond.* **172**, 92–106.

GANONG, W. F. (1974). Brain mechanisms regulating the secretion of the pituitary gland. In *The neurosciences: third study program.* (ed. F. O. Schmitt and F. G. Worden), pp. 549–63. M.I.T. Press, Cambridge, Massachusetts and London.

—and MARTINI, L. (ed.) (1969). *Frontiers in neuroendocrinology,1969.* Oxford University Press, London.

GESCHWIND, I. I. (1969). Mechanism of action of releasing factors. In *Frontiers in neuroendocrinology, 1969* (ed. W. F. Ganong and L. Martini), pp. 389–431. Oxford University Press, London.

GREEN, J. D. (1951). The comparative anatomy of the hypophysis, with special reference to its blood supply and innervation. *Am. J. Anat.* **88**, 225–311.

HARRIS, G. W. and DONOVAN, B. T. (ed.) (1966). *The pituitary gland.* Vols 1–3. Butterworths, London.

HAYMAKER, W. (1969). Hypothalamo-pituitary neural pathways and the circulatory system of the pituitary. In *The hypothalamus* (ed. W. Haymaker, E. Anderson, and W. J. H. Nauta), pp. 219–50. Thomas, Springfield, Illinois.

—ANDERSON, E., and NAUTA, W. J. H. (ed.) (1969). *The hypothalamus.* Thomas, Springfield, Illinois.

LI, C. H. (1972). Recent knowledge of the chemistry of lactogenic hormones. In *Lactogenic hormones* (ed. G. E. W. Wolstenholme and J. Knight). Churchill Livingstone, Edinburgh and London.

—DIXON, J. S., and CHUNG, D. (1961). Adrenocorticotropins. XXI The amino acid sequence of bovine adrenocorticotropin. *Biochem. biophys. acta* **46**, 324–44.

MCCANN, S. McD. and PORTER, J. C. (1969). Hypothalamic pituitary stimulating and inhibiting hormones. *Physiol. Rev.* **49**, 240–84.

NICOLL, C. S. and BERN, H. A. (1972). On the actions of prolactin among the vertebrates: is there a common denominator? In *Lactogenic hormones* (ed. G. E. W. Wolstenholme and J. Knight), pp. 299–317. Churchill Livingstone, Edinburgh and London.

OLIVER, G. and SCHÄFER, E. A. (1895). On the physiological action of extracts of pituitary body and certain other glandular organs. *J. Physiol., Lond.* **18**, 277–9.

SAWYER, W. H. (1966). Neurohypophysial principles of vertebrates. In *The pituitary gland* (ed. G. W. Harris and B. T. Donovan), Vol. 3: *Pars intermedia and neurohypophysis,* pp. 307–29. Butterworths, London.

SCHALLY, A. V. and KASTIN, A. J. (1970). The role of sex steroids, hypothalamic LH-releasing hor-

mone and FSH-releasing hormone in the regulation of gonadotropin secretion from the anterior pituitary gland. *Adv. Steroid Biochem. Pharmacol.* **2**, 41–69.

SCHWYZER, R. and SIEBER, P. (1963). Total synthesis of adrenocorticotropic hormone. *Nature.* **199**, 172–4.

SHARE, L. (1969). Extracellular fluid volume and vasopressin secretion. In *Frontiers in neuroendocrinology, 1969* (ed. W. F. Ganong and L. Martini), pp. 183–210. Oxford University Press, London.

SLOPER, J. C. (1966). The experimental and cytopathological investigation of neurosecretion in the hypothalamus and pituitary. In *The pituitary gland* (ed. G. W. Harris and B. T. Donovan), Vol. 3: *Pars intermedia and neurohypophysis,* pp. 131–239. Butterworth, London.

SMITH, P. E. (1926). Ablation and transplantation of the hypophysis in the rat. *Anat. Rec.* **32**, 221.

TURKINGTON, R. W. (1972). Molecular biological aspects of prolactin. In *Lactogenic hormones* (ed. G. E. W. Wolstenholme and J. Knight), pp. 111–27. Churchill Livingstone, Edinburgh and London.

WINGSTRAND, K. G. (1966). Microscopic anatomy, nerve supply and blood supply of the pars intermedia. In *The pituitary gland* (ed. G. W. Harris and B. T. Donovan), Vol. 3: *Pars intermedia and neurohypophysis,* pp. 1–27. Butterworths London.

WOLSTENHOLME, G. E. W. and KNIGHT, J. (ed.) (1972). *Lactogenic hormones.* Ciba Foundation Symposium. Churchill Livingstone, Edinburgh and London.

XEUREB, G. P., PRITCHARD, M. M. L., and DANIEL, P. M. (1954*a*). The arterial supply and venous drainage of the human hypophysis cerebri. *Q. Jl. exp. physiol.* **39**, 199–217.

——and—(1954*b*). The hypophysial portal system of vessels in man. *Q. Jl. exp. physiol.* **39**, 219–30.

YATES, F. E., RUSSELL, S. M., and MARAN, J. W. (1971). Brain–adenohypophysial communication in mammals. *A. Rev. Physiol.* **33**, 393–444.

Chapter 49. The pineal gland

KELLY, D. E. (1962). Pineal organs: photoreception, secretion and development. *Am. Scient.* **50**, 597–625.

LERNER, A. B., CASE, J. D., TAKAHASHI, Y., LEE, T. H., and MORI, W. (1958). Isolation of melatonin, the pineal gland factor that lightens melanocytes. *J. Am. chem. Soc.* **80**, 2587.

STEBBINS, R. C. and EAKIN, R. M. (1958). The role of the "third eye" in reptilian behavior. *Am. Mus. Novit.* **1870**, 1–40.

WOLSTENHOLME, G. E. W. and KNIGHT, J. (ed.) (1971). *The pineal gland.* Ciba Foundation symposium. Churchill Livingstone, Edinburgh and London.

WURTMAN, R. J., AXELROD, J., and KELLY, D. E. (1968). *The pineal.* Academic Press, New York and London.

YOUNG, J. Z. (1973). The pineal gland. *Philosophy* **48**, 70–4.

Chapter 50. Internal secretions and homeostasis

BARRINGTON, E. J. W. (1963). *An introduction to general and comparative endocrinology.* Clarendon Press, Oxford.

— (1964). *Hormones and evolution.* English University Press, London.

BURNS, R. K. (1961). Role of hormones in the differentiation of sex. In *Sex and internal secretions* (3rd edn, ed. W. C. Young), Vol. 1, pp. 76–158. Baillière, Tindall & Cox, London.

FINAR, I. L. (1968). *Organic Chemistry,* Vol. 2, *Stereochemistry and the chemistry of natural products* (4th edn). Longman, London.

JOST, J.-P. and RICKENBERG, H. V. (1971). Cyclic AMP. *A. Rev. Biochem.* **40**, 741–74.

LOEWI, O. (1921). Über humorale Übertragbarkeit der Nernerven-wirkung. *Pflüg. Arch. ges. Physiol.* **189**, 239–42.

LOEWENSTEIN, W. R. and KANNO, Y. (1967). Intercellular communication and tissue growth. I. Cancerous growth. *J. cell. Biol.* **33**, 225–34.

— and PENN, R. D. (1967). Intercellular communication and tissue growth. II. Tissue regeneration. *J. cell. Biol.* **33**, 235–42.

PARKER, G. H. (1932). *Humoral agents in nervous activity.* Cambridge University Press.

RASMUSSEN, H., GOODMAN, D. B. P., and TENENHOUSE, A. (1972). The role of cyclic AMP and calcium in cell activation. *Crit. Rev. Biochem.* **1**, 95–148.

STARLING, E. H. (1905). The Croonian lectures on the chemical correlation of the functions of the body. *Lancet* (2), 339.

WITSCHI, E. (1939). Modification of development of sex in lower vertebrates and in mammals. In *Sex and internal secretions* (2nd edn, ed. E. Allen), pp. 145–226. Baillière, Tindall & Cox, London.

YOUNG, J. Z. (1934). Hormones and chemical correlation. *Sch. Sci. Rev.* **60**, 502–8.

Chapter 51. The reproductive tracts of mammals

AUSTIN, C. R. and SHORT, R. V. (ed.) (1972). *Hormones in reproduction.* (*Reproduction in mammals,* Book 3.) Cambridge University Press.

BISHOP, D. W. (ed.) (1962). *Spermatozoan motility.* American Association for the Advancement of Science, Washington, D.C.

FAWCETT, D. W. (1962). Sperm tail structure in relation to the mechanism of movement. In *Spermatozoan motility* (ed. D. W. Bishop), pp. 147–69. American Association for the Advancement of Science, Washington, D.C.

HAFEZ, E. S. E. (ed.) (1971). *Comparative reproduction of non-human primates.* Thomas, Springfield, Illinois.

LAMMING, G. E. and AMOROSO, E. C. (ed.) (1967). *Reproduction in the female mammal.* Butterworth, London.

METZ, C. B. and MONROY, A. (ed.) (1967). *Fertilization: comparative morphology, biochemistry and immunology,* Vols 1 and 2. Academic Press, New York and London.

MONROY, A. (1965). *Chemistry and physiology of fertilization.* Holt, Rinehart and Winston, New York.

Chapter 52. The oestrous cycle of mammals

ARIMURA, A. and FINDLAY, A. (1971). Hypothalamic map for the regulation of gonadotropin release. *Res. Reprod.* **3**, (1).

ASDELL, S. A. (1964). *Patterns of mammalian reproduction* (2nd edn). Constable, London.

BULLOUGH, W. S. (1961). *Vertebrate reproductive cycles* (2nd edn). Methuen, London.

MARKEE, J. E. (1940). Menstruation in intraocular endometrial transplants in the Rhesus monkey. *Contr. Embryol.* **28**, 219–308.

PARKES, A. S. (ed.) (1960). *Marshall's Physiology of reproduction* (3rd edn), Vol. 2. Longmans, London.

PERRY, J. S. (1971). *The ovarian cycle of mammals.* Oliver & Boyd, Edinburgh.

ROSS, G. T., CARGILLE, C. M., LIPSETT, M. B., RAYFORD, P. L., MARSHALL, J. R., STROTT, C. A., and RODBARD, D. (1970). Pituitary and gonadal hormones in women during spontaneous and induced ovulatory cycles. *Recent Prog. Horm. Res.* **26**, 1–48.

SHORT, R. V. (1972). Role of hormones in sex cycles. In *Hormones in reproduction. Reproduction in mammals,* Book 3 (ed. C. R. Austin and R. V. Short), pp. 42–72. Cambridge University Press.

VANDE WIELE, R. L., BOGUMIL, J., DYRENFURTH, I., FERIN, M., JEWELEWICZ, R., WARREN, M., RIZKALLAH, T., and MIKHAIL, G. (1970). Mechanisms regulating the menstrual cycle in women. *Recent Prog. Horm. Res.* **26**, 63–95.

ZUCKERMAN, S. (1932). *The social life of monkeys and apes.* Kegan Paul, London.

Chapter 53. The development of mammals

AMOROSO, E. C. (1952). Placentation. In *Marshall's Physiology of reproduction* (3rd edn, ed. A. S. Parkes), Vol. 2, pp. 127–311. Longmans, London.

ASSALI, N. S. (ed.) (1968). *Biology of gestation* Vols. 1 and 2. Academic Press, New York and London.

ASSALI, N. S. DILTS, P. V., Jr, PLENTL, A. A., KIRSCHBAUM, T. H., and GROSS, S. J. (1968). Physiology of the placenta. In *Biology of gestation,* Vol. 1: *The maternal organism* (ed. N. S. Assali), pp. 185–289. Academic Press, New York and London.

BELLAIRS, R. (1971). *Developmental processes in higher vertebrates.* Logos Press, London.

BOYD, J. D. and HAMILTON, W. J. (1952). Cleavage, early development and implantation of the egg. In *Marshall's Physiology of reproduction* (3rd edn, ed. A. S. Parkes), Vol. 2, pp. 1–126. Longmans, London.

DAVIDSON, E. H. (1968). *Gene activity in early development.* Academic Press, New York and London.

DE HAAN, R. L. and URSPRUNG, H. (ed.) (1965). *Organogenesis.* Holt,

Rinehart and Winston, New York.

HARRISON, R. J. (1961). Some aspects of mammalian embryology. In *Recent advances in anatomy* (second series, ed. F. Goldby and R. J. Harrison), pp. 129–74. J. & A. Churchill, London.

HERTIG, A. T. and ROCK, J. (1941). Two human ova of the pre-villous stage, having an ovulation age of about eleven and twelve days respectively. *Contr. Embryol.* **29**, 127–56.

— and — (1945). Two human ova of the pre-villous stage, having a developmental age of about seven and nine days respectively. *Contr. Embryol.* **31**, 65–84.

HUGGETT, A. ST. G. (1961). Carbohydrate metabolism in the placenta and foetus. *Br. med. Bull.* **17**, 122–6.

KIRBY, D. R. S. (1970). Immunological aspects of implantation. In *Ovo-implantation, human gonadotropins, aad prolactin* (ed. P. O. Hubinot, F. Leroy, C. Robyn, and P. Leleux), pp. 86–97. Second International Seminar on Reproductive Physiology and Sexual Endocrinology, 1968. S. Karger, Basel.

LEWIS, W. H. and WRIGHT, E. S. (1935). On the early development of the mouse egg. *Contr. Embryol.* **25**, 113–43.

RUNNER, M. N. (1947). Development of mouse eggs in the anterior chamber of the eye. *Anat. Rec.* **98**, 1–17.

SIMMER, H. H. (1968). Placental hormones. In *Biology of gestation*, Vol. 1: *The maternal organism* (ed. N. S. Assali), pp. 290–354. Academic Press, New York and London.

STREETER, G. L. (1942). Developmental horizons in human embryos. *Contr. Embryol.* **30**, 211–45.

— (1945). *Contr. Embryol.* **31**, 27–63.

— (1948). *Contr. Embryol.* **32**, 133–85.

WOLSTENHOLME, G. E. W. and KNIGHT, J. (ed.) (1972). *Lactogenic hormones*. Ciba Foundation Symposium. Churchill Livingstone, Edinburgh and London.

WYNN, R. M. (1968). Morphology of the placenta. In *Biology of gestation* (ed. N. S. Assali), Vol. 1, *The maternal organism*, pp. 93–184. Academic Press, New York and London.

References added in proof

KEIDEL, W. D. and NEFF, W. D. (ed.) (1974). *Handbook of sensory physiology*, Vol. 5, part 1: *Auditory system: anatomy, physiology* (*ear*). Springer-Verlag, Berlin.

KORNHUBER, H. H. (ed.) (1974a). *Handbook of sensory physiology*, Vol. 6, part 1: *Vestibular system: basic mechanisms*. Springer-Verlag, Berlin.

KORNHUBER, H. H. (ed.) (1974b). *Handbook of sensory physiology*, Vol. 6, part 2: *Psychophysics, applied aspects and general interpretations*. Springer-Verlag, Berlin.

The following articles from the above-named books should be used in conjunction with the bibliography.

Chapter 31
From Kornhuber (1974a):

BRODAL, A. Anatomy of the vestibular nuclei and their connections (pp. 239–352).

GACEK, R. R. Morphological aspects of the efferent vestibular system (pp. 213–20).

PRECHT, W. The physiology of the vestibular nuclei (pp. 353–416).

From Kornhuber (1974b):

KORNHUBER, H. H. The vestibular system and the general motor system (pp. 581–620).

Chapter 35
From Keidel and Neff (1974):

ADES, H. W. and ENGSTRÖM, H. Anatomy of the inner ear (pp. 125–58).

From Kornhuber (1974a):

FREDRICKSON, J. M., KORNHUBER, H. H., and SCHWARZ, D. W. F. Cortical projections of the vestibular nerve (pp. 565–82).

Chapter 43
From Keidel and Neff (1974):

ELDRIDGE, D. H. Inner ear—cochlear mechanics and cochlear potentials (pp. 549–84).

FEX, J. Neural excitatory processes of the inner-ear (pp. 585–646).

HARRISON, J. M. and HOWE, M. E. Anatomy of the efferent auditory nervous system of mammals (pp. 283–336).

—and— Anatomy of the descending auditory system (mammalian) (pp. 363–88).

MØLLER, A. R. Function of the middle ear (pp. 491–518).

RAUCH, S. and RAUCH, I. Physicochemical properties of the inner ear especially ion transport (pp. 647–82).

From Kornhuber (1974a):

LOWENSTEIN, O. E. Comparative morphology and physiology (pp. 75–120).

WERSÄLL, J. and BAGGER-SJÖBÄCK, D. Morphology of the vestibular sense organs (pp. 123–70).

Author Index

The names appearing in the Bibliography are indexed only in the case of general references for which there is no corresponding mention in the text. In such cases the page number is followed by an italic chapter number. (*gen.*) indicates that the entry is in the list of general works.

The names appearing in the Bibliography are indexed only in the case of general references for which there is no corresponding mention in the text. In such cases the page number is followed by an italic chapter number. (*gen.*) indicates that the entry is in the list of general works.

The names appearing in the Bibliography are indexed only in the case of general references for which there is no corresponding mention in the text. In such cases the page number is followed by an italic chapter number. (*gen.*) indicates that the entry is in the list of general works.

The names appearing in the Bibliography are indexed only in the case of general references for which there is no corresponding mention in the text. In such cases the page number is followed by an italic chapter number. (*gen.*) indicates that the entry is in the list of general works.

Subject Index